Thieme

Vascular Diagnosis with Ultrasound

Clinical Reference with Case Studies

Volume 1: Cerebral and Peripheral Vessels

Michael G. Hennerici, M.D., Ph.D.
Professor and Chairman
Department of Neurology
University of Heidelberg
Klinikum Mannheim
Mannheim, Germany

Doris Neuerburg-Heusler, M.D.
Former Director of the Department
of Noninvasive Diagnostics
Aggertalklinik, Engelskirchen
Cologne, Germany

With contributions by

Michael Daffertshofer, M.D, Ph.D.
Associate Professor
Department of Neurology
University of Heidelberg
Klinikum Mannheim
Mannheim, Germany

Thomas Karasch, M.D.
Department of
Cardiology/Angiology
University of Cologne
Cologne, Germany

Stephen Meairs, M.D., Ph.D.
Associate Professor
Department of Neurology
University of Heidelberg
Klinikum Mannheim
Mannheim, Germany

2nd revised edition

532 illustrations
49 tables

Thieme
Stuttgart · New York

Library of Congress Cataloging-in-Publication Data
Hennerici, M. (Michael) G.
[Gefässdiagnostik mit Ultraschall. English]
Vascular diagnosis with ultrasound : clinical references with case studies / Michael G. Hennerici, Doris Neuerburg-Heusler ; with contributions by Michael Daffertshofer, Thomas Karasch, Stephen Meairs.
p. cm.
Rev. translation of: Gefässdiagnostik mit Ultraschall. 2. Aufl. 1995.
Includes bibliographical references and index.
ISBN 3-13-103832-2 (alk. paper) -- ISBN 1-58890-144-0 (alk. paper)
1. Blood-vessels--Ultrasonic imaging. 2. Blood-vessels--Ultrasonic imaging--Case studies.
[DNLM: 1. Vascular Diseases--ultrasonography. WG 500 H515 g 2005a] I. Neuerburg-Heusler, Doris. II. Hennerici, M. (Michael) Gefässdiagnostik mit Ultraschall. English III. Title.
RC691.6.U47H4613 2005
616.1'307543--dc22
2005010530

1st English edition 1998

Important note: Medicine is an ever-changing science undergoing continual development. Research and clinical experience are continually expanding our knowledge, in particular our knowledge of proper treatment and drug therapy. Insofar as this book mentions any dosage or application, readers may rest assured that the authors, editors, and publishers have made every effort to ensure that such references are in accordance with **the state of knowledge at the time of production of the book.** Nevertheless, this does not involve, imply, or express any guarantee or responsibility on the part of the publishers in respect to any dosage instructions and forms of applications stated in the book. **Every user is requested to examine carefully** the manufacturers' leaflets accompanying each drug and to check, if necessary in consultation with a physician or specialist, whether the dosage schedules mentioned therein or the contraindications stated by the manufacturers differ from the statements made in the present book. Such examination is particularly important with drugs that are either rarely used or have been newly released on the market. Every dosage schedule or every form of application used is entirely at the user's own risk and responsibility. The authors and publishers request every user to report to the publishers any discrepancies or inaccuracies noticed. If errors in this work are found after publication, errata will be posted at www.thieme.com on the product description page.

Some of the product names, patents, and registered designs referred to in this book are in fact registered trademarks or proprietary names even though specific reference to this fact is not always made in the text. Therefore, the appearance of a name without designation as proprietary is not to be construed as a representation by the publisher that it is in the public domain.

Rüdigerstrasse 14, 70469 Stuttgart, Germany
http://www.thieme.de
Thieme New York, 333 Seventh Avenue,
New York, NY 10001 USA
http://www.thieme.com

Cover design: Martina Berge, Erbach
Typesetting by primustype Hurler GmbH, Notzingen
Printed in Germany by Grammlich, Pliezhausen

ISBN 3-13-103832-2 (GTV)
ISBN 1-58890-144-0 (TNY) 1 2 3 4 5

Preface

Diagnosis and treatment of vascular diseases have made tremendous progress since the introduction of non-invasive ultrasound technologies in clinical practice in the early 1970s. Today ultrasound is capable of monitoring the early silent stages of atherogenesis as well as the morphological features of advanced atherosclerosis during its progression and regression in all major arteries of the body. The ability of ultrasound to visualize both arterial and venous thrombus formation has been expanded to include detection and quantification of circulating microemboli. Vascular ultrasound studies are important tools in individual patient diagnosis and during follow-up, and notably in randomized, prospective clinical trials that serve to strengthen evidence-based medicine for clinical practice. The recent use of state-of-the-art ultrasound monitoring in such studies has helped to introduce new pathways for the management of our patients, both in modern industrialized societies and in developing countries. Increasing age combined with the unfortunately remaining high prevalences and incidences of myocardial infarction, stroke, and peripheral vascular disease underline the need for early identification of subjects at high risk for treatment in a yet asymptomatic period with potential therapeutic impact in primary prevention, and for means to improve monitoring for secondary prevention. Ultrasound will continue to play a major role in realizing these important goals of preventive medicine; as it is non-invasive, always available, and economically viable, it has distinct advantages over all other vascular diagnostic tools.

The first German edition of this book (1988) was welcomed by both novice and experienced sonographers due to its strict illustrative composition. A major revision published in 1994 was necessary for inclusion of rapidly developing technologies. This edition first introduced a collection of individual case histories and vascular findings, which became of major interest to many readers of the book. The atlas illustrated from the very beginning the combined use of ultrasound technologies with clinical data and other methods applied in clinical practice and was particularly useful and well accepted both by the specialized collaborations in the vascular laboratory, as well as by clinicians unfamiliar with specific ultrasound tests but using their reports. In 1997 when the first English edition was published, rare findings and specific problems were only occasionally included and selected repetitions were intentional for educational purposes. This edition was also a complete revision of the second German monograph and included advanced ultrasound applications as well as data from large study trials and basic research. It also introduced experimental techniques just investigated in research laboratories at that time (harmonic imaging, 3-D and 4-D imaging, flow volume measurements, intra-arterial and interventional applications, functional and monitoring studies, etc.). A third German edition was published very soon thereafter in 1999; due to the fast development of ultrasound, this edition included new data on imaging of perfusion in different organs and investigations of small vessel networks by means of new echocontrast media.

Since the amount of material to be included in the new second English edition of "Vascular Diagnosis with Ultrasound" has been growing so rapidly and because we did not want to miss the broad spectrum of vascular ultrasound applications addressing a large community of both investigators and clinicians (who are now confronted with results of ultrasound studies from throughout the world using different instruments, different technologies, and different economic restrictions), we decided to completely revise the text and split the book into two volumes. The first volume deals with cerebral and peripheral vessels. The second volume addresses the abdominal vessels, small parts vessels, and topics such as tumor vascularization, which has become a fascinating area of vascular ultrasound. The complete revision includes both diagnostic and therapeutic aspects of ultrasound that address, particularly in the atlas, the advantages of recent neurosonologic technologies and applications. This edition also provides new information on ultrasound procedures in peripheral angiology, reflecting the results of recent ultrasound studies that provide normative parameters for therapeutic decisions. Newer horizons in vascular sonography, such as sonothrombolysis in stroke, peripheral artery thrombolysis using intra-arterial ultrasound to cause micro-fragmentation of thrombi, and molecular imaging for non-invasive detection of diseases using microbubbles targeted to disease-associated molecular structures, are only a few of the fascinating perspectives addressed in this new volume.

The first volume of this new second edition has continued to be written by M.G. Hennerici and D. Neuerburg-Heusler. However, this would not have been possible without the help and cooperation of new coworkers, such as Michael Daffertshofer and Stephen Meairs from the Department of Neurology, University of Heidelberg, Universitätsklinikum Mannheim and ongoing cooperation with Thomas Karasch, Department of Internal Medicine (Cardiology/Angiology), University of

Cologne, whose expertise was so well acknowledged already for previous editions. We are grateful to our vascular technicians and secretaries who helped us with the enormous work on the manuscript, including the continuous preparation of the abundant literature in this exciting and developing field. It took us longer than expected but we finally managed to finish the first volume, with the support of our team from Thieme publishers headed by Cliff Bergman. We are especially grateful for the continuous support of our families, Marion Hennerici, and the late Helmut Neuerburg; without their love and support we would have failed.

Michael G. Hennerici and Doris Neuerburg-Heusler

Table of Contents

1 Physics and Technology of Ultrasound 1

2 Extracranial Cerebral Arteries 27

3 Intracranial Cerebral Arteries 89

5 Peripheral Arteries 149

6 Peripheral Veins 211

7 Case Histories 257

Important Abbreviations and Symbols

Examination Methods

A-mode	amplitude mode
B-mode	brightness mode
CW	continuous wave duplex sonography
dB	decibel
DSA	digital subtraction angiography
FFT	fast Fourier transformation
CBFV	cerebral blood flow velocity
CFDS	color flow duplex sonography
HITS	high intensity transient signals
Hz	hertz
I	intensity
ICP	intracranial pressure
kHz	kilohertz
MHz	megahertz
M-mode	motion mode
MRA	magnetic resonance angiography
MRI	magnetic resonance imaging
PET	positron emission tomography
PRF	pulse repetition frequency
PW	pulsed wave Doppler sonography
PVR	peak velocity ratio
SAP	systolic arterial pressure
SB	spectral broadening
TAMX	time averaged maximum
TCD	transcranial Doppler sonography
TCCD	transcranial color coded duplex
TCM	transcranial Doppler monitoring
TEE	transesophageal echocardiography
TGC	time gain compensation
TTE	transthoracic echocardiography
UDS	ultrasound Doppler sonography
v	velocity
vdlast	enddiastolic velocity
vmax	maximum systolic velocity
vmean	mean velocity

Calibration and Tests

ordinate:	signal amplification (if not otherwise noted)
abscissa:	time 5 mm/s, 25 mm/s, or 50 mm/s as indicated
→[flow toward the probe	Doppler signal above baseline
←[flow away from the probe	Doppler signal above baseline
▬	duration of a compression test
↓ ↓ ↓	repeated tapping
E	expiration
I	inspiration
K = C	compression
PI	pulsatility index (Gosling)
R	reflux
RI	resistance index (Pourcelot)

Clinical Terms

CAD	coronary artery disease
PAOD	peripheral arterial occlusive disease
PRIND	prolonged reversible ischemic neurological deficit
PTA	percutaneous transluminal angioplasty
RIND	reversible ischemic neurological deficit
TIA	transient ischemic attack

1 Physics and Technology of Ultrasound

Basic Ultrasound Physics

Sound is mechanical energy that is transmitted through a medium such as air. Periodic changes in air pressure are created by forces acting on air molecules, causing them to oscillate. A pressure wave is transmitted from one location to another when vibrating molecules interact with neighboring molecules. This molecular motion is necessary for the transmission of sound and explains why sound cannot be transmitted in a vacuum.

Sound waves above a frequency of 20 kHz are termed "ultrasound." Like all sound waves, ultrasound propagates through various media in the form of a pulsating pressure wave. Waves are basically of two types—longitudinal and transverse. Longitudinal waves are those in which particle motion is along the direction of propagation of the wave energy. Sound waves are longitudinal. Transverse waves are perpendicular to the direction of propagation of the wave energy. Wave motion resulting from a stone being thrown into water is an example of a transverse wave. Bone is the only biological tissue that can cause the production of transverse waves, which are also referred to as "shear waves" or "stress waves."

Properties of Waves

When particle displacement is plotted against distance, the *wavelength* (λ) of a wave is the distance from crest to crest, or from trough to trough. A wave cycle is a sequence of changes in the amplitude that recur at regular intervals. The *frequency* (*f*) of a wave is the number of cycles passing a given point in one unit of time (usually one second). The unit of frequency is the hertz (Hz; one cycle per second) (Fig. 1.**1**).

The speed of wave propagation through a medium is known as the acoustic velocity, *c*. This speed depends on the density and compressibility of a medium. For sound to propagate, it is essential for a medium to be present. In addition, the medium also has to be compressible—that is, it must be able to deform temporarily and then return back to its original shape.

The velocity of sound in a medium is inversely proportional to the square root of the medium's density, $\sqrt{\varrho}$. Therefore, the denser the medium, the slower the velocity of sound.

The relationship between sound velocity and compressibility is also inversely proportional to the square root of the medium's compressibility, $\sqrt{\beta}$. Since dense materials such as bone have low compressibility, sound propagates through them at a high velocity. In contrast, since gas molecules in air are far apart and easily compressible, the velocity of sound in air is slower.

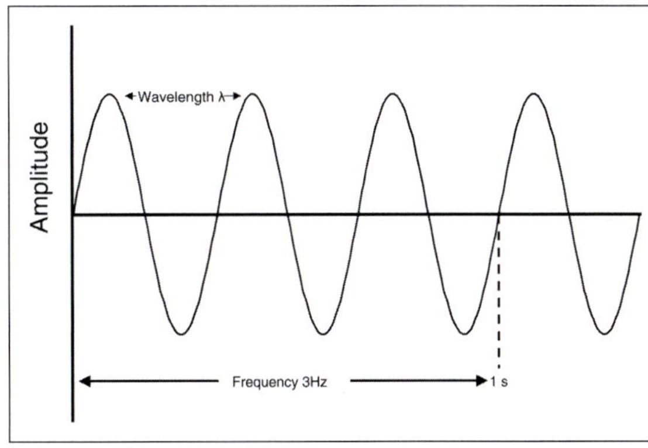

Fig. 1.**1** Schematic drawing of ultrasound wave

When the effects of compressibility and density are combined into a single equation, acoustic velocity can be defined as:

$$c = 1/\sqrt{\varrho\beta}$$

Compressibility and density are interdependent—changes in density are accompanied by opposing changes in compressibility. However, since compressibility varies more rapidly than density does, it becomes the dominant factor in determining the relative acoustic velocity through a medium.

Reflection

Ultrasound reflections occur when there is an interface between two media with differing levels of resistance to ultrasound. Such interfaces are known as "specular reflectors." Resistance to ultrasound is termed *acoustic impedance*. Acoustic impedance depends on the speed of ultrasound propagation in tissue and on the density of the tissue concerned; the greater the difference in impedance, the stronger the reflection. Mathematically, acoustic impedance (*Z*) is the product of the density of the tissue (ϱ) and the velocity of sound in the medium (*c*):

$$Z = \varrho - c$$

Acoustic impedance is thus is a measure of the resistance to sound passing through a medium. High-density materials are associated with high sound velocities and therefore high acoustic impedances. Similarly, low-density media such as gases have low acoustic impedances. It is not important which impedance is larger or smaller for two media forming an interface; the same reflection occurs whether sound is propagating from high-impedance tissue to low-impedance tissue or vice versa.

The angle of reflection of a sound beam is equal to the angle of incidence of the sound beam. To obtain maximum detection of the reflected ultrasound signal, the transducer, which both sends and receives signals, has to be oriented in such a way that the ultrasound beam generated strikes the interface perpendicularly (normal incidence).

Refraction

If a sound beam strikes an interface at an angle other than 90°, the part of the beam that is transmitted further is refracted or bent away from the straight path that would have been expected. This refraction takes place in accordance with Snell's law of optics, which relates the angle of transmission to the relative velocities of sound in the two media. The bending occurs because the portion of the wavefront in the second medium travels at a different velocity compared with that in the first medium. As may be expected, the amount of deviation from the expected path changes with the angle of incidence and with the velocities in the associated media. If the velocity of sound is the same in both media, then there will be no refraction, even though there may be different acoustic impedances. In diagnostic vascular ultrasound, problems of refraction are encountered in transcranial applications, in which refraction occurs at both bone–tissue and tissue–bone interfaces when insonating through the transtemporal bone window. This may result in a significant reduction of the image quality.

Scattering and Diffraction

Tissue particles that are relatively small in relation to the wavelength (e.g., blood cells), and particles with differing impedance that lie very close to one another, cause *scattering* or *speckling*. Reflecting media that lie at an angle to the ultrasound propagation axis can only be recognized due to these scattering phenomena, which are accompanied by an attenuation in the echo intensity. *Speckling effects* result from the extinction of reflexes from structures lying adjacent to each other or close behind each another at a distance of $\lambda/4$.

Table 1.**1** Attenuation of various human tissues at 1 MHz

Tissue	Np/cm
Blood	0.021
Fat	0.069
Brain	0.098
Liver	0.103
Kidney	0.115
Skull bone	2.3
Lung	4.6

Np: neper.

Attenuation

The stronger the reflections are at the interfaces, the less ultrasound energy is available to reach deeper tissue. If a total reflection occurs at an interface to air, bone, or calcium-containing tissue, then an *ultrasound shadow* results. Most of the ultrasound energy is converted into kinetic energy within the tissue. The term *attenuation* characterizes the reduction in the amplitude of an ultrasound wave as it propagates through a medium, due to both scattering and absorption. Attenuation is described by an exponential function. The attenuation coefficient is given by the sum of the scattering coefficient a_s and the absorption coefficient α:

$$a = a_s + \alpha$$

The coefficients quantitate the respective fractional loss in amplitude per unit length from absorption, scattering, and both processes together. The special unit used for these coefficients is the neper (Np) per centimeter. The attenuation coefficients at 1 MHz for various tissue types are shown in Table 1.**1.**

Intensity and Power

The *intensity* of the ultrasonic beam is a physical parameter that describes the amount of energy flowing through a unit of cross-sectional area each second. Acoustic intensity is expressed in mixed units of watts per centimeter squared. One watt is equal to one joule per second. The intensity of an ultrasonic beam is proportional to the square of the pressure amplitude. The instantaneous intensity (i) is given by

$$i = p^2/\varrho c$$

where p is the acoustic pressure, c the velocity of sound and ϱ the density. The intensity of the beam decreases exponentially with distance, the magnitude of which depends on the *amplitude attenuation coefficient.*

The Doppler Effect

The flow velocity of corpuscular elements in the blood can be detected with ultrasound by using a principle named after Christian Andreas Doppler (1803–1852). The physical context of the *Doppler effect* may well be familiar from everyday experience: when a truck blowing its horn approaches a stationary observer on a highway, the observer will hear the sound of the horn rising in pitch until the truck is directly in front of him. When the truck has passed and drives away from the observer with its horn still blowing, the pitch of the horn starts to fall. When we use this principle to investigate blood flow, we refer correspondingly to a change in frequency called the *Doppler shift*, Δf (Hz), which is proportional to the flow velocity of blood, v (cm/s), and to the transmission frequency of ultrasound, f (MHz).

Exact measurement of the Doppler shift requires that the angle α between the ultrasound beam and the longitudinal axis of the blood vessel is known:

$$\Delta f = 2f \cdot v \cdot \cos \alpha / c$$

(where c is the flow velocity through tissue, 1540 m/s).

When blood vessels are examined, the reflection caused by the blood's corpuscular elements, mainly erythrocytes, plays a major role. Since the flow velocity through the different areas of any given cross-section of a blood vessel varies, the Doppler signal itself does not correspond to a single frequency, but contains a broad *frequency spectrum* (in the normal internal carotid artery, this ranges from 0.5 kHz to 3.5 kHz and < 120 cm/s, respectively, if a transmission frequency of 4 MHz is used).

Reflections at the blood vessel wall appear as low-frequency artifacts with an intensity up to 30 times higher than that of moving blood elements. They can be eliminated using *high-pass filters* (100–400 Hz). However, it should be noted that slow velocities that are naturally part of the flow phenomena in a particular blood vessel will also be filtered out. The larger the number of moving blood cells, the higher the amplitude of the Doppler signal will be.

Ultrasound Technology

Ultrasound waves are produced by a vibrating crystal using a *piezoelectric effect*. This effect involves the characteristic that certain materials have of changing their form when subjected to an electric current—or vice versa, of producing an electric current due to changes in their form (depending on whether they are used as the transmitter or receiver). These piezoelectric crystals are placed in a specific arrangement *(array)* within the ultrasound probe.

In order to improve the *acoustic coupling* (impedance match), a gel is applied between the crystal and the surface of the skin.

Doppler Systems

Two types of Doppler system are used in vascular diagnosis:

- Continuous-wave Doppler systems
- Pulsed-wave Doppler systems

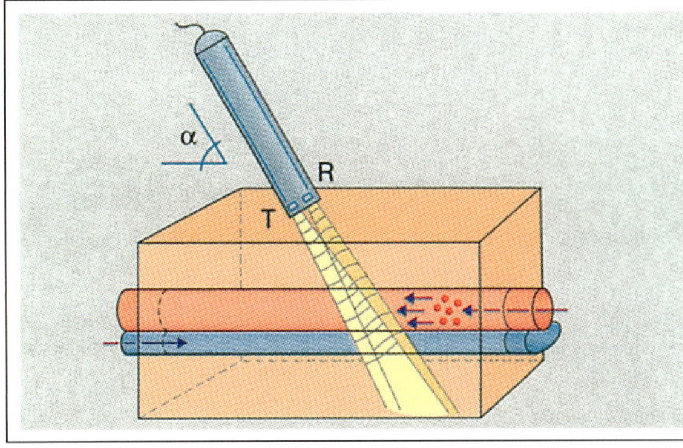

Fig. 1.**2** The principle of continuous-wave Doppler sonography. The arrow shows the direction of flow, and angle α is formed by the ultrasound and vascular axes. T: transmission crystal, R: receiving crystal

Continuous-Wave Doppler Sonography

Continuous-wave (CW) Doppler equipment uses two *piezoelectric elements* with overlapping ultrasound fields, which serve as the transmitter and the receiver (Fig. 1.**2**). Due to the continuous and simultaneous emission and reflection of ultrasound waves, no exact information about depth can be obtained with this equipment. However, specific areas of interest can be selected using appropriate focusing.

Pulsed-Wave Doppler Sonography

Pulsed-wave Doppler systems are used to detect blood flow at a specifically defined depth. Using a single piezoelectric element, these systems function alternatingly as the transmitter and receiver (Fig. 1.**3**). By electronically selecting the length of the time interval between the transmitting and receiving pulses, it is possible to localize signals at different depths using the transit time of ultrasound through the tissue. This region is called the *sample volume*, and it can be shifted along the ultrasound axis. Its size is variable and is determined both by the characteristics of the ultrasound beam and by the duration of the pulse, which is controlled by the oscillator. Shorter ultrasound packets during the transmission phase (duration approximately 0.5–2.0 µs) result in a smaller sample volume, with a corresponding increase in the axial resolution.

At the receiving end, changing the time delay causes displacement of the sample volume along the ultrasound axis (sample volume localization).

To ensure unequivocal depth measurements, a new pulse can only be sent after the ultrasound packet has returned. The *pulse repetition frequency* (PRF) is therefore defined as the transit time required for the pulse to travel to the investigation area and back again at a given ultrasound propagation velocity (c), with E referring to the penetration depth:

$$PRF_{max} = c/2 \cdot E$$

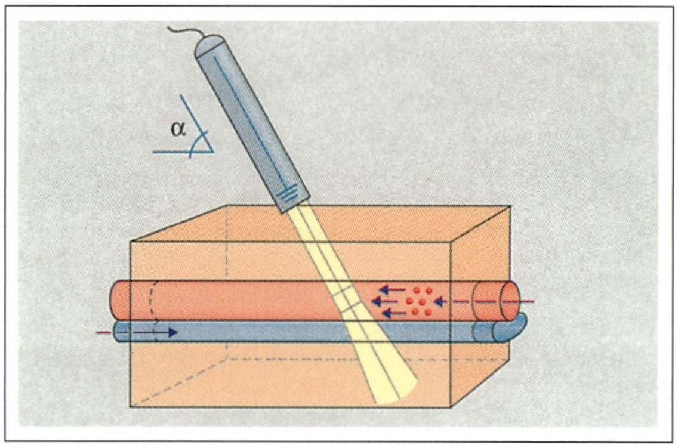

Fig. 1.**3** The principle of pulsed-wave Doppler sonography. The arrow shows the direction of flow, and angle α is formed by the ultrasound and vascular axes

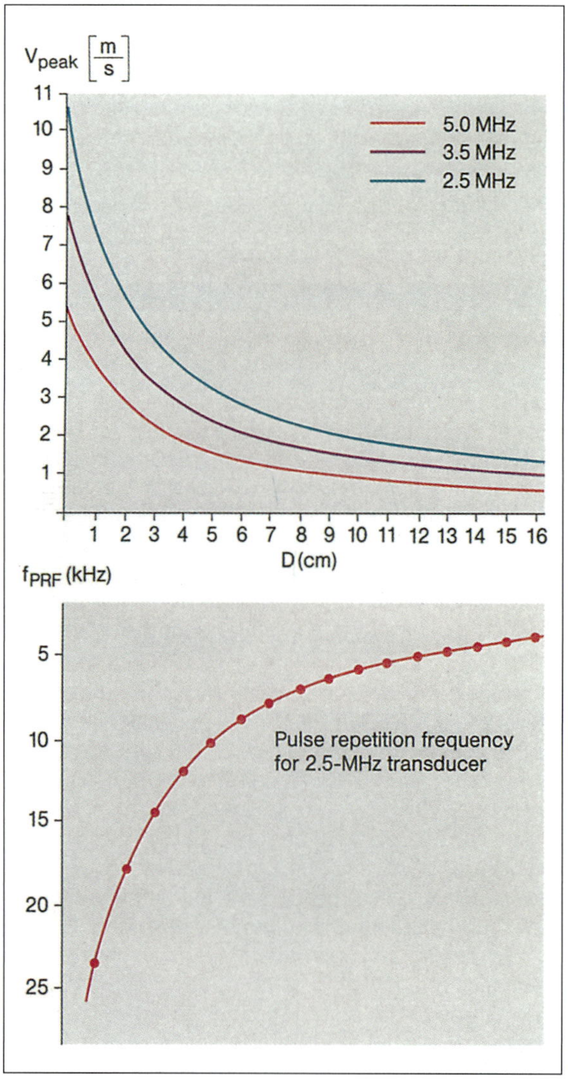

Fig. 1.**4** The relationship between the positioning of the sample volume D (depth), the pulse repetition frequency (f_{PRF}), and the peak flow velocity (V_{peak}) for different ultrasound frequencies (2.5–5.0 MHz)

The frequency at which individual ultrasound packets are emitted restricts the maximum Doppler shift that can be measured. The pulse repetition frequency has to be at least twice as large as the Doppler shift frequency in order to be recorded. For example, a PRF of 10 kHz allows detection of a Doppler shift of a maximum of 5 kHz.

Referred to as the *Nyquist theorem,* this rule applies to all events that cannot be observed continuously. For example, in a movie, the rotation of a spoked wheel can be perceived correctly when the wheel begins to move. With further acceleration, the movement of the wheel increases up to a maximum velocity, beyond which it incorrectly appears that the wheels start to rotate backward, against the direction in which the vehicle is moving. This phenomenon is due to the pulse effect deriving from the camera technology: the higher a camera's pulse repetition speed, the higher the maximum velocity that can still be clearly depicted. Beyond this limit, errors occur. Different ultrasound transmission frequencies lead to different upper limits for the pulse repetition frequencies, and as a result, different threshold values for the flow velocities that can still be detected in different investigation areas. This is illustrated in Fig. 1.**4**.

With regard to the pulsed-wave process at a given depth, the application of low emission frequencies is preferable if high flow velocities are to be recorded. While these statements with regard to the maximum Doppler frequency are correct, it should also be noted that the standardized Doppler frequencies have to be much lower than PRF/2 to ensure that they can still be analyzed accurately enough. The PRF depends on the ultrasound propagation velocity in tissue and the distance between the transducer and the measuring point. It has to be selected at a frequency low enough to allow the ultrasound packet to return before the next one is sent. Applying the Nyquist theorem, measurement difficulties can arise particularly in relation to stenoses involving the deeper-lying blood vessels. (For example, if a PRF upper limit value of 15 kHz is given at a depth of 5 cm, this restricts the maximum ascertainable Doppler shift to 7.5 kHz.)

Doppler shifts can be registered even above the Nyquist limit. This measurement situation is termed *aliasing,* and it describes a degree of uncertainty in determining the velocity (Fig. 1.**5**). In the above example, the impression was that the wheel was turning backward at almost maximum speed just above the Nyquist limit. The faster the wheel turns, the further the spokes tend to drift back to their original position, eventually coming to a standstill at twice the Nyquist speed. In fast Fourier transform (FFT) spectral analysis, the velocity is plotted in the opposite direction. Using a *baseline shift* and deleting the counter-directional indication, the recordable velocity can be increased to a maximum value of twice the Nyquist limit. Multiple aliasing, however, can produce ambiguous diagrams. Alternatively, the pulse repetition rate can be raised *(high PRF),* so that instead of waiting for a transmitted pulse to return, a second pulse can be sent within a defined interval. When this procedure is extended, a transition into the continuous-wave mode takes place. In principle, blood flow velocities can be analyzed at different points within the cross-section of a blood vessel, provided a suitable pulsed-wave Doppler system is available. Using a *single-channel system,* these recordings would have to be carried out sequentially during different heart cycles. Due to the variability of the blood flow velocity from one heartbeat to the next, however, this method involves potential errors.

It may be difficult with a single-channel system to determine the exact location of the sample volume. This is particularly important when trying to demonstrate altered blood flow velocities near the vessel wall while excluding artifacts. *Multichannel pulsed-wave Doppler systems,* using a parallel arrangement of several measuring channels, can simultaneously measure the velocity in various neighboring areas of a blood vessel and display the velocity distribution as a function of time for each heart cycle (Fig. 1.**6**). Provided the angle between the ultrasound beam and the vascular axis is known, an algorithm can then be used to estimate the flow volume.

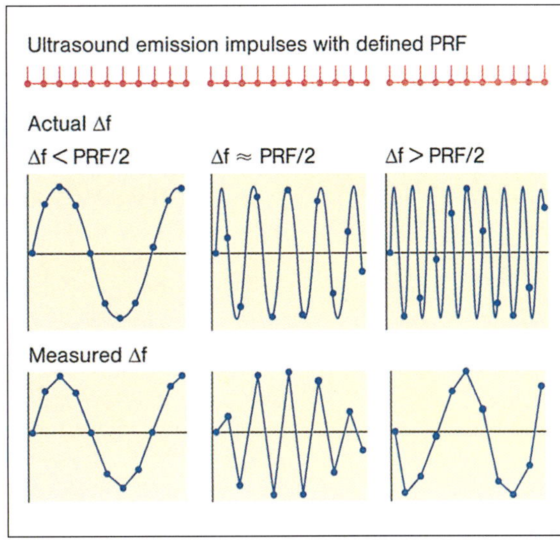

Fig. 1.**5** Reconstruction of Doppler oscillations at various frequencies from individual high-frequency ultrasound emission impulses (represented by the dots on the lines) with a defined pulse repetition frequency (PRF). The benchmark points for the reconstruction were the amplitude values of the sinus oscillation at the moment of each emitted ultrasound impulse. If the Doppler frequency shift μf exceeds half of the PRF *(right),* then the measured Doppler frequency will be interpreted incorrectly ("aliasing" effect) (adapted from Widder 1995)

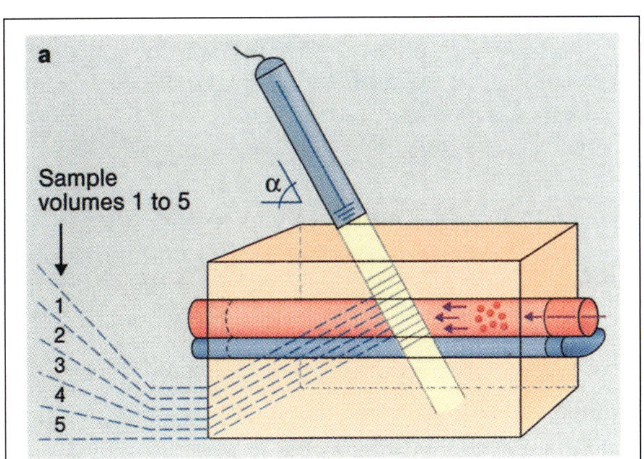

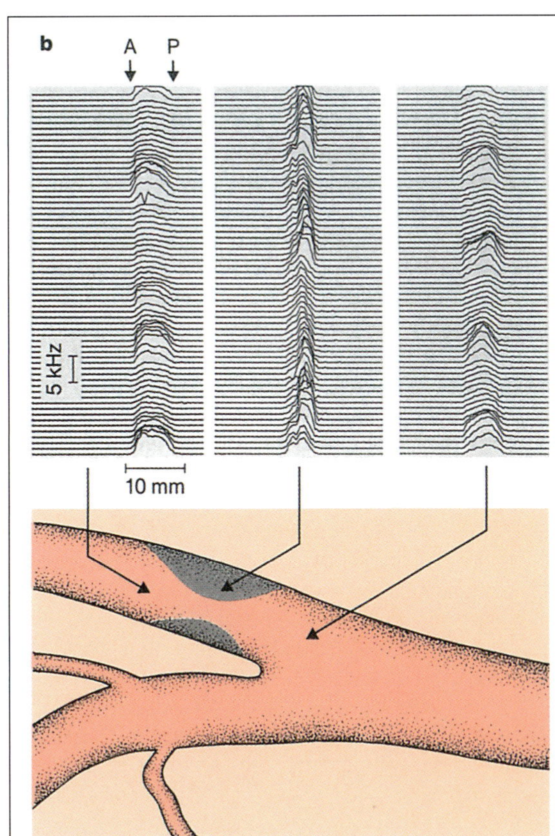

Fig. 1.**6 a** The principle used in a multichannel pulsed-wave Doppler system. **b** The display from a medium-grade stenosis of the internal carotid artery at the bifurcation. A: anterior wall, B: posterior wall

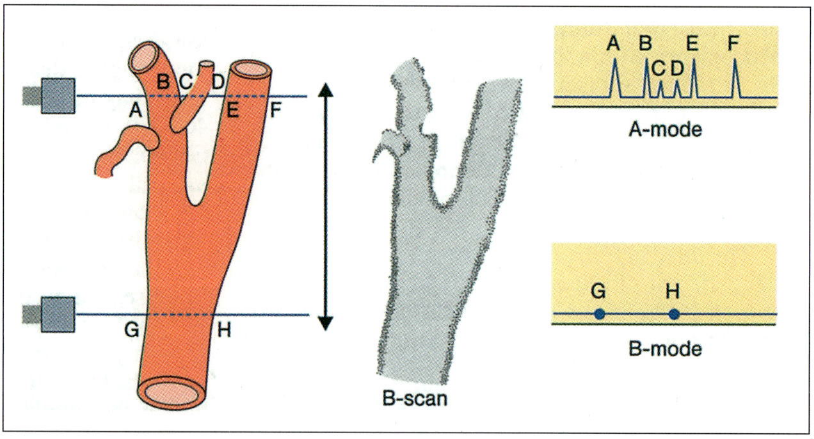

Fig. 1.7 Schematic drawing illustrating examination conditions using the echo-impulse technique: A-mode and B-mode (echotomogram)

B-Mode Imaging

When the speed of ultrasound propagation through tissue is known, the depth of the echo's reflection can be measured (Fig. 1.7) by:

- Amplitude modulation, *A-mode;* or
- Brightness modulation, *B-mode.*

When the ultrasound axes are shifted sequentially and laterally—which is technically possible in various ways with different arrangements and settings in the transducer components—it is possible to produce a two-dimensional ultrasound image in real time. With appropriate selection of different planes, blood vessel segments of interest can be displayed, both in longitudinal sections and cross-sections.

The quality of the image produced depends on the *image resolution.* Resolution is defined as the smallest distance between two points at which they can still be depicted as separate. The *axial resolution* depends exclusively on the length of the ultrasound impulse. Since for B-mode ultrasound the ultrasound pulses consist of only one to two sinus wavelengths, the axial resolution lies in the range of the ultrasound wavelength λ (0.2–1.0 mm). Since this value is reciprocal to the ultrasound frequency ($\lambda = c/f$), the *axial resolution* improves with increasing ultrasound frequency.

The *lateral resolution* refers to the ability to depict as separate entities two points that lie at right angles to the direction of ultrasound propagation. This is primarily dependent on the width of the ultrasound beam. To be able to resolve points that lie close together, the width of the ultrasound beam has to be kept reasonably small. This is possible when the wavelength is kept small and the diameter of the transducer is kept as large as possible (i.e., small phased-array transducers have a poorer lateral resolution than large linear or curved-array transducers). For physical reasons, secondary wavefronts also form in addition to the main wavefront and in suboptimal conditions may cause interference. The intensity of the reflected echo is assigned to a specific brightness level along a gray scale, providing a finely shaded image.

Transmission frequencies used for investigation of blood vessels range between 1 and 10 MHz, depending on the specific area of application (for arteries located close to the skin > 7.5 MHz, for arteries located deeper within the body between 3 and 5 MHz, and for transcranial application usually ≤ 2 MHz).

When selecting a transmission frequency, one should also note the following:

- The *axial resolution* is directly proportional to the ultrasound frequency.
- The *signal intensity,* however, depends on the attenuation of the ultrasound as it passes through tissue: the higher the ultrasound frequency is, the stronger the attenuation of the signal intensity will be.

The selection of a specific frequency for various diagnostic procedures always involves a compromise between the optimal ultrasound penetration depth and the ability to identify relevant tissue changes.

The B-mode procedure serves to identify arteries and veins and to position the sample volume for the Doppler mode accurately. Although it is possible to use CW Doppler sonography alone to evaluate blood vessels that lie close to the body's surface, such as the extracranial cerebral arteries and the accessible areas of the extremities, continuous tracing of the vasculature of the extremities is only feasible with the B-mode. Because of the deep location of the abdominal and pelvic blood vessels, they can only be examined using duplex sonography.

The advantage of the B-mode system is the information it provides about morphological structures within the vascular lumen and the surrounding tissue, such as:

- Anatomical topography
- Spatial relationship to neighboring structures
- Vascular caliber
- Vascular wall morphology

- Structures inside the lumen (plaque, thrombi)
- Compressibility (of the veins)

It can also be used to evaluate deep venous thrombosis, since it has been shown that an absence of venous compressibility by the transducer, widening of the lumen, and absent responses to breathing maneuvers are reliable parameters for establishing the diagnosis. In contrast, the sole use of B-mode sonography for the examination of arterial disease is unsuitable and potentially misleading. However, in combination with functional Doppler sonography, the B-mode component of duplex systems provides valuable information.

Color Doppler Flow Imaging

Color Doppler flow imaging (CDFI) is the principal technique used in clinical diagnosis (Fig. 1.**8**). Two technically different methods are commercially available. The *"frequency-domain"* method uses Doppler shifts and phase shifts of ultrasound echoes to display flow velocity superimposed on the B-mode image—i.e., this is a Doppler-related method. The *"time-domain"* method analyzes the alteration of high-frequency echo signals to display movement patterns—i.e., this method is independent of Doppler-shifted signals. Due to the larger computing capacity it requires, this method is not in widespread use, and its advantages (it is said to improve interpretation at low flow velocities) have not yet been demonstrated (Herment et al. 1996). Using the frequency-domain method, real-time spectral analysis of the individual Doppler signals cannot be performed due to the extremely high demands on data-processing facilities. Instead, color-coded two-dimensional flow velocity information is registered separately from the image information, although both types of data are displayed quasi-simultaneously on the screen. Black-and-white encoding is usually used to depict structures, while the flow velocity is presented in color-coded form. Technically, linear and phased-array systems can register a combination of both Doppler and image beams in a single pass, while annular-array systems combine a quick succession of images in one direction with a slower Doppler component. Instead of spectral analysis, the *autocorrelation principle* is used to determine three Doppler measurement parameters from the individual signal pattern, and the mean of these is displayed. In this process, it is not the frequency change in the individual ultrasound pulses that is displayed, but the phase change between two successive signals. If the velocity (v) of a blood component examined by ultrasound during the analysis time (t) remains constant between two ultrasound pulses, then the following relationship describes the phase change in the reflected ultrasound wave and the distance (d) it has covered:

$$v = d/t$$

If the interval between two ultrasound pulses is known, then simply measuring the phase change is

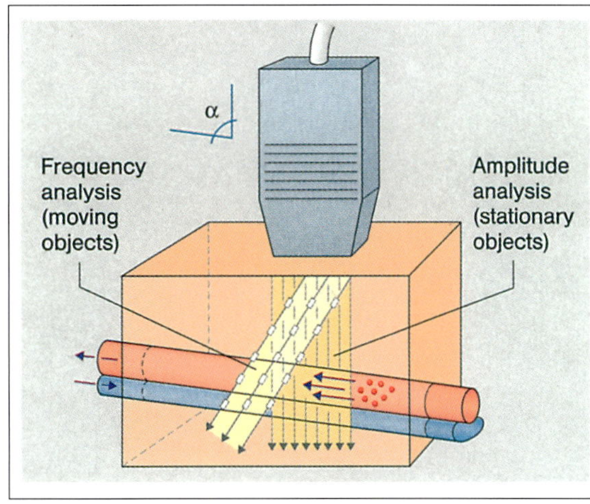

Fig. 1.**8** Color flow duplex sonography is a procedure that allows nearly simultaneous display of flow signals and tissue structures. In contrast to conventional duplex sonography, two ultrasound processes are used to construct the B-mode and color-mode images: frequency-domain and amplitude-domain analyses (for details, see text)

sufficient to determine the velocity. Usually, the following measurement parameters are determined:

- The mean velocity from the frequency distribution of all the frequencies, independent of their signal intensities *(velocity mode)*.
- The scatter range of frequencies around the mean *(variance mode)*.
- The total amplitude of the Doppler reflection within a sample volume *(amplitude mode* or *power–energy mode)*.

Only a single value is calculated for each pixel with a color-coded Doppler signal (Mitchell 1990). This represents a significant limitation on the amount of available information in comparison with what is possible with FFT signal intensity-weighted frequency analysis. It is possible to underestimate the true maximum velocity significantly, and the variability in flow velocities can also be significantly underestimated in turbulent flow conditions, when the velocity values within a sample volume are averaged out. In addition, artifacts due to wall movements are included in the calculated signal, and this can lead to serious displacement of the autocorrelation relationship (Fig. 1.**9**), since there is no objective method of noise reduction for individual frequencies.

The data are analyzed by a phase detector (Wells 2001). The phase change is given a digital value, and the value calculated for a specific sample volume over time is then displayed in color. The boundaries of the color scale used can be variably defined by the investigator, based on the expected flow velocities in relation to the Nyquist theorem. Equipment manufacturers all use the basic colors red and blue, but the representation of the direction of flow varies. In the illustrations shown in the present book, arterial blood flow is coded

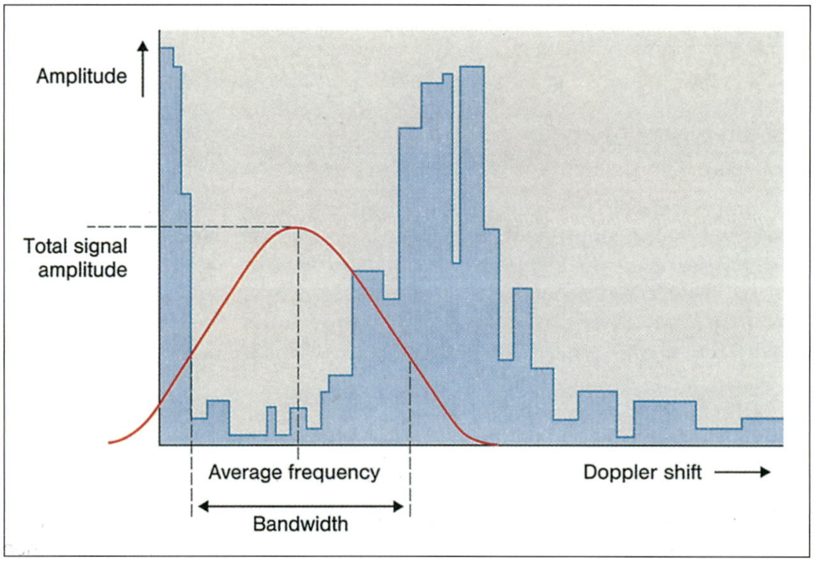

Fig. 1.**9** Comparing the parameters from an amplitude-weighted spectral analysis (fast Fourier transform [FFT] analysis) and a signal distribution calculated using autocorrelation. In the FFT spectrum, documentation of amplitude-weighted low and medium Doppler frequencies is distinct (e.g., wall movements and laminar flow within a blood vessel). In contrast, in color-coded duplex instruments using the frequency-domain analysis, the individual frequency intensities are disregarded due to the overall signal statistics (amplitude, bandwidth, and average frequency). This should be borne in mind when interpreting color-coded flow signals (adapted from Fehske 1988)

red, and venous blood flow is coded blue. The color saturation level increases with increasing flow velocities, and the boundaries between red and blue meet during aliasing. The intensity of the Doppler signal is represented by the color saturation level, and variable scales are available for high-amplitude and low-amplitude Doppler signals, so that better differentiation is possible in depicting the amplitude fluctuations often seen in higher-intensity signals. A simpler grid is sufficient for lower-amplitude signals. Application of the variance mode uses data from the frequency distribution per time interval analyzed subsequently. If the broadening of the Doppler spectrum increases at low intensities with an increasing degree of stenosis, the variance of the color-coded digital values increases too. This condition is often displayed by green signals scattered throughout the color-coded signals.

The advent of the color flow Doppler mode, with its visual and dynamic representation of blood flow, made it much easier to detect blood vessels, and in some regions made this type of analysis possible for the first time. In particular, it was not possible to identify the deep-lying vasculature or small-caliber arteries and veins within a reasonable amount of time until color-coded depiction of blood flow became available. Technically, color flow duplex sonography exploits all the potential of conventional duplex systems, and in addition makes it possible to represent ultrasound frequency shifts in different colors when moving corpuscular blood components are encountered. This information is superimposed on the B-mode image in a selectable display section. When the equipment is given a known setting that is appropriate to the area being examined and the inquiry being pursued, the various colors displayed can provide relevant information about the direction of the blood flow and the relative changes in its average velocity within the arteries. It is best to select the color setting of the duplex equipment to display the physiological flow within a blood vessel as a constant and strong color—i.e., without any aliasing. In straight blood vessels—i.e., those that maintain a constant angle to the Doppler axis—the color representation throughout the course of the blood vessel remains constant. In optimally accessible arterial segments, one can even detect in color-coded form the different average blood flow velocities, which present in a longitudinal profile with a fast component at the center of the blood vessel and slower components on its periphery. The color coding parameters have to be adjusted properly, and equipment-dependent factors have to be appropriately preselected for the requirements of a specific analysis and the specific inquiry. In order to obtain optimal arterial color representation in segments of interest, the following color adjustments need to be selected:

- Color
- Gain or power
- Pulse repetition frequency (PRF)
- Filter
- Angle
- Baseline
- Scale
- Variance

Zones of pathologically elevated blood flow velocities can be quickly detected due to the distinctly lighter shade of the basic color that has been selected, or due to a color change produced by aliasing. However, it should be noted that a change in the vascular axis relative to the transducer, as in the naturally curving course of an artery, will also lead to a change in the color flow Doppler information (Fig. 1.**10**).

Power Doppler Imaging

Power Doppler imaging (PDI) is a technique that displays the amplitude of Doppler signals. The color and brightness of the signals are related to the number of blood cells producing the Doppler shift. The greater sensitivity of power Doppler imaging for detecting blood flow in comparison with color Doppler flow imaging is due to several factors. Noise can be assigned to a homogeneous background, thus allowing the gain to be increased over the level of color Doppler flow imaging. Moreover, in power Doppler imaging more of the dynamic range of the Doppler signal can be used to increase sensitivity. Power Doppler is also less angle-dependent than CDFI, thus allowing better display of curving or tortuous vessels. Finally, relying on the Doppler amplitude means there is no aliasing; this improves the display of vessel wall pathology in areas of turbulent flow. Reports on the clinical value of power Doppler imaging (Steinke et al. 1996) have demonstrated distinct advantages with this technique in assessing the surface structures of atherosclerotic plaque (Fig. 1.**11**) (Griewing et al. 1996).

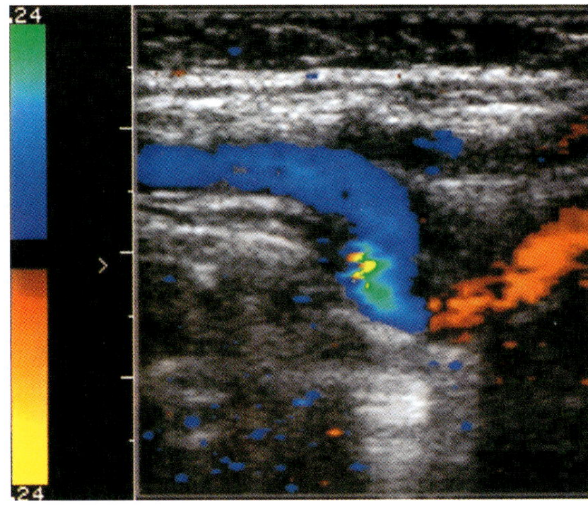

Fig. 1.**10** Kinking of the internal carotid artery as displayed in a color flow duplex system. Note the change of color at the top of the arterial loop due to the change of flow direction relative to the transducer, despite physiological flow conditions

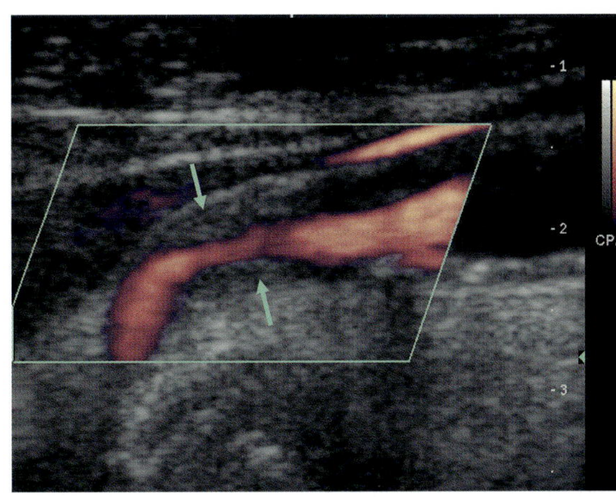

Fig. 1.**11** Power Doppler display of a stenosis in the left internal carotid artery. Simultaneous B-mode imaging (duplex mode) allows depiction of homogeneous plaque material (arrows).

Duplex Systems

In duplex systems, a pulsed wave Doppler system (single-channel or multi-channel) is combined with an echo-impulse system (*Arbeille et al. 1984, Barber et al. 1974, Blackshear, Jr. et al. 1979*) (*Hennerici and Freund 1984*). This type of equipment primarily uses linear-array, curved-array and sector transducers. In most systems, the angle between the Doppler ultrasound beam and the vascular axis can be modified continuously or incrementally. This allows one to locate the Doppler sample volume within the specific area of interest in a B-scan. Simultaneous use of echo-impulse and Doppler sonography techniques may cause quality reductions in both procedures. However, the technique is very useful for purposes of anatomical orientation.

Due to the split between the image and the Doppler pulses, a reduction in the pulse repetition rate can occur. This results in a reduction of the maximally detectable flow velocity. Under visual control in two dimensions, the sample volume is usually first positioned on an electronic beam; with the image in freeze-frame, a Doppler investigation is then carried out. If the Doppler transducer is not integrated with the B-mode transducer, as is usually the case with linear and phased-array transducers, differences between the two transducer transmission frequencies can lead to artifacts in positioning the sample volume. Unnoticed changes in the transducer position or in the area investigated between measurements are additional error sources.

In the conventional duplex procedure, Doppler analysis is carried out using a sample volume selected on the basis of its position in the B-mode scan. Attempts to use *multigate procedures*, however, did not prove to be practicable. Although this method allows evaluation of Doppler sample volumes at different depths, producing a wealth of flow velocity graphics, the information proved to be confusing and ambiguous for diagnostic purposes (*Hennerici and Freund 1984, Keller et al. 1976*), (*Reneman et al. 1985, Reneman et al. 1986*). However, because their measurements are so precise, these systems are still of significance for scientific purposes.

Conventional duplex-system equipment consists of a B-mode scanner that incorporates a pulsed wave Doppler mode. In *color flow duplex sonography*, it is also possible to depict the velocity and the direction of flow in blood vessels with almost simultaneous superimposition on the B-mode image.

Duplex-system transducers usually operate with a pulsed wave Doppler mode. Using a spatially well-defined *sample volume*, the pulsed wave Doppler mode is able to detect local flow velocity phenomena that are mainly free of any superimposed artifacts. To examine small-caliber arteries and veins (foot, lower arm, penis and scrotum), the pulsed wave Doppler ultrasound beam should be narrowly adjusted and precisely focused. By analyzing the Doppler frequency spectrum, one can determine the hemodynamic effect of wall deposits that are shown in B-mode images, calibrate arterial stenoses, and confirm any proximal and distal changes shown by the Doppler flow curve.

Duplex-system examinations are useful for all types of vascular disease affecting both arteries and veins. Measurements made using the Doppler mode to produce qualitative and quantitative assessments of vascular stenoses and to localize and measure occlusions require very careful arterial scanning, which is time-consuming and—especially in small-caliber blood vessels—often unsuccessful. Usually, a screening test using continuous-wave Doppler sonography is carried out to provide general orientation, especially when dealing with the cerebral arteries and arteries in the extremities, in order to identify a blood vessel segment requiring closer investigation.

Compound Imaging

Speckle artifact can be described as noise caused by interference from echoes returning from tissue. Speckle is an inherent artifact and reduces image contrast and detail resolution.

The results of in-vitro B-mode imaging studies have shown that combining scanning information from more than one angle can lead to a significant reduction in the amount of speckle (Barry et al. 1997, Jespersen et al. 1998). This technique, known as compounding, has been used experimentally to increase the signal-to-noise ratio in three-dimensional (3D) ultrasound imaging (Barry et al. 1997, Rohling et al. 1998). Compound imaging has also been applied in an in-vivo 3D system registering irregularly sampled ultrasound images obtained from different perspectives to maximize vessel wall information in the carotid artery (Meairs et al. 2000a). The major obstacle to the implementation of compound imaging in routine clinical practice, however, has been the time-consuming offline processing of image data required. Recently, real-time spatial compound imaging has been made possible through substantial improvements in the computational power of ultrasound equipment. Real-time compound imaging uses computed beam-steering technology to steer ultrasound beams "off-axis," providing multiple transmission angles in the course of a single scan. Pipeline signal-processing architecture is used to render these steered frames accurately into the appropriate display geometry and to update the compound image in real time as each new frame is acquired. Initial experience with this technique in the field of vascular ultrasound imaging (Jespersen et al. 2000) demonstrated its ability to improve the visualization of atherosclerotic plaques in vitro. These experiments have been complemented by studies in patients with carotid artery disease to show better intraobserver and interobserver reproducibility (Kofoed et al. 2001) and artifact reduction, with improved assessment of the plaque surface and texture (Kern et al. 2004) (Fig. 1.**12**).

B-Flow Imaging

A new method of detecting blood flow with ultrasound (B-flow) was developed in 2000. With this technology, direct visualization of blood reflectors has made B-mode flow imaging possible without limitations of Doppler technology such as aliasing, signal dropout at orthogonal detection angles, and wall filter limitations (Henri and Tranquart 2000, Weskott 2000). B-flow visualizes real-time hemodynamic flow relative to stationary tissue. These attributes of B-flow imaging make this technique an important additional tool in the evaluation of vascular disease.

B-flow images are generated by using digitally encoded ultrasound technology consisting of a transmission encoder and a reception decoder in a digital beam former that provides electronic array focusing. A small number of digitally encoded wideband pulses are transmitted into the body for each scan line. Unlike color imaging techniques, in which the typical packet size is 10–12, B-flow imaging uses a packet size as small as two to four. Directly after receiving the reversed pulses, the decoder performs pulse-length compression ("coded excitation") on the acoustic data and then carries out clutter suppression filtering. The rest of the processing is essentially the same as in the conventional B-flow mode.

Coded excitation is a technique that increases the transmission energy by as much as one order of magnitude without compromising transverse resolution and is therefore especially suited for high spatial resolution and high temporal resolution imaging of echo sources that are normally weak (such as red blood cells). Through the digital encoder, the scanner transmits not one, but a sequence of *n* wideband pulses in accordance with a specific pattern referred to as a *code*; a decoder on the receiving side is used to compress the returning echo effectively back into a single pulse that has nearly the same resolution but *n* times more energy. If the coded sequence received and the code sent are exactly matched, the response is a pulse of amplitude *n* times greater than a single uncoded pulse. During in-vivo scanning, the returning signal represents the sum of reflections from multiple sources in body tissue, so that the output of the sum should equal

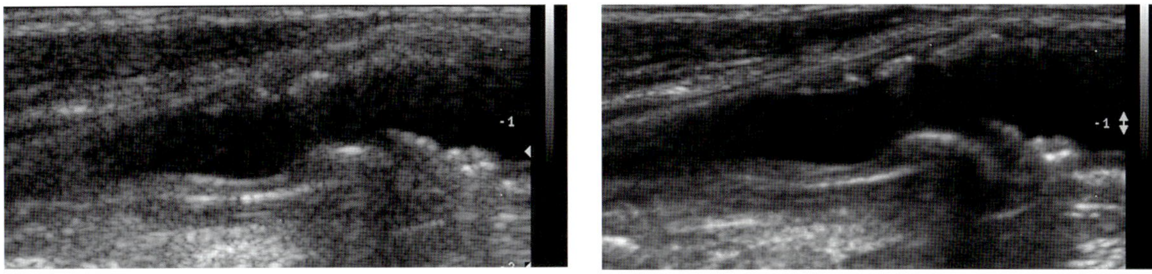

a b

Fig. 1.**12** **a** A high-resolution B-mode scan of a heterogeneous plaque on the wall of the internal carotid artery. **b** A real-time compound image of the plaque shown in **a**, from the same scanning plane. Note the considerable reduction in artifacts, allowing improved assessment of the surface and texture of the plaque

the sum of the outputs from individual contributing reflectors.

Clutter suppression filtering. For each of the coded transmissions (e.g., two), a stream of acoustic backscatter data from the insonated anatomy is received and coded. These data are stored in a buffer in the equalization filter, which then subtracts a fraction of the second transmission from the first transmission. This process suppresses any large and slow-moving tissue clutter component relative to any moving blood-echo component.

B-flow has been shown to be very effective in visualizing hemodynamic flow and in detecting stenotic lesions in the cervical carotid artery (Umemura and Yamada 2001) (Fig. 1.**13**). Moreover, B-flow can provide detailed hemodynamic imaging of phenomena such as bloodstream swirls that is not possible with color Doppler flow imaging or power Doppler imaging. B-flow also visualizes bloodstream flow patterns at the site of wall ulcerations or vascular tortuousness. The exact nature of these visualized flow patterns remains to be elucidated. Although it has been suspected that the form and intensity of poststenotic jets as demonstrated by B-flow imaging may allow characterization of carotid stenosis, this is not supported by recent data (Bucek et al. 2002).

One limitation of B-flow imaging is that excessive vessel pulsations can lead to movement of the surrounding structures, so that the vessel wall is sometimes ill-defined (Weskott 2000). A further disadvantage is decreased sensitivity of B-flow imaging with increasing depth, because of the strong dependence on signal strength.

Transducers

A wide range of ultrasonic probes are available for neurovascular imaging. These are generally electronic phased-array or linear-array transducers, which have replaced early rotating or oscillating mechanical transducers. Common to these transducers is the B-mode image construction used, which consists of individual scan lines that are adjacent to one another in either a parallel (linear) or sector (radial) arrangement. The image quality increases with the density (number of scan lines) and also with the frame rate (number of image repetitions). However, both parameters cannot be increased indefinitely, since their product yields the pulse repetition frequency (PRF), which in turn is decreasing with the image depth.

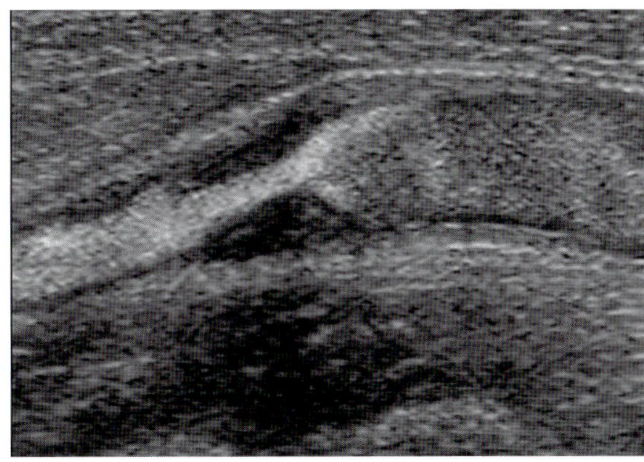

Fig. 1.**13** A B-flow image of carotid stenosis

$$\text{PRF (Hz)} = \text{frame rate x image line density}$$

Electronic transducers (linear-array, curved-array) incorporate multiple small converters that can be excited either sequentially or segmentally, creating an ultrasound image of a rectangular or sector- shaped composition (Fig. 1.**14**).

Adjustable time-gain compensation (TGC) provides selective amplification of attenuated reflections from tissue cross-sections at various depths. Also, the total amplification (gain) can usually be continuously regulated. The selected image can be enlarged, and is displayed on the monitor in real time. A scale, consisting of electronically produced lines that indicate distance, allows measurement of the echo structures depicted.

The advantage of transducers that produce a sector image (using the phased-array principle) is a small

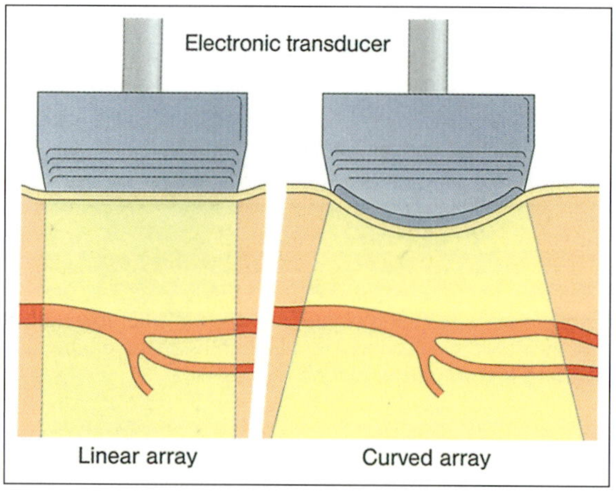

Fig. 1.**14** The linear-array (linear) and curved-array (concave) transducers, and their ultrasound fields. The linear transducer's contact area is relatively long, while that of the concave transducer matches the body surface better. Despite their sector-shaped ultrasound field, the resolution of curved-array transducers is nearly as good as that of linear scanners. This type of concave transducer is becoming increasingly popular

contact area; however, they have the disadvantage that they can only assess the vascular structure orthogonally on a limited and selective basis. Transducers with a larger surface contact area (linear array) allow examination of a larger vascular segment under the optimal 90° angle. The „curved-array" transducer has a geometric structure in between that of a linear and a sector-shaped ultrasound field. The near field is broader and the depth resolution is also better than that of linear scanners.

The choice between the various transducers depends on the vascular segment to be examined. Blood vessels that are parallel to the body surface are preferably examined with linear scanners. On the other hand, if there is only a small ultrasound window available and a larger and deeper vascular segment needs to be examined, then sector transducers are necessary – e.g., to examine the abdomen, the pelvic vasculature, or the bifurcation of the carotid artery when it lies in a far cranial position.

As a supplementary information source, intravasal ultrasound probes are also used in vascular diagnostics. These provide a better display of plaque structures using an additional image emanating from the center of the blood vessel (Fig. 1.**15**). Recent advances include development of 2D electronic matrix arrays capable of displaying 3D images in real time.

Signal Analysis

Audio Signal Analysis

The simplest way of evaluating a Doppler signal is to interpret the audio impression reproduced on a loudspeaker (Barnes et al. 1981). The direction of flow can also be represented acoustically using two speakers, with flow away from the probe and flow toward the probe being distributed separately to the two speakers. Even with these simple methods, an experienced investigator is able to identify factors that are significant in evaluating the condition of a blood vessel. Audio analysis is indispensable for selecting and optimizing the signal that is subsequently to be investigated.

Zero-Crossing Counter

Using the directional Doppler technique, directions of flow "toward the probe" and "away from the probe" are registered on separate channels. These Doppler signals are depicted graphically as a two-dimensional diagram, with the value of the Doppler shift on the y-axis plotted against time on the x-axis. When the incident angle α is known and a constant baseline is maintained, both the Doppler shift itself and its corresponding velocity can be read off.

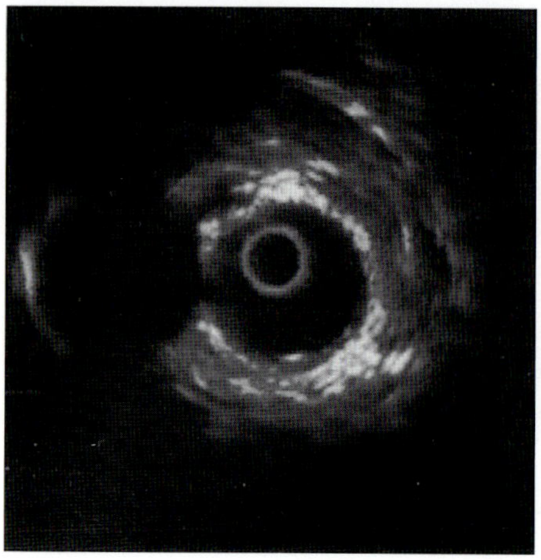

a

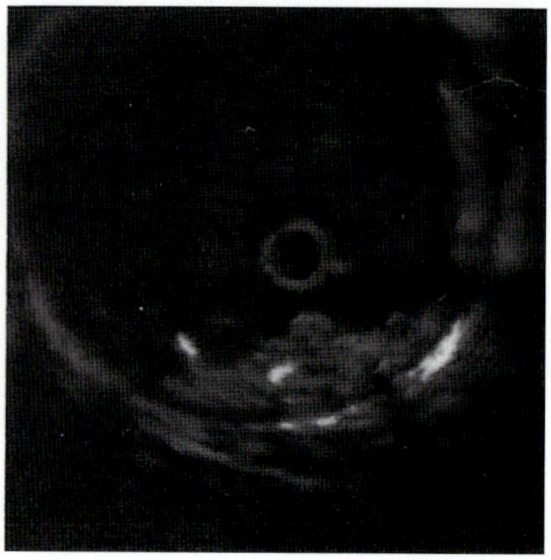

b

Fig. 1.**15** Intravascular sonography, showing (**a**) a normal common carotid artery and (**b**) a heterogeneous plaque deposit

In fact, however, using what are termed zero-crossing counters *(zero-crossers)*, which display less information than the envelope curve, only an average frequency for the Doppler spectrum is documented. Many studies have shown that this simple procedure does not even enable one to determine the average frequency of the Doppler spectrum corresponding to the average flow velocity (Reneman et al. 1973). Instead, a systematic error develops, which depends on the form of the received signal and increases proportionally to the true average frequency (Fig. 1.**16**). This is especially detrimental when stenoses are present. In the low-frequency range, high-amplitude signals—movements of the vessel walls, for example—are noticeable sources of error, and using suitable high-pass filters also excludes important information provided by the Doppler spectrum (e.g., areas of turbulence). For a long time, these disadvantages were outweighed by the advantages of having a simple and cost-effective technology. The envelope curve recording the analogue signal of the Doppler spectrum (known as the *analogue procedure)* provides an adequate overview of changes in the average flow velocity. This procedure is still widely used to provide general orientation.

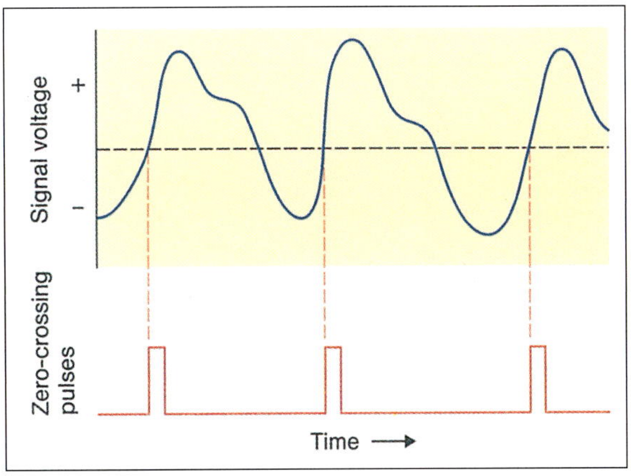

Fig. 1.**16** A zero-crossing histogram (adapted from Zagzebski and Madsen 1982)

Time-Interval Histograms

The time-interval histogram, which preceded amplitude-weighted spectral analysis up to the 1980s, is also based on the zero-crossing counter principle. This method analyzes not only the time at which the Doppler signal crosses the baseline in a particular direction, but also the time interval to the previous crossing (Fig. 1.**17**). This produces an image that contains more information than the envelope curve.

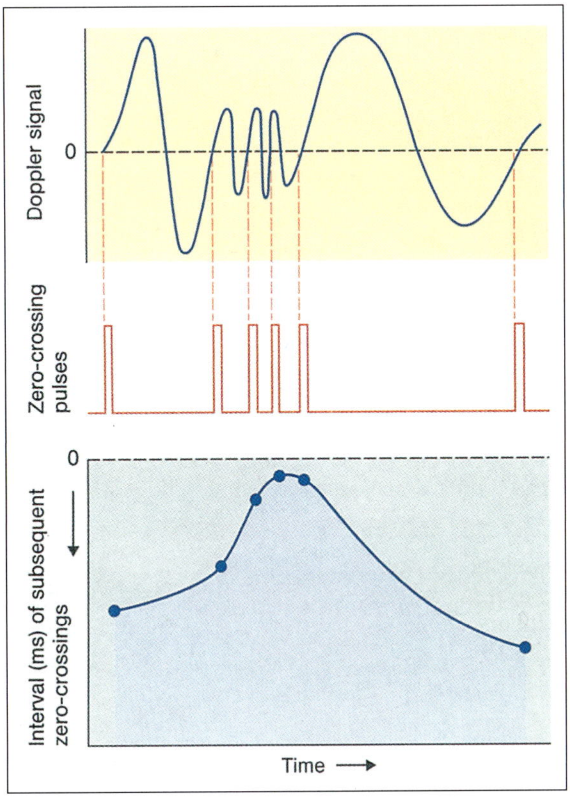

Fig. 1.**17** Time interval histograms. An increase in the Doppler frequency leads to shorter time intervals between the zero-crossings (adapted from Zagzebski and Madsen 1982)

Spectral Analysis

The frequencies contained within a Doppler signal, the time at which they occur, and their intensity can be calculated and depicted using a *fast Fourier transform* (FFT) (Felix et al. 1976, Lewis et al. 1978, Spencer and Reid 1979) (Fig. 1.**18**).

The basis of the Fourier transform is the Fourier theorem, according to which a periodically repeating event can be almost exactly defined by superimposing various sinus waveforms. This technique is used both for continuous-wave and pulsed-wave Doppler investigations. The frequency of the FFT analysis is plotted as the ordinate, the time as the abscissa, and the amplitude of the signal is documented as a gray scale, or, alternatively, in color-coded format. The flow direction is indicated as a positive signal (above the baseline) or a negative signal (below the baseline).

To analyze complex flow phenomena precisely, various parameters are calculated and displayed digitally, including:

- Maximum Doppler frequency *(peak frequency)*
- Average Doppler frequency *(mean frequency)*
- Doppler frequency with the maximum amplitude *(mode frequency)*
- Bandwidth *(spectral broadening)*
- Spectral window

As can be seen from the individual *power spectra*, all of these parameters vary from one moment of analysis to

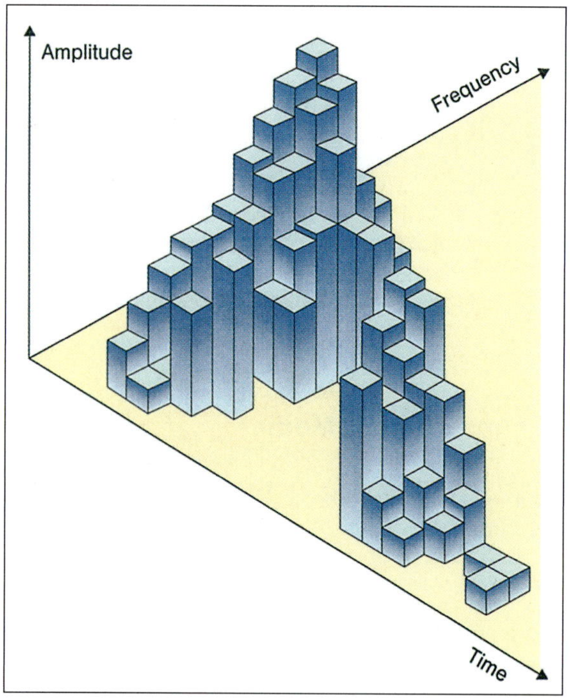

Fig. 1.**18** A three-dimensional diagram showing the different components in a fast Fourier transform spectrum (adapted from Seitz and Kubale 1988)

the next. Each power spectrum depicts the frequency distribution (on the abscissa) at a particular time, plotted against its amplitude (on the ordinate). The amplitudes are raised to powers of ten to provide clearer differentiation between the individual components.

It should be noted that the spectrum is particularly influenced by the area measured, as well as a variety of other factors, so that it is subject to significant variations, especially in pulsed-wave Doppler systems. Due to laminar flow, a selected sample volume that is too small can incorrectly enlarge the bandwidth or reduce the spectral window, especially when different velocities are being analyzed (this also applies to vascular branching points). In addition, a clear distinction between background noise and signal should be ensured by selecting an appropriate noise reduction filter—otherwise, low-amplitude but high-frequency signals reflecting significant stenosis may be missed. If the signal is weak, it can still be successfully differentiated in small steps by changing the amplitude scale *(compression)*. In contrast, the global amplification *(gain)* works like a loudspeaker's volume control, without any amplitude weighting of the Doppler spectra. Since the signal-to-noise ratio deteriorates with increasing signal amplification, care should be taken to make the settings as precise as possible—taking full advantage of global amplification to detect any remaining flow signals only makes sense in particular circumstances such as differentiating between subtotal stenosis and complete occlusion.

In the same way that certain growth processes are best documented using time-delay photography and rapid movements are best represented by slow-motion photography, various analysis times are also required for individual spectral components. With increasing analysis time (T), a Doppler shift (Δf) can therefore be precisely determined:

$$\Delta f = 1/T \text{ or } T = 1/\Delta f$$

A signal observation time of 5 ms would therefore be sufficient to define a Doppler shift of 200 Hz (Δf) precisely:

$$T = 1 : 200\,\text{Hz} = 5\,\text{ms}$$

When calculating the spectrum, care should be taken to select adequate time intervals for the analysis (between 4 ms and 40 ms). It is important to note that the analysis time is inversely proportional to the frequency resolution. This means that with an analysis time of 4 ms, the discrimination capability is 280 Hz, and with 16 ms it is 70 Hz.

It should be noted that the precision of the measurement, which depends on the analysis time, is further affected by the ultrasound frequency and the transit distance (defined by the sample volume length). Imprecision in measurement that occurs during the observation time therefore has to be balanced out by the *transit time effect*—i.e., in regions of increasing velocity, the observation time also has to increase. Nevertheless, the analysis time cannot be extended indefinitely, since flowing blood can only be considered constant for a limited time (approximately 10–15 ms). An excessive analysis time would reduce the precision of the velocity measurement.

Statistical analyses of Doppler signals are complex, and with the exception of specific investigations, have not had any diagnostic significance (e.g., Laplace transforms).

Three-Dimensional Ultrasound

Three-dimensional imaging with ultrasound was first described in 1956 for stereoscopic observation of body structures (Howry et al. 1956). Several years later, a 3D ultrasonographic system for display of the human orbit was reported (Baum and Greenwood 1961). This technique obtained serial parallel ultrasound images and then created a 3D display by stacking sequential photographic plates of the images. Later, Dekker and coworkers reported 3D ultrasound imaging of the heart with a mechanical tracking device to register images for volume reconstruction (Dekker et al. 1974). Although rapid developments in ultrasound scanning equipment and in computer hardware and software have occurred since then, the basic requirements for producing 3D ultrasonographic reconstructions have not changed: adequate two-dimensional images, a known spatial relationship, and orientation of each two-dimensional image to a common external reference point, as well as techniques for volume reconstruction and visualization.

Image Acquisition

Many multidimensional ultrasound systems use frame-grabbers to digitize analog data from ultrasound machines. Data acquisition with video recorders has also been proposed (Barry et al. 1997), although this technique can result in loss of image quality on redigitization of captured video data. Some manufacturers have implemented "internal" 3D systems that have the distinct advantage of using raw digital ultrasound data to reconstruct 3D ultrasound images. However, these systems may be limited by a lack of true image registration (see below).

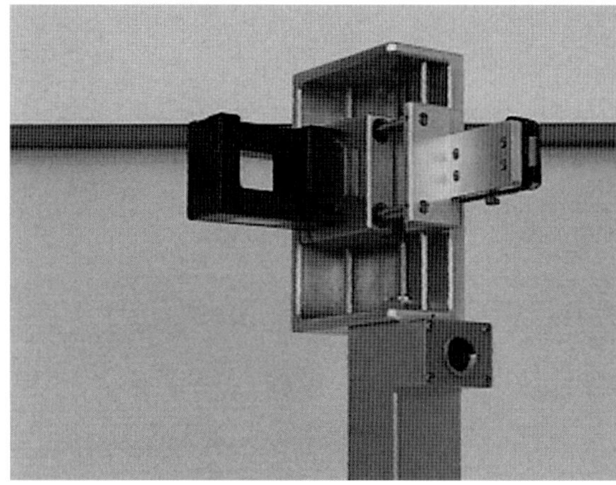

Fig. 1.**19** A high-precision parallel motor system used for three-dimensional ultrasound data acquisition. The ultrasound transducer (not shown) is mounted on the motor system. The system is adaptable for a wide variety of transducers from different manufacturers

Monitoring Spatial Position

A number of methods have been developed for monitoring the spatial position of the ultrasound transducer during volume data acquisition. These have included mechanical arms (Geiser et al. 1982), multiposition tripods (Steinke and Hennerici 1989), catheter pull-back (Mintz et al. 1993), acoustical spatial locating systems (King et al. 1990, Rosenfield et al. 1991, von Birgelen et al. 1997), and motor-driven rotational systems (Eiho et al. 1987) and parallel systems (Franceschi et al. 1992, Hell et al. 1993). Sweep volume systems (Deng et al. 1996, Steiner et al. 1994) employ a specially constructed probe to mechanically sweep the B-scans through a volume of interest and have been successfully applied in a variety of clinical settings, such as imaging of the fetus (Merz et al. 1995), gallbladder (Fine et al. 1991), breast (Moskalik et al. 1995), and heart (Salustri and Roelandt 1995). Dedicated tomographic transducers (Hamper et al. 1994) have also been reported, but have only had limited application. In semiregistration techniques, it is assumed that the transducer follows a constant planar sweep, and an attempt is made to estimate the distance between the individual image slices for 3D volume reconstruction.

Three-dimensional ultrasound image acquisition using position and orientation measurement (POM) devices capable of tracking scanheads in six degrees of freedom (6-DOF) has received considerable attention (Barry et al. 1997, Detmer et al. 1994, Hodges et al. 1994, Leotta et al. 1997). This technique allows "freehand" scanning, with image data being collected from different perspectives, and potentially offers the ability to maximize tissue information that is not readily available from one imaging plane alone. Among the various methods of monitoring the spatial position of images, computerized motor-driven systems and POM scanhead tracking have been successfully implemented for 3D ultrasonographic assessment of vascular disease. Interesting results have also been obtained with semiregistration techniques.

Computerized Motor-Driven Systems

Computerized motor-driven systems are equipped with both electrocardiographic (ECG) and respiratory gating, allowing exact temporal data acquisition from arteries. An example of a computerized motor for parallel image acquisition is shown in (Fig. 1.**19**). It has good precision and is capable of delivering ECG-gated ultrasound slices as thin as 0.2 mm, thus approaching the theoretical limits of ultrasound resolution. ECG-gated imaging of the carotid arteries using this 3D ultrasound technique has shown a high level of accuracy with little translational error. Since the ultrasonographic data obtained lie along one orthogonal plane, reconstruction techniques are identical to those used in 3D applications for computed tomography and magnetic resonance imaging.

Semiregistration Techniques

Three-dimensional applications include those that do not use a method of registering the transducer coordinates relative to the acquired ultrasound images. Reconstruction of volume data assumes an evenly distributed fan movement of the transducer along one imaging plane. Assuming an equidistant slice thickness between each acquired image, the reconstruction algorithm depends on both the image frame rate and the speed of the fan movement. Since the true voxel size is unknown, measurements of volume data are not possible. However, excellent 3D visualizations of arteries from various tissues have been achieved with this semi-3D technique (Postert et al. 1998), and these often compare well with those obtained with magnetic reso-

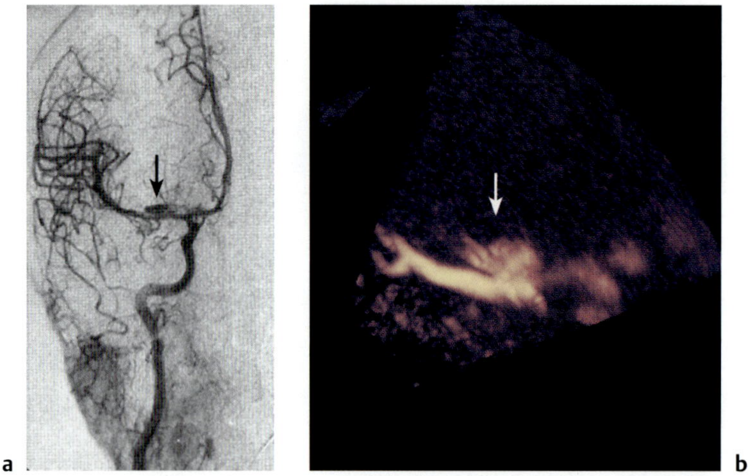

Fig. 1.**20** Radiographic angiography (**a**) and three-dimensional transcranial power Doppler imaging (**b**) of an aneurysm in the middle cerebral artery (arrows). The three-dimensional ultrasound reconstruction provides excellent visualization of the aneurysm. Note the rough surface contour of aneurysm on the power Doppler image, reflecting thrombotic material

nance angiography. The main advantage of this approach is speed, as the entire process of image acquisition, volume reconstruction, and volume rendering can take less than a minute to accomplish. This can be valuable in critical care situations. For example, in patients with acute subarachnoid hemorrhage, 3D renderings of a bleeding aneurysm can be achieved rapidly in the emergency room (Perren et al. 2004) (Fig. 1.**20**). New approaches for estimating the probe trajectory based on decorrelation algorithms of B-scan images (Prager et al. 2003) may be of particular value in improving 3D ultrasound applications using semiregistration techniques.

POM Scanhead Tracking

The precise spatial location of fixed feature points can be identified using a 3D ultrasound imaging system based on magnetic position and orientation measurement (Detmer et al. 1994). Since one of the key issues in 3D ultrasonography using POM devices is accurate registration of ultrasound images to a regular 3D volume, extensive work has been undertaken to achieve optimal calibration of 6-DOF systems (Barry et al. 1997, Meairs et al. 2000a, Prager et al. 1998). Sophisticated automated methods of 6-DOF calibration have also been described (Rohling et al. 1998).

Reconstruction Techniques

Reconstruction of multidimensional ultrasound data acquired with mechanical devices is straightforward, since each image lies along one plane and the distance between the individual slices is constant. In this case, the reconstruction involved is comparable to that used in magnetic resonance imaging applications. Reconstruction of irregular, arbitrary-plane 6-DOF ultrasound data, however, can be quite complex. Since images are acquired in an arbitrary orientation and position with 6-DOF scanning techniques, the individual slices must be inserted at various positions and rotations within the volume data set. This can lead to regions of missing data if a blood vessel has not been completely scanned. Moreover, since ultrasound reflection and backscatter intensities are angle-dependent (Picano et al. 1985), a further problem arises when handling overlapping data obtained from different insonation angles. Since these data occupy nearly identical voxel coordinates, there may be uncertainty concerning which intensity best represents the tissue structure under investigation.

Several approaches have been presented for volume reconstruction of ultrasound data acquired with a POM device. One group addressing this problem has proposed that individual voxel intensities can be calculated through distance-weighted averaging of the sum of pixels registered within the voxel radius (Barry et al. 1997). Another technique uses ellipsoid Gaussian convolution kernels to handle arbitrary irregularly sampled ultrasound data (Meairs et al. 2000a).

Visualization Techniques

The quality of 3D volume rendering in vascular applications will be judged according to the ability of the rendered images to depict pathological conditions in comparison with established techniques such as magnetic resonance angiography and computed tomographic (CT) angiography. This will ultimately determine whether 3D ultrasound technology will establish a place for itself in routine clinical practice. There are three basic techniques for visualizing 3D ultrasound data: planar slicing, surface fitting, and volume rendering.

Planar slicing. Orthogonal slicing of 3D ultrasound volume data produces two-dimensional images in the axial, sagittal, and coronary planes—i.e., it samples data along one of the three x, y, or z coordinate axes. Arbitrary slices at oblique angles require resampling of the volume data, with some form of interpolation of pixel values that may lie between individual voxels. The quality of such oblique images will depend on the type of interpolation used, as well as on the way in which the 3D volume was acquired and reconstructed. Data slicing can be very useful for quickly reviewing a series of two-dimensional images to estimate the accuracy of an acquisition technique. Two-dimensional slicing may also allow visualization of planes not accessible during the ultrasound examination. Moreover, individual planes can be selected for purposes of standardization and for better comparison in follow-up studies.

Surface fitting. A common technique used in 3D data visualization is to display a constant density surface, also known as an "isosurface." Given a 3D grid of density measurements, these can be contoured to display the surfaces of anatomical structures. Polygons are used to describe the surface of a contour, from which a shaded surface representation is calculated. Shading values can be determined from a light source model, vertex shading values, or polygon shade values. Using a scan line algorithm, the shading model usually combines diffuse reflection and depth cueing. Polygons are shaded using either constant shading, in which each polygon is given a constant intensity, or with Gouraud shading, in which the intensity is computed at each vertex and then interpolated over the polygon. Surface fitting is used in 3D vascular applications (Meairs et al. 2000b), particularly for visualizing the extracranial and intracranial arteries (Fig. 1.**21**).

Fig. 1.**21** A three-dimensional reconstruction of an atherosclerotic plaque in the internal carotid artery, scanned with compound B-mode imaging. A grid on the plaque surface demonstrates a surface defect, corresponding to a traumatic plaque rupture

Volume rendering. Volume rendering generates visualizations of volumetric data by computing two-dimensional projections of a colored, semitransparent volume. Parallel rays from any given direction are cast through the volume onto a viewing plane. A maximum-intensity projection, commonly used in 3D magnetic resonance angiography, is a simple form of volume rendering in which the value of each pixel in the viewing plane is the maximum data value along the corresponding ray. Volume rendering can also incorporate depth cueing and opacity to affect the appearance of the 3D image. In addition, colors and opacities can be assigned to arbitrary data ranges, simulating the appearance of materials contained within a volume (Fig. 1.**22**).

Fig. 1.**22** Volume rendering of a high-grade carotid stenosis. The data were acquired using scanhead tracking with six degrees of freedom (6-DOF). Irregularly sampled power Doppler images served as the raw data for the reconstruction. Note the excellent definition of the stenosis and the surrounding plaque material. The walls of the carotid artery are well delineated

Four-Dimensional Applications

Four-dimensional ultrasound is a temporal extension of 3D ultrasound to assess moving tissue. In addition to the requirements for information regarding the position of each image in space, four-dimensional imaging also requires exact data concerning the time of each image acquisition. This allows, for example, reconstruction of vessel walls or vessel flow characteristics during a cardiac cycle. There are several areas in the field of vascular diagnostics in which four-dimensional ultrasound data may provide information to help elucidate the pathophysiology of arterial disease.

Analysis of atherosclerotic plaque motion may provide insights into plaque modeling, as well as into the mechanisms of plaque rupture with subsequent embolism. It has been postulated that plaque surface movement may be attributable to deformations resulting from crack propagation of multiple local internal tears in the plaque. The identification of local variations in surface deformability with four-dimensional ultrasound may provide information concerning relative vulnerability to plaque fissuring or rupture (Meairs and Hennerici 1999).

Another promising four-dimensional ultrasound technique is quantitative four-dimensional color Doppler flow imaging (4D-CDFI). Although it is still in the research phase, 4D-CDFI could have potential applications for multidimensional analysis of vessel wall shear stress, quantification of blood flow, and acquisition of quantitative in-vivo flow data for numerical studies of carotid plaque instability.

Contrast Imaging

The ability of intravenous contrast media to increase the echogenicity of flowing blood has been known for some time (Ophir and Parker 1989). However, it is only in the last few years that increasing interest has developed in the use of echo-enhancing agents in the assessment of vascular disease.

Commercially available contrast agents consist of microbubbles with average diameters ranging from 3 µm to 6 µm in concentrations typically on the order of 10^8 microbubbles/mL. The microbubbles are normally stabilized against dissolution by surfactants, phospholipids, or a surface layer of partially denatured albumin. Current contrast agents can enhance the ultrasound signal by 10–30 dB (Burns 1997), allowing flow to be detected in deeper and smaller vessels (Table 1.**2**).

The first generation of ultrasound contrast agents consisted of air-filled microbubbles. Examples of such agents are Albunex (Molecular Biosystems, Inc., San Diego, California, USA), which is produced by controlled sonication of a 5% human serum albumin solution, and Levovist (Schering, Inc., Berlin, Germany), which is a galactose-based agent stabilized by 0.01% palmitic acid. Albunex has been approved by the Food and Drug Administration (FDA) for use in the United States, and Levovist is approved for use in Europe. However, because of the low concentration of air gases in the systemic circulation, the air contained inside these agents quickly diffuses out of the microbubbles into the body. The type of gas inside the bubble determines the persistence time in the circulation, and can also affect the backscattered signal. A second generation of ultrasonic contrast agents was therefore developed, consisting of microbubbles containing less soluble gases, such as perfluorocarbons (Optison) or sulfur hexafluoride (SonoVue). These agents have now also been approved for use in Europe.

Nonlinear Characteristics of Microbubbles

An ultrasound acoustic wave consists of alternating high and low pressures at frequencies of 1.5–10 MHz. When an acoustic wave encounters a microbubble, it alternately compresses the microbubble under positive pressure and expands it under negative pressure. In the positive portion of the wave, the microbubbles are compressed in a different way from the way in which they expand during the negative portion. This results in asymmetric, nonlinear bubble oscillation. This asymmetry produces harmonics, which can be used to enhance the signals from bubbles (Fig. 1.**23**).

Microbubble Destruction

Bubbles in a liquid tend to diffuse and disappear unless they are stabilized by some type of shell. Once the shell is disrupted, the gas inside will diffuse into the surrounding fluid. The mechanical index (MI), originally defined to predict the onset of cavitation in fluids, also gives an indication of the likelihood of bubble destruction. The MI is defined as:

$$\text{MI} = \text{peak negative pressure}/\sqrt{(\text{ultrasound frequency})}$$

or

$$\text{MI} = \text{peak negative pressure} \cdot \sqrt{(\text{period of ultrasound wavelength})}$$

The more the bubble is expanded (peak negative pressure) and the longer it is expanded (period of ultra-

Table 1.**2** Commercially available ultrasound contrast agents

Name	Manufacturer	Gas	Shell
Optison	Mallinckrodt Medical, St. Louis, MO, USA	Air and perfluoropropane	Denatured albumin
Definity	Dupont Pharmaceuticals, North Billerica, MA, USA	Air and perfluoropropane	Lipid
SonoVue	Bracco Imaging, Milan, Italy	Air and sulfur hexafluoride	Lipid
Levovist	Schering, Berlin, Germany	Air	Stabilized with surfactant
Imagent	Schering, Berlin, Germany	Air and perfluorohexane	Stabilized with surfactant
Sonazoid	Nycomed Amersham, Amersham, UK	Perfluorocarbon	Lipid

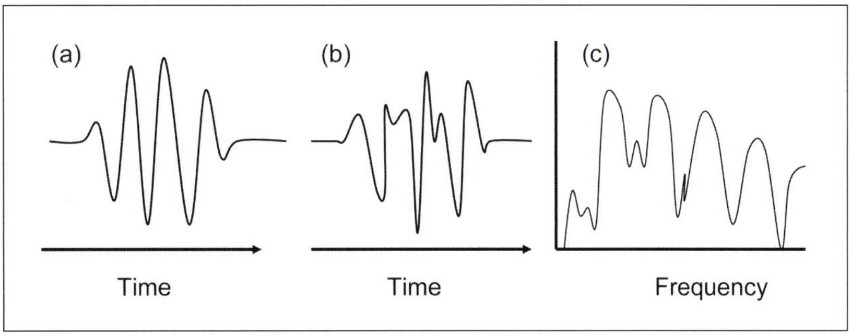

Fig. 1.**23** Ultrasound characteristics of microbubbles: (**a**) an incident acoustic wave; (**b**) nonlinear echoes from bubbles; and (**c**) the frequency spectrum of nonlinear echoes. The first major hump reflects the fundamental echo, while the following ones are the second, third, and fourth harmonics

sound wavelength), the more likely it is to break. It is well established that the acoustic power level used during routine examinations destroys contrast microbubbles.

Blood flow in a normal capillary bed is on the order of 1 mm/s, and a typical capillary is about 1 mm long. Thus, if the contrast within a capillary is destroyed, it will take about a second or more to refill the capillary. Given the branching structure of the microvasculature and the thickness of a typical scan plane, it can take several seconds to replenish the contrast in the scan plane, depending on the flow rate to the organ.

During real-time ultrasound scanning at normal diagnostic output power levels, the contrast agent never has a chance to fill the microvasculature. This was first observed by Porter when he found that intermittent triggered imaging (e.g., one pulse every four or more cardiac cycles) allows much better visualization of contrast within the myocardium. Microbubble destruction imaging has recently been developed for depiction of acute brain infarction. This technique uses a rapid, real-time pulse sequence with high acoustic energy to destroy all microbubbles within a field of view in less than 0.5 seconds. First results with this advanced method indicate that brain perfusion deficits can be rapidly identified in the territory of the middle cerebral artery (*Kern et al. 2004*).

Low Mechanical Index Imaging

Until recently, visualization of flow in the microcirculation required intermittent triggering or low frame rate imaging. To detect the bubbles, the ultrasound energy had to be sufficient to cause their destruction. However, the increased sensitivity provided by newer imaging techniques (see below) now allows imaging of contrast microbubbles nondestructively in real time at very low acoustic pressures.

In low MI scanning, bubble destruction is avoided. At an MI of about 0.15, microbubbles with different shell compositions are not significantly destroyed, but still provide a good harmonic contrast signal. Low MI scanning also reduces the harmonic component in the tissue echoes relative to bubble echoes. Because tissue is less nonlinear than bubbles, it requires a higher MI than the contrast microbubbles for a certain harmonic response. Therefore, at low MI, the contrast-to-tissue ratio is higher than at high MI, helping to remove the tissue signal and leave only the contrast agent's "signature."

Stimulated Acoustic Emission

Stimulated acoustic emission, also known as contrast burst imaging, involves color or power Doppler imaging with the transmission power set high enough to ensure contrast agent bubble disruption on the first pulse. Certain ultrasound contrast agents, such as those with thin polymer coatings, are durable linear scatterers at low acoustic pressures, but undergo destruction, fusion, or splitting at higher acoustic pressures (Postert et al. 2000) that are within the range of diagnostic ultrasound. This causes a transient high-amplitude broadband response (Uhlendorf and Hoffmann 1994). Since ultrasound Doppler systems correlate the signals backscattered from a target within a number of successive pulses, the loss of signal correlation caused by the transient bubble collapse is interpreted by the machine as a random Doppler shift, resulting in a mosaic of colors at the location of the microbubbles, even without flow. This technique has been shown to be useful in detecting bleeding sites (Goldberg et al. 1998) and liver tumors (Forsberg et al. 1999, Moriyasu et al. 1998). It has been demonstrated that stimulated acoustic emission is capable of visualizing brain perfusion (Postert et al. 2000).

Harmonic Imaging

The nonlinear behavior of bubbles can be exploited to enhance the contrast relative to tissue. Harmonic imaging transmits at a fundamental frequency of f_0 and forms an image from the second harmonic component $2f_0$ of the backscattered echoes by using filters to remove the fundamental component. Although harmonic imaging can enhance microbubbles relative to tissue, the bandwidth available for imaging to ensure that the received harmonic signal can be separated from the fundamental signal is restricted. If the bandwidth of the fundamental signal overlaps with that of the second harmonic, the received signals cannot be completely separated. A narrower transmission bandwidth is therefore used in harmonic imaging. Harmonic imaging has traditionally been used as a high MI technique, requiring intermittent imaging to allow sufficient time for new bubbles to refill the region of interest.

It was expected that harmonic imaging would allow complete separation of contrast from tissue, as it was assumed that tissue was completely linear. While it has long been known that tissue does produce nonlinear energy, it was believed that the higher-frequency harmonics would be eliminated by attenuation. However, it was soon found that tissue did produce significant harmonic energy and that the high sensitivity and bandwidth of modern ultrasound equipment was able to detect it. In fact, the harmonic image produced by tissue alone has beneficial qualities such as reduced clutter in the image and improved resolution. A tissue image is therefore present even in the absence of a contrast agent, so that perfect separation was not achieved.

Pulse Inversion Harmonic Imaging

Pulse inversion harmonic (PIH) imaging avoids the narrow bandwidth limitations of conventional harmonic imaging by subtracting rather than filtering out the fundamental signals. PIH is therefore able to separate the fundamental component of the bubble echoes from the harmonic even when they overlap. This allows the use of broader transmission and reception bandwidths for improved resolution, and increased sensitivity to contrast agents. In pulse inversion harmonic imaging, two pulses are transmitted down each ray line. The first is a normal pulse and the second is an inverted replica of the first, so that wherever there is positive pressure on the first pulse there is an equal negative pressure on the second. Any linear target that responds equally to positive and negative pressures will reflect back to the transducer with equal but opposite echoes. The summation of these linear echoes results in a cancellation of the corresponding signals—i.e., no signal (Fig. 1.**24**).

However, microbubbles respond differently to positive and negative pressures and do not reflect identical inverted waveforms. This is illustrated in Fig. 1.**25**. When the echoes are added from the positive and negative portions of nonlinear reflectors, they do not cancel out completely. The harmonic components add together, thus giving twice the harmonic level of a single echo. An example of pulse inversion contrast imaging of the brain is shown in Fig. 1.**26**.

Power Pulse Inversion Imaging

To achieve greater sensitivity than PIH, additional pulses of alternating polarity can be used. This also improves separation between tissue and contrast, and reduces motion sensitivity. Since PIH uses two pulses to form each image line, anything that moves between the two pulses is not completely canceled, leading to incomplete tissue removal. This is similar to color Doppler motion artifacts, but since PIH is a gray-scale mode, the effect is to brighten the gray-scale image slightly. To remove tissue motion artifacts while increasing sensitivity, a longer sequence of inverted pulses is used.

Microvascular Imaging

Steady improvements in methods of imaging microbubbles without destroying them have led to the ability to image individual bubbles in small vessels with very low blood flow rates. In some of these vessels, the flow rate is so low that a bubble may pass through only every few seconds. A novel technique for microvascular imaging has recently been introduced. This approach uses specially designed image processing software to capture and track the bubbles as they pass through these small vessels. The software measures changes in the image from frame to frame, suppressing any back-

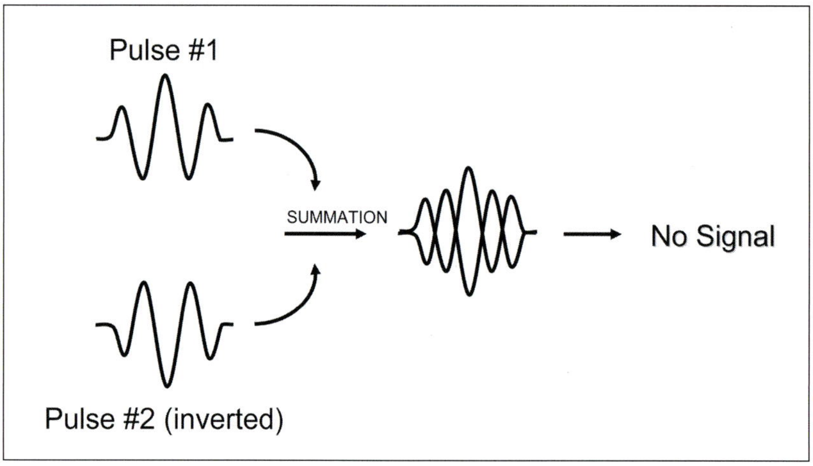

Fig. 1.**24** The summation of two consecutive echoes from inverted pulses. Pulse inversion cancels out the fundamental echoes without filtering

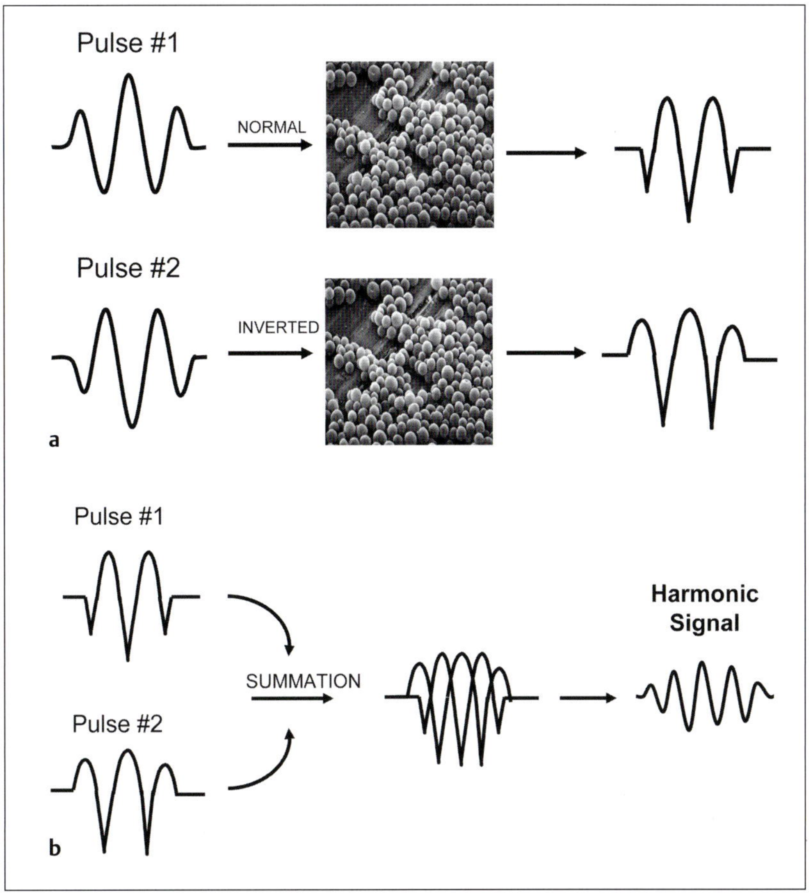

Fig. 1.**25a** Two consecutive nonlinear bubble echoes from inverted pulses. Note the differences in the response to positive and negative pressures between the two signals. **b** Summation of these bubble echoes results in stronger harmonic signals with the pulse inversion technique

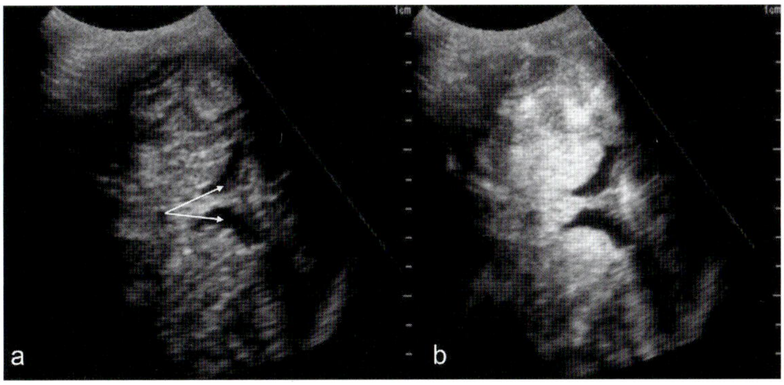

Fig. 1.**26a, b** Excellent contrast enhancement of the brain following injection of Levovist is seen in **b**, using pulse inversion contrast harmonic imaging. The arrows indicate the frontal horns of the lateral ventricles

ground tissue signal and capturing the bubbles as they pass through the vasculature. This dramatically enhances vessel conspicuity, showing tracks of single bubbles flowing through the microvasculature.

Microbubble Refill Kinetics

The level of contrast enhancement in an image represents the volume of contrast within the tissue, not the flow rate. Blood volume can be fairly constant distal to a stenosis even when the flow rate is reduced. Accordingly, once a vascular bed has filled with contrast agent, it becomes difficult to differentiate altered flow rates. Traditionally, indicator dilution techniques with determination of the time to peak after a bolus injection have been used to assess the flow rate. However, the absolute concentration of an ultrasound contrast agent is unknown, and the contrast agent bolus spreads after an intravenous injection. By destroying the contrast within the scan plane, a “negative bolus” of contrast is created locally. The time it takes for contrast to refill the scan plane is an indicator of the local blood flow velocity.

Flash contrast imaging provides the tools required to assess microbubble refill kinetics. By “flashing” the tissue with a short series of high MI frames and then switching back to low MI, the replenishment of contrast agent in the tissue can be observed in real time. A microbubble replenishment curve after bubble destruction is shown in Fig. 1.**27**. The data correspond to the equation

$$\gamma = A(1 - e^{-\beta t})$$

where γ is the acoustic intensity measured at the pulse interval t, A is the plateau of acoustic intensity, and β is the rate constant that determines the rate of increase in the acoustic intensity.

The technique of microbubble refill kinetics has been used to measure blood flow to various tissues, including the myocardium (Wei et al. 1998), skeletal muscle (Vincent et al. 2002) and kidney (Wei et al. 2001). In animal experiments, it has also been shown to be suitable for assessing cerebral blood flow (CBF) (Seidel et al. 2001). In dog brains, Rim and co-workers have demonstrated that CBF can be measured using real-time pulse inversion harmonic imaging (Rim et al. 2001). Emerging extensions of this approach to real-time power pulse inversion harmonic imaging may allow significant advances in both ultrasonographic visualization of parenchymal infarctions and in the assessment of tissue blood flow. An example of the use of microbubble refill kinetics with power pulse inversion technology to assess tissue perfusion following cerebral infarction is shown in Fig. 1.**28**.

References

Arbeille P, Lapierre F, Patat F, Benhamou AC, Alison D, Dusorbier C, et al. [Evaluation of the degree of carotid stenosis by spectral analysis of the Doppler signal: comparison of the results of spectral analysis, angiography and anatomo-pathology; in French.] Arch Mal Coeur Vaiss 1984;77:1097–1107.

Barber, FE, Baker DW, Nation AW, Strandness DE Jr, Reid JM. Ultrasonic duplex echo-Doppler scanner. IEEE Trans Biomed Eng 1974;21:109–13.

Barnes RW, Nix L, Rittgers SE. Audible interpretation of carotid Doppler signals: an improved technique to define carotid artery disease. Arch Surg 1981;116:1185–9.

Barry CD, Allott CP, John NW, Mellor PM, Arundel PA, Thomson DS, et al. Three-dimensional freehand ultrasound: image reconstruction and volume analysis. Ultrasound Med Biol 1997;23:1209–24.

Baum G, Greenwood I. Orbital lesion localization by three dimensional ultrasonography. NY State J Med 1961;61:4149–57.

Blackshear WM Jr, Phillips DJ, Thiele BL, Hirsch JH, Chikos PM, Marinelli MR, et al. Detection of carotid occlusive disease by ultrasonic imaging and pulsed Doppler spectrum analysis. Surgery 1979;86:698–706.

Bucek RA, Reiter M, Koppensteiner I, Ahmadi R, Minar E, Lammer J. B-flow evaluation of carotid arterial stenosis: initial experience. Radiology 2002;225:295–9.

Burns PN. Overview of echo-enhanced vascular ultrasound imaging for clinical diagnosis in neurosonology. J Neuroimaging 1997;7 Suppl 1:S2–14.

Dekker DL, Piziali RL, Dong E. A system for ultrasonically imaging the human heart in three dimensions. Comput Biomed Res 1974;7:544–53.

Deng J, Gardener JE, Rodeck CH, Lees WR. Fetal echocardiography in three and four dimensions. Ultrasound Med Biol 1996;22:979–86.

Detmer PR, Bashein G, Hodges T, Beach KW, Filer EP, Burns DH, et al. 3D ultrasonic image feature localization based on magnetic scanhead tracking: in vitro calibration and validation. Ultrasound Med Biol 1994;20:923–36.

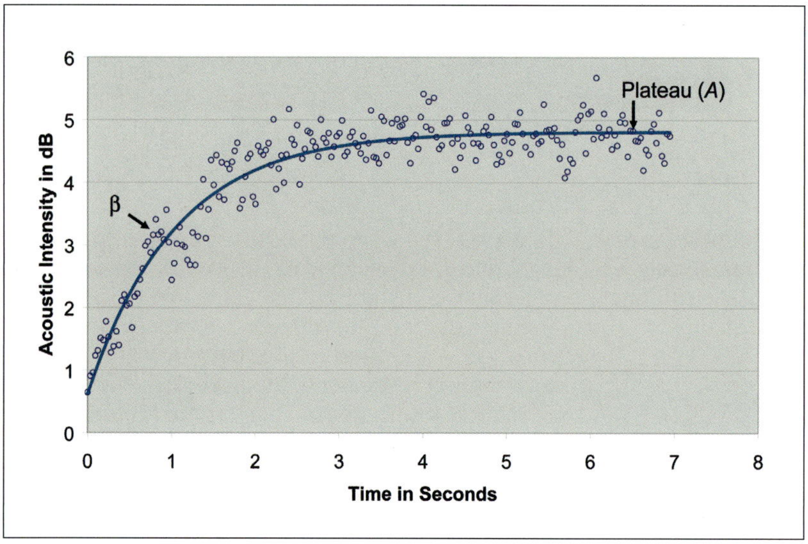

Fig. 1.**27** The microbubble replenishment curve following bubble destruction. The data correspond to the equation $\gamma = A(1 - e^{-\beta t})$, where γ is the acoustic intensity measured at the pulse interval t, A is the plateau of acoustic intensity, and β is the rate constant that determines the rate of increase in the acoustic intensity

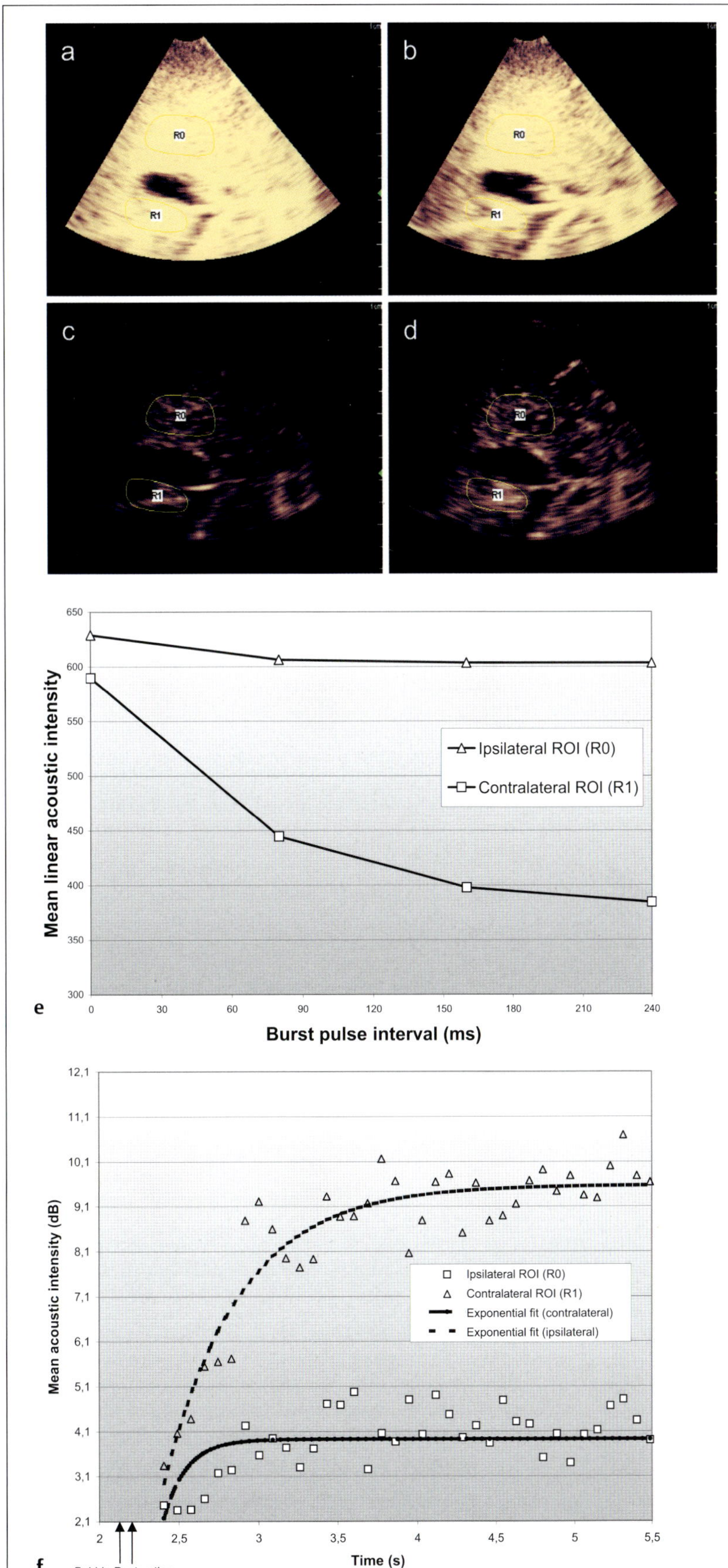

Fig. 1.**28 a–f** Real-time power pulse inversion contrast imaging of ipsilateral cerebral infarction of the middle cerebral artery territory after partial recanalization. **a**, **b** Real-time destruction of the contrast agent SonoVue using a high mechanical index of 1.3. Discrete color desaturation is seen in **b** after bubble destruction. **c, d** A low mechanical index of 0.1 is used to observe contrast agent replenishment in real time. The left image (**c**) was taken immediately after bubble destruction (in **b**), while the right image (**d**) is several seconds later. Two regions of interest (ROI) have been set to evaluate bubble destruction (**e**) and refill kinetics (**f**) in the two hemispheres. In **e**, it should be noted that there is only marginal bubble destruction in the ipsilateral hemisphere after tissue infarction (i.e., low flow conditions). The replenishment curves in **f** clearly show better perfusion of the contralateral hemisphere

Eiho S, Kuwahara M, Asada N. Left ventricular image processing. Med Prog Technol 1987;12:101–15.
Fehske W. Praxis der konventionellen und farbcodierten Doppler-Echokardiographie. Berne: Huber, 1988
Felix RW Jr, Sigel B, Gibson RJ, Williams J, Popky GL, Edelstein AL, et al. Pulsed Doppler ultrasound detection of flow disturbances in arteriosclerosis. J Clin Ultrasound 1976;4:275–82.
Fine D, Perring S, Herbetko J, Hacking CN, Fleming JS, Dewbury KC. Three-dimensional (3D) ultrasound imaging of the gallbladder and dilated biliary tree: reconstruction from real-time B-scans. Br J Radiol 1991;64:1056–7.
Forsberg F, Goldberg BB, Liu JB, Merton DA, Rawool NM, Shi WT. Tissue-specific US contrast agent for evaluation of hepatic and splenic parenchyma. Radiology 1999;210:125–32.
Franceschi D, Bondi JA, Rubin JR. A new approach for three-dimensional reconstruction of arterial ultrasonography. J Vasc Surg 1992;15:800–4.
Geiser EA, Christie LGJ, Conetta DA, Conti CR, Gossman GS. A mechanical arm for spatial registration of two-dimensional echocardiographic sections. Catheter Cardiovasc Diagn 1982;8:89–101.
Goldberg BB, Merton DA, Liu JB, Forsberg F. Evaluation of bleeding sites with a tissue-specific sonographic contrast agent. J Ultrasound Med 1998;17:609–16.
Griewing B, Morgenstern C, Driesner F, Kallwellis G, Walker ML, Kessler C. Cerebrovascular disease assessed by color-flow and power Doppler ultrasonography: comparison with digital subtraction angiography in internal carotid artery stenosis. Stroke 1996;27:95–100.
Hamper UM, Trapanotto V, Sheth S, DeJong MR, Caskey CI. Three-dimensional US: preliminary clinical experience. Radiology 1994;191:397–401.
Hell B, Walter FA, Schreiber S, Blase H, Bielke G, Meindl S, et al. Three-dimensional ultrasonography in maxillofacial surgery: a new diagnostic tool. Int J Oral Maxillofac Surg 1993;22:173–7.
Hennerici M, Freund HJ. Efficacy of CW-Doppler and duplex system examinations for the evaluation of extracranial carotid disease. J Clin Ultrasound 1984;1:155–61.
Henri P, Tranquart F. [B-flow ultrasonographic imaging of circulating blood; in French.] J Radiol 2000;81:465–7.
Herment A, Demoment G, Dumee P. Improved estimation of low velocities in color Doppler imaging by adapting the mean frequency estimator to the clutter rejection filter. IEEE Trans Biomed Eng 1996;43:919–27.
Hodges TC, Detmer PR, Burns DH, Beach KW, Strandness DE Jr. Ultrasonic three-dimensional reconstruction: in vitro and in vivo volume and area measurement. Ultrasound Med Biol 1994;20:719–29.
Howry DH, Posakony G, Cuschman R, Holmes JH. Three dimensional and stereoscopic observation of body structures by ultrasound. J Appl Physiol 1956;9:304–6.
Jespersen SK, Wilhjelm JE, Sillesen H. Multi-angle compound imaging. Ultrason Imaging 1998;20:81–102.
Jespersen SK, Wilhjelm JE, Sillesen H. In vitro spatial compound scanning for improved visualization of atherosclerosis. Ultrasound Med Biol 2000;26:1357–62.
Keller HM, Meier WE, Anliker M, Kumpe DA. Noninvasive measurement of velocity profiles and blood flow in the common carotid artery by pulsed Doppler ultrasound. Stroke 1976;7:370–7.
Kern, R., F. Perren, K. Schoeneberger, A. Gass, M. Hennerici, S. Meairs: Ultrasound microbubble destruction imaging in acute middle cerebral artery stroke. Stroke 35 (2004) 1665–1670
Kern R, Szabo K, Hennerici M, Meairs S. Characterization of carotid artery plaques using real-time compound B-mode ultrasound. Stroke 2004;35:870–5.
King DL, King DL Jr, Shao MY. Three-dimensional spatial registration and interactive display of position and orientation of real-time ultrasound images. J Ultrasound Med 1990;9:525–32.
Kofoed SC, Gronholdt ML, Wilhjelm JE, Bismuth J, Sillesen H. Real-time spatial compound imaging improves reproducibility in the evaluation of atherosclerotic carotid plaques. Ultrasound Med Biol 2001;27:1311–7.
Leotta DF, Detmer PR, Martin RW. Performance of a miniature magnetic position sensor for three-dimensional ultrasound imaging. Ultrasound Med Biol 1997;23:597–609.
Lewis RR, Beasley MG, Hyams DE, Gosling RG. Imaging the carotid bifurcation using continuous-wave Doppler-shift ultrasound and spectral analysis. Stroke 1978;9:465–71.
Meairs S, Hennerici M. Four-dimensional ultrasonographic characterization of plaque surface motion in patients with symptomatic and asymptomatic carotid artery stenosis. Stroke 1999;30:1807–13.
Meairs S, Beyer J, Hennerici M. Reconstruction and visualization of irregularly sampled three- and four-dimensional ultrasound data for cerebrovascular applications. Ultrasound Med Biol 2000a;26:263–72.
Meairs S, Timpe L, Beyer J, Hennerici M. Acute aphasia and hemiplegia during karate training. Lancet 2000b;356:40.
Merz E, Bahlmann F, Weber G, Macchiella D. Three-dimensional ultrasonography in prenatal diagnosis. J Perinat Med 1995;23:213–22.
Mintz GS, Pichard AD, Satler LF, Popma JJ, Kent KM, Leon MB. Three-dimensional intravascular ultrasonography: reconstruction of endovascular stents in vitro and in vivo. J Clin Ultrasound 1993;21:609–15.
Mitchell DG. Color Doppler imaging: principles, limitations, and artifacts. Radiology 1990;177:1–10.
Moriyasu F, Kono Y, Kamiyama N, Bruns D, Matsumura T, Suginoshita Y, et al. Flash echo imaging of liver tumors using ultrasound contrast agent and intermittent color Doppler scanning. J Ultrasound Med 1998;17:S63.
Moskalik A, Carson PL, Meyer CR, Fowlkes JB, Rubin JM, Roubidoux MA. Registration of three-dimensional compound ultrasound scans of the breast for refraction and motion correction. Ultrasound Med Biol 1995;21:769–78.
Ophir J, Parker KJ. Contrast agents in diagnostic ultrasound. Ultrasound Med Biol 1989;15:319–33.
Perren F, Horn P, Kern R, Bueltmann E, Hennerici M, Meairs S. A rapid, noninvasive method to visualize ruptured aneurysms in the emergency room: three-dimensional power Doppler imaging. J Neurosurg 2004;100:619–22.
Picano E, Landini L, Distante A, Salvadori M, Lattanzi F. Angle dependence of ultrasonic backscatter in arterial tissues: a study in vitro. Circulation 1985;72:573–6.
Postert T, Braun B, Pfundtner N, Sprengelmeyer R, Meves S, Przuntek H, et al. Echo contrast-enhanced three-dimensional power Doppler of intracranial arteries. Ultrasound Med Biol 1998;24:953–62.
Postert T, Hoppe P, Federlein J, Helbeck S, Ermert H, Przuntek H, et al. Contrast agent specific imaging modes for the ultrasonic assessment of parenchymal cerebral echo contrast enhancement. J Cereb Blood Flow Metab 2000;20:1709–16.

Prager RW, Rohling RN, Gee AH, Berman L. Rapid calibration of 3-D freehand ultrasound. Ultrasound Med Biol 1998;24:855–69.

Prager RW, Gee AH, Treece GM, Cash CJ, Berman LH. Sensorless freehand 3-D ultrasound using regression of the echo intensity. Ultrasound Med Biol 2003;29:437–46.

Reneman RS, Clarke HF, Simmons N, Spencer MP. In vivo comparison of electromagnetic and Doppler flowmeters: with special attention to the processing of the analogue Doppler flow signal. Cardiovasc Res 1973;7:557–66.

Reneman RS, van Merode T, Hick P, Hoeks AP. Flow velocity patterns in and distensibility of the carotid artery bulb in subjects of various ages. Circulation 1985;71:500–9.

Reneman RS, van Merode T, Hick P, Hoeks AP. Cardiovascular applications of multi-gate pulsed Doppler systems. Ultrasound Med Biol 1986;12:357–70.

Rim SJ, Leong-Poi H, Lindner JR, Couture D, Ellegala D, Mason H, et al. Quantification of cerebral perfusion with "real-time" contrast-enhanced ultrasound. Circulation 2001;104:2582–7.

Rohling RN, Gee AH, Berman L. Automatic registration of 3-D ultrasound images. Ultrasound Med Biol 1998;24:841–54.

Rosenfield K, Losordo DW, Ramaswamy K, Pastore JO, Langevin RE, Razvi S, et al. Three-dimensional reconstruction of human coronary and peripheral arteries from images recorded during two-dimensional intravascular ultrasound examination. Circulation 1991;84:1938–56.

Salustri A, Roelandt JR. Ultrasonic three-dimensional reconstruction of the heart. Ultrasound Med Biol 1995;21:281–93.

Seidel G, Claassen L, Meyer K, Vidal-Langwasser M. Evaluation of blood flow in the cerebral microcirculation: analysis of the refill kinetics during ultrasound contrast agent infusion. Ultrasound Med Biol 2001;27:1059–1064.

Seitz K, Kubale R. Duplexsonographie der abdominellen und retroperitonealen Gefässe. Weinheim: Verlag Chemie, 1988.

Spencer MP, Reid JM. Quantitation of carotid stenosis with continuous-wave (C-W) Doppler ultrasound. Stroke 1979;10:326–30.

Steiner H, Staudach A, Spitzer D, Schaffer H. Three-dimensional ultrasound in obstetrics and gynaecology: technique, possibilities and limitations. Hum Reprod 1994;9:1773–8.

Steinke W, Hennerici M. Three-dimensional ultrasound imaging of carotid artery plaques. J Cardiovasc Technol 1989;8:15–22.

Steinke W, Meairs S, Ries S, Hennerici M. Sonographic assessment of carotid artery stenosis: comparison of power Doppler imaging and color Doppler flow imaging. Stroke 1996;27:91–4.

Uhlendorf V, Hoffman C. Nonlinear acoustic response of coated microbubbles in diagnostic ultrasound. Proceedings of the IEEE International Ultrasonics Symposium. Piscataway, NJ: IEEE, 1994;1559–62.

Umemura A, Yamada K. B-mode flow imaging of the carotid artery. Stroke 2001;32:2055–7.

Vincent MA, Dawson D, Clark AD, Lindner JR, Rattigan S, Clark MG, et al. Skeletal muscle microvascular recruitment by physiological hyperinsulinemia precedes increases in total blood flow. Diabetes 2002;51:42–8.

Von Birgelen C, de Vrey EA, Mintz GS, Nicosia A, Bruining N, Li W, et al. ECG-gated three-dimensional intravascular ultrasound: feasibility and reproducibility of the automated analysis of coronary lumen and atherosclerotic plaque dimensions in humans. Circulation 1997;96:2944–52.

Wei K, Jayaweera AR, Firoozan S, Linka A, Skyba DM, Kaul S. Quantification of myocardial blood flow with ultrasound-induced destruction of microbubbles administered as a constant venous infusion. Circulation 1998;97:473–83.

Wei K, Le E, Bin JP, Coggins M, Thorpe J, Kaul S. Quantification of renal blood flow with contrast-enhanced ultrasound. J Am Coll Cardiol 2001;37:1135–40.

Wells PNT. Duplex and color flow imaging. In: Hennerici M, Meairs S, editors. Cerebrovascular ultrasound: theory, practice and future developments. Cambridge: Cambridge University Press, 2001: 88–100.

Weskott HP. [B-flow: a new method for detecting blood flow; in German.] Ultraschall Med 2000;21:59–65.

Widder B. Doppler- und Duplexsonographie der hirnversorgenden Arterien. Berlin: Springer, 1995.

Zagzebski JA, Madsen EL. Physics and instrumentation in Doppler and B-mode ultrasonography. New York: Grune and Stratton, 1982.

2 Extracranial Cerebral Arteries

Examination

Special Equipment and Documentation

Color Flow Duplex Sonography

Various systems allow a combination of structural and functional arterial analysis; B-mode and color mode with frequency and amplitude pattern display are used. Special types of equipment are provided by different manufacturers, as described in Chapter 1. For optimal image quality, transducers with transmission frequencies of between 7.5 and 15 MHz should be used when examining the extracranial cerebral arteries and veins.

Continuous-Wave Doppler Sonography

Continuous-wave Doppler sonography is still used for first-line screening investigations in some centers—it combines rapid accessibility, using inexpensive instruments, with the advantages of mobile application even outside the hospital environment. Transmission frequencies of 4–5 MHz are used to examine the extracranial arteries supplying the brain, while frequencies of between 8 and 10 MHz produce better results when investigating the orbital arteries (the indirect examination method).

Examination Conditions

Patient and Examiner

The *patient* should sit in a relaxed position on a comfortable seat or a tilted examination chair. It is best to place the patient in an examination chair with a backrest that can be lowered diagonally. The patient's head should be resting on an easily adjustable support, and should be firmly secured during compression maneuvers.

The position of the patient's head needs to be adjusted depending on requirements in the region being examined, and the position has to be corrected appropriately when any variations in the normal courses of the blood vessels are encountered.

The *examiner* does not have a set position. In addition to the seated position behind the patient, seated or standing access to the patient from the side or front is also common. Steady movement of the probe and fixation of the patient's head during compression maneuvers should always be ensured. Ultrasound examination carried out with the patient lying in bed has disadvantages for the examiner, as it does not provide adequate fixation of the patient's head.

For three-dimensional reconstructions, the position of the probe has to be displayed precisely relative to the position of the vessels. This is commonly achieved using a separate ultrasound positional sensor adapted to the transducer. Computerized data analysis is necessary; separate online or offline systems are provided by different manufacturers, but many laboratories use their own algorithms. Electrocardiography-triggered studies are mandatory for time control measurements (four-dimensional analysis).

Conducting the Examination

Color Flow Duplex Sonography

When carrying out ultrasound imaging procedures, it should be ensured that the blood vessels being examined are depicted in at least three planes (Fig. 2.**1**). A standardized examination sequence is recommended, e.g.:

- Transverse (cross-section)
- Longitudinal (longitudinal section)
  - Anterior
  - Posterolateral

These planes can only be obtained consistently in the extracranial system (Fig. 2.**2**).

Continuous-Wave Doppler Sonography

The probe is guided by hand. The angle between the ultrasound beam and the blood vessel varies over a wide range and needs to be optimized by monitoring the audio signal and the Doppler waveform or spectrum that is registered. Only the best signal should be recorded.

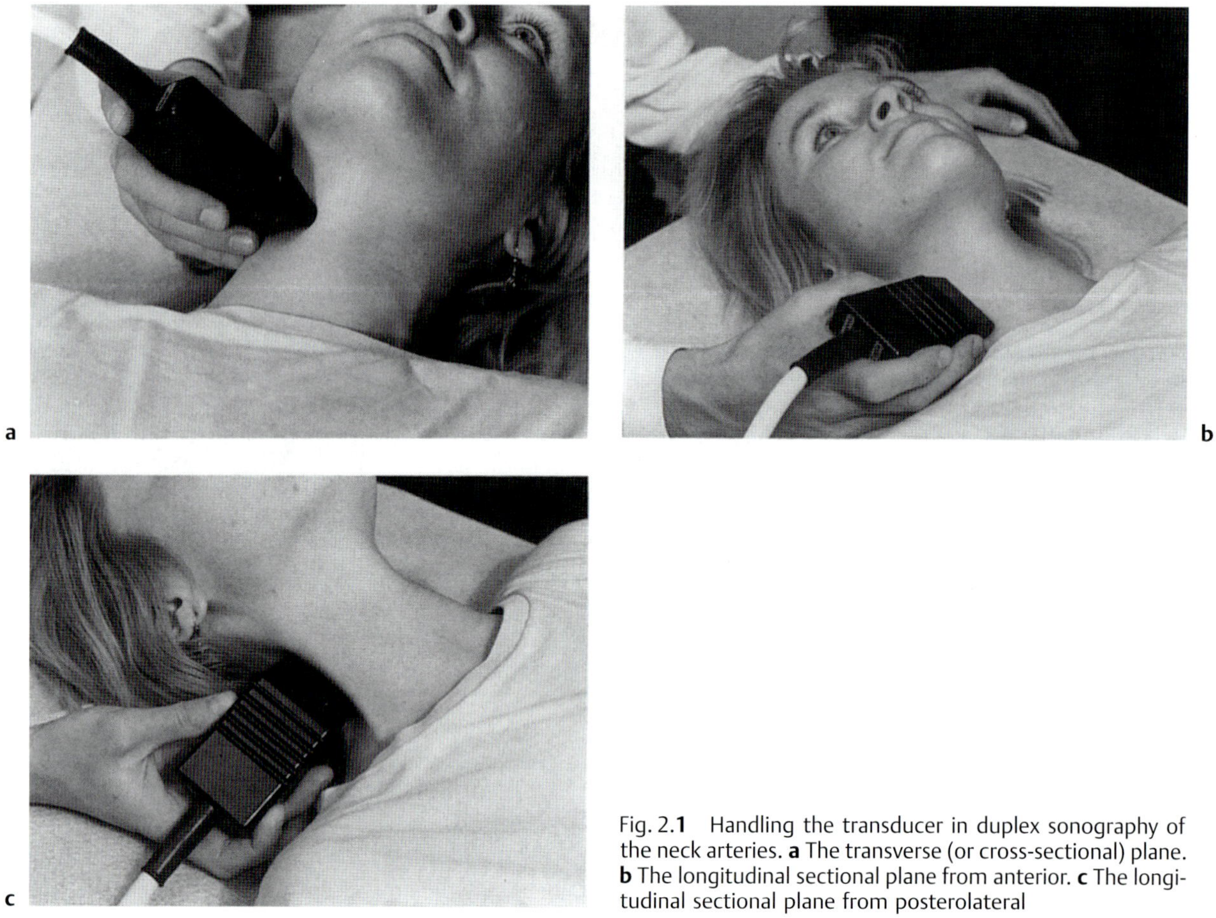

Fig. 2.1 Handling the transducer in duplex sonography of the neck arteries. **a** The transverse (or cross-sectional) plane. **b** The longitudinal sectional plane from anterior. **c** The longitudinal sectional plane from posterolateral

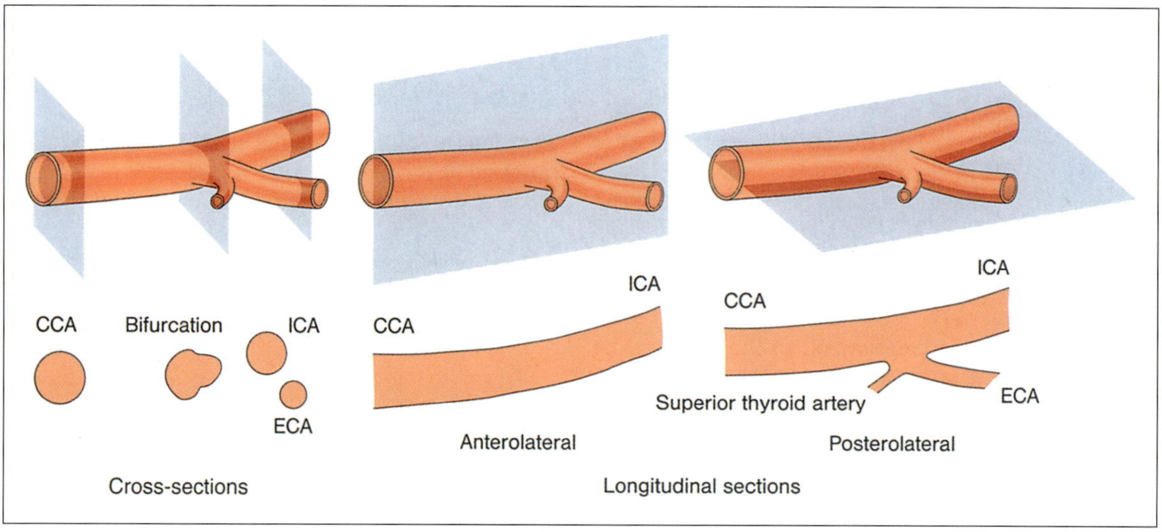

Fig. 2.2 Sectional planes used when examining the carotid system in the neck with duplex sonography

Examination Sequence

Using side-to-side comparisons, information should be gathered from the individual examination areas, following a set laboratory sequence. A logical sequence of examination steps is best—e.g., first taking recordings that yield only overview results (indirect Doppler sonography), or proceeding anatomically along the neck from caudal to cranial.

Orbital Arteries

Color Flow Duplex Sonography

Ophthalmologists recommend using color flow duplex sonography to visualize the ophthalmic, central retinal, and posterior ciliary arteries. A variety of conditions affecting the orbital blood vessels have been studied, and it has been suggested that embolism in particular represents a challenge for future investigations (Hedges 1995).

Continuous-Wave Doppler Sonography

The continuous-wave Doppler method is still preferable when examining the orbital arteries. This is the traditional procedure used to detect obstructions in the carotid system.

The terminal branches of the ophthalmic artery are the supratrochlear artery and the supraorbital artery. Ultrasound evaluation of the supratrochlear artery takes place at the nasal canthus, while the supraorbital artery is assessed at the upper orbital margin.

The probe is placed on the nasal canthus, slightly above the eyeball, without applying any pressure and using only a small amount of contact gel. The examiner may choose to support the ball of his or her thumb on the patient's forehead, or to "rein in" the probe by using the cable to hold it gently in place (Figs. 2.**3**, 2.**4**). Proceeding in a mediocranial direction, and using only slight variations in the angle of the probe, a search is made for the optimal acoustic signal, or the one with the highest amplitude on the spectrum. Often, a slight change in position or a circular probe movement is necessary to "thread" the ultrasound beam through the thin blood vessels.

Compression tests. The temporal artery or facial artery, or both, are usually compressed ipsilaterally at the same time. Figure 2.**4** shows the typical compression points.

A little practice is needed to master the technique, first palpating the pulsating arteries with the thumb and middle finger without applying any pressure, and then exerting firm pressure, moving neither the head nor the probe held in the other hand above the orbit. It may sometimes be necessary to apply pressure to the contralateral arteries; an additional examiner can be helpful here.

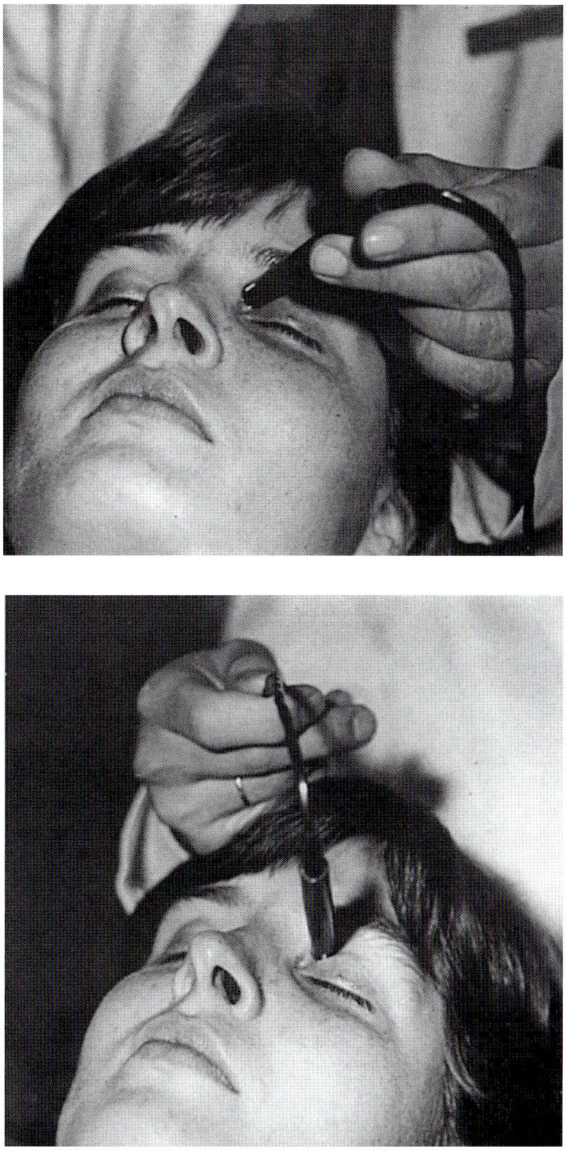

a

b

Fig. 2.**3** Insonation of the orbital arteries: **a** with support from the hand, and **b** "reining in" the probe

Neck Arteries

Common Carotid Artery

Color Flow Duplex Sonography

To display a longitudinal section through the common carotid artery, the probe is placed above the clavicle, usually in a mediocranial direction, lateral to the proximal part of the sternocleidomastoid muscle (Fig. 2.**1**). From this point, a search is made for the bifurcation, which can be used as the topmost border. One can then examine the wall structure of the entire vascular segment that can be directly imaged above the clavicle. It is important to examine different longitudinal and cross-sectional images, in order to obtain a three-di-

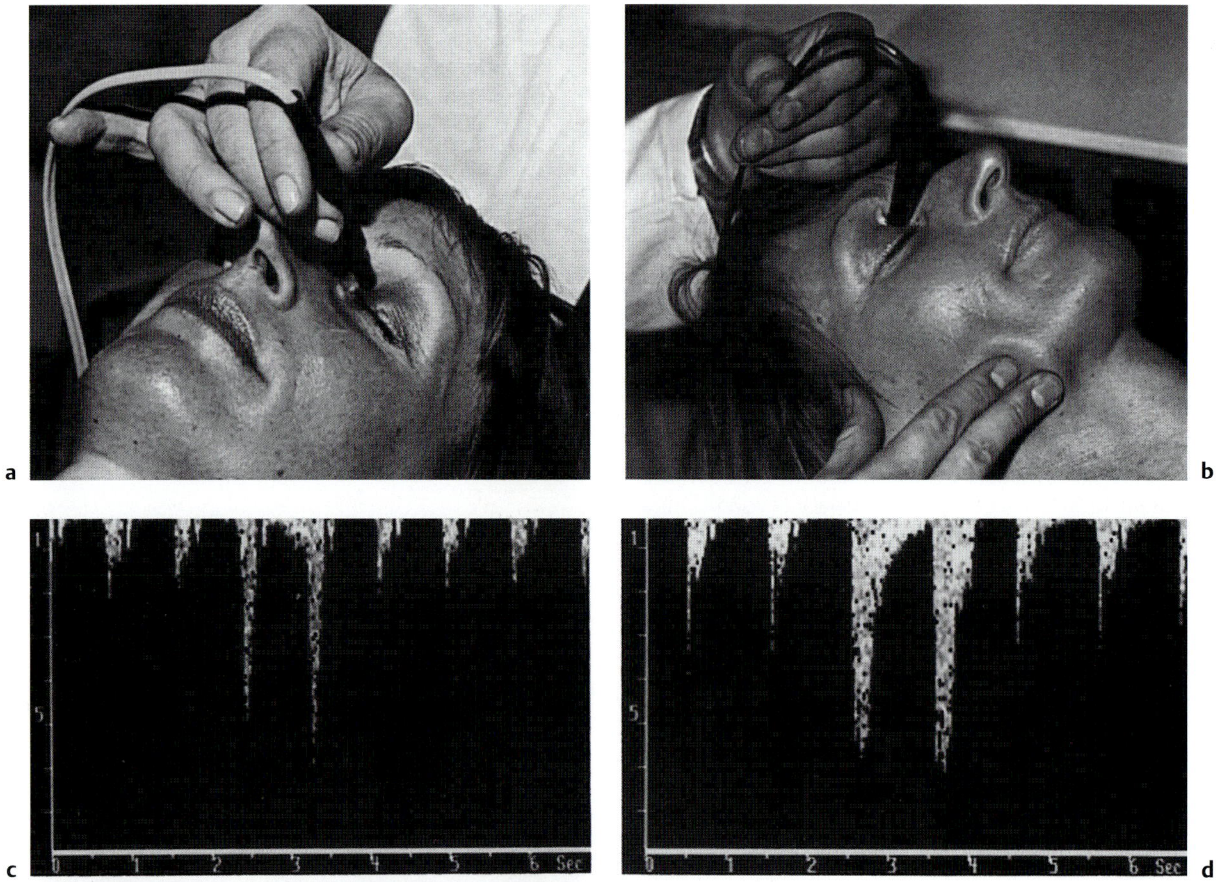

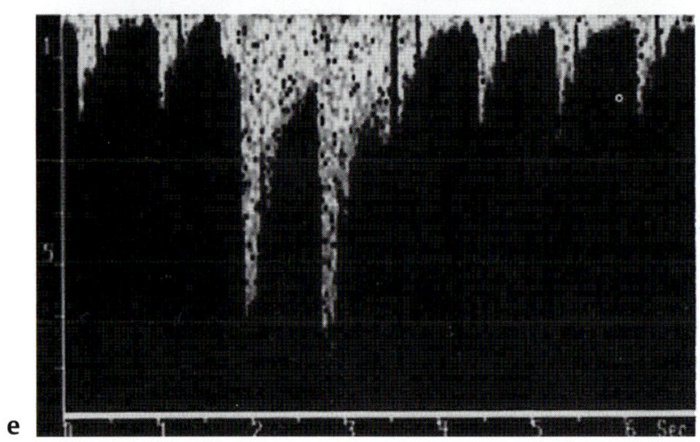

Fig. 2.**4** Compression tests of the facial artery and superficial temporal artery while simultaneously recording signals from the orbital arteries. **a, b** Reining in the probe. **c–e** Doppler spectra of the supratrochlear artery while compressing the superficial temporal artery (**c**), the facial artery (**d**), and simultaneously compressing both blood vessels (**e**) for two heart cycles

mensional overview of the blood vessel's course that is as complete as possible (Fig. 2.**2**). In addition, the color mode can then be added to identify blood flow velocity patterns either in an amplitude or frequency mode (see Fig. 2.**6**). Finally, the sample volume of the pulsed-wave Doppler is placed within the lumen, and after the angle has been optimized, a Doppler spectrum should be documented.

Continuous-Wave Doppler Sonography

The examination of the extracranial carotid system usually starts with the proximal common carotid artery. The probe is placed flat above the clavicle, in a mediocranial direction, lateral to the proximal part of the sternocleidomastoid muscle. Usually, the artery can be identified immediately at this location and can be differentiated from the branches of the external carotid artery and the branches of the thyrocervical trunk by the deep, slightly hollow sound it produces (Fig. 2.**5**).

Very occasionally, the artery needs to be located further cranially and medially to the sternocleidomastoid muscle. It can sometimes be displaced into an extremely lateral position. If the signals received are asymmetric (different incident angles for applying the ultrasound on each side, due to differences in the course of the vessels), then side-to-side comparison should be repeated more distally. To assess the proximal intrathoracic part of the blood vessel, the probe has to be turned upward (flow toward the probe).

Internal and External Carotid Arteries

Color Flow Duplex Sonography

Displaying the carotid bifurcation together with both dividing arteries in a single longitudinal plane is possible in only 40–60 % of cases. Particularly when this is not possible, it is sometimes difficult to distinguish clearly between the end of the common carotid artery (CCA) and the start of the internal carotid artery (ICA). The ICA is more difficult to display than the CCA, since the ICA is more curved and—particularly in older patients—it is more often slightly elongated. With a state-of-the-art duplex device, displaying the bifurcation and the transition from the CCA to ICA and external carotid artery (ECA) is usually straightforward when scanning from the proximal CCA to the distal ICA in cross-section and with color. Scanning should be conducted in a standardized sequence. Again, one should attempt to obtain a good overview of the anatomic and flow relationships using longitudinal sections and cross-sections (Griewing et al. 1996, Steinke et al. 1996). One possible sequence might be to obtain an overview of the carotid system by scanning in a cross-sectional approach and then switching to the longitudinal plane, starting with a B-mode display of the bifurcation and ending with flow analysis using color and pulsed-wave Doppler recordings.

It is usually possible to distinguish the internal carotid artery from the external carotid artery when a blood vessel branching off from the external carotid artery is visible (Fig. 2.**6**). Differences in pulsatility between the ICA and ECA (the ECA normally shows higher pulsatility) can also be useful for distinguishing between segments of the ICA and ECA. Individual compression tests in the external carotid artery are helpful for identification purposes not only in CW Doppler but also in duplex sonography, and should be used when necessary, especially in pathological conditions (Fig. 2.**7**).

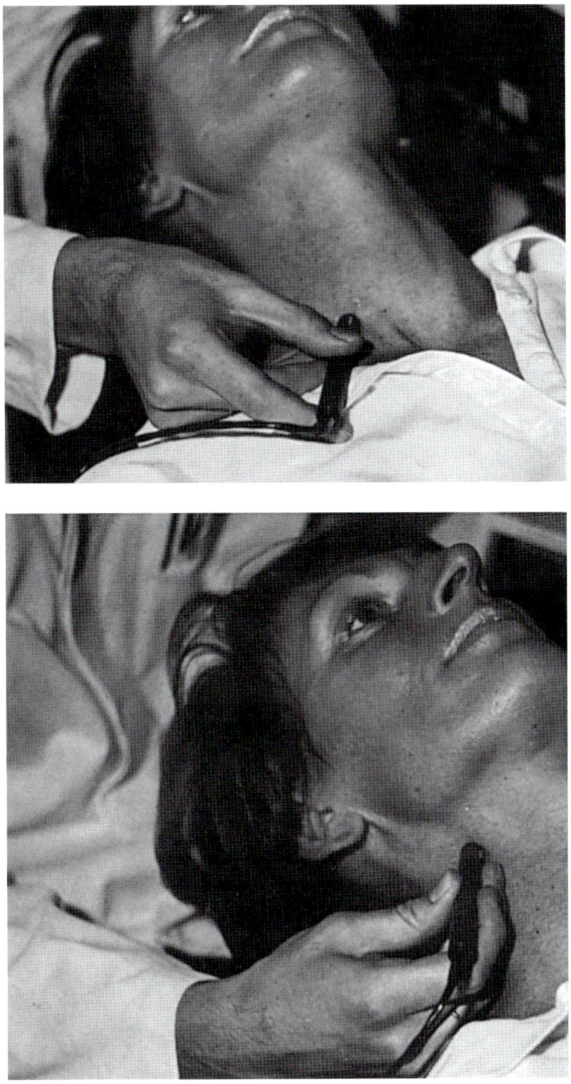
a
b

Fig. 2.**5** Continuous-wave Doppler sonography of the common carotid artery (**a**) and carotid bifurcation (**b**)

Fig. 2.**6** Color flow duplex sonogram of the carotid bifurcation; **a** power mode, **b** frequency mode. The origin of the superior thyroid artery as it branches off from the external carotid artery is shown (arrow)

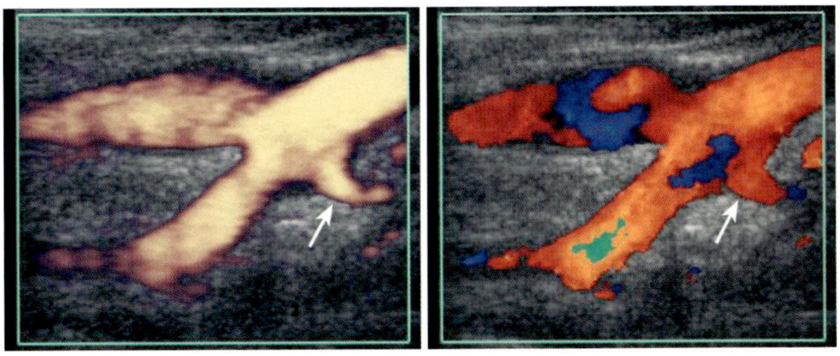
a b

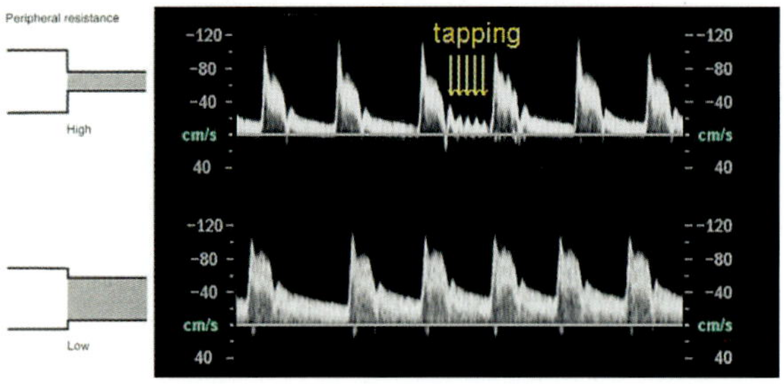

Fig. 2.**7** The effect of the peripheral resistance on the Doppler spectrum. *Upper row:* The external carotid artery (ECA) shows high pulsatility with low (sometimes zero) end-diastolic flow. This is due to the high peripheral resistance of the muscular territory supplied by the ECA. Note the oscillating flow components in one cardiac cycle, resulting from tapping(arrows) of the temporal artery (compression test for identifying the ECA). *Lower row:* The normal internal carotid artery (ICA) shows higher end-diastolic flow and hence less pulsatility. This is due to the low peripheral resistance of brain tissue

Continuous-Wave Doppler Sonography

Usually, it is possible to glide continuously with the probe from the common carotid artery in a cranial direction to the bifurcation, where a transition can be heard to the whipping, systolically weighted signal of the external carotid artery, or the soft low sound of the internal carotid artery. In a wide sinus, the sound of the internal carotid artery is not yet typical, but the velocity of the blood flow is usually noticeably slower than that of the common carotid artery. In addition, the sinus can be recognized as the center of a triangle formed by the common carotid artery (caudally), the internal carotid artery (laterally), and the external carotid artery (medially). In doubtful cases, it may be much simpler to start by locating the submandibular component of the internal carotid artery and then proceed from there, following it caudally to the carotid sinus, which is clearly recognized by its "bubbling," low-frequency sound and its low flow velocity.

The positional relationship between the internal and external carotid arteries varies considerably. Most commonly (49%), the internal carotid artery lies directly dorsolateral to the external carotid artery. However, the internal and external carotid arteries are often superposed significantly at the bifurcation, so that a mixed signal is heard. If the internal carotid artery lies dorsally (39%), a more lateral approach is recommended. Occasionally (6%), the internal carotid artery is found medial to the external carotid artery (Tismer and Boehlke 1986) (Fig. 2.**8**).

Identifying branches of the external carotid artery (Fig. 2.**9**) can be facilitated by repeated manual compression or tapping e.g. of the superficial temporal artery, occipital artery, or facial artery. This produces retrograde pulse waves that are transmitted acoustically, and can be registered as small oscillations (Fig. 2.**7**). Increased peripheral resistance may also be registered, as well as the hyperemic phase after releasing the compression of the blood vessel.

Identifying the individual branches of the external carotid artery is only useful when pathological findings are present. Using compression tests (von Reutern et al. 1976b), it is possible to identify:

- The superior thyroid artery: manual compression of the thyroid gland.
- The lingual artery: pressing the tongue against the palate.
- The facial artery: manual compression against the mandible.
- The ascending pharyngeal artery. This is difficult to perform, but acoustic modification of the Doppler signal may occur during swallowing.
- The occipital artery: manual compression at the mastoid process. It is important to distinguish this from the vertebral artery.
- The superficial temporal artery: preauricular manual compression.
- The maxillary artery: clenching the teeth is only occasionally effective, due to superposition of the superficial temporal artery.

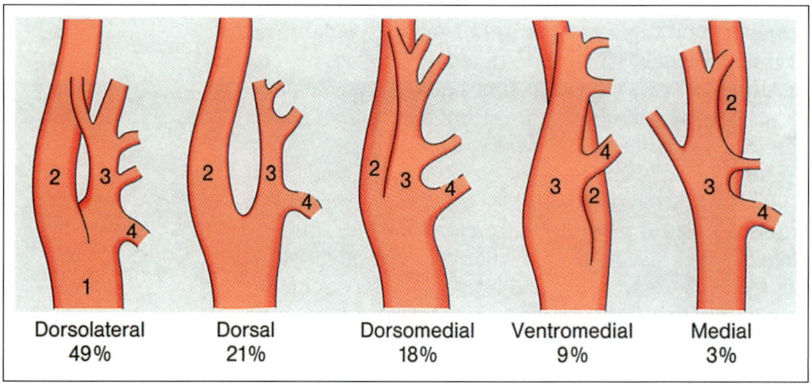

Fig. 2.**8** Variations in the location of the internal carotid artery relative to the external carotid artery
1 Common carotid artery
2 Internal carotid artery
3 External carotid artery
4 Superior thyroid artery

Fig. 2.**9** Anatomy of the large arteries supplying the brain
1 Innominate artery
2 Subclavian artery
3 common carotid artery
4 Internal carotid artery
5 External carotid artery
6 Superior thyroid artery
7 Lingual artery
8 Facial artery
9 Maxillary artery
10 Superficial temporal artery
11 Occipital artery
12 Ophthalmic artery
13 Supraorbital artery
14 Supratrochlear artery
15 Angular artery
16 Vertebral artery
17 Basilar artery
18 Carotid siphon

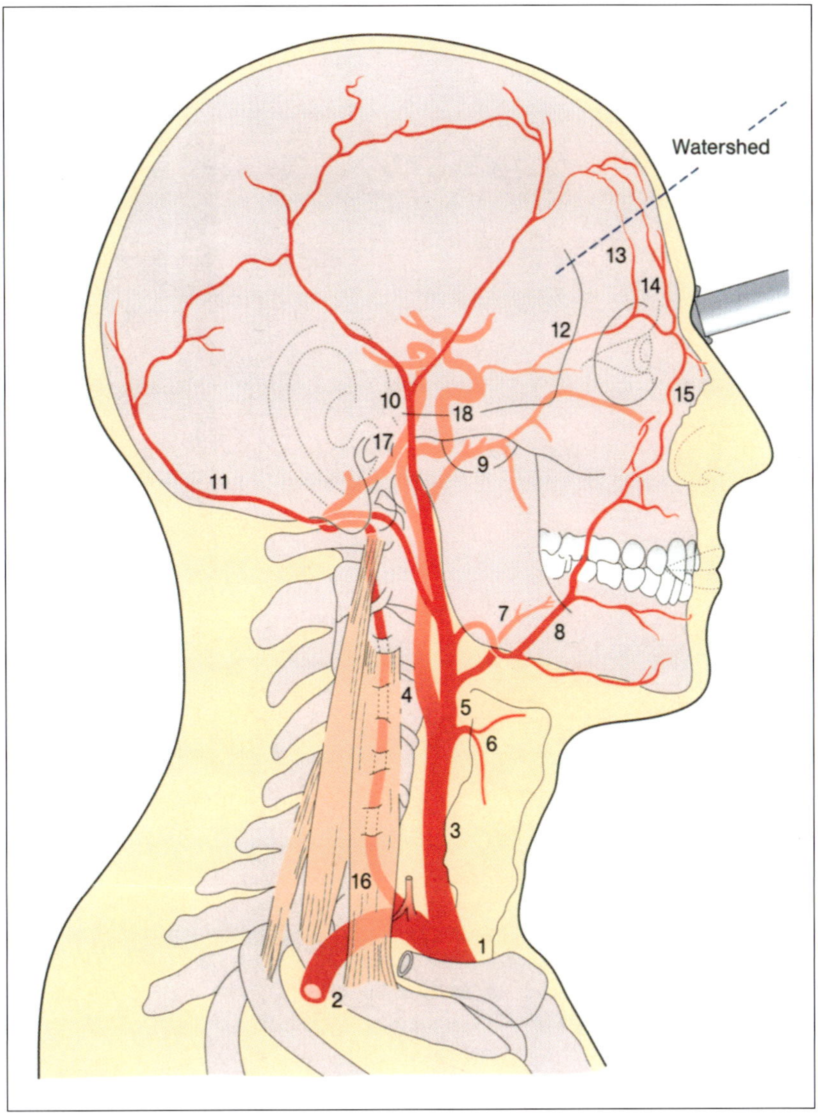

Vertebral–Subclavian System

Vertebral Artery

Color Flow Duplex Sonography

The vertebral artery can be recorded from its origin to the atlas loop (Figs. 2.**10**, 2.**11**). The course in the neck and between the transverse processes of the cervical vertebral bodies can be examined particularly well using color flow duplex sonography (Fig. 2.**12**). By tipping the transducer slightly from anteromedial to lateral, the main bodies of the cervical vertebrae can be displayed in longitudinal section with the corresponding ultrasound shadow. The vertebral artery is seen in the intervertebral spaces. Depending on the specific anatomic configuration, the proximal segment can be depicted all the way to the origin. At the origin, as in the intertransverse segment, Doppler spectra are documented after positioning the sample volume (Ackerstaff et al. 1984, Touboul et al. 1988, Trattnig et al. 1990, Bartels and Flügel 1993, Ries et al. 1996). The flow direction can easily be assessed even in the presence of a subclavian steal phenomenon, based on color patterns from the adjacent vertebral and common carotid arteries. This can further be strengthened by compression tests using the following procedure (Figs. 2.**13**, 2.**14**):

1. Place the blood pressure cuff around the upper arm.
2. Locate the vertebral artery at the atlas loop, or at its origin from the subclavian artery, and register the flow waveform (the characteristic sound made by the vertebral artery may be different when its function has been altered to supplying the extremities instead of the brain—i.e., when it has a low diastolic flow velocity).

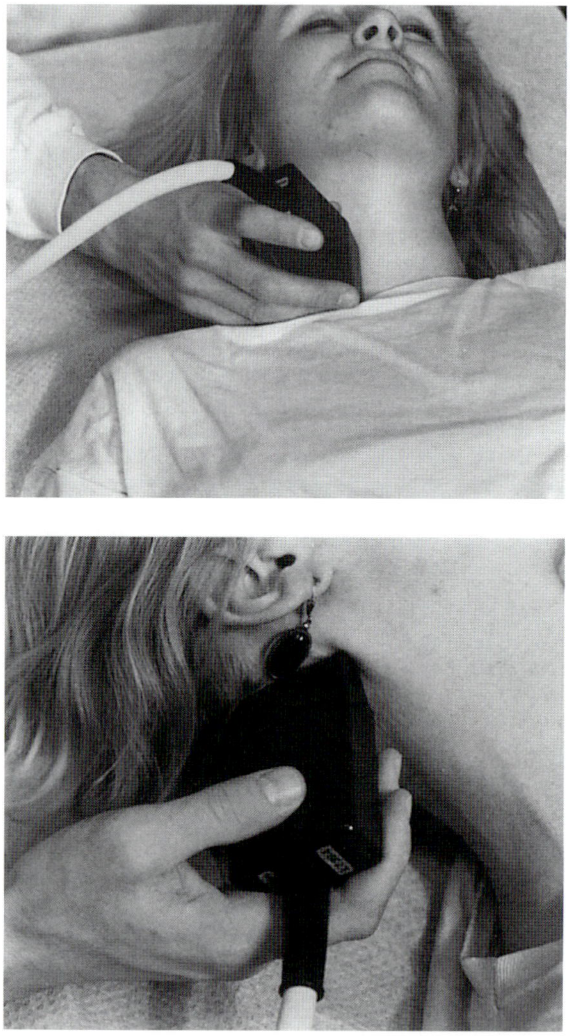

a

b

Fig. 2.**10** Handling the transducer in duplex sonography of the vertebral arteries. Insonation in the neck section (**a**) and at the atlas loop (**b**)

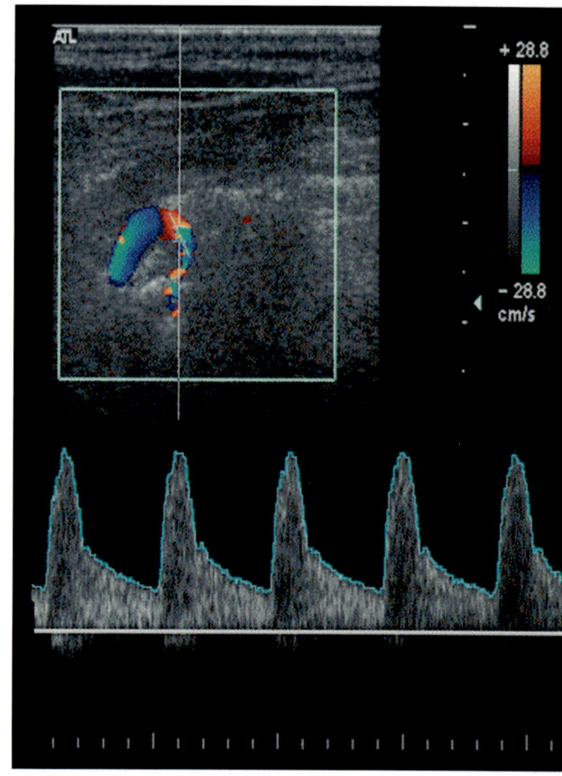

Fig. 2.**11** Vertebral artery. Insonation in a cross-sectional plane in the neck, showing the atlas loop with flow toward (right leg of the bow) and away from the probe (left leg of the bow)

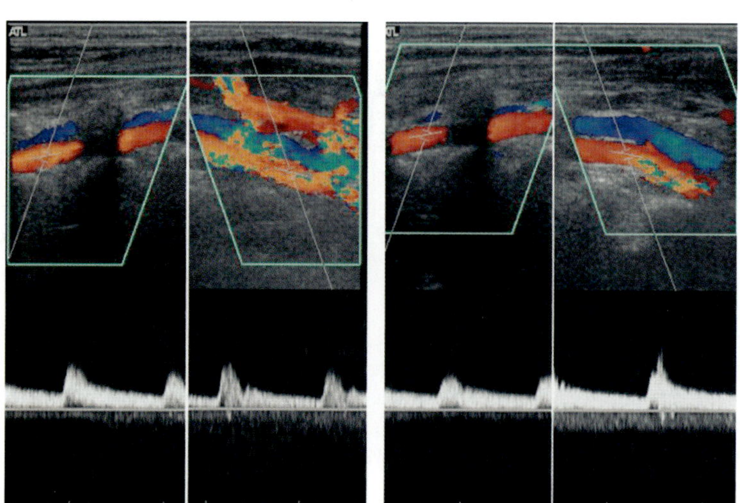

a b c d

Fig. 2.**12** The normal results of a color-mode (frequency-mode) examination with pulsed-wave Doppler registration. **a** Color display of the left vertebral artery at the V_2 segment. **b** Color display of the left vertebral artery at the V_1 segment, just distal to the origin. **c** Color display of the right vertebral artery at the V_2 segment. **d** Color display of the right vertebral artery at the proximal V_1 segment (due to its deep intrathoracic location its origin is often missed)

3. Interrupt the blood flow to the arm by exerting suprasystolic pressure by means of a cuff (if necessary, with help from an assistant). Alternatively, the patient can form a tightly clenched fist, which reduces the diastolic flow.
4. After the blood pressure cuff is released, register the hyperemic reaction, with an increase in the diastolic blood flow component in the vertebral artery. This phenomenon produces a marked acoustic impression lasting for several cardiac cycles.

Continuous-Wave Doppler Sonography

Examination of the atlas loop at the mastoid. With the patient's head turned to the side, the probe is placed below the mastoid process, directly lateral to the sternocleidomastoid muscle and in the direction of the contralateral orbit or ear (Fig. 2.**13**). The sound characteristics of the signal are the same as in the internal carotid artery, but the volume is lower due to the smaller vascular caliber, the unfavorable angle for the investigation, and the greater distance from the transducer. An experienced investigator can distinguish the vertebral artery from the occipital artery purely acoustically, due to the differences in the sound qualities produced by blood vessels supplying the brain and those supplying skin and muscle. In addition, the occipital artery can be compressed by applying pressure broadly in the occipital region.

Examination at the origin. Examining the vertebral artery at its origin is not an easy task. It can be mistaken for branches of the thyrocervical trunk or of the common carotid artery. The direction of flow is cranial.

Identification of the proper blood vessel should be ensured by repeatedly tapping the vertebral artery where it loops round the atlas.

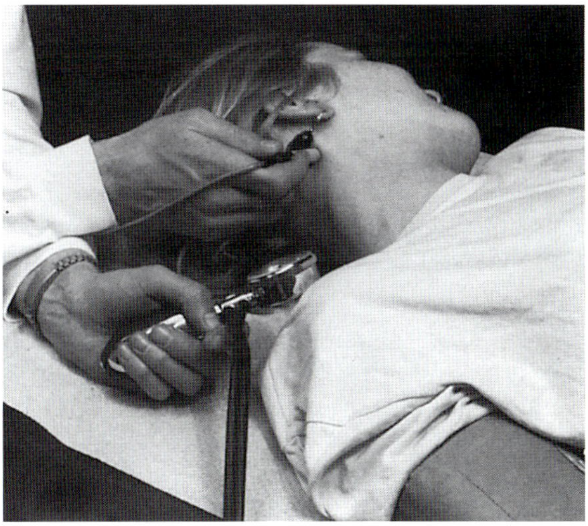

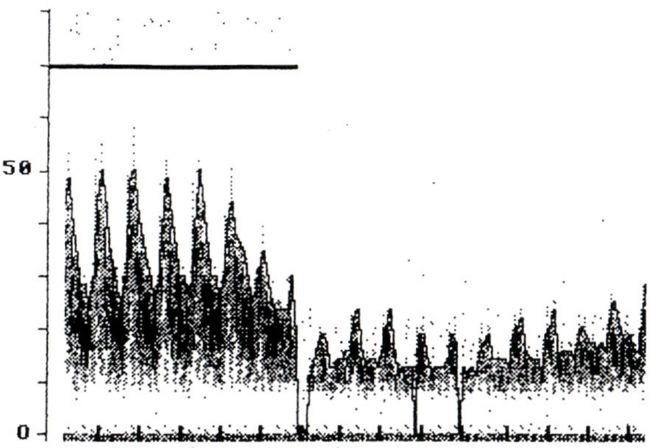

Fig. 2.**13** Testing for the subclavian steal phenomenon: suprasystolic pressure in the upper arm produced by a blood-pressure cuff causes an increase in flow velocity signals (–) from the vertebral artery at the mastoid process. After the pressure is released, a sudden short retrograde flow signal is followed by gradual restoration of the pretest situation

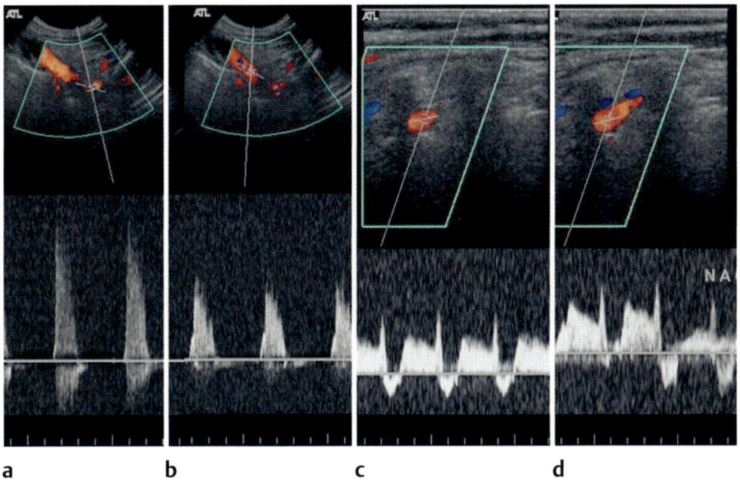

Fig. 2.**14** Display of a subclavian artery stenosis with subclavian steal phenomenon (call subclavian steal *syndrome* only when the patient is symptomatic).
a Color (frequency-mode) display and pulsed-wave Doppler registration from the stenotic lesion, showing increased flow velocities. **b** Post-stenotic registration, showing loss of back-flow (closure of the aortic valve) in early diastole.
c Color display and Doppler spectrum registration of the ipsilateral vertebral artery at the V_2 segment, showing intermediate flow: the flow has a paradoxical pattern, with back-flow in early diastole.
d After release of compression (from the blood pressure cuff around the ipsilateral arm after two heart beats), there is almost complete reversal of the flow direction in the left vertebral artery from the brain towards the arm

Subclavian Artery

Color Flow Duplex Sonography

The position used to examine the subclavian artery is the same in both Doppler sonography and duplex sonography. It is not possible to examine the vessel in selected cross-sections or in longitudinal sections rather than in diagonal sections, due to the limited overview resulting from the anatomic relationships and the large size of the transducer (Fig. 2.**15**).

Continuous-Wave Doppler Sonography

A search is made for the subclavian artery in the supraclavicular fossa. Flow direction may be toward the probe (proximal segment), or away from the probe (distal segment).

The triphasic forward and backward components of the pulse can usually be displayed. The spectrum resembles that of an artery supplying the extremities, with high peripheral resistance: systolic forward flow, early diastolic back-flow, and late diastolic forward flow. In doubtful cases, one can identify the artery by interrupting the blood flow in the upper arm, or using repeated compressions proximal to the elbow.

Innominate Artery (Brachiocephalic Trunk)

Color Flow Duplex Sonography

In addition to the location previously described, one can also attempt to depict the innominate artery in longitudinal section directly above the jugular fossa, using color flow Doppler sonography to record the flow signal; this is achievable in very few patients only.

Continuous-Wave Doppler Sonography

The innominate artery can also be located, with the probe being applied in a mediocaudal direction within the right supraclavicular fossa. The Doppler signal in this large, wide blood vessel that branches off into the common carotid artery (supplying the brain) and the subclavian artery (supplying the extremities) often sounds rather hollow and turbulent.

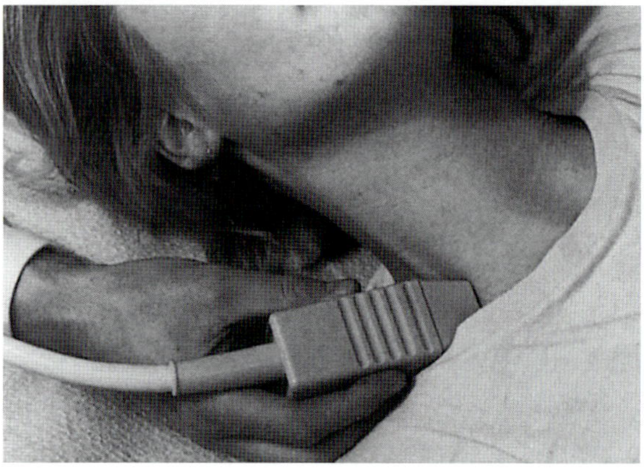

Fig. 2.**15** Handling the transducer in duplex sonography of the subclavian artery in the supraclavicular fossa

Thoracic Aorta

Color Flow Duplex Sonography and Transthoracic or Transesophageal Echocardiography

The duplex procedure is usually used for this vascular segment, especially for the branching, proximal arteries supplying the brain, either from a suprasternal position using a low-frequency probe or from transthoracic or transesophageal probe positions known from echocardiography (Seward et al. 1990, Rauh et al. 1996). The aorta cannot be examined using continuous-wave Doppler sonography. However, using a 1–2 MHz small PW Doppler probe it can often be performed much better than with the larger duplex probe.

Normal Findings

Orbital Arteries

Principle

The direction of flow in the fronto-orbital terminal branches of the ophthalmic artery is determined using bidirectional CW or PW Doppler equipment. Known as the *indirect method,* this technique is used to obtain information about patency of the internal carotid artery and reflects the flow relationships within the *circulatory system of the ophthalmic artery* (Fig. 2.**9**).

Physiologically, pressure in the two branches of the anastomosis of the ophthalmic artery—the internal carotid artery and the external carotid artery—is balanced. The two branches form this important collateral, via the supratrochlear artery and the supraorbital artery, with the watershed above the orbits. If the Doppler probe is placed on the medial canthus, it is usually possible to record blood flow signals from the carotid siphon directed toward the probe.

Even in a physiological case, it is sometimes found that the watershed has been displaced to a location within the orbital cavities, so that it is not possible to document a definite signal, or only a very weak one in comparison with the contralateral side.

Anatomy and Findings

Anatomy

The ophthalmic artery originates from the frontal convexity of the carotid siphon (C_2), and then proceeds through the dura and the optic canal, entering the orbit. It follows a course around the optic nerve until it divides into different and very variable branches (Hayreh and Dass 1962a, b):

- Laterally, through the lacrimal artery to an anastomosis with the middle meningeal artery
- Medially, through the anterior and posterior ethmoidal arteries to an anastomosis with the maxillary artery
- Rostrally, via the supratrochlear and supraorbital arteries to an anastomosis with the superficial temporal artery

The lumen of the ophthalmic artery is approximately 1 mm.

Findings

It is usually sufficient to record the supratrochlear artery. When the results are unclear, or the Doppler signal is poor, obtaining additional data from the supraorbital artery is recommended. The supraorbital artery is often hypoplastic, but rarely (5%), when the supratrochlear artery is hypoplastic, it can take on a role as the terminal vascular system. The blood flow is directed toward the probe, and is usually registered as a negative deflection in relation to the baseline. Although physiological signals are usually recorded as positive deflections, this convention has become common, since it avoids any need to change the polarity during a given series of measurements. The diastolic component of the blood flow is relatively low, since the artery supplies both the skin and musculature. During *compression tests* of the branches of the external carotid artery (the superficial temporal and facial artery, or both), the velocity of the flow normally increases, since the counterpressure in the external carotid artery is reduced. This confirms the *physiological* (orthograde) *flow direction.*

Evaluation

Only the direction of flow is relevant; formally analyzing the Doppler spectrum is of secondary importance. The diastolic flow component, in particular, has different characteristics in each individual, and depends on the peripheral resistance in the skin and musculature.

The significance of any differences in the flow velocity spectrum between the two sides of the body is limited. Such differences depend on various factors (e.g., the position of the probe in relation to the blood vessel, variability in the vascular caliber, pressure relationships in the circulatory system of the external and internal carotid arteries), and they cannot be properly evaluated as isolated phenomena. Compression tests are therefore indispensable in clearly determining the direction of flow in the ophthalmic artery.

Sources of Error

When examining a vascular loop (similar to evaluating the vertebral artery at the atlas loop), it is possible to misinterpret the blood flow direction if compression tests are not performed (Fig. 2.**16**). They are necessary here to determine whether or not blood is in fact flowing through the ophthalmic artery into the cranium—i.e., whether the blood flow originates in the branches

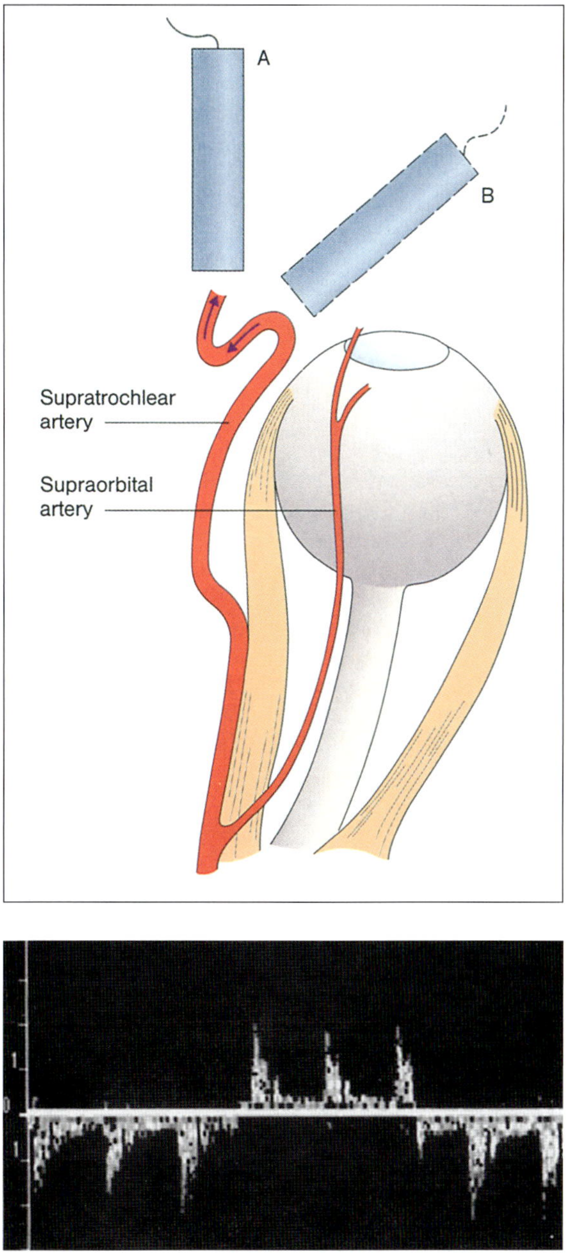

Fig. 2.**16 a** Loop formation in terminal branches of the orbital artery. **b** Doppler spectra from the supratrochlear artery with changing flow direction, depending on the probe position. Despite orthograde perfusion, blood flow away from the probe is seen in position B. Flow runs toward the probe only in A

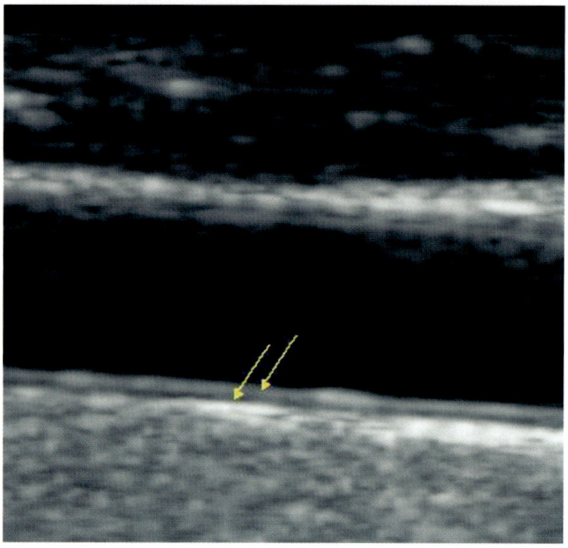

Fig. 2.**17** The normal intima–media complex in the common carotid artery, perfectly illustrating the typical hyperechoic–hypoechoic–hyperechoic structure in the far wall, while the display quality in the near wall is much poorer. The intima–media thickness (measured between the two arrows) averages 0.7 mm

of the external carotid artery. If this is the case, the blood flow should either decrease, cease, or change its direction when branches of the external carotid artery are compressed (i. e., retrograde perfusion). If the flow increases under compression, this is evidence of physiological or orthograde flow. If compression has no effect on the blood flow, no certain statement concerning the flow direction can be made. The latter is almost always the case when an anastomosis has formed from the ophthalmic artery through the maxillary artery.

Neck Arteries

Principle

Color Flow Duplex Sonography

Initially, Doppler sonography made functional and hemodynamic assessment of flow relationships possible in the neck arteries (Planiol et al. 1972, Büdingen et al. 1976). Since the 1980s, blood vessels that are close to the surface have also been imaged using echotomography (Hennerici 1983, Comerota et al. 1984, Marosi and Ehringer 1984).

Nowadays, duplex system analysis combining structural and functional vascular imaging is the standard. B-mode imaging provides morphological information about the location and texture of the blood vessel and its contents. It is possible to display the characteristics of the arterial wall using the intima–media thickness (IMT) (Fig. 2.**17**) and to assess the extent of plaques and stenoses (Hennerici et al. 2004). When this method is combined with the color mode, it is possible to distinguish the various arterial segments. Computer-assisted image processing can now provide a three-dimensional display of the different sectional planes (Fig. 2.**18**). This facilitates the evaluation and quantification of findings encountered during the examination, particularly when there are unusual anatomical relationships and pathological conditions (Steinke and Hennerici 1989, Meairs et al. 1995, Guo and Fenster 1996). Spectral analysis is useful for analyzing Doppler signals and provides semi-quantitative numerical documentation of various parameters for describing the Doppler signal (Spencer and Reid 1979, Rittgers et al. 1983).

Continuous-Wave Doppler Sonography

All flow velocity spectra from the neck arteries have to be differentiated acoustically and formally. Their shape is mainly affected by the varying peripheral resistances in the vascular system (Fig. 2.**7**).

a

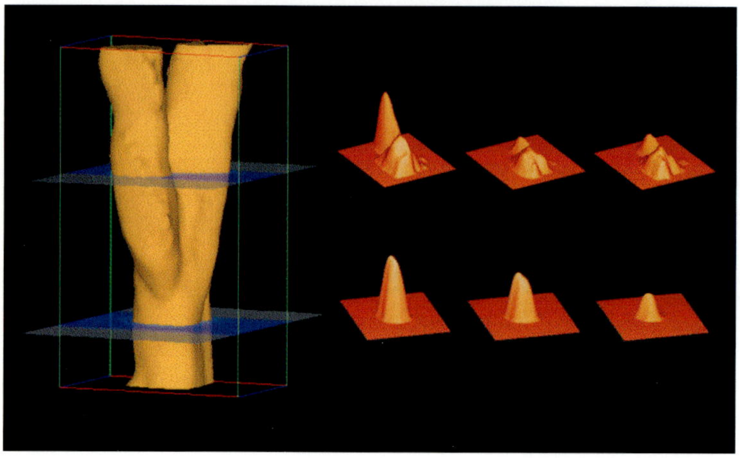

b

Fig. 2.**18** Three-dimensional display of flow information from the carotid bifurcation. **a** Surface reconstruction from a three-dimensional flow data set. **b** Reconstruction of systolic, early diastolic, and end-diastolic flow in the two cross-sections indicated on the left, showing the time-related changes flow profile

- *Arteries supplying a parenchymatous organ* such as the brain have a low peripheral resistance: the internal carotid artery, vertebral artery, superior thyroid artery, and (with qualifications) the common carotid artery (to the extent that it supplies the internal carotid artery).
- *Arteries supplying muscle and skin* have a high peripheral resistance: the branches of the external carotid artery (facial artery, maxillary artery, superficial temporal artery), the subclavian artery, and (with qualifications) the common carotid artery (to the extent that it supplies the external carotid artery).

Anatomy and Findings

Carotid System

Anatomy

The extracranial *common carotid artery* originates from the aortic arch or the innominate artery, and divides into the internal carotid artery and the external carotid artery at different levels, usually at vertebrae C4/C5 (48%) or C3/C4 (34%), and only very rarely at a higher level, at C1, or significantly lower, at C2 (Krayenbühl et al. 1979). The structure of the *carotid bifurcation* is unique in the human vascular system. While the tunica media of the common carotid artery consists of elastic fibers in different layers and muscle cells, it is the muscle cells that predominate in the internal carotid artery. At the carotid bifurcation, the transition zone, which is not present in 6% of individuals, the hybrid structure of the media decreases, and the elastic fiber structure predominates. Four different sections of the *internal carotid artery* can be distinguished: cervical, petrous, cavernous, and cerebral. After entering the cranium, the course of the internal carotid artery is initially vertical for more than 1 cm, then horizontal until the apex of the petrous bone is reached. Since the artery is reduced to at least half of its diameter (3.0–3.5 mm) in the bony canal here, its pulsatile wall movements alter over a length of 25–35 mm. Stereomicroscopic investigations have shown that there is a close relationship between the carotid artery and the transbasal veins. The artery in the proximal section of the canal, mainly fixated by longitudinal fibers, cannot move in an extracranial direction, and it is only supplied by a few accompanying veins. Further distally, there is an extensive venous drainage system into the cavernous sinus, which shows clear changes in the volume and wall movements that are pulse-dependent. At the apex of the petrous bone, the internal carotid artery gives off branches to the middle ear (caroticotympanic artery) and to the pterygoid canal. When there is an occlusion at the external carotid bifurcation, these branches often keep the intracranial segment open to this point (Paullus et al. 1977). The C_3–C_5 segments are between 30 mm and 50 mm long, and are very variable in their diameters. Subsequently, the diameter narrows conically to 6 mm (C_5/C_6 segments) and then to 4 mm (C_1/C_2 segments) (Gabrielsen and Greitz 1970). From proximal to distal, the most important branches of the internal carotid artery are (Hennerici et al. 1988a):

- The ophthalmic artery
- The posterior communicating artery
- The anterior choroidal artery

The *external carotid artery* supplies the viscerocranium and, above the carotid bifurcation, divides into several branches (Salamon et al. 1968, Merland 1973). Its anastomoses with the branches of the internal carotid artery are very important in pathological conditions, and play a vital role in ultrasound diagnosis (on the anatomy of the ophthalmic artery, see pp. 36–37 above). In addition, there are many anastomoses between the branches of the external carotid arteries on both sides and also, in pathological conditions, between the deep neck arteries and the vertebral arteries. The latter are significant as collaterals, especially when there are proximal vascular occlusions of the common carotid artery and the subclavian artery. The main external branches of the carotid artery are the superficial temporal artery and the maxillary artery. Anastomoses formed with the middle meningeal artery and the fronto-orbital terminal branches of the ophthalmic artery are very important. Caroticobasilar anastomoses, which are persistent developmental disturbances of the fetal brain circulation, are rarely encountered. The best known of these is the primitive trigeminal artery.

Common Carotid Artery

Color Flow Duplex Sonography

Due to its course, which is extensive and close to the surface, it is very easy to image the common carotid artery. The wall structure has the characteristic double reflection, with a hypoechoic zone lying in between (Fig. 2.**17**). This is termed the *intima–media complex* and the measurement of this parameter (*intima–media thickness, IMT*), is used in multicenter trials and in many standardized protocols as an indicator and prognostic parameter of atherosclerosis (progression vs. regression). Changes amounting to about 100 µm can be resolved with certainty (Hennerici and Steinke 1994). Normal values lie definitely below 1.5 mm (Riley et al. 1992) and, according to large series, lie in the range 0.96 ± 0.19 mm in women and 1.04 ± 0.22 mm in men with risk factors for atherosclerosis (Fig. 2.**19**). The American Heart Association consensus is that IMT measurements represent independent risk factors for atherosclerosis (Smith et al. 2000). The Mannheim IMT consensus adresses both IMT and plaque measurements as important parameters to evaluate vascular disease progression and regression (Touboul et al. 2004). IMT also characterizes a silent period of atherogenesis from infancy to senescence (Hennerici and

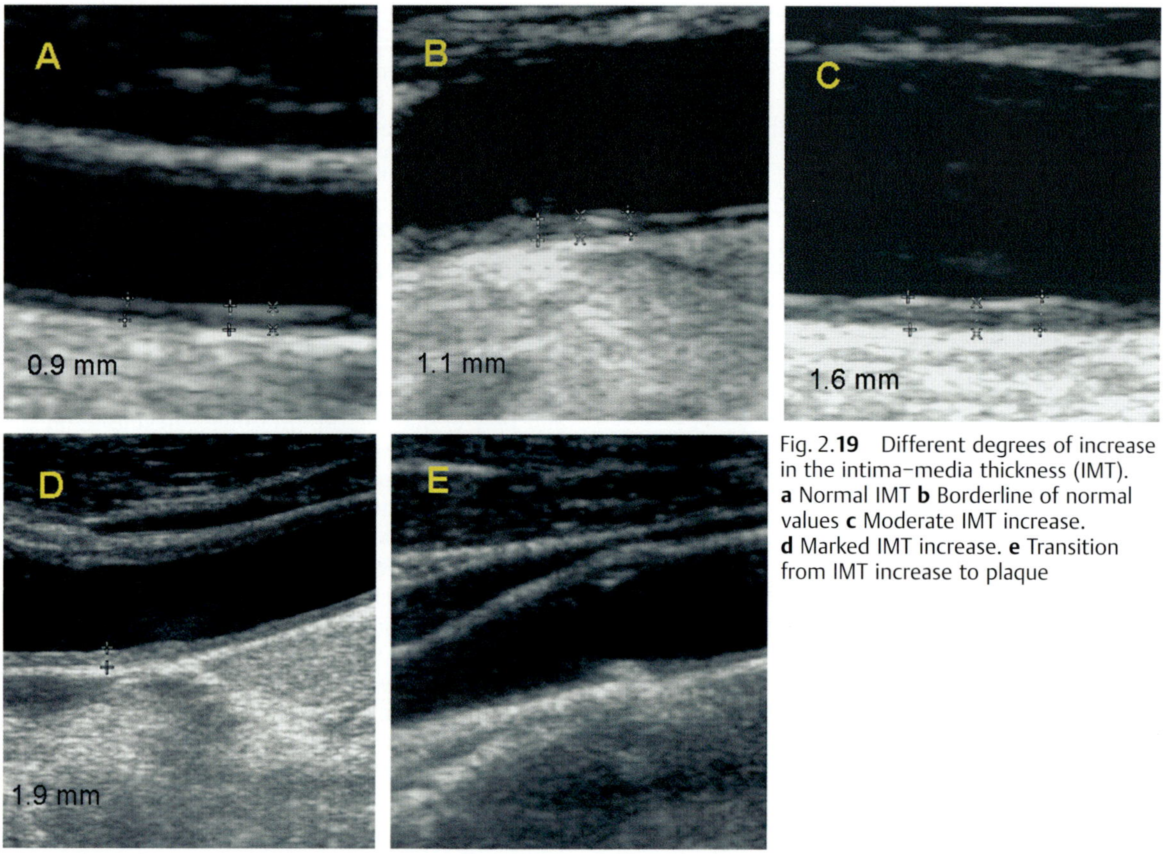

Fig. 2.**19** Different degrees of increase in the intima-media thickness (IMT). **a** Normal IMT **b** Borderline of normal values **c** Moderate IMT increase. **d** Marked IMT increase. **e** Transition from IMT increase to plaque

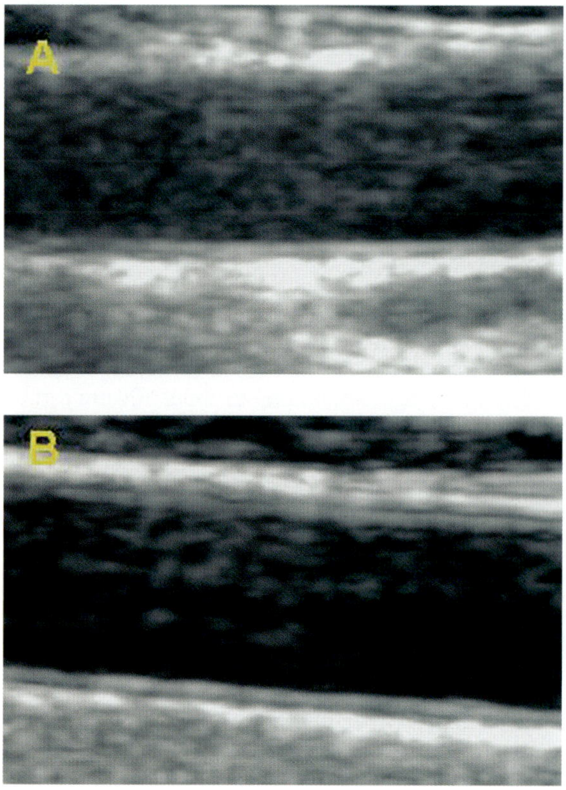

Fig. 2.**20 a** The display of the intima-media thickness (IMT) at the far wall is better than that at the near wall. **b** The display of the IMT can be improved by insonating through an interposed vein (in this case the jugular vein)

Meairs 2000). If there is associated coronary heart disease (CHD) or clinically manifest atherosclerosis, the IMT increases (O'Leary et al. 1996). The anterior wall is usually more difficult to assess than the posterior wall (Figs. 2.**17** and 2.**20**), due to image distortion in the near field; to prevent image distortion, it helps to have a vein interposed between the artery and the transducer. Normally, the width of the blood vessel is between 6.3 mm and 7.0 mm (Marosi and Ehringer 1984).

Different methods of making quantitative measurements of the *flow volume* on the basis of duplex sonography have produced very contradictory results (Licht et al. 1999, Li et al. 2001, Starmans-Kool et al. 2002). Without exactly reconstructing the ultrasound planes from several pulsed-wave Doppler components, it is not possible to distinguish correctly between the laminar flow that takes place within the common carotid artery and the plug distortions near the arterial wall. Measurements using quantitative flow volumetry therefore have an error rate of between

Table 2.**1** Blood flow volume in different age groups (adapted from Scheel et al. 2000)

Age group (y)	n	VA bilateral	ICA bilateral	ECA bilateral	CCA bilateral
20–85	78	158 ± 48 (24 ± 7)	499 ± 108 (76 ± 7)	328 ± 111	816 ± 198
20–39	24	173 ± 41 (24 ± 6)	554 ± 99 (76 ± 6)	290 ± 63	853 ± 197
40–59	24	147 ± 36 (23 ± 6)	508 ± 114 (77 ± 6)	350 ± 146	868 ± 223
60–85	30	155 ± 58 (25 ± 8)	448 ± 85 (75 ± 8)	340 ± 103	745 ± 160

CCA: common carotid artery; ECA: external carotid artery; ICA: internal carotid artery; VA: vertebral artery.

10 % and 40 % (Müller et al. 1985, Eicke et al. 1994, Licht et al. 1999) (Table 2.**1**). The same also applies to the widely used angle-corrected color-coded duplex sonography measurements, various versions of which are provided by several manufacturers. Using standard techniques, flow volume measurements based on color velocity imaging appear to be more accurate than those based on spectral Doppler imaging (Ho and Metreweli 2000, Tan et al. 2002). Sophisticated techniques using three-dimensional algorithms and gray-scale decorrelation had been introduced (Li et al. 2001, Rubin et al. 2001) and may facilitate volume flow measurements.

Using color flow duplex sonography, it is possible to improve the assessment of turbulence phenomena, flow separation zones, and movement disturbances at the vascular wall, and variations in the hemodynamics, throughout two-dimensional sections of the blood vessel during each cardiac cycle. This has diagnostic significance in patients with suspected artery-to-artery embolism (Hennerici et al. 1995, Meairs et al. 1995, Hoskins 2002).

Continuous-Wave Doppler Sonography

The Doppler spectrum of the common carotid artery is similar both to that of the internal carotid artery, as a brain-supplying blood vessel, and to that of the external carotid artery, in its role as a vessel supplying muscle and skin. Compared to the internal carotid artery, the amplitude of its systolic signal is greater, and the diastolic flow velocity is smaller. In contrast to the external carotid artery, the common carotid artery has a smaller signal amplitude and a similar diastolic flow velocity. The flow signal from the common carotid artery is acoustically associated with a rough sound (Reneman et al. 1985, Zbornikova and Lassvik 1986).

Signal frequency analysis is not as important here as it is with the internal and external carotid arteries. Stenoses usually lie at the origin, which is difficult to examine with continuous-wave Doppler ultrasound but may be possible with pulsed-wave Doppler studies. Stenoses occur much less frequently than they do at the bifurcation.

Internal Carotid Artery

Color Flow Duplex Sonography

It is usually possible to image the internal carotid artery at its origin, since the widened sinus is clearly recognizable (Figs. 2.**21**, 2.**22**). According to Marosi and Ehringer (1984), the width of the blood vessel at the sinus is 6.5–7.5 mm, and distal from this location it is 4.3–5.3 mm. IMT values lie in the range 1.35 ± 0.64 mm and 1.57 ± 0.67 mm in women and men, respectively, with risk factors of atherosclerosis, and 1.56 ± 0.70 and 1.87 ± 0.73 mm in women and men, respectively, with

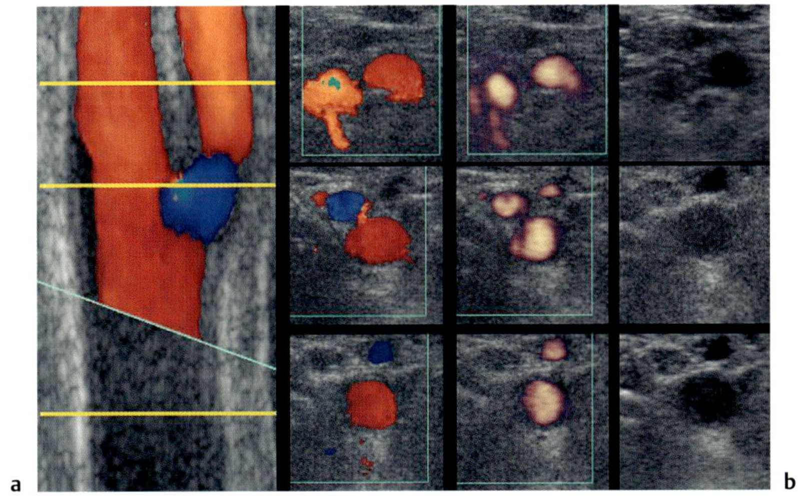

Fig. 2.**21** Display of a normal carotid artery in longitudinal section (**a**) and cross-sections at several levels (**b**). The levels are indicated by the yellow lines in the longitudinal section

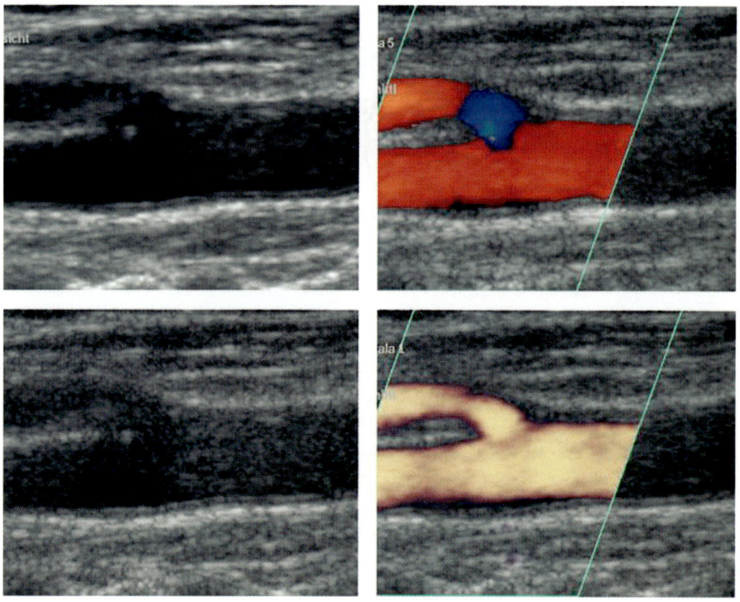

Fig. 2.**22** A normal carotid bifurcation in B-mode (**a**, **b**), frequency-coded color mode (**c**), and amplitude-coded color/power mode (**d**). Note the sensitivity for flow direction of the frequency-coded color mode—showing flow irregularities with relative back-flow components (blue areas) in the bifurcation

CHD and clinically manifest atherosclerosis, which is significantly associated with abnormal IMT values (O'Leary et al. 1996).

Dorsolateral access usually provides an adequate display of the course of the internal carotid artery. Although it is usually possible to depict its origin at the common carotid artery in a single longitudinal section, simultaneous display with the external carotid artery should be attempted when both dividing arteries are on the same plane (Fig. 2.**22**).

Depiction of the complex flow changes within the carotid sinus and in the proximal segment of the internal carotid artery is considerably better using the frequency mode of color flow duplex sonography (Schmid-Schönbein and Perktold 1995). Flow separation zones, variations in circulation, and cardiac cycle changes can be imaged in relation to the structural findings in at least two planes (Fig. 2.**23**). It is only when the carotid sinus is absent that no flow separation zones are seen. Flow separation zones can usually be detected at the proximal origin of the internal carotid artery, and in almost one in two cases they are also found at the origin of the external carotid artery and, during systole, they can be seen as a zone of clearly altered flow direction (against the axial or wall-directed flow) (Steinke et al. 1990a). A color signal is also sometimes missing if there is a clear decrease in the flow velocity. Experimental results show that changes in flow are usually not found directly at the location that causes the flow separation, but inside the carotid sinus at the origin of the external carotid artery, and that they are distributed in a horseshoe shape around the flow separation point (Karino and Goldsmith 1985, Zierler et al. 1987, Steinke et al. 1990b). The extent of the distribution of the flow separation (Fig. 2.**24**) is highly variable (Middleton et al. 1988, Steinke et al. 1990b). Recently, more sophisticated methods using multiple insonation angles to analyze the flow profile have been developed (Hoskins 2002, Starmans-Kool et al. 2002).

Continuous-Wave Doppler Sonography

Usually, the artery can be identified in its typical location by a soft, high-frequency sound and corresponding spectrum characteristics: the systolic signal amplitude is less steep, but the diastolic flow is significantly greater than that of the common carotid artery.

The transition from the common carotid artery to the internal carotid artery is almost always marked by a clear decrease in the signal amplitude and in the remaining diastolic flow in particular. This may be due to a widening of the blood vessel within the carotid sinus, or may be caused by a change in the angle between the transducer and the blood vessel. In individuals with hereditary absence of the carotid sinus (approximately 6–8% of cases), these phenomena may be absent, but this should not be interpreted as early arteriosclerotic plaque formation (Reneman et al. 1985, Harward et al. 1986, Steinke et al. 1990a, Hennerici and Steinke 1991, Steinke et al. 1992a). Continuous-wave Doppler sonography does not provide any information about the complex hemodynamics of this region.

Depending on the specific equipment and the transducer used, the upper normal value for the systolic peak frequency lies in the range of 3000–4000 Hz at a transmission frequency of 4 MHz and a flow velocity of < 120 cm/s.

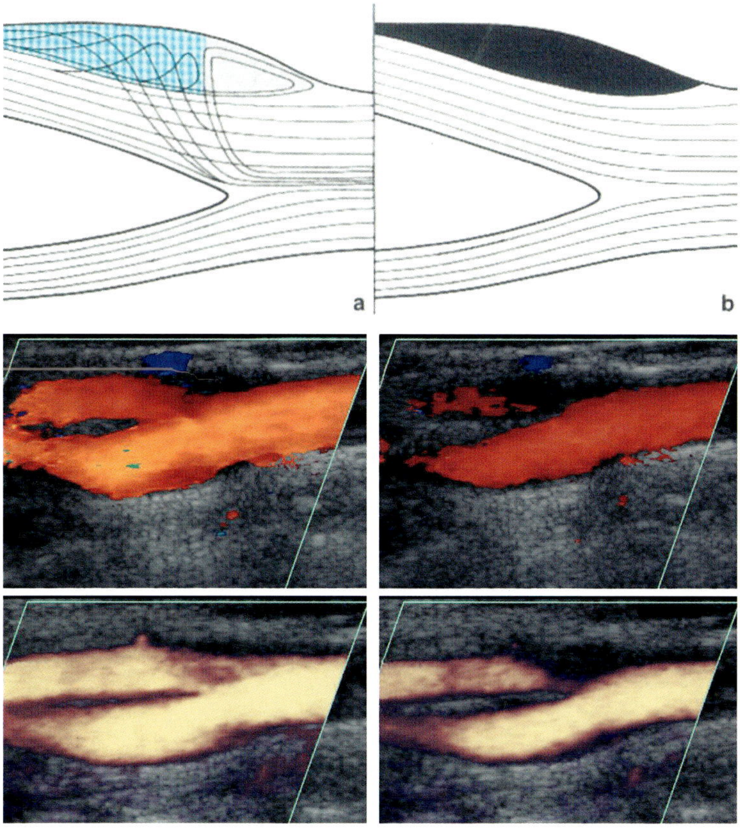

Fig. 2.**23** Diagrams and color flow duplex images (frequency mode and power mode) in systole (**a**) and diastole (**b**), showing different flow patterns in the internal carotid artery, with very low flow (no color display) at the end of diastole

External Carotid Artery

Color Flow Duplex Sonography

In B-mode, there is often no difference between the origin of the external carotid artery and that of the internal carotid artery, although it is occasionally possible to use the origin of the superior thyroid artery to assist in the identification. Display of the typical flow signals in either color-coded mode is more reliable. Both branches of the common carotid artery can be identified immediately (Figs. 2.**6**, 2.**8**, 2.**22**). Further differentiation may be assisted by evaluating the flow dynamics in the two arteries (Fig. 2.**7**). While the external carotid artery is usually displayed very briefly as a saturated color signal during systole, the internal carotid artery has more constant flow characteristics during both systole and diastole. This reflects the different vascular peripheral resistances in the two distribution areas. Using amplitude mode makes this distinction easier (Steinke et al. 1996).

In contrast to what might be expected from experimental results, significant flow separation and even flow reversal zones are also found at the origin of the external carotid artery. This is due to the configuration of the carotid sinus and the variable branching of the carotid bifurcation (Steinke et al. 1990b).

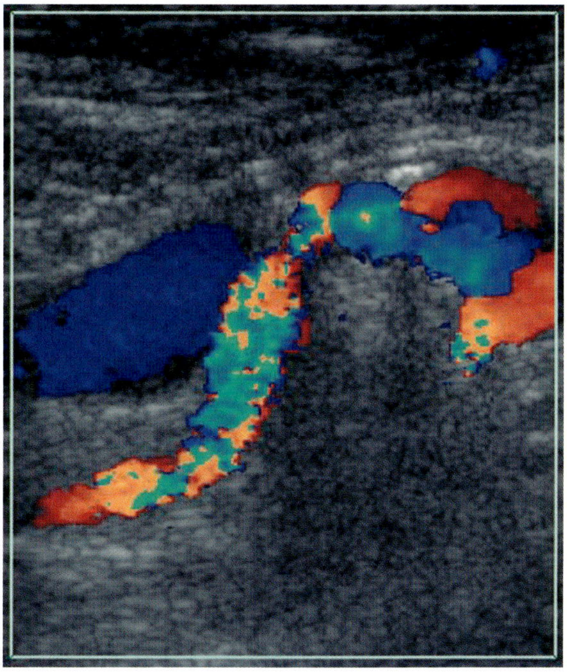

Fig. 2.**24** Color flow duplex sonography of the carotid bifurcation and an internal carotid artery (ICA) with a kink but normal laminar flow. At the proximal origin of the ICA, there are flow reversal zones (areas with black transition zones between red and blue zones) as well as aliasing (to detect slow flow components, which causes aliasing the pulse repetition frequency was set to 1000). Color changes due to aliasing show direct changes from blue to red, without zero-flow (black) transition zones

Continuous-Wave Doppler Sonography

At its origin in the carotid bifurcation, the external carotid artery can be conclusively identified due to its characteristic whiplike sound, which is accompanied by a characteristic flow velocity curve. Since the external carotid artery supplies the muscle and skin, most of its branches show a high systolic flow, with an almost complete absence of residual flow in diastole due to the high peripheral resistance in their vascular territories. Diastolic back-flow, found in proximal arteries of the extremities, is not observed. The superior thyroid artery, which proceeds in a caudal direction shortly after its origin, is an exception to these rules. As a vessel supplying a parenchymal organ, the superior thyroid artery has a flow signal that is similar to that of the internal carotid artery, due to the low vascular resistance within the thyroid gland. Thus, even in a completely normal situation, and particularly when there are loop formations, confusion with the internal carotid artery may occur. Using compression tests can clarify the situation in doubtful cases.

Vertebral–Subclavian System

Anatomy

The vertebral artery forms the first branch of the subclavian artery. It rarely originates directly from the aortic arch. Its course proceeds in a cranial direction like a curve (V_1, prevertebral part). Near vertebra C5 or C6, it enters the costotransverse foramen and follows a perpendicular path toward the cranium (V_2, transverse part). After the C2 vertebra, it proceeds laterally and winds behind the lateral mass of the atlas (V_3, atlantic part). At this point, muscular branches originate and form anastomoses with the branches of the external carotid artery. Continuing from the atlanto-occipital joint, it moves further cranially and proceeds into its subarachnoid part (V_4).

The anatomy of the subclavian artery and innominate artery is described in Chapter 5. Due to the bilateral origin of the vertebral artery from the subclavian artery, the proximal segment of the subclavian artery also counts as an artery that supplies the brain.

Vertebral Artery

Color Flow Duplex Sonography

The entire course of the artery can reliably be displayed using color flow duplex sonography, which has led to a significant improvement (Trattnig et al. 1990, Bartels et al. 1992, Trattnig et al. 1993, Ries and Steinke 1997, Ries et al. 1998, Johnson et al. 2000, Kuhl et al. 2000). After the flow signal has been detected, identification of the vessel walls of the artery and its accompanying vein is quick and direct (Fig. 2.**12**). Starting with its proximal segment in the neck, the blood vessel can be followed over a considerable distance cranially, even when its caliber is hypoplastic and when it has a curving course, and it can be readily identified between the vertebral bodies of the cervical vertebrae up to the loop formation in the mastoid region (Ries et al. 1996). The flow direction within the blood vessel can be determined with certainty without any need for compression tests. It is occasionally difficult to depict the blood vessel at its origin, due to the superimposition of other vessels or a division of the blood vessel that continues far intrathoracically. The width of the blood vessel varies between 1 mm and 5 mm, with lateral asymmetry being quite common. The atlas bow can be easily displayed using a cross-sectional approach (Fig. 2.**11**).

Continuous-Wave Doppler Sonography

It is difficult to follow the entire course of the artery using Doppler sonography alone. At best, both the afferent and efferent branches near the mastoid process can be detected (Fig. 2.**25a**, **b**). Acoustically, the flow signal shows little frequency modulation, and is therefore clearly distinguishable from the occipital artery, which at this point is little more than a crossing terminal branch of the external carotid artery, or forms anastomoses with cervical collateral arteries (Fig. 2.**25c**, **d**). Compressing the occipital artery proximally at the mastoid process helps avoid erroneous identification. It can sometimes be difficult to document the Doppler signal from the vertebral artery when the neck is thick, the blood vessel is quite deep-lying, or in the presence of variations in the caliber of the artery (hypoplasia). To determine the flow direction in the vertebral artery, a compression maneuver involving the upper arm is necessary. Bilateral differences in the flow velocity occur quite often. Color-coded duplex sonography is the preferred method for distinguishing between hypoplasia and aplasia or vertebral artery obstructions.

At its origin, the artery can be examined for several centimeters proximally (Fig. 2.**25e**). The proximity of the vertebral artery to the common carotid artery, the thyrocervical trunk, the proximal subclavian artery, and numerous veins makes it difficult to identify, especially in normal conditions (Fig. 2.**25f**). Stenoses are usually easy to detect, due to their distinctive bruit. Proceeding from the origin of the vertebral artery at the subclavian artery, with the probe directed ventromedially and caudally, it is usually possible to record signals from the proximal segment (V_1). The flow signal is directed toward the probe, and the flow sound is identical to that obtained at the mastoid process (Fig. 2.**25b**, **e**). With repeated tapping at the mastoid process, it is also possible to further identify the proximal section as a segment of the vertebral artery. Depending on the position of the loop of the vertebral artery round the atlas and the investigator's technique, the effect produced may vary. One should beware of artifacts caused by compression tests that are too vigorous, or resulting from simultaneous compression of other terminal branches of the neck arteries.

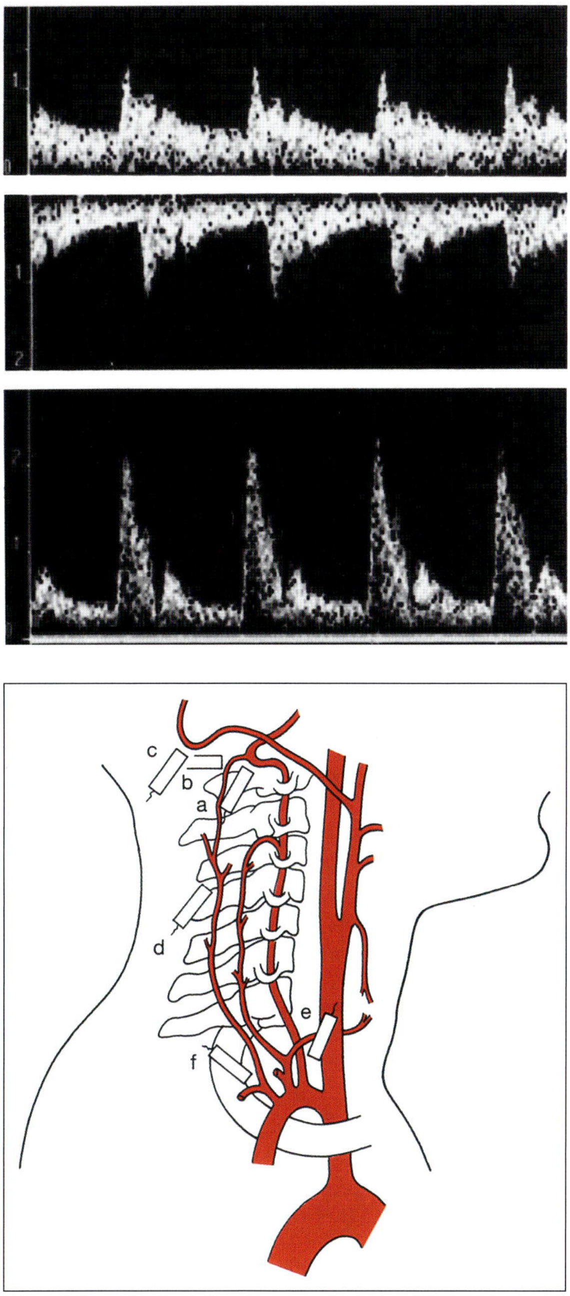

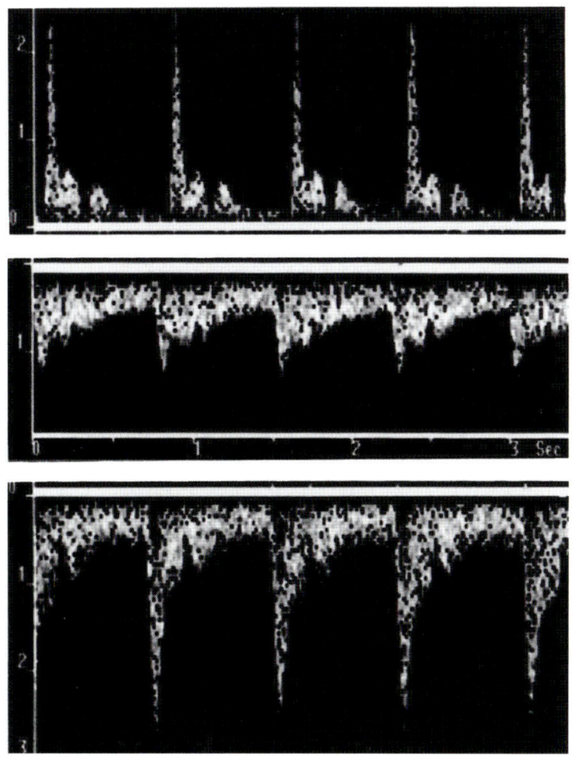

Fig. 2.**25 a–g** Doppler spectra (**a–f**) and diagram of probe positions (**g**) from an examination of the extracranial posterior circulation using continuous-wave Doppler sonography
a Vertebral artery: efferent segment of the atlas loop
b Vertebral artery, afferent segment of the atlas loop
c Occipital artery
d Cervical collaterals
e Vertebral artery at the origin
f Thyrocervical trunk
g Probe positions

Subclavian Artery

Color Flow Duplex Sonography

Due to the crooked course of the vessel, ultrasound assessment can be difficult, and is sometimes only possible on a section-by-section basis. The origin of the subclavian artery from the aortic arch is occasionally difficult to display on the left side. On the right, the origin from the innominate artery is easier to evaluate. Classifying the artery and differentiating it from the vein is easier with a properly adjusted Doppler signal.

Continuous-Wave Doppler Sonography

The Doppler signal shows the typical triphasic components of proximal arteries supplying the extremities. A high-amplitude systolic component is recorded, which increases steeply and decreases quickly. Early in the diastole, this is followed by a negative signal, and in the late diastole a brief forward flow is seen.

Acoustically, the whiplike sound of the artery's signal is even more marked than that of external carotid artery. Due to superimposed venous signals or an unfavorable ultrasound application angle, it is sometimes difficult to document flow phenomena.

Measuring the different parameters in the frequency spectrum is a secondary consideration, since quantitative values from bilateral blood pressure comparisons are usually sufficient.

Innominate Artery–Aortic Arch

Color Flow Duplex Sonography

It is sometimes possible to image the innominate artery using the supraclavicular or the suprasternal access points. The display of structural and flow-dynamic findings is better at low transmission frequencies. This vascular segment, along with the ascending aorta, can usually be displayed well using transesophageal access. Interpretation of the results is also improved when three-dimensional sectional planes are used (Rauh et al. 1996).

Continuous-Wave Doppler Sonography

Since both the origin of the innominate artery and its overall course are located in a deep intrathoracic position, direct assessment using continuous-wave Doppler sonography is not possible. Reliable evaluation of this region is possible with pulsed-wave Doppler systems, and the overall course can be displayed in a bilateral comparison similar to the left common carotid artery. Alternatively, it is possible to examine this intrathoracic segment using transesophageal echocardiography.

At a low ultrasound transmission frequency (2 MHz), flow velocities around 54.6 ± 16.9 cm/s are physiological when a supraclavicular access point for ultrasound application is chosen (Rautenberg and Hennerici 1988).

Evaluation

Optimal interpretation of the results of the examination is only possible when one observes the procedure oneself, taking into account the immediate acoustic and visual information and analyzing the documented findings carefully. Any interpretation that is made without knowledge of the specific circumstances associated with the examination of a patient can lead to significant misinterpretations. Graphic documentation provides a fair indication of the care exercised by the investigator and the extent of the examination, but it cannot replace the subtle and direct experience of the individual investigator.

Color Flow Duplex Sonography

Duplex sonography has clearly simplified the evaluation of the results and made it easier to distinguish normal findings and variants within the vertebral artery system. This was made possible by introducing improved imaging capabilities for structural wall relationships, and especially after the introduction of simultaneous flow analysis into the color flow procedure. In general, systematic evaluation of all of the important parameters is carried out in normal conditions, and the findings are carefully described and documented.

Continuous-Wave Doppler Sonography

The *analogue graphs* are usually only evaluated qualitatively, comparing the two sides. The end-diastolic flow within the common carotid artery is particularly important in bilateral comparisons. With optimal ultrasound application, variations of more than 15 % in the values obtained are abnormal. Due to the slightly different courses of both common carotid arteries, it may therefore be necessary to repeat the side-to-side comparison at two or three levels, in order to avoid misinterpretations. By contrast, bilateral comparison of the measurements is less informative in the internal carotid artery, due to the more variable vascular course and the less standardized probe position in relation to the arterial axis. Asymmetries can be quite normal here. However, hypoplasia is extremely rare, in contrast to the vertebral system, where flow asymmetries are common (25–30 %). In this context, it should also be noted that the flow velocity in a hypoplastically narrow blood vessel can be higher than that in a comparably normal or even pathologically narrow contralateral vertebral artery, while the flow velocity is more likely to be lower in a dilated vascular segment (Ries et al. 1996). It should also be noted that a hypoplastic vertebral artery often has a limited vascular territory—e.g., supplying the posterior inferior cerebellar artery instead of the basilar artery. In such cases, the flow is low even if the caliber of the blood vessel is normal. Better interpretation of the often difficult and inadequately classifiable hemodynamic findings can only be achieved using color flow duplex sonography (Trattnig et al. 1990, 1993, Ries and Steinke 1997).

Formal analysis of the Doppler spectrum in the carotid and vertebral systems often makes it easy to determine which of the arteries in the neck is being depicted, particularly when compression tests and vibration maneuvers (i.e., tapping) are registered at the same time.

Only two parameters in the Doppler spectrum are usually evaluated:

- The systolic peak frequency
- The average mean frequency

In addition, the diastolic frequency is sometimes important for side-to-side comparison. Asymmetries in the spectrum have so far been difficult to detect.

Many algorithms have been proposed for calculating the width of the spectrum or the systolic window, particularly in the carotid system, including comparisons of preceding and subsequent vascular segments.

These investigations mostly date from the period when the quality of echotomographic imaging was still relatively poor. The aim was to detect early forms of arteriosclerosis by calculating this parameter. However, systematic comparison of the structural and hemodynamic parameters shows that, even in completely normal situations, and particularly in younger patients with physiological flow displacements and separation phenomena, there may be limitations. In doubtful cases, an unusual result obtained when analyzing the spectrum should subsequently be checked using duplex sonography, which is the ultimate diagnostic arbiter.

Sources of Error

Ultrasound Incident Angle

Color Flow Duplex Sonography

Flow irregularity may be incorrectly represented by a change in color if the orientation of the blood vessel is not parallel to the probe. Actually, only the direction of the vascular axis in relation to the transducer position has changed when this happens. This is a frequent source of misinterpretations, especially in loop formations (Fig. 2.**26**).

Continuous-Wave Doppler Sonography

In Doppler sonography, whether in analogue recording or spectral analysis, the most important source of error is still the unknown incident angle of ultrasound application.

Since the arterial course is very variable and bilateral symmetry is not an absolutely reliable parameter, applying the ultrasound at too flat angles may erroneously simulate a pathological flow acceleration or a decrease in flow if the angle used is unfavorable (too steep or too flat) (Fig. 2.**27**). When the patient has a long, slender neck, this is easily possible in the submandibular region of the internal carotid artery.

Due to the almost perpendicular angle at which ultrasound is applied where the vertebral artery loops round the atlas, a lower flow velocity is registered here than is actually the case.

Cardiac Arrhythmia

Due to the varying filling phases of the cardiac cycle, absolute arrhythmia makes it more difficult to evaluate the flow spectra.

Vascular Width

At the same flow intensity, the flow velocity will be lower in a wider vessel, and higher in a narrower vessel. Therefore, the flow in a wide sinus can easily be disturbed, and may incorrectly suggest a pathological finding. Particularly in older patients, aneurysmal dilations are difficult to distinguish from dilative arteriopathy without duplex sonography.

Anomalies in the Vascular Course

An atypical arterial course can make it more difficult to locate the blood vessels and can simulate stenosis, kinking, or coiling, or an occlusion. This is most often the case when the internal carotid artery has a medial origin (Fig. 2.**24**).

Venous Superimposition

Venous superimposition can disturb the assessment with continuous-wave ultrasound and distort the signals as a result of opposed deflections. The venous noise often mimics a stenotic signal. By contrast, duplex sonography may be facilitated by venous superimposition.

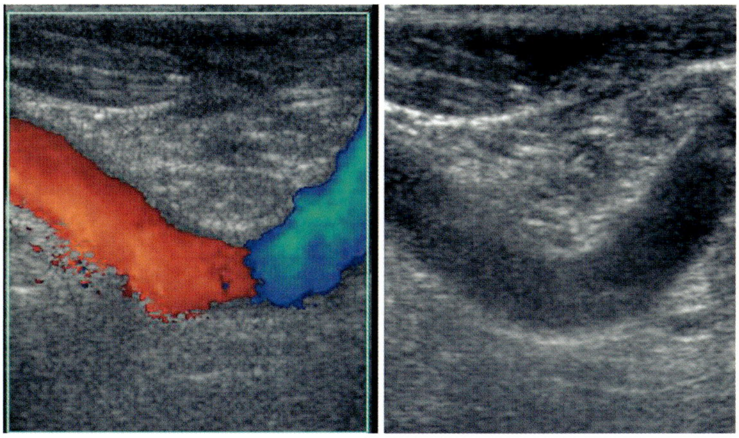

a b

Fig. 2.**26** Kinking of the internal carotid artery in color mode. Note the color changes relative to the ultrasound probe in the absence of flow abnormalities

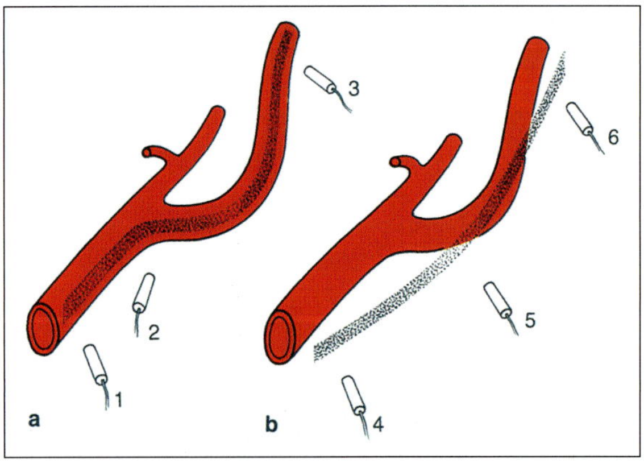

Fig. 2.**27** Differences in the Doppler waveform during examinations of the extracranial carotid system, **a** when following the course of the common carotid artery (CCA) and internal carotid artery (ICA) with the probe position optimized, and **b** when holding the probe rigidly adjusted

a1 a2 a3

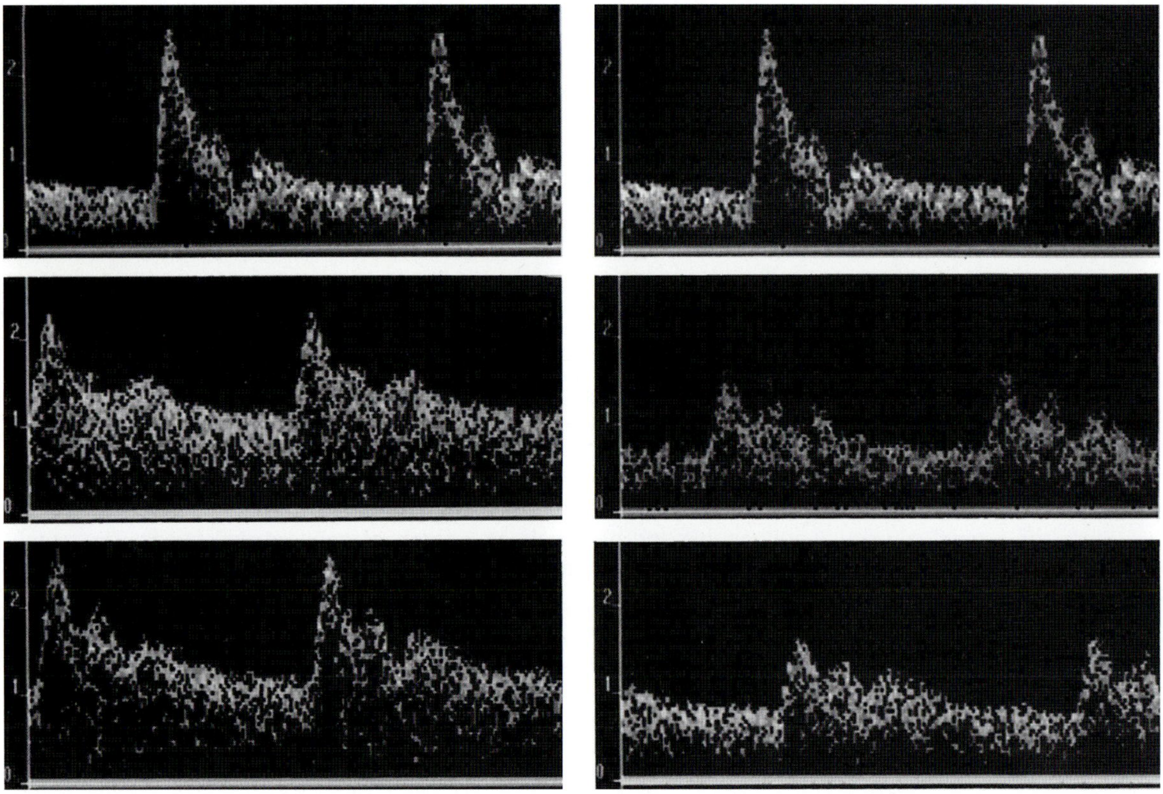

b4 b5 b6

Echogenicity

Fresh thrombotic material in blood vessels has the same acoustic impedance as blood. In the B-mode image, the blood vessel appears to be open. It is only with the Doppler signal that it is possible to determine whether the lumen of the blood vessel has narrowed or, when the signal is missing, has closed (Zwiebel et al. 1983). Missing pulsatility in the B-mode image and omission of color flow Doppler signals are characteristic.

If the sectional plane is selected unfavorably, a plaque may be bypassed, and the blood vessel be misdiagnosed as completely open. However, the opposite is also possible—plaque formation reaching into the lumen can be mimicked if wall segments of tortuous arteries are imaged diagonally. Examinations should therefore always be carried out in several planes, including both longitudinal sections and cross-sections, in order to avoid misinterpretations.

Postintervention

Depending on the specific type of procedure carried out, there may be alterations in the Doppler signal and the arterial wall interface after surgery (Fig. 2.**28**), as well as after angioplasty and stenting (Fig. 2.**29**).

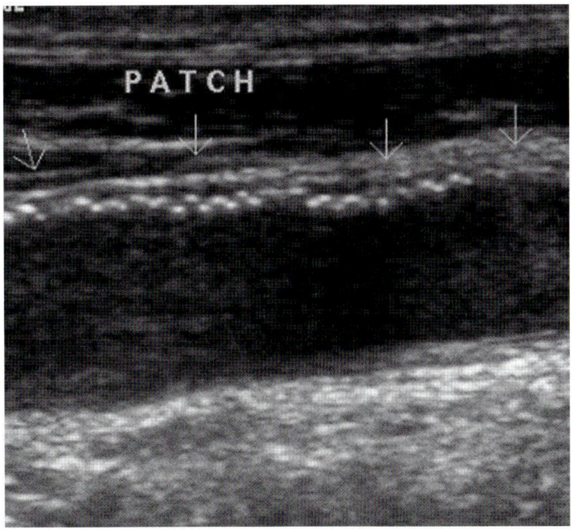

Fig. 2.**28** An ultrasound examination 6 days after carotid surgery. Parts of the suture or patch material are sometimes prominent in the acute period after surgery (arrows). The findings vary, depending on the individual surgeon's technique

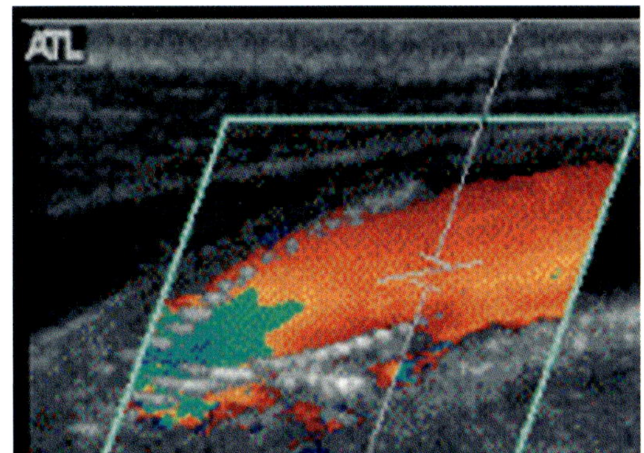
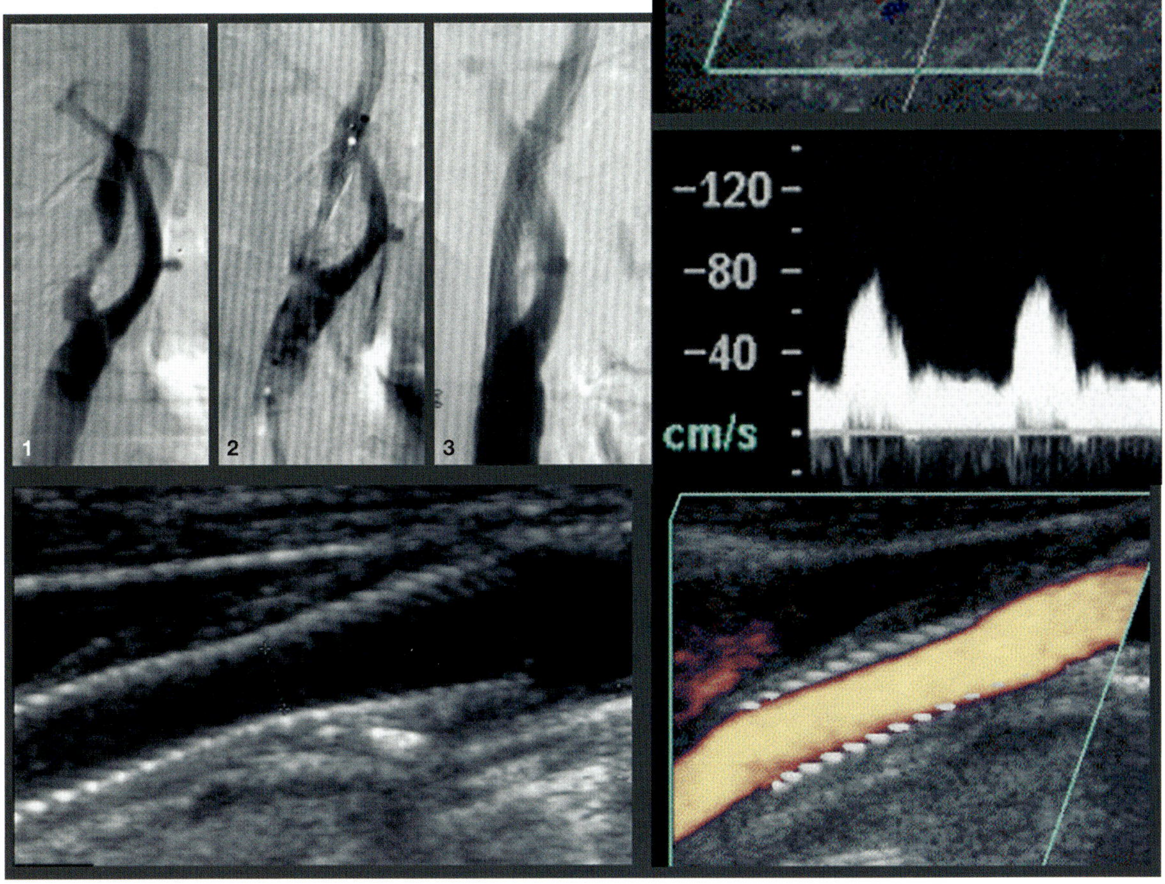

Fig. 2.**29** Stenting in high-grade carotid stenosis. The angiograms (**a**) show, from left to right: 1, high-grade carotid stenosis; 2, the stent catheter in place; and 3, the situation after angioplasty and stent implantation. The ultrasound images show a regular flow profile (**b**); regular stent placement (**c**); and complete filling of the intrastent lumen (**d**). Note the slight residual stenosis (approximately 30 %) evident in the B-mode image

Diagnostic Effectiveness

Color Flow Duplex Sonography

The *specificity* of duplex investigations lies between 54% and 100% (Tables 2.**3**, 2.**4**) (de Bray et al. 1995, de Bray and Glatt 1995). B-mode examinations on their own have a low accuracy level and are inadequate. Angiographic examination, which for a long time used to be the accepted reference method for such measurements, do no longer serve as the gold standard (de Bray and Glatt 1995). For example, digital subtraction angiography, even using selective intra-arterial contrast medium administration, often fails to detect small changes in the walls of blood vessels. Comparative studies of duplex sonography and magnetic resonance angiography have repeatedly been published (Mattle et al. 1991, Riles et al. 1992), and show a good correlation with normal findings. Flow irregularities that appear physiologically in the carotid bifurcation can produce artifacts in magnetic resonance angiography, which can be identified by color flow duplex sonography if structural wall changes are absent.

The predictive value of the ultrasound method is significantly dependent on the quality of the examination procedure. Unfavorable conditions (e.g., an inability to study the arterial course in several sectional planes) can influence the specificity in individual cases. Also, isolated echotomographic imaging of the arteries of the neck, without additional examination of the hemodynamic findings using Doppler sonography, has a negative influence on the diagnostic predictive value (Ricotta et al. 1987), and should therefore be avoided. The advantage of color flow duplex sonography lies in the easy and quick simultaneous display of morphological and hemodynamic parameters. The specificity of the method ranges between 86% and 100% (Steinke et al. 1990a, Sitzer et al. 1993, Görtler et al. 1994), similar to that of continuous-wave and pulsed-wave Doppler versus angiography in the detection of carotid stenoses > 50% (92–99% and 78–95%, respectively) in experienced hands. It is improved by the combined use of color flow velocity and signal amplitude analysis (*power imaging*) (Steinke et al. 1996) (Tables 2.**3**, 2.**4**).

Few investigations show that the specificity for evaluating the vertebral arteries lies at about 83% (Ackerstaff et al. 1984). For color flow duplex sonography, these figures may well turn out to be higher.

Table 2.**2** Calculated reliability of CW/PW Doppler sonography versus intra-arterial angiography in internal carotid artery stenoses and occlusions

			Specificity	**Sensitivity**								
First author	Year	Arteries (n)	(%)	Degree of stenosis (%)					Occlusion (%)	Validity (%)	F	D
				0–20	20–40	40–60	60–80	80–100				
Barnes	1981	199	91		30		85		96	61	+	CW
Hennerici	1981	488	97		–	87	96	95	93	96	–	CW
Doorly	1982	190	79		85		91		90	84	+	PW
Rittgers	1983	123	89	28	62	77	88	63	–		+	CW
Neuerburg	1984	209	98		68		90		97	91	–	CW
Trockel	1984	431	96		34	81	93	86	96	85	–	CW
Hennerici	1985	28	89		68		–		–		+	CW
Fischer	1985	229	97	52	91		100		97	92	+	CW
Caes	1987	48	96		79		100	33	100	83	+	CW
Sillesen	1988	128	90		83		87		100		+	PW
Bornstein	1988	662	–			99			86	98	+	CW
Perry	1985	155	92		79		72	87	94	85	+	CW
Bandyk	1985	243	90		85		87	72	90	88	+	CW
Lindegaard	1984	432	94			96			97	94	+	PW
Floriani	1987	469	89		75		73	87	88	85	+	CW
Brown	1982	205	92				91		96		+	CW
Hames	1985	137	93				91		–		+	CW
Harward	1986	130	93				93		–	92	+	CW
Widder	1987	65	94				90		94	76	+	CW
Blackshear	1987	48	88				87		–		+	PW

F frequency analysis (+ yes, – no); D Doppler technique (CW continuous wave, PW pulsed wave Doppler)

Table 2.**3** Calculated reliability of conventional duplex sonography versus intra-arterial angiography in patients with internal artery stenoses and occlusions

<table>
<tr><th></th><th></th><th></th><th>Specificity</th><th colspan="6">Sensitivity</th><th></th><th></th><th></th></tr>
<tr><th>First author</th><th>Year</th><th>Arteries (n)</th><th>(%)</th><th colspan="5">Degree of stenosis (%)</th><th>Occlusion (%)</th><th>Validity (%)</th><th>Ulceration (%)</th><th>T</th></tr>
<tr><th></th><th></th><th></th><th></th><th>0–20</th><th>20–40</th><th>40–60</th><th>60–80</th><th>80–100</th><th></th><th></th><th></th><th></th></tr>
<tr><td>Abu Rahma</td><td>1987</td><td>200</td><td>71</td><td colspan="2">51</td><td>81</td><td colspan="2">79</td><td>94</td><td>71</td><td></td><td>D</td></tr>
<tr><td>Moore</td><td>1988</td><td>170</td><td>–</td><td colspan="2">94</td><td colspan="3">98</td><td>100</td><td>93,5</td><td>64</td><td>D</td></tr>
<tr><td>Mattle</td><td>1991</td><td>39</td><td>–</td><td colspan="3">88</td><td colspan="2">86</td><td></td><td>85</td><td></td><td>B</td></tr>
<tr><td>Comerota</td><td>1984</td><td>1723</td><td>–</td><td colspan="2">87</td><td colspan="2">72</td><td>66</td><td>64</td><td>80</td><td></td><td>B</td></tr>
<tr><td>Jones</td><td>1982</td><td>100</td><td>85</td><td colspan="5">98</td><td>75</td><td>–</td><td></td><td>B</td></tr>
<tr><td>Terwey</td><td>1981</td><td>200</td><td>75</td><td colspan="2">99</td><td colspan="3">74</td><td>62</td><td>–</td><td></td><td>B</td></tr>
<tr><td>Wolverson</td><td>1983</td><td>97</td><td>47</td><td colspan="2">83</td><td colspan="3">78</td><td>36</td><td>69</td><td></td><td>B</td></tr>
<tr><td>Zwiebel</td><td>1983</td><td>393</td><td>95</td><td colspan="2">72</td><td>52</td><td colspan="2">21</td><td>91</td><td>62</td><td>44</td><td>B</td></tr>
<tr><td>Daiss</td><td>1984</td><td>488</td><td>89</td><td colspan="2">50</td><td colspan="3">85</td><td>87</td><td>66</td><td></td><td>D</td></tr>
<tr><td>Hames</td><td>1985</td><td>85</td><td>58</td><td colspan="2">85</td><td colspan="2">76</td><td>85</td><td>92</td><td>79</td><td></td><td>D</td></tr>
<tr><td>Hennerici</td><td>1984</td><td>193</td><td>76</td><td colspan="2">88</td><td colspan="3">100</td><td>71</td><td>83</td><td></td><td>B</td></tr>
<tr><td>Ratlif</td><td>1985</td><td>39</td><td>–</td><td>75</td><td>67</td><td colspan="2">69</td><td>79</td><td>–</td><td>74</td><td></td><td>D</td></tr>
<tr><td>Zbornikova</td><td>1985</td><td>249</td><td>54</td><td>80</td><td colspan="2">77</td><td colspan="2">92</td><td>94</td><td>76</td><td></td><td>D</td></tr>
<tr><td>Ricotta</td><td>1987</td><td>1578</td><td>50</td><td colspan="2">43</td><td colspan="2">39</td><td>36</td><td>41</td><td>43</td><td>32</td><td>B</td></tr>
<tr><td>Langlois</td><td>1983</td><td>77</td><td>50</td><td>69</td><td colspan="2">70</td><td colspan="2">89</td><td>100</td><td>77</td><td></td><td>D</td></tr>
<tr><td>Cape</td><td>1984</td><td>586</td><td>93</td><td>64</td><td colspan="2">59</td><td>71</td><td>68</td><td>54</td><td>84</td><td></td><td>B</td></tr>
<tr><td>Blackshear</td><td>1979</td><td>66</td><td>100</td><td>44</td><td colspan="2">74</td><td colspan="2">95</td><td>92</td><td>80</td><td></td><td>D</td></tr>
<tr><td>Rush</td><td>1985</td><td>20</td><td colspan="2">81</td><td colspan="2">75</td><td colspan="2">92</td><td>100</td><td>–</td><td></td><td>D</td></tr>
<tr><td>Dreisbach</td><td>1983</td><td>101</td><td colspan="2">94</td><td>84</td><td>76</td><td colspan="2">83</td><td>94</td><td>–</td><td></td><td>D</td></tr>
<tr><td>Humphrey</td><td>1990</td><td>186</td><td colspan="2">84</td><td colspan="2">83</td><td>94</td><td>68</td><td>100</td><td>–</td><td></td><td>D</td></tr>
<tr><td>Burns</td><td>1985</td><td>500</td><td colspan="3">71</td><td>38</td><td>26</td><td>6</td><td>83</td><td>–</td><td></td><td>D</td></tr>
<tr><td>Withers</td><td>1990</td><td>180</td><td colspan="4">94</td><td>80</td><td>98</td><td>96</td><td>94</td><td></td><td>D</td></tr>
<tr><td>Lane</td><td>1982</td><td>304</td><td colspan="4">80</td><td colspan="2">76</td><td>–</td><td>–</td><td></td><td>D</td></tr>
<tr><td>Keagy</td><td>1982</td><td>171</td><td colspan="4">94</td><td colspan="2">89</td><td>94</td><td>93</td><td></td><td>D</td></tr>
<tr><td>Bornstein</td><td>1988</td><td>124</td><td colspan="4">96</td><td colspan="2">–</td><td>96</td><td>96</td><td></td><td>D</td></tr>
</table>

T: technique used (D: duplex mode, B: B–mode)

Table 2.**4** Calculated reliability of color flow duplex sonography versus intra-arterial angiography in patients with internal carotid artery stenoses and occlusions

<table>
<tr><th></th><th></th><th></th><th>Specificity</th><th colspan="6">Sensitivity</th><th></th></tr>
<tr><th>First author</th><th>Year</th><th>Arteries (n)</th><th>(%)</th><th colspan="5">Degree of stenosis (%)</th><th>Occlusion (%)</th><th>Validity (%)</th></tr>
<tr><th></th><th></th><th></th><th></th><th>0–20</th><th>20–40</th><th>40–60</th><th>60–80</th><th>80–100</th><th></th><th></th></tr>
<tr><td>Steinke</td><td>1990</td><td>60</td><td>88</td><td colspan="2">69</td><td>78</td><td>100</td><td>88</td><td>100</td><td>80</td></tr>
<tr><td>Polak</td><td>1992</td><td>41</td><td colspan="4">91</td><td colspan="2">85</td><td>100</td><td>88</td></tr>
</table>

Continuous-Wave Doppler Sonography

Directly assessing the arteries of the neck using ultrasound has led to a significant improvement in noninvasive vascular diagnosis in this region. It is now possible to differentiate the various vascular segments and confirm whether or not they are open.

Using direct, continuous-wave Doppler sonography, stenoses exceeding 50% and occlusions in the carotid system can be excluded with a high degree of certainty. In the hands of experienced investigators, the *specificity* of the technique lies between 88% and 98% (Table 2.**2**) (de Bray and Glatt 1995). There is no increase in the reliability of diagnostic statements based on spectral analysis with regard to excluding a pathological finding. Depending on which parameters are used to evaluate the spectrum, the number of false-positive results may vary (Sheldon et al. 1983, Lally et al. 1984, Neuerburg-Heusler et al. 1985) (Table 2.**2**).

Fewer case numbers are available for the vertebral arteries. Due to the frequent variants, the specificity is lower, at 75–90% (von Reutern and Clarenbach 1980, Winter et al. 1988).

Pathological Findings

Orbital Arteries

Principle

The ophthalmic artery and its terminal branches, the supraorbital and supratrochlear arteries (the lateral or medial frontal arteries), form a watershed with branches of the external carotid artery. Impediments to normal blood flow in this circulatory system, also known as the *ophthalmic artery anastomosis*, can lead to distortions in the flow equilibrium when they are accompanied by changes in the pressure relationships (Fig. 2.**30**) (Melis-Kisman and Mol 1970, Müller 1971, Planiol et al. 1972, Büdingen et al. 1976). Using Doppler sonography, a variety of observations can be made via the supratrochlear, supraorbital, and ophthalmic arteries.

Findings

Color Flow Duplex Sonography

Recent attempts to examine branches of the ophthalmic artery in the orbit to identify small embolic events have not been useful for clinical diagnosis. If emboli travel into the retinal blood vessels, they may be demonstrated—e.g., by absent flow signals and hyperintense reflections, sometimes producing echo shadows in the central retinal artery (Hedges et al. 1993, Hedges 1995).

Continuous-Wave Doppler Sonography

Flow reversal. The direction of flow can change due to a decrease in pressure within the internal carotid artery (occlusion or high-grade stenosis). In the supratrochlear and supraorbital arteries, blood can be observed to move away from the probe—through the branches of the external carotid artery and the ophthalmic artery, blood flows into the carotid siphon and subsequently into the intracranial cerebral arteries. When the branches of the external carotid artery are compressed, this retrograde inflow is reduced or even reversed into the normal, physiological flow direction.

Since the peripheral vascular resistance in cerebral arteries is lower than that in the vasculature supplying skin and muscle, the shape of the Doppler waveform also changes—the diastolic residual blood flow, in particular, often remains very high in the orbital arteries when flow reversal occurs.

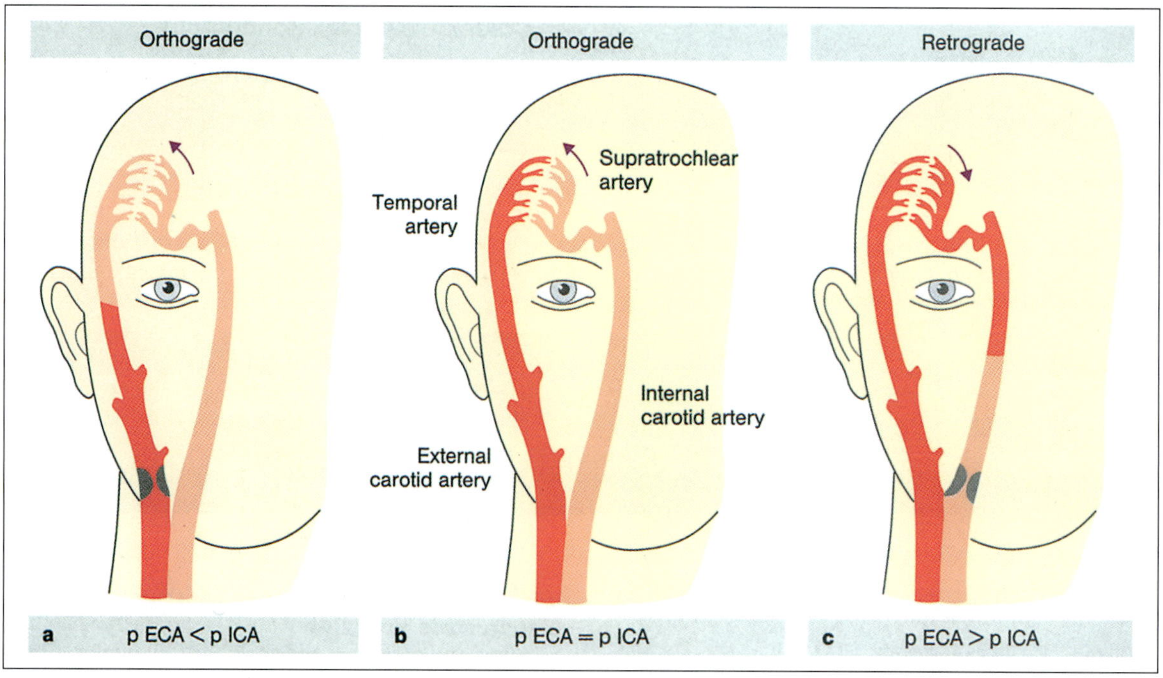

Fig. 2.**30 a–c** The collateral system and resulting blood flow directions in the ophthalmic artery. **a** A pressure decrease in the external carotid artery = orthograde blood flow. **b** Normal findings (pressure in the internal and external carotid arteries is at equilibrium, indicating orthograde blood flow in the orbital arteries. **c** A pressure decrease in the internal carotid artery = retrograde blood flow

No blood flow. Occasionally, it is not possible to record a signal spontaneously, and acoustic detection is often not possible either. When the branches of the external carotid artery that form the anastomosis are compressed, physiological blood flow may result. This finding is often caused by a hemodynamically significant obstruction in the internal carotid artery (high-grade stenosis or occlusion), with insufficient collateralization through the ophthalmic artery anastomosis—e.g., in the presence of a stenosis of the extracranial carotid bifurcation, when pressure is reduced in both the internal and the external carotid arteries.

Alternating forward and backward (pendular) blood flow. When the fronto-orbital terminal branches of the ophthalmic artery are being examined, a rare observation is a breath-like audio signal, corresponding to alternating forward and backward blood flow around the baseline on the recorded graph. Compression of branches of the external carotid artery leads to a gradual return of blood flow to normal (watershed displacement). A high-grade obstruction in the innominate artery with significant reduction of pressure in the entire extracranial carotid system is often responsible for this rare finding (Hennerici et al. 1981a). Occasionally, a stenosis at the carotid bifurcation in the neck may also cause this phenomenon, and in rare instances, hypoplastic orbital arteries (Keller 1961) or a variant in the origin of the ophthalmic artery may be present (Vogelsang 1961).

Reduced physiological blood flow. A wide variety of reasons for asymmetric amplitudes (systolic or diastolic, or both) are observed in side-to-side comparisons, with an increase in the flow signal when a branch of the external carotid artery participating in the collateral blood flow anastomosis is compressed; these differences are not significant in diagnosis. Flow obstructions in the extracranial and intracranial arteries may lead to such findings, as well as variations in the caliber of the ophthalmic arteries (Decker and Schlegel 1957, Di Chiro 1961). More often, these lateral differences are caused artificially by variations in the angle between the ultrasound probe and the vasculature.

Pathologically elevated nonpulsatile flow. Rarely, a marked, continuous sibilant audio signal may be heard—e.g., in the vicinity of a significant stenosis of the neck arteries (Büdingen and von Reutern 1993). If the flow direction is toward the probe, venous drainage through the orbital veins should be considered, as in the case of a cavernous sinus fistula involving the carotid artery. Increased pressure in the cavernous sinus can be caused by retrograde blood flow through the orbital veins, which subsequently passes through the jugular vein. The differential diagnosis should include large intracranial arteriovenous malformations, which generally produce a pathological retrograde nonpulsatile flow signal in the orbital arteries (von Reutern et al. 1977).

Evaluation

The most reliable indirect criteria for assessing flow obstructions with hemodynamic significance in the internal carotid artery are, in decreasing order of importance:

- Retrograde blood flow
- Alternating forward and backward (pendular) blood flow
- No blood flow

in the main branch of the ophthalmic artery (usually the supratrochlear artery).

Asymmetries noted during side-to-side comparisons when the flow direction is orthograde can have several causes, and cannot be properly assessed until results are obtained from direct Doppler sonography of the neck arteries (Büdingen et al. 1976, Ackermann 1979, Trockel et al. 1984).

Sources of Error

Loop formation. Since, in a loop formation, compression tests in the external carotid artery branches affected are necessary to determine the actual direction of flow within the orbital arteries, a signal directed toward the probe can indicate both orthograde and retrograde flow (Fig. 2.**16**, Table 2.**5**). A positive compression effect with an accompanying *decrease* or *reversal* in the registered blood flow confirms *retrograde flow* in the orbital artery. Any increase in the flow velocity observed while compressing the branches of the external carotid artery indicates a physiological *orthograde* flow direction. If only one of the two components of the loop

Tabelle 2.**5** Interpreting the Doppler-sonographic findings in the orbital arteries

Flow direction related to the probe	Compression effect on flow	Verified flow direction
→ [ ] ~ ← [Loop] ~	Increase	Orthograde
← [ ] ~ → [Loop] ~	Decrease/ Reversal	Retrograde
← [ ] ~ → [Loop] ~	No reaction	None

can be assessed, the compression test is indispensable to determine the flow direction.

Variants. Usually, the supratrochlear artery is a terminal branch of the ophthalmic artery, whereas the supraorbital artery is frequently hypoplastic. More than 80% of the time, the supratrochlear artery and the dorsal artery of the nose form the two terminal branches of the ophthalmic artery. In approximately 3% of cases each, the supratrochlear and supraorbital arteries, or all three of the above arteries, form the terminal branches. Only in about 5% of cases each does the supratrochlear or supraorbital artery alone form the terminal branch (Krayenbühl et al. 1979). High-grade obstructions to blood flow in the internal carotid artery, with an accompanying decrease in pressure, usually produce a clear flow change within the supratrochlear artery (*retrograde flow,* or *no flow*). In individual cases, however, the supraorbital artery is the main terminal branch of the ophthalmic artery, and the supratrochlear artery is hypoplastic. Retrograde blood flow within the hypoplastic supraorbital artery does not reliably indicate a significant carotid obstruction, but can also reflect altered flow relationships within the orbit in an intact carotid system. Therefore, only the orbital blood vessel that has the *strongest Doppler signal* supports the interpretation of any flow obstruction in the extracranial carotid system.

No signal. If no flow is observed for the supratrochlear or supraorbital arteries in the typical probe position, it may be due to a variant location, and the positioning of the probe may have to be changed. In this case, it is best to extend the search for the blood vessel by constant compression of the facial artery and/or the superficial temporal artery bilaterally, and to observe the effect when the vascular system of the external carotid artery is released again.

No compression effect. *Retrograde flow with a negative compression effect* can be caused by:

- A combined stenosis of both the internal and external carotid arteries
- A more proximal obstruction in the common carotid artery

Compression maneuvers in the branches of the contralateral external carotid artery are required, and these often confirm retrograde flow in the orbital arteries.

A negative compression effect can also be present:

- When the ophthalmic artery is collateralized via the maxillary artery and its largest terminal branch, the middle meningeal artery, which cannot be selectively compressed

A positive compression effect may also be absent in:

- Bilateral obstructions of the external carotid arteries

Obstruction of the ophthalmic artery and its terminal branches. Rarely, proximal obstructions of these arteries can cause changes in the Doppler signals, in the absence of any obstructions within the extracranial carotid system. In such cases, it is best, in addition to the supratrochlear and supraorbital arteries, to locate and record the ophthalmic artery by placing the probe directly on the eyeball over the closed eyelid, or to evaluate its origin transorbitally using the pulsed-wave Doppler (Spencer and Whisler 1986).

Diagnostic Effectiveness

The specificity of indirect Doppler sonography lies between 88% and 97%, and as it is simple and straightforward to use, it makes a useful rapid screening test (Büdingen et al. 1976, Neuerburg-Heusler 1984) instead of more complex ultrasound tests (Howard et al. 1996). In pathological cases, it serves as an additional way of evaluating the hemodynamic significance of an extracranial stenosis of the ICA. In addition, in cases of intracranial internal carotid artery obstruction, it may help to establish whether the process is proximal or distal to the ophthalmic artery.

However, examination of the orbital arteries alone when diagnosing obstructive lesions of the carotid artery is obsolete. Indirect test results support the final diagnosis only if combined with direct evaluation of the neck arteries, and if the findings are definitively pathological.

False-negative results—i.e., orthograde blood flow in the orbital arteries—also occur in hemodynamically marked flow obstructions involving the internal carotid artery if:

- There is good collateralization through the circle of Willis. In spite of a high-grade ICA obstruction, no retrograde flow or flow incidents occur (approximately 20% of cases).
- The anastomosis of the ophthalmic artery is not open—e.g., there is a high-grade external carotid artery stenosis ipsilaterally and the contralateral external carotid artery does not contribute to collateral blood flow in the ophthalmic anastomosis. The pressure gradient remains unchanged in this case, since the pressure in both the internal and external carotid arteries decreases.
- The flow obstruction lies in the neck region proximal to the carotid bifurcation (common carotid artery or innominate artery). Here, the pressure gradient also remains unaffected.

False-positive results are extremely rare, provided that:

- Anatomical variants are taken into account.
- Compression maneuvers are performed completely and correctly. The most frequent mistake involves an artificial head movement while Doppler signals are being recorded, resulting in a simulated reduction in flow velocity.
- The possibility of an ophthalmic artery stenosis is taken into account.
- Intracranial arteriovenous malformations are not forgotten.

There is *no* reliable correlation between a change in flow relationships within the orbital arteries and the degree of stenosis in the internal carotid artery. Rather, the opening of the ophthalmic artery anastomosis and the resulting flow relationships reflect only the potential and effectiveness of collateralization to compensate for the obstruction in the carotid arteries.

Hemodynamic compensation through the anastomosis of the ophthalmic artery is only relevant to the perfusion of the brain in rare instances. Much more often, it is of no significance at all. In some cases, after a transcranial examination has been carried out, it may be possible to ascertain whether or not the ipsilateral middle or anterior cerebral arteries are being supplied through the collateral pathways of the ophthalmic artery.

In summary, the verification of retrograde flow in the orbital arteries is of considerable diagnostic significance. It indicates a marked flow obstruction (severe stenosis or occlusion) in the preceding carotid system.

Vice versa, verified orthograde flow in the orbital arteries is conceivable with accompanying direct findings in the neck arteries that suggest a high-grade stenosis or occlusion in the distribution area of the internal carotid artery. In order to avoid misinterpretation, however, attempts should always be made to document the underlying collateralization pathways via the contralateral carotid system or the posterior circulation and the circle of Willis—i.e., supplementary intracranial ultrasound diagnostic procedures are advisable.

Apparently abnormal but not necessarily pathological findings (e.g., alternating forward and backward blood flow, no flow, as well as clearly asymmetric orthograde flow) can be an important aid in locating an intracranial flow obstruction or pathological vascularization (e.g., combined with asymmetries in the common or internal carotid artery). In these cases, supplementary intracranial Doppler studies are also useful.

Table 2.**6** Criteria for evaluating B-mode image findings in the carotid artery, based on the consensus conference (de Bray et al. 1997)

Form and course
- Regular
- Variation/anomaly
- Dilated
- Tortuosity
- Kinking/coiling

Vessel size
- External diameter
- Internal diameter
- Stenosis diameter
- Pulsatility changes (systolic/diastolic)
- Intima–media thickness
- Vessel motion
  - Transverse
  - Axial
  - Systolic–diastolic

Pathological changes
Location
- Common carotid artery (anterior wall/posterior wall, lateral or medial)
- Internal carotid artery (anterior wall/posterior wall, lateral or medial)
- External carotid artery (anterior wall/posterior wall, lateral or medial)

Form
- Circular
- Semicircular

Size
- Plaque diameter in longitudinal and cross-section at the broadest point

Surface
- Regular/irregular
- Flat, smooth
- Ulcerated
- Cavitation

Echo pattern
- Strength (strong/weak)
- Size (progressive/regressive)
- Echogenicity (anechoic/echogenic)
- Texture (homogeneous/heterogeneous)
- Ultrasound shadow (present/absent)

Plaque motion (uniform/discrepant)
Plaque volume
Embolic activity

Carotid System

Principle

Color Flow Duplex Sonography

B-mode. Various studies have described the validity of ultrasound for the display of arteriosclerotic vascular processes (Comerota et al. 1981, Reilly et al. 1983, Cape et al. 1984, Hennerici and Freund 1984, Hennerici 1987, Ricotta et al. 1987). It is useful to evaluate these arteriosclerotic changes in B-mode imaging, using suitable parameters (Table 2.**6**), and it is important to display plaques in both longitudinal sections as well as cross-sections (Fig. 2.**31**). B-mode studies of the cerebral circulation alone, however, are often misleading and should thus be combined with Doppler analysis. This should also be considered if IMT measurements are included in multicenter trial protocols (see p. 39) (Touboul et al. 2004) in order not to overlook remote severe obstructive lesions. Furthermore, without reliable differentiation between the ICA and ECA (in Doppler mode or color mode signal analysis), B-mode studies alone can cause significant errors.

Sonographic characteristics of plaques are:
- *Echogenicity* (from anechoic to hyperechoic or echodense); this is a reliable, sufficiently reproducible parameter, as evidenced by three independent studies (Geroulkos et al. 1994, de Bray et al. 1998). "Echoic" can refer to anatomic structures ("isoechoic" relative to the sternocleidomastoid muscle

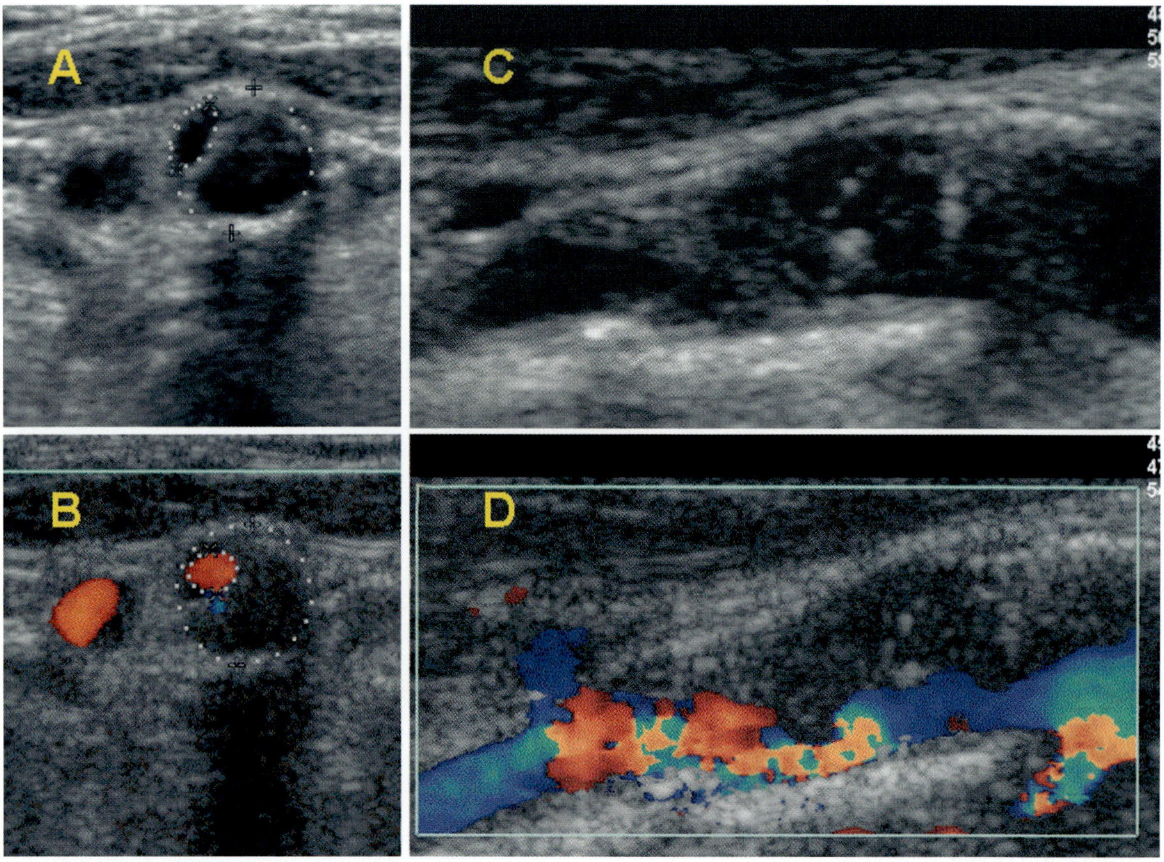

Fig. 2.**31** High-grade stenosis with heterogeneous plaque formation, in cross-sections (**a**, **b**) and longitudinal sections (**c**, **d**). Color Doppler information allows better delineation of the plaque surface structures than B-mode alone

and "hyperechoic" relative to the cervical vertebrae).

- *Texture* (from homogeneous to heterogeneous); classification schemes have been proposed, but are not fully established (de Bray et al. 1998).
- *Surface characteristics* (from irregular to excavated forms) are difficult to evaluate and often a matter of interpreter disagreement; the accuracy of diagnosis can be improved by additional color flow duplex sonography: recesses and vortices (> 2 mm in depth and length) are characteristic features of ulcers (Sitzer et al. 1994).
- *Evidence of embolic activity* (indirect signs such as high-intensity transient signals, HITS); several studies have shown a significant association with such findings and plaque activity as a potential predictor for stroke and transient ischemic attack (TIA) (Siebler et al. 1993, Markus and Brown 1994, Valton et al. 1995, Ries et al. 1996).
- *Changes in plaque appearance* over time; this is currently the most valid predictor of associated stroke risks, apart from the degree of stenosis (> 70% obstruction) (Hennerici and Steinke 1991, Sillesen 1995).
- *Plaque motion* is an interesting parameter; it is hypothesized that plaque motion correlates with the risk of plaque distortion and rupture, giving rise to cerebral embolism (Meairs et al. 1995).

Real-time compound imaging is a new modality, which may be able to improve the ultrasonographic visualization and characterization of carotid artery plaques (Kern et al. 2004). This technique acquires ultrasound beams that are steered off-axis from the orthogonal beams used in conventional ultrasound. The number of frames and steering angles varies depending on the transducer characteristics. Frames acquired from sufficiently different angles contain independent random speckle patterns, which are averaged to reduce speckle and improve tissue differentiation. Figure 2.**32** presents a comparison between this new technology and conventional B-mode imaging. Compound imaging allows suppression of edge shadowing and improves contrast resolution, and it can therefore improve the characterization of arterial walls and plaques (see Fig. 1.**12**).

Doppler mode. In continuous-wave and pulsed-wave Doppler procedures—e.g., in duplex systems—similar parameters can be used to describe the Doppler spec-

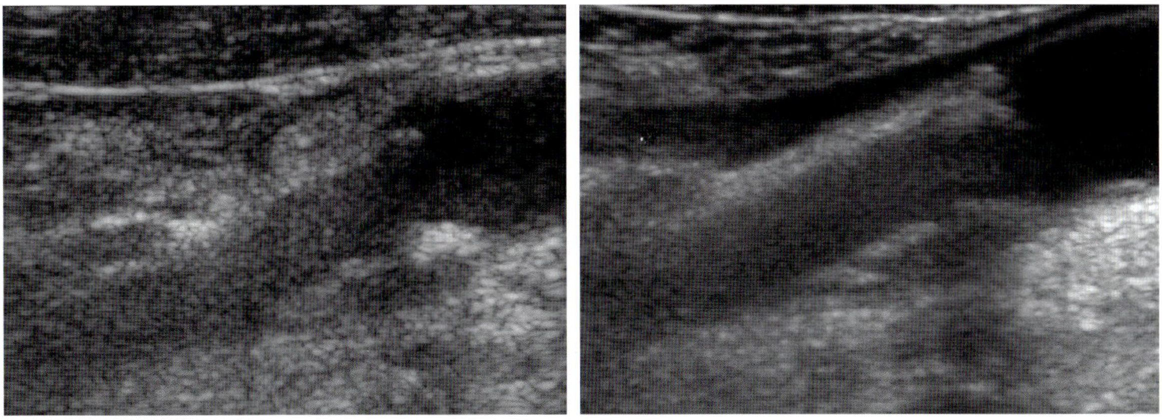

a b

Fig. 2.**32** Acute thrombosis of the internal carotid artery. **a** The conventional B-mode image. **b** B-mode with compound imaging technology. Note the significant increase in the image quality, with better delineation of the proximal thrombus formation.

trum (Blackshear et al. 1980, Roederer et al. 1982, Jacobs et al. 1985, Robinson et al. 1988, Withers et al. 1990, Moneta et al. 1993). The application of quantitative parameters from the Doppler spectra is particularly useful in patients with evidence of dynamic changes in the degree of stenosis. The criteria used are mainly values for maximum and average frequencies. The spectral width is a less reliable indicator of minimal changes in the vascular wall.

Interpreting spectral parameters is carried out in the same way as in continuous-wave and conventional pulsed-wave Doppler sonography. Parameters reflecting hemodynamic compromise induced by severe ICA obstructions are considered to be relevant (de Bray et al. 1995). The systolic carotid ratio should be recorded between the site of the stenosis and the CCA, obtained 3 cm below the bifurcation: a value > 1.5 determines stenoses > 50 %, and a threshold > 4 determines stenoses of > 70 %.

Color flow mode. Since the early 1990s, studies have increasingly used color flow duplex sonography (velocity-guided and amplitude-guided analysis) to assess the value of a combined analysis of the echotomographic and associated hemodynamic changes in the presence of varying degrees of carotid stenosis (Polak et al. 1989, Steinke et al. 1990b, Londrey et al. 1991, Anderson et al. 1992, Steinke et al. 1992b, de Bray and Glatt 1995). Adequate use of structural (B-mode) and hemodynamic information (Doppler mode) support a highly reliable classification of diffent degrees of stenosis of the carotid system in the neck (Fig. 2.**33**).

Systematic analysis of high-grade carotid stenoses (Erickson et al. 1989, Steinke et al. 1992b, Sitzer et al. 1993, de Bray et al. 1995) in comparison with angiographic imaging yielded agreement in over 70 % of cases in relation to the surface characteristics of plaques. Measurements of the degree of stenosis and cross-sectional measurements corresponded in 85 % of cases. So far, evidence of specific hemodynamic patterns associated with various structural wall changes

Table 2.**7**

- Changes in the audio signal (“turbulences”)
- Increase in the diastolic or systolic flow velocity, or both
- Decrease or loss in pulsatility, with diastolic flow acceleration and systolic deceleration
- Decrease in the diastolic or systolic flow velocity (proximal and distal to the obstruction)
- Absent Doppler signal

has not yet allowed sufficient clinical correlations to be made in evaluating the malignant potential of such vascular processes. Using Doppler mode, analyses have been made of the intrastenotic and post-stenotic maximum peak frequencies, of the extent and location, as well as the spread, of the color saturation zone in correlation with “jet flow,” and the spread and expression of the color reversal and color mixture zones (Table 2.**7**).

Comparison with the angiographic findings in the literature is puzzling, due to heterogeneous criteria in estimating the degree of stenosis from image planes (Croft et al. 1980, Eikelboom et al. 1983, Langlois et al. 1983, Thomas et al. 1986, Widder et al. 1986, Vanninen et al. 1995), and the variable procedures used to calculate the hemodynamic compromise from Doppler studies (Bladin et al. 1994, de Bray and Glatt 1995). Also, comparing Doppler findings with measurements of the degree of stenosis obtained from endarterectomy preparations (Arbeille et al. 1985), or from estimates of the original lumen using echotomographic findings (Terwey et al. 1984), has not provided a satisfactory solution. The same applies to interpretations of the degree of stenosis that are based on angiographic findings, which usually involves an examination in two planes at best, representing a sketchy image of the real three-dimensional relationships. Similar problems are also encountered when evaluating the results with magnetic resonance angiography (MRA), which simulates a morphological image based on flow-sensitive relaxation phenomena (Huston et al. 1993, Laster et al. 1993). Even with a pathoanatomical approach, quantitative

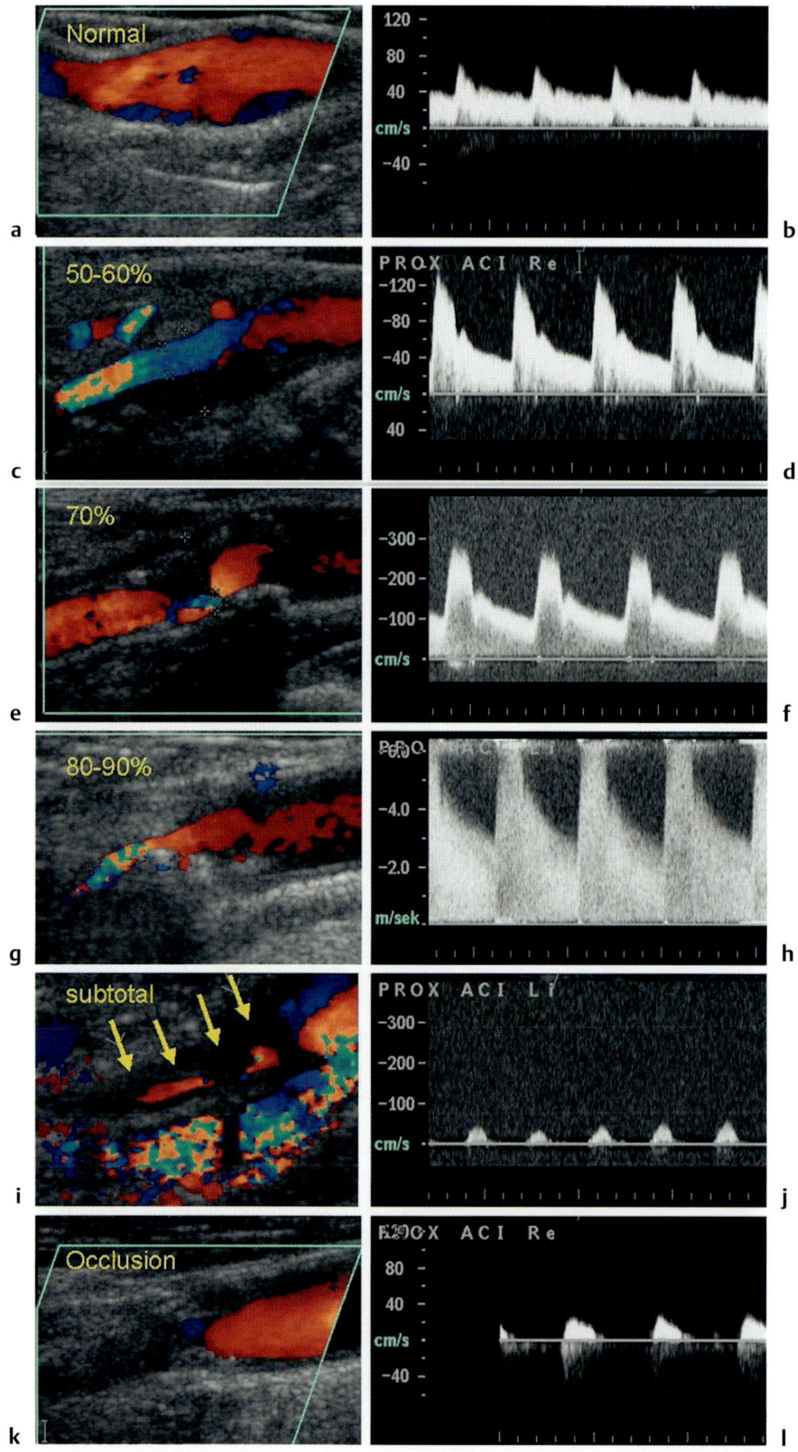

Fig. 2.**33** A template of typical color flow duplex findings for different degrees of stenosis. *Left:* color duplex images; *right:* pulsed-wave Doppler measurements. **a**, **b** Normal findings. **c**, **d** Mild stenosis. **e**, **f** Moderate stenosis. **g**, **h** Severe stenosis. **i**, **j** Subtotal stenosis. **k**, **l** Occlusion

and reliable estimation of the degree of stenosis is difficult due to the heterogeneous nature of arteriosclerotic processes and the structural artifacts that result from the processing of tissue (Picot et al. 1993).

Since the completion of trials that proved the effectiveness of surgery in symptomatic patients with stenoses of > 70% in the carotid system (Barnett et al. 1998), and again after presentation of ACST (Halliday et al. 2004) the appropriate method of estimating the degree of stenosis has become a matter of controversy. When interpreting the extent of the stenosis angiographically, both the *local* and *distal degree of the stenosis* were measured (Fig. 2.**34**). In addition, evaluating the proximal degree of stenosis seemed to be useful, provided this calculation related to the aforementioned criteria used in the multicenter trials. In individual cases, these results can also be further complemented with appropriate MRA and ultrasound diagnostic evaluations to determine the degree of stenosis. All of these methods have a similar predictive value, provided that appropriate care is taken when interpreting the results. Today clinical decisions are taken on the basis of noninvasive findings without the help of conventional angiography in the majority of patients. Hemodynamic values from ultrasound and MRA studies have their own classification, and should not be considered as synonymous with the "morphological degree of stenosis." An adequate safety margin is recommended for individual values (e.g., 50–60 %, 60–80 %) (Glagov and Zarins 1983, de Bray and Glatt 1995, Erdoes et al. 1996).

Comparison between angiographic measurements and ultrasound studies (Fig. 2.**35**) to classify degrees of stenosis have shown that in radiography, the local degree of stenosis overestimates and the distal degree of stenosis underestimates the true values for obstruction, while ultrasonography similarly underestimates higher degrees of stenosis and overestimates smaller ones (Steinke et al. 1996). It has been suggested that ultrasound contrast agents can improve the diagnosis in patients in whom conventional methods are inconclusive (Hennerici et al. 2004). In particular, in patients with complex calcified plaques, a very high-grade stenosis may be mistaken for complete occlusion (about 15–20 % of patients in large neurovascular centers). It was hoped that such limitations might be overcome using contrast media—with the addition of contrast agents, the compounds acquire improved ultrasound reflection characteristics, so that the rate of false-negative results when depicting high-grade stenoses ought to be reduced (Fig. 2.**36**). However, this hope was not confirmed in clinical studies. When the issue of whether there is subtotal stenosis or total occlusion is still unclear after ultrasound imaging alone, it is satisfactorily resolved in fewer than 10 % of cases when ultrasound contrast agents are additionally administered (CEDAS—Hennerici et al. 2004). Since ultrasound contrast agents are comparatively expensive, their use in the examination of the extracranial domain is therefore currently very limited, even in countries in which such agents have received approval.

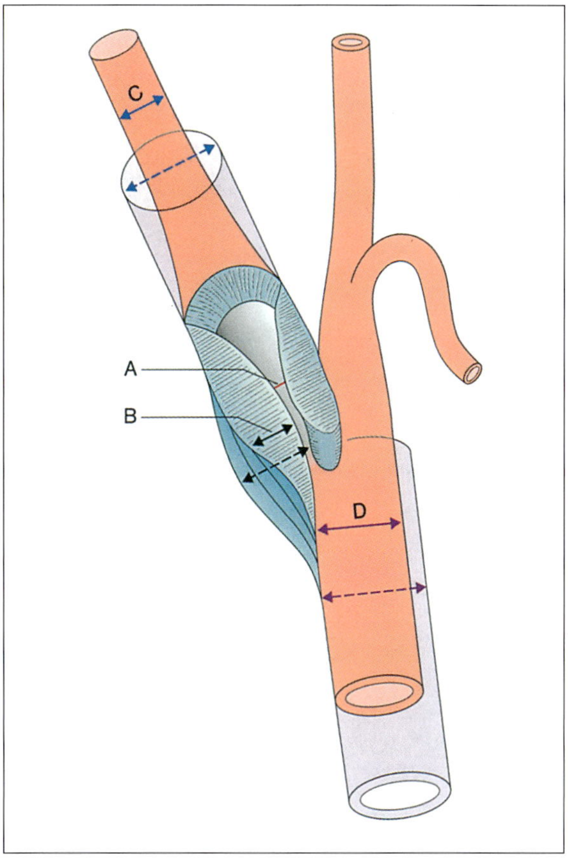

Fig. 2.**34** Parameters used to calculate the degree of stenosis.
A, residual lumen; B, plaque; C, distal vascular diameter; D, standard lumen 3 cm proximal to the bifurcation
Local degree of stenosis (1 − A/B) × 100 %
Distal degree of stenosis (1 − A/C) × 100 %
Proximal degree of stenosis (1 − A/D) × 100 %

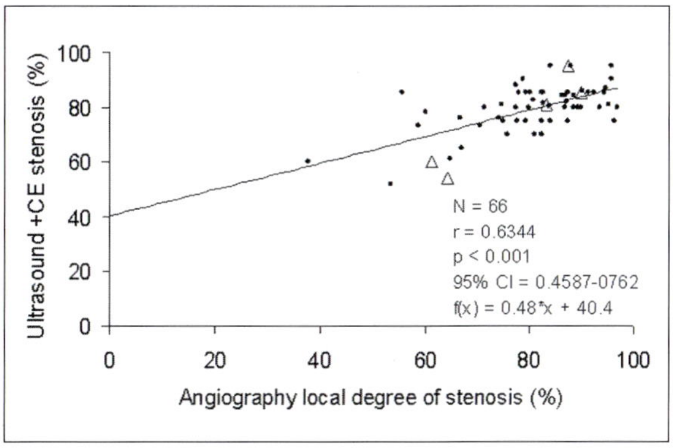

Fig. 2.**35** Ultrasound plus CE (contrast enhancement) versus angiographic quantification of high-grade stenoses. There is a significant positive correlation, indicating that ultrasound has a high level of diagnostic accuracy in comparison with angiography as the gold standard

Continuous-Wave Doppler Sonography

Obstructions within the carotid system can be detected by continuous-wave Doppler sonography when the narrowing of the lumen is at least 40 % and it is accessible to direct ultrasound application (Miyazaki and Kato 1965a, 1965b, Keller et al. 1976a, von Reutern et al. 1976, Renemann and Spencer 1979, Spencer and Reid 1979, Blackshear et al. 1980, Humphrey and Bradbury 1984, Trockel et al. 1984, Beach et al. 1989).

Using five fundamental Doppler criteria (Table 2.**7**), it is possible to classify carotid obstructions according to the degree of stenosis and to compare them with the angiographic results (Hennerici et al. 1981b, Trockel et al. 1984, von Reutern and Büdingen 1993) (Table 2.**8**):

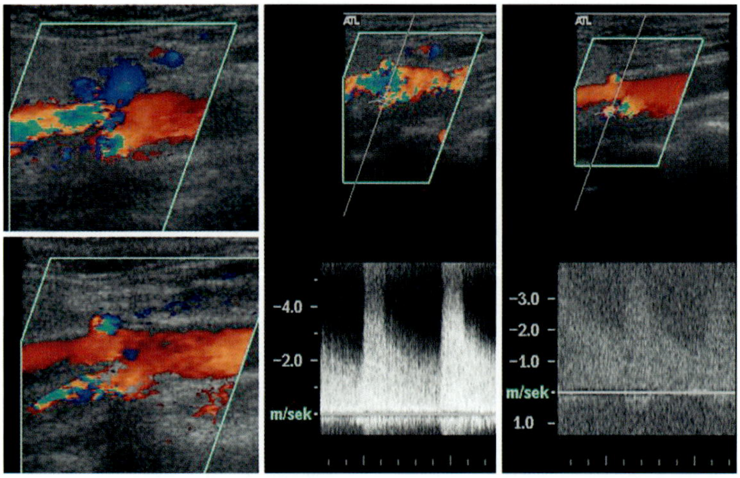

a b c d

Fig. 2.**36** High-grade carotid stenoses of the proximal internal carotid artery (ICA) (**a**) and external carotid artery (ECA) (**b**), with aliasing in the frequency-modulated color-mode images and high flow velocities (approximately 5 m/s) in the Doppler-mode analysis (**c** and **d**)

Table 2.**8** Structural and hemodynamic criteria for the evaluation of the degree of stenosis in the internal carotid artery (based on Widder 1985); frequency data refer to a 4-MHz transmission frequency

	I **Nonstenotic plaques**	II **Low-grade stenosis**	III **Medium-grade stenosis**	IV **High-grade stenosis**	V **Subtotal stenosis**
Angiographic estimates					
Local degree of stenosis	< 40%	40–60%	60–70%	ca. 80%	> 90%
Distal degree of stenosis	0	< 30%	≈50%	≈70%	> 90%
Dopplersonographic parameters					
Indirect criteria	No sign of flow obstruction			Ophthalmic artery: no flow or retrograde flow Common carotid artery: reduced flow	
Direct criteria					
Analogue waveform					
Near the stenosis	Normal	Audiosignal change, slight increase in local flow	Clear flow increase, pulsatility loss and systolic deceleration	Strong local flow increase with systolic deceleration	Variable stenosis signal with decreased intensity
Post-stenotic	Normal	Normal	Normal	Decreased systolic flow velocity	Difficult to find, strongly reduced signal
Spectrum analysis					
Near the stenosis	Normal	Spectral broadening	Spectral broadening with increasing intensity of the low-frequency component	Inverse frequency components in a broadened spectrum	Inverse frequency components in a broadened spectrum
Systolic peak frequency	< 4 kHz	> 4 kHz	4–8 kHz	> 8 kHz	Variable
End diastolic frequency	< 1.3 kHz	< 1.3 kHz	> 1.3 kHz	> 3.3 kHz	Variable
Systolic peak velocity (cm/s)	< 120	> 120	> 120	> 240	Variable
End diastolic velocity (cm/s)	< 40	< 40	> 40	> 100	Variable
Systolic ratio (ICA/CCA)	< 1.5	< 1.8	> 1.8	> 3.7	Variable

Table 2.**8** Structural and hemodynamic criteria for the evaluation of the degree of stenosis in the internal carotid artery (based on Widder 1985); frequency data refer to a 4-MHz transmission frequency (continued)

	I Nonstenotic plaques	II Low-grade stenosis	III Medium-grade stenosis	IV High-grade stenosis	V Subtotal stenosis
B-mode					
Proof quality	+ + +	+ + +	+ +	+	+
Findings	Small plaque extension		Medium-grade lumen constriction	High-grade lumen constriction	Maximum lumen constriction
Color flow mode					
Findings	No or local turbulences only	Long segmental systolic flow acceleration Color fading in systole only	Localized segmental systolic flow acceleration Color fading Turbulences Increased flow velocity in diastole	Extremely localized segmental high-grade flow acceleration, poststenotic backflow components (mosaic pattern), secondary vortices Short segment of marked color fading or aliasing	

CCA: common carotid artery; ICA: internal carotid artery.

- Low-grade stenoses (40–60%)
- Medium-grade stenoses (60–70%)
- High-grade stenoses (around 80%)
- Subtotal stenoses (over 90%)
- Occlusions

This classification scheme only applies to unilateral, isolated obstructions of the internal carotid artery. Modifications are necessary for:

- Common carotid artery obstructions
- External carotid artery obstructions
- Processes involving several blood vessels
- Unilateral multiple stenoses at specific levels of the circulatory system ("tandem stenosis"), especially when these cannot be directly assessed with ultrasound (e.g., those in the proximal segment close to the aorta, or extending submandibularly until they reach an intracranial location)

Spectral analysis can be used to determine the degree of stenosis from the parameters summarized in Table 2.**9**:

- The systolic and diastolic peak frequency can easily be measured in normal flow conditions; in pathological cases, problems can be encountered here. The peak frequency is usually used as the primary criterion for classifying the extent of a stenosis (Spencer and Reid 1979, Hennerici et al. 1981b, Norrving and Cronqvist 1981, Humphrey and Bradbury 1984, Trockel et al. 1984, Johnson et al. 1986).
- The average Doppler frequency reflects a calculated variable (at least 30 different formulas are used in commercial instruments), which is insensitive to local flow changes near wall processes, and insensitive to variable intervals of sequential heartbeats.
- The distribution of amplitude-intensive frequency bands differs at different degrees of stenosis; with increasing obstruction, a displacement toward the baseline results.
- The systolic frequency window describes the relationship between amplitude-weighted low frequencies and the frequencies of the overall spectrum. With an increase in the degree of flow obstruction, the percentage value of the frequency window decreases (Brown et al. 1982, Rittgers et al. 1983, Bandyk et al. 1985).
- Negative frequencies indicate flow components toward the probe, and reflect turbulence phenomena.

Table 2.**9** Important spectral analysis parameters for evaluating the degree of stenosis in the extracranial cerebral arteries

- Systolic and diastolic peak frequency
- Average mean Doppler frequency
- Mode frequency band with the highest signal amplitude
- Systolic frequency window (spectral broadening)
- Negative frequencies

In an international consensus conference (de Bray and Glatt 1995), systolic peak Doppler shifts > 4 kHz (f_0 = 4 MHz, > 120 cm/s) to identify most ICA stenoses > 50% in local diameter reduction and end-diastolic values > 4.5 kHz (f_0 = 4 MHz, > 135 cm/s) to identify ICA stenoses > 80% were considered to be the most reliable parameters for classification.

Findings

Common Carotid Artery

Color Flow Duplex Sonography

B-mode. The course of the common carotid artery in the neck region can be ideally demonstrated using a B-mode image, since this vascular segment can be displayed regularly and very reliably in several planes, even when a large transducer is being used. Local wall changes in this region can therefore be detected, and their morphological structure can be described with more certainty than with angiography (Croft et al. 1980, Eikelboom et al. 1983, Hennerici and Freund 1984).

Intima–media thickness (IMT). The common carotid artery is the standard vascular segment for measuring the IMT, which is considered to be an independent risk factor for atherosclerosis. Measurement of the IMT is therefore not only useful in the setting of cerebrovascular examinations, but can also be used as an indicator for the risk of coronary arteriosclerotic disease. The normal finding in the vessel wall is an anechoic lumen, bordered by two light reflections. The echo line facing the lumen (intimal and subintimal portions) is separated from a second exterior interface reflection (medial and adventitial components) by a hypoechoic middle layer—i.e., the "intima–media contour" (Fig. 2.**17**) (Pignoli et al. 1986, Poli et al. 1988). B-mode and color-coded duplex sonography of the CCA is increasingly being used to evaluate the IMT (Glagov et al. 1995, O'Leary et al. 1996, Young and Humphrey 1996). These values are used for the analysis of early atherosclerosis, its natural history, and changes induced by reduction of risk factors and pharmacological agents—e.g., in patients treated with special diets and antihypertensive or antilipid drug regimens. According to the literature, this parameter is rather useful in multicenter trials and can be applied reliably if the protocol exactly indicates where and when to measure the extracranial carotid system (e.g., the proximal and distal common carotid artery, the carotid bulb, the proximal and distal internal carotid artery bilaterally), etc. In addition, the diameter of the vessels and pulsatility variations have to be noted and be adapted for standardized values (Touboul et al. 2004).

Arteriosclerotic disease—plaque development. Common carotid artery plaques are rarer than those involving the internal carotid artery. They can also be detected using the criteria mentioned above (Fig. 2.**37**).

The results obtained are often the same as those described for the internal carotid artery (p. 65). The same applies to changes associated with dilative arteriopathy, which is characterized by an increased vascular diameter, with decreased pulsatility and flow velocity. In cases of acute occlusion, pulsatility is completely absent. In the chronic stage, when the scarlike changes have produced a complete restructuring of the vascular lumen, it can be difficult even to identify the lumen of the common carotid artery.

Doppler mode. Flow changes in the common carotid artery can also be detected easily and regularly under visual control using conventional duplex sonography. By moving the sample volume, flow accelerations can be detected in the direct vicinity of plaque using sequential ultrasound application. Multi-channel pulsed-wave systems have mainly been used in scientific studies and to calculate the blood volume because of their complex application modalities. Such systems have not acquired any significant diagnostic relevance in clinical use.

Severe obstructions in vascular segments remote from the area of examinations challenge the examinor's experience. When there is obstruction within the innominate artery, duplex sonography can be particularly helpful if to identify structural and hemodynamic changes in the carotid system Doppler sonography has registered only weak signals or signals that are difficult to classify (Fig. 2.**38**). If there is a clear reduction in the flow signal caused by a vascular process in the artery and there are simultaneous changes in the right common or internal carotid artery, then it is almost always missed by continuous-wave Doppler sonography. In this case, identifying any structural wall changes in B-mode and locating the sample volume in this particular vascular segment can be helpful in registering the mostly intermediate signals. When appropriate compression maneuvers are used, the hemodynamics of the compensatory collateral circulatory systems, which are often extremely complex, can be comprehended. Depending on the location of the obstruction with respect to the watershed, various results may be obtained (Fig. 2.**39**).

Color flow mode. This procedure is particularly suitable for detecting abnormalities within the middle and distal common carotid artery. If the internal carotid artery is open, but has low flow velocities, a search for a flow signal under visual control is particularly important here. Often, the vertebro-occipital collateral—which is usually effective—is directly detectable, running via the ipsilateral vertebral artery, with retrograde perfusion of the external carotid branch, and ultimately passing through the bifurcation distal to the occluded common carotid artery into the internal carotid artery.

Plaque formations occur particularly often in the common carotid artery. Fresh wall adhesions, which are marked by an absent flow signal and hypoechoic or absent structural wall changes in B-mode imaging, are by no means rare. Changes in the color signal can only occasionally be predicted from the structural image. Bizarre configurations may be accompanied by nearly laminar flow, while minimal plaque formations with a smooth surface structure can produce significant separation zones and flow turbulences. The extent to which clinically relevant information might be provided by these findings (e.g., differentiating between threatening and innocuous plaque formation) is not yet clear (Steinke et al. 1992b, Hennerici et al. 1995, Meairs et al. 1995).

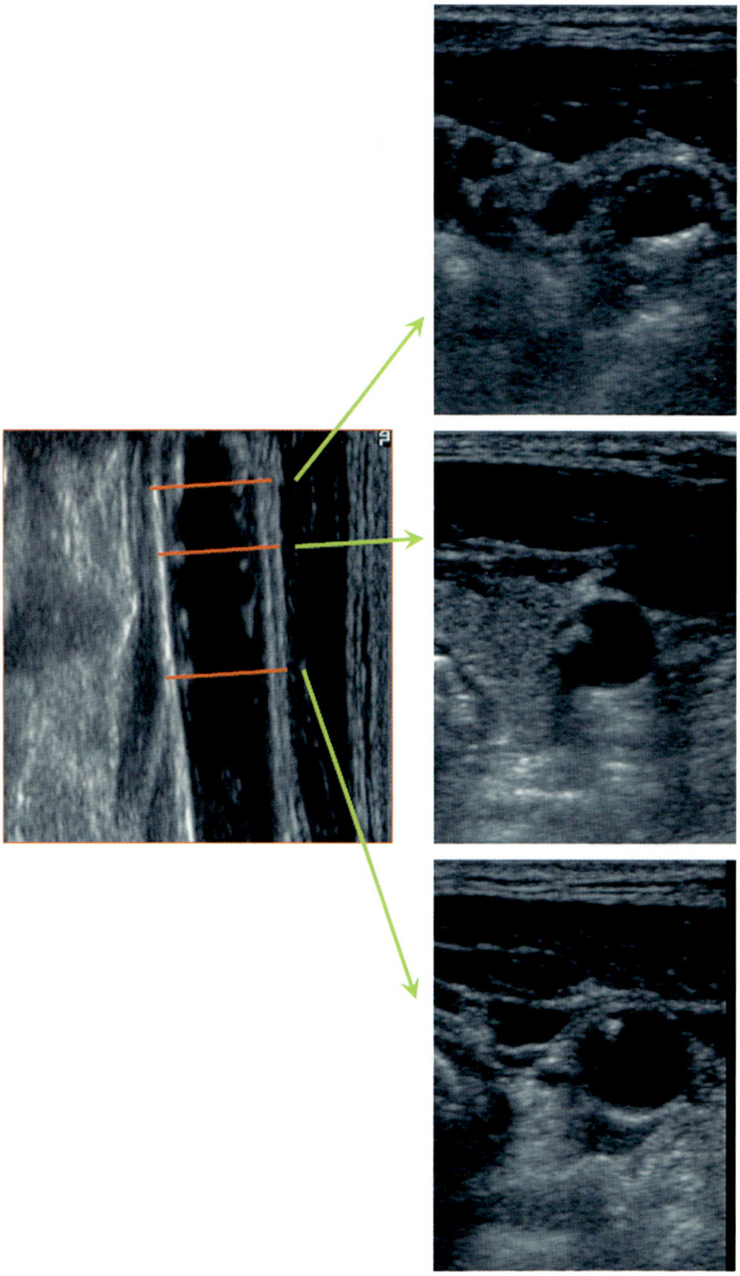

Fig. 2.**37** B-mode image of multiple small plaques within the common carotid artery. Longitudinal view (left side). Cross sectional views (on the right side). The combination of both is needed to delineate and uncover plaques which may also be located outside a single longitudinal section

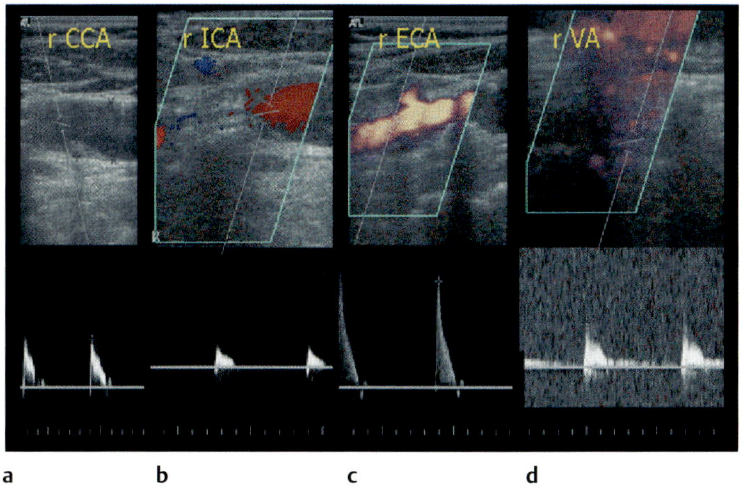

Fig. 2.**38** Occlusion of the innominate artery. Unusual pathological flow changes are seen in the CCA in the right common carotid artery (**a**), right internal carotid artery (**b**), right external carotid artery (**c**), and particularly in the right vertebral artery (**d**) (as well as the right middle cerebral artery and anterior cerebral artery). In this case, high pulsatility and zero flow during diastole are evident; other flow patterns are also possible

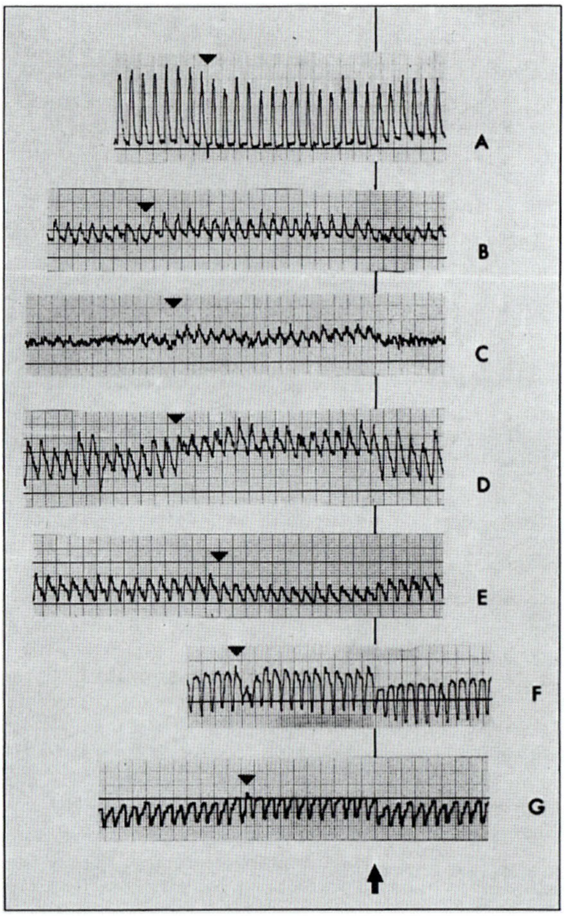

Fig. 2.**39** Doppler waveforms in the common carotid artery in seven patients with high-grade obstructions of the innominate artery. The arrows mark compression (↑) and release (↓) of the right brachial artery in the upper arm. Physiological blood flow, directed toward the brain, corresponds to an upward curve deflection, and retrograde flow to a downward deflection. Consistent with the hemodynamic relationships in the carotid artery system, the following variations are found, as well as transitional forms:

A An apparently normal flow profile, the diastolic flow velocity of which changes during a compression test, reflecting a hemodynamic connection between the cerebral and the peripheral arterial systems

B Flow signal with reduced amplitude, **C** with only slightly modulated amplitude, and **D** with an exaggerated modulated signal, with a significantly delayed diastolic decrease in the flow velocity. All of these have in common an increase in the orthograde flow pattern during compression

E, F Early systolic retrograde flow, with a late systolic physiological flow signal and a renewed diastolic decrease in the flow velocity (intermediate flow); when the upper arm is compressed, physiological blood flow directed toward the brain can result (**E**), or, during the reactive hyperemic phase, a "carotid steal" phenomenon may appear (alternating flow) (**F**)

G A permanent "carotid steal" syndrome, with retrograde blood flow

Continuous-Wave Doppler Sonography

Stenoses of the common carotid artery show similar changes in the Doppler spectrum parameters to those listed in Tables 2.**7** and 2.**9**.

Increased Flow Velocity

Stenoses at the origin cannot be detected directly using continuous-wave Doppler alone. Nevertheless, in individual cases, their presence may be assumed from changes in the audio signal. Using pulsed-wave Doppler and an appropriate depth selection to record the Doppler signals, a flow velocity acceleration can be detected as positive proof. A proximal stenosis at the common carotid artery can be differentiated from the innominate artery by demonstration of intact flow relationships in the ipsilateral and proximal subclavian and vertebral arteries.

Continuously elevated flow. In contrast to the internal carotid artery, a significant compensatory increase in flow within the common carotid artery cannot usually be detected when contralateral occlusions are present. However, when there are extensive intracranial shunt volumes (fistula formation, arteriovenous blood vessel malformation), an increase in the flow velocity can be detected within the entire neck region accessible to examination. In this case, it is quite possible to superimpose flow signal irregularities, both acoustically and in the spectrum waveform itself. Rarely, flow accelerations may be found, resulting from circular constrictions of the common carotid artery (e.g., in inflammatory vascular diseases).

Decreased Flow Velocity

1. The most frequent and most significant pathological finding in the common carotid artery is decreased flow velocity caused by a *distal high-grade internal carotid artery obstruction* (internal carotid artery stenosis exceeding 80%, internal carotid artery occlusion, hemodynamically significant intracranial ipsilateral vascular obstruction).
2. The differential diagnosis should include a *dilative arteriopathy* of the common carotid artery), which is often encountered (unilaterally or bilaterally) in patients at an advanced age, and causes a reduction in the flow signal similar to that seen in distal vascular processes with hemodynamic significance.
3. In a few cases, an acute obstruction of the middle cerebral artery, with poor collateralization through leptomeningeal anastomoses and near the circle of Willis (M_1 segment), can lead to a flow velocity reduction in the ipsilateral common carotid artery.
4. Rarely, a high-grade common carotid artery stenosis at the origin of the artery is also present.

Absent Signal

When the common carotid artery is occluded, no signal can be recorded. In contrast, signals from the external and internal carotid arteries may still be recorded distal to the bifurcation, because the most common collateral fills other branches of the external carotid artery from the ipsilateral vertebral artery via the occipital artery, which shows retrograde flow. If the internal carotid artery is open, intracranial circulation can also be maintained through this vertebral–occipital anastomosis. The important search for a flow signal within the internal carotid artery with intact circulation is much more successful using the duplex system under "visual control" than with the conventional Doppler technique. This is important, since in these cases surgical reconstruction of the carotid system is possible.

Internal Carotid Artery

Color Flow Duplex Sonography

B-mode. Investigations of different plaques and obstructive lesions depend on a careful analysis of structural patterns at the carotid bifurcation and proximal ICA.

- *Flat plaques* are characterized by an increased broadening and restructuring of the normally hypoechoic middle layer. The inner or exterior interface reflections can also appear thickened. Although flat plaques represent the earliest detectable form of arteriosclerotic wall change, they are not always exclusively signs of a progressive development, but can also persist as the final stage of a regressive development (Hennerici et al. 1985, Hennerici and Steinke 1991).
- If the wall change expands into the vascular lumen with increasing size, then a homogeneous echo structure is usually retained initially; the *surface structure* and the location have to be precisely evaluated by suitable longitudinal sections and cross-sections.
- With increasing size and morphological changes, *complicated plaque formations* with a nonhomogeneous echo structure and surface irregularities can form. Hypoechoic and hyperechoic sections may alternate. Hypoechoic plaques at the carotid bifurcation are associated with a significantly higher risk of TIA and stroke (Sterpetti et al. 1988, Bock et al. 1993, de Bray et al. 1998). Sequential longitudinal sections or cross-sections then often show only individual aspects of a complex morphology. With advancing development, clear narrowing of the lumen and broadening of the atheromatous bed occur. During this process, bleeding (*"hemorrhagic plaques"*) and calcium deposits (*"hard plaques"*) may appear, which can cause an ultrasound shadow with significant energy loss. The validity of the test for identifying hemorrhages within the plaque is still controversial: sensitivities and specificities reported range between 72% and 94% (79% and 89% respectively) (Reilly et al. 1983, Bluth et al. 1986, Widder et al. 1990).
- Ruptures in the plaque cap and *ulcerations* associated with them can appear at any stage of arteriosclerotic development, and cannot always definitely be detected in B-mode (Comerota et al. 1981, Sitzer et al. 1995, Valton et al. 1995). The presence of an ulcer in a tight stenosis (> 70% lumen narrowing) is associated with an increased risk of cerebral ischemia (Eliasziw et al. 1994, Sitzer et al. 1995). Recent attempts to improve the image quality and to display three-dimensional and even four-dimensional plaque reconstructions may support the diagnosis of stable or friable plaques (Hennerici 2004).

Doppler mode. It is usually easy to interpret flow changes in the vicinity of low-grade stenoses, and to assign flow accelerations and also sometimes flow irregularities. With medium and high-grade stenoses, however, the primary concern for the Doppler mode in duplex sonography is to interpret hemodynamic criteria, since the image quality may be limited due to ultrasound shadows (Fig. 2.**40**).

Color flow Doppler mode. Even in difficult examination conditions, color coding makes it easier to detect and grade medium to higher-grade stenoses in the carotid artery region than it is with the single-channel Doppler mode in echotomography. Usually, the flow acceleration, depicted as a color saturation, is immediately recognizable from the color-coded signal. Local or extensive sudden color changes, with artificially high flow velocities caused by aliasing, can be directly detected. The B-mode image only has a secondary role here, providing additional information by defining structural wall changes when these vascular processes, which are often calcified, are being examined. With low-grade plaque formations, there are usually changes near the carotid sinus, such as limited or completely absent flow separation phenomena, irregular color distributions near a detectable structural wall change in B-mode, or isolated color changes without any detectable morphological correlation. This is especially important as an indication of a fresh thrombosis, possibly with a high embolization potential. Cross-sectional imaging provides additional information about the actual degree of stenosis and is particularly useful if performed prior to surgery in the absence of conventional angiograms (de Bray and Glatt 1995, Ringelstein 1995, Carpenter et al. 1996, Steinke et al. 1996).

Color flow duplex sonography also provides better visualization of changes within a plaque (Fig. 2.**31**), such as niche formation and surface irregularities, with or without accompanying flow changes, hypoechoic plaque formations, and ulcerations. Ulcerated lesions show both recess (≥ 2 mm in depth and ≥ 2 mm in length, with a well-defined back wall at its base) and flow vortices (i.e., reversed flow without aliasing) within or at the level of the recess. Only one prospec-

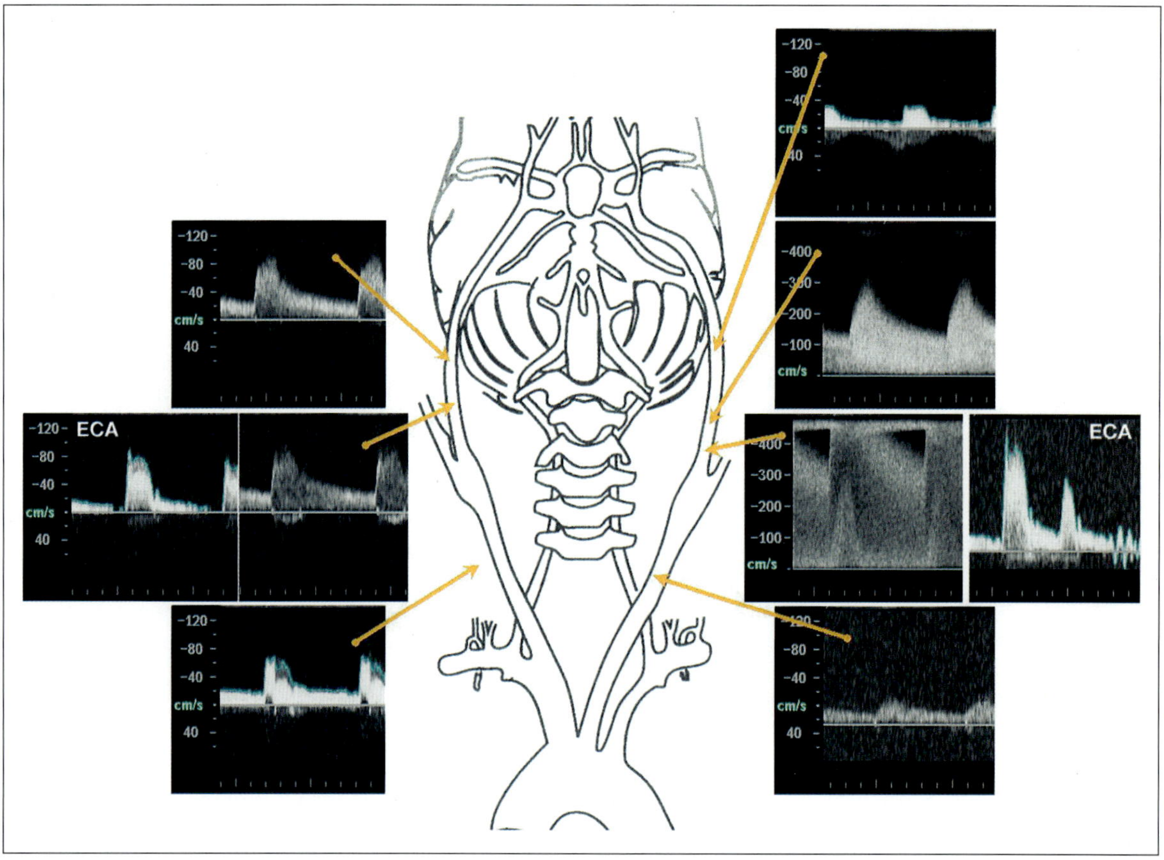

Fig. 2.**40** Examples of the Doppler spectra in different arterial segments, from a patient with high-grade carotid stenosis on the right side. *Left:* normal spectra from the common carotid artery (CCA) and proximal and distal internal carotid artery (ICA). Note the low pulsatility, low frequency, irregular components in the proximal part near the bifurcation, and the slightly higher systolic velocities (the distal ICA diameter is lower) in the distal ICA. *Right:* low velocity in the CCA (lower than on the normal left side). Increased flow velocities in the external carotid artery (ECA) (identified by compression test and tapping on the temporal artery). There are very high flow velocities in the proximal ICA intrastenotic Doppler recording and very low flow in the distal ICA

tive study has so far been published comparing color flow duplex imaging with surgical specimens. It reported rather low sensitivity and specificity values (0.33 and 0.76, respectively) (Kessler et al. 1991), which could be improved in better study conditions.

If the system is accurate enough to detect slow flow velocities, then the color flow Doppler mode is also helpful in evaluating subtotal stenoses (Fig. 2.**41**), pseudo-occlusions, and dissections (Steinke et al. 1990a, Wernz et al. 1994) (see Fig. 2.**43**).

Several methods using special transmitter arrangements, emission characteristics, or postprocessing algorithms have been introduced to enhance the imaging quality in B-mode and color mode imaging. These techniques and methods are highly dependent on the type of ultrasound equipment used. The current Siemens machines use "photopic ultrasound imaging," while Acuson uses "chirp-coded excitation" and Phillips employs "compound imaging." In addition to the basic methodological information given in Chapter 1, these technologies can substantially enhance the resolution and quality of the ultrasound image and quality; the compound imaging technique is illustrated in Fig. 2.**32** (Kern et al. 2004).

Stents are increasingly being used with percutaneous transluminal angioplasty (PTA) in interventional procedures. Although there is at present a lack of evidence-based data on the efficacy and risks associated with this experimental treatment, increasing numbers of patients are undergoing PTA and stenting procedures. The location and patency are easily displayed during and after procedure, but the insonation capacity appears to deteriorate, with increasing delays in follow-up studies (Fig. 2.**29**).

Continuous-Wave Doppler Sonography

Increased Flow Velocity

Low-grade stenosis (40–60% luminal narrowing). A local increase in flow velocity is observed, with an acoustically intensified Doppler signal. The systolic–diastolic amplitude modulation is retained; prestenotically and post-stenotically, there are no changes in the

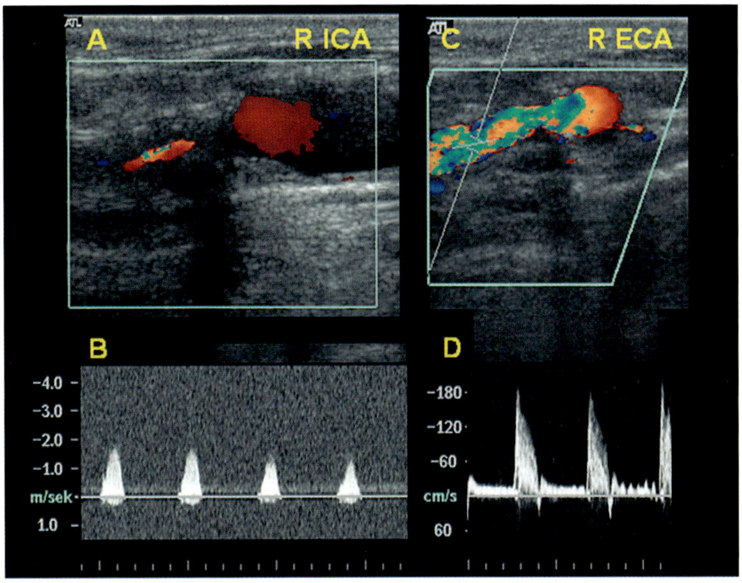

Fig. 2.**41** Combined stenosis of the internal (ICA) and external carotid arteries (ELA) with subtotal obstruction of the ICA. Power mode (**a**, **c**) and corresponding Doppler mode (**b**, **d**) registrations

Doppler waveforms; there are sometimes marked irregularities in the audio signal (turbulence phenomena), comparable to fine or medium crepitating rales (Fig. 2.**42**).

Compensatory elevated flow. In the presence of hemodynamically significant contralateral flow obstructions, a similar Doppler signal intensification may possibly be recorded ipsilaterally, characterized by a *continuous* increase mainly in the diastolic flow velocity (rarely also in the systolic) throughout the internal carotid artery after the bifurcation. If the internal carotid artery can only be followed for a short distance (short neck, bifurcation located at a high level), it is impossible to distinguish between low-grade stenosis and compensatory elevated flow at the origin of the internal carotid artery.

Dilative arteriopathy of the common carotid artery or hypoplasia of the internal carotid artery. Abnormal differences in the vascular caliber between the common and internal carotid arteries can cause similar changes in the Doppler signal, mimicking compensatory collateralization, or a mild stenosis at the origin of the internal carotid artery.

Intracranial arteriovenous malformation. Intracranial angiomas with a high shunt volume are sometimes recognizable due to a distinct increase in flow ipsilateral to the malformation, with a retained—or in rare cases also disturbed—amplitude modulation in the feeding extracranial cerebral arteries (i.e., the common and internal carotid arteries). In such cases, the Doppler signal is often reduced in the terminal branches of the ophthalmic artery, or even becomes retrograde when the blood is drained through these vessels into the malformation.

Loss of Pulsatile Amplitude Modulation and Systolic Deceleration

Medium to high-grade stenosis (60–90% luminal narrowing). In medium to high-grade stenoses, with a lumen reduction exceeding 60%, not only does the diastolic flow velocity at the stenosis increase, but the pulsatile amplitude modulation is also lost as well. When the transducer is moved along the obstruction, the analogue registration shows a systolic deceleration, also known as a "systolic peak reversal," with the average flow signal directed toward rather than away from the probe. Post-stenotically, the average flow velocity again decreases, and significant turbulences in the audio signal appear (comparable to medium to coarse crepitating rales, crepitations, and rhonchi) (Fig. 2.**42**).

Extensive wall pulsations. A systolic deceleration may be imitated by extensive wall pulsations in vessels with abnormally high flow feed in the absence of any structural stenosis (e.g., "functional stenosis" in patients with intracranial arteriovenous malformations).

Kinking/coiling. Loop formations can also simulate a systolic deceleration. Misdiagnosis can be prevented when the exact physiological waveform is watched, particularly noting the normal audio signal, which does not show any change in association with the changing flow direction.

Decreased Flow Velocity

High-grade and subtotal stenosis (more than 80% luminal narrowing). In high-grade and subtotal stenoses, the diastolic and systolic flow velocities decrease in both segments proximal and distal to the narrowing of the lumen. Post-stenotically, pulsatile amplitude modulation may be completely absent, can be replaced by

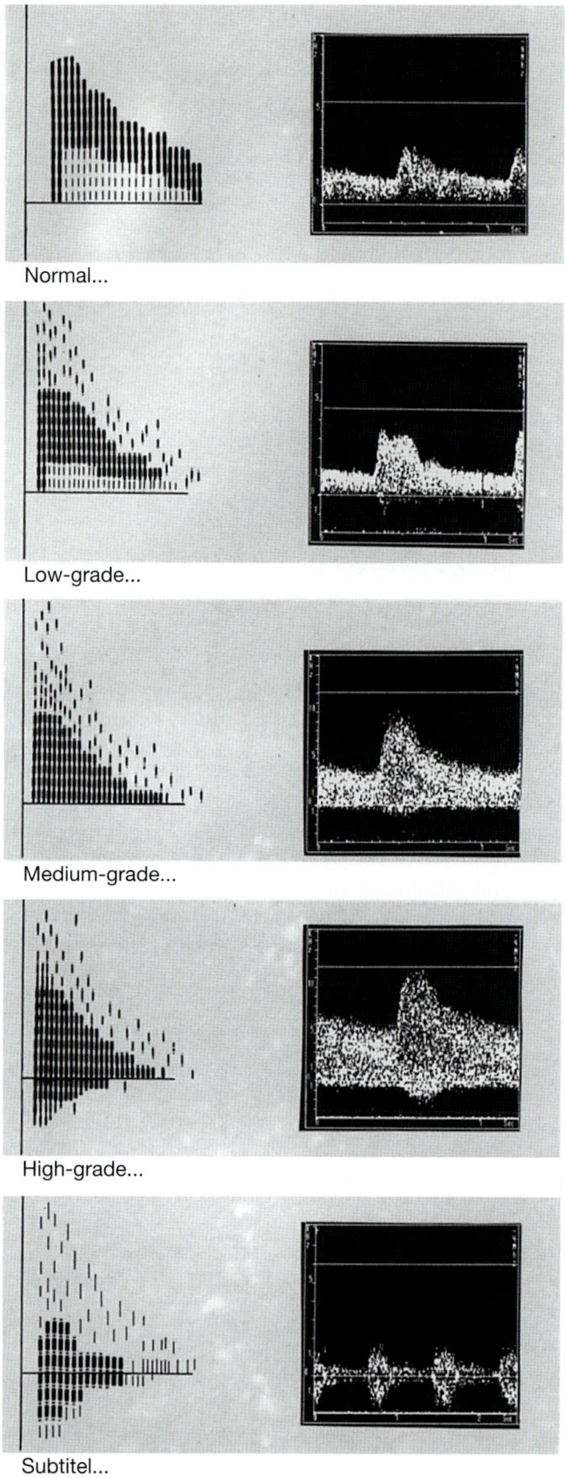

Fig. 2.**42** Diagrams and fast Fourier transform (FFT) spectra showing continuous-wave Doppler signals from the internal carotid artery for different degrees of stenosis (adapted from Arbeille et al. 1985)

an unmodulated, turbulent flow with pronounced audio signal changes, and often reveals a marked systolic deceleration zone. High-grade stenoses show increased flow within the tightest vascular segment. In subtotal stenoses, these changes are frequently absent and the Doppler signal is sometimes hardly audible, characterized by a quiet, high-frequency sound that is difficult to register because of the poor signal-to-noise ratio caused by the low ultrasound energy reflected by the few corpuscles passing (an acoustically fine, chirping sound) (Fig. 2.**42**).

Proximal and distal high-grade stenosis. Luminal narrowing in vascular segments near the aorta or in the submandibular or intracranial areas can be detected by a reduced flow velocity, with a retained or slightly reduced amplitude modulation. These are indirect criteria for the presence of a proximal or distal stenosis.

Slosh Phenomenon

A "carotid dissection" is usually characterized by a filiform stenosis, which originates submandibularly at the entrance of the internal carotid artery into the base of the skull, and reaches down into the bifurcation (Fig. 2.**43**) (Fisher et al. 1978, Biller et al. 1986, Mokri et al. 1986, Steinke and Hennerici 1989). Using Doppler sonography, a highly typical but non-pathognomonic signal is found, which the present authors were the first to describe as the "slosh" phenomenon in the internal carotid artery (Hennerici et al. 1989). In contrast to an intermediate flow signal, systolic flow components can be detected that are simultaneously directed in both orthograde and retrograde directions, indicating a significant increase in the peripheral resistance due to the pronounced distal stenosis in the absence of any flow velocity during the diastole. Some of these are pulsatile movement-induced artifacts caused by hemorrhage within the vascular wall, superimposed on a very low flow velocity signal. A similar flow signal, although usually without this characteristic systolic forward-and-backward (or "to-and-fro") blood flow, is seen in rare instances in which the main branch of the middle cerebral artery is acutely and completely occluded—in serious and extensive vasculitis, and in incomplete retrograde thrombosis of the internal carotid artery due to a siphon stenosis. Depending on the progression of the dissection, which is often followed by spontaneous recanalization after an average of 4–6 weeks, the slosh phenomenon shifts cranially, and the lumen opens again. During this process, various stenotic signs can be detected with Doppler sonography. Sometimes does the blood vessel remain completely occluded (Steinke et al. 1994).

Absent Signal

When the internal carotid artery is occluded at its origin, the flow signal is absent. In contrast to the changes described above, there is no positive criterion that can be used as an indicator of a flow obstruction.

Indirect criteria are therefore particularly important:

- All detectable Doppler signals have to be identified (e.g., most of the branches of the ECA by applying compression tests).
- Confusion with the vertebral artery, which often functions as a collateral, has to be avoided by following its course. If the signal is interrupted or cannot be traced back to the bifurcation, suspicion arises as to whether the ICA might be occluded.
- Possible variations in location have to be noted (medial location, loop formation).
- A filiform stenosis (i.e., pseudo-occlusion), often with only local flow acceleration, has to be looked for meticulously (Wernz et al. 1994). It is only rarely that the proximal stump is still open, or that an alternating flow signal extending in a submandibular direction can be detected in this location; this finding is characteristic of a dissection, as described above (Fig. 2.**43**).

The parameters given in Table 2.**9** vary depending on the degree of stenosis in internal carotid artery obstructions:

- The *peak frequency and velocity* (systolic and diastolic) increases continuously (from a lumen narrowing of 30 % to around 90 %) and then decreases again, corresponding to the transition from high-grade to subtotal stenosis.
- The average flow velocity also increases with the degree of stenosis (from approximately 30 % to 90 %), but is subject to less pronounced fluctuations.
- In high-grade stenoses (> 80 %), it is useful to combine the values measured at different registration points (e.g., intrastenotic and prestenotic or poststenotic) in order to detect the change in the vascular resistance induced by the stenosis (Rittgers et al. 1983).
- The frequency associated with the highest signal amplitude decreases with an increasing degree of stenosis, and has a *negative frequency component*, especially in high-grade stenoses (over 75 %).
- The *systolic window* decreases with increasing stenosis (from 30 % to subtotal stenosis).

It is therefore not recommended to use single parameters to classify the degree of stenosis although most studies so far published on the topic have either used peak or mean frequency values (for review, see de Bray and Glatt 1995). Appropriate differentiation between normal results and evaluation of different degrees of stenosis (Table 2.**9**), rather than a widely ranging classification (e.g., < 50 %, 50–80 %, > 80), requires the analysis of appropriate combinations of parameters for reliable estimation (Hennerici et al. 1985, Arbeille et al. 1985). This is especially important for follow-up studies (Daigle et al. 1988, Daffertshofer and Hennerici 1990, Ringelstein 1995, Young and Humphrey 1996).

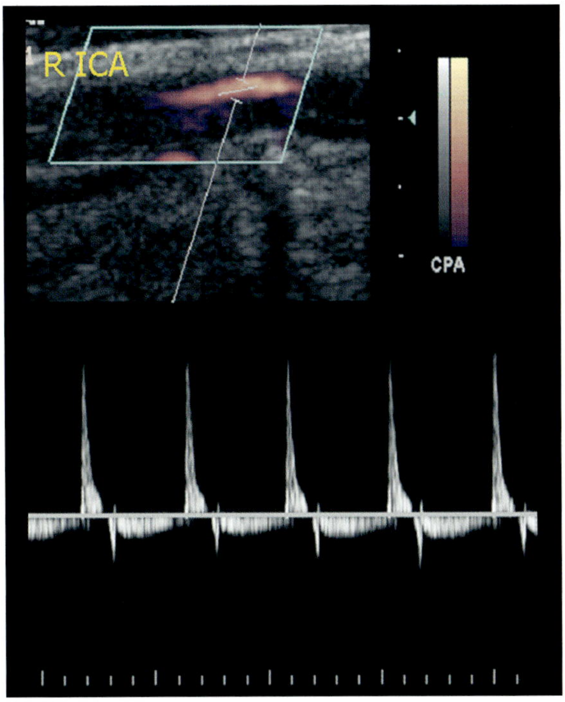

Fig. 2.**43** Dissection of the right internal carotid artery (ICA), leading to functional occlusion. There is very narrow filling of the right ICA in power mode. The pulsed-wave Doppler measurements show an oscillating “to-and-fro” signal, with zero net flow

External Carotid Artery

Color Flow Duplex Sonography

Isolated external carotid artery stenoses and occlusions are rare findings. They are more often seen in the context of bifurcation stenoses. Although the same Doppler-sonographic criteria apply as in the internal carotid artery, grading of the hemodynamic relevance is different.

B-mode. All proximal branches of the external carotid artery can be displayed in B-mode and color-mode. Most distal branches can only be poorly imaged, due to:

- An unfavorable imaging angle
- Thin caliber
- Frequent loop formations
- Very variable anatomy

Additional identification of the flow signal using duplex system analysis is recommended. The wall changes in the external carotid artery are similar in structure to those described above for the internal carotid and common carotid arteries.

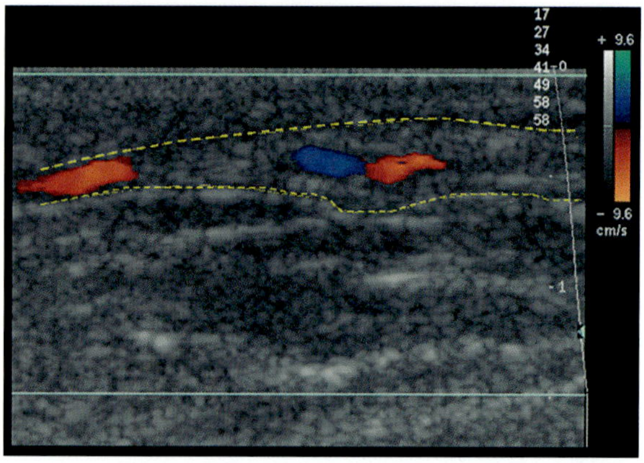

Fig. 2.**44** A color duplex image of the superficial temporal artery in a patient with temporal arteritis. There is normal flow, but an increased, slightly hyperechoic wall thickness. The outer wall of the artery is outlined in yellow

Color flow Doppler mode. It is easier to identify small, thin-caliber branches and loop formations with color flow duplex sonography than with conventional duplex procedures. Individual branches at their origin from the bifurcation are often displayed simultaneously (Fig. 2.**8**). In addition, flow signal changes can be visually recognized—e.g., saturation and separation phenomena, as well as local signal obliterations by thrombotic wall adhesions. This is especially significant when a stenosis is present at the bifurcation. With both continuous-wave and conventional duplex sonography, it may be difficult to distinguish terminal branches and differentiate the degree of stenosis in the external and internal carotid arteries independently. Retrograde blood flow through the external carotid artery, as in a proximal occlusion of the common carotid artery (Chang et al. 1995), can also be easily recognized with color flow duplex sonography. In addition, complex processes involving several blood vessels can be depicted, and the collateral vascular network in the neck region can be identified (e.g., innominate artery obstruction with collateralization through the contralateral carotid and vertebral vascular systems). Local external carotid artery stenoses can easily be missed in conventional duplex and Doppler sonography in patients with "low-flow" conditions—e.g., dilative arteriopathies and subsequent arteriosclerotic processes, or high-grade stenoses near the aortic arch. If a vascular reconstruction has been carried out ("stump syndrome"; Barnett 1978) in a case of occlusion of the internal carotid artery with an embolizing external carotid artery stenosis, or when new symptoms associated with a stenosis of the external carotid artery appear in patients with an extracranial–intracranial bypass, postoperative follow-up observations should include this method.

In patients with temporal arteritis, color mode imaging has become useful for identifying thickened vessel wall structures with associated flow pattern abnormalities (Salvarani et al. 2002) (Fig. 2.**44**).

It is less common nowadays to examine patients with external–internal carotid bypass conditions; however, assessing patency and alterations in flow with increasing age and atherosclerosis in these patients may occasionally be useful (Arakawa et al. 2003). For research reasons, it may also be interesting to identify supportive vessels and to monitor the outcome in highly selected patients in whom this procedure may still be appropriate.

Continuous-Wave Doppler Sonography

Whether it is possible to assess ECA pathology using Doppler sonography alone has long been a controversial issue in vascular clinics. Such studies often have more research value than clinical relevance, and nowadays require duplex sonography analysis. However, if Doppler instruments are available, it is useful to identify the ECA branches from the ICA and CCA, particularly if more than one vessel is obstructed.

Increased Flow Velocity

Only stenoses with extensive luminal narrowing (> 70%) can be detected in the audio signal.

Due to the reduced peripheral vascular resistance, dural fistulas and angiomas with an extracranial blood supply can cause significant increases in flow velocity at the origin and along the course of the feeding branches. This imitates all the criteria for high-grade stenosis in the ICA (loss of pulsatility, systolic deceleration, extreme turbulence). Pronounced pulsations in the neck region are usually also found, which may cause artifacts in the Doppler signal of the ECA.

Exceptions to this are conditions such as *extracranial–intracranial bypass* and, to a lesser extent, ECA branches collateralizing occlusion of the internal carotid artery, in which the function of the artery changes.

Absent Signal

As in the internal carotid artery, an occlusion of the external carotid artery or its individual branches is characterized by an absent Doppler signal. It is easy to overlook occlusions of the branches of the external carotid artery, since there are no indirect criteria for the diagnosis, in contrast to the situation in the internal carotid artery (no change of flow direction in the orbital, common, or internal carotid arteries). Often, only certain branches of the external carotid artery can be evaluated with Doppler studies.

Spectral analysis is particularly helpful when external carotid artery stenoses are accompanied by internal carotid artery stenoses in bifurcation processes. Interpreting the audio signal in such cases is difficult, and classification of different degrees of stenoses in the two branching arteries is sometimes impossible. In contrast, identification of the individual branches of the external carotid artery with pathological flow patterns is easily accomplished by compression tests.

Evaluation

Common pathological and ultrasound situation (diagnostic effectiveness and importance for various aspects of ultrasound analysis: +++ very important; ++ supportive; + helpful):

Vascular Degeneration and Atherogenesis

Direct criteria

- Imaging (+++):
  - Increase in the IMT measured on defined positions in the CCA and ICA.
  - Plaque formation in longitudinal and cross-sectional views (without increased blood flow velocity).
  - Differentiation between various pathological processes and distinguishing these from normal variants (e.g., wide or congenitally absent carotid sinus, caliber variations between the common and internal carotid arteries, "low-flow phenomenon" in an extracranial carotid system aneurysm, and dilative arteriopathy.
- Flow analysis. Color flow and power mode (++):
  - Occasionally, pathological flow irregularities with increased color changes.
- Spectrum analysis: continuous-wave Doppler or pulsed-wave Doppler (duplex mode) (+): no valid pathological changes. Flow irregularities sometimes cause broadening of the spectrum, with closing of the spectral systolic window. In older systems, sometimes only evident in the audio signal.

Indirect criteria

- None.

Internal Carotid Artery Stenosis

(Figs. 7.**1**–**2** pp. 263–264)

Direct criteria

- Imaging: (+++)
  - Wall changes in B-mode and display of plaque formation
  - Degree of stenosis in longitudinal and cross-sectional view
- Flow analysis. Color flow and power mode (++):
  - Increase of blood flow velocity associated color level
  - Post-stenotic flow irregularities with changes in flow direction
  - Aliasing in color-mode in high grade stenosis
- Spectrum analysis: continuous-wave Doppler or pulsed-wave Doppler (duplex mode) (+++):
  - Increase of blood flow velocity depending on degree of stenosis, within a range of ~ 40–90% stenosis
  - Abnormal spectrum pattern within the stenosis, with a shift of the highest intensities within the spectrum toward lower frequencies, despite an increase in peak blood flow velocities
  - In hemodynamically relevant stenoses (ca. 70% stenosis or more), post-stenotic flow irregularities with increase of lower frequencies and sometimes back-flow phenomena during the cardiac cycle

Indirect criteria (+++)

- In hemodynamically relevant stenosis (ca. 70% stenosis or more), decreased blood flow velocity in the proximal CCA and the distal (submandibular or intracranial, ICA) arterial segments
  - Increased pulsatility in the proximal CCA
  - Decreased pulsatility in the distal ICA (due to intracranial arteriolar vasodilation; vasomotor reserve capacity)
  - Flow velocity changes in the orbital artery (when there is collateralization)

The degree of obstruction is assessed using the combined parameters given in Table 2.**8**. Subtotal stenoses are often difficult to detect, and there is a risk of misdiagnosing an occlusion of the internal carotid artery. Examination using Duplex systems is more successful than with exclusively continuous-wave Doppler sonography for detecting and classifying carotid stenosis (Steinke et al. 1990a, Wernz et al. 1994, Steinke et al. 1996, Filis et al. 2002, Knudsen et al. 2002).

Internal Carotid Artery Occlusion

Direct criteria

- **Imaging (++):**
  - Confirmed wall change, occluded lumen of the ICA in B-mode analysis—e.g., hypoechoic material indicating fresh thrombus, arteriosclerotic plaque occluding the ICA
- Flow analysis. Color flow and power mode (++):
  - Absent flow even when there is contrast enhancement
  - "To-and-fro" colors alternating regularly in an ICA stump
- Spectrum analysis: continuous-wave Doppler or pulsed-wave Doppler (duplex mode) (++):
  - Absent flow even with contrast enhancement "To-and-fro" colors alternating regularly in an ICA stump.

Indirect criteria (+++)

- Decreased blood flow velocity in the proximal CCA, particularly decreased diastolic blood flow velocity
  - Increase of pulsatility in the proximal CCA (the CCA spectrum adapts to the ECA spectral pattern)
  - The ECA compression test is clearly positive in the CCA
  - If there is collateralization in the intracranial vessels, ECA pulsatility decreases and the Doppler spectrum pattern of the ECA changes to that of an ICA
  - Flow velocity changes in the orbital artery (in case of collateralization)

Carotid Bifurcation Stenosis
(Fig. 7.**5** pp. 268–270)

Direct criteria
If both the internal and external carotid arteries are involved, it is necessary to classify the registered signals separately in the "carotid triangle" by:

- Carefully moving the ultrasound probe
- Identifying the external carotid artery by compressing or tapping ECA branches in order to separately:
- Evaluate the degree of stenosis in each artery
- Identify the vascular processes

Duplex systems have significantly better diagnostic accuracy here than exclusive use of continuous-wave Doppler.

- Imaging (+++):
  - Wall changes in B-mode and display of plaque formation
  - Degree of stenosis in longitudinal and cross-sectional view
- Flow analysis. Color flow and power mode (++):
  - Increase in blood flow velocity–associated color level
  - Post-stenotic flow irregularities with changes in flow direction
  - Aliasing in color-mode in high-grade stenosis
- Spectrum analysis: continuous-wave Doppler or pulsed-wave Doppler (duplex mode) (+++):
  - Increase in blood flow velocity depending on the degree of stenosis, within a range of ca. 40–90% stenosis
  - Abnormal spectral pattern within the stenosis, with shifting of the highest intensities within the spectrum toward lower frequencies despite an increase in peak blood flow velocities
  - In hemodynamically relevant stenosis (ca. 70% stenosis or more), post-stenotic flow irregularities with increase in lower frequencies and sometimes back-flow phenomena during the cardiac cycle

Indirect criteria
- Post-stenotically reduced flow, which can be assigned to the external or internal carotid arteries using compression tests
- Changes in the flow relationships of the orbital arteries (in higher-grade stenoses), but these can be even normal if the degree of stenosis is equibalanced (from a hemodynamic standpoint)
- Prestenotically decreased flow velocity in the common carotid artery is seen in high-grade internal carotid artery stenoses, but not in external carotid artery stenoses

Sources of error
- Local variants—e.g., internal carotid artery medial to the external carotid artery
- Superposition of two blood vessels (changing the neck and head position sometimes allows better assignment); color flow duplex analysis facilitates isolated Doppler signal registration
- Loop formation in the internal carotid artery and individual branches of the external carotid artery

Internal Carotid Artery Occlusion and Ipsilateral External Carotid Artery Stenosis

This situation often leads to misdiagnoses, as it is difficult to establish whether an internal carotid artery signal is absent in the vicinity of an external carotid artery that has pronounced stenotic signs.

Color flow duplex sonography (including power amplitude analysis) helps: in a series of cross-sections proceeding from the common carotid artery into the internal carotid artery, no flow signal can be detected. If the bifurcation is located high in the submandibular region, or if there is a stenosis over an extended distance, the examination is sometimes subject to technical limitations. Occasionally, absence of a signal from the submandibular ICA may support the findings.

External Carotid Artery Occlusion and Internal Carotid Artery Stenosis

In general, this situation does not present any diagnostic difficulties. To evaluate the degree of stenosis in the internal carotid artery, the criteria given above apply; there is no external carotid artery signal. Pathological flow changes in the branches of the ophthalmic artery are rare, due to the failure of ophthalmic artery to collateralize via the ipsilateral external carotid artery. Compressing the branches of the contralateral external carotid artery while recording Doppler signals from the ophthalmic artery and its branches is important here, because collateralization usually proceeds via the contralateral side.

Sequential Ipsilateral Stenoses in the Carotid System (Fig. 7.**7** pp. 272–273)

These conditions are often easily detectable when duplex systems are used, but they may be difficult to diagnose with continuous-wave Doppler alone, since sequential stenoses may affect the flow velocity measurements.

- If both stenoses lie in an area directly accessible to the Doppler probe, it is often possible to detect them.
- If the higher-grade stenosis is more proximal, then it is easy to miss the distal stenosis, because the Doppler-sonographic findings can be explained just by the changes in the flow relationships near the high-grade stenosis. This applies in all cases in which the distal stenosis is no longer directly accessible.
- If the higher-grade stenosis lies distally, and both stenoses lie in the directly accessible neck region, then they can usually be reliably detected. Even when the higher-grade stenosis is not in the neck

region that is directly accessible, indirect criteria often provide clues to the existence of a second stenosis.

Multivessel Disease (Fig. 7.**5** pp. 268–270)

Obstructions involving several arteries on both sides of the neck are more difficult to identify, due to:

- The possible presence of many different collateral mechanisms
- The greater difficulty of carrying out side-to-side comparisons (e.g., mild reductions in the blood flow velocity in both CCAs are easily overlooked)
- The required modification of the criteria used to classify the degree of stenosis

Carotid Artery Dissection (Fig. 7.**3** p. 265)

There is a characteristic Doppler signal when a carotid artery dissection causes carotid lumen obstruction. This "slosh phenomenon" can be found along the internal carotid artery in the neck region, depending on the extent to which intramural hematoma formation, which usually starts submandibularly, reaches toward the carotid bifurcation in the neck and constricts the lumen. Depending on the extent to which the lumen is constricted and on the distance from the dissection area, there may be a sequence of stenotic signs, ranging from high-grade obstruction to medium-grade or low-grade stenosis, all the way to completely normal findings. In a few cases, a complete occlusion after the bifurcation can be detected, which in B-mode typically resembles a long, pointed cap. In color flow Doppler mode, a flow signal is seen that reaches as far the tip of the cap, sometimes with a superimposed component showing retrograde flow. Carotid dissections that do not cause lumen obstruction—i.e., with wall hematoma and/or aneurysm alone—can be missed with ultrasound, particularly if such dissections are located very distally.

Postoperative Findings

Check-up assessments are a frequent indication for ultrasound diagnosis, both in the early postoperative period to document the effectiveness of a procedure, and also later, to detect any re-stenoses, particularly in young patients (Valentine et al. 1996). Shortly after an operation, the examination can be difficult due to swelling of the neck, but continuous-wave Doppler sonography can usually provide a reliable evaluation of the hemodynamic situation. The surgical technique used (e.g., patch) will determine whether a wide bifurcation can be displayed in B-mode. In color flow duplex sonography, well-developed flow separation zones occur (Steinke et al. 1991); occasionally, larger hypoechoic intravascular structures can be detected.

Postinterventional Findings (Stenting) (Fig. 7.**4** pp. 266–267)

Ultrasound examinations after interventional procedures (angioplasty and particularly stenting) are not obstructed by swelling or other surface alterations. The stent (material) can easily be displayed directly after the procedures. The stent continues to be visible for several weeks or months, with perfect display of stent placement as well as any residual stenosis. After reendothelialization has taken place, it can become much more difficult or sometimes almost impossible to display the stent.

Sources of Error

Although there is usually no problem in the differential diagnosis of isolated low, medium, and high-grade stenoses of the carotid arteries, mistakes in addition to those mentioned above are frequently caused by the following:

- Pseudo-occlusion versus total occlusion
- Proximal or distal stenoses outside the directly accessible area
- Multiple obstructions of varying degree at several levels of the carotid system, both extracranial and intracranial

Misinterpretations can be avoided if the plausibility of findings from *all* arteries examined is carefully considered, with particular attention to the following:

- Nonstenotic wall changes can be suspected rarely by Doppler sonography alone when they are directly accessible to ultrasound. Differential diagnosis from normal variants is only possible with supplementary duplex system studies.
- Flow signal changes in juveniles.
- Dilative arteriopathy.
- Intracranial vascular processes, obstructions, fistulas, and arteriovenous malformations.
- Analysis of altered collateral circulation.
- Aneurysmal dilations.
- Variations in location and course (a medial internal carotid artery, coiling, and kinking).
- Status after vascular surgery in the neck (scar formation, hematoma, edema, synthetic patch), or insertion of an extracranial–intracranial bypass.

Overestimation of the degree of stenosis (flow increase and audio signal irregularities) may result from the presence of increased collateral flow, which is usually caused by a contralateral hemodynamically marked flow obstruction (80% stenosis or more, or occlusion). Since duplex sonography is able to provide structural confirmation in multiple planes of a slight or medium constriction of the lumen and depict the associated alterations in the vascular flow, appropriate classification of the degree of stenosis is usually possible in these cases. However, assessing the degree of the stenosis in high-grade stenoses of the internal carotid arteries is more difficult because of loss of image quality associated with hemodynamic alterations. In these cases, evaluation of the higher-grade stenosis should take into account not only the local degree of stenosis and the circumscribed maximum constriction at one location, but should also consider:

- The extent of the vascular process
- Collateral function in the vertebrobasilar system
- Possible cumulative effects due to stenoses that follow or precede the stenosis in question
- The current intracranial collateralization through the often (50%) insufficiently developed circle of Willis (Padget 1944)
- Collateralization through the external carotid artery anastomosis

If the vertebral artery collateralizes an occlusion in the ipsilateral carotid system (common or internal carotid artery), then the signal from the vertebral artery may be mistakenly interpreted as coming from the internal carotid artery, particularly in slender patients, since the blood vessels are projected on top of one other. Comparing this signal near the bifurcation to the vertebral artery signal at the posterior arch of the atlas or the V_2 segment usually allows the correct diagnosis to be made. Here too, duplex sonography is diagnostic.

Proximal common carotid artery stenoses at the origin of the vessel from the aortic arch can also cause changes in the Doppler curves in the neck region, preventing correct assessment of the degree of stenosis of the internal carotid artery at the bifurcation. It should be noted that in these cases, the ophthalmic collateral is not usually open, in spite of an intact external carotid artery. In the differential diagnosis, it is important to evaluate the collateral function in the remaining arteries supplying the brain: if there is no elevated blood flow in the contralateral carotid system or in either vertebral artery, a careful search needs to be made for a proximal stenosis at the origin as an explanation. Intrathoracic pulsed-wave Doppler/duplex sonography is often helpful here.

Diagnostic Effectiveness

Color Flow Duplex sonography

The specificity of a duplex investigation ranges from 47% to 100%, depending on methodology, technical equipment, and expertise (Tables 2.**3** and 2.**4**). B-mode examinations alone are less valid, and no longer meet today's examination standards. Radiographic angiography, which has usually been regarded as the accepted reference method in making such assessments, can no longer be considered as the gold standard without further qualification (de Bray and Glatt 1995, Patel et al. 1995, Erdoes et al. 1996). Digital subtraction angiography, for example, often fails to detect small changes in the vessel walls, even when selective intra-arterial contrast medium is used (Eikelboom et al. 1983). Flow irregularities that appear physiologically in the carotid bifurcation can produce artifacts in magnetic resonance angiography. Duplex sonography using the color flow examination method is able to show that this apparent pathology represents normal flow separation phenomena, without any structural wall changes. Thus, the two methods can complement each other.

The main advantage of color flow duplex sonography is that it can simultaneously display both the morphological and hemodynamic parameters. In both in-vitro and in-vivo examinations, the sensitivity and specificity of high-resolution B-mode imaging (> 7.5 MHz) are very good for *nonstenotic plaques* (< 40% obstruction). Adequate differentiation of the various plaque forms, according to the criteria described above, is also possible. Using new methods that allow three-dimensional reconstruction from a series of sections (Steinke and Hennerici 1989, Delcker et al. 1995, Hennerici et al. 1995, Meairs et al. 1995), an outstanding image can be obtained that demonstrates the total extent and shape of the plaque formation. However, equipment manufacturers have been slow to provide the software required for this. As an atheroma expands and causes increasing stenosis, the quality of the image from various angles becomes poorer due to energy loss, even with sufficiently precise ultrasound. Uncomplicated plaques, usually consisting of an atheromatous seed, often still provide a good image, and the B-mode image is not significantly altered by fibrous caps covering a smooth-muscle cell bed, or by collagen, elastin, or intracellular and extracellular lipids. However, *calcium deposits* and *surface irregularities*, which indicate progressive, complicated plaques, are unfavorable conditions for imaging in a high-resolution system. Whether this association with patients' individual prognoses can actually be scientifically established remains to be seen.

Plaque-associated flow changes and plaque movements can also be analyzed using color flow duplex sonography, and three-dimensional analysis is possible in hemodynamic investigations, although it is still somewhat experimental (Hennerici et al. 1995, Meairs et al. 1995). With appropriate clinical indications, a supplementary duplex-sonographic examination is therefore mandatory when the conventional Doppler sonogram or angiogram have normal or doubtful appearances—particularly signs of embolizing wall changes, since fresh thrombotic deposits that may have increased embolic potential, as well as niche and ulcer formation, can be detected fairly reliably using the color flow Doppler mode. The same applies to the detection of aneurysmal wall changes. Follow-up observations after obliterating carotid artery occlusions, and differentiation between dilative vascular processes and obstructive ones, are further important issues (Steinke et al. 1991, Bonithon-Kopp et al. 1996).

Color flow duplex sonography increases the diagnostic efficacy and validity of the classification of extracranial carotid disease (Vanninen et al. 1995, Carpenter et al. 1996, Howard et al. 1996). The advantage of the procedure, especially for the less experienced investigator, lies in the fact that it is easy to learn and provides both visual and acoustic access, in contrast to the traditional, rather one-sided acoustic investigation method. Significantly shorter examination times are possible, particularly in complicated high-grade vascular processes in which it is difficult to display the anatomy. Standardized research to determine the validity of this method using defined stenosis criteria in a reference procedure has not really been carried out and probably never will be, since even without conventional angiography it is possible to establish the diagnosis of a carotid stenosis of more than 70 % using the appropriate criteria for surgery (de Bray et al. 1995, Patel et al. 1995, Barnett et al. 1998, Rothwell et al. 2003).

In *high-grade and subtotal flow obstructions,* all of the available procedures may fail. It is difficult to differentiate a pseudo-occlusion from a complete occlusion, even when all possible hemodynamic and imaging methods, including conventional angiographic procedures, are used. It can only be done when a weak flow signal is positively confirmed in an almost occluded blood vessel. Color flow duplex sonography is superior to the other methods in this instance as well, especially once algorithms for displaying slow flow signals, combined with new echo contrast media have been developed (Wernz et al. 1994, Steinke et al. 1996).

According to recent results from the Asymptomatic Carotid Surgery Trial (Halliday et al. 2004), asymptomatic patients with high-grade carotid stenoses may occasionally benefit from surgery, provided that low (< 3 %) and carefully documented carotid endarterectomy complication rates are established and life expectancy is not limited by known diseases. Similar regimen applies to high-risk patients with severe symptomatic carotid stenoses (Rothwell et al. 2003). In both groups, however, angiography should be avoided in order to minimize the risk of side effects (ACAS, Executive Committee 1995). Thus, provided that Doppler results from certified vascular laboratories are straightforward, surgeons should operate without conventional radiographic studies, but may occasionally use further noninvasive imaging studies (MRA) if the ultrasound results are inconclusive.

Continuous-Wave Doppler Sonography

Using direct and continuous-wave Doppler sonography, it is possible to exclude stenoses of over 50 % and occlusions of the carotid vascular system with a high degree of certainty. With experienced investigators, the specificity lies between 88 % and 98 % (Table 2.**2**). Spectral analysis does not improve the diagnostic and predictive value of the method for excluding pathological findings.

In addition to reliable detection of obstructions in the extracranial carotid system, continuous-wave and pulsed-wave Doppler sonography can provide reliable estimates of the degree of stenosis. However, the ultrasound investigation criteria and the reference methods used vary so widely in different studies that it is not possible to say whether the method can reliably detect a carotid stenosis of less than 70 % in general. Combined use of Doppler and imaging ultrasound techniques is therefore the current practice (de Bray et al. 1995).

Vertebral Artery System

Principle

There are fundamental differences in the diagnosis of vertebral artery obstructions in comparison with the extracranial carotid system:

- The course of the vertebral artery cannot be examined continuously
- Flow obstructions are preferentially located at the origin, at the atlas loop, or in the intracranial segment
- Variations in vessel caliber and hereditary anomalies occur more frequently here than they do in the carotid system
- Subclavian artery flow obstructions affect Doppler signals from the vertebral artery

As in the carotid system, the findings at any single point only provide an overview. Examinations need to be carried out in several vascular segments when there are pathological conditions, or in the context of specific problems.

Findings

If the vertebral artery segment affected is accessible to direct examination with ultrasound, then the assessment of the degree of stenosis follows the criteria discussed in detail for the carotid artery system (pp. 55–70). However, there are more problems here, because the origin of the vertebral artery can usually only be depicted in a single projection, and the degree of stenosis in the vascular course where it loops round the atlas cannot be reliably evaluated. Further distally, and up to the point at which both vertebral arteries join the basilar artery, there are fluctuations in the vascular caliber. These variations can lead to evaluation difficulties when segmental or extended stenoses are involved.

Certain unique characteristics result from problems in interpreting processes that involve several blood vessels in:

- Unilateral tandem stenoses
- Proximal occlusions with distal collateralization
- Contralateral and ipsilateral decreases in flow
- Intermediate flow patterns
- Retrograde flow waveforms

Depending on the development of collateral arteries, various steal forms are encountered (Vollmar 1975) (Fig. 2.**45**):

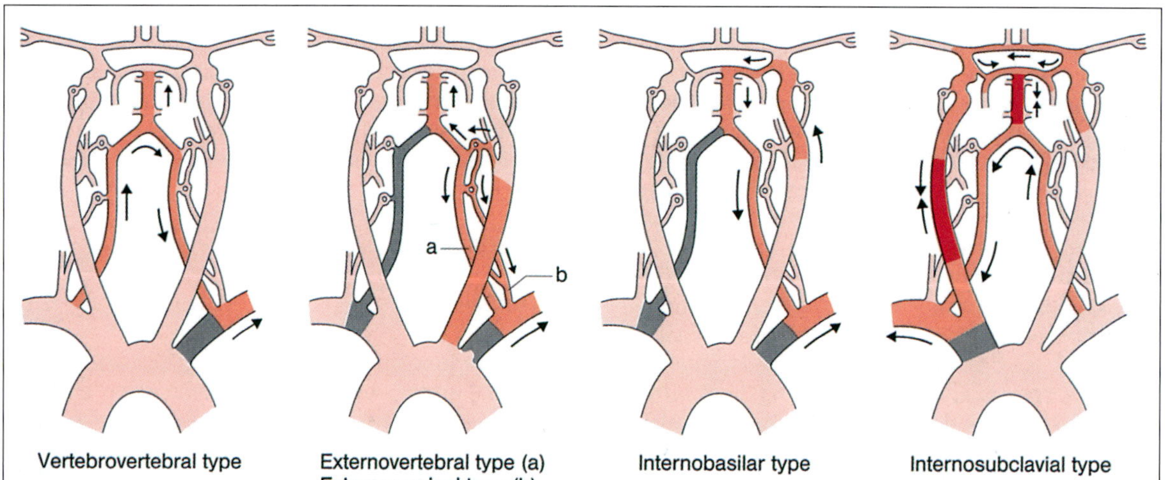

Fig. 2.**45** Various forms of steal phenomenon associated with proximal supra-aortic flow obstruction

- Vertebrovertebral
- Caroticobasilar
- Externovertebral
- Caroticosubclavian

An additional unique trait in the vertebral artery system is the existence of many collateral circulatory pathways in the neck, which usually continue intracranially even when the proximal or distal segment of either or both vertebral arteries is occluded. This is a significant difference in comparison with the carotid artery system, in which stenoses at the origin of the internal carotid artery are compensated for by collateral pathways via the ophthalmic anastomosis or the contralateral carotid system.

Color Flow Duplex Sonography

In contrast to previous studies mainly involving Doppler investigations, duplex sonography is now the method of choice for evaluating abnormalities and obstructions in the vertebrobasilar system. As in the carotid system, appropriate examinations can no longer be restricted to the extracranial vasculature, but should always use a combined extracranial and intracranial test approach, which is facilitated by duplex system analysis. Hemodynamic parameters already identified in this way may be further substantiated by MRA imaging in selected cases.

Increased Flow Velocities

Stenoses. As in the carotid artery, *stenoses at the origin of the vertebral artery* (Fig. 2.**46**), the intravertebral segments, or around the atlas loop can be diagnosed by flow velocity and imaging studies. Depending on the location of the stenosis, flow changes can be found in the proximal or distal segments, or in both vascular loop segments around the atlas. When there is physiological vascular perfusion, the efferent segment is usually the distal vertebral artery, and in retrograde perfusion it may also involve the proximal vascular segment.

Decreased Flow Velocities

Collaterals. There may also be an increase in vascular flow velocity when:

- The contralateral vertebral artery is hypoplastic, aplastic, contains a high-grade stenosis, or is occluded.
- The contralateral vertebral artery is perfused in a retrograde direction.
- There is a significant carotid system obstruction.
  Normal Doppler waveforms with clear side-to-side asymmetries (> 75 % of the residual diastolic flow) suggest various potential causes, assuming that the examination has excluded abnormal flow acceleration *at least at the origin and near the vertebral atlas loop*:
- Unfavorable transducer location. This finding does not imply any pathology. It is the most common cause of a misdiagnosis if Doppler studies alone are performed.
- Vertebral artery hypoplasia or dilation.
- Flow obstruction in the intracranial segment of the vertebral artery (when there is unilaterally reduced flow velocity) or in the basilar artery (when there is bilateral flow velocity reduction) can cause decreased flow conditions in the extracranial vertebral artery. *Recommendation:* supplementary transcranial examination.
- Far less often than might be thought, a decrease in flow velocity results from changes in the neck and head position caused by partial or complete compression of the vertebral artery (neurological symp-

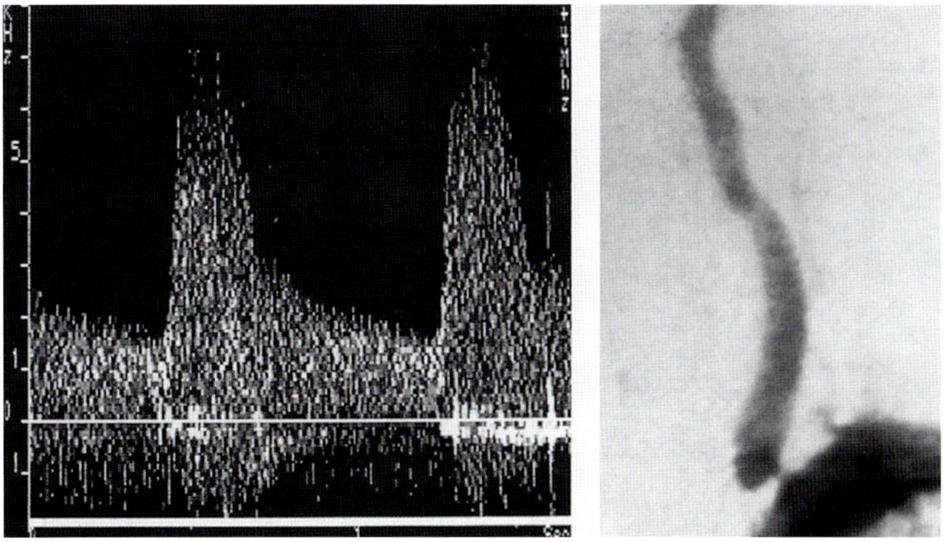

Fig. 2.**46** Doppler spectrum (**a**) and corresponding angiogram (**b**) of a high-grade stenosis at the origin of the vertebral artery

toms or subjective complaints are associated with this finding even more rarely), but misdiagnosis is frequent. Among thousands of examinations carried out in our vascular laboratories, only three patients were found who were suffering from sustained neurological deficits in which both vertebral arteries were functionally obstructed (e.g., when reclining or inclining) in the absence of collateralization through the circle of Willis from the carotid artery system.

Slosh Phenomenon

As in the internal carotid artery, an abnormal, usually low-amplitude flow signal, with a narrow systolic peak and at the same time a shorter back-flow phase and a significantly delayed diastolic flow component, is a typical highly sensitive sign in vertebral artery dissections. This is not caused by intermediate blood flow, but is probably due to a combined intraluminal flow signal and secondary pulse waves that are superimposed on the Doppler signal by intramural hematoma formation. In contrast to carotid dissection, this phenomenon is encountered much less frequently in vertebral artery dissection (in approximately one-third of the cases). The diagnosis can usually be confirmed during the follow-up examination by a normalized flow signal after an interval of days to a few weeks. More often, unspecified findings suggesting wall bleedings are encountered in arterial imaging when it is used to evaluate a vertebral artery dissection (Sliwka and Rautenberg 1998, Pfadenhauer and Müller 1995, Bartels and Flügel 1996). Since angiographic confirmation of this clinical picture can be difficult, and experience shows that the intramural bleeding can be missed even with magnetic resonance imaging (MRI) and MRA, this condition is probably not diagnosed often enough (Steinke et al. 1994).

Intermediate Blood Flow

Different intermediate forms of flow velocity, also called pendular flow (alternating forward and backward flow) or systolic flow reduction (von Reutern and Pourcelot 1978), are found in the vertebral artery when:

- A flow obstruction in the proximal subclavian artery or the innominate artery is detected. This is due to what is known as *latent subclavian steal phenomenon*, which can be a *temporary* phenomenon in hyperemia of the ipsilateral upper extremity. A medium-grade stenosis is often found to be the cause, and less frequently proximal subclavian artery occlusion.
- There are completely normal flow relationships in the proximal subclavian artery. This situation suggests *vertebral artery hypoplasia* and can be confirmed by color flow imaging.

Absent Signal

This finding can point in several diagnostic directions:

- *Systematic error* (a frequent occurrence)—e.g., due to incorrect or unfavorable probe placement or anatomically unfavorable conditions for conducting the examination.
- A *pathological finding*—e.g., an occlusion or segmental occlusion (aplasia) in the course of the vertebral artery. To differentiate this, an examination in all three locations (proximal, in the neck region, and at the posterior atlas arch) should be performed. When there is a segmental occlusion, distal filling of the vertebral artery through collateral vasculature is frequently found, so that even when pulsed-wave Doppler sonography is being used, evaluation of the intracranial segment of the vertebral artery from a

transnuchal direction has to be carried out in order to ensure that an extensive extracranial occlusion is not missed. In contrast to proximal occlusion of the vertebral artery, a reduced flow signal is generally found in a distal occlusion. Stenosis of the vertebral artery is only rarely accompanied by an absent signal—at most, an extremely high-grade stenosis at the origin can cause a signal near the mastoid process, which can only be detected with great difficulty. When there is an absent signal, as in vascular hypoplasia, a compensatory increase in flow through the contralateral vertebral artery is the most reliable indirect indicator.
- A *flow signal* that is only detectable acoustically. This is often due to poor ultrasound examination conditions (unfavorable signal-to-noise ratio, deep location of the loop of the vertebral artery around the atlas, large distance between the examination point and the probe position). In both conditions imaging is a potential guide to identify flow patterns in small caliper vessels.

Retrograde Blood Flow

Significant obstructions affecting the proximal subclavian artery (over 80% stenosis or occlusion) or the innominate artery often form a permanent subclavian steal phenomenon, with the arteries of the arm being supplied by the ipsilateral vertebral artery. The presence of retrograde flow can be demonstrated by a compression test (p. 35), which is accompanied by a clear change in the Doppler signal recorded from the vertebral artery. This effect is often clearer in the audio signal than in the documented waveform changes. In normal conditions, no change—or at most only a slight flow reduction—may occur during compression, followed by an increase in flow after the blood vessel is reopened. Comparisons with the angiographic findings have confirmed the excellent reliability of the method (100%) for diagnosing a permanent subclavian steal phenomenon (Keller et al. 1976b, Pourcelot et al. 1977, von Reutern and Pourcelot 1978, Hennerici and Aulich 1979, Büdingen and von Reutern 1993).

The various types of steal pathways are summarized in Fig. 2.**45**.

Continuous-Wave Doppler Sonography

All hemodynamic aspect of increased/decresed flow velocities, intermediate and retrograde flow patterns can be identified in CW Doppler studies similar to CFDS examinations. However, due to the more pronounced variability of the anatomical structure in the vertebrobasilar system rather than in the carotid system final interpretation of best results is limited.

Evaluation

Combining different examination techniques and gathering data from several probe positions, evaluation mainly involves:

- Registering flow obstructions.
- Distinguishing between the different forms: proximal stenosis at the origin, proximal segmental occlusion, stenosis in the neck region, stenosis at the atlas loop, distal high-grade stenosis, distal segmental occlusion, complete occlusion, hypoplasia, aplasia, dissection (Table 2.**10**).
- A permanent or temporary subclavian steal phenomenon can be detected with a high degree of reliability if the flow relationships in the subclavian or innominate arteries are observed and appropriate compression procedures are applied.
- When assessing global circulation in the brain in processes involving blood vessels, in the anterior circulation, it is very important to be aware of the collateral function of the vertebrobasilar arteries.

Further areas of interest are

- Segmental occlusions of the neck region.
- Stenoses before the vessel enters the intervertebral foramina.
- Quantitative measurement of the diameter and direct evaluation of intra-arterial flow relationships (i.e., diagnosis of hypoplasia, stenosis, dilative arteriopathy).
- Proximal dissections (V_1) of the vertebral artery in the neck.

Sources of Error

Systematic errors occur when there is:

- A weak or missing signal
- A poor signal-to-noise ratio
- Unfavorable anatomy
- Venous superimposition
- Stenosis in hypoplastic segments with extremely slow flow velocity

Confusion is possible with:

- The inferior thyroid artery or the thyrocervical trunk and the subclavian artery, in proximal ultrasound application.
- The internal carotid artery in a slender neck (when ultrasound is applied at the arch of the atlas).
- External carotid artery branches (e.g., the occipital artery and other cervical arteries, when ultrasound is applied at the arch of the atlas).
- Other cervical arteries or open collaterals in rare types of steal phenomenon. The flow velocity in the cervical collateral vasculature can be so pronounced that the normal Doppler signals from the distal vertebral artery may be missed.

Table 2.**10** Summary of findings at various vertebral artery recording sites for different pathological conditions

	Doppler findings in the vertebral artery system				Special information (B-mode, CFDS)
	Origin	Neck area	Atlas loop	Intra-cranial	
Stenosis at the origin – High-grade – Low-grade	 >2 ss <2 ss	 ↓ n	 ↓ n	 ↓ n	 Detection is possible at the origin Detection at the origin is difficult
Stenosis at the atlas loop – High-grade – Low-grade	 ↓ n	 ↓ n	 >2 ss <2 ss	 ↓ n	 Detection is possible at the atlas loop Detection is possible at the atlas loop
Stenosis – Intracranial, high-grade – Intracranial, low grade	 ↓ n	 ↓ n	 ↓ n	 >2 ss <2 ss	 Normal lumen, possibly with a reduced color-coded signal
Occlusion (V0–V4)	∅	∅	∅	∅	Absence of color signal
Segmental occlusion (V0–V2, e.g., proximal dissection)	∅	∅	n	n	Cervical collaterals are detectable, occasionally wall lesion detectable at the site of dissection
Distal occlusion (V3, V4, e.g., distal dissection)	↓	↓	↓	∅	Normal lumen, reduced color signal in the neck region
Hypoplasia	↓	↓	↓	↓	Narrow lumen
Functional collaterals in carotid occlusion	↑	↑	↑	↑	Lumen often widened

n Normal flow signal ∅ No signal
↓ Reduced flow signal ss: signs of stenosis (Tables 2.**8**, 2.**9**)
↑ Increased flow signal

When only a selective examination is carried out at a single registration point:

- Stenoses at the origin, and *proximal* occlusions, may be collateralized via:
  - Thyrocervical trunk anastomoses; a normal signal results at the atlas arch.
  - External carotid artery branches into the distal vertebral artery; a normal signal is obtained at the atlas loop.
- *Distal occlusion* of the vertebral artery: collaterals of the cervical arteries and the external carotid artery produce a normal signal proximally, and in some cases also distally (Table 2.**10**).

Various combinations of findings from pathological vascular processes in the distribution area of the vertebral artery are summarized in Table 2.**10**. Comparisons between various vascular examination methods are given in Table 2.**11**.

Diagnostic Effectiveness

Color Flow Duplex Sonography

Detailed studies have been carried out, mainly involving vertebral artery stenosis, dissection, and vessel variants. The sensitivity for stenoses of more than 50% and a specificity of > 90% has been reported for color flow imaging devices (Sliwka and Rautenberg 1998, Pfadenhauer and Müller 1995, Bartels and Flügel 1996).

Continuous-Wave Doppler Sonography

Studies have shown that in comparison with angiography, the diagnostic value of Doppler findings from the vertebral arteries is good if evaluation of differentiated flow obstruction is not required (e.g., hypoplasia, stenosis, or occlusion). The sensitivity lies between 70% and 90%, and the specificity is reported to be over 95% (Keller et al. 1976b, von Reutern and Pourcelot 1978, von Reutern and Clarenbach 1980, Ackerstaff et al. 1984, Winter et al. 1988, Büdingen and von Reutern 1993).

Table 2.**11** Comparison of vascular examination methods

Method	Potential findings	Limitations and disadvantages
Pulse palpation	– Subclavian artery occlusion – Common carotid artery occlusion – External carotid artery occlusion	– Internal carotid artery not distinguishable – Side-to-side differences in the carotid pulse are often a methodological artifact
Auscultation	– Subclavian artery stenosis – Medium-grade carotid stenosis – Arteriovenous and dural fistula – Angioma	– The internal/external carotid artery cannot be differentiated – Flow sounds propagated from the heart or the subclavian artery may be misleading – Highest-grade and low-grade stenoses are often misdiagnosed – Collateral flow creates bruits in normal vasculature
Doppler sonography (indirect methods)	– Occlusion or stenosis > 80% of the common and internal carotid arteries	– Locating the obstructions in the vascular system is not possible – Occlusion and stenosis cannot be differentiated – Stenoses < 80% and bifurcation processes cannot be detected – False-negative results in approx. 20% of carotid obstructions > 80%
Doppler sonography (indirect and direct methods) with spectrum analysis	– Carotid processes with lumen constriction > 40% – Differentiation between occlusion and high-grade stenosis is often possible – Evaluating the vertebral artery system – Subclavian steal phenomenon – Quantitative evaluation of acoustic spectral phenomena	– Carotid stenoses < 40% cannot be detected with certainty – Difficult to differentiate between vertebral artery aplasia-occlusion, and stenosis-hypoplasia – Only in a few cases, intracranial processes may be suspected based on indirect criteria
Color flow duplex sonography	– Low-grade carotid stenoses, plaques and large ulcerations – Evaluating vascular width, wall movement, and plaque structure – Isodense plaques/thrombi vs. flowing blood – Subtotal stenosis vs. occlusion (i.e., pseudo-occlusion) – Small ulcers and dissections	– Occlusions and subtotal stenoses are sometimes difficult to separate – Variants in anatomy of the carotid bifurcation and complex plaque structure reduces validity – Lesions outside of the area directly examined can be easily missed – Artifacts and aliasing may cause misinterpretation

Supra-Aortic System

Principle

The arterial segments that supply the brain close to the aortic arch (the proximal subclavian artery, innominate artery, and proximal common carotid artery) can only be evaluated to a limited degree with Doppler sonography, since they are poorly accessible and beyond the scope of vascular ultrasound. Usually, early atherosclerotic plaques in the proximal section are not detectable. Mild and moderate stenoses can be recorded using a pulsed-wave Doppler instrument, provided the sample volume is applicable intrathoracically to the origin of the arteries from the aorta. High-grade stenoses and occlusions lead to changes in the Doppler signals in the neck arteries and can thus be identified indirectly by continuous-wave and pulsed-wave Doppler methods.

Findings

Subclavian Artery

Color Flow Duplex Sonography

The wall changes can sometimes be structurally detected from a supraclavicular direction using duplex sonography. Aneurysmal dilations can be well documented. Diagnosing the subclavian artery in the supraclavicular fossa is easier using color flow duplex sonography, as it allows the artery to be identified more quickly in contrast to the neighboring vasculature, particularly venous structures.

Continuous-Wave Doppler Sonography

Obstruction Proximal to the Origin of the Vertebral Artery

Medium and high-grade stenoses cause a flow acceleration, with changes in the audio signal and possibly a systolic deceleration, comparable to the changes associated with external carotid artery stenosis. Distal to the stenosis, a pathological waveform can be observed, with loss of the early diastolic back-flow phenomenon.

If the stenosis cannot be directly evaluated, or an *occlusion* is present, a reduced flow signal and an absent early diastolic back-flow phenomenon is characteristic. Since blood then flows through the vertebral artery in a retrograde direction and drains into the distal segment of the subclavian artery, the curves from the two blood vessels are similar in form and in the resulting audio signal. It is only rarely that a temporary or permanent *subclavian steal phenomenon* (i.e., the hemodynamic pattern in the absence of typical signs or symptoms; Hennerici et al. 1988b) fails to form—for example, when the vertebral artery originates directly from the aorta.

Even if no direct signs of stenosis are detected in the continuous-wave Doppler sonogram, a high-grade stenosis may still be present, due to the inaccessibility of the site of stenosis far away from the close-up focus of the ultrasound area. This can usually be distinguished from occlusion by a careful supplementary examination using the pulsed-wave method, with selective depth-gating.

Obstruction Distal to the Origin of the Vertebral Artery

- No steal phenomenon in the ipsilateral vertebral artery.
- If located in a segment directly accessible to ultrasound, reliable differentiation between stenosis and occlusion is possible.
- If located distally, there is often no change in the waveform of the proximal subclavian artery, or only indirect signs.

With spectral analysis, quite reliable quantification of the maximum systolic frequency is usually possible. It is only in high-grade stenoses that the signal is sometimes difficult to document.

Innominate Artery

Color Flow Duplex Sonography

Due to the significant reduction in flow in the vascular territory affected, duplex sonography is extremely helpful in detecting structural changes that would otherwise have been missed because of the lack of associated flow disturbances—e.g., in the absence of any detectable velocity at rest, even a significant stenosis in the watershed region can remain undiscovered when using Doppler sonography alone. An additional B-mode examination is therefore particularly important if surgery is planned.

The innominate artery and its branches can sometimes be documented, proceeding from a supraclavicular direction and using low-frequency probes such as those generally used in echocardiography. By simultaneously displaying hemodynamic and structural parameters in the distal neck arteries, one can improve the results of both Doppler mode and B-mode studies.

Continuous-Wave Doppler Sonography

As a result of compensatory collateral circulation, hemodynamically significant obstructions of the innominate artery cause a variety of changes in the extracranial system (Hennerici et al. 1981a) (see Fig. 2.**39**):

- In the orbital arteries, there is often a characteristic finding of a double-peaked physiological flow during systole, corresponding to a very characteristic breath-like audio signal acoustically. When the arteries of the right upper arm are compressed, diastolic flow sometimes increases.
- In the ipsilateral internal and common carotid arteries, quite a variety of flow changes are seen in individual cases and even in the same patient on different occasions, depending on the location and lability of the watershed. *All* of these changes undergo modulation during the compression test.
- The external carotid artery also often shows altered, but slightly less diverse, flow patterns, which are modulated in the compression test.
- In the ipsilateral vertebral artery, all physiological (rare), intermediate, or retrograde flow variations can be present.

As a peculiarity, the carotid and vertebral systems show rather *unstable hemodynamic behavior,* depending on systemic blood pressure, muscular activity, and changes in the arterial flow resistance of the right arm. Neurological signs and symptoms are seldom reported; rarely, symptoms may be provoked by physical stress, since a variety of compensatory mechanisms are available:

- The contralateral carotid artery
- The contralateral vertebral artery
- The circle of Willis

Aortic Arch

Transesophageal Echocardiography

Changes in the aortic arch and in the heart, which may be a significant source of cerebral emboli, can be readily detected using transesophageal echocardiography (TEE) (Amarenco et al. 1992, 1994, 1995). Relevant works on echocardiography should be consulted in connection with the method itself and the interpretation of its findings (Amarenco et al. 1992, Horowitz et al. 1992, Rubin et al. 1992, Jones et al. 1995, Rauh et al. 1996).

Evaluation

Continuous-wave Doppler sonography is only of limited value in assessing vascular segments that are located intrathoracically (subclavian artery, innominate artery, and proximal common carotid artery), but they can usually be depicted using the pulsed-wave Doppler method. Associated flow velocity changes in the distal territories of the extracranial carotid and vertebral artery systems can be well documented using all the available methods. They involve many hemodynamic changes.

Sources of Error

If severe flow reduction is present in obstructive lesions involving the innominate artery, it is often difficult to detect and assess additional structural changes in the distal territory. Occasionally, arterial signals are so distorted that they can be mistaken for venous signals. In the "low flow" watershed area, low and medium-grade stenoses, as well as high-grade ones, then fail to cause any abnormal changes in the Doppler signal, and are easily overlooked during the diagnosis. Even when angiography is used, such processes can be missed due to limited documentation in one single projection or inadequate concentration of contrast medium in the post-stenotic vascular segments. It is therefore mandatory to search deliberately for any associated atherosclerotic signs using a supplementary duplex system examination. Since MRA and helical computed tomography (CT) share the same disadvantages, TEE of the aortic arch and the proximal branching cerebral arteries is often necessary (Amarenco et al. 1992, 1994, 1995, Rauh et al. 1996). In addition, otherwise unidentified important sources of cardioembolic stroke may be diagnosed in patients *without* evidence of artery-to-artery embolism or hemodynamic sources of cerebral ischemia (i.e., left atrial thrombus, atrial septal aneurysm with or without patent foramen ovale, ventricular aneurysm, global left-ventricular wall motion abnormalities, etc.).

Diagnostic Effectiveness

High-grade flow obstructions in the innominate artery are rarer than stenoses or occlusions in the extracranial carotid artery system, the left subclavian artery, or the vertebral arteries—approximately 0.6% in the series examined by Hass et al. (1968) and also in our own study (Hennerici et al. 1981a). Clinically, they are usually asymptomatic, just like subclavian artery stenoses and occlusions, but they can lead to significant changes in the extracranial and intracranial circulation, including the subclavian steal phenomenon, and create a complex collateral circulatory network that can be analyzed sonographically in detail. More often, such lesions cause ischemic or embolic disturbances in the blood supply to the upper extremity. During a clinical examination in which blood pressure in both upper arms should always be measured, such conditions can already be suspected. However, since measuring the blood pressure does not allow precise localization of the vascular process (e.g., location proximal versus distal to the origin of the vertebral artery), a Doppler-sonographic examination is indicated. In addition, coexisting flow obstructions in the carotid artery territory are sometimes difficult to assess, both with Doppler sonography and angiography. Duplex sonography is the most useful additional method here. Clinically, such vascular processes become active (transient or permanent neurological deficits) once the capacity of the collateral vasculature is limited by additional obstructions in other extracranial or intracranial arteries or when embolism occurs.

Due to the rarity of these findings and the limited angiographic display that is possible, hardly any comparative studies exist concerning the sensitivity and specificity of ultrasound diagnosis (Grosveld et al. 1988, Rautenberg and Hennerici 1988).

Local wall changes and low-grade or medium-grade stenoses are almost always missed in routine ultrasound studies if the examiner is not specifically informed about the reasons for suspicion. Studies in the intrathoracic zone are time-consuming and demanding and are therefore not included in routine services, although they may be useful on request.

References

Ackermann RH. Perspective on noninvasive diagnosis of carotid disease. Neurology 1979;29:615–22.

Ackerstaff RG, Hoeneveld H, Slowikowski JM, Moll FL, Eikelboom BC, Ludwig JW. Ultrasonic duplex scanning in atherosclerotic disease of the innominate, subclavian and vertebral arteries: a comparative study with angiography. Ultrasound Med Biol 1984;10:409–18.

Alexandrov AV, Bladin CF, Maggisano R, Norris JW. Measuring carotid stenosis: time for a reappraisal. Stroke 1993;24:1292–6.

Amarenco P, Cohen A, Baudrimont M, Bousser MG. Transesophageal echocardiographic detection of aortic arch disease in patients with cerebral infarction. Stroke 1992; 23: 1005–9.

Amarenco P, Cohen A, Tzourio C, Bertrand B, Hommel M, Besson G, et al. Atherosclerotic disease of the aortic arch and the risk of ischemic stroke. N Engl J Med 1994;331:1474–9.

Amarenco P, Heinzlef O, Lucas C. The risk of cerebral infarct in patients with aortic arch plaques: a prospective study [abstract]. Stroke 1995; 26: 184.

Anderson CM, Saloner D, Lee RE, Griswold VJ, Shapeero LG, Rapp JH, et al. Assessment of carotid artery stenosis by

MR angiography: comparison with x-ray angiography and color-coded Doppler ultrasound. AJNR Am J Neuroradiol 1992;13:989–1003; discussion 1005–8.

Arakawa S, Kamouchi M, Okada Y, Kishikawa K, Omae T, Inoue T, et al. Ultrasonographically predicting the extent of collateral flow through superficial temporal artery-to-middle cerebral artery anastomosis. AJNR Am J Neuroradiol 2003;24:886–91.

Arbeille P, Lapierre F, Pourcelot L. Evaluation des sténoses carotidiennes par les ultrasons. Encyclop Méd Chir 1985;322:1–8.

Bandyk DF, Levine AW, Pohl L, Towne JB. Classification of carotid bifurcation disease using quantitative Doppler spectrum analysis. Arch Surg 1985;120:306–14.

Barnett HJ. Delayed cerebral ischemic episodes distal to occlusion of major cerebral arteries. Neurology 1978;28:769–74.

Barnett HJ, Warlow CP. Carotid endarterectomy and the measurement of stenosis. Stroke 1993;24:1281–4.

Barnett HJ, Taylor DW, Eliasziw M, Fox AJ, Ferguson GG, Haynes RB, et al. Benefit of carotid endarterectomy in patients with symptomatic moderate or severe stenosis. North American Symptomatic Carotid Endarterectomy Trial Collaborators. N Engl J Med 1998;339:1415–25.

Bartels E, Flügel KA. Advantages of color Doppler imaging for the evaluation of vertebral arteries. J Neuroimaging 1993;3:229–33.

Bartels E, Flügel KA. Evaluation of extracranial vertebral artery dissection with duplex color-flow imaging. Stroke 1996;27:290–5.

Bartels E, Fuchs HH, Flügel KA. Duplex ultrasonography of vertebral arteries: examination, technique, normal values, and clinical applications. Angiology 1992;43(3 Pt 1):169–80.

Beach K, Lawrence R, Phillips D, Primozich E. The systolic velocity criterion for diagnosing significant internal carotid stenoses. J Vasc Technol 1989;13:65–8.

Biller J, Hingtgen WL, Adams HP Jr, Smoker WR, Godersky JC, Toffol GJ. Cervicocephalic arterial dissections: a ten-year experience. Arch Neurol 1986;43:1234–8.

Blackshear WM, Phillips DJ, Chikos PM, Harley JD, Thiele BL, Strandness DE Jr. Carotid artery velocity patterns in normal and stenotic vessels. Stroke 1980;11:67–71.

Bladin CF, Alexandrov AV, Norris JW. How should we measure carotid stenosis? Lancet 1994;344:69.

Bluth EI, Kay D, Merritt CR, Sullivan M, Farr G, Mills NL, et al. Sonographic characterization of carotid plaque: detection of hemorrhage. AJR Am J Roentgenol 1986;146:1061–5.

Bock RW, Gray-Weale AC, Mock PA, Robinson DA, Irwig L, Lusby RJ. The natural history of asymptomatic carotid artery disease. J Vasc Surg 1993;17:160–9; discussion 170–1.

Bonithon-Kopp C, Touboul PJ, Berr C, Magne C, Ducimetière P. Factors of carotid arterial enlargement in a population aged 59 to 71 years: the EVA study. Stroke 1996;27:654–60.

Brown PM, Johnston KW, Douville Y. Detection of occlusive disease of the carotid artery with continuous wave Doppler spectral analysis. Surg Gynecol Obstet 1982;155:183–6.

Büdingen HJ, von Reutern GM. Ultraschalldiagnostik der hirnversorgenden Arterien. Stuttgart: Thieme, 1993.

Büdingen HJ, Hennerici M, Voigt K, Kendel K, Freund HJ. Die Diagnostik von Stenosen oder Verschlüssen der A. carotis interna mit der direktionellen Ultraschall-Doppler-Sonographie der A. supratrochlearis. Dtsch Med Wochenschr 1976;101:269–75.

Cape CA, DeSaussure RL, Nixon J. Carotid ultrasonography in carotid artery disease. South Med J 1984;77:183–6.

Carpenter JP, Lexa FJ, Davis JT. Determination of duplex Doppler ultrasound criteria appropriate to the North American Symptomatic Carotid Endarterectomy Trial. Stroke 1996;27:695–9.

Chang YJ, Lin SK, Ryu SJ, Wai YY. Common carotid artery occlusion: evaluation with duplex sonography. AJNR Am J Neuroradiol 1995;16:1099–105.

Comerota AJ, Cranley JJ, Cook SE. Real-time B-mode carotid imaging in diagnosis of cerebrovascular disease. Surgery 1981;89:718–29.

Comerota AJ, Cranley JJ, Katz ML. Real-time B-mode carotid imaging: a three-year multicenter experience. J Vasc Surg 1984;1:85–95.

Croft RJ, Ellam LD, Harrison MJ. Accuracy of carotid angiography in the assessment of atheroma of the internal carotid artery. Lancet 1980;i:997–1000.

Daffertshofer M, Hennerici M. [Spectrum analysis of Doppler signals; in German.] Ultraschall Med 1990;11:219–26.

Daigle RJ, Stavros AT, Platon R, Anderst DB, Nurre PD. Velocity criteria for differentiation of 60%–79% carotid stenoses from 80% or greater stenoses. J Vasc Technol 1988;7:176–83.

de Bray JM, Glatt B. Quantification of atheromatous stenosis in the extracranial internal carotid artery. International Consensus Conference. Cerebrovasc Dis 1995; 5: 414–26.

de Bray JM, Galland F, Lhoste P, Nicolau S, Dubas F, Emile J, et al. Colour Doppler and duplex sonography and angiography of the carotid artery bifurcations: prospective, double-blind study. Neuroradiology 1995;37:219–24.

de Bray JM, Baud JM, Delanoy P, Camuzat JP, Dehans V, Descamp-Le Chevoir J, et al. Reproducibility in ultrasonic characterization of carotid plaques. Cerebrovasc Dis 1998;8:273–7.

de Bray, JM, Baud JM, Dauzat M on behalf of the Consensus Conference. Concerning the morphology and the risk of carotid plaques. Cerebrovascular Diseases 7(1997)289–296.

Decker K, Schlegel HJ. Normbilder und Normvarianten der A. ophthalmica im Röntgenbild. Albrecht von Graefes Arch Ophthalmol 1957;159:302–10.

Delcker A, Diener HC, Wilhelm H. Influence of vascular risk factors for atherosclerotic carotid artery plaque progression. Stroke 1995;26:2016–22.

Di Chiro G. Ophthalmic arteriography. Radiology 1961;77:948–57.

Easton JD. Accuracy of high resolution ultrasound imaging for quantitative assessment for early carotid atherosclerosis. Cerebrovasc Dis 1994;4:109–13.

ECST [no authors listed]. Randomised trial of endarterectomy for recently symptomatic carotid stenosis: final results of the MRC European Carotid Surgery Trial (ECST). Lancet 1998; 351: 1379–87.

Eicke BM, Tegeler CH, Dalley G, Myers LG. Angle correction in transcranial Doppler sonography. J Neuroimaging 1994;4: 29–33.

Eikelboom BC, Riles TR, Mintzer R, Baumann FG, DeFillip G, Lin J, et al. Inaccuracy of angiography in the diagnosis of carotid ulceration. Stroke 1983;14:882–5.

Eliasziw M, Streifler JY, Fox AJ, Hachinski VC, Ferguson GG, Barnett HJ. Significance of plaque ulceration in symptomatic patients with high-grade carotid stenosis. North American Symptomatic Carotid Endarterectomy Trial. Stroke 1994;25:304–8.

Erdoes LS, Marek JM, Mills JL, Berman SS, Whitehill T, Hunter GC, et al. The relative contributions of carotid duplex scanning, magnetic resonance angiography, and cerebral arteriography to clinical decision-making: a prospective study in patients with carotid occlusive disease. J Vasc Surg 1996;23:950–6.

Erickson SJ, Mewissen MW, Foley WD, Lawson TL, Middleton WD, Quiroz FA, et al. Stenosis of the internal carotid artery: assessment using color Doppler imaging compared with angiography. AJR Am J Roentgenol 1989;152:1299–305.

Executive Committee for the Asymptomatik Carotid Atherosclerosis Stenosis Study. Endarterectomy for asymptomatic carotid artery stenosis. *JAMA* 1995;273: 1421–1428.

Filis KA, Arko FR, Johnson BL, Pipinos II, Harris EJ, Olcott C 4th, et al. Duplex ultrasound criteria for defining the severity of carotid stenosis. Ann Vasc Surg 2002;16:413–21.

Fisher CM, Ojemann RG, Roberson GH. Spontaneous dissection of cervico-cerebral arteries. Can J Neurol Sci 1978;5:9–19.

Gabrielsen TO, Greitz T. Normal size of the internal carotid, middle cerebral and anterior cerebral arteries. Acta Radiol Diagn (Stockh) 1970;10:1–10.

Geroulakos G, Domjan J, Nicolaides A, Stevens J, Labropoulos N, Ramaswami G, et al. Ultrasonic carotid artery plaque structure and the risk of cerebral infarction on computed tomography. J Vasc Surg 1994;20:263–6.

Glagov S, Zarins CK. Quantitating atherosclerosis: problems of definition. In: Bond MG, Insull W, Glagov S, Chandler AB, Cornhill JF, editors. Clinical diagnosis of atherosclerosis. Berlin: Springer, 1983.

Glagov S, Bassiouny HS, Giddens DP, Zarins CK. Intimal thickening: morphogenesis, functional significance and detection. J Vasc Investig 1995;1:2–14.

Görtler M, Niethammer R, Widder B. Differentiating subtotal carotid artery stenoses from occlusions by colour-coded duplex sonography. J Neurol 1994;241:301–5.

Griewing B, Morgenstern C, Driesner F, Kallwellis G, Walker ML, Kessler C. Cerebrovascular disease assessed by color-flow and power Doppler ultrasonography: comparison with digital subtraction angiography in internal carotid artery stenosis. Stroke 1996;27:95–100.

Grosveld WJ, Lawson JA, Eikelboom BC, vd Windt JM, Ackerstaff RG. Clinical and hemodynamic significance of innominate artery lesions evaluated by ultrasonography and digital angiography. Stroke 1988;19:958–62.

Guo Z, Fenster A. Three-dimensional power Doppler imaging: a phantom study to quantify vessel stenosis. Ultrasound Med Biol 1996;22:1059–69.

Halliday A, Mansfield A, Marro J et al. MRC Asymptomatic Carotid Surgery Trial (ACST) Collaborative Group. Prevention of disabling and fatal strokes by successful carotid endarterectomy in patients without recent neurological symptoms: randomised controlled trial. Lancet 2004;363:1491–1502.

Harward TR, Bernstein EF, Fronek A. Continuous-wave versus range-gated pulsed Doppler power frequency spectrum analysis in the detection of carotid arterial occlusive disease. Ann Surg 1986;204:32–7.

Hass WK, Fields WS, North RR, Kircheff II, Chase NE, Bauer RB. Joint study of extracranial arterial occlusion, 2: arteriography, techniques, sites, and complications. JAMA J Am Med Assoc 1968;203:961–8.

Hayreh SS, Dass R. The ophthalmic artery, 1: origin and intracranial and intracanalicular course. Br J Ophthalmol 1962a;46:46.

Hayreh SS, Dass R. The ophthalmic artery, 2: intraorbital course. Br J Ophthalmol 1962b;46:165.

Hedges TR. Ocular ischemia. In: Caplan LR, editor. Brain ischemia: basic concepts and clinical relevance. Berlin: Springer, 1995:61–73.

Hedges TR, Reichel E, Duker JS, Puliafito CA, Heggerick PA. Color Doppler imaging identifies different mechanisms of central retinal artery occlusion. Invest Ophthalmol Vis Sci 1993;34:842.

Hennerici MG. Nicht-invasive Diagnostik des Frühstadiums arteriosklerotischer Karotisprozesse mit dem Duplex-System. Vasa 1983;12:228–32.

Hennerici MG. Hochauflösende Ultraschall-Duplexsystemanalyse der extrakraniellen Karotisstrombahn. In: Hartmann A, Wassmann H, editors. Hirninfarkt. Munich: Urban & Schwarzenberg, 1987:228–253.

Hennerici MG. The unstable plaque. Cerebrovase Dis. 2004; 17 Suppl 3:17–22

Hennerici MG, Aulich A. Ultraschall-Doppler-Sonographie. Mediuz 1979;3:168–79.

Hennerici M, Freund HJ. Efficacy of CW-Doppler and duplex system examinations for the evaluation of extracranial carotid disease. J Clin Ultrasound 1984;12:155–61.

Hennerici M, Meairs S. Imaging arterial wall disease. Cerebrovasc Dis 2000;10(Suppl 5):9–20.

Hennerici MG, Steinke W. Carotid plaque developments: aspects of hemodynamic and vessel wall platelet interaction. Cerebrovasc Dis 1991;1:142–8.

Hennerici MG, Steinke W. Accuracy of high-resolution ultrasound imaging for quantitative assessment of early carotid atherosclerosis. Cerebrovasc Dis 1994;4:109–13.

Hennerici M, Aulich A, Sandmann W, Freund HJ. Incidence of asymptomatic extracranial arterial disease. Stroke 1981a;12:750–8.

Hennerici MG, Aulich A, Sandmann W, Lerut J. Stenosen und Verschlüsse des Truncus brachiocephalicus. Dtsch Med Wochenschr 1981b;106:1697–703.

Hennerici M, Rautenberg W, Trockel U, Kladetzky RG. Spontaneous progression and regression of small carotid atheroma. Lancet 1985;i:1415–9.

Hennerici MG, Aulich A, Freund JH. Carotid system syndromes. In: Vinken PJ, Bruyn GW, Klawans HL, editors. Handbook of clinical neurology. Amsterdam: Elsevier, 1988a: 291–337.

Hennerici M, Klemm C, Rautenberg W. The subclavian steal phenomenon: a common vascular disorder with rare neurologic deficits. Neurology 1988b;38:669–73.

Hennerici M, Steinke W, Rautenberg W. High-resistance Doppler flow pattern in extracranial carotid dissection. Arch Neurol 1989;46:670–2.

Hennerici MG, Meairs S, Daffertshofer M, Neff W, Schminke U, Schwartz A. Investigation of wall motion, local hemodynamics, and atherosclerotic plaque geometry in carotid arteries using 4-D ultrasound and high resolution magnetic resonance angiography. In: Moskowitz MA, Caplan LR, editors. Cerebrovascular diseases: 19th Princeton Stroke Conference. London: Butterworth-Heinemann, 1995:123–133.

Hennerici M, Baezner H, Daffertshofer M. Ultrasound and arterial wall disease. Cerebrovasc Dis 2004;17 (Suppl 1): 19–33.

Ho SS, Metreweli C. Preferred technique for blood flow volume measurement in cerebrovascular disease. Stroke 2000;31:1342–5.

Horowitz DR, Tuhrim S, Budd J, Goldman ME. Aortic plaque in patients with brain ischemia: diagnosis by transesophageal echocardiography. Neurology 1992;42: 1602–4.

Hoskins PR. Ultrasound techniques for measurement of blood flow and tissue motion. Biorheology 2002;39:451–9.

Howard G, Baker WH, Chambless LE, Howard VJ, Jones AM, Toole JF. An approach for the use of Doppler ultrasound as a screening tool for hemodynamically significant stenosis (despite heterogeneity of Doppler performance). A multicenter experience. Asymptomatic Carotid Atherosclerosis Study Investigators. Stroke 1996;27:1951–7.

Humphrey PR, Bradbury PG. Continuous wave Doppler ultrasonography in the detection of carotid stenosis and occlusion. J Neurol Neurosurg Psychiatry 1984;47:1128–30.

Huston J 3rd, Lewis BD, Wiebers DO, Meyer FB, Riederer SJ, Weaver AL. Carotid artery: prospective blinded comparison of two-dimensional time-of-flight MR angiography with conventional angiography and duplex US. Radiology 1993;186:339–44.

Jacobs NM, Grant EG, Schellinger D, Byrd MC, Richardson JD, Cohan SL. Duplex carotid sonography: criteria for stenosis, accuracy, and pitfalls. Radiology 1985;154:385–91.

Johnson C, Grant R, Dansie B, Taylor J, Spyropolous P. Measurement of blood flow in the vertebral artery using colour duplex Doppler ultrasound: establishment of the reliability of selected parameters. Man Ther 2000;5:21–9.

Johnston KW, Baker WH, Burnham SJ, Hayes AC, Kupper CA, Poole MA. Quantitative analysis of continuous-wave Doppler spectral broadening for the diagnosis of carotid disease: results of a multicenter study. J Vasc Surg 1986;4:493–504.

Jones EF, Kalman JM, Calafiore P, Tonkin AM, Donnan GA. Proximal aortic atheroma: an independent risk factor for cerebral ischemia. Stroke 1995;26:218–24.

Karino T, Goldsmith HL. Particle flow behavior in models of branching vessels, 2: effects of branching angle and diameter ratio on flow patterns. Biorheology 1985;22:87–104.

Keller HL. Varianten der Arteria carotis interna, der A. meningea media und der A. ophthalmica im Carotisangiogramm. Fortschr Röntgenstr 1961;95:472–82.

Keller HM, Meier W, Yonekawa Y, Kumpe D. Noninvasive angiography for the diagnosis of carotid artery disease using Doppler ultrasound (carotid artery Doppler). Stroke 1976a;7:354–63.

Keller HM, Meier WE, Kumpe DA. Noninvasive angiography for the diagnosis of vertebral artery disease using Doppler ultrasound (vertebral artery Doppler). Stroke 1976b;7:364–9.

Kern R, Szabo K, Hennerici M, Meairs S. Characterization of carotid artery plaques using real-time compound B-mode ultrasound. Stroke 2004 Apr;35(4):870–5.

Kessler C, Maravic CV, Marravic MV, Kömpf D. Color Doppler flow imaging of the carotid arteries. Neuroradiology 1991;3:114–7.

Knudsen L, Johansen A, Justesen P, Jorgensen HB. Accuracy of duplex scan of internal carotid arteries. Eur J Vasc Endovasc Surg 2002;24:86–7.

Krayenbühl H, Yasargil MG, Huber P. Zerebrale Angiographie für Klinik und Praxis. Stuttgart: Thieme, 1979.

Kuhl V, Tettenborn B, Eicke BM, Visbeck A, Meckes S. Color-coded duplex ultrasonography of the origin of the vertebral artery: normal values of flow velocities. J Neuroimaging 2000;10:17–21.

Lally M, Johnston KW, Cobbold RS. Limitations in the accuracy of peak frequency measurements in the diagnosis of carotid disease. J Clin Ultrasound 1984;12:403–9.

Langlois Y, Roederer GO, Chan A, Phillips DJ, Beach KW, Martin D, et al. Evaluating carotid artery disease: the concordance between pulsed Doppler/spectrum analysis and angiography. Ultrasound Med Biol 1983;9:51–63.

Laster RE Jr, Acker JD, Halford HH 3rd, Nauert TC. Assessment of MR angiography versus arteriography for evaluation of cervical carotid bifurcation disease. AJNR Am J Neuroradiol 1993;14:681–8.

Lee VS, Hertzberg BS, Kliewer MA, Carroll BA. Assessment of stenosis: implications of variability of Doppler measurements in normal-appearing carotid arteries. Radiology 1999;212:493–8.

Li J, Li X, Mori Y, Rusk RA, Lee JS, Davies CH, et al. Quantification of flow volume with a new digital three-dimensional color Doppler flow approach: an in vitro study. J Ultrasound Med 2001;20:1303–11.

Licht PB, Christensen HW, Roder O, Hoilund-Carlsen PF. Volume flow estimation by colour duplex. Eur J Vasc Endovasc Surg 1999;17:219–24.

Londrey GL, Spadone DP, Hodgson KJ, Ramsey DE, Barkmeier LD, Sumner DS. Does color-flow imaging improve the accuracy of duplex carotid evaluation? J Vasc Surg 1991;13:659–63.

Markus HS, Brown MM. Asymptomatic cerebral embolic signals in carotid artery stenosis [abstract]. Cerebrovasc Dis 1994;4:235.

Marosi L, Ehringer H. [The extracranial carotid artery in a high-resolution real-time ultrasonic imaging system: the morphological findings in healthy young adults; in German.] Ultraschall Med 1984;5:174–81.

Mattle HP, Kent KC, Edelman RR, Atkinson DJ, Skillman JJ. Evaluation of the extracranial carotid arteries: correlation of magnetic resonance angiography, duplex ultrasonography, and conventional angiography. J Vasc Surg 1991;13:838–44; discussion 844–5.

Meairs S, Röther J, Neff W, Hennerici M. New and future developments in cerebrovascular ultrasound, magnetic resonance angiography, and related techniques. J Clin Ultrasound 1995;23:139–49.

Melis-Kisman E, Mol JM. [Application of the Doppler effect to cerebrovascular exploration (preliminary report); in French.] Rev Neurol (Paris) 1970;122:470–2.

Merland JJ. Artériographie supersélective de la carotide externe [dissertation]. Paris: University of Paris, 1973.

Middleton WD, Foley WD, Lawson TL. Flow reversal in the normal carotid bifurcation: color Doppler flow imaging analysis. Radiology 1988;167:207–10.

Miyazaki M, Kato K. Measurement of cerebral blood flow by ultrasonic Doppler technique. Jpn Circ J 1965a;29:375–82.

Miyazaki M, Kato K. Measurement of cerebral blood flow by ultrasonic Doppler technique: hemodynamic comparison of right and left carotid artery in patients with hemiplegia. Jpn Circ J 1965b;29:383–6.

Mokri B, Sundt TM Jr, Houser OW, Piepgras DG. Spontaneous dissection of the cervical internal carotid artery. Ann Neurol 1986;19:126–38.

Moneta GL, Edwards JM, Chitwood RW, Taylor LM Jr, Lee RW, Cummings CA, et al. Correlation of North American Symptomatic Carotid Endarterectomy Trial (NASCET) angiographic definition of 70% to 99% internal carotid artery stenosis with duplex scanning. J Vasc Surg 1993;17:152–7; discussion 157–9.

Müller HR. Direktionelle Doppler-Sonographie der Arteria frontalis medialis. Z EEG EMG 1971; 2: 24–32.

Müller HR, Radue EW, Saia A, Pallotti C, Buser M. Carotid blood flow measurement by means of ultrasonic techniques: limitations and clinical use. In: Hartmann A, Hoyer S, editors. Cerebral blood flow and metabolism measurement. Berlin: Springer, 1985.

Neuerburg-Heusler D. [Doppler sonographic diagnosis of extracranial occlusive disease: validity of indirect, direct and combined procedures; in German.] Vasa Suppl 1984;12:59–70.

Neuerburg-Heusler D, Todt M, Roth FJ. Quantitative computergestützte Frequenzenanalyse bei Karotisstenosen. In: Maurer HJ, editor. Berichtsband Deutsch–Japanischer Kongress für Angiologie. Gräfelfing 1985.

Norrving B, Cronqvist S. Doppler examination of the carotid arteries: a comparative study with angiography. Acta Neurol Scand 1981;64:241–52.

O'Leary DH, Polak JF, Kronmal RA, Savage PJ, Borhani NO, Kittner SJ, et al. Thickening of the carotid wall: a marker for atherosclerosis in the elderly? Cardiovascular Health Study Collaborative Research Group. Stroke 1996;27:224–31.

Padget DH. The circle of Willis: its embryology and anatomy. Ithaca, NY: Comestock, 1944.

Patel MR, Kuntz KM, Klufas RA, Kim D, Kramer J, Polak JF, et al. Preoperative assessment of the carotid bifurcation: can magnetic resonance angiography and duplex ultrasonography replace contrast arteriography? Stroke 1995;26:1753–8.

Paullus WS, Pait TG, Rhoton Al Jr. Microsurgical exposure of the petrous portion of the carotid artery. J Neurosurg 1977;47:713–26.

Pfadenhauer K, Müller H. [Color-coded duplex ultrasound of the vertebral artery: normal findings and pathologic findings in obstruction of the vertebral artery and remaining cerebral arteries; in German.] Ultraschall Med 1995;16:228–33.

Picot PA, Rickey DW, Mitchell R, Rankin RN, Fenster A. Three-dimensional colour Doppler imaging. Ultrasound Med Biol 1993;19:95–104.

Pignoli P, Tremoli E, Poli A, Oreste P, Paoletti R. Intimal plus medial thickness of the arterial wall: a direct measurement with ultrasound imaging. Circulation 1986;74:1399–406.

Planiol T, Pourcelot L, Pottier JM, Degiovanni E. [Study of carotid circulation by means of ultrasonic methods and thermography; in French.] Rev Neurol (Paris) 1972;126:127–41.

Polak JF, Dobkin GR, O'Leary DH, Wang AM, Cutler SS. Internal carotid artery stenosis: accuracy and reproducibility of color-Doppler-assisted duplex imaging. Radiology 1989;173:793–8.

Poli A, Tremoli E, Colombo A, Sirtori M, Pignoli P, Paoletti R. Ultrasonographic measurement of the common carotid artery wall thickness in hypercholesterolemic patients: a new model for the quantitation and follow-up of preclinical atherosclerosis in living human subjects. Atherosclerosis 1988;70:253–61.

Pourcelot L, Ribadeau-Dumas JL, Fagret D, Planiol T. [Contribution of the Doppler examination to the diagnosis of subclavian steal syndrome; in French.] Rev Neurol (Paris) 1977;133:309–23.

Rauh R, Fischereder M, Spengel FA. Transesophageal echocardiography in patients with focal cerebral ischemia of unknown cause. Stroke 1996;27:691–4.

Rautenberg W, Hennerici M. Pulsed Doppler assessment of innominate artery obstructive diseases. Stroke 1988;19:1514–20.

Reilly LM, Lusby RJ, Hughes L, Ferrell LD, Stoney RJ, Ehrenfeld WK. Carotid plaque histology using real-time ultrasonography: clinical and therapeutic implications. Am J Surg 1983;146:188–93.

Reneman RS, Spencer MP. Local Doppler audio spectra in normal and stenosed carotid arteries in man. Ultrasound Med Biol 1979;5:1–11.

Reneman RS, van Merode T, Hick P, Hoeks AP. Flow velocity patterns in and distensibility of the carotid artery bulb in subjects of various ages. Circulation 1985;71:500–9.

Ricotta JJ, Bryan FA, Bond MG, Kurtz A, O'Leary DH, Raines JK, et al. Multicenter validation study of real-time (B-mode) ultrasound, arteriography, and pathologic examination. J Vasc Surg 1987;6:512–20.

Ries S, Steinke W. Amplitude modulated and frequency modulated color duplex flow imaging for the evaluation of normal and pathological vertebral arteries. Cerebrovasc Dis 1997;7:63.

Ries S, Daffertshofer M, Steinke W, Hennerici MG. Power amplitude duplex ultrasound imaging of vertebral artery dolichoectasia. Cerebrovasc Dis 1996;6:374.

Ries S, Steinke W, Devuyst G, Artemis N, Valikovics A, Hennerici M. Power Doppler imaging and color Doppler flow imaging for the evaluation of normal and pathological vertebral arteries. J Neuroimaging 1998;8:71–4.

Riles TS, Eidelman EM, Litt AW, Pinto RS, Oldford F, Schwartzenberg GW. Comparison of magnetic resonance angiography, conventional angiography, and duplex scanning. Stroke 1992;23:341–6.

Riley WA, Barnes RW, Applegate WB, Dempsey R, Hartwell T, Davis VG, et al. Reproducibility of noninvasive ultrasonic measurement of carotid atherosclerosis. The Asymptomatic Carotid Artery Plaque Study. Stroke 1992;23:1062–8.

Ringelstein EB. Skepticism toward carotid ultrasonography: a virtue, an attitude, or fanaticism? Stroke 1995;26:1743–6.

Rittgers SE, Thornhill BM, Barnes RW. Quantitative analysis of carotid artery Doppler spectral waveforms: diagnostic value of parameters. Ultrasound Med Biol 1983;9:255–64.

Robinson ML, Sacks D, Perlmutter GS, Marinelli DL. Diagnostic criteria for carotid duplex sonography. AJR Am J Roentgenol 1988;151:1045–9.

Roederer GO, Langlois YE, Chan AW, Primozich J, Lawrence RA, Chikos PM, et al. Ultrasonic duplex scanning of extracranial carotid arteries: improved accuracy using new features from the common carotid artery. J Cardiovasc Ultrasound 1982;1:373–8.

Rothwell PM, Eliasziw M, Gutnikov SA, Fox AJ, Taylor DW, Mayberg MR, et al. Analysis of pooled data from the randomised controlled trials of endarterectomy for symptomatic carotid stenosis. Lancet 2003;361:107–16.

Rubin DC, Plotnick GD, Hawke MW. Intraaortic debris as a potential source of embolic stroke. Am J Cardiol 1992;69:819–20.

Rubin JM, Tuthill TA, Fowlkes JB. Volume flow measurement using Doppler and grey-scale decorrelation. Ultrasound Med Biol 2001;27:101–9.

Salamon G, Guerinel G, Demard F. [Radioanatomical study of the external carotid artery; in French.] Ann Radiol (Paris) 1968;11:199–215.

Salvarani C, Silingardi M, Ghirarduzzi A, Lo Scocco G, Macchioni P, Bajocchi G, et al. Is duplex ultrasonography useful for the diagnosis of giant-cell arteritis? Ann Intern Med 2002;137:232–8.

Sauve JS, Thorpe KE, Sackett DL, Taylor W, Barnett HJ, Haynes RB, et al. Can bruits distinguish high-grade from moderate symptomatic carotid stenosis? The North American Symptomatic Carotid Endarterectomy Trial. Ann Intern Med 1994;120:633–7.

Scheel P, Ruge C, Petruch UR, Schoning M. Color duplex measurement of cerebral blood flow volume in healthy adults. Stroke 2000;31:147–50.

Schmid-Schönbein H, Perktold K. Physical factors in the pathogenesis of atheroma formation. In: Caplan LR, editor. Brain ischemia: basic concepts and clinical relevance. Berlin: Springer, 1995:185–214.

Seward JB, Khandheria BK, Edwards WD, Oh JK, Freeman WK, Tajik AJ. Biplanar transesophageal echocardiography: anatomic correlations, image orientation, and clinical applications. Mayo Clin Proc 1990;65:1193–213.

Sheldon CD, Murie JA, Quin RO. Ultrasonic Doppler spectral broadening in the diagnosis of internal carotid artery stenosis. Ultrasound Med Biol 1983;9:575–80.

Siebler M, Sitzer M, Rose G, Bendfeldt D, Steinmetz H. Silent cerebral embolism caused by neurologically symptomatic high-grade carotid stenosis; event rates before and after carotid endarterectomy. Brain 1993;116:1005–15.

Sillesen H. Carotid artery plaque composition: relationship to clinical presentation and ultrasound B-mode imaging. European Carotid Plaque Study Group. Eur J Vasc Endovasc Surg 1995;10:23–30.

Sitzer M, Fürst G, Fischer H, Siebler M, Fehlings T, Kleinschmidt A, et al. Between-method correlation in quantifying internal carotid stenosis. Stroke 1993;24:1513–8.

Sitzer M, Fürst G, Siebler M, Steinmetz H. Usefulness of an intravenous contrast medium in the characterization of high-grade internal carotid stenosis with color Doppler-assisted duplex imaging. Stroke 1994;25:385–9.

Sitzer M, Müller W, Siebler M, Hort W, Kniemeyer HW, Jancke L, et al. Plaque ulceration and lumen thrombus are the main sources of cerebral microemboli in high-grade internal carotid artery stenosis. Stroke 1995;26:1231–3.

Sliwka U, Rautenberg W. Multimodal ultrasound versus intra-arterial angiography for imaging the vertebrobasilar circulation [letter]. J Neuroimaging 1998;8:182.

Smith SC Jr, Greenland P, Grundy SM. AHA Conference Proceedings. Prevention conference V. Beyond secondary prevention: identifying the high-risk patient for primary prevention: executive summary. American Heart Association. Circulation 2000;101:111–6.

Spencer MP, Reid JM. Quantitation of carotid stenosis with continuous-wave (C-W) Doppler ultrasound. Stroke 1979;10:326–30.

Spencer MP, Whisler D. Transorbital Doppler diagnosis of intracranial arterial stenosis. Stroke 1986;17:916–21.

Starmans-Kool MJ, Stanton AV, Zhao S, Xu XY, Thom SA, Hughes AD. Measurement of hemodynamics in human carotid artery using ultrasound and computational fluid dynamics. J Appl Physiol 2002;92:957–61.

Steinke W, Hennerici MG. Three-dimensional ultrasound imaging of carotid artery plaques. J Cardiovasc Technol 1989;8:15–22.

Steinke W, Kloetzsch C, Hennerici M. Carotid artery disease assessed by color Doppler flow imaging: correlation with standard Doppler sonography and angiography. AJNR Am J Neuroradiol 1990a;11:259–66.

Steinke W, Kloetzsch C, Hennerici M. Variability of flow patterns in the normal carotid bifurcation. Atherosclerosis 1990b;84:121–7.

Steinke W, Hennerici M, Kloetzsch C, Sandmann W. Doppler colour flow imaging after carotid endarterectomy. Eur J Vasc Surg 1991;5:527–34.

Steinke W, Els T, Hennerici M. Comparison of flow disturbances in small carotid atheroma using a multi-gate pulsed Doppler system and Doppler color flow imaging. Ultrasound Med Biol 1992a;18:11–8.

Steinke W, Hennerici M, Rautenberg W, Mohr JP. Symptomatic and asymptomatic high-grade carotid stenoses in Doppler color-flow imaging. Neurology 1992b;42:131–8.

Steinke W, Rautenberg W, Schwartz A, Hennerici M. Noninvasive monitoring of internal carotid artery dissection. Stroke 1994;25:998–1005.

Steinke W, Meairs S, Ries S, Hennerici M. Sonographic assessment of carotid artery stenosis: comparison of power Doppler imaging and color Doppler flow imaging. Stroke 1996;27:91–4.

Sterpetti AV, Schultz RD, Feldhaus RJ, Davenport KL, Richardson M, Farina C, et al. Ultrasonographic features of carotid plaque and the risk of subsequent neurologic deficits. Surgery 1988;104:652–60.

Tan TY, met LM, Schminke U, Tesh P, Reynolds PS, Tegeler CH. Hemodynamic effects of innominate artery occlusive disease on anterior cerebral artery. J Neuroimaging 2002;12:59–62.

Terwey B, Gahbauer H, Montemayor M, Proussalis A, Zollner G. [B-image sonography of carotid bifurcation; in German.] Ultraschall Med 1984;5:190–201.

Thomas AC, Davies MJ, Dilly S, Dilly N, Franc F. Potential errors in the estimation of coronary arterial stenosis from clinical arteriography with reference to the shape of the coronary arterial lumen. Br Heart J 1986;55:129–39.

Tismer R, Boehlke J. Die Anatomie der Carotisgabel. Ein Beitrag zur Real-Time-Sonographie der extracraniellen A. carotis. Ultraschall Klin Prax 1986;1:86.

Touboul PJ, Mas JL, Bousser MG, Laplane D. Duplex scanning in extracranial vertebral artery dissection. Stroke 1988;19:116–21.

Touboul PJ et al. Mannheim intima-media thickness consensus. Cerebrovase Dis. 2004;18(4):346–9.

Trattnig S, Hübsch P, Schuster H, Polzleitner D. Color-coded Doppler imaging of normal vertebral arteries. Stroke 1990;21:1222–5.

Trattnig S, Matula C, Karnel F, Daha K, Tschabitscher M, Schwaighofer B. Difficulties in examination of the origin of the vertebral artery by duplex and colour-coded Doppler sonography: anatomical considerations. Neuroradiology 1993;35:296–9.

Trockel U, Hennerici M, Aulich A, Sandmann W. The superiority of combined continuous wave Doppler examination over periorbital Doppler for the detection of extracranial carotid disease. J Neurol Neurosurg Psychiatry 1984;47:43–50.

Valentine RJ, Myers SI, Hagino RT, Clagett GP. Late outcome of patients with premature carotid atherosclerosis after carotid endarterectomy. Stroke 1996;27:1502–6.

Valton L, Larrue V, Arrue P, Geraud G, Bes A. Asymptomatic cerebral embolic signals in patients with carotid stenosis: correlation with appearance of plaque ulceration on angiography. Stroke 1995;26:813–5.

Vanninen R, Manninen H, Soimakallio S. Imaging of carotid artery stenosis: clinical efficacy and cost-effectiveness. AJNR Am J Neuroradiol 1995;16:1875–83.

Vogelsang H. Über eine angiographisch selten nachzuweisende Anastomose zwischen dem A. carotis interna und dem A. carotis externa-Kreislauf. Nervenarzt 1961;32:518–20.

Vollmar J. Rekonstruktive Chirurgie der Arterien. Stuttgart: Thieme, 1975.

von Reutern GM, Büdingen HJ. Ultrasound diagnosis of cerebrovascular disease: Doppler sonography of the extra- and intracranial arteries, duplex scanning. Stuttgart: Thieme, 1993.

von Reutern GM, Clarenbach P. Valeur de l'exploration Doppler des collatérales cervicales et de l'ostium vertébral dans le diagnostic des sténoses et occlusions de l'artère vertébrale. Ultrasonics 1980;1:153–62.

von Reutern GM, Pourcelot L. Cardiac cycle-dependent alternating flow in vertebral arteries with subclavian artery stenoses. Stroke 1978;9:229–36.

von Reutern GM, Büdingen HJ, Hennerici M, Freund HJ. [The diagnosis of stenoses and occlusions of the vertebral arteries and subclavian steal syndrome using directional Doppler sonography; in German.] Arch Psychiatr Nervenkr 1976a;222:191–207.

von Reutern GM, Voigt K, Ortega-Suhrkamp E, Büdingen HJ. [Doppler findings in intracranial vascular disorders: differential diagnosis of extracranial and intracranial vascular occlusions; in German.] Arch Psychiatr Nervenkr 1977;223:181–96.

Wernz MG, Hetzel A, Eckenweber B, Beyersdorf F. Pseudoocclusion (PO) of the internal carotid artery (ICA): carotid endarterectomy (CEA) without angiography? Cerebrovasc Dis 1994;7(Suppl 4):1–88.

Withers CE, Gosink BB, Keightley AM, Casola G, Lee AA, van Sonnenberg E, et al. Duplex carotid sonography: peak systolic velocity in quantifying internal carotid artery stenosis. J Ultrasound Med 1990;9:345–9.

Widder B, von Reutern GM, Neuerburg-Heusler D. [Morphologic and Doppler sonographic criteria for determining the degree of stenosis of the internal carotid artery; in German.] Ultraschall Med 1986;7:70–5.

Widder B, Berger G, Hackspacher J, Horz R, Nippe A, Paulat K, et al. [Reproducibility of ultrasound criteria for characterizing carotid artery stenoses; in German.] Ultraschall Med 1990;11:56–61.

Winter R, Biedert S, Staudacher T, Betz H, Reuther R. Vertebral artery Doppler sonography. Eur Arch Psychiatry Neurol Sci 1987;237:21–8. Erratum in: Eur Arch Psychiatry Neurol Sci 1988;237:124.

Young GR, Humphrey PR. Skepticism and carotid ultrasonography. Stroke 1996;27:768–70.

Zbornikova V, Lassvik C. Duplex scanning in presumably normal persons of different ages. Ultrasound Med Biol 1986;12:371–8.

Zierler RE, Phillips DJ, Beach KW, Primozich JF, Strandness DE Jr. Noninvasive assessment of normal carotid bifurcation hemodynamics with color-flow ultrasound imaging. Ultrasound Med Biol 1987;13:471–6.

Zwiebel WJ, Austin CW, Sackett JF, Strother CM. Correlation of high-resolution, B-mode and continuous-wave Doppler sonography with arteriography in the diagnosis of carotid stenosis. Radiology 1983;149:523–32.

3 Intracranial Cerebral Arteries

Examination

Special Equipment and Documentation

Low-frequency (1.5–3.0 MHz) pulsed-wave Doppler sonography with a high energy output is needed for examinations of the intracranial basal cerebral arteries in adults. Continuous-wave Doppler sonography cannot be used for the purpose. Transcranial Color flow *duplex systems* have now widely replaced the purely *Doppler systems* that were long in use.

Normally, a *hand-held probe* is used at various access points to examine the arteries at the base of the skull. Data from successive vascular segments are recorded with depth control. A display of the anatomical landmarks in biplanar B-mode is useful to identify

1) the patency of the bone window
2) the optimal section to start searching for the cerebral arteries with Doppler
3) several anatomical structures—e.g., the brainstem, the midbrain and the temporal lobe in transtemporal approaches (Figs. 3.**1a**, 3.**2**) (Bogdahn et al. 1993, Rosenkranz et al. 1993, Seidel et al. 1995, Baumgartner and Baumgartner 1996, Ries et al. 1997).

In addition, frequency—(flow velocity and amplitude-modulated Doppler signals are used in color flow Duplex sonograms to superimpose the course of arteries

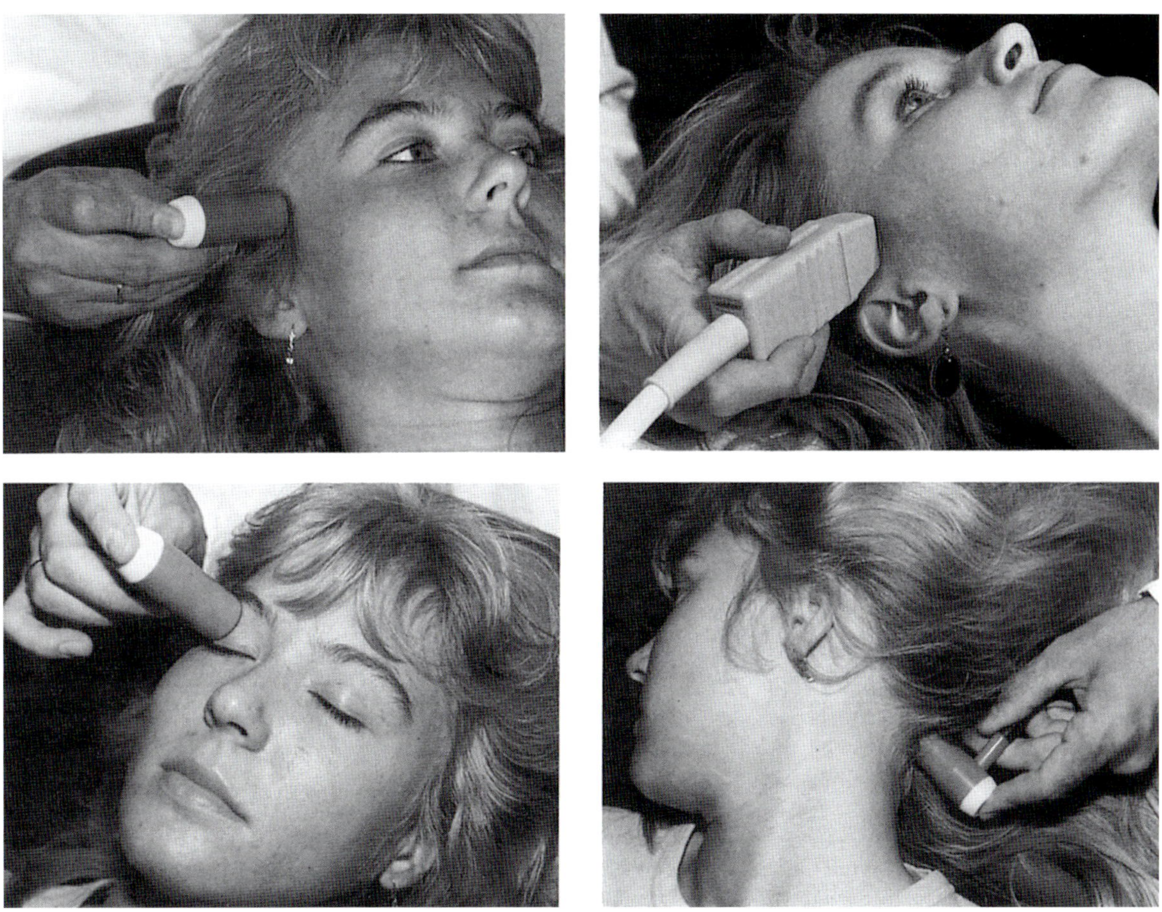

Fig. 3.**1 a** The probe position used in transcranial Doppler sonography to examine the anterior section of the circle of Willis. **b** The probe position in transtemporal duplex sonography. **c** Demonstrating the probe position for transorbital ultrasound application. **d** Transnuchal ultrasound application

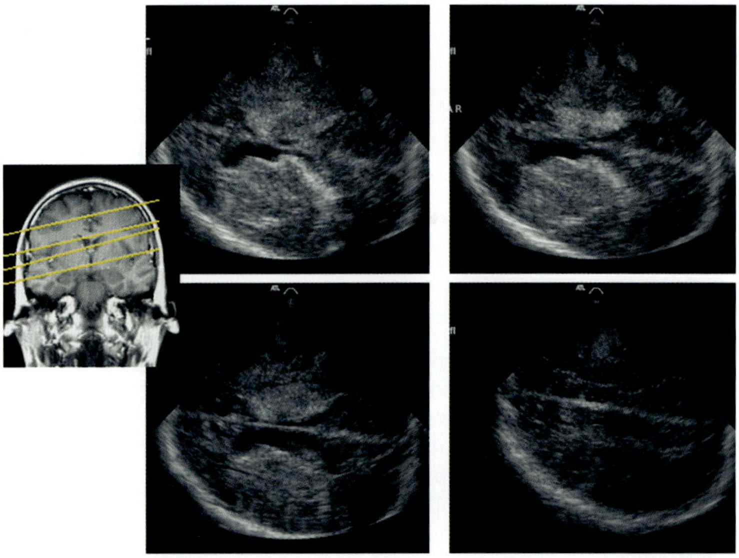

Fig. 3.**2** Four axial transtemporal ultrasound sections in a patient with a particularly good transtemporal bone window. The section levels are projected here onto a coronary magnetic resonance image (MRI) serving as a localizer. This makes it clear that the axial transtemporal levels obtained with ultrasound B-imaging do not correspond to the axial levels obtained with MRI, but instead represent a slope from the ipsilateral caudal temporal region to a substantially more cranial area in the contralateral hemisphere

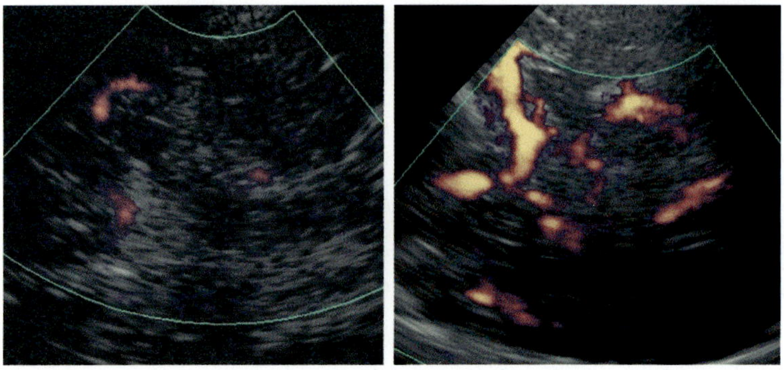

Fig. 3.**3** The effect of administering ultrasound contrast media. **a** Power mode depiction of the basal afferent cerebral arteries from the left transtemporal area. Continuous depiction of the cerebral arteries is not possible. **b** After contrast administration (SonoVue®), there is a clear improvement in the signal-to-noise ratio, and the arteries are well displayed.

and veins. Pulsed-wave Doppler displays from selected positions and vessel volumes can be used to classify normal and pathological waveforms) and enable comparison to analysis from pure Doppler system measures.

Ultrasound contrast agents are useful for improving the signal-to-noise ratio in patients with insufficient acoustic bone windows and in the presence of slow flow signals and tortuous slopes of arteries and veins. In contrast to extracranial examinations, ultrasound contrast agents are of potential valve in transcranial investigations (Fig. 3.**3**). More recent attempts to study brain tissue perfusion using harmonic imaging also require the use of ultrasound contrast agents (see Chapter 1).

For special purposes, such as emergency conditions, Doppler systems are still useful in the hands of experienced examiners. In addition, several areas of insonation are best with either one or the other technique—e.g., the distal basilar segment (better with Doppler systems) and internal carotid segment (better with duplex systems). Simultaneous availability of both devices is recommended. Stable fixation of the probe for an extended period is required for recordings to assess function and progress over time (e.g., for monitoring purposes). In addition, recent developments aimed at speeding up recombinant tissue plasminogen activator (rt-PA) thrombolysis by using ultrasound at the same time when the drug is being administered require rigid probe positioning and a special ultrasound configuration for more than 60–90 min. In principle, simultaneous studies from both sides are useful, allowing a comparative analysis of flow conditions in the right and left middle cerebral arteries.

Examination Conditions

Patient and Examiner

When the intracranial arteries are being examined, it is necessary for the patient to be evaluated in controlled conditions with regard to cardiovascular function and respiration. The patient should therefore lie in a relaxed position in the examining room until a steady

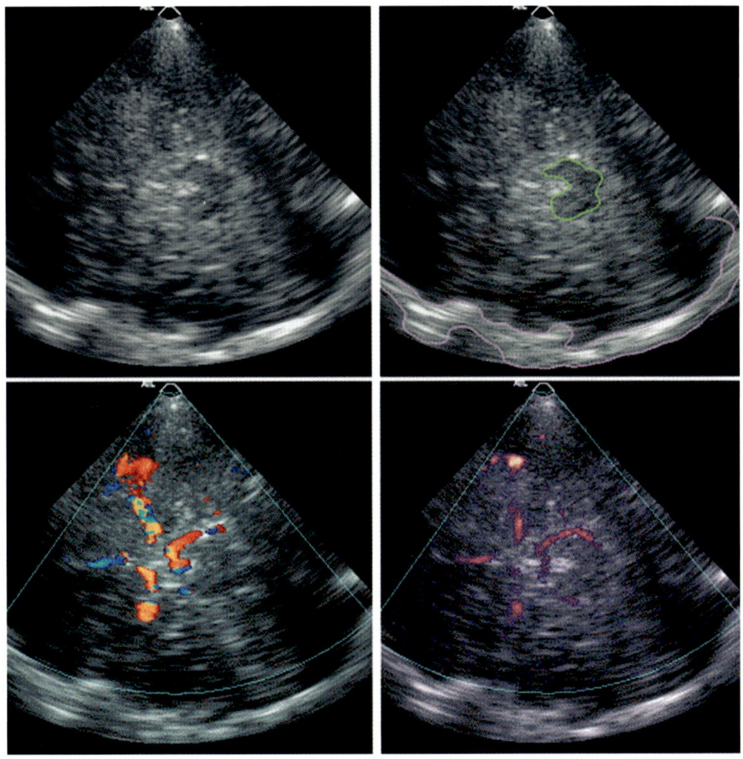

Fig. 3.**4** The typical basic settings in transcranial duplex sonography. *Upper row:* setting the base level in the B-image (left side, B-image alone; right side, with contouring of the main structures). The basic settings show what is known as a "butterfly" (green outline, corresponding to the mesencephalon, with the crus cerebri) and the contralateral calvaria (purple outline, from the inion to the frontal bone). In addition, when there is a good transtemporal bone window, additional midline structures (the third ventricle) can be seen at this level, as well as the cerebellum or tentorium. The basal ganglia are also sometimes visible. *Lower row:* Frequency-mode (left side) and amplitude-mode (right side) of the circle of Willis

state is reached. For quantitative and sequential velocity measurements and functional tests, constant cardiovascular parameters should be obtained (blood pressure, heart rate, and end-expiratory P_{CO_2} values).

Examinations in the emergency room, intensive-care and stroke units require standardized conditions with regard to the systems being used, examiners' skills, and documentation. Attention also needs to be given to these requirements so that different conditions can be compared during interventional and surgical procedures.

Conducting the Examination

Examining the intracranial cerebral arteries requires assessment and documentation of the Doppler signals and a display of the flow velocity distribution and adjacent tissue at defined ultrasound access points. Anatomical orientation at the base of the brain is possible using the B-mode (for example, the midbrain can be depicted as a butterfly-shaped structure) (Figs. 3.**2**, 3.**4**), and flow within the arteries at the base of the skull can be depicted using color. After the sample volume has been placed in the color flow image, additional evaluation of the Doppler spectrum signals should follow. The methods of application, probe position, and identification of the individual blood vessels are the same in Doppler and duplex system studies, as described below (Baumgartner et al. 1996a, Klötzsch et al. 1996).

Ultrasound Application and Probe Position

Transtemporal Ultrasound

After contact gel has been applied, the probe is placed on the temporal region (Figs. 3.**1**, 3.**5**). The proper access point through the bone window (a thin-walled section of the skull) can be identified using slight, incremental changes in the incident angle and displacement (Fig. 3.**6**). The required point is the one at which good probe coupling allows maximum flow signals to be recorded. In general, ultrasound energies of between 75 and 100 mW/cm^2 are required.

Examination area. *Arteries:* Anterior cerebral artery (ACA), middle cerebral artery (MCA), posterior cerebral artery (PCA), intracranial internal carotid artery (ICA), anterior communicating artery, posterior communicating artery, superior cerebellar artery, top of the basilar artery (BA). *Brain structures:* Mesencephalon, lateral and third ventricles, thalamus and midline structures, temporal lobe.

Transorbital Ultrasound

With the patient's eyes closed and after contact gel has been applied, the probe is placed without pressure above the eyeball and above the ophthalmic artery (Fig. 3.**1 c**). Alternatively, it can be placed in the lateral canthus, where the contralateral anterior cerebral artery can also be assessed. The examination should proceed at a low intensity level (10–25 mW/cm^2). Since

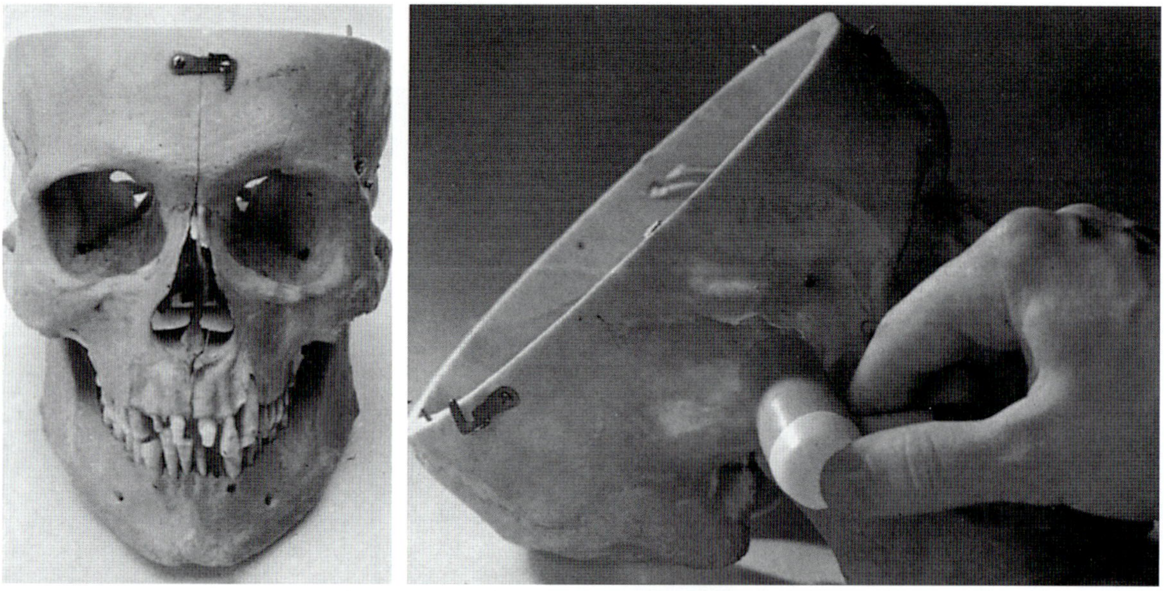

a b

Fig. 3.**5 a** Location of ultrasound windows for transorbital application, and **b** in the temporal region for transtemporal application (for ACA, MCA, and PCA recordings)

the angle between the ultrasound beam and the vascular axis is much more variable when the carotid siphon is being examined transorbitally than transtemporally, quantitative interpretation of the data obtained is limited.

Examination area. Siphon of the internal carotid artery (C_1, C_2 segments), ophthalmic artery, anterior cerebral artery middle cerebral artery in part.

Transnuchal Ultrasound

The probe is placed in the midline or laterally underneath the inion, causing the ultrasound to enter mainly through the great foramen (Fig. 3.**1 d**). The patient's head is inclined to the side; occasionally, the examination may be more successful with the patient seated than recumbent. Depending on the placement of the sample volume, an ultrasound energy of between 50 and 100 mW/cm^2 should be selected.

Examination area. Vertebral artery (V_4 segment), proximal and middle basilar artery (posterior inferior cerebellar artery, PICA; anterior inferior cerebellar artery, AICA).

Ophthalmic Artery and Internal Carotid Artery (Siphon)

Probe position. Proceeding transorbitally, the flow signal from the ophthalmic artery is searched for at a depth of between 25 mm and 45 mm, and the ultrasound beam is adjusted along the course of the blood vessel to the carotid siphon (Fig. 3.**1 c**).

Identification. When the sample volume is continuously displayed through the superior orbital fissure (Fig. 3.**5 a**), the characteristic audio signal from the internal carotid artery siphon can be obtained after approximately 60 mm (Fig. 3.**6 c**). By slightly changing the incident angle of the applied ultrasound, it is usually possible to assess both the proximal part of the siphon (C_4 segment, with flow toward the probe) and the distal part (C_2 segment, with flow away from the probe). It should be noted that the flow direction changes as this is done.

Middle Cerebral Artery (MCA)

Probe position. To apply transtemporal ultrasound through the frontal bone window, the probe is placed at the lateral canthus, approximately 1 cm in front of the external acoustic meatus and above the zygomatic bone (Fig. 3.**5 b**). The ultrasound beam is aimed in a slightly orbital direction (Fig. 3.**1 a**). A slight frontal shift over the tragus is sometimes helpful. If color flow duplex systems are used the bone window can be identified by B-mode imaging with sector depth of approximately 16 cm displaying the contralateral skull and at least some intracranial structure.

Fig. 3.**6** A diagram showing the relationship between the ▷ ultrasound axis and the vascular axis, for examining the large arteries at the base of the skull

a Probe position and sample volume depth location for examining the MCA (left) or the ACA (right)

b Probe position for examining the proximal segment (P_1 segment, left) and the distal segment (P_2 segment, right) of the PCA, with transtemporal ultrasound application, and the BA using transnuchal access

c Application of transorbital ultrasound to the ophthalmic artery

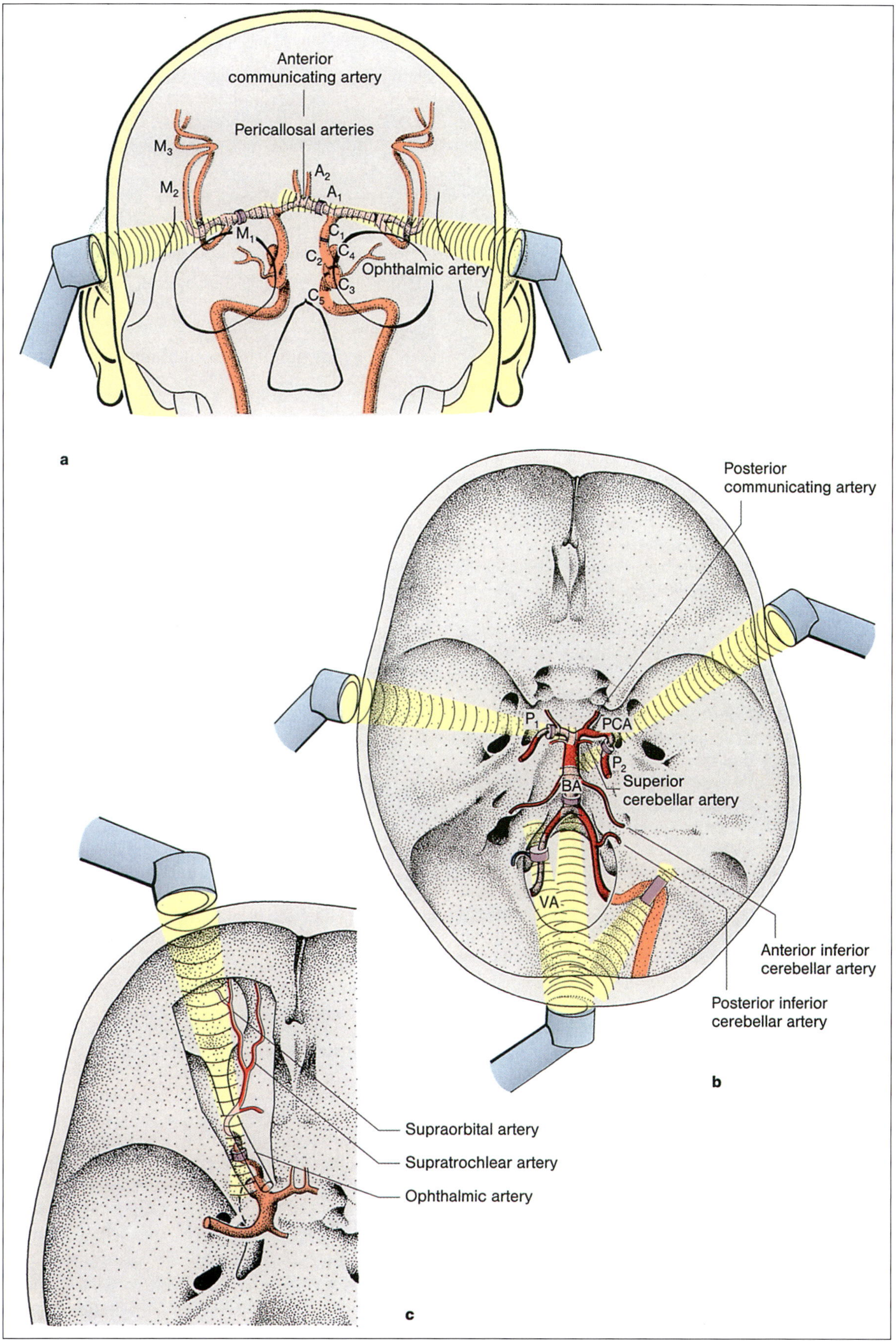
Anterior communicating artery
Pericallosal arteries
M3
M2
A2
A1
M1
C1
C4
C2
Ophthalmic artery
C3
C5
a
Posterior communicating artery
P1
PCA
P2
Superior cerebellar artery
BA
VA
Anterior inferior cerebellar artery
Posterior inferior cerebellar artery
b
Supraorbital artery
Supratrochlear artery
Ophthalmic artery
c

Identification. At a depth of between 30 mm and 55 mm, the main trunk of the MCA (the M_1 segment) can be identified from the flow directed toward the probe, and the trunk can be followed for approximately 30 mm (Fig. 3.**6a**). Since the course of the distal branches (the M_2 segment) is almost perpendicular to the ultrasound axis, these blood vessels cannot be reliably evaluated using the pure Doppler system approach. However, this area can often be displayed using a duplex system investigation. Color coding is useful in the MCA, to distinguish its stem from other segments of the circle of Willis and to improve the identification of branches of the M_2 segment. This is particularly facilitated by additional use of ultrasound contrast agents (Otis et al. 1995, Postert et al. 1997). A practical guideline is ultrasound contrast agents being valid in case of poor or no display of the MCA but at least some structure or the contralateral skull visible in B-mode. If B-mode fails to show anything even with ultrasound contrast agents the MCA is barely visible.

Anterior Cerebral Artery (ACA)

Probe position. To obtain the best possible signal, the transducer is placed for transtemporal ultrasound application through the frontal bone window, and it usually has to be inclined forward slightly. The direction of the ultrasound beam should be adjusted very flexibly, since there are many variants in this section of the circle of Willis (Padget 1944, Fisher 1965).

Identification. The anterior cerebral artery can be assessed at a depth of between 55 mm and 80 mm; the blood flow is normally directed away from the probe (Fig. 3.**6a**).When data are being recorded in deeper segments, it is sometimes possible to detect the contralateral anterior cerebral artery, showing blood flow directed toward the probe. Color flow information is useful for the identification of the ACA, although the course of the ACA is sometimes more difficult to follow than that of the MCA. After administration of ultrasound contrast media, it may be possible to display both A_1 segments and the anterior communicating artery (Postert et al. 1997).

Intracranial Internal Carotid Artery and T-Junction

Probe position. The transducer is placed at the frontal bone window to allow a transtemporal ultrasound examination. The probe is usually orientated in an axial plane.

Identification. The intracranial internal carotid artery is easily detected by adjusting the probe to the MCA and then tilting the probe slightly caudally. The MCA image disappears and the intracranial internal carotid artery (ICA) appears in cross-section (in the C_1 segment). With further tilting, the cross-section changes into a longitudinal arterial color image showing the C_2 segment. The C_1 and C_2 segments can almost always be displayed using modern color duplex systems if the entire main stem of the MCA can be visualized. In some patients, tilting the probe even further sometimes reveals another longitudinal arterial color sign showing flow in the opposite direction (the C_4 segment). Imaging of the knee of the siphon (the C_3 segment) is a very rare finding with the temporal approach.

Posterior Cerebral Artery (PCA)

Probe position. The probe is placed directly above the ear, with slight dorsal and caudal displacement, so that ultrasound can be applied transtemporally at the posterior bone window.

Identification. The posterior cerebral artery is usually detected at a depth of between 55 mm and 75 mm. In contrast to the MCA, the signal cannot be followed in regions closer to the skull. There are two sections that can be distinguished from one another (Fig. 3.**6b**):

- The *proximal section* (P_1) usually has blood flow signals directed toward the probe, and can be detected at a depth of between 60 mm and 80 mm.
- The *distal section* (P_2) usually has the blood flow going away from the probe, and is found closer to the skull when the ultrasound probe is placed at a flatter angle (55–70 mm). With regard to the location of the butterfly-shaped mesencephalon, the P_1 segment, with blood flow toward the probe, and the P_2 segment can usually be depicted using color flow duplex sonography and distinguished from contralateral PCA segments (Fig. 3.**4**) (Bogdahn et al. 1993, Baumgartner et al. 1996b).

As with the ACA, it is possible to detect the contralateral PCA at greater depth when passing over the midline. It is sometimes possible to identify other blood vessels in this region, especially when using instruments with a smaller ultrasound beam or with color flow imaging—e.g., the superior cerebellar artery and the posterior communicating artery (Hennerici et al. 1987a, b, Klötzsch et al. 1996). This is facilitated by administering ultrasound contrast media (Postert et al. 1997).

Vertebral Artery (VA)

Probe position. When ultrasound is applied nuchally, the VAs can be registered along their intracranial course in an adult at a depth of between 40 mm and 65 mm, and sometimes even deeper. To achieve this, the incident ultrasound angle has to be optimally adapted to the varying directions of these vascular segments (Figs. 3.**1d**, Fig. 3.**6**b). Occasionally, it may also be useful to displace the probe laterally away from the midline, especially when the VAs have loop formations.

Identification. There is a major difference in the sensitivity and specificity of insonation between Doppler and duplex system analysis. Generally, duplex imaging is much easier to apply and often provides better im-

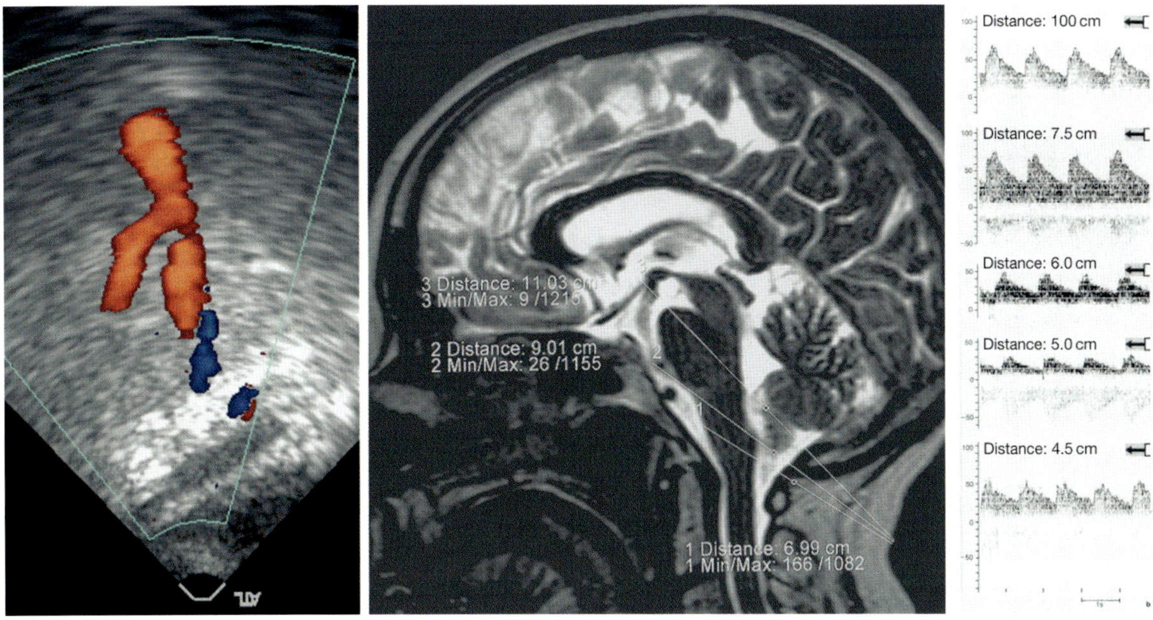

Fig. 3.**7** Insonation of the vertebrobasilar flow region using transcranial color duplex imaging. *Left:* duplex image, inverted to provide greater clarity (the bottom is caudal in the illustration and the top is cranial). Insonation was applied from the transnuchal direction. The color has been adjusted here to show all of the vessels in which the flow is away from the heart with red coding if possible. The two vertebral arteries and the BA are displayed. In contrast to the blind method of transcranial Doppler (TCD) sonography, it is easy to distinguish here between the vertebral artery and the BA, by visualizing the vertebral confluence. In the center, a sagittal magnetic resonance image shows the distances from the transnuchal insonation point to the vertebral confluence or the level of the proximal basilar artery (1), to the level of the medial basilar artery (2), and to the level of the distal basilar artery (3). On the far right there is a recording using simple TCD in which the distance information serves as a fundamental criterion for identifying the vessels

ages (Fig. 3.**7**) than Doppler. This is particularly important, since in the posterior cerebral circulation there is a high incidence of unilateral hypoplasia (10–30%) and course of the vasculature is very varied (Krayenbühl and Yasargil 1957). Duplex imaging is also better when there is pathology obstructing the blood flow and when carotid artery obstructions result in the collateral circulation being taken over by the vertebrobasilar system. Uncertainties and errors in interpreting the flow signals can be compensated for using additional magnetic resonance imaging (MRI) or magnetic resonance angiography (MRA) studies. Progress has been made in displaying the full course and junction of both VAs from extracranial to intracranial (Fig. 3.**7**). With today's ultrasound systems, conclusive judgments can often be made in these situations (Mull et al. 1990, Röther et al. 1993).

Basilar Artery (BA)

Probe position. Using nuchal access, the origin of the basilar artery can almost always be detected at a depth of between 70 mm and 90 mm (Figs. 3.**1 d**, 3.**6 b**), but it can be displayed along its full length in a minority of cases even if color flow duplex systems with amplitude or power energy mode are used. Due to the variable position of the confluence of the two vertebral arteries, the proximal segment is likely to be encountered at various depths easily displayed with the duplex system approach (Fig. 3.**7**). However, since the BA has an average length of 33 mm, the top of the artery is often not detectable (Hennerici et al. 1987a, b, Mull et al. 1990). In this case, transtemporal access is necessary.

Identification. It may be difficult to distinguish between the BA and the intracranial segments of the vertebral arteries and neighboring blood vessels using the pure Doppler technique:

- Due to unilateral hypoplasia (10–30%).
- Due to the variable distance between the basilar artery and the clivus (in 80% of cases this is less than 90 mm (Krayenbühl and Yasargil 1957, Busch 1966).
- Due to variations in the distance to the transducer caused by the musculature of the nape of the neck.
- Due to the course of the basilar artery, which is rich in variants, sometimes horizontal or curved, and the other arteries in the posterior cranial fossa.

Depth alone is therefore not a reliable parameter for identifying the BA. Signals that are located deeper than 80 mm from the transducer are usually from the BA, but occasionally the confluence of the two vertebral arteries is not found until 90–110 mm, or cannot be determined at all.

Classifying the bidirectional signals that sometimes occur is often possible with color-coded system analysis. Such signals originate from the large branches of the vertebral artery (posterior inferior cerebellar artery, PICA) or the basilar artery (anterior inferior cerebellar artery, AICA) (Fig. 3.**6b**). Definitive classification of the signals requires familiarity with the anatomical situation at the vertebral confluence. A comparative examination using both MRA (or computed-tomographic angiography, CTA) is useful (Röther et al. 1993).

Functional Tests

Compression Tests

Reliable compression tests can be carried out on the CCA. Compressing the extracranial segments of the *vertebral arteries* at the mastoid process is difficult, and is often not complete due to the deep location of the blood vessels. Due to their intracranial confluence, only bilateral compression would make sense, but even with assistance from a second investigator, this is only occasionally satisfying and it can only be recommended for research purposes. Ischemic complications during compression tests have been reported (Mast et al. 1993). Compression tests should therefore be restricted to cases in which alternative noninvasive methods do not solve the question.

Before compressing the common carotid artery, one should be certain:

- That light pressure on the blood vessel will not cause a significant decrease in the heart rate.
- That the compression will *definitely* take place below the carotid bifurcation (extracranial duplex examination beforehand).
- That an area in which wall changes are evident will not be compressed and that there are no significant arteriosclerotic lesions at the compression site (extracranial duplex).
- That compression is carried out only for single cardiac cycles.

When the ipsilateral CCA is compressed:

- The signal from the ipsilateral MCA is reduced (Fig. 3.**8**a).
- The signal from the ipsilateral ACA will reverse its flow direction (Fig. 3.**8**b) or will be reduced it the anterior communicating artery is patent.

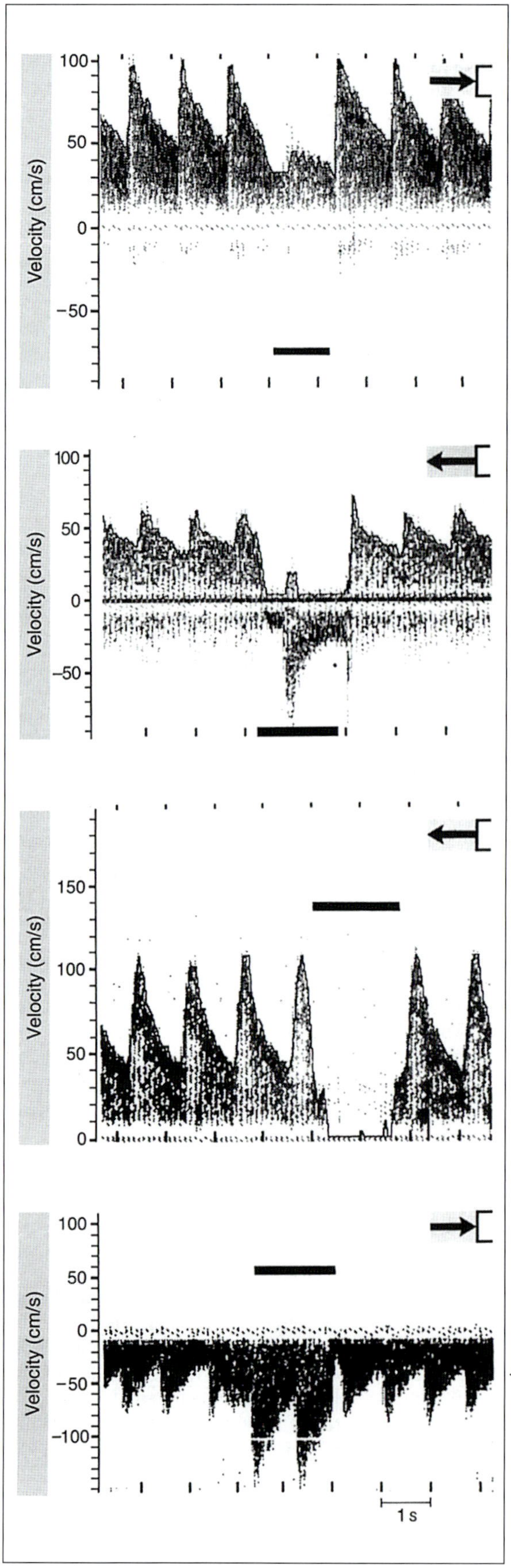

Fig. 3.**8** Doppler spectrum of the cerebral arteries near the base of the skull after ipsilateral compression of the common carotid artery (indicated by the bar)

a Flow velocity reduction in the MCA
b Flow signal reversal in the ACA
c Signal reduction in the PCA if originating from the carotid artery
d Signal increase in the PCA or the posterior communicating artery if originating from the BA

- The signal from the P1 will be reduced if this vessel originates directly from the carotid artery and behaves like the MCA (Fig. 3.**8**c).
- The signal from the P1-Segment of the PCA will be increased if the vessel does not originate from the carotid artery and the posterior communicating artery is intact (Fig. 3.**8**d).
- The signal from the posterior communicating artery will increase if good collateral function is possible (Fig. 3.**8**d).

Carotid compression tests may also be useful for assessing the posterior circulation.

When the upper arm arteries are compressed with a blood-pressure cuff in the subclavian steal phenomenon, the blood flow in the vertebral arteries, as well as in the BA and occasionally the PCA, may be either *reduced* or *elevated*, depending on whether or not the artery being examined is contributing to blood circulation in the brain.

During the reactive hyperemic phase after the blood-pressure cuff has been released again, antagonistic effects may be observed, depending on the prevailing hemodynamics.

Normally, neither the vertebral artery nor the BA show any reaction to this compression test. Corresponding findings that can even be more complex; involving also how velocity changes after compression (or compression release) in the right CCA, ECA, ICA and sometimes the MCA, can be found in obstructions of the innominate artery.

Vasomotor Reactivity Tests

Vasomotor reactivity (VMR) is defined as the range between maximal arteriolar constriction and dilation, and it forms part of the cerebral autoregulation system. In physiological conditions, vascular tone is at a medium level between maximal constriction and maximal dilation. Vasomotor reactivity can be measured by applying a selective dilative or constrictive vasoactive stimulus. Most commonly, cerebral VMR is assessed using a potent vasodilating stimulus such as CO_2 or acetazolamide. Tissue P_{CO_2} is directly related to pH, which is an important factor in determining vascular tone, and it therefore influences cerebrovascular resistance.

The CO_2 reactivity test measures the change in cerebral blood flow velocity (CBFV) that takes place in response to hypercapnia induced by inhaling a gas mixture with a raised carbon dioxide content. The resulting CO_2 effect is mediated by vasodilation of the cerebral arterioles, whereas the diameter of the proximal arteries remains constant (Kleiser et al. 1995). Several studies have shown a good correlation between the changes in the CBFV induced by acetazolamide or CO_2 measured using transcranial Doppler sonography and changes in cerebral blood flow measured with xenon inhalation methods or single photon emission computed tomography (SPECT) (Piepgras et al. 1990, Dahl et al. 1992a, Brauer et al. 1998).

CO_2 Reactivity Test

CO_2 inhalation. The measurement of CO_2-induced CBFV changes is the most established transcranial Doppler ultrasound test for analyzing VMR. Many variations in the test have been published; some of the most commonly used methods are described here.

Cerebral blood flow velocity (CBFV) is assessed bilaterally by simultaneous monitoring of both MCAs while the patient is lying on the back with eyes closed. The MCAs are insonated at a depth of about 50 mm via the transtemporal approach. The probe should be fixed to the insonation position (headband, metal frame, etc.). The patient inhale a gas mixture of e.g. 5 % CO_2 and 95 % oxygen (carbogen) through a mouthpiece connected to a respiratory balloon. A nose clip ensures correct inhalation of carbogen. The CO_2 content of the inhaled gas has to be continuously monitored e.g. with an infrared gas analyzer. After a baseline period of 2–5 min, patients inhale carbogen for at least 2 min, or better until a steady-state is reached. The maximum or mean CBFV, blood pressure, and CO_2 content (end-expiratory or transcutaneous) have to be continuously monitored.

The vasomotor reactivity can be calculated using the following equation:

$$VMR = 100 \cdot \frac{CBFV_{hyper} - CBFV_{normo}}{CBFV_{normo} \cdot (P_{CO_{2[hyper]}} - P_{CO_{2[normo]}})} \; [\%/mmHg]$$

where VMR = vasomotor reactivity; $CBFV_{hyper}$ = the cerebral blood flow velocity value after a constant has been reached while breathing CO_2-enriched air; and $CBFV_{normo}$ = mean value for 1 min at baseline.

Rebreathing. After the ultrasound probes have been positioned and fixed, patients are connected to a rebreathing system. Initially, of the beginning the system is open and the patient breathes normal air. Baseline CBFV and expiratory CO_2 are monitored for at least 1 min, after which the patient has to hold a deep breath. One or two of the empty breathing bags (at least 5–10 L) are connected to the mouthpiece, and the patient has to breath into the bag or bags for 60 s with a constant breathing frequency and depth. The bag or bags are then disconnected and the patient breathes normal air again. Vasomotor reactivity can be calculated on analogy with the CO_2 inhalation method. Since CO_2 and CBFV saturation do not occur using this method, the maximum CBFV and CO_2 are used as stimulation parameters.

Breath-holding test. The breath-holding method has potential as a convenient screening method of assessing carbon dioxide reactivity that does not require administration of carbon dioxide. The CBFV has to be monitored for at least 1 min at baseline. After baseline registration, the patient has to stop breathing. The apnea time is measured during continuous recording of the CBFV. The apnea time should be at least 30 s, as the validity of the results increases with a longer duration of apnea (Fig. 3.**9**). After apnea, CBFV registration

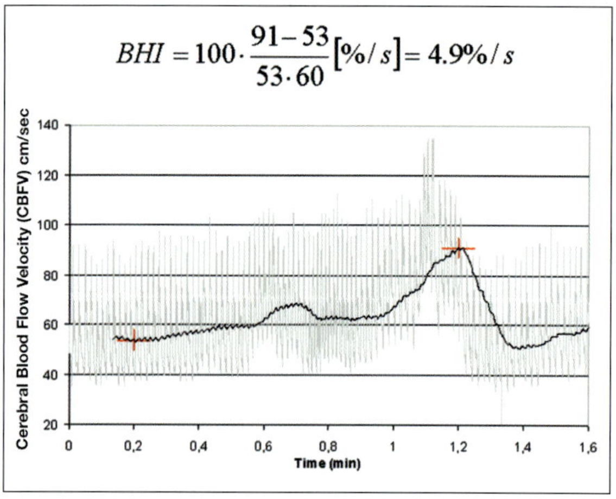

Fig. 3.**9** A normal CO_2 reaction after apnea. This curve in a normal volunteer shows the course of CBFV after 1 min of apnea. The curve provides the basis for the example calculation of the breath-holding index given in the text

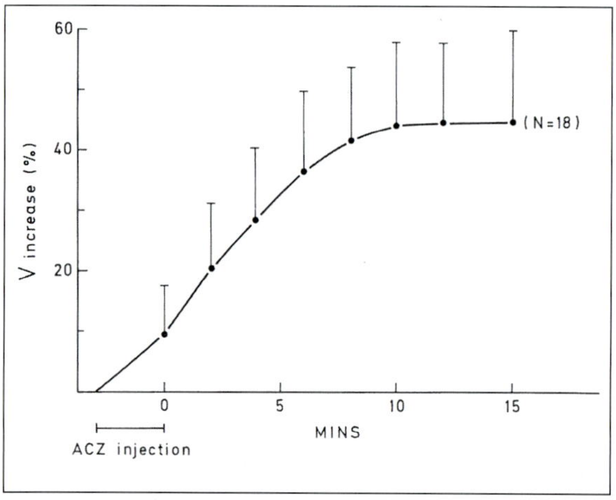

Fig. 3.**10** The normal range of increase in CBFV after an intravenous bolus administration of 100 mg acetazolamide. Data from 18 normal volunteers

should be continued for approximately 30 s. The mean velocity during baseline is taken as $CBFV_{normo}$. The maximum CBFV occurs approximately 10 s after the end of the apnea test. $CBFV_{apnea}$ is the maximum CBFV average over two respiratory cycles. The induced CBFV can be quantified using the breath-holding index (BHI):

$$BHI = 100 \cdot \frac{CBFV_{apnea} - CBFV_{normo}}{CBFV_{normo} \cdot T_{apnea}} [\%/s]$$

where $CBFV_{normo}$ = average CBFV (1 min) at baseline; $CBFV_{apnea}$ = maximum CBFV during apnea; and T_{apnea} = time of apnea.

Markus and Harrison (1992) compared a breath-holding method with two different CO_2 stimulation tests. All three methods provided results that correlated well with the degree of carotid stenosis, although the correlation was best when the full vasodilatory range was measured. The CO_2 inhalation method was adopted as the gold standard, and the other methods were compared with it. The breath-holding method correlated at least as well ($\varrho = 0.67$) as the 5% CO_2 method ($\varrho = 0.64$). It identified a similar group of low reactors to the gold standard method.

Acetazolamide test. Intravenous application of acetazolamide leads to arterial acidosis and hypercapnia, initiating vasodilation (Ehrenreich et al. 1961, Posner and Plum 1960). When testing VMR, 500–1000 mg acetazolamide can be administered intravenously during TCD monitoring.

Acetazolamide is an inhibitor of carboanhydrase, an enzyme that catalyzes bicarbonate from CO_2 and water. Intravenous application of acetazolamide almost completely blocks carboanhydrase in the erythrocytes. This results in an increase in dissolved CO_2 in the blood and also in the intracranial extracellular CO_2, causing arteriolar dilation comparable to that observed with the direct CO_2 stimulation test.

The CBFV can be monitored from the MCA or PCA (depending on where vasomotor reactivity is being measured). Most published studies have used unilateral measurements. With today's Doppler technology, bilateral registration is more useful, as it allows direct and more valid comparison of the two sides (e.g., when demonstrating a unilateral decrease in VMR due to proximal high-grade carotid obstruction). The patient should be in a constant position (usually recumbent). In addition, blood pressure and CO_2 (e.g., end-expiratory CO_2) should be recorded. The baseline CBFV should be measured for at least 5 min. Acetazolamide (500–1000 mg, usually dissolved in 10 mL water) is administered intravenously by bolus injection within approximately 1 min. The CBFV typically increases after approximately 1 min (Fig. 3.**10**), and the increase lasts for a minimum of 15–20 min. The VMR can be estimated by dividing the difference between the maximum increased CBFV and the baseline CBFV by the baseline CBFV:

$$VMR = 100 \cdot \frac{CBFV_{ace} - CBFV_{normo}}{CBFV_{normo}} [\%]$$

where VMR = vasomotor reactivity; $CBFV_{ace}$ = maximum CBFV after acetazolamide administration; $CBFV_{normo}$ = baseline CBFV (before acetazolamide administration).

The acetazolamide test has the advantage that it does not require cooperation from the patient, but it has the disadvantage of being associated with certain risks due to other pharmacological actions of acetazolamide (e.g., hypertension).

Autoregulation Tests

Static and dynamic autoregulation. Cerebral autoregulation can be evaluated by measuring the total CBFV change in response to a constant change in the blood pressure—*static autoregulation;* or by measuring the time until the CBFV reaches a new equilibrium after rapid blood-pressure alteration—*dynamic autoregulation.* Measuring dynamic autoregulation yields results similar to those with static testing of intact and impaired autoregulation. Tiecks et al. compared methods of measuring both static and dynamic autoregulation with TCD (Tiecks et al. 1995). The changes in cerebrovascular resistance were estimated on the basis of changes in cerebral blood flow velocity and arterial blood pressure in response to manipulations of blood pressure. Static autoregulation was determined by analyzing the response to a phenylephrine-induced rise in blood pressure, while rapid deflation of a blood-pressure cuff around the thighs served as a stimulus for testing dynamic autoregulation. Measurements were carried out in 10 volunteers, first with intact autoregulation and again in the same patients after autoregulation had been impaired by administering high-dose isoflurane. Reduced autoregulatory capacity after the administration of high-dose isoflurane was demonstrated using static ($P < 0.0001$) and dynamic ($P < 0.0001$) methods. The correlation between static, or steady-state, and dynamic autoregulation measurements was highly significant ($r = 0.93$, $P < 0.0001$).

Constant alterations in blood pressure are needed to measure static autoregulation. This type of continuous blood-pressure alteration has usually been achieved in clinical autoregulation studies using pharmacological agents that affect blood pressure. In a study investigating autoregulation and vasomotor reactivity during propofol-induced electrical silence on the electroencephalogram, cerebral autoregulation was tested by increasing blood pressure by 24 ± 5 mmHg from the baseline with an infusion of phenylephrine. No significant change in the CBFV was observed in the MCA (34 ± 2 cm/s and 35 ± 2 cm/s). Autoregulation remained intact during propofol-induced isoelectric electroencephalography (Matta et al. 1995).

Short-term blood-pressure alterations can be achieved without using pharmacological agents. The best-established technique is the cuff method, in which a sudden decrease in blood pressure is achieved by deflating blood-pressure cuffs at the thighs. Alternative methods include using an orthostasis paradigm to reduce blood pressure or to measure spontaneous blood-pressure fluctuations (Fig. 3.**11**).

Aaslid et al. (1989) induced a rapid decrease in blood pressure of approximately 20 mmHg by rapidly deflating thigh blood-pressure cuffs following a 2-min period of inflation above systolic blood pressure. Instantaneous arterial blood pressure was measured, and CBFV changes were assessed by transcranial Doppler recording of the blood flow velocity in the MCA. The results showed that in hypocapnia, full restoration of blood flow to the pretest level was seen as early as 4.1 s after the steep decrease in blood pressure, while the response was slower in normocapnia and hypercapnia. The rates of regulation were 0.38/s, 0.20/s, and 0.11/s in hypocapnia, normocapnia, and hypercapnia, respectively. There was a highly significant inverse correlation between the rate of regulation and P_{CO_2} ($P < 0.001$), indicating that the cerebral autoregulation response rate in conscious normal humans is extremely dependent on vascular tone. As an alternative to the cuff method, carotid compression can be used to achieve an intracranial reduction in blood pressure (the *transient hyperemic response test*) (Smielewski et al. 1996). Smielewski published a study comparing autoregulation measurements using the leg-cuff test with those using carotid compression (lasting 3, 4, 5, 7, and 9 s) at different levels of CO_2. The transient hyperemic response ratio, calculated as the maximum increase in CBFV to baseline values after release of the carotid compression, was taken as the autoregulation index. There was a linear correlation between the transient hyperemic response ratio and the autoregulation index derived by the leg-cuff test ($r = 0.86$). However, there are several restrictions with both the leg-cuff test and measurement of autoregulation via carotid compression. Particularly in patients with arteriosclerotic disease, compressing an arterial segment (either the carotid artery or femoral artery) is associated with a significant risk for arterial ischemia.

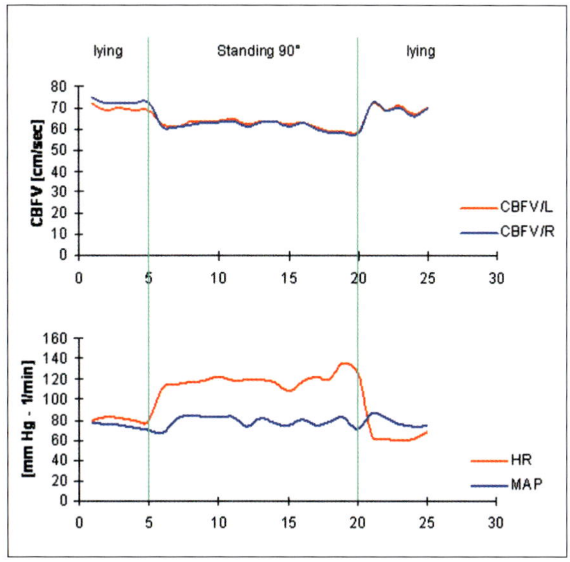

Fig. 3.**11** A polygraphic recording of the cerebral blood flow velocity (CBFV) and circulatory parameters during a tilting-table paradigm with 15 min in the upright position (90°). Normally, there is a slight decline in CBFV (maximum 20 %), even though blood pressure remains perfectly constant (at the level of the upper arm) and there is good orthostatic regulation, mainly with an increased heart rate in this case. CO_2 remained constant during the recording and should always be investigated as well in order to exclude ventilation-related alterations in CBFV

The most sophisticated technique for assessing cerebral autoregulation makes use of spontaneous changes in cardiovascular parameters—blood pressure and CBFV (Steinmeier et al. 2002, Puppo et al. 2002). Both the blood pressure and the CBFV oscillate at slow frequencies caused by vagal oscillations affecting heart

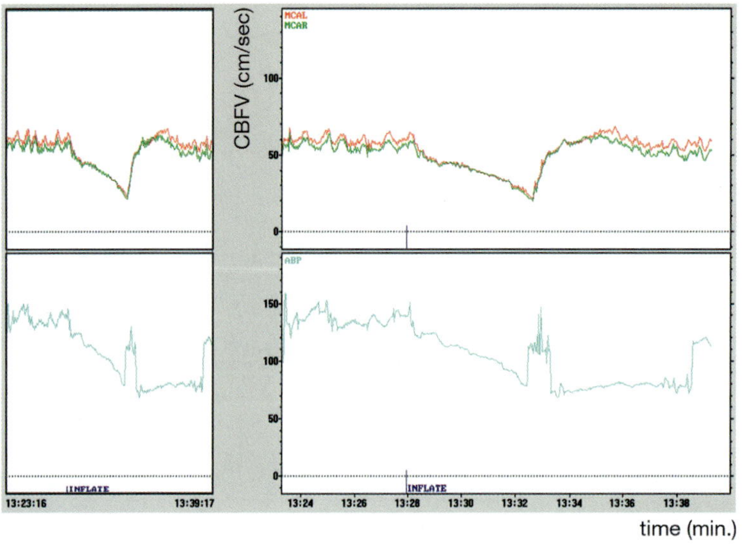

Fig. 3.**12** A pathological orthostatic reaction in a patient with neurocardiogenic syncope. When there is a reduction in the mean arterial blood pressure (light blue curve: lower trace), it results in reduced cerebral blood flow velocity (CBFV) in both MCA's. With presyncopal symptoms, the patient is returned to the horizontal position. Although blood pressure quickly returns to normal, the CBFV shows initially slightly higher values in comparison with the baseline situation, but then returns to normal as well

and respiration. Autoregulation can be measured using phase-shift angle analysis (coherence) between oscillations in the CBFV and blood pressure during deep breathing. In a study of 50 healthy volunteers, 20 patients with occlusive cerebrovascular diseases, and 10 patients with arteriovenous malformations, measurements of CBFV in the MCA and blood pressure during deep breathing at a rate of 6/min, autoregulation was quantified as a phase-shift angle between oscillations in CBFV and ambulatory blood pressure (ABP) at a frequency of 6/min (Diehl et al. 1995). A phase-shift angle of 0° indicated a total absence of autoregulation, while 90° can be regarded as optimal autoregulation. All normal individuals showed positive phase-shift angles between CBFV and ABP (70.5 ± 29.8°). Patients with occlusive cerebrovascular diseases showed significantly decreased phase-shift angles for the MCA exclusively on the side of carotid obstruction (51.7 ± 35.1°; $P < 0.05$). Patients with arteriovenous malformations showed significantly reduced phase-shift angles on both the affected side (26.8 ± 13.5°; $P < 0.001$) and the unaffected side (40.6 ± 26.6°; $P < 0.01$). Other studies confirmed the usefulness of this method in normal individuals (Birch et al. 1995) as well as in patients with subarachnoid hemorrhage (Giller and Iacopino 1997) and carotid disease (Panerai et al. 1998). However, the measurement of cerebral autoregulation on the basis of spontaneous slow oscillations in blood pressure and CBFV is based on continuous blood-pressure measurement; since intra-arterial measurements are invasive procedures, the monitoring data recorded represent qualitative rather than quantitative measures.

Orthostatic regulation. Cerebral autoregulation can also be estimated using an orthostasis paradigm (Brunhölzl and Müller 1986, Stoll et al. 1999b). Blood-pressure alterations are completed by a sudden tilt to the vertical position. However, the CBFV adaptation to the upright position depends not only on cerebral autoregulation, but is also influenced by other variables (e.g., cerebral pressure, venous constrictions, etc.). Orthostatic regulation tests changes and variability in the pure cerebral autoregulation. If the test is appropriately conducted and interpreted with caution, it can provide relevant estimates of autoregulation that have the advantage of being simple and reproducible (Stoll et al. 1999b). Clinically, testing orthostasis is often useful for diagnosing sources of syncope, and it may indicate threatened collapse of hemodynamic compensation in severe large-artery disease in the brain (Grubb et al. 1991, Daffertshofer and Hennerici 1993, Ladwig et al. 1997, Hsu et al. 1999). A drop in CBFV and blood pressure of < 20% from individual baseline values during orthostasis is normal. On the basis of these reference values, four different types of CBFV responses were observed:

- Type 1. Normal systemic blood pressure and heart rate regulation and constant CBFV.
- Type 2. A pathological pressure drop of more than 25 mmHg from the baseline blood pressure while the CBFV remains constant, indicating adequate compensation by the local cerebral autoregulation.
- Type 3. A significant decrease in the CBFV by more than 20% from the initial CBFV values following a pathological fall in blood pressure indicates exhausted cerebral autoregulation (Fig. 3.**12**).
- Type 4. Normal regulation of blood pressure and heart rate, while the CBFV decreases significantly by more than 20% from the initial CBFV values, suggests exclusive disturbance of cerebrovascular autoregulation.

Testing cerebrovascular orthostatic regulation is a simple method. A TCD monitoring device is needed. Generally, monitoring a single intracranial arterial segment (e.g., one MCA segment) is sufficient. Bilateral monitoring is recommended, particularly in the presence of small-artery disease, which is otherwise difficult to identify, or unilateral large-vessel disease. Recordings of heart rate and blood pressure should be automated and carried out in a polygraphic fashion. For most purposes, conventional blood pressure recording using a discontinuous blood-pressure measuring device (the arm-cuff method) is adequate. A more sophisticated method is continuous blood pressure and heart rate recording. Since some patients tend to change their breathing pattern when tilting to an upright position recording CO_2 is helpful for confirming a constant CO_2 level.

The examination should start with the patient in a supine position on a tilting table. Doppler probes, as well as blood pressure and heart rate devices, have to be attached, and monitoring should be registered for 5 min to allow reliable baseline analysis. After baseline registration, the table is tilted upwards to approximately 80° (angles of 60–90° have been reported). The tilting should take place quickly, within 1–2 s, to test early vascular dysregulation. The duration for which the upright position should be maintained depends on the clinical issues being investigated; for example, if neurocardiogenic syncopes (vasovagal syncopes) are being considered, the tilt-up phase should be extended up to 45 min. Changes in blood pressure and heart rate during the upright position are expressed as absolute values, while changes in CBFV are expressed as relative values. The estimate of systemic vascular regulation is based on the analysis of blood pressure and heart rate. Most authors report pathological blood-pressure regulation if the blood pressure decreases by more than 20–25 mmHg. For more detailed classification, heart rate and diastolic blood pressure also have to be taken into account. A more sophisticated approach is to establish individual normal values. Ways of interpreting blood pressure and heart rate changes are well described in textbooks on autonomic disorders.

During analysis of the maximum CBFV changes occurring in the upright position, diagnosis of systemic circulatory regulation can be enhanced by assessing whether or not there is sufficient cerebrovascular compensation. The normal range of reduced CBFV during tilt-up is reported as 20–30%. When the fall in CBFV is less than the normal range, it can be interpreted as representing sufficient cerebrovascular compensation; conversely, when the fall in CBFV is more than the normal range, it can be interpreted as representing insufficient cerebrovascular compensation.

Vasoneural Coupling

Brain activity, metabolism, and blood flow are closely related. Assuming that vessels have a constant diameter (see above), velocity changes are mainly related to changes in the volume of flow. Consequently, changes in the diameter of small resistance vessels, caused by metabolic changes that produce changes in the volume of blood flow in a proximal artery, will also cause an increase in CBFV in the large arteries that supply the brain. Transcranial Doppler sonography therefore makes it possible to measure vasoneural coupling. The introduction of bilateral continuous TCD monitoring has led to a variety of sophisticated applications. Moreover, if vasoneural coupling is normal, different stimuli can be applied to investigate brain function. TCD is therefore complementary to PET and fMRI, with the advantage of real-time measurements but the disadvantage of low spatial resolution. Applications reported so far include various studies of visually evoked responses and evaluations of functional recovery after stroke (Silvestrini et al. 1998a); investigation of perfusion asymmetries during complex spatial tasks (Klingelhofer et al. 1997) and melody recognition (Matteis et al. 1997); assessment of hemispheric dominance in candidates for epilepsy surgery (Knecht et al. 1997); and elucidation of the temporal patterns of regional neuronal activity (Tiecks et al. 1998).

Posterior Cerebral Artery Stimulation

TCD measurements of vasoneural coupling within the vascular territory of the PCA are commonly made using visual stimulation. Applying a complex, colored, and mobile visual stimulus to activate the large striate and extrastriate neuronal networks within the PCA territory at constant time intervals (usually 20 s) is the most common paradigm used to measure visual evoked CBFV responses in the PCA (Fig. 3.**13**).

The test has been shown to be useful for diagnosis in patients with migraine (Thie et al. 1990a, b, Silvestrini et al. 1996). CBFV changes were originally observed during the migraine attack (Thie et al. 1990b), but inconsistent results have been reported in the interval between attacks (Olesen 1991). In our own study of patients with migraine (26 with and 10 without aura) during the interictal state, we found that Δ CBFV was increased (42%) in migraineurs with aura vs. normal subjects ($P < 0.05$). If adequate treatment is administered, this difference in CBFV values often normalizes.

Since vasoneural coupling depends on the integrity of the cerebrovascular autoregulation, the visual stimulation test results decline as the cerebrovascular reserve capacity becomes exhausted (Sitzer et al. 1992). When there is a relevant decrease in the local cerebral perfusion pressure—e.g., distal to a hemodynamically relevant stenosis—the flow increase after visual stimulation is diminished. In addition, in conditions of hemodynamic compromise in the MCA (e.g., distal to a carotid occlusion), the CBFV response in the PCA to visual stimulation may be diminished if the PCA serves as a collateral pathway. Whereas CBFV is diminished in cases of neuronal damage, patients with an ischemic lesion within the PCA territory show a significant decrease in the CBFV response. In patients with normal cerebrovascular autoregulation, the visual stimulation test can be used to test vasoneural coupling.

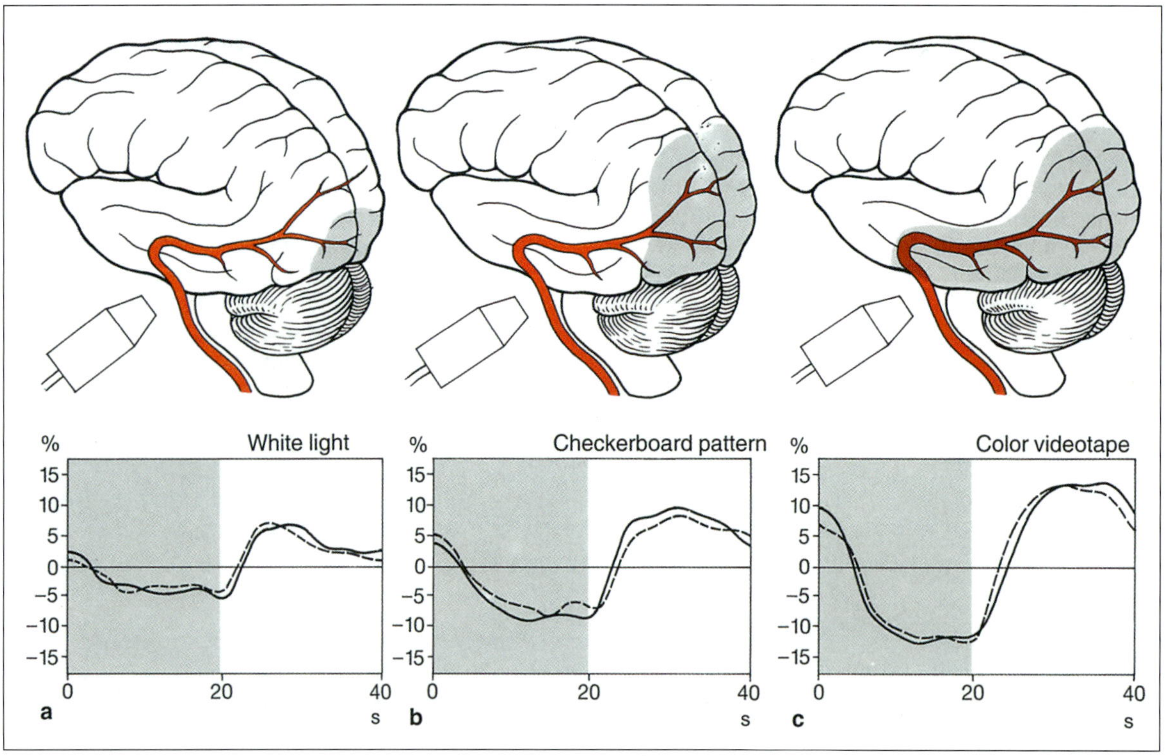

Fig. 3.**13** The principle of vasoneural coupling. Flow velocity in the PCA after visual stimulation increases with the complexity of the visual stimulus (**a–c**). Complex stimuli activate larger brain areas than when white light alone is presented

Middle Cerebral Artery Stimulation

The MCA supplies a wide variety of territories including sensorimotor and higher cortical functions (Fig. 3.**14**). The reported CBFV responses to MCA stimulation are usually smaller than the CBFV responses within the PCA territory after full visual stimulation (Diehl et al. 1990, Kelley et al. 1992). This depends on the complexity of the stimulation paradigms, and in particular on the relationship between the functionally responding brain territory and the total vascular territory measured. Reproducible CBFV responses in the MCA territory have been demonstrated using motor stimulation (Sander et al. 1995), sensory stimulation (Knecht et al. 1997), cognitive stimulation (Droste et al. 1989a, b), and emotional stimulation (Troisi et al. 1999).

Increased flow velocities in *both* the contralateral and ipsilateral MCA during motor tasks have been demonstrated in stroke patients, suggesting that areas of the healthy hemisphere can be activated soon after a focal ischemic injury and can contribute to the positive evolution of a functional deficit (Silvestrini et al. 1998a). This phenomenon of ipsilateral activation may be long-lasting, as it has also been demonstrated several months after the onset of stroke (Silvestrini et al. 1998b). CBFV changes after magnetic stimulation of motor cortex areas in stroke patients with a poor outcome were significantly lower (4.2 % ± 1.1 % increase in CBFV) than in stroke patients who had a good recovery (8.0 % ± 2.5 % increase of CBFV) (Daffertshofer and Hennerici 1995). It has therefore been suggested that this technique could be useful as a prognostic test.

In normal individuals, CBFV responses after magnetic stimulation of the motor cortex were slightly larger in the left MCA territory in right-handed persons, reflecting the larger motor fields in the dominant hemisphere (motor evoked potential stimulation was nonfocal, and the stimulation procedure was carried out with equal stimulation of both hemispheres). This was taken as potentially indicative of hemispheric dominance, as was confirmed by subsequent Wada testing (Knecht et al. 1998). Side-to-side differences in CBFV responses during cued word generation tasks showed perfect concordance in determining activated lateralized language areas when compared with intracarotid amobarbital anesthesia (Wada test) in patients being evaluated for epilepsy surgery. Another group (Cupini et al. 1996) showed lateralized activation during working memory tasks; Klingelhöfer et al. (1997) also reported noninvasive assessment of hemispheric language dominance, based on perfusion asymmetries during complex spatial tasks. Finally, TCD vasoneural testing for predicting hemispheric dominance was compared with similar fMRI studies and provided perfect matching (Knecht et al. 1999, Deppe et al. 2000).

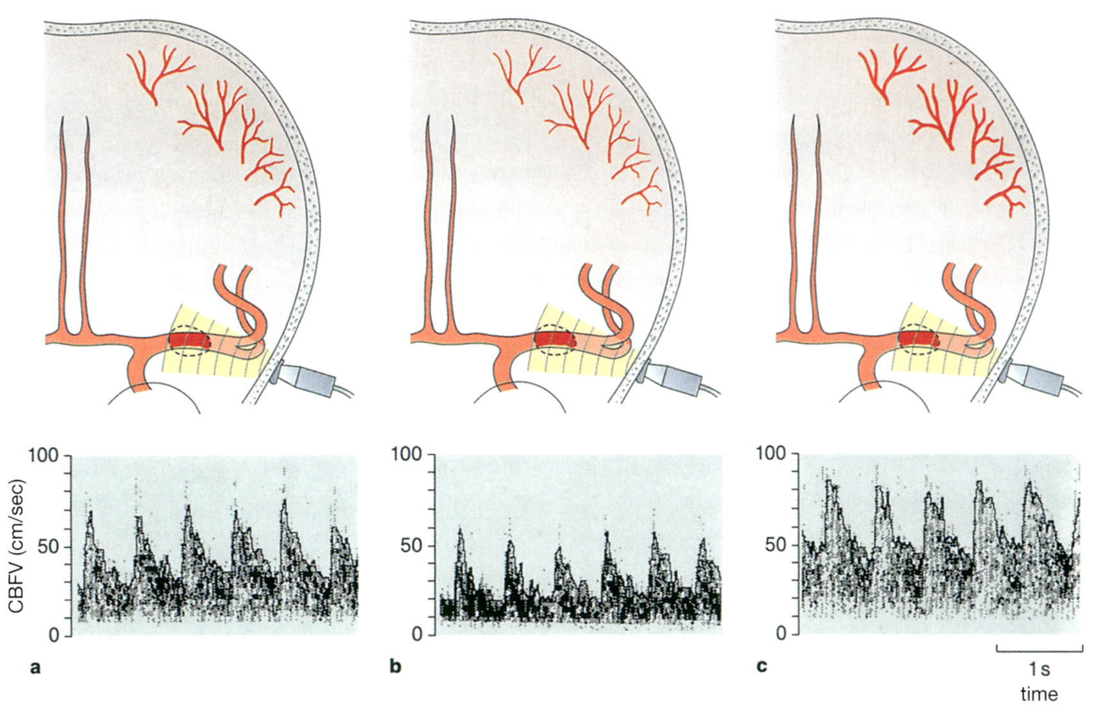

Fig. 3.**14** A diagram representing the CO_2 reactivity principle (adapted from Hassler 1986), with the Doppler spectra of the MCA in the typical position (50 mm deep, with flow toward the probe). Data obtained from a healthy control individual for different end-expiratory P_{CO_2}. **a** 3.8 % (normocapnia). **b** 3.3 % (hypocapnia). **c** 4.8 % (hypercapnia)

Techniques

Bilateral simultaneous TCD registration should be used to test vasoneural coupling. For visual stimulation, the PCA should be insonated whenever possible at the P_2 segment that mainly cover the brain areas involved in visual processing. In hemispheric stimulation the insonated MCA segment depends on the stimulation paradigm (e.g., more proximal MCA insonation should be used for the speech production test in the Broca center, while more distal insonation is optimal for language reception tasks in the Wernicke center). In general, insonation should be located in the most distal point available in the artery supplying the cortical area stimulated by the paradigm.

The position of the patient should be adjusted to the paradigm. After the TCD monitoring probes have been fixed and the registration has been started, CBFV should be monitored for at least 5 min to allow adaptation to the test situation and to ensure stable baseline conditions. After this, stimulation is started. Since stimulation-related CBFV changes are small in comparison with systolic–diastolic cycles and spontaneous oscillations (about 5–10 % in the MCA and 20–30 % in the PCA), averaging is useful. The most commonly used procedure is to present the stimuli for approximately 20 s, followed by a 20-s off-phase. On and off cycles should be repeated at least five times and should not be extended to more than 20 trials, because of response habituation.

Normalization of CBFV changes can be assessed using values relative to a figure of 100 % set at the end of the 20-s off-phase or in the last minute of the baseline registration. The evoked CBFV response is represented by the change from baseline values occurring after a maximum increase in CBFV (off stimulation, usually 10–15 s after stimulus presentation).

Transcranial Monitoring (TCM)

Due to its noninvasive nature and very high time resolution, TCD is also used as a *monitoring procedure* during acute stroke treatment, surgery, and neuroradiological interventions (e.g., examining the cranial circulation carotid and cardiopulmonary bypass surgery or angioma embolization). Care should be taken to ensure that parameters such as blood pressure, pulse, P_{CO_2}, and metabolism (blood sugar, hematocrit, or heart rates during surgery), which can significantly influence flow velocity, remain constant; they should be monitored and kept under control. Spontaneous variations of up to 30 % are physiological (Diehl et al. 1991). Monitoring is attracting increasing interest in stroke and intensive-care units, but caution should be exercised, as it is not known whether many hours of monitoring at may have bio effect with potentially inherent risks. The individual examination technique depends on the specific clinical or scientific issue being investigated. Discontinuous monitoring—e.g., monitor-

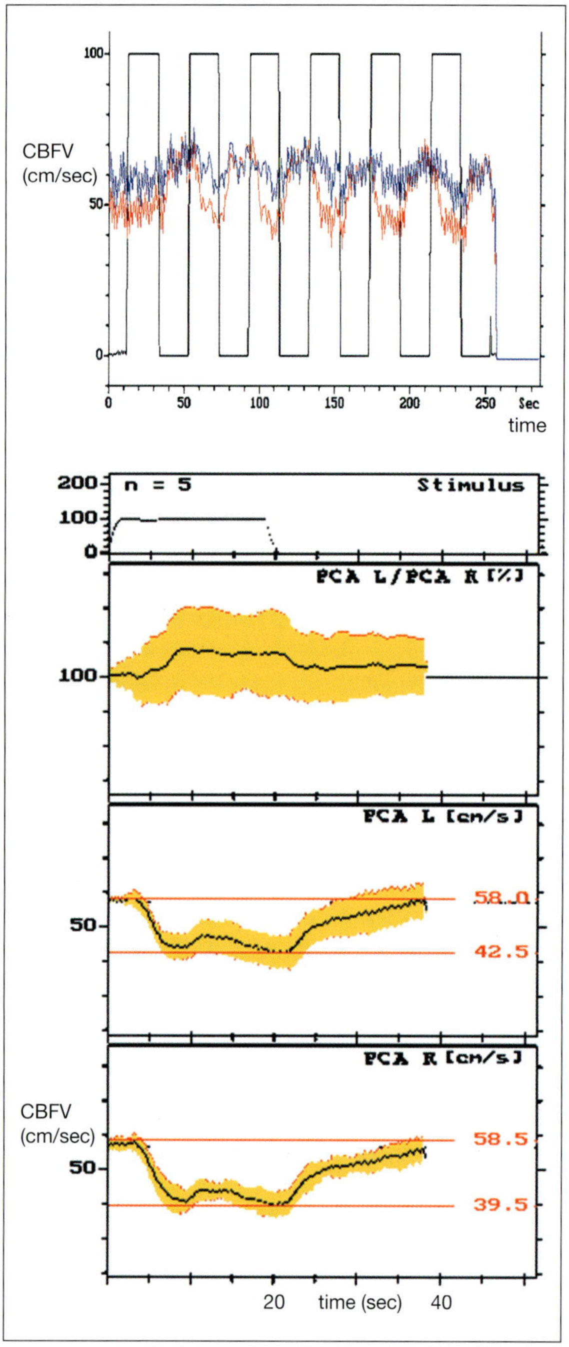

Fig. 3.**15** (A) CBFV Registration of both PCA's during visual stimulation (20 sec eyes open and stimulation with a movie vs 20 sec eyes closed) in a patient with migraine in the interictal interval. (B) Mean average curve (black) and SD (orange) of 5 cycles with visual stimulation on/off. Upper tray ratio between CBFV of the PCA right to left. Middle tray left PCA, lower tray right PCA. Mean CBFV increase as well as SD of the single responses are increased on the left side. The ratio (left/right) showed a high SD.

ing for vasospasm in subarachnoid hemorrhage (SAH) or inflammation—can be carried out with a handheld device. Long-term monitoring to measure flow changes, autoregulation, microemboli, etc., means that the probe have to be fixed using different types of head band or head frame. In principle, monitoring can be carried out with either Doppler systems or duplex systems. Currently, long-term monitoring is mainly conducted using Doppler systems (as they are much more practicable). All vessels that can be identified using transcranial ultrasound can theoretically also be monitored, but monitoring mainly focuses on the MCA, as MCA insonation is simplest and the situation is particularly stable. The need for stability when monitoring is the most important limitation. Stability mainly depends on the bone window (insonation quality) and the fixation device. Bilateral insonation is recommended for most applications (e.g., monitoring for recanalization).

Originally, during intravenous and intra-arterial injection of microscopic air bubbles (Rautenberg et al. 1987, Gerraty et al. 1996) and surgical procedures, as well as in patients with artificial heart valves, high-amplitude signals were detected (high-intensity transient signals, HITS) (Hennerici 1995) that were initially interpreted as signs of cerebral emboli (Spencer et al. 1992). There is clear evidence that HITS do indeed often reflect microembolic signals (MES), but they may also be due to artifacts or sources that do not cause cerebral ischemia, such as microcavitations in artificial heart valves (i.e., transient irregularities in blood flow without formation of structural emboli, which can be distinguished from nongaseous emboli by means of oxygen inhalation (Droste et al. 1997). The true clinical significance of findings suspected to be associated with MES, such as spontaneous HITS within intracranial blood vessels distal to carotid stenoses, is at present still unclear, although large numbers of reports have been published in recent years (Hennerici 1994, Daffertshofer et al. 1996, Nabavi et al. 1996) and despite many attempts to increase the validity of TCD spectral analysis to distinguish between artefacts and genuine microembolic signals (Cullinane et al. 2000, Mess et al. 2000, Moehring and Spencer 2002, Russell and Brucher 2002).

Ultrasound Contrast Administration

Two types of ultrasound contrast medium are available, based on different principles: agents that pass the pulmonary capillaries and those that do not.

Ultrasound contrast that does not pass the lung capillaries (right-to-left shunt testing). Administering ultrasound contrast media that do not pass the lung capillaries can illuminate potential routes of paradoxical embolism. For example, paradoxical embolization through a patent foramen ovale of the heart, which may cause cerebral ischemia (Lechat et al. 1988, Schminke et al. 1995, Zanette et al. 1996) can be imi-

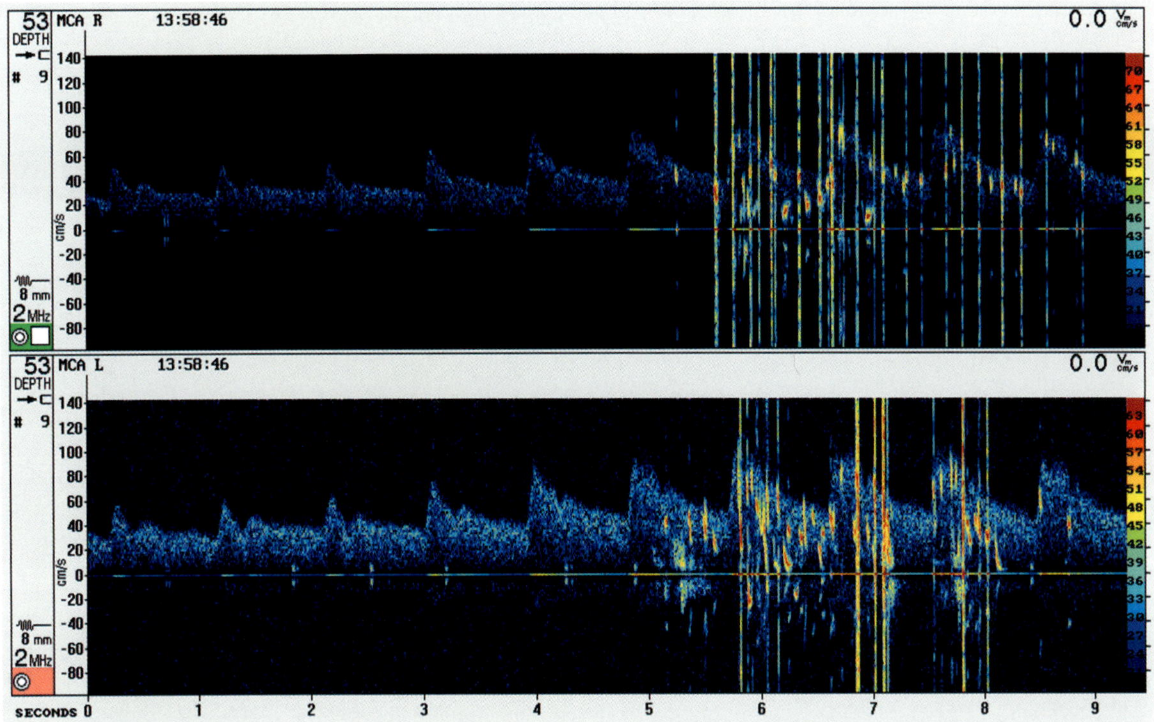

Fig. 3.**16** Bilateral recording of the Doppler spectrum over the middle cerebral artery (MCA) after administration of an ultrasound contrast agent that does not pass the lung capillaries (Echovist 300®). Six seconds after the injection, high-intensity transient signals (HITS) are seen over both MCAs, providing evidence of a right–left shunt

tated during transesophageal echocardiography (TEE) and/or TCD studies if ultrasound contrast media are administered to follow this pathway with a high degree of sensitivity (Teague and Sharma 1991, Job et al. 1993, Nikutta et al. 1993) (Fig. 3.**16**). Continuous bilateral TCD monitoring is recommended when testing for paradoxical embolism. The contrast should be injected into a peripheral vein, usually the cubital vein. If there is a right-to-left shunt, HITS can be observed after a few seconds. After a short time—e.g., 1 min—the procedure should be repeated while the patient does some form of exercise to increase thoracic (venous) pressure (e.g., a Valsalva maneuver or coughing).

Ultrasound contrast that passes the lung capillaries. Ultrasound contrast agents that pass the lung capillaries are at present mainly used to enhance the image quality (Droste et al. 2000). In addition, small encapsulated microbubbles can be used to measure brain perfusion in special settings and may be able to provide carriers for local administration of drugs or genetic material (see Chapter 1). In contrast to other vascular territories—e.g., the extracranial arteries—ultrasound contrast application at the transcranial level has a substantial impact, particularly when there is a suboptimal or poor transcranial bone window. Contrast application often allows analysis of both color flow imaging and Doppler signal analysis of intracranial arteries that are otherwise not visible (Hansberg et al. 2002, Zunker et al. 2002).

Normal Findings

Principle

In *pulsed-wave Doppler sonography,* the sample volume is usually positioned between 25 mm and 140 mm, proceeding in 1–5 mm increments (sometimes variable). Normal Doppler spectrum values refer to a healthy control group and vary for different arteries and selected vascular segments in different studies. Our own data (Table 3.**1**) correspond well with the results obtained by other groups (Aaslid et al. 1982, Arnolds and von Reutern 1986, Harders 1986, von Reutern and Büdingen 1993). Overall, they indicate a certain age dependence of flow velocities, which sometimes reach significance in the MCA and show clearly lower average values for the posterior circulation in comparison with the intracranial carotid region. It should be noted that the mean values used to calculate the standard data may have been obtained from different arterial segments, due to interindividual variations in arterial length and width. Also, it is not always possible to take into account the optimal flow signal in groups while maintaining a consistent location for the measuring point. Finally, topical asymmetries due to different angles between the ultrasound and vascular axes also play a role and can be misleading if falsely interpreted from multigate pulsed Doppler instruments (Bäzner et al. 1996, Mess et al. 1996). Such problems,

Table 3.**1** Mean values and standard deviations of Doppler parameters from spectral waveforms recorded in 50 adult control individuals

Arteries		Peak systolic velocity (cm/s)	Mean velocity (cm/s)	Range of normal values (mean velocity) (cm/s)
ICA	Siphon (60 mm) Transorbital	81.0 ± 16.1	52.3 ± 11.4	20–77
ACA	(70 mm) Transtemporal	78.7 ± 19.1	48.6 ± 12.5	18–82
MCA	(50 mm) Transtemporal			
	< 60 years of age	92.7 ± 15.3	58.1 ± 10.0	32–82
	≥ 60 years of age	78.1 ± 15.0	44.7 ± 11.1	18–64
PCA	(60 mm) Transtemporal	54.8 ± 14.6	33.6 ± 8.9	16–58
BA/VA	(75 mm) Transtemporal	55.6 ± 14.5	33.9 ± 10.6	12–66

ACA: anterior cerebral artery; BA: basilar artery; ICA: internal carotid artery; MCA: middle cerebral artery; PCA: Posterior cerebral artery; VA: vertebral artery.

Table 3.**2** Position of the sample volume and flow direction for identifying the cerebral arteries

Depth	Flow direction		Artery
60–70	←[	C_2 segment	ICA
	→[	C_4 segment	
60–75	←[		ACA
30–60	→[		MCA
55–80	→[	P_2 segment	PCA
	←[	P_1 segment	
70–110	←[		VA/BA

ACA: anterior cerebral artery; BA: basilar artery; ICA: internal carotid artery; MCA: middle cerebral artery; PCA: Posterior cerebral artery; VA: vertebral artery.

which were particularly inherent in pure Doppler studies, can be minimized with modern duplex systems. Angle correction alone, however, is probably unlikely and in many segments is inadequate. That is because of the inability to display most of the intracranial arterial segments in the long axis.

Technical problems and artifacts caused by the methodology used also play a role (see Chapter 1). For the sake of clarity, only *the systolic peak velocity and the mean velocity* with their calculated averages and standard deviations are given here. Whereas the peak flow velocity is best depicted with the Doppler fast Fourier transform technique, mean flow velocity has been shown to mirror regional CBF more closely than peak or modal velocities (see Glossary, pp. 349–351). The lower and upper boundary values observed for the average flow velocity are also listed (Hennerici et al. 1987a).

The position and depth of the sample volume and the direction of the Doppler signal relative to the ultrasound probe in the large basal arteries of the adult brain are given in Table 3.**2**. Due to the differing size relationships, significantly varying values can be expected in infants and small children.

Anatomy and Findings

Carotid Siphon—Ophthalmic Artery

Anatomy

After leaving the cavernous sinus, the ophthalmic artery is the first main branch of the ICA. Before the origin of the ophthalmic artery, the diameter of the ICA is 3.3–5.4 mm, and distal to this it is 2.4–4.1 mm; this is a physiological narrowing of the lumen, and does not represent an obstruction. The ophthalmic artery has a diameter of 0.7–1.4 mm, and follows a course some 7.9 mm long that is intracranial, intracanalicular, and intraorbital (Krayenbühl et al. 1979) (Fig. 3.**6**).

Findings

Transorbital ultrasound. Applying ultrasound to the ophthalmic artery, which is located at a depth of between 20 mm and 60 mm at the roof of the orbits, usually shows a low Doppler flow signal directed toward the probe.

At its origin in the ICA one can sometimes register mixed signals in both directions (toward the ultrasound probe and away from it). By moving the ultrasound probe slightly, or correcting the incident angle, a characteristic flow curve is recorded, with a corresponding audio signal in the C_3/C_4 segment of the ICA. The audio signal is recognizable by its high diastolic flow velocity, with a slight systolic amplitude modulation. Color flow duplex sonography can be used for

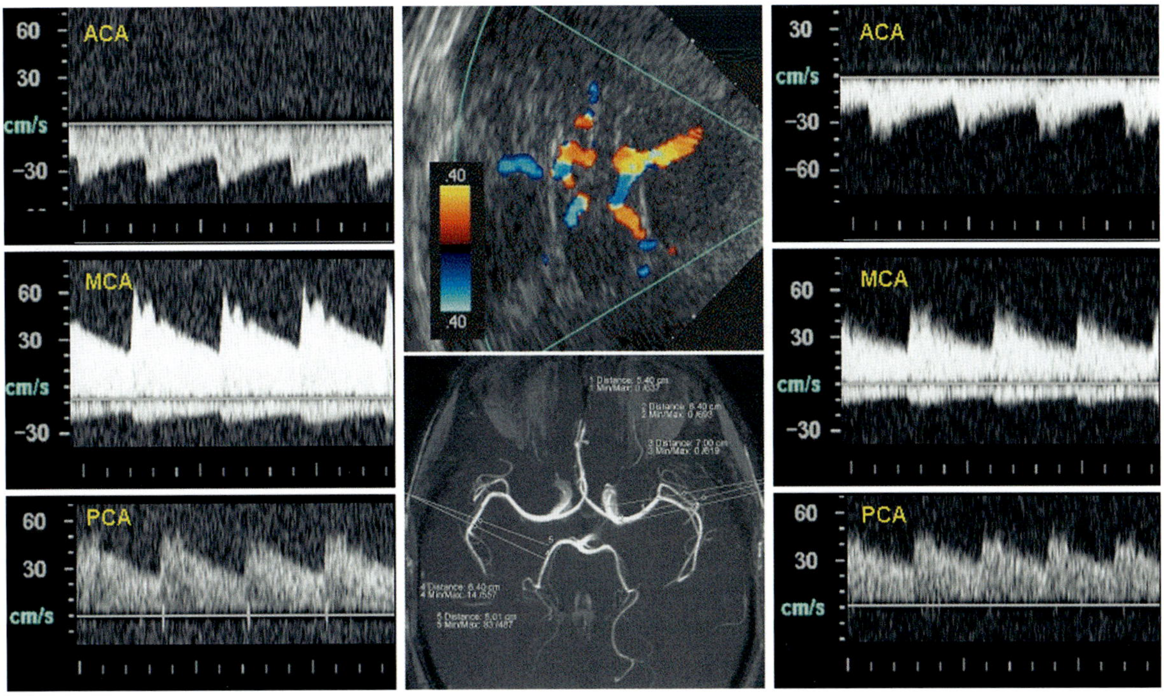

Fig. 3.**17** Display of the circle of Willis using transcranial duplex sonography. *Central column:* a color map of the circle of Willis (top) from a right temporal insonation position (flow to the probe is shown in red and flow away from the probe is shown in blue). Magnetic resonance angiography (MRA; bottom) is shown for comparison of the anatomical situation. Duplex scanning delineates the communicating posterior arteries on both sides; these segments are not displayed on the MRA, but MRA provides additional information regarding the distal branches. *Left column:* flow velocity recordings from the main proximal brain-supplying arteries on the left. *Right column:* flow velocity recordings from the main proximal brain supplying arteries on the right

selected studies in patients with suspected retinal embolism (Hedges 1995).

Transtemporal ultrasound. The carotid siphon and the T-junction (Figs. 3.**2**, 3.**17**, 3.**18**) can also be displayed using the transtemporal bone window in cross-section, orienting the probe caudally.

Middle Cerebral Artery

Anatomy

After its origin from the internal carotid artery, the MCA runs laterally along the wing of the sphenoid bone into the Sylvian fossa over a path 5–30 mm long (average 14–16 mm) (Herman et al. 1963, Jain 1964). The interior lumen of the blood vessel is 3–5 mm in diameter (M_1 segment: sphenoidal part). Usually, a large number of small arteries originate in this segment (the lenticulostriate arteries). Branching off in two or three segments at the level, distal long superficial arteries support the superficial cortical and deep perforating arteries.

Findings

Using the temporal bone window, the main branch of the MCA (M_1 segment) can be examined step by step with ultrasound over a length of 30–60 mm (Fig. 3.**18**). Since the artery usually divides into two or three main branches, or into a double bifurcation at a distance less than 30 mm from the ultrasound probe, and since its origin from the ICA is variable (usually 50–65 mm deep), normal values for the flow signals refer to a specific vascular segment (around 50 mm) (Table 3.**1**). It should be noted that significant changes in the Doppler frequencies are often only an expression of a loop formation in this region. Also, due to bending or crossing vascular segments, an impression of bidirectional flow may result. As a means of differentiating retrograde perfusion through leptomeningeal anastomoses, it is helpful to examine this region from several sample volume positions. The distal vascular segments (M_2 segment: insular part; M_3 segment: opercular part; M_4 segment: terminal part) are sometimes difficult and often even impossible to examine, due to the unfavorable incident angle of insonation. *Color flow duplex imaging* is helpful for identifying the individual intracerebral course of the MCA and improving the accessibility of the M_2, and even M_3 branches, particularly if ultrasound contrast media and/or amplitude energy

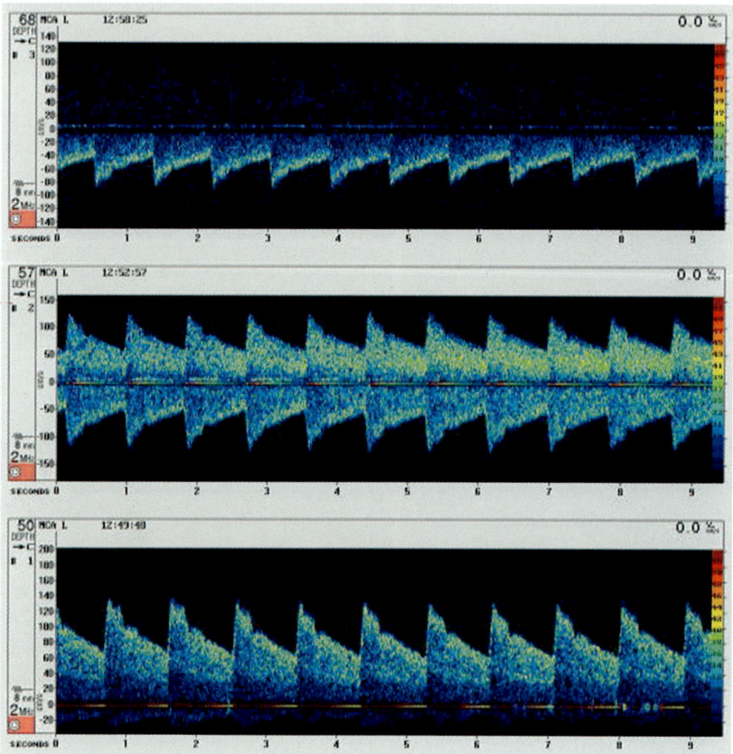

a b c

Fig. 3.**18** Doppler spectra from transtemporal examinations of the anterior cerebral artery (**a**), the T-junction of the internal carotid artery (**b**), and the middle cerebral artery (**c**). All of these spectra were obtained from a healthy control individual. At a typical sample volume position, the anterior cerebral artery shows flow away from the probe, while the middle cerebral artery shows flow toward the probe . A flow signal in both directions is seen near the T-junction of the internal carotid artery

a Anterior cerebral artery (depth 68 mm)
b Internal carotid artery, T-junction (depth 57 mm)
c Middle cerebral artery (depth 50 mm)

Table 3.**3** The calculated mean pulsatility indices (PI; ± one standard deviation), based on 50 control individuals, for comparing Doppler spectra intra-individually and side to side

Arteries		Average PI	(± SD)
MCA	Right	0.90	(0.24)
	Left	0.94	(0.27)
ACA	Right	0.78	(0.15)
	Left	0.83	(0.17)
PCA	Right	0.88	(0.23)
	Left	0.88	(0.20)

ACA: anterior cerebral artery; MCA: middle cerebral artery; PCA: Posterior cerebral artery.

modes are used (Kenton et al. 1996, Postert et al. 1997). In addition, the T-junction, the ICA and the ACA can often be displayed simultaneously, facilitating the evaluation of the anterior circle of Willis and avoiding misinterpretation (Fig. 3.**18**).

Systolic diastolic pulsatility is mainly influenced by the peripheral resistance, which is an important mechanism in compensating for stenosis or other hemodynamic compromise but may also be altered in case of inflammation, intracranial pressure (ICP) changes, or metabolic disorders. Several indices (see Glossary, pp. 351–352) have been developed to quantify systolic diastolic ratio. Table 3.**3** shows our normal values for the pulsatility index.

Anterior Cerebral Artery and Anterior Communicating Artery

Anatomy

In the bifurcation of the ICA, the ACA initially follows a horizontal path (A_1 segment = precommunical part = horizontal part) before it turns in a cranial direction as the pericallosal artery (A_2 segment) (Fischer 1938, Krayenbühl et al. 1979). The internal diameter of the ACA is 1–3 mm; a diameter of less than 1 mm indicates hypoplasia (Wollschläger and Wollschläger 1966):

- *Unilateral* in 4.1–14.0 %
- *Bilateral* in 3.2 % (Wollschläger and Wollschläger 1974, Krayenbühl et al. 1979)
- *Aplasia* in 0.7–1.1 % (Tönnis and Schiefer 1959, Krayenbühl et al. 1979).

As the connecting link between both ACAs (anterior circle of Willis), the anterior communicating artery is the shortest cerebral artery (0.1–3.0 mm) and has significant variants (Figs. 3.**17**, 3.**18**). The frequencies of hypoplasia and aplasia are 0.5–1 % and 6–8 %, respectively (Sedzimir 1959, Krayenbühl et al. 1979).

Findings

After the T-junction of the internal carotid artery has been identified (Fig. 3.**18**), the ACA can be detected at 60–75 mm, with flow moving away from the probe. By increasing the depth adjustment, the contralateral ACA can sometimes be registered for a distance of 70–

75 mm. It can be recognized by a change in the flow direction (toward the probe).

Flow signals from the *anterior communicating artery* cannot usually be documented in isolation. They are similar to those obtained from the posterior communicating artery, and reflect an intermediate position between the two A_1 segments, at a low signal amplitude.

Color flow duplex sonography facilitates the depiction of the anterior circle of Willis (Fig. 3.**17**). It simplifies identification of the ACA and significantly enhances the display of anatomical variants (e.g. A_1 segment aplasia). However, due to the often curving and winding arterial course of the vessel, continuous imaging in a single sectional plane is more difficult than in the MCA if velocity modes alone are used. Using amplitude modes (power imaging) improves the validity of the display considerably (Kenton et al. 1996, Postert et al. 1997). Selective display of distal segments of this blood vessel that are not detectable with pulsed-wave sonography alone is also possible. Display of the A_2 segment is often poor due to the unfavorable insonation angle when using the transtemporal bone window. Particularly with color duplex systems, a frontal window can sometimes be used, which provides a better display of the A_2 segment (Stolz et al. 1999b).

Normal Values

Due to physiologically occurring variants in this segment, in particular, the normal values at a distance of about 70 mm from the probe (Tables 3.**1**, 3.**3**) fluctuate, and show no significant age dependence.

Posterior Cerebral Artery and Posterior Communicating Artery

Anatomy

Phylogenetically and ontogenetically, the PCA originates from the ICA. Its connection with the basilar artery is a later development. Anatomical examinations in humans have shown that in 10–30% of cases, the PCA originates directly from the ICA (Sunderland 1948, von Mitterwallner 1955). Normally both PCAs course as the terminal branches of the basilar artery around the brainstem (0.5–1 cm; P_1 segment) until they join the *posterior communicating artery*, which is a branch of the internal carotid artery. In their further course (P_2 segment), they lead into the calcarine and parieto-occipital arteries.

Findings

The supratentorial segment of the PCA can be detected at a depth of 60–80 mm. The proximal segment (P_1 segment) usually lies slightly deeper, and its signal is directed toward the probe (Figs. 3.**17**, 3.**19**). In the distal section (P_2 segment), the direction of flow is away from the probe (Fig. 3.**19**).

Doppler signals recorded at favorable ultrasound application angles should be compared with the calculated normal values (Tables 3.**1**, 3.**3**). The contralateral PCA can be evaluated at considerable depth (usually beyond 80 mm) while maintaining the specific angle used to examine the ipsilateral PCA. The flow direction reverses in relation to the ipsilateral P_1 segment, and is oriented away from the probe. Flow signals from the *posterior communicating artery* cannot usually be documented in isolation. They show an intermediate direction with low amplitude. This is firstly because in normal conditions, flow in the posterior communicating artery may be very low or even absent; and secondly, because the insonation angle is very poor with transtemporal insonation. If the frontal bone window is open, the angle for the posterior communicating artery is better (Stolz et al. 1999b).

In normal conditions, the *posterior communicating artery* can often only displayed during compression tests.

It is possible for the superior cerebellar artery to be confused with the PCA. The latter lies cranial to the superior cerebellar artery, and often shows signal modulation after the eyes are opened, due to changes in neuronal activation in the occipital lobe (Fig. 3.**19**).

Vertebral Artery and Basilar Artery

Anatomy

The BA, which has an average diameter of 4.1 mm and a length of 33.3 mm (Busch 1966), runs cranially from the junction of the two vertebral arteries at the clivus. It has many variants (Krayenbühl and Yasargil 1957, Busch 1966), due to embryological factors in its development and also, in older patients, as a result of arteriosclerotic, ectatic changes. Both the cranial bifurcation and the caudal junction are highly variable in location (Lang 1991). The course of the artery winds repeatedly, and it is sometimes markedly elongated (the so-called megadolichobasilar artery). The vertebral arteries and the origins of the cerebellar arteries have a wide range of variations (Krayenbühl and Yasargil 1957).

Findings

Approaching from the occipital direction, one can assess the BA in the midline or paramedially, and follow it for a longer distance by applying ultrasound through the great foramen or a thin bony access point at the posterior cranium (Fig. 3.**7**). Near the loop formation of the vertebral arteries around the atlas, opposed signals similar to those made by crossing arteries, can be registered (PICA, AICA, etc.).

The diagnostic accuracy for identifying the vertebral artery and distinguishing it from the BA is much better when using a color duplex system, as the junction of the vertebral arteries forming the basilar artery can often be displayed from a transnuchal approach (Fig. 3.**7**). Particularly at the vertebrobasilar level, diag-

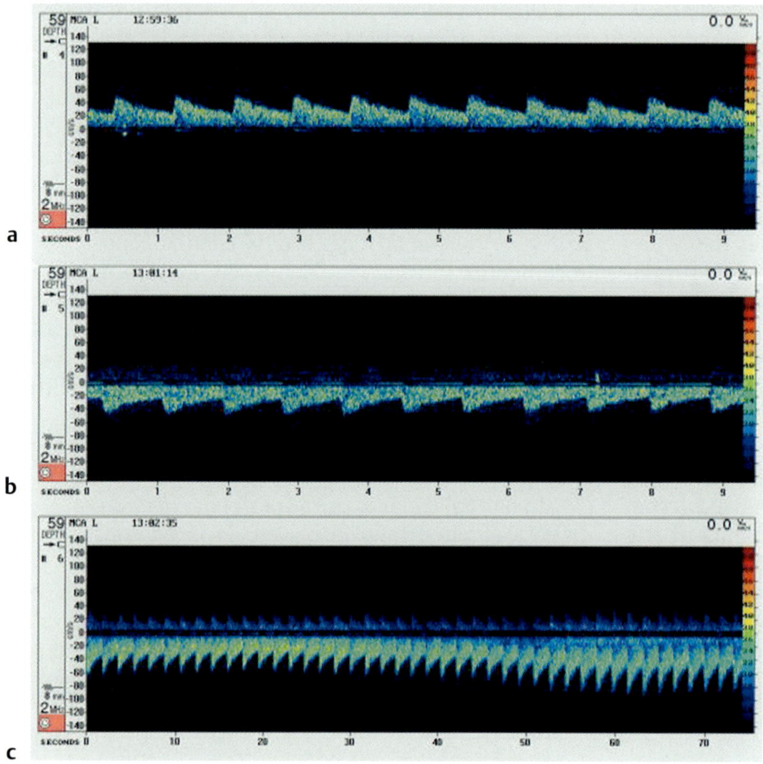

Fig. 3.**19** Doppler spectra from transtemporal examinations of the posterior cerebral artery, P_1 segment with flow towards the probe (**a**) and P_2 segment (**b**) with flow away from the probe. (**c**) Increase in cerebral blood flow velocity (CBFV) of more than 5 % during visualstimulation (open the eyes) identifies the PCA; the maximum CBFV increase is seen in the P2 segment

nostic confidence can be markedly increased using ultrasound contrast agents, which provide adequate diagnosis with duplex examinations in more than 90 % of cases (Stolz et al. 2002b).

Functional Tests

Compression Tests

Compression tests are helpful for identifying the insonated arterial segment in TCD examinations. Compression tests can be used to illuminate patent collateral pathways and are particularly helpful in unmasking feeding arteries in cerebral arteriovenous malformations. Routine use of these tests has become less important since the introduction of color flow duplex sonography and studies using noninvasive imaging methods (MRA/CTA). In addition, since ischemic episodes during compression tests have been reported (Mast et al. 1993), the indication for compression tests should be restricted.

A compression test involving both the ipsilateral and the contralateral CCA is important in the functional assessment of the ACA, but can almost always be dispensed if color flow duplex imaging provides good visualization. Compression may be helpful for analyzing the patency of the collateral pathways even when using color-coded duplex systems, since flow velocities can often be obscured in the communicating arteries in normal hemodynamic conditions. During ipsilateral compression of the CCA, the flow direction in the ACA typically reverses, while contralateral compression leads to increasing flow velocities if the anterior communicating artery is intact. If a change in flow direction is absent or only a decrease is observed, then collateral blood flow is either being maintained through the posterior communicating artery or the anterior communicating artery is missing or hypoplastic. Flow signals in the MCA remain stable or decrease depending on the collateral capacity of the communicating arteries. During compression, signs of *functional stenosis* often appear when the Doppler sample volume is located within the communicating arteries.

When the ipsilateral CCA is compressed, it is often possible to clarify whether the PCA originates directly from the ICA or from the BA. In the first case, the flow velocity is reduced. The examination is easiest in the P_2 segment. If the PCA is supplied through the basilar artery and the posterior communicating artery is intact, an increase in flow velocity in the P_1 segment occurs. In this instance, care should be taken to examine the P_1 segment in order not to miss this collateralization (Fig. 3.**5**).

Vasomotor Reactivity Test

Many authors report normal findings for vasomotor reactivity using CO_2 and acetazolamide (Bishop et al. 1986, Ringelstein et al. 1986, Piepgras et al. 1990, Widder et al. 1994, Kastrup et al. 1998).

When the blood supply to the brain changes due to hypocapnia (decreased CO_2), the flow velocity in the intracranial cerebral arteries decreases correspondingly. In hypercapnia (increased P_{CO_2}), the flow velocity increases. From very low end-expiratory P_{CO_2} values (1.0–3.5%), the cerebral tissue perfusion increases only slowly to return to normocapnia. In the middle range (3.5–5.5%), there is an almost linear dependence, and in the upper P_{CO_2} zone (above 5.5%), a flattening of the overall sigmoidal relationship results (Sitzer et al. 1992). The pathophysiological cause of this phenomenon is related to increasing P_{CO_2} values during hypoxia, which in turn lead to dilation of the brain's arteriolar blood vessels. This is combined with a further increase in the cross-sectional area of the large cerebral arteries (Poulin and Robbins 1996). Within the preceding larger cerebral arteries that are examined with Doppler sonography, the CBFV increases disproportionately, reflecting the reduction in the peripheral vascular resistance and thus compensating for the hypoxic situation. When hyperventilation takes place, the blood's P_{CO_2} value falls, and the small cerebral arteries in the periphery counterregulate this by constricting. The preceding large intracranial cerebral arteries then show a reduction in their flow velocities, with reduced diastolic phases and low end-systolic flow velocities, as happens when there is increased intracranial pressure. Similar phenomena are found in cerebral autoregulation as a response to changes in the systemic blood pressure. However, the terms for these two mechanisms, which are qualitatively and quantitatively different, should not be used interchangeably as if they were truly identical processes.

There is the relatively broad variability between individuals in the flow velocity observed during different P_{CO_2} changes.

Reported normal values (see below) can only serve as an approximation and are useful for validating normal ranges in the individual laboratory. The importance of developing laboratory-specific normal ranges, rather than relying on data from other laboratories for this method, is mainly due to the complexity of the methodological set-up, which requires not only TCD but also CO_2 measurements—which of course crucially depend on the individual methods and devices used.

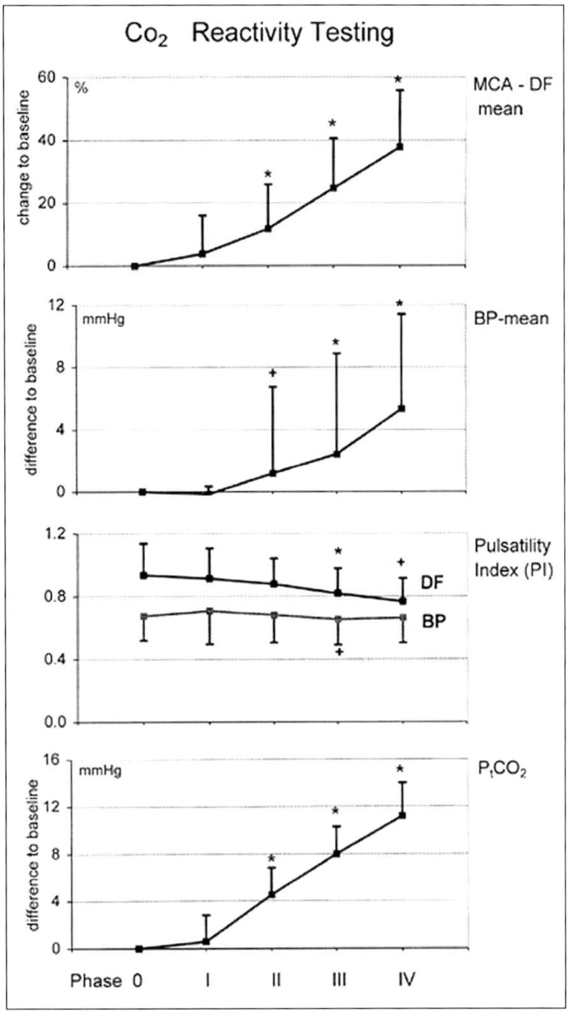

Fig. 3.**20** Multimodal monitoring during CO_2 reactivity testing with simultaneously recorded blood pressure (BP) (in mmHg), Doppler frequency over the MCA MCA-DF (kHz), heart rate (bpm) and PETCO2 (mmHg). Relative changes compared with baseline values are calculated for all parameters. Data are displayed as mean ± SD The symbols indicate significant differences between phases in Wilcoxon's test for paired samples (+P<0.01, *P<0.001).
(from Hetzel et al. Stroke, 1999; 30:398–401

CO_2 Reactivity Tests

CO_2 inhalation. There is a linear correlation between P_{CO_2} and MCA flow velocity (Markwalder et al. 1984, Bishop et al. 1986, Ringelstein et al. 1988, Widder 1989, Widder et al. 1994). It can be difficult to distinguish normal from pathological findings, due to large ranges of standard deviation in changes in blood pressure and CO_2 during CO_2 reactivity testing (Fig. 3.**20**). The common Doppler CO_2 test should therefore be interpreted very carefully, as recommended by Widder et al. (1994). In their opinion, only exhausted CO_2 reactivity is a significant finding.

The high variance in the MCA flow velocity increase to hypercapnia is based on the fact that CO_2 also affects blood pressure, which in turn affect flow velocity (Kastrup et al. 1998, Hetzel et al. 1999). The parallel increase in blood pressure, more than the heart rate, is induced by activation of the central sympathetic nervous system (Braune et al. 1997). In some individuals, the effect of blood pressure reaches the level of CO_2-induced changes. The problem of variable CO_2 effects on flow velocity and blood pressure can lead to misinterpretation of the CO_2 test results if not all relevant parameters are monitored. The effect of blood pressure on the MCA Doppler shift is not linear. It might be explained by the high-pass filter properties of cerebral autoregulation. Depending on the frequency of blood-pressure oscillation, phase displacement and potentially a rise in the amplitude of flow velocity oscillations was observed by Diehl et al. (1991). There have been attempts to understand the effects of CO_2-related blood-pressure changes using mathematical models (Piechnik et al. 1999). However, due to the complex relationships involved, there is some concern that simple mathematical methods may not be able to account for the irregular variations in blood flow velocities caused by variations in blood pressure.

Rebreathing. The considerations regarding normal values discussed above in relation to CO_2 inhalation techniques also apply to all methods of measuring VMR capacity with CO_2-stimulated blood flow velocities. With regard to the rebreathing technique, Diehl et al. (Diehl and Berlit 1996) report a normal VMR of 5.26 ± 1.61 %/mmHg and a pathological threshold of < 2.15 %/mmHg. A side-to-side difference of > 2.90 %/mmHg was also shown to indicate pathological VMR.

Apnea test. The normal range of the breath-holding method was 1.2 ± 0.6 %/s. This means that even normal individuals may have very low breath-holding indices. The specificity of the test is probably not very high. Breath-holding indices of less than 0.5 %/s can be classified as suspicious results. Because of the low specificity, however, all authors recognize that this test is a screening method that should be followed by the standard CO_2 stimulation testing if there are any abnormal results.

Measuring vasomotor reactivity with acetazolamide. In healthy individuals, the CBFV increases after the administration of acetazolamide (Dahl et al. 1992a, b). Studies in normal individuals have shown a uniform increase in the CBFV in both MCAs. The increases in the mean MCA blood flow velocity were 23.2 ± 1.8 cm/s and 24.7 ± 1.9 cm/s on the right and left sides, respectively (Dahl et al. 1994). There was no significant side-to-side asymmetry in the increase in CBFV in the MCA after acetazolamide administration. These normal ranges refer to a dosage of 1 g acetazolamide given intravenously, with an observation interval of at least 20 min. Normal values for the ACA, PCA, and BA have not yet been reported.

Autoregulation Tests

Static and dynamic autoregulation. There are many different approaches to the measurement of autoregulation using transcranial Doppler sonography. Normal ranges for static and dynamic autoregulation crucially depend on the paradigm used and should therefore be established on the basis of the individual situation.

Analyzing autoregulation by applying blood-pressure cuffs adjusted to suprasystolic values on both lower extremities is a method that has been used since the very start of autoregulation measurements with TCD (Aaslid et al. 1989). The disadvantage of this method is that it can only be used in the setting of an intensive-care unit or in experimental studies with cooperative and healthy patients. After a steady-state situation has been reached with the patient recumbent, a more or less noticeable reduction in the flow velocity of the middle cerebral artery follows any change in the equilibrium of the systemic blood pressure. Usually, the MCA is used as a representative of all the cerebral arteries. Normally, the reduction in the average CBFV in this phase does not exceed 20 % of the resting value. In general, changes in CBFV in the MCA correlate well with systemic blood-pressure values.

The interpretation of autoregulation results is affected by a significant limitation, however. The assumption that the vascular diameter of the cerebral arteries examined remains constant during such changes in the systemic blood pressure is disputed. Angiographic studies that failed to demonstrate any significant changes, for reasons probably related to the limited accuracy of the method (Huber and Handa 1967), contradict experimental results suggesting delayed dilation (Fujii et al. 1991). With the help of transcranial duplex measurements, this may be investigated further if appropriate conditions are selected.

Testing orthostatic regulation with TCD. Normal ranges in the CBFV decrease during tilt-up are reported to range from 20 % to 30 %. Autoregulation can be estimated by analyzing the relationship between the relative CBFV and blood-pressure decrease (e.g., CBFV reduction = 25 %, blood-pressure reduction = 15 %; quotient = 1.67). If this quotient is greater than the 97.5 % percentile of normal values, a disturbed autoregulation can be estimated. In a recent study including 50 normal individuals, the cut-off-value for this quotient was calculated as 1.40 (Diehl et al. 1998).

Vasoneural Coupling

Various studies of the link between perfusion and neuronal metabolism have shown that, with specific stimuli, changes can be measured in the CBFV within the large arteries of the brain. This type of investigation was previously only possible using nuclear medicine techniques such as PET or fMRI, which demonstrate perfusion of the brain or changes in metabolism with reasonable and good spatial resolution, but with poor temporal resolution in active and resting conditions. Examinations conducted while the subject is reading, counting, writing, undergoing emotional activation, etc., show an increase in the cerebral blood flow in the MCA (Klingelhofer et al. 1997, Knecht et al. 1997, Schmidt P et al. 1999, Stoll et al. 1999a, Troisi et al. 1999, Knecht et al. 2000, Vingerhoets and Luppens 2001, Drager and Knecht 2002). The amplitude of this increase is critically dependent on the stimulation paradigm, so that individual laboratory-specific and stimulation paradigm–specific normal values need to generated instead of using general normal values. Significant side-to-side differences can be registered depending on the stimulated brain volume and depend-

Table 3.**4** Normal Values of Vasoneural Coupling in the PCA after Visual Stimulation* in Different Age Groups

Age group (y)	VEFR (%)	S_{inc} (cm/s²)	S_{dec} (cm/s²)	$T_{L\ on}$ (s)	$T_{L\ off}$ (s)	PI_{off}	PI_{on}	dPI (%)
20–40	41.7 ± 10.5	3.04 ± 0.70	3.26 ± 1.07	2.6 ± 0.9	3.0 ± 1.3	0.94 ± 0.13	0.77 ± 0.11	24.3 ± 3.2
40–60	35 ± 9.2	2.69 ± 0.72	2.37 ± 0.55	3.2 ± 0.5	4.1 ± 1.2	0.97 ± 0.26	0.77 ± 0.18	19.7 ± 7.3
60–80	33.9 ± 8.6	2.03 ± 0.83	1.81 ± 0.94	3.2 ± 0.8	3.8 ± 1.4	1.25 ± 0.32	1.09 ± 0.33	12.9 ± 11.2

Values are mean ± SD

* Video movie stimulation with 20-s on/off phases; P2 segment insonated

ing on the hemispheric lateralization. Taking this information into account, vasoneural testing has been shown to be a valid tool for diagnosing which hemisphere of the brain is dominant (Knecht et al. 1998, Deppe et al. 2000).

Measurements made from the PCA have been successful, provided that an adequate experimental paradigm is used; for example, the individual characteristics of the stimulus may involve basic cardiovascular parameters, age (Niehaus et al. 2001), and in particular increasing complexity (uniform background, structured black and white patterns, motion pictures) (Fig. 3.**13**) (Sitzer et al. 1992). Individual exposures to the stimuli (eyes open/eyes closed) are repeated several times, and the CBFV change calculated are standardized, in each case lasting more than 10–20 s (Conrad and Klingelhöfer 1989, Sitzer et al. 1992, Hennerici and Daffertshofer 1993). Using the following equation (D):

$$D = \frac{V_s - V_0}{V_0} \cdot 100$$

—where V_s is the maximal CBFV during stimulation. V_0 is the mean CBFV during stimulation off phase (Fig. 3.**13**). This result corresponds well with similar functional MRI and PET investigations. This test is not yet suitable for use in acute ischemic cases and in patients with severe hemodynamically relevant obstructive lesions, in whom the elimination of the vasoneural correlation leads to misinterpretations.

Normal values. Normal values for vasoneural coupling tests are crucially dependent on the stimulation paradigm, which defines the amount of stimulated brain tissue. This is particularly true when measuring vasoneural-related CBFV increases in the MCA territory, since standardized stimulation paradigms in this area are lacking and since there is a widely varying range of stimulation. Some published data can be used to provide an estimate. On the other hand, normal values for visual stimulation and PCA-evoked CBFV response has been established and validated by several groups (Sitzer et al. 1992, Panczel et al. 1999) (Table 3.**4**). Nevertheless, it is strongly recommended that laboratory-specific normal values should be established when performing vasoneural studies.

Monitoring

During normal TCD monitoring, there may be oscillations other than the systolic–diastolic flow velocity oscillations, particularly at different slow frequencies. Such oscillations have been identified as reflecting breath-induced R waves, the autonomically induced M (Mayer) waves, and what are known as B waves. It is still a matter of debate whether B waves reflect spontaneous fluctuations in arteriolar tone or are secondary phenomena due to fluctuations in intracranial pressure. Oscillations should not be interpreted as changes in brain perfusion or in the diameters of large arteries.

MESs are absent in normal circumstances, but HITS may be detectable, as they often reflect artifacts and may be difficult to distinguish from microemboli. In particular, gaseous bubbles due to high-pressure phenomena, which often occur in patients with mechanical artificial heart valves (but also in other conditions), have been observed.

Evaluation

Basic routine assessment of the normal Doppler spectrum is qualitative, following established formal criteria (Fig. 3.**21**). Semiquantitative and quantitative parameters increase the precision of diagnostic information, especially when pathological findings are reported. Various indices have been developed for measuring the flow pulses. Among these, the pulsatility index (PI) has gained widespread acceptance (see Glossary, pp. 351–352).

The advantage of the PI is that it is independent of the angle between the ultrasound axis and the vascular axis, since a potential error affects both the numerator and the denominator to the same extent. An index of around one indicates a normal value. In intracranial examinations, the interpretation of such indices is influenced by proximal, local, and distal arteriolar conditions and is therefore sometimes difficult.

Only two parameters in the Doppler spectrum are usually evaluated:

- The systolic peak flow velocity
- The mean flow velocity

In addition, the following are occasionally analyzed:

- The end-diastolic flow velocity
- The width of the spectral window

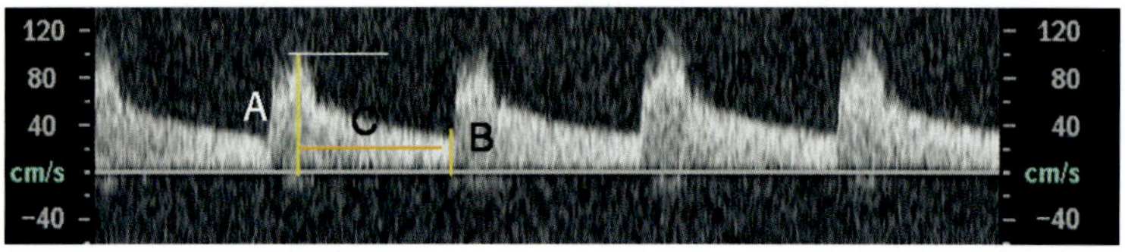

Fig. 3.**21** A normal Doppler flow velocity recording in the M_1 segment of the middle cerebral artery (MCA). The various parameters for analysis are indicated. A = peak systolic velocity, B = peak diastolic velocity, C = mean velocity. The pulsatility index is calculated (A-B)/C and the resistence index = (A-B)/A

There is a wide range of variation in the individual parameters, and the confidence interval is also very large. This applies in particular to examinations comparing individuals, and in some respects also to intra-individual follow-up checks and bilateral comparisons with different indications (Hennerici et al. 1987a).

Sources of Error

Weak or Absent Signals

A poor signal-to-noise ratio occurs when there is:

- Ultrasound energy loss at the skull
- A narrow ultrasound window
- Frontal or parietal displacement of the ultrasound bone window

In individual cases (less than 15% of patients under the age of 60, but up to 25% of those over the age of 70, and even higher in blacks and women), no signals can be registered transtemporally, despite high energy levels and good coupling (Gerriets et al. 2000). Studies by Kenton et al. (1996) and Postert et al. (1997) have shown that amplitude mode color flow imaging may facilitate the identification of intracranial arteries: A_1 (83–93%), A_2 (–90%), M_1 (89–100%), M_2 (54–96%), M_3 (–76%), P_1 (83–100%), P_2 (75–96%), P_3 (–45%), BA (–46%) (successfully studied vessels in both studies).

The use of ultrasound contrast media can increase the display quality if there is a weak or absent signal. In duplex system analysis, contrast media allow signal analysis if the B-mode image shows any structures in the ultrasound sector (e.g., bone structures, tissue, etc.). Ultrasound contrast media are therefore useful in about 10% of patients (Gerriets et al. 1999, 2000, Stolz et al. 2002b, Zunker et al. 2002). In this and comparable indications, ultrasound contrast media allow additional diagnosis in approximately 80% of cases with insufficient bone windows (Fig. 3.**3**). If B-mode image only shows uniform noise, it indicates total closure of the bone window; in these cases, even administering ultrasound contrast media will not produce an adequate signal.

Anatomical Reasons for Error

- There may be variants in the anterior and posterior segment of the circle of Willis.
- Asymmetries and loops in the course of a blood vessel may prevent adequate optimization of the ultrasound probe position.
- There may be asymmetrical skull configuration and bone density (hyperostosis, especially in older women); here, the varying configuration of the diploë may be mainly responsible for the loss of ultrasound energy.

Technical Artifacts

- Due to large sample volumes, different CBFV values may be recorded, some of which may run in opposite directions simultaneously and may originate from several blood vessels.
- Many changes in the course of the small-caliber arteries at the base of the brain reduce the signal amplitude in relation to a large thickness (5–10 mm) provided by the instrument for the section being examined, especially in color flow duplex sonography.
- A poor signal-to-noise ratio can be misleading, especially for the calculated values of the average flow velocity.
- There may be disproportionate underrepresentation or overrepresentation of high-frequency or low-frequency areas.
- There may be unrecognized changes in the transducer or probe position during follow-up examinations (monitoring) or functional tests.

Hemodynamic Causes

- Inhomogeneous areas within the vascular wall that disturb the flow.
- Collateral circulatory systems supporting signs of "functional stenosis."
- Extracranial obstructions that are unknown or unclear.
- Changing heart rates (extrasystoles, absolute arrhythmia) can cause disturbances, which particularly affect the maximum systolic velocity.

Pathophysiological Causes

- Rarely, spontaneous Doppler signal fluctuations.
- Significant changes in the hematocrit, and possibly also in the viscosity.
- Metabolic parameters, e.g., hypoglycemia and abnormal P_{CO_2} values.
- Elevated intracranial pressure.

- Arteriovenous malformations and shunt/steal conditions.

Identification Problems

- Several problems mainly affect the TCD method only and can be solved using the color-coded duplex technique.
- The PCA may be interpreted as the MCA. This mistake can be avoided by following both MCAs deeply until the ACA is reached, and also near the skull (20–30 mm). Compression tests, tests for vasoneural coupling, and additional use of color flow duplex imaging are recommended.
- The T-junction of the carotid artery on the side facing the probe may be confused with the anterior communicating artery and the contralateral ACA. This happens particularly when there are anomalies involving the anterior section of the circle of Willis. The individual vascular segments can be identified reliably when the extracranial parts of the ipsilateral and contralateral common carotid arteries are compressed.
- The distal part of the carotid siphon may be confused with the ACA when the probe is inclined too steeply in relation to the base of the brain. Compressing the ipsilateral carotid artery would change the flow direction in the ACA, but not in the siphon.
- The superior cerebellar artery may be confused with the PCA. To avoid this, vasoneural coupling tests using exposure to light make sense, and particular note should be taken of the relative level of the arteries.
- The intracranial section of the vertebral arteries can be mistaken for the BA. Although this error cannot be completely excluded, it is helpful to follow the blood vessel for the maximum penetration depth (the patient's position should be changed if necessary), and in doubtful cases to attempt a compression test (a second examiner is required for this).

Diagnostic Effectiveness

Only a few references in the literature have reported on the specificity of the procedure, since intracranial stenoses are much rarer than extracranial ones. There is very little information about the specificity in the posterior circulatory system, but this will eventually become available when MRA and helical computed tomography (CT) are more widely used. According to our own investigations (Rautenberg et al. 1990, Röther et al. 1993), the specificity for the brain's anterior and posterior circulatory systems ranges between 99 % and 100 % (Table 3.**5**). Similar results for the anterior circulatory system have been reported by Arnolds et al. (1986) (98 %) and Ley-Pozo and Ringelstein (1990).

Table 3.**5** The sensitivity, specificity, and validity of transcranial Doppler sonography in comparison with intra-arterial angiography for detecting intracranial stenoses in 467 patients (adapted from Rautenberg 1990)

	MCA	ICA	PCA	VA	BA
True positive	24	21	10	19	9
True negative	438	445	455	442	453
False positive	3	–	–	1	–
False negative	2	1	2	5	5
Sensitivity (%)	92	91	83	79	64
Specificity (%)	99	100	100	99	100
Validity	99	100	100	99	99
Positive predictive value (%)	89	100	100	99	100
Negative predictive value (%)	99	100	99	98	99

BA: basilar artery; ICA: internal carotid artery; MCA: middle cerebral artery; PCA: posterior cerebral artery; VA: vertebral artery.

Pathological Findings

Principle

Several parameters are selected from the spectra in transcranial pulsed-wave Doppler sonography to evaluate pathological changes:

- Peak systolic (and possibly diastolic) velocity (Table 3.**1**)
- Mean flow velocity
- Spectral waveform (window width and signal amplitude distribution)

In addition, the following indices can be used:
- Pulsatility index (PI) (Table 3.**3**)
- Pulsatility transmission index (PTI):

$$PTI = \frac{PI_{ipsilat}}{PI_{contralat}} \cdot 100$$

The criteria listed below are used to classify the degree of stenosis.

Low-grade stenoses:
- A local, circumscribed increase in peak systolic and, if applicable, normal diastolic velocity.
- Local increase in the CBFV.
- Possibly a slight signal amplitude increase in the lower flow velocity range.

Medium and high-grade stenoses:
- A clear increase in the peak systolic and diastolic velocity.
- A clear mean CBFV increase.
- Significant disturbance of the spectrum, with increased systolic back-flow components.

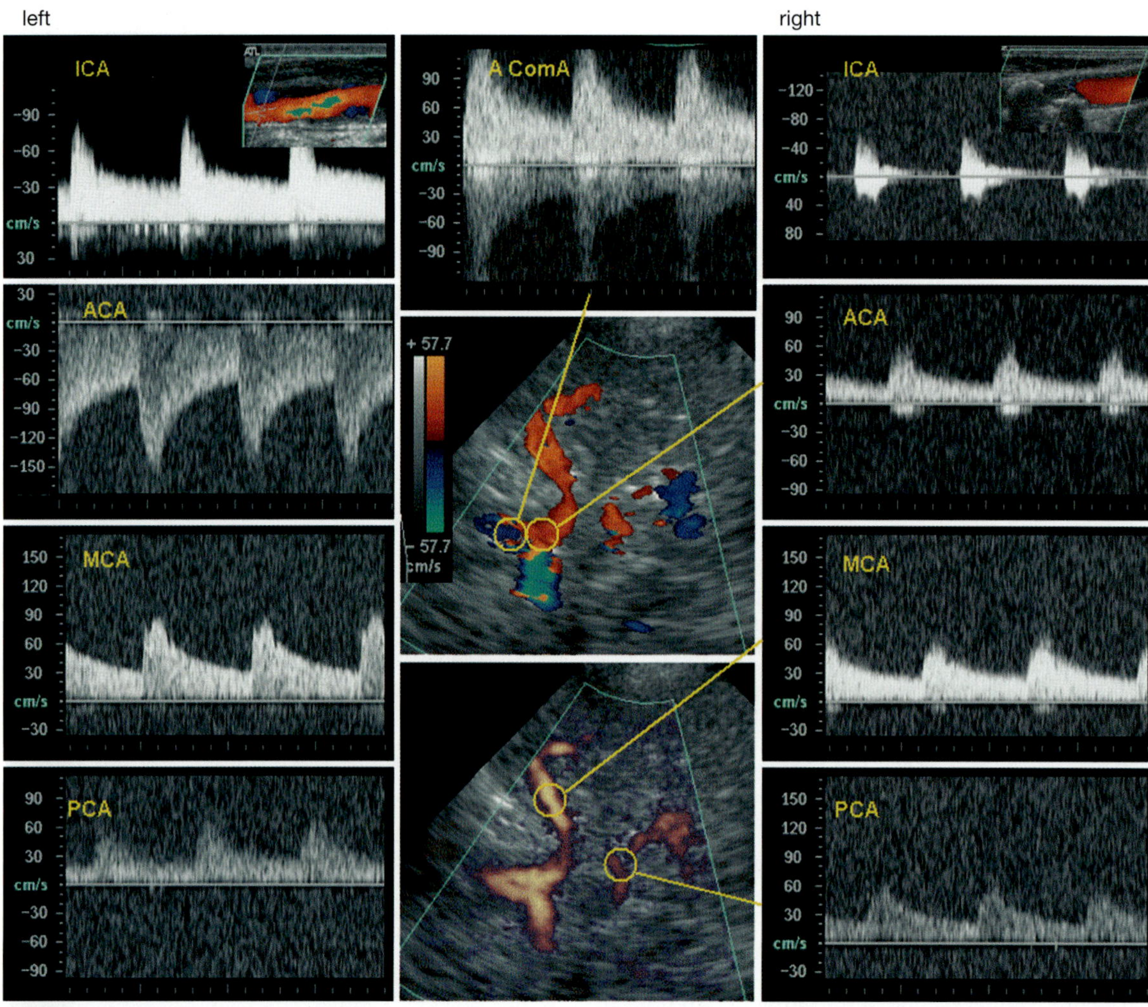

Fig. 3.**22** Bilateral Doppler spectrum waveforms of all basal cerebral arteries in a patient with a 90% right extracranial carotid stenosis. Middelrow: color duplex maps of frequency-mode (middle) and amplitude-mode (bottom) display arteries recorded (yellow arrows). The mean flow velocities and peak flow velocities on the right side (ipsilateral to the stenosis) are slightly decreased on (ACA, MCA, PCA) with flow reversal in the ACA. Due to the intracranial collateralization from the left internal carotid artery, signs of functional stenosis are found in projection to the left ACA (flow acceleration, bidirectional flow signals) and more clearly in the communicating artery (flow increase, bi-directional signals and increase in low frequency, high intensity components) as well as in the right ACA. Normal flow conditions in the left MCA and PCA. Asymmetry of flow in the left ICA > right ICA (post stenotic flow reduction)

- "Musical murmurs" (i.e., bands of discrete velocity at high signal amplitudes) may appear.
- CBFV in arterial segments distal to the stenosis. Reduction in the peak or mean.

The criteria listed above do not allow differentiation between a structural and a functional arterial stenosis at the base of the brain: the latter is quite frequent in communicating arteries in case of high-grade extracranial vascular processes with active intracranial collateralization through the anterior or posterior circle of Willis (Fig. 3.**22**).

Studies analyzing transcranial Doppler findings with angiographic and morphological results have been reported in the literature (Arnolds and von Reutern 1986, Lindegaard et al. 1986, Hennerici et al. 1987a, 1992, Röther et al. 1994, Postert et al. 1997). These demonstrate, as do our own data summarized in Table 3.**5**, that the reliability of the method for detecting (i.e., sensitivity) or excluding (i.e., specificity) arterial processes at the base of the brain is comparable to that obtained with conventional angiography or MRA/spiral CTA, with some overestimation and underestimation in low-degree and high-degree lesions. Therefore, hardly any results are yet available with regard to the validity of these methods in evaluating different degrees of stenosis. This is due to some extent to the difficulty of evaluating arterial stenoses, which involve projection problems and loop formation and are often limited by being able to depict the

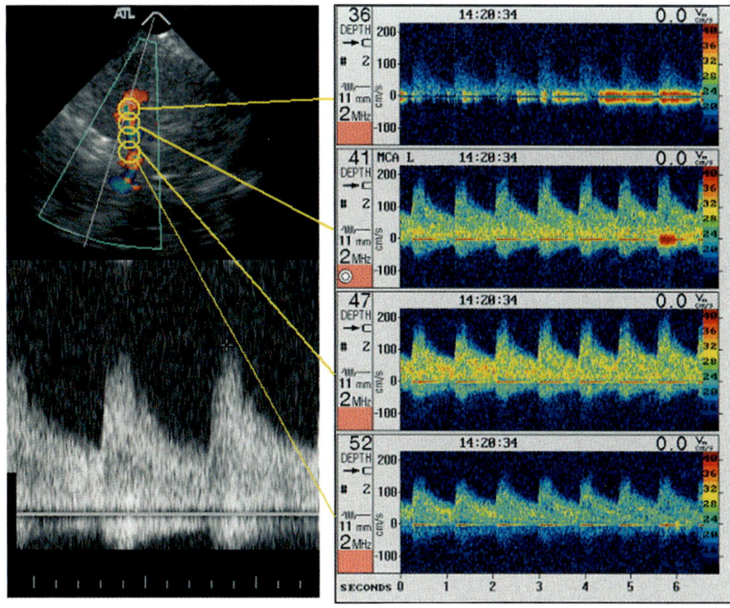

Fig. 3.**23** *Left column:* color duplex scanning and duplex spectral analysis of a distal vertebral artery stenosis, showing aliasing at the stenosis in color map and increased flow velocity in Doppler spectrum. *Right column:* transcranial Doppler recordings at various depths (top to bottom: from proximal to distal), showing flow velocity increases and decreases from pre-stenosis to post- stenosis

stenosis in only a single plane. Conventional angiography and MRA/spiral CTA share the same problems. These procedures are all mutually complementary (Fig. 3.**23**).

Findings

Stenoses

Criteria. Changes in the Doppler spectrum can be used to describe flow obstructions in the extracranial and intracranial cerebral arteries. In transcranial color flow duplex sonography, intracranial stenoses can sometimes be detected by a change in the color signal ("aliasing") or only from guided spectrum analysis (Fig. 3.**24**). For routine use, a classification proposed by Röther et al. (1994) may be of value (Table 3.**6**).

- The *peak flow velocity* (systolic and diastolic) initially increases with increasing stenosis, and decreases again in high-grade and subtotal stenoses.
- The *mean flow velocity* also increases with increasing stenosis, but is subject to less pronounced fluctuations.
- In significant high-grade stenoses, a clear mean flow velocity reduction may result, both prestenotically and post-stenotically.
- Medium and high-grade stenoses have bidirectional signals—i.e., negative flow components.
- With increasing stenosis, the low-frequency components increase (during systole in low-grade stenosis, during the entire cardiac cycle in high-grade stenosis) and the systolic window becomes smaller.

Although peak flow velocity has the highest value in the diagnosis of stenosis, no single parameter seems to

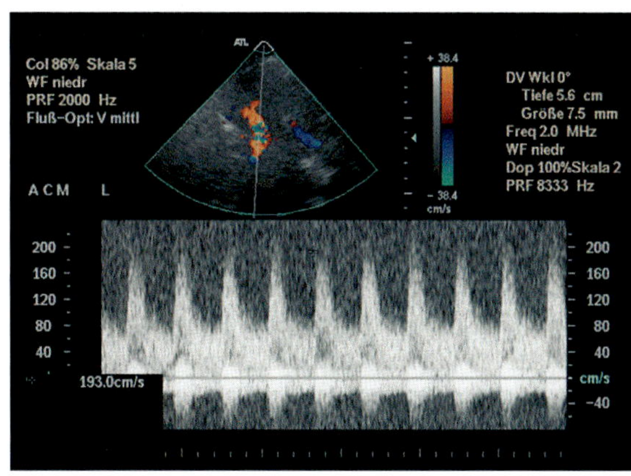

Fig. 3.**24** Color duplex recording of a moderate obstruction of the M_1 segment of the middle cerebral artery (MCA). The color image shows localized aliasing in the mid-segment of the MCA main stem, where the CBFV has increased locally to a peak systolic velocity of approximately 200 cm/s. The low-frequency components further support the diagnosis of stenosis

be able on its own to provide a reliable assessment of the degree of stenosis. Which combination of parameters correlates best with the morphological result has not yet been determined. Since the establishment of a "gold standard" (i.e., postmortem specimens) is unlikely to be achieved, correlations between different test procedures should be established on the basis of standardized studies using a priori criteria, such as those already available for extracranial measurements (de Bray and Glatt 1995).

Location. Stenoses can be followed along the horizontal course of the MCA—i.e., they can be examined and documented prestenotically, intrastenotically, and post-stenotically. Moreover, quantitative flow velocity

Table 3.**6** Proposed classification for stenosis of the middle cerebral artery (adapted from Röther et al. 1994). 0 = normal, I = mild stenosis, II = moderate stenosis, III = severe stenosis

Magnetic resonance angiography (MRA)		
0	No signal loss	
I	Signal reduction	a = < 50 % b = > 50 %
II	Complete signal loss	a = limited to stenosis b = including poststenotic
III	Signal loss including post-stenotic with signal rarefaction in distal vessels	
Transcranial Doppler (TCD) sonography		
0	< 140 cm/s	
I	140–209 cm/s	
II	210–280 cm/s	
III	> 280 cm/s	
Digital subtraction angiography (DSA)		
0	No lumen reduction	
I	Pallor of contrast medium	
II	Lumen reduction	a = < 50 % b = ≥ 50 %
III	Subtotal stenosis with delayed distal filling and/or border zone shift	

ranges for different degrees of stenosis have been established for stenosis of the MCA (Felberg et al. 2002, Gao et al. 2002). In obstructions of the intracranial ICA, the vertebral artery, the ACA, and the PCA, the examinations are more difficult to assess and particularly to quantify, as the vessels have a tortuous course, and only short vascular segments are optimally accessible to ultrasound. This is particularly the case when obstructive processes need to be followed.

Changes in the ascending arteries (e.g., the M_2 segment of the MCA) can be detected only if color flow imaging and spectrum analysis are used (Fig. 3.**24**).

Stenoses of the basilar and vertebral arteries can usually be detected quite well, provided they can be continuously examined prestenotically, intrastenotically, and post-stenotically (de Bray et al. 1997, Stolz et al. 2002b). Due to the unique vascular course and specific positioning of the sample volume in relation to the transducer, only parts of the lesion can be assessed, and classification may not be possible. It is also difficult to differentiate between the two vertebral arteries intracranially, and to define the boundaries between the distribution areas of the vertebral artery and the BA, which do not show any hemodynamic difference (Mull et al. 1990). Unless additional diagnostic methods are used, definite identification of flow obstructions is not possible.

Vascular changes within the intracranial segment of the carotid artery can be detected with combined transtemporal and transorbital ultrasound. Locating and detecting an intracranial carotid artery stenosis is more reliable than with continuous-wave conventional Doppler sonography, which only indicates pathological findings associated with high-grade obstructions indirectly. Stenoses at the origin of the ophthalmic artery, which are extremely difficult to verify using extracranial continuous-wave Doppler sonography, can be directly detected.

Using supplementary examination from the neck, it is sometimes possible to detect extracranial flow obstructions that are inaccessible to continuous-wave Doppler sonography (Schwartz and Hennerici 1986b, Ley-Pozo and Ringelstein 1990).

Occlusions

Occlusion criteria. Changes in the Doppler spectrum in intracranial arterial occlusions are characterized by:
- A locally absent signal, in spite of good study conditions (normal bony access point for ultrasound) and correct Doppler probe and sample volume position (clear signals from neighboring arteries with the identical probe position).
- Absent signal even after ultrasound contrast administration.
- Reduced flow velocity values in proximal vascular segments.
- Detection of collaterals and anastomoses with a flow velocity increase or flow reversal. (Kaps et al. 1990, Burgin et al. 2000, Gerriets et al. 2000).

Location. Occlusions, often caused by emboli, are mainly found in the MCA. During acute ischemia, embolic material in the M_1 segment can fragment due to *spontaneous thrombolysis,* and subsequently cause distal branch occlusions (M_2 and M_3 segments). The following findings are therefore seen:

- In side-to-side comparisons, there is a clearly lower flow velocity at the transition point from the bifurcation of the internal carotid artery into the MCA.
- There is an absent signal toward the superior part of the cranium in the distal course of the vessel; at best, smaller anastomoses can sometimes be detected here, with reversed blood flow.
- A flow velocity increase in the ipsilateral ACA is sometimes also seen bilaterally.

When there is a *branch occlusion of the middle cerebral artery,* the following are also found (Fig. 3.**25**):

- A reduction in the maximum systolic CBFV and also in the average flow velocity, with a normal flow direction in the M_1 segment.
- Occasionally, retrogradely perfused distal anastomoses.
- No collateral flow increase in the ACA.
- No direct indication of the occluded branch of the MCA.

Emboli in the posterior distribution area are also quite common; they predominantly affect the PCA, and sometimes also VA branches or, in isolated cases, the BA.

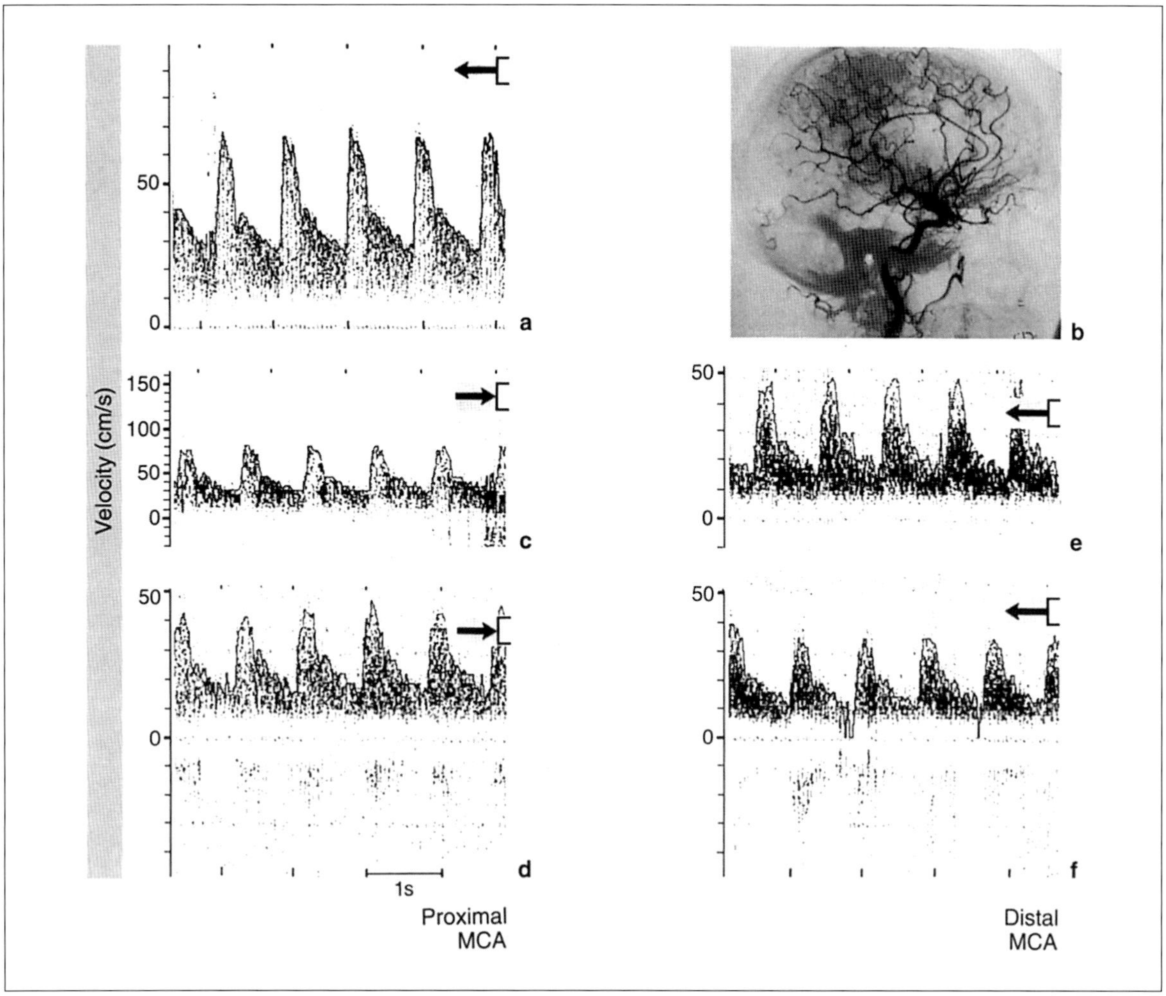

Fig. 3.**25** Selective angiogram of the internal carotid artery (**b**) and Doppler spectrum waveforms of the ACA (**a**) and MCA (**c**–**f**) in an occluded branch of the left middle cerebral artery. In the ipsilateral ACA, the Doppler spectrum waveform is normal (**a**), while in the proximal MCA (**c**, **d**), the systolic peak and mean flow velocities are reduced. Close to the brain surface (**e**, **f**), only arteries with inverse flow directions (away from the probe) and low flow velocities can be identified in leptomeningeal anastomoses here (from Hennerici et al., 1987)

a ACA (depth 70 mm, systolic peak flow velocity 66 cm/s, mean flow velocity 30 cm/s)
c Proximal MCA (depth 65 mm, systolic peak flow velocity 82 cm/s, mean flow velocity 40 cm/s)
d Proximal MCA (depth 50 mm, systolic peak flow velocity 48 cm/s, mean flow velocity 24 cm/s)
e Distal MCA (depth 40 mm, systolic peak flow velocity 48 cm/s, mean flow velocity 22 cm/s)
f Distal MCA (depth 30 mm, systolic peak flow velocity 36 cm/s, mean flow velocity 20 cm/s)

Proximal PCA occlusions (P_1 segment) are characterized by:

- An absent PCA signal despite a reasonable ultrasound window.
- No clearly detectable collaterals or anastomoses.

When there are occlusions in the P_2 segment of the PCA:

- A reduced flow velocity can be detected in the P_1 segment and in the P_2 segment proximal to the flow obstruction, in comparison with the contralateral side.
- Sonographic display of anastomoses or collateral vasculature is not possible.

VA and BA occlusions cannot be diagnosed with certainty using Doppler systems alone, due to the significant variability in the vascular course and the many collateral blood vessels that are present. Even complete occlusion of the basilar artery may be misdiagnosed when the collateral vasculature shows a high flow amplitude alongside the occluded vessel (Mull et al. 1990). The diagnostic accuracy is substantially improved with duplex systems, particularly when ultrasound contrast media are used. In particular, the specificity increases, and a basilar thrombosis is unlikely when there is normal display of the vertebral junction and origin of the basilar artery. Duplex systems minimize the limitation of interfering neighboring vessel segments and allow much better localization (de Bray et al. 1997, Stolz et al. 2002b). Individual parameters are helpful in providing sonographic support for the suspected diagnosis:

- Detection of collateral blood vessels or anastomoses—e.g., retrograde perfusion at the top of the basilar artery, in the posterior communicating artery, or in the P_1 segment of the PCA.
- A reduction in the peak and mean flow velocity in both vertebral arteries.
- Intracranial Doppler-sonographic findings always need to be supplemented by results obtained from an extracranial examination; if the results are divergent, further investigations need to be considered.

The *ability to distinguish a stenosis from an occlusion* with transcranial Doppler in the posterior circulation is also limited by the potential sources error described earlier. High-grade basilar artery stenoses are difficult to record along their whole length, and flow signals from collateral vessels—e.g., the AICA or PICA—may be missed or misinterpreted. If they are located distally at the top of the basilar artery, they can completely escape detection. In addition, the flow velocity is so low that it can no longer be registered if the angle of the applied ultrasound is unfavorable. Color flow imaging improve the validity of these tests substantially because of the better anatomical allocation.

Occlusions of the ACA are relatively infrequent. The diagnosis is difficult, as there are many variations in the course of the artery and its origin, such as hypoplasias and aplasias. Due to compensatory elevation of the blood flow in collateral blood vessels via the anterior communicating artery, misdiagnosis of a stenosis is likely.

Intracranial ICA occlusions can be diagnosed indirectly by examining the carotid system and the orbital arteries extracranially. Depending on the extent of the occlusion and the involvement out of either the ACA, MCA, or both, intracranial cross-flow from the contralateral side or a completely absent signal are common findings. As yet, there have only been anecdotal reports on the use of transcranial color flow duplex sonography and ultrasound contrast to identify flow downstream from high-grade stenoses and occlusions (Baumgartner et al. 1997).

Collateralization of Extracranial Carotid Artery Obstructions

Examination criteria. The interpretation of Doppler signals from intracranial arteries can be misleading without knowledge of the extracranial vascular situation. Hemodynamically significant extracranial obstructions, in particular, lead to flow alterations in the intracranial circulation due to collateral compensation (Fig. 3.**22**). In small-caliber arteries, the increase in hemodynamics reflects *functional stenoses.* They can show almost all the signs of a structural lesion:

- Flow velocity increase
- Bidirectional flow signals
- Increase in the low-frequency, high-amplitude components of the systolic spectrum

Signs by which *morphological stenoses* can be differentiated from functional ones are:

- In functional stenoses, the flow increase persists over the entire vascular segment—i.e., there is no Doppler spectrum change with gradually altering sample volume positions.
- In functional stenoses, there is no flow velocity reduction in vascular segments proximal or distal to the maximum flow velocity area.

Collateralization pathways. Diagnostically, the collateralization pathways can be well documented (Stolz et al. 2002a).

An *ophthalmic anastomosis* can be documented by retrograde flow in the orbital arteries. Registering the MCA signal while compressing the feeding branch of the ECA allows assessment of the significance of the contribution of the external carotid artery to brain circulation. Even when there are highly retrograde signals through the branches of the ophthalmic artery, a significant contribution to the MCA flow occurs only rarely. In contrast to what was previously assumed on the basis of injection-induced pressure displacement in angiography, opening of the ophthalmic artery collateral does not appear to play a significant functional role in the perfusion of the brain.

Collateralization through the *anterior communicating artery* is characterized by:

- Retrograde flow direction in the ipsilateral ACA.
- Elevated peak and mean flow velocity in the ACA bilaterally.
- Optionally, signs of functional stenosis.
- When the carotid artery is compressed contralaterally, there is a velocity reduction in the MCA and the ACA ipsilateral to the flow obstruction.

The collateralization path through the *posterior communicating artery* is characterized by the following findings:

- Increased velocity in the P_1 segment of the ipsilateral PCA and BA.
- Signs of functional stenosis at an ultrasound application depth of between 60 mm and 70 mm.
- Flow increase in the PCA when the carotid artery is compressed.
- Display of the posterior communicating artery with transcranial color-coded ultrasound.

Collateralization through *leptomeningeal anastomoses* is located far distally. In addition, the blood vessels involved are often poorly developed in caliber. Usually, collateralization via this path is therefore quite difficult to prove if amplitude-mode color imaging or contrast enhanced studies are not available. Instead, indirect parameters have to be relied on:

- Increased velocities in the proximal and distal vascular segments reflect collateral blood flow moving distally (in the P_1 and P_2 segments of the PCA, in the A_1 segment of the ACA without crossflow).
- Retrograde flow in the proximal arteries at the base of the brain—e.g., retrograde flow in the distal MCA segments (M_2/M_3) to compensate for a proximal (M_1) occlusion of the main branch of the middle cerebral artery.

Collateralization via a *persistent primitive trigeminal artery* is difficult to detect. Theoretically, the signs of functional stenosis mentioned above could be found in the area between the anterior and posterior circulation. However, it is not possible to delineate any collateralization in the posterior communicating artery properly unless the morphological relationships are known.

Collateral vascular capacity. Although valid parameters are not yet available to estimate the hemodynamic capacity of the collateral vasculature, it can be approximated by comparing the pulsatility indices of the Doppler signals obtained from the bilateral corresponding cerebral arteries (known as PTI values) (Lindegaard et al. 1985). Vasomotor reactivity (Widder et al. 1986) can be determined when there is a change in the end-expiratory P_{CO_2} or after administering acetazolamide, and also by evaluating the link between perfusion and neuronal metabolism (i.e., vasoneural coupling). These methods are based on the assumption that a vascular process has caused maximum vasodilation in the terminal (arteriolar) system, so that a further increase in the flow velocity of the corresponding intracranial cerebral arteries can neither be attained by additional hypercapnia, nor by activating the vasoneural reserve. If there are no disturbances in neurological function, it can be assumed that the hemodynamic reserve capacity mechanisms have been exhausted, and that function is being maintained by metabolic compensation (e.g., increased oxygen extraction).

Evaluation

Dilative Arteriopathy

With increasing age, arteriosclerosis may also show fusiform dilations, the wall consistency of which may be difficult to distinguish histologically from sacciform aneurysms (Schwartz et al. 1993). Dilated and significantly elongated blood vessels can be found in all large arteries at the base of the brain, especially in the ICA and the basilar artery (known as the megadolichobasilar artery) (Boeri and Passerini 1964). The structural changes in the vascular wall and associated losses of elasticity lead to vascular dilations (ectasias) that are accompanied by a reduced flow velocity. The following findings are characteristic (Figs. 3.**19**):

- A significantly reduced mean flow velocity (less than the calculated average –2 SD)
- A reduced peak systolic velocity

Carotid Cavernous Fistula

Even with continuous-wave Doppler sonography in the orbital arteries, a high-frequency flow signal can regularly be observed with a marked loss of pulsatility and flow directed toward the probe, reflecting the venous drainage of the cavernous sinus fistula through the orbital veins. Using pulsed-wave Doppler sonography transorbitally or transtemporally, the fistula can sometimes be evaluated directly, and high-amplitude, high-frequency, nonpulsatile flow signals can be demonstrated. In addition, there are sometimes significant low-frequency components (machine noise), which decrease when the ipsilateral carotid artery is compressed. Depending on the shunt volume, abnormal results can also be detected above the neck arteries; however, these may also be completely absent (Fig. 3.**26**). During and after interventional therapy, closure of the fistula can be well documented with Doppler sonography and color flow duplex sonography.

Arteriovenous Malformation

Arteriovenous malformations have to be investigated using standard neuroradiological methods (CT, MRI, angiography). Ultrasound can only serve as a supplementary technique, with a few unique capabilities for measuring hemodynamic effects (Kilic et al. 1998, Uggowitzer et al. 1999). The following examinations using transcranial duplex sonography appear to be important:

- Examining hemodynamics in the arteries supplying the malformation.
- Analyzing collateralization and the shunt size.
- Recognizing blood supply in the neighboring vasculature in time.
- Observing the path of blood flow before and after therapy (Hassler 1986, Schwartz and Hennerici 1986a, Hennerici et al. 1992).

Characteristic changes in the vessels supplying the angioma are:

- Increased flow velocity
- Increased signal amplitude affecting slow flow velocities
- Bidirectional flow signals
- Significant audio signal changes
- Clearly reduced pulsatility indices
- Disturbed autoregulation in hypercapnia and hypocapnia

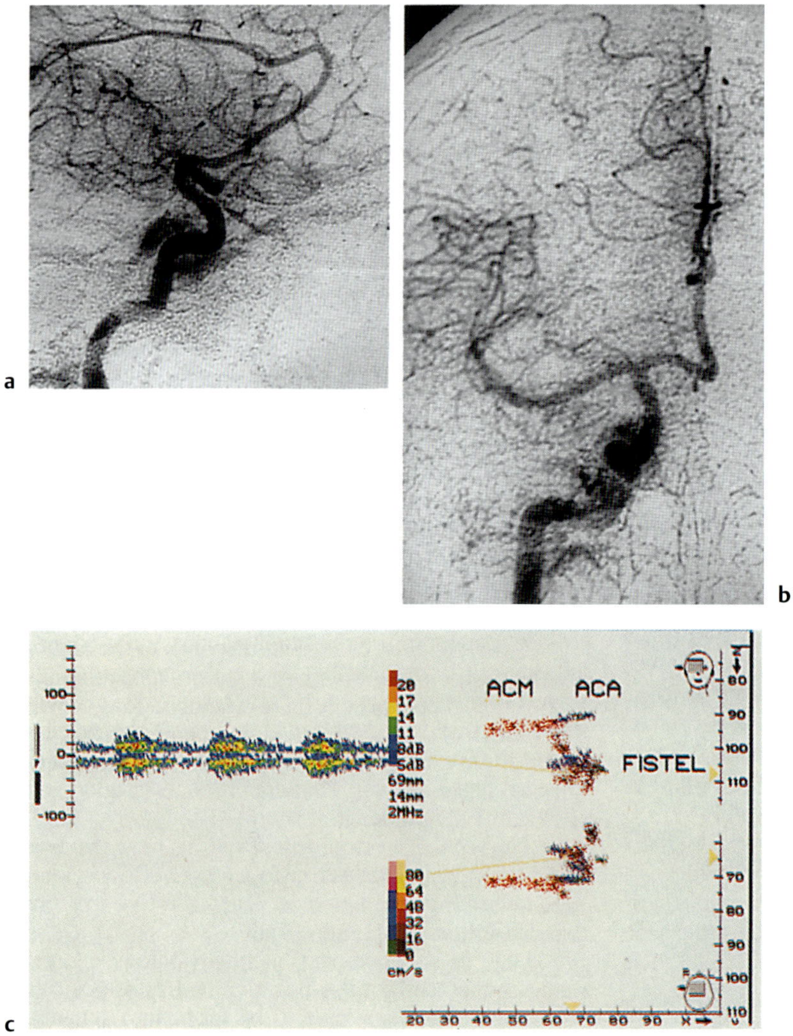

Fig. 3.**26** Angiograms in **a** lateral and **b** anteroposterior projections, and **c** transcranial Doppler sonography findings in a patient with a carotid cavernous fistula, using a two-dimensional scanning system (frontal and horizontal scans on the right). Ultrasound recorded from a depth of 69 mm demonstrates a pathological flow signal with low-frequency flow components (left)

In the neighboring arteries, the following are found:

- Decreased flow velocities
- Simultaneous registrations containing overlapping vascular spectra

Preoperatively, it is possible to differentiate the draining arteries using the criteria given above; postoperatively, reduced flow velocities are found in the vascular segments, which are usually still dilated and show disturbed CO_2 reactivity. Both parameters normalize with time.

Spasms and Aneurysms

One early use of intracranial sonography was to document and follow the clinical course of spasms (Fig. 3.**27**) in patients suffering from subarachnoid hemorrhage (Aaslid et al. 1984, Harders 1986, Seiler et al. 1986). TCD has become a wideley used examination procedure for detecting, quantifying, and following up vasospasms after subarachnoid hemorrhage (SAH) (Mariak et al. 2002, Suarez et al. 2002, Topcuoglu et al. 2002). Vasospasms generally occur on the fourth day after SAH, while peak flow velocities can be observed between the 11th and 18th days. Normalization of flow velocities occurs within the third or fourth week after SAH. A rapid increase in velocities 4–8 days after SAH is associated with an increased risk of ischemic stroke.

Due to the complex pathophysiology of vasospasms, potential false-negative and false-positive sonographic findings have to be taken into account. Despite angiographically detectable spasms, elevated intracranial pressure due to subarachnoid bleeding can cause increased peripheral resistance in both the arteriolar and capillary vascular systems, accompanied by a corresponding flow velocity reduction in the preceding arteries consequencing normalisation of CBFV even in case of vasospasm. This may lead to false-negative sonographic findings.

However, in other cases, substantial sonographic changes may be found without angiographic confirmation. These apparently false-positive results are generally interpreted as being caused by arteriolar dilation that can arise when inflammation processes take place. They cause a clear in peripheral resistance, and decrease induce increased flow velocity. There have been a few attempts to improve the diagnosis of spasm using continuous monitoring with TCD (Venkatesh et al. 2002).

In addition, it should be noted that routine use of calcium blockers in patients with subarachnoid hemorrhage may have an influence on the value of sonographic findings. Laumer et al. (1993) detected a significant increase in intracranial flow velocity prior to secondary cerebral ischemia, which appeared in only three of their 11 treated patients. Elevated CBFV, apart from local stenosis or spasm in SAH patients, can also be found in patients with meningitis or encephalitis, in female patients with eclampsia, and also in cases of fever, hypercapnia [Author – check! hyper- or hypo-], anemia or young age (child).

Finally, although they are rarely significant, abnormally elevated flow velocities may be found in patients who have not been diagnosed as suffering from subarachnoid hemorrhage. The extent to which functional spasms may be present (e.g., migraine)—as in Prinzmetal's angina of the coronary arteries—remains unclear.

Apart from rare giant aneurysms, direct detection of an aneurysm as the bleeding source is not possible using TCD. In individual cases, transcranial color flow duplex sonography has been reported to facilitate direct imaging of an aneurysm (Wardlaw and Cannon 1996, White et al. 2001). Baumgartner et al. (1996a) were able to detect 13 of 18 aneurysms sonographically, ranging in size between 6 mm and 28 mm, which had previously been demonstrated by angiography. Two thrombosed aneurysms and three ranging in size from 4 mm to 5 mm escaped sonographic detection.

Extracranial–Intracranial Bypass

It has been shown that carrying out an extracranial–intracranial bypass procedure to treat significant hemodynamic compromise in an intracranial arterial territory (usually the MCA territory) does not have a beneficial effect (EC/IC Bypass Study Group 1985). There have recently been several attempts to analyze the use of extracranial–intracranial bypass techniques with new and improved study designs, with stricter indications. Several ultrasound techniques are useful in protocols of this type for confirming the obstructive process and analyzing hemodynamic failure (e.g., vasomotor tests, perfusion measurements). The special aspect of transcranial ultrasound is that it allows very easy screening for the patency of extracranial–intracranial bypasses (Umemura et al. 2002).

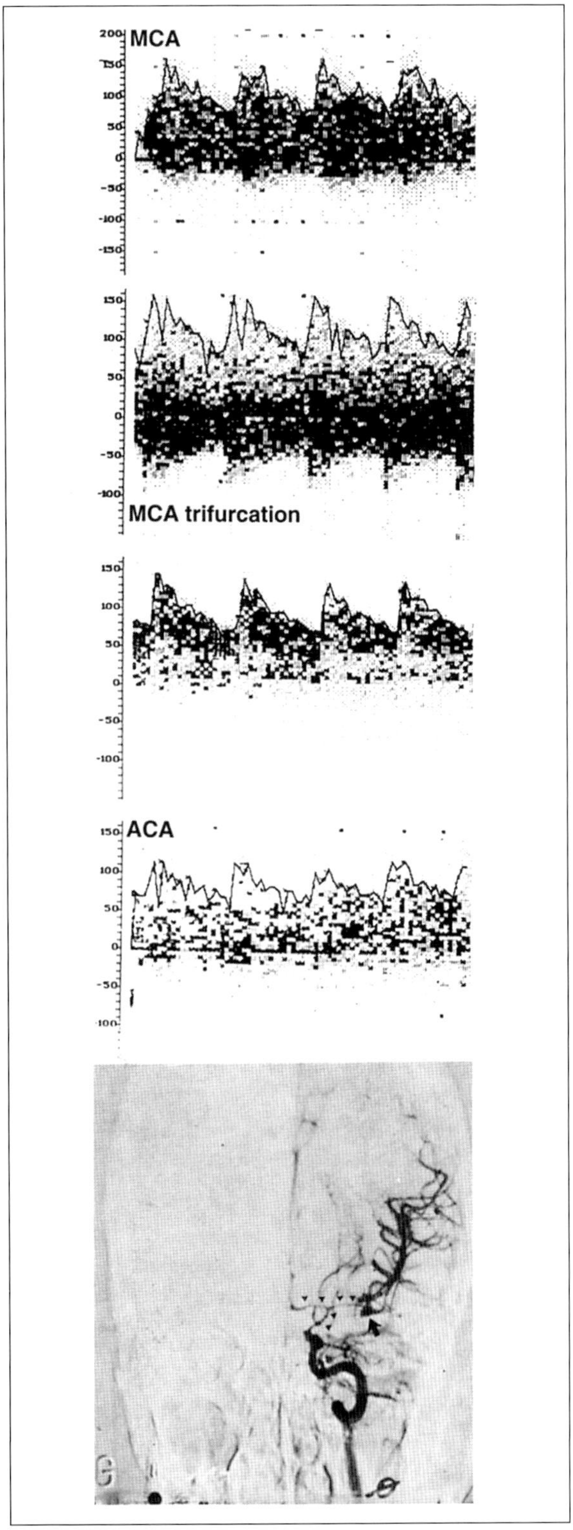

Fig. 3.**27** Doppler spectrum waveforms and the corresponding carotid angiogram in a patient with a marked spasm (arrowheads) in the intracranial T-junction of the carotid artery, the anterior cerebral artery (ACA), and the middle cerebral artery (MCA), after subarachnoid hemorrhage from a middle cerebral artery aneurysm (arrow). There is a flow acceleration in the entire area of both the MCA and the ACA at depths of 40–60 mm

Increased Intracranial Pressure and Cerebral Circulatory Arrest

An increase in intracranial pressure results after serious trauma, global hypoxia, and pronounced cerebral edema due to various causes. When the systemic blood pressure is constant, this leads to a decrease in intracranial circulation, with increasing peripheral resistance. As mentioned above, CBFV depends on cerebral perfusion pressure (CPP), which results from the blood pressure and the ICP (Terwey et al. 1981, Erickson et al. 1989). Taking this into account and assuming constant blood pressure, CBFV measurements directly correlate with the ICP (Withers et al. 1990, Martin et al. 1994). Several attempts have been made to measure ICP quantitatively using TCD. However, since CBFV values do not depend on ICP alone, but also on the blood pressure, all of these methods require continuous recording of blood pressure (Hennerici 1987, Geroulakos et al. 1994, Schmidt B et al. 1997). Moreover, the equations given above are simplifications to demonstrate the principle, and CBFV is also affected by other factors such as venous pressure and arteriolar dilation. In addition, the assumption that an insonated arterial segment has a constant diameter is usually misleading in conditions with variable ICP values (e.g., SAH, meningitis, or space-occupying lesions). Vasospasm may also lead to alterations in vessel diameter, causing uncertain shifts in ICP measurements. More valid data have been obtained in cases of brain injury, in which CBFV measurements have been shown to improve intensive-care monitoring documentation (Hennerici et al. 1988). In a study including 96 patients, arterial pressure, ICP, and transcranial Doppler (TCD) blood flow velocity were studied to monitor for episodes of plateau waves of ICP, which were found in eight patients (Nabavi et al. 1996). The dramatic increase in ICP was followed by a profound drop in CPP (by 45 %). In contrast, flow velocity fell by only 20 %. Autoregulation was documented to be intact both before and after the plateau, but was disturbed during the wave ($P < 0.05$). The pressure–volume compensatory reserve was always depleted before the wave. Cerebrovascular resistance decreased during the wave by 60 % ($P < 0.05$), and TCD pulsatility increased ($P < 0.05$). CBFV measurements can be helpful for demonstrating preserved cerebral autoregulation but a reduced pressure–volume compensatory reserve in these patients. In this group of patients, TCD monitoring may be most helpful for noninvasive prediction of drops and rises in intracranial pressure. However, for quantitative ICP measurements, highly elaborate algorithms for ICP have been based on blood pressure and CBFV (Schmidt B et al. 1999, Schmidt EA et al. 2000, Michaeli and Rappaport 2002). Qualitatively, it has been known for years that the Doppler flow pattern shows specific changes (Fig. 3.**28**) when ICP increases (Perry et al. 1985, Klingelhofer et al. 1987, Becker et al. 1995, Görtler et al. 1994, Glagov et al. 1995). The ICP leads to increased peripheral resistance with a reduction in CBF and also CBFV. Decreases in CBFV affect the systolic and diastolic velocity profiles in a different way, resulting in an increased pulsatility index (Withers et al. 1990). Thus, decreased mean flow velocities and increased pulsatility indices reflect increased ICP in patients without interfering vascular conditions such as brain trauma or tumors (Busch 1966, Klingelhofer et al. 1988).

However, CBFV is not only affected by ICP, and flow pattern changes therefore always have to be interpreted in conjunction with the clinical diagnosis. There is a characteristic alteration in CBFV when ICP levels reach mean arterial pressure levels and autoregulation is overridden. Increased pulsatility leads to diastolic zero flow, with short systolic velocity peaks then pulsating around zero (Fig. 3.**28**), and finally with a complete zero flow (Hass et al. 1968). Feri et al. (1994) investigated 37 patients with Glasgow coma scores of < 7. TCD waveforms showed high resistance profiles, with low, zero, and then reversed diastolic flow velocity. Only three waveform patterns—diastolic reverse flow without diastolic forward flow; brief systolic forward flow; and undetectable flow in the MCA—were recorded in the 22 brain-dead patients, but in none of the other comatose patients (Geroulakos et al. 1994). Several other groups have also reported this pattern, which is observed uniformly in all of these patients (Gabrielsen and Greitz 1970). TCD has therefore been introduced for the diagnosis of brain death in children (Gabrielsen and Greitz 1970) and adults (Glagov and Zarins 1983, Hames et al. 1985, Guo and Fenster 1996). It has been shown that brain death can be diagnosed with perfect sensitivity and specificity if flow in the MCA was found initially and changed to the typical flow profile described above or absent flow (Harward et al. 1986, Griewing et al. 1996, Yamamoto et al. 1998). TCD can therefore be used as a supporting criterion for determining brain death if flow signals from both the extracranial and the intracranial cerebral arteries are modified in their course and if these changes are documented by an experienced investigator (von Reutern 1991). The criteria for brain death vary from country to country, and applicable national guidelines should be consulted; the position taken by the Scientific Advisory Council of the Federal Council of Physicians in Germany requires at least two examinations that are at least 30 min apart to determine brain death. Similarly, according to the President's Commission on guidelines for determination of brain death in the United States, confirmatory tests (such as TCD, blood flow studies, and radioisotope scanning) may be used in patients who meet the clinical criteria for brain death, in order to shorten the observation period before organ donation.

Intracranial Findings in Steal Phenomena

In the majority of patients with subclavian steal phenomenon, the direction of blood flow through the basilar artery remains normal (Huang et al. 1997). Approximately one-third of the patients show intermediate flow phenomena, and retrograde perfusion is a rarity, almost exclusively found in patients with multivessel disease, or anomalies in the circle of Willis, or both. About 40 % of patients with subclavian steal phenomena show changes in the flow relationships within the basilar artery when the affected arm is compressed. In patients with innominate artery obstructions, complex hemodynamic changes can be detected in the intracranial vasculature, similar to those reported extracranially. The abnormal flow profiles in these patients are characterized by a muffled spectrum during the early systole, which is accompanied by a breath-like flow signal predominantly caused by a delayed systolic increase (Rautenberg and Hennerici 1988). A latent steal phenomenon can be detected in various intracranial blood vessels when the right upper extremity is compressed (Hennerici et al. 1988). Currently available studies suggest that latent and intermediate arterial flow changes at the base of the brain may act as a possible source of emboli. This revises the earlier pathophysiological view, according to which a hemodynamic perfusion disturbance was seen as the primary mechanism in this condition.

Functional Disturbances

Various functional disturbances can have a general influence on Doppler signals in the cerebral arteries, and their role as potential pathological mechanisms in neurological illnesses is under discussion.

Cerebral autoregulation disturbances. Examinations using a tilting table have been used in the diagnosis of syncope and cerebral ischemia of undetermined origin:

- In patients with serious peripheral or central autonomic nervous dysfunction (e.g., diabetic neuropathy, dysautonomia, multisystem degeneration).
- In patients without mechanisms of compensation for a significant reduction in flow parameters registered in the arteries of the brain.
- Due to endocrine/metabolic failure of systemic blood-pressure control.

A different patient group, in whom this type of reduction is observable as a sign of isolated cerebral autoregulation disturbance, is also of interest:

- Despite stable systemic blood-pressure regulation
- With compensatory tachycardia (Daffertshofer et al. 1991, Hennerici and Daffertshofer 1993)

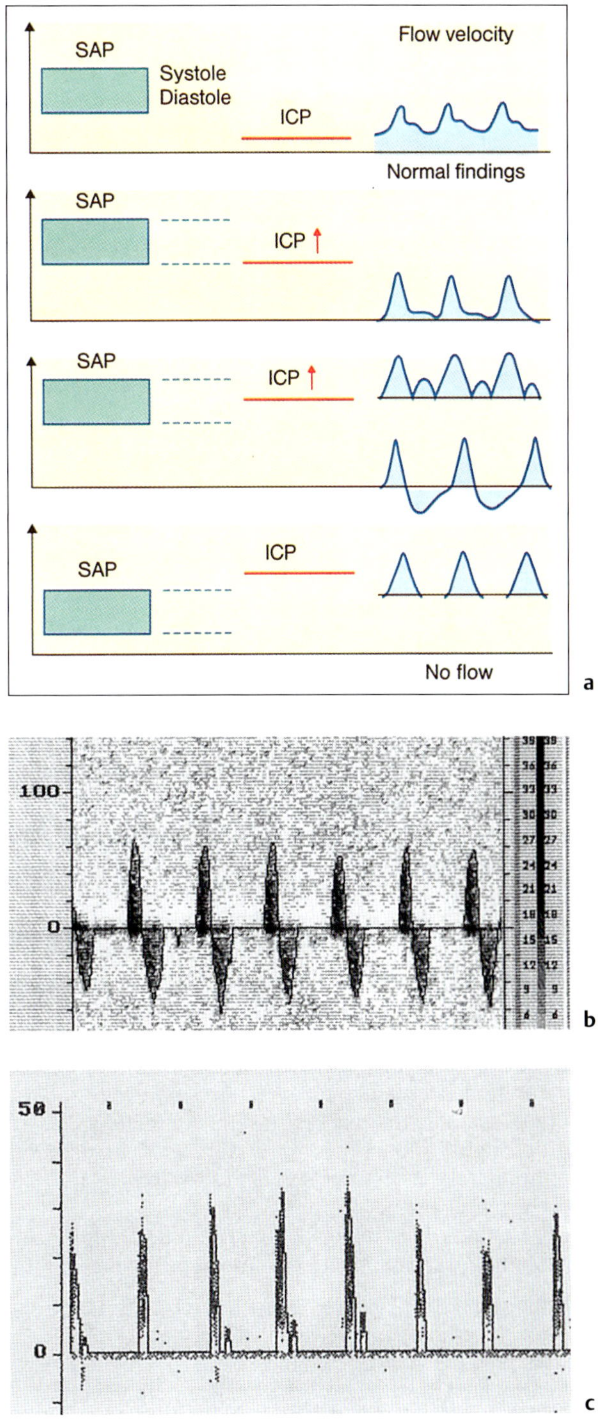

Fig. 3.**28** Schematic illustration (**a**) of the relationship between systolic arterial pressure (SAP), intracranial pressure (ICP), and Doppler spectrum waveforms recorded from the intracranial cerebral arteries (adapted from Hassler et al. 1988). If the intracranial pressure exceeds the systolic blood pressure, flow velocity ceases, first during diastole and then completely. Examples of biphasic flow (**b**) and early systolic peaks (**c**) in a patient with intracranial arrest of cerebral perfusion

Monitoring

Acute Stroke

TCD can accurately predict complete recanalization (positive predictive value 91%) and complete occlusion (positive predictive value 100%) (Burgin et al. 2000). Findings of partial occlusion on TCD are relatively sensitive but not highly specific in comparison with angiography, with partial occlusion representing complete angiographic occlusion in 44% of cases, but rarely representing complete recanalization. Therefore, nonresponders to tissue plasminogen activator (t-PA) with TCD findings of complete or partial occlusion are likely to have persistent occlusion on angiography and thus may be potential candidates for further intervention. Conversely, the remaining 27% of patients treated with intravenous t-PA who have persistent neurological deficit despite recanalization might be spared the risk and expense of an angiogram that is unlikely to reveal a treatable thrombus.

Information about MCA recanalization obtained using TCD is of clinical importance. Recovery after intravenous t-PA is associated with recanalization and resumption of flow. Many studies (Demchuk et al. 1999, Alexandrov et al. 2000) have shown that recanalization corresponds to clinical improvement in some patients during or shortly after intravenous t-PA infusion. The question remains of how patients who do not experience early recovery should be treated. Whether or not the continued neurological deficit is due to persistent occlusion has important clinical implications. A lack of clinical recovery or worsening of a neurological deficit is often associated with persistent occlusion or reocclusion in 50% of patients undergoing intravenous t-PA (Alexandrov et al. 2000). These patients would be potential candidates for intra-arterial therapy used in a bridging protocol (Lewandowski et al. 1999).

Several studies have shown that both Doppler systems (Fig. 3.**29**) and duplex systems (Fig. 3.**30**) are valid methods of assessing the presence of MCA occlusion and for monitoring MCA recanalization in acute stroke patients (Burgin et al. 2000, Akopov and Whitman 2002, Suwanwela et al. 2002, Wardlaw et al. 2002). Burgin et al. (2000) have reported that abnormal MCA waveforms correlate with the angiographic presence of occlusion (Fig. 3.**31**). TCD evidence of partial occlusion (blunted or dampened flow signals) should be interpreted as predicting persistent MCA occlusion at angiography. Complete MCA recanalization on TCD can be assumed if low-resistance flow is observed with a velocity of 70% of that in the contralateral MCA. TCD cannot reliably distinguish between residual stenosis and hyperemia after reperfusion. The simple criteria published by the Houston Stroke Center, however, allow accurate prediction of thrombolysis in brain infarction (TIBI) grade III, since low-resistance flow, regardless of the velocity increase, is predictive of rapid opacification of the distal vessels. Repeat TCD examinations may be needed in order to distinguish hyperemia (decreasing velocities) from residual stenosis (a persistent focal velocity increase).

The intra-arterial Prourokinase for Acute Ischemic Stroke Trial (PROACT II; Furlan et al. 1999) showed a benefit in acute ischemic stroke patients who received intra-arterial thrombolysis. However, of the 474 patients who received diagnostic angiograms, only 180 were eligible for treatment on the basis of the PROACT II criteria. Bedside diagnosis of MCA flow status can minimize the number of diagnostic angiograms required to find a treatable MCA occlusion.

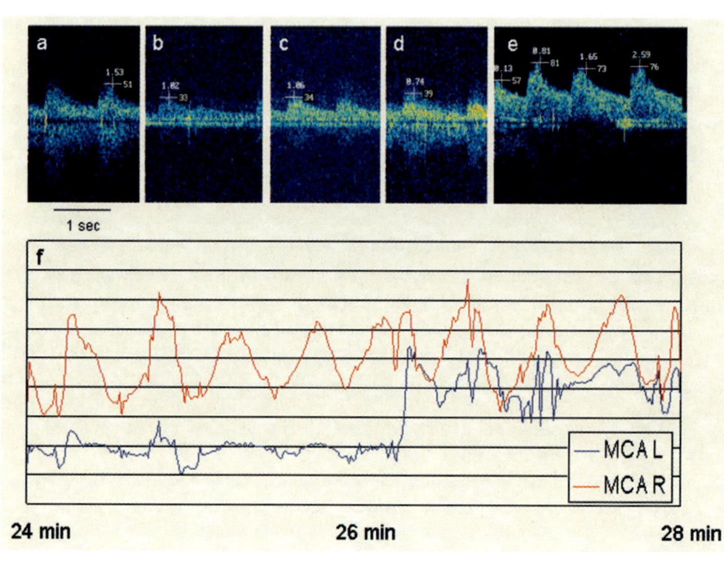

Fig. 3.**29** Online monitoring of the middle cerebral artery (MCA) on both sides in a patient with left-hemispheric stroke. *Upper row:* Spectra from the left MCA, showing reduced and sometimes oscillating flow with progressive amplitude recovery in the left MCA (**a**-**d**) until abrupt normalization (**e**) 26 min after start of intravenous recombinant tissue plasminogen activator (rt-PA) administration. *Lower row:* bilateral monitoring with high time resolution, showing the significant side difference in cerebral blood flow velocity, with abrupt normalizing 26 min after onset of treatment (i.e. clot lysis in this case)

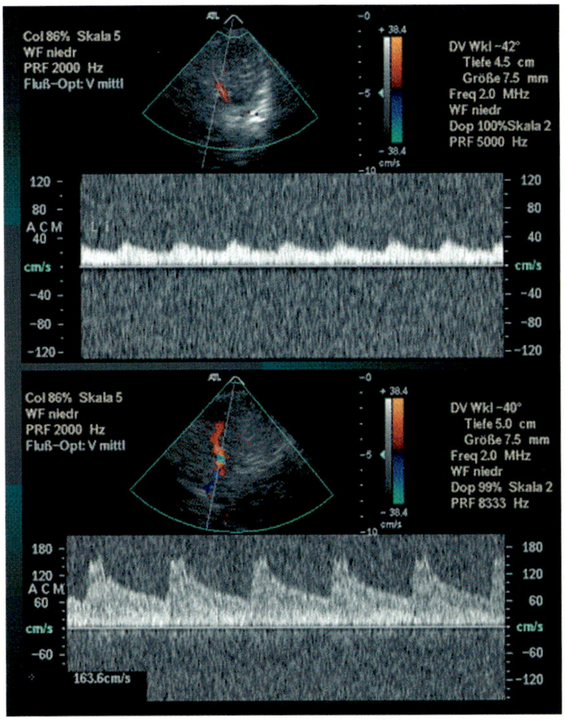

Fig. 3.**30** Color duplex sonography of the left middle cerebral artery (MCA) in a 63-year-old patient with acute left-hemispheric stroke (right-sided hemiparesis and aphasia). *Upper row:* a recording 2 h after onset of the stroke. There is a reduced CBFV (duller signal) in the M_1 segment of the MCA, indicating occlusion of a distal MCA branch. *Lower row:* a recording 3 h 20 min after the stroke and 20 min after intravenous recombinant tissue plasminogen activator (rt-PA) administration. There is complete normalization of the Doppler flow, indicating recanalization.

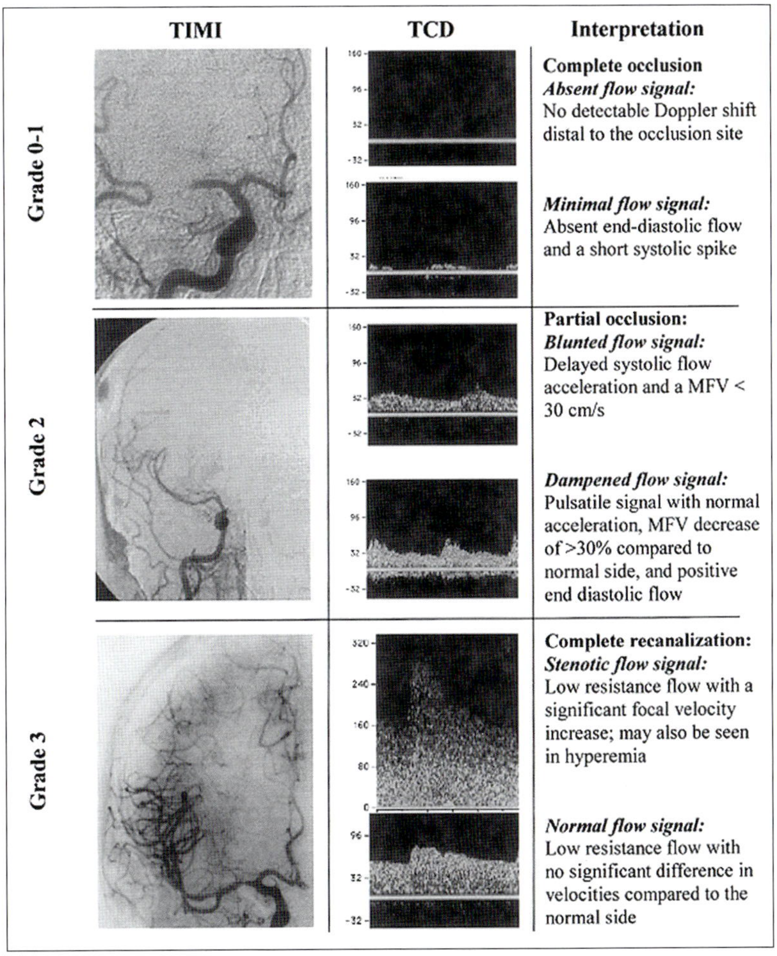

Fig. 3.**31** Grading of MCA signals during recanalization treatment (thrombolysis) to allow Doppler results to be converted into angiographic findings, and particularly to adapt them to cardiological gradings for obstructive coronary disease (from Alexandrov et al. 2000)

Spontaneous Microemboli (MES) and High-Intensity Signals (HITS)

High-intensity transient signals (HITS) were first observed during carotid endarterectomy. Similar signals have also been detected during cerebral angiography of the intracranial arteries at the base of the brain (Rautenberg et al. 1987). HITS are short, high-amplitude, unidirectional signals associated with a typical chirping sound, which can appear at any time in the cardiac cycle (Fig. 3.**32**). In-vitro examinations have shown that HITS sometimes reflect the size and velocity of corpuscular intra-arterial particles (Microemboli, MES). However, reliable conclusions about the varying composition of the reflecting particles (e.g., atheromatous material, platelet aggregate) from signals recorded are not currently possible in the individual patient. (Markus and Brown 1993). HITS or MES have been reported in patients who are potentially at risk from embolism (e.g., patients with asymptomatic or symptomatic carotid artery stenoses and especially patients with artificial heart valves) (Hennerici 1994). In these patients with artificial heart valves, 40 000–50 000 HITS occur per day without any clinically detectable neurological deficits. These HITS are most likely to be due to harmless microcavitations. Since there is growing evidence for HITS generated from arteriosclerotic lesions or cardiac arrhythmia or coagulation disorders being real [Author – check!] MES, HITS generated from artificial heart valves (Droste et al. 1997, Telman et al. 2002).

Intracranial emboli from extracranial vascular processes or cardiac sources may fragment during arterial passage or even after closure of distal arteries through spontaneous lysis. To what extent this is of any pathogenetic significance remains unclear. Automatic embolus detection software devices (Cullinane et al. 2000, Mess et al. 2000, Moehring and Spencer 2002) have shown less valid results in comparison with assessment by experienced observers (Daffertshofer et al. 1996, van Zuilen et al. 1996). Emboli detection is a highly attractive measure, but despite many years of research, evidence for clinical significance is still low. There have been proposals suggesting that detection of emboli can be used as a surrogate marker for treatment monitoring (Junghans and Siebler 2003), but even this application has not yet become fully established (Goertler et al. 2001, 2002). Very recent studies indicate that detection of emboli can be used as a predictive measure of reocclusion after carotid endarterectomy (Muller et al. 1998, Georgiadis et al. 2002).

Monitoring During Interventional Procedures

Continuous monitoring of the flow relationships in the MCA has been carried out during carotid endarterectomy (Ackerstaff et al. 1995, van Zuilen et al. 1995) and during cardiopulmonary bypass surgery. In addition to flow alterations, particularly reduced CBFV as a measure of hemodynamic compromise, the focus of intraoperative TCD monitoring has been on the occurrence of HITS as indicators of microembolic or even macroembolic events. In surgical circumstances, the occurrence of HITS may correlate with postoperative neuropsychological symptoms (Russell 2002). The occurrence of HITS after closure of the artery has also been suspected to be an indicator of early postoperative reocclusion. However, intraoperative hemodynamic disturbances and ischemia are best monitored by clinical observation when local anesthesia is being used. In patients with early postoperative occlusion, early duplex scanning of the carotid artery itself appears to be the gold standard. It is basically still not clear whether the complication rate associated with surgery can be reduced using this type of TCD monitoring (Hennerici 1993).

Right-to-Left Shunt Detection

In TCD with contrast administration, the signal changes that appear can be used diagnostically in the presence of a patent foramen ovale (Lechat et al. 1988, Schminke et al. 1995, Angeli et al. 2001). After an intravenous injection of an ultrasound contrast medium, paradoxical embolization paths can be spontaneously detected by observing microbubbles in the cerebral arteries, or after using the Valsalva maneuver to induce a right-to-left atrial shunt. Depending on the particle size used, the reported sensitivity of contrast TCD may differ from that seen with contrast TEE. With a particle diameter of less than 5 µm, contrast TCD also detects small in-

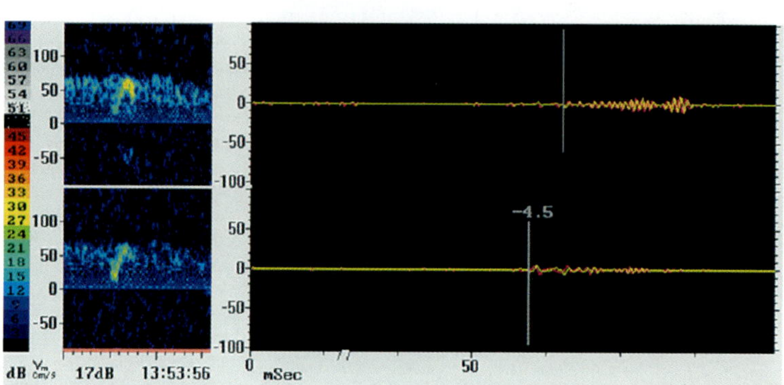

Fig. 3.**32** Microembolic signals (MES) in the left MCA. A multigate recording showing a delay in the signal between the proximal gate (lower row) and the distal gate (upper row), allowing some reduction in artefacts. The left column shows a segment of the Doppler spectrum, with high-intensity transient signals (HITS). The right side shows the electrical input signal, clearly demonstrating a difference in appearance between the two gates. However, identifying artifacts is still an issue

trapulmonary shunt formations; this error is lower when contrast medium particles larger than 20 µm are used (Fig. 3.**16**).

TCD and Migraine

Follow-up investigations during, and in the intervals between, migraine attacks have not so far yielded any convincing and consistent examination results. The presence of local increases in flow velocities (Abernathy et al. 1994), as well as decreases, has contributed nothing to the diagnosis. More recent findings (Panczel et al. 1999), however, obtained using complex stimulus patterns that stimulate almost the entire vascular territory of the PCA, including both the primary and secondary visual projection areas, indicate significant abnormal vasoneural coupling in almost all patients with migraine (with aura) who were examined, even when there was no current migraine attack (Fig. 3.**15**). These results are modified when the patients are undergoing treatment. Confirmation of the findings and evaluations of these studies is not yet complete.

TCD and Epilepsy

The monitoring capability of TCD, its perfect temporal resolution, and data from vasoreactivity and vasoneural coupling studies have led to studies investigating CBFV behavior during epilepsy or during the interictal phase. It has been shown that CBFV increases during generalized seizures (Vanninen et al. 1994). Whether the flow also increases during partial seizures, however, is unclear; some reports have suggested that there is an increase in flow in the MCA during focal motor seizures, while others found a decrease during amnestic episodes (Vanninen et al. 1994) and complex partial seizures. In contrast to other techniques (PET, MRI, SPECT), TCD allows online perfusion studies, with excellent monitoring capabilities even when the patient is noncooperative. Vasoneural coupling appears to be altered at least in some types of epilepsy. TCD may therefore become even more useful for epilepsy monitoring in the future.

Sources of Error

Incorrect Anatomic Identification of Flow Accelerations

- “Functional stenosis” in a hypoplastic blood vessel
- A high shunt volume (arteriovenous malformation)
- Extracranial collateralization due to obstructive processes
- Intracranial collateralization due to obstructive processes
- Hyperperfusion after ischemia induced by arterial occlusion, cardiac arrest, hypoxia, spreading depression in migraine, meningitis, head trauma, etc.

Absent Signal

- When the sample volume is incorrectly placed, or there is a poor signal-to-noise ratio or unfavorable ultrasound window.

Inadequate Control

- Of orthopnea, hematocrit, hypoglycemia

Normal Variants

- Direct origin of the PCA from the carotid artery
- Pseudostenosis signs (e.g., collateral flow in the communicating arteries of the circle of Willis)
- Reduced flow velocity (e.g., basilar hypoplasia)

Special Applications

Brain Structure Imaging

Today’s transcranial Duplex systems allow B-mode display of the cerebral parenchyma via transtemporal insonation in the majority of patients (80–90 % of cases). In patients with an intact skull, the display of anatomical structures is limited by restricted insonation routes and ultrasound principles. Some brain structures can be displayed perfectly, while other structures are only depictable in selected patients (Table 3.7). The display and image quality of ultrasound brain imaging is substantially better in patients who have undergone a craniectomy and in younger patients. In contrast to other imaging modalities, there is currently no stand-

Table 3.**7** Feasability of display of anatomical structures in 2D Ultrasound

Anatomical structure	
Mesencephalic brainstem	+++
Aqueduct	++
Red nucleus	++
Basal cistern	+++
Hypophysis, chiasm	– –
Bones of the skull base	+++
Vermis cerebelli	+
Cerebral hemisphere	++
Third ventricle	+++
Pineal body	+++
Plexus of the triangle	+++
Lateral ventricles	++
Septum pellucidum	++
Thalamus	+++
Internal capsule	+
Lenticular nuclei	– –
Lateral sulcus (Sylvian fissure)	+
Corpus callosum	+
Sulci on the brain surface	++
Tentorium	++
Falx	++

a 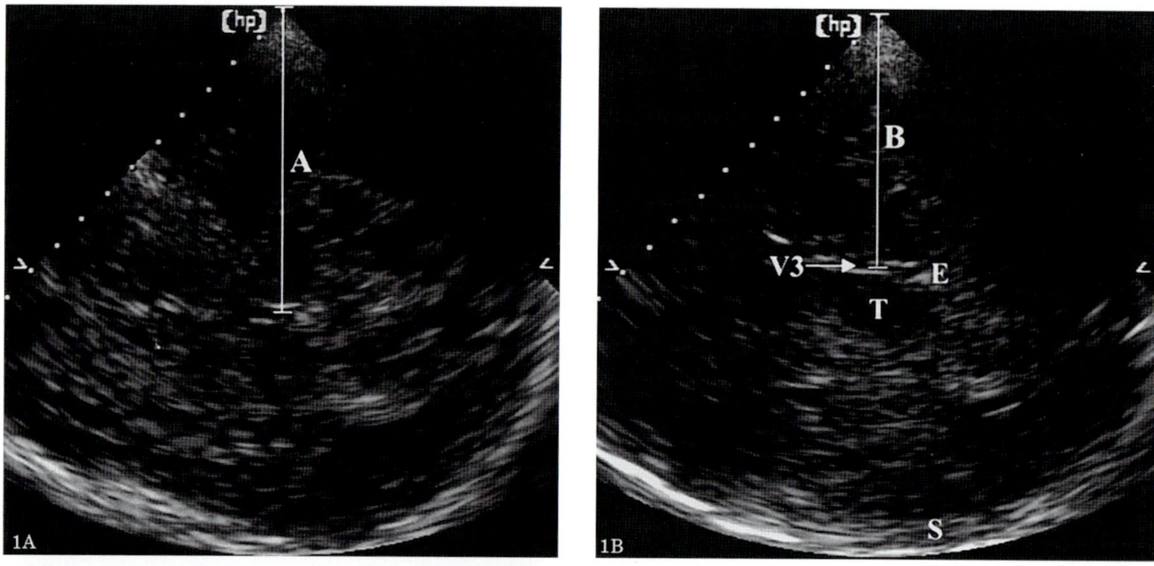 b

c 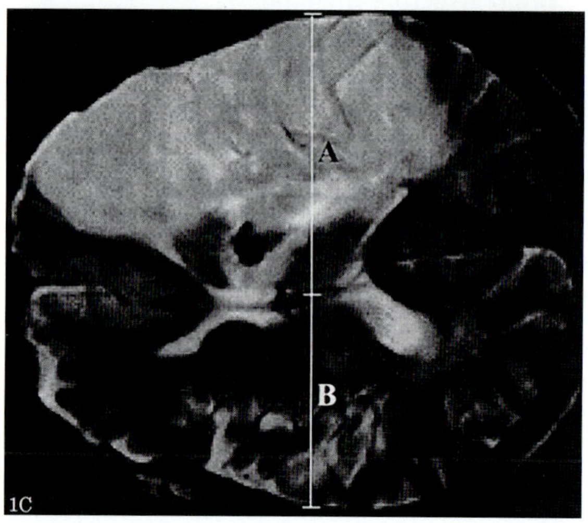

Fig. 3.**33** Assessing midline shift, using transcranial sonography and magnetic resonance imaging, in a 63-year-old man with a large MCA infarction. **a**, **b** Transcranial color-coded sonography scans 3.5 days after the onset of the stroke. Insonation was applied from the left temporal bone window (**a**) and right temporal bone window (**b**). A and B indicate the distances between the probe and the center of the third ventricle from the symptomatic side (A) and asymptomatic side (B). V_3: third ventricle; E: pineal gland; T: thalamus; S: contralateral skull. **c** A magnetic resonance image 4.5 days after the onset of the stroke. A and B indicate the distances between the cortex and the center of the third ventricle from the symptomatic side (A) and the asymptomatic side (B).

ardization of the image sections used. In addition, ultrasound brain imaging cannot currently be performed using a standard tomographic design, and the sections used are highly variable and subjective.

Hemorrhage, Midline Shift

Transcranial color-coded sonography (TCCS) is a noninvasive, easily reproducible and reliable method of monitoring midline dislocation in the third ventricle in stroke patients (Stolz et al. 1999a). It is well suited for monitoring the space-occupying effect of supratentorial strokes (Fig. 3.**33**) during treatment in critical-care and stroke units (Bertram et al. 2000). The technique can also be used to facilitate the identification of patients who are unlikely to survive without decompressive craniectomy (Gerriets et al. 2001).

Several groups have reported that duplex system analysis makes it possible to display intracerebral hemorrhage (Seidel et al. 1993, Krejza et al. 2000). However, current display technologies allow a diagnosis of intracerebral hemorrhage in only a minority of cases, and are far from providing an alternative to the standard radiographic methods (CT and MRI). In particular, it is not possible to definately exclude intracerebral bleeding. This may be due to limitations in the imaging quality of transcranial B-mode examinations or the nature of the ultrasound display of hemorrhage. Intracerebral hemorrhage has only been studied in small numbers of patients, and in particular, the ultrasound appearance of hemorrhage in relation to time has not been well described.

Tissue Perfusion

Since perfusion studies may be able to detect ischemic lesions earlier than CT and can distinguish the stroke subtype and severity of cerebral ischemia, there is growing interest in the use of perfusion imaging to predict recovery, to identify the pathogenesis of a stroke, and to monitor therapy. Validated proportional indicators of cerebral blood flow and potential diagnostic tools in stroke include technetium hexamethyl-propyleneamine oxine (Tc-HMPAO) SPECT, PET, xenon CT and perfusion-weighted MRI. The main disadvantages of these methods are that they are time-consuming, require the use of radioactive tracers, are expensive, or are poorly tolerated by critically ill or restless patients. Noninvasive and easily available methods of studying perfusion are clearly needed. In this regard, contrast harmonic imaging (CHI) may represent a useful bedside tool to provide a reliable proxy of brain perfusion.

Contrast harmonic imaging. Physiological and pathological myocardial perfusion has been assessed with contrast harmonic imaging (CHI) (Porter et al. 1996, Linka et al. 1998). More recently, several studies on CHI have shown that this technique may allow identification of physiological parenchymal cerebral ultrasound contrast enhancement in different brain areas and is related to brain perfusion. (Postert et al. 1998, Seidel et al. 1999, Postert et al. 2000b, Seidel et al. 2000, Wiesmann and Seidel 2000, Postert et al. 2001). Initial applications of CHI in patients with cerebrovascular disease have demonstrated its potential to detect perfusion deficits (Federlein et al. 2000).

Although CHI has been shown to be capable of imaging brain tissue perfusion, there remain important limitations. Significant energy loss, signal reverberations, and aberrations occur when insonating through the transtemporal bone window. Moreover, in dual-frequency harmonic imaging, the bandwidths have to be narrow to avoid overlapping between the fundamental and second harmonic frequencies. This leads to an inherent trade-off in image resolution, accentuating the problem of the transtemporal bone window.

Stimulated acoustic emission. Stimulated acoustic emission, also known as contrast burst imaging, involves color or power Doppler imaging with the transmitting power set high enough to ensure disruption of the contrast bubbles on the first pulse. Some ultrasound contrast agents, such as those with thin polymer coatings (e.g., SHU 563A, Schering, Inc., Berlin, Germany), are durable linear scatterers at low acoustic pressures but undergo destruction, fusion, or splitting at higher acoustic pressures (Postert et al. 2000a) that are within the range of diagnostic ultrasound—causing a transient high-amplitude broadband response (Uhlendorf et al. 2000). This technique has been shown to be useful for detecting bleeding sites (Goldberg et al. 1998) and liver tumors (Forsberg et al. 1999). The ability of stimulated acoustic emission to visualize brain perfusion has recently been demonstrated (Postert et al. 2000b). Current research is focusing on the ability of this technique to depict acute cerebral infarction.

Pulse inversion contrast harmonic imaging. Pulse inversion CHI (PICHI) is an ultrasound technique that minimizes the shortcomings of CHI. It uses a two-pulse sequence with a 180° phase difference to cancel out the effects of transmitted second harmonics on the received signal (Seidel et al. 1999). By preserving axial resolution and avoiding harmonic frequency overlaps, PICHI may open up new opportunities for qualitative, as well as quantitative, evaluation of the cerebral circulation.

Early results with PICHI have demonstrated excellent ultrasonographic visualization of adult brain tissue (Federlein et al. 2000). Exceptional depth penetration allows simultaneous measurement of harmonic microbubble contrast enhancement in both ipsilateral and contralateral temporal lobes, providing a basis for qualitative comparison of perfusion characteristics with a single bolus injection of contrast (see Fig. 3.**11**).

Diagnostic Effectiveness

The stenosis criteria available do not allow differentiation between *functional* and *structural* arterial stenosis in the intracranial cerebral arteries. In high-grade extracranial vascular processes, intracranial collateralization through the anterior or posterior circle of Willis often occurs. The extracranial vascular findings should always be known in order to avoid misinterpretations in such circumstances.

For methodological reasons, there have so far been hardly any reports on the validity of the evaluation of different degrees of stenosis, as is possible in extracranial Doppler sonography. As in other vascular regions, while a definite assessment of the degree of stenosis is quite possible on the basis of hemodynamic principles using Doppler sonography, angiographic evaluation of arterial stenoses of the intracranial cerebral arteries is severely restricted by projection problems, loop formation, and the fact that the stenosis can often only be depicted on a single plane.

The literature only provides isolated information about the specificity of the procedure, since intracranial stenoses occur far more rarely than extracranial ones in Western communities. Information about the specificity in the posterior circulatory system is rare. According to our own investigations (Rautenberg et al. 1990), the specificity for the anterior and posterior circulatory systems of the brain lies between 99% and 100% (Table 3.**5**). Similar results (98%) were reported by Arnolds et al. (1986) and Ley-Pozo and Ringelstein (1990) in the anterior circulation.

In the anterior circulation, and when the normal Doppler spectrum values reported agree, reliable detection of vascular processes due to pathological phenomena is possible, as long as the vascular topography remains relatively standardized. In the posterior region, vascular variants and anomalies in the vessel course predominate; continuously applying ultrasound at the specific depths that are of interest is often difficult. Using transcranial Doppler sonography and MRA, helical CTA in combination with the other techniques (Röther et al. 1993, Neff et al. 1997) appears to be promising. With its ability to display the vascular anatomy, especially in the posterior circulatory system, MRA, helical CTA provides significant additional information. The detailed additions to the static examination technique described are useful in documenting functional changes. The diagnostic usefulness of transcranial color flow Doppler sonography is attractive, but cannot yet be conclusively evaluated (Seidel et al. 1995).

References

Aaslid R, Markwalder TM, Nornes H. Noninvasive transcranial Doppler ultrasound recording of flow velocity in basal cerebral arteries. J Neurosurg 1982;57:769–74.

Aaslid R, Huber P, Nornes H. Evaluation of cerebrovascular spasm with transcranial Doppler ultrasound. J Neurosurg 1984;60:37–41.

Aaslid R, Lindegaard KF, Sorteberg W, Nornes H. Cerebral autoregulation dynamics in humans. Stroke 1989;20:45–52.

Abernathy M, Donnelly G, Kay G, Wieneke J, Morris S, Bergeson S, et al. Transcranial Doppler sonography in headache-free migraineurs. Headache 1994;34:198–203.

Ackerstaff RG, Jansen C, Moll FL, Vermeulen FE, Hamerlijnck RP, Mauser HW. The significance of microemboli detection by means of transcranial Doppler ultrasonography monitoring in carotid endarterectomy. J Vasc Surg 1995;21:963–9.

Akopov S, Whitman GT. Hemodynamic studies in early ischemic stroke: serial transcranial Doppler and magnetic resonance angiography evaluation. Stroke 2002;33:1274–9.

Alexandrov AV, Demchuk AM, Felberg RA, Christou I, Barber PA, Burgin WS, et al. High rate of complete recanalization and dramatic clinical recovery during tPA infusion when continuously monitored with 2-MHz transcranial Doppler monitoring. Stroke 2000;31:610–4.

Angeli S, Del Sette M, Beelke M, Anzola GP, Zanette E. Transcranial Doppler in the diagnosis of cardiac patent foramen ovale. Neurol Sci 2001;22:353–6.

Arnolds BJ, von Reutern GM. Transcranial Doppler sonography: examination technique and normal reference values. Ultrasound Med Biol 1986;12:115–23.

Arnolds BJ, Ochme M, von Reutern GM. Detection of intracranial stenosis and occlusion with transcranial Doppler sonography [abstract]. J Cardiovasc Ultrasonogr 1986;5:4.

Babikian VL, Hyde C, Pochay V, Winter MR. Clinical correlates of high-intensity transient signals detected on transcranial Doppler sonography in patients with cerebrovascular disease. Stroke 1994;25:1570–3.

Baumgartner RW, Baumgartner IB. [Transcranial Doppler and color duplex ultrasound: familiar and new uses; in German.] Ultraschall Med 1996;17:50–4.

Baumgartner RW, Mattle HP, Schroth G. Transcranial colour-coded duplex sonography of cerebral arteriovenous malformations. Neuroradiology 1996a;38:734–7.

Baumgartner RW, Schmid C, Baumgartner I. Comparative study of power-based versus mean frequency-based transcranial color-coded duplex sonography in normal adults. Stroke 1996b;27:101–4.

Baumgartner RW, Mattle HP, Aaslid R, Kaps M. Transcranial color-coded duplex sonography in arterial cerebrovascular disease. Cerebrovasc Dis 1997;7:57–63.

Bäzner H, Ries S, Daffertshofer M, Hennerici M. Localizing the emboligenic focus: documentation of microemboli generated across an MCA stenosis by a new multi-gate technology. Cerebrovasc Dis 1996;6(Suppl 3):62.

Becker G, Bogdahn U, Gehlberg C, Frohlich T, Hofmann E, Schlief MD. Transcranial color-coded real-time sonography of intracranial veins: normal values of blood flow velocities and findings in superior sagittal sinus thrombosis. J Neuroimaging 1995;5:87–94.

Bertram M, Khoja W, Ringleb P, Schwab S. Transcranial colour-coded sonography for the bedside evaluation of mass effect after stroke. Eur J Neurol 2000;7:639–46.

Birch AA, Dirnhuber MJ, Hartley-Davies R, Iannotti F, Neil-Dwyer G. Assessment of autoregulation by means of periodic changes in blood pressure. Stroke 1995;26:834–7.

Bishop CC, Powell S, Insall M, Rutt D, Browse NL. Effect of internal carotid artery occlusion on middle cerebral artery blood flow at rest and in response to hypercapnia. Lancet 1986;i:710–2.

Boeri R, Passerini A. The megadolichobasilar anomaly. J Neurol Sci 1964;1:475–84.

Bogdahn U, Becker G, Schlief R, Reddig J, Hassel W. Contrast-enhanced transcranial color-coded real-time sonography: results of a phase-two study. Stroke 1993;24:676–84.

Brauer P, Kochs E, Werner C, Bloom M, Policare R, Pentheny S, et al. Correlation of transcranial Doppler sonography mean flow velocity with cerebral blood flow in patients with intracranial pathology. J Neurosurg Anesthesiol 1998;10:80–5.

Braune S, Hetzel A, Prasse A, Dohms K, Guschlbauer B, Lucking CH. Stimulation of sympathetic activity by carbon dioxide in patients with autonomic failure compared to normal subjects. Clin Auton Res 1997;7:327–32.

Brunholzl C, Muller HR. [Transcranial Doppler sonography in orthostasis; in German.] Ultraschall Med 1986;7:248–52.

Burgin WS, Malkoff M, Felberg RA, Demchuk AM, Christou I, Grotta JC, et al. Transcranial Doppler ultrasound criteria for recanalization after thrombolysis for middle cerebral artery stroke. Stroke 2000;31:1128–32.

Busch W. Beitrag zur Morphologie und Pathologie der Arteria basilaris (Untersuchungsergebnisse bei 1000 Gehirnen). Arch Psychiatr Nervenkr 1966;208:326–44.

Conrad B, Klingelhöfer J. Dynamics of regional cerebral blood flow for various visual stimuli. Exp Brain Res 1989;77:437–41.

Cullinane M, Reid G, Dittrich R, Kaposzta Z, Ackerstaff R, Babikian V, et al. Evaluation of new online automated embolic signal detection algorithm, including comparison with panel of international experts. Stroke 2000;31:1335–41.

Cupini LM, Matteis M, Troisi E, Sabbadini M, Bernardi G, Caltagirone C, et al. Bilateral simultaneous transcranial Doppler monitoring of flow velocity changes during visuospatial and verbal working memory tasks. Brain 1996;119(Pt 4):1249–53.

Daffertshofer M, Hennerici M. [Syncope in neurologic diseases; in German.] Herz 1993;18:187–201.

Daffertshofer M, Hennerici M. Cerebrovascular regulation and vasoneuronal coupling. J Clin Ultrasound 1995;23:125–38.

Daffertshofer M, Diehl RR, Ziems GU, Hennerici M. Orthostatic changes of cerebral blood flow velocity in patients with autonomic dysfunction. J Neurol Sci 1991;104:32–8.

Daffertshofer M, Ries S, Schminke U, Hennerici M. High-intensity transient signals in patients with cerebral ischemia. Stroke 1996;27:1844–9.

Dahl A, Lindegaard KF, Russell D, Nyberg-Hansen R, Rootwelt K, Sorteberg W, et al. A comparison of transcranial Doppler and cerebral blood flow studies to assess cerebral vasoreactivity. Stroke 1992a;23:15–9.

Dahl A, Russell D, Nyberg-Hansen R, Rootwelt K. A comparison of regional cerebral blood flow and middle cerebral artery blood flow velocities: simultaneous measurements in healthy subjects. J Cereb Blood Flow Metab 1992b;12:1049–54.

Dahl A, Russell D, Nyberg-Hansen R, Rootwelt K, Mowinckel P. Simultaneous assessment of vasoreactivity using transcranial Doppler ultrasound and cerebral blood flow in healthy subjects. J Cereb Blood Flow Metab 1994;14:974–81.

de Bray JM, Glatt B. Quantification of atheromatous stenosis in the extracranial internal carotid artery. International Consensus Conference. Cerebrovasc Dis 1995; 5: 414–26.

de Bray JM, Missoum A, Dubas F, Emile J, Lhoste P. Detection of vertebrobasilar intracranial stenoses: transcranial Doppler sonography versus angiography. J Ultrasound Med 1997;16:213–8.

Demchuk AM, Felburg RA, Alexandrov AV. Clinical recovery from acute ischemic stroke after early reperfusion of the brain with intravenous thrombolysis. N Engl J Med 1999;340:894–5.

Deppe M, Knecht S, Papke K, Lohmann H, Fleischer H, Heindel W, et al. Assessment of hemispheric language lateralization: a comparison between fMRI and fTCD. J Cereb Blood Flow Metab 2000;20:263–8.

Diehl RR, Berlit P. Dopplerfunktionstests. In: Diehl RR, Berlit P. Funktionelle Dopplersonographie in der Neurologie. Berlin: Springer, 1996:23–46.

Diehl RR, Linden D, Lucke D, Berlit P. Spontaneous blood pressure oscillations and cerebral autoregulation, Clin Anton Res. 1998 Feb; 8(1):7–12.

Diehl RR, Sitzer M, Hennerici M. Changes of cerebral blood flow velocity during cognitive activity [letter]. Stroke 1990;21:1236–7.

Diehl RR, Diehl B, Sitzer M, Hennerici M. Spontaneous oscillations in cerebral blood flow velocity in normal humans and in patients with carotid artery disease. Neurosci Lett 1991;127:5–8.

Diehl RR, Linden D, Lucke D, Berlit P. Phase relationship between cerebral blood flow velocity and blood pressure: a clinical test of autoregulation. Stroke 1995;26:1801–4.

Drager B, Knecht S. When finding words becomes difficult: is there activation of the subdominant hemisphere? Neuroimage 2002;16(3 Pt 1):794–800.

Droste DW, Harders AG, Rastogi E. A transcranial Doppler study of blood flow velocity in the middle cerebral arteries performed at rest and during mental activities. Stroke 1989a;20:1005–11.

Droste DW, Harders AG, Rastogi E. Two transcranial Doppler studies on blood flow velocity in both middle cerebral arteries during rest and the performance of cognitive tasks. Neuropsychologia 1989b;27:1221–30.

Droste DW, Hansberg T, Kemeny V, Hammel D, Schulte-Altedorneburg G, Nabavi DG, et al. Oxygen inhalation can differentiate gaseous from nongaseous microemboli detected by transcranial Doppler ultrasound. Stroke 1997;28:2453–6.

Droste DW, Kaps M, Navabi DG, Ringelstein EB. Ultrasound contrast enhancing agents in neurosonology: principles, methods, future possibilities. Acta Neurol Scand 2000;102:1–10.

EC/IC Bypass Study Group. The International Cooperative Study of Extracranial/Intracranial Arterial Anastomosis (EC/IC Bypass Study): methodology and entry characteristics. Stroke 1985;16:397–406.

Ehrenreich DL, Burns RA, Alman RW, Fazekas JF. Influence of acetazolamide on cerebral blood flow. Arch Neurol 1961;5:227–32.

Erickson SJ, Mewissen MW, Foley WD, Lawson TL, Middleton WD, Quiroz FA, et la. Stenosis of the internal carotid artery: assessment using color Doppler imaging compared with angiography. AJR Am J Roentgenol 1989;152:1299–305.

Federlein J, Postert T, Meves S, Weber S, Przuntek H, Buttner T. Ultrasonic evaluation of pathological brain perfusion in acute stroke using second harmonic imaging. J Neurol Neurosurg Psychiatry 2000;69:616–22.

Felberg RA, Christou I, Demchuk AM, Malkoff M, Alexandrov AV. Screening for intracranial stenosis with transcranial Doppler: the accuracy of mean flow velocity thresholds. J Neuroimaging 2002;12:9–14.

Feri M, Ralli L, Felici M, Vanni D, Capria V. Transcranial Doppler and brain death diagnosis. Crit Care Med 1994;22:1120–6.

Fischer E. Die Lageabweichung der vorderen Hirnarterie im Gefässbild. Zentralbl Neurochir 1938;3:300–13.

Fisher CM. The circle of Willis: anatomical variations. Vasc Dis 1965;2:99–105.

Forsberg F, Goldberg BB, Liu JB, Merton DA, Rawool NM, Shi WT. Tissue-specific US contrast agent for evaluation of hepatic and splenic parenchyma. Radiology 1999;210:125–32.

Fujii K, Heistad DD, Faraci FM. Flow-mediated dilatation of the basilar artery in vivo. Circ Res 1991;69:697–705.

Furlan A, Higashida R, Wechsler L, Gent M, Rowley H, Kase C, et al. Intra-arterial prourokinase for acute ischemic stroke. The PROACT II study: a randomized controlled trial. Prolyse in acute cerebral thromboembolism. JAMA 1999;282:2003–11.

Gabrielsen TO, Greitz T. Normal size of the internal carotid, middle cerebral and anterior cerebral arteries. Acta Radiol 1970;10:1–10.

Gao S, Lam WW, Chan YL, Liu JY, Wong KS. Optimal values of flow velocity on transcranial Doppler in grading middle cerebral artery stenosis in comparison with magnetic resonance angiography. J Neuroimaging 2002;12:213–8.

Georgiadis D, Schwab S, Baumgartner RW. [Clinical relevance of detection of microembolism signals with transcranial Doppler ultrasound diagnosis; in German.] Nervenarzt 2002;73:125–32.

Geroulakos G, Domjan J, Nicolaides A, Stevens J, Labropoulos N, Ramaswami G, et al. Ultrasonic carotid artery plaque structure and the risk of cerebral infarction on computed tomography. J Vasc Surg 1994;20:263–6.

Gerraty RP, Bowser DN, Infeld B, Mitchell PJ, Davis SM. Microemboli during carotid angiography: association with stroke risk factors or subsequent magnetic resonance imaging changes? Stroke 1996;27:1543–7.

Gerriets T, Seidel G, Fiss I, Modrau B, Kaps M. Contrast-enhanced transcranial color-coded duplex sonography: efficiency and validity. Neurology 1999;52:1133–7.

Gerriets T, Postert T, Goertler M, Stolz E, Schlachetzki F, Sliwka U, et al. DIAS I: duplex-sonographic assessment of the cerebrovascular status in acute stroke. A useful tool for future stroke trials. Stroke 2000;31:2342–5.

Gerriets T, Stolz E, Konig S, Babacan S, Fiss I, Jauss M, et al. Sonographic monitoring of midline shift in space-occupying stroke: an early outcome predictor. Stroke 2001;32:442–7.
Giller CA, Iacopino DG. Use of middle cerebral velocity and blood pressure for the analysis of cerebral autoregulation at various frequencies: the coherence index. Neurol Res 1997;19:634–40.
Glagov S, Bassiouny HS, Giddens DP, Zarins CK. Intimal thickening: morphogenesis, functional significance and detection. J Vasc Investig 1995;1:2–14.
Goertler M, Niethammer R, Widder B. Differentiating subtotal carotid artery stenoses from occlusions by colour-coded duplex sonography. J Neurol 1994;241:301–5.
Goertler M, Blaser T, Krueger S, Lutze G, Wallesch CW. Acetylsalicylic acid and microembolic events detected by transcranial Doppler in symptomatic arterial stenoses. Cerebrovasc Dis 2001;11:324–9.
Goertler M, Blaser T, Krueger S, Hofmann K, Baeumer M, Wallesch CW. Cessation of embolic signals after antithrombotic prevention is related to reduced risk of recurrent arterioembolic transient ischaemic attack and stroke. J Neurol Neurosurg Psychiatry 2002;72:338–42.
Goldberg BB, Merton DA, Liu JB, Forsberg F. Evaluation of bleeding sites with a tissue-specific sonographic contrast agent: preliminary experiences in an animal model. J Ultrasound Med 1998;17:609–16; quiz 617–8.
Griewing B, Morgenstern C, Driesner F, Kallwellis G, Walker ML, Kessler C. Cerebrovascular disease assessed by color-flow and power Doppler ultrasonography: comparison with digital subtraction angiography in internal carotid artery stenosis. Stroke 1996;27:95–100.
Grubb BP, Gerard G, Roush K, Temesy-Armos P, Montford P, Elliott L, et al. Cerebral vasoconstriction during head-upright tilt-induced vasovagal syncope: a paradoxic and unexpected response. Circulation 1991;84:1157–64.
Guo Z, Fenster A. Three-dimensional power Doppler imaging: a phantom study to quantify vessel stenosis. Ultrasound Med Biol 1996;22:1059–69.
Hames TK, Humphries KN, Gazzard VM, Powell TV, McLellan DL. The role of continuous wave Doppler imaging in a vascular unit. Cardiovasc Res 1985;19:631–5.
Hansberg T, Wong KS, Droste DW, Ringelstein EB, Kay R. Effects of the ultrasound contrast-enhancing agent Levovist on the detection of intracranial arteries and stenoses in chinese by transcranial Doppler ultrasound. Cerebrovasc Dis 2002;14:105–8.
Harders A. Neurosurgical applications of transcranial Doppler sonography. Vienna: Springer, 1986.
Harward TR, Bernstein EF, Fronek A. Continuous-wave versus range-gated pulsed Doppler power frequency spectrum analysis in the detection of carotid arterial occlusive disease. Ann Surg 1986;204:32–7.
Hass WK, Fields WS, North RR, Kircheff II, Chase NE, Bauer RB. Joint study of extracranial arterial occlusion, 2: arteriography, techniques, sites, and complications. JAMA 1968;203:961–8.
Hassler W. Hemodynamic aspects of cerebral angiomas. Vienna: Springer, 1986.
Hedges TR. Ocular ischemia. In: Caplan LR, editor. Brain ischemia: from basic science to treatment. London: Springer, 1995: 61–73.
Hennerici MG. Can carotid endarterectomy be improved by neurovascular monitoring? Stroke 1993;24:637–8.
Hennerici MG. High intensity transcranial signals (HITS): a questionable "jackpot" for the prediction of stroke risk. J Heart Valve Dis 1994;3:124–5.
Hennerici M. High-intensity transient signals: evolution or revolution in understanding cerebral embolism? Eur Neurol 1995;35:249–53.
Hennerici MG, Daffertshofer M. Noninvasive vascular testing. In: Fisher M, Bogousslavsky J, editors. Current review of cerebrovascular disease. Philadelphia: Current Medicine, 1993: 121–37.
Hennerici M, Rautenberg W, Sitzer G, Schwartz A. Transcranial Doppler ultrasound for the assessment of intracranial arterial flow velocity, 1: examination technique and normal values. Surg Neurol 1987a;27:439–48.
Hennerici M, Rautenberg W, Schwartz A. Transcranial Doppler ultrasound for the assessment of intracranial arterial flow velocity, 2: evaluation of intracranial arterial disease. Surg Neurol 1987b;27:523–32.
Hennerici M, Klemm C, Rautenberg W. The subclavian steal phenomenon: a common vascular disorder with rare neurologic deficits. Neurology 1988;38:669–73.
Herman LH, Ostrowski AZ, Gurdjian ES. Perforating branches of the middle cerebral artery: an anatomical study. Arch Neurol 1963;8:32–4.
Hetzel A, Braune S, Guschlbauer B, Dohms K. CO_2 reactivity testing without blood pressure monitoring? Stroke 1999;30:398–401.
Hsu LC, Chern CM, Sheng WY, Wong WJ, Luk YO, Hu HH. Transcranial Doppler monitoring with head-upright tilting in patients with syncope. Zhonghua Yi Xue Za Zhi (Taipei) 1999;62:544–9.
Huang Y, Gao S, Wang B, Li S. The evaluation of intra- and extra-cranial circulation in subclavian steal syndrome. Chin Med J (Engl) 1997;110:286–8.
Huber P, Handa J. Effect of contrast material, hypercapnia, hyperventilation, hypertonic glucose and papaverine on the diameter of the cerebral arteries: angiographic determination in man. Invest Radiol 1967;2:17–32.
Jain KK. Some observations on the anatomy of the middle cerebral artery. Can J Surg 1964;7:134–9.
Job FP, Grafen A, Flachskampf FA, Ringelstein EB, Hanrath P. Stellenwert der transkraniellen Kontrastdopplersonographie in der Diagnostik des klinisch relevanten offenen Foramen ovale (PFO) bei der kardialen Emboliequellensuche [abstract]. Z Kardiol 1993;82(Suppl 1):123.
Junghans U, Siebler M. Cerebral microembolism is blocked by tirofiban, a selective nonpeptide platelet glycoprotein IIb/IIIa receptor antagonist. Circulation. 2003 Jun 3;107(21):2717–21.
Kaps M, Damian MS, Teschendorf U, Dorndorf W. Transcranial Doppler ultrasound findings in middle cerebral artery occlusion. Stroke 1990;21:532–7.
Kastrup A, Dichgans J, Niemeier M, Schabet M. Changes of cerebrovascular CO_2 reactivity during normal aging. Stroke 1998;29:1311–4.
Kelley RE, Chang JY, Scheinman NJ, Levin BE, Duncan RC, Lee SC. Transcranial Doppler assessment of cerebral flow velocity during cognitive tasks. Stroke 1992;23:9–14.
Kenton AR, Martin PJ, Evans DH. Power Doppler: an advance over colour Doppler for transcranial imaging? Ultrasound Med Biol 1996;22:313–7.
Kilic T, Pamir MN, Budd S, Ozek MM, Erzen C. Grading and hemodynamic follow-up study of arteriovenous malformations with transcranial Doppler ultrasonography. J Ultrasound Med 1998;17):729–38.
Kleiser B, Scholl D, ▪▪, et al. Doppler CO_2 and Diamox test: decreased reliability by changes of vessel diameter? Cerebrovasc Dis 1995;5:397–402.

Klingelhofer J, Conrad B, Benecke R, Sander D. Intracranial flow patterns at increasing intracranial pressure. Klin Wochenschr 1987;65:542–5.

Klingelhofer J, Conrad B, Benecke R, Sander D, Markakis E. Evaluation of intracranial pressure from transcranial Doppler studies in cerebral disease. J Neurol 1988;235:159–62.

Klingelhofer J, Matzander G, Sander D, Schwarze J, Boecker H, Bischoff C. Assessment of functional hemispheric asymmetry by bilateral simultaneous cerebral blood flow velocity monitoring. J Cereb Blood Flow Metab 1997;17:577–85.

Klotzsch C, Popescu O, Berlit P. Assessment of the posterior communicating artery by transcranial color-coded duplex sonography. Stroke 1996;27:486–9.

Knecht S, Deppe M, Backer M, Ringelstein EB, Henningsen H. Regional cerebral blood flow increases during preparation for and processing of sensory stimuli. Exp Brain Res 1997;116:309–14.

Knecht S, Deppe M, Ebner A, Henningsen H, Huber T, Jokeit H, et al. Noninvasive determination of language lateralization by functional transcranial Doppler sonography: a comparison with the Wada test. Stroke 1998;29:82–6.

Knecht S, Deppe M, Ringelstein EB. Determination of cognitive hemispheric lateralization by "functional" transcranial Doppler cross-validated by functional MRI [letter]. Stroke 1999;30:2491–2.

Knecht S, Drager B, Deppe M, Bobe L, Lohmann H, Floel A, et al. Handedness and hemispheric language dominance in healthy humans. Brain 2000;123:2512–8.

Krayenbühl H, Yasargil MG. Die vaskulären Erkrankungen im Gebiet der Arteria vertebralis und Arteria basilaris. Stuttgart: Thieme, 1957.

Krayenbühl H, Yasargil MG, Huber P. Zerebrale Angiographie für Klinik und Praxis. Stuttgart: Thieme, 1979.

Krejza J, Mariak Z, Bert RJ. Transcranial colour Doppler sonography in emergency management of intracerebral haemorrhage caused by an arteriovenous malformation: case report. Neuroradiology 2000;42:900–4.

Ladwig S, Ries S, Henning O, Valikovics A, Daffertshofer M, Pohlmann-Eden B. Combined electroencephalography and measurements of transcranial blood flow velocity during orthostatic testing: a new approach to assess syncope of unknown origin? Clin Auton Res 1997;7:305–9.

Lang J. Clinical anatomy of the posterior cranial fossa and its foramina. Stuttgart: Thieme, 1991.

Laumer R, Steinmeier R, Gonner F, Vogtmann T, Priem R, Fahlbusch R. Cerebral hemodynamics in subarachnoid hemorrhage evaluated by transcranial Doppler sonography. Part 1. Reliability of flow velocities in clinical management. Neurosurgery 1993;33:1–8; discussion 8–9.

Lechat P, Mas JL, Lascault G, Loron P, Theard M, Klimczac M, et al. Prevalence of patent foramen ovale in patients with stroke. N Engl J Med 1988;318:1148–52.

Lewandowski CA, Frankel M, Tomsick TA, Broderick J, Frey J, Clark W, et al. Combined intravenous and intra-arterial r-TPA versus intra-arterial therapy of acute ischemic stroke: Emergency Management of Stroke (EMS) Bridging Trial. Stroke 1999;30:2598–605.

Ley-Pozo J, Ringelstein EB. Noninvasive detection of occlusive disease of the carotid siphon and middle cerebral artery. Ann Neurol 1990;28:640–7.

Lindegaard KF, Bakke SJ, Grolimund P, Aaslid R, Huber P, Nornes H. Assessment of intracranial hemodynamics in carotid artery disease by transcranial Doppler ultrasound. J Neurosurg 1985;63:890–8.

Lindegaard KF, Bakke SJ, Aaslid R, Nornes H. Doppler diagnosis of intracranial artery occlusive disorders. J Neurol Neurosurg Psychiatry 1986;49:510–8.

Linka AZ, Sklenar J, Wei K, Jayaweera AR, Skyba DM, Kaul S. Assessment of transmural distribution of myocardial perfusion with contrast echocardiography. Circulation 1998;98:1912–20.

Mariak Z, Krejza J, Swiercz M, Kordecki K, Lewko J. Accuracy of transcranial color Doppler ultrasonography in the diagnosis of middle cerebral artery spasm determined by receiver operating characteristic analysis. J Neurosurg 2002;96:323–30.

Markus HS, Brown MM. Differentiation between different pathological cerebral embolic materials using transcranial Doppler in an in vitro model. Stroke 1993;24:1–5; erratum in: Stroke 1993;24:913.

Markus HS, Harrison MJ. Estimation of cerebrovascular reactivity using transcranial Doppler, including the use of breath-holding as the vasodilatory stimulus. Stroke 1992;23:668–73.

Markwalder TM, Grolimund P, Seiler RW, Roth F, Aaslid R. Dependency of blood flow velocity in the middle cerebral artery on end-tidal carbon dioxide partial pressure–a transcranial ultrasound Doppler study. J Cereb Blood Flow Metab 1984;4:368–72.

Martin PJ, Gaunt ME, Naylor AR, Hope DT, Orpe V, Evans DH. Intracranial aneurysms and arteriovenous malformations: transcranial colour-coded sonography as a diagnostic aid. Ultrasound Med Biol 1994;20:689–98.

Mast H, Ecker S, Marx P. Cerebral ischemia induced by compression tests during transcranial Doppler sonography. Clin Investig 1993;71:46–8.

Matta BF, Lam AM, Strebel S, Mayberg TS. Cerebral pressure autoregulation and carbon dioxide reactivity during propofol-induced EEG suppression. Br J Anaesth 1995;74:159–63.

Matteis M, Silvestrini M, Troisi E, Cupini LM, Caltagirone C. Transcranial Doppler assessment of cerebral flow velocity during perception and recognition of melodies. J Neurol Sci 1997;149:57–61.

Mess WH, Titulaer BM, Ackerstaff RGA. An in vivo model to detect microemboli with multidepth technique preliminary results [abstract]. Cerebrovasc Dis 1996;6(Suppl 3):60.

Mess WH, Titulaer BM, Ackerstaff RG. A new algorithm for off-line automated emboli detection based on the pseudo-Wigner power distribution and the dual gate TCD technique. Ultrasound Med Biol 2000;26:413–8.

Michaeli D, Rappaport ZH. Tissue resonance analysis: a novel method for noninvasive monitoring of intracranial pressure. J Neurosurg 2002;96:1132–7.

Moehring MA, Spencer MP. Power M-mode Doppler (PMD) for observing cerebral blood flow and tracking emboli. Ultrasound Med Biol 2002;28:49–57.

Mull M, Aulich A, Hennerici M. Transcranial Doppler ultrasonography versus arteriography for assessment of the vertebrobasilar circulation. J Clin Ultrasound 1990;18:539–49.

Muller M, Behnke S, Walter P, Omlor G, Schimrigk K. Microembolic signals and intraoperative stroke in carotid endarterectomy. Acta Neurol Scand 1998;97:110–7.

Nabavi DG, Georgiadis D, Mumme T, Schmid C, Mackay TG, et al. Clinical relevance of intracranial microembolic signals in patients with left ventricular assist devices: a prospective study. Stroke 1996;27:891–6.

Neff KW, Lehmann KJ, Ries S, Sommer A, Steinke W, Schwartz A, et al. CTA of middle cerebral artery stenosis and occlusion: is CTA as valid as MRA and TCD? [abstract]. Cerebrovasc Dis 1997;7(Suppl 4):14.

Niehaus L, Lehmann R, Roricht S, Meyer BU. Age-related reduction in visually evoked cerebral blood flow responses. Neurobiol Aging 2001;22:35–8.

Nikutta P, Schneider M, Claus G, Schellong SM, Kühn H, Hausmann D, et al. Wie zuverlässig ist der transkranielle Dopplerultraschall in der Diagnostik des offenen Foramen ovale? [abstract] Z Kardiol 1993;82(Suppl 1):123.

Olesen J. Cerebral and extracranial circulatory disturbances in migraine: pathophysiological implications. Cerebrovasc Brain Metab Rev 1991;3:1–28.

Otis S, Rush M, Boyajian R. Contrast-enhanced transcranial imaging: results of an American phase-two study. Stroke 1995;26:203–9.

Padget DH. The circle of Willis: its embryology and anatomy. Ithaca, NY: Comestock, 1944.

Panczel G, Daffertshofer M, Ries S, Spiegel D, Hennerici M. Age and stimulus dependency of visually evoked cerebral blood flow responses. Stroke 1999;30:619–23.

Panerai RB, White RP, Markus HS, Evans DH. Grading of cerebral dynamic autoregulation from spontaneous fluctuations in arterial blood pressure. Stroke 1998;29:2341–6.

Perry MO, Silane MF, Calcagno D, Spradlin S. Quantitative Doppler spectrum analysis of extracranial carotid artery stenosis. N Y State J Med 1985;85:577–80.

Piechnik SK, Yang X, Czosnyka M, Smielewski P, Fletcher SH, Jones AL, et al. The continuous assessment of cerebrovascular reactivity: a validation of the method in healthy volunteers. Anesth Analg 1999;89:944–9.

Piepgras A, Schmiedek P, Leinsinger G, Haberl RL, Kirsch CM, Einhaupl KM. A simple test to assess cerebrovascular reserve capacity using transcranial Doppler sonography and acetazolamide. Stroke 1990;21:1306–11.

Porter TR, Xie F, Kricsfeld D, Armbruster RW. Improved myocardial contrast with second harmonic transient ultrasound response imaging in humans using intravenous perfluorocarbon-exposed sonicated dextrose albumin. J Am Coll Cardiol 1996;27:1497–501.

Posner JB, Plum F. The toxic effects of carbon dioxide and acetazolamide in hepatic encephalopathy. J Clin Invest 1960;39:1246–58.

Postert T, Meves S, Bornke C, Przuntek H, Büttner T. Power Doppler compared to color-coded duplex sonography in the assessment of the basal cerebral circulation. J Neuroimaging 1997;7:221–6.

Postert T, Muhs A, Meves S, Federlein J, Przuntek H, Büttner T. Transient response harmonic imaging: an ultrasound technique related to brain perfusion. Stroke 1998;29:1901–7.

Postert T, Hoppe P, Federlein J, Helbeck S, Ermert H, Przuntek H, et al. Contrast agent specific imaging modes for the ultrasonic assessment of parenchymal cerebral echo contrast enhancement. J Cereb Blood Flow Metab 2000a;20:1709–16.

Postert T, Hoppe P, Federlein J, Przuntek H, Buttner T, Helbeck S, et al. Ultrasonic assessment of brain perfusion. Stroke 2000b;31:1460–2.

Postert T, Federlein J, Rose J, Przuntek H, Weber S, Büttner T. Ultrasonic assessment of physiological echo-contrast agent distribution in brain parenchyma with transient response second harmonic imaging. J Neuroimaging 2001;11:18–24.

Poulin MJ, Robbins PA. Indexes of flow and cross-sectional area of the middle cerebral artery using Doppler ultrasound during hypoxia and hypercapnia in humans. Stroke 1996;27:2244–50.

Puppo C, Lopez L, Panzardo H, Caragna E, Mesa P, Biestro A. Comparison between two static autoregulation evaluation methods. Acta Neurochir Suppl 2002;81:129–32.

Rautenberg W, Hennerici M. Pulsed Doppler assessment of innominate artery obstructive diseases. Stroke 1988;19:1514–20.

Rautenberg W, Schwartz A, Mull M, Aulich A, Hennerici M. Noninvasive detection of intracranial stenoses and occlusions. Stroke 1990;21:49.

Rim SJ, Leong-Poi H, Lindner JR, Couture D, Ellegala D, Mason H, et al. Quantification of cerebral perfusion with "real-time" contrast-enhanced ultrasound. Circulation 2001;104:2582–7.

Ringelstein EB, Grosse W, Matentzoglu S, Glockner WM. Non-invasive assessment of the cerebral vasomotor reactivity by means of transcranial Doppler sonography during hyper- and hypocapnea. Klin Wochenschr 1986;64:194–5.

Ringelstein EB, Sievers C, Ecker S, Schneider PA, Otis SM. Noninvasive assessment of CO_2-induced cerebral vasomotor response in normal individuals and patients with internal carotid artery occlusions. Stroke 1988;19:963–9.

Rosenkranz K, Zendel W, Langer R, Heim T, Schubeus P, Scholz A, et al. Contrast-enhanced transcranial Doppler US with a new transpulmonary echo contrast agent based on saccharide microparticles. Radiology 1993;187:439–43.

Röther J, Wentz KU, Rautenberg W, Schwartz A, Hennerici M. Magnetic resonance angiography in vertebrobasilar ischemia. Stroke 1993;24:1310–5.

Röther J, Schwartz A, Wentz KU, Rautenberg W, Hennerici M. Middle cerebral artery stenoses: assessment by MRA and TDU. Cerebrovasc Dis 1994;4:273–9.

Russell D. Cerebral microemboli and cognitive impairment. J Neurol Sci 2002;203–204:211–4.

Russell D, Brucher R. Online automatic discrimination between solid and gaseous cerebral microemboli with the first multifrequency transcranial Doppler. Stroke 2002;33:1975–80.

Sander D, Meyer BU, Roricht S, Klingelhofer J. Effect of hemisphere-selective repetitive magnetic brain stimulation on middle cerebral artery blood flow velocity. Electroencephalogr Clin Neurophysiol 1995;97:43–8.

Schmidt B, Klingelhofer J, Schwarze JJ, Sander D, Wittich I. Noninvasive prediction of intracranial pressure curves using transcranial Doppler ultrasonography and blood pressure curves. Stroke 1997;28:2465–72.

Schmidt B, Czosnyka M, Schwarze JJ, Sander D, Gerstner W, Lumenta CB, et al. Cerebral vasodilatation causing acute intracranial hypertension: a method for noninvasive assessment. J Cereb Blood Flow Metab 1999;19:990–6.

Schmidt EA, Czosnyka M, Matta BF, Gooskens I, Piechnik S, Pickard JD. Non-invasive cerebral perfusion pressure (nCPP): evaluation of the monitoring methodology in head injured patients. Acta Neurochir Suppl 2000;76:451–2.

Schmidt P, Krings T, Willmes K, Roessler F, Reul J, Thron A. Determination of cognitive hemispheric lateralization by "functional" transcranial Doppler cross-validated by functional MRI. Stroke 1999;30:939–45.

Schminke U, Ries S, Daffertshofer M, Staedt U, Hennerici M. Patent foramen ovale: a potential source of cerebral embolism? Cerebrovasc Dis 1995;5:133–8.

Schwartz A, Hennerici M. Noninvasive transcranial Doppler ultrasound in intracranial angiomas. Neurology 1986a;36:626–35.

Schwartz A, Rautenberg W, Hennerici M. Dolichoectatic intracranial arteries: review of selected aspects. Cerebrovasc Dis 1993;3:273–9.

Sedzimir CB. An angiographic test of collateral circulation through the anterior segment of the circle of Willis. J Neurol Neurosurg Psychiatry 1959;22:64–8.

Seidel G, Kaps M, Dorndorf W. Transcranial color-coded duplex sonography of intracerebral hematomas in adults. Stroke 1993;24:1519–27.

Seidel G, Kaps M, Gerriets T. Potential and limitations of transcranial color-coded sonography in stroke patients. Stroke 1995;26:2061–6.

Seidel G, Greis C, Sonne J, Kaps M. Harmonic grey scale imaging of the human brain. J Neuroimaging 1999;9:171–4; erratum in: J Neuroimaging 1999;9:247.

Seidel G, Algermissen C, Christoph A, Claassen L, Vidal-Langwasser M, Katzer T. Harmonic imaging of the human brain: visualization of brain perfusion with ultrasound. Stroke 2000;31:151–4.

Seidel G, Claassen L, Meyer K, Vidal-Langwasser M. Evaluation of blood flow in the cerebral microcirculation: analysis of the refill kinetics during ultrasound contrast agent infusion. Ultrasound Med Biol 2001;27:1059–64.

Seiler RW, Grolimund P, Aaslid R, Huber P, Nornes H. Cerebral vasospasm evaluated by transcranial ultrasound correlated with clinical grade and CT-visualized subarachnoid hemorrhage. J Neurosurg 1986;64:594–600.

Silvestrini M, Matteis M, Troisi E, Cupini LM, Bernardi G. Cerebrovascular reactivity in migraine with and without aura. Headache 1996;36:37–40.

Silvestrini M, Cupini LM, Placidi F, Diomedi M, Bernardi G. Bilateral hemispheric activation in the early recovery of motor function after stroke. Stroke 1998a;29:1305–10.

Silvestrini M, Troisi E, Matteis M, Razzano C, Caltagirone C. Correlations of flow velocity changes during mental activity and recovery from aphasia in ischemic stroke. Neurology 1998b;50:191–5.

Sitzer M, Diehl RR, Hennerici M. Visually evoked cerebral blood flow responses: normal and pathological conditions. J Neuroimaging 1992;2:65–70.

Smielewski P, Czosnyka M, Kirkpatrick P, McEroy H, Rutkowska H, Pickard JD. Assessment of cerebral autoregulation using carotid artery compression. Stroke 1996;27:2197–203.

Spencer MP, Thomas GI, Moehring MA. Relation between middle cerebral artery blood flow velocity and stump pressure during carotid endarterectomy. Stroke 1992;23:1439–45.

Steinmeier R, Bauhuf C, Hubner U, Hofmann RP, Fahlbusch R. Continuous cerebral autoregulation monitoring by cross-correlation analysis: evaluation in healthy volunteers. Crit Care Med 2002;30:1969–75.

Stoll M, Hamann GF, Mangold R, Huf O, Winterhoff-Spurk P. Emotionally evoked changes in cerebral hemodynamics measured by transcranial Doppler sonography. J Neurol 1999a;246:127–33.

Stoll M, Seidel A, Schimrigk K, Hamann GF. Orthostasis as a test for cerebral autoregulation in normal persons and patients with carotid artery disease. J Neuroimaging 1999b;9:113–7.

Stolz E, Gerriets T, Fiss I, Babacan SS, Seidel G, Kaps M. Comparison of transcranial color-coded duplex sonography and cranial CT measurements for determining third ventricle midline shift in space-occupying stroke. AJNR Am J Neuroradiol 1999a;20:1567–71.

Stolz E, Kaps M, Kern A, Dorndorf W. Frontal bone windows for transcranial color-coded duplex sonography. Stroke 1999b;30:814–20.

Stolz E, Mendes I, Gerriets T, Kaps M. Assessment of intracranial collateral flow by transcranial color-coded duplex sonography using a temporal and frontal axial insonation plane. J Neuroimaging 2002a;12:136–43.

Stolz E, Nuckel M, Mendes I, Gerriets T, Kaps M. Vertebrobasilar transcranial color-coded duplex ultrasonography: improvement with echo enhancement. AJNR Am J Neuroradiol 2002b;23:1051–4.

Suarez JI, Qureshi AI, Yahia AB, Parekh PD, Tamargo RJ, Williams MA, et al. Symptomatic vasospasm diagnosis after subarachnoid hemorrhage: evaluation of transcranial Doppler ultrasound and cerebral angiography as related to compromised vascular distribution. Crit Care Med 2002;30:1348–55.

Sunderland S. Neurovascular relations and anomalies at the base of the brain. J Neurol Neurosurg Psychiatry 1948;11:243–54.

Suwanwela NC, Phanthumchinda K, Suwanwela N. Transcranial Doppler sonography and CT angiography in patients with atherothrombotic middle cerebral artery stroke. AJNR Am J Neuroradiol 2002;231352–5.

Teague SM, Sharma MK. Detection of paradoxical cerebral echo contrast embolization by transcranial Doppler ultrasound. Stroke 1991;22:740–5.

Telman G, Kouperberg E, Sprecher E, Yarnitsky D. The nature of microemboli in patients with artificial heart valves. J Neuroimaging 2002;12:15–8.

Terwey B, Krier C, Gerhardt P. [The demonstration of jugular venous thrombosis by high resolution ultrasound; in German.] RöFo Fortschr Geb Rontgenstr Nuklearmed 1981;134:557–9.

Thie A, Fuhlendorf A, Spitzer K, Kunze K. Transcranial Doppler evaluation of common and classic migraine, 1: ultrasonic features during the headache-free period. Headache 1990a;30:201–8.

Thie A, Fuhlendorf A, Spitzer K, Kunze K. Transcranial Doppler evaluation of common and classic migraine, 2: ultrasonic features during attacks. Headache 1990b;30:209–15.

Tiecks FP, Lam AM, Aaslid R, Newell DW. Comparison of static and dynamic cerebral autoregulation measurements. Stroke 1995;26:1014–9.

Tiecks FP, Haberl RL, Newell DW. Temporal patterns of evoked cerebral blood flow during reading. J Cereb Blood Flow Metab 1998;18:735–41.

Tönnis W, Schiefer W. Zirkulationsstärungen des Gehirns im Serienangiogramm. Berlin: Spinger, 1959.

Topcuoglu MA, Pryor JC, Ogilvy CS, Kistler JP. Cerebral vasospasm following subarachnoid hemorrhage. Curr Treat Options Cardiovasc Med 2002;4:373–84.

Troisi E, Silvestrini M, Matteis M, Monaldo BC, Vernieri F, Caltagirone C. Emotion-related cerebral asymmetry: hemodynamics measured by functional ultrasound. J Neurol 1999;246:1172–6; erratum in: J Neurol 2000;247:157.

Uggowitzer MM, Kugler C, Riccabona M, Klein GE, Leber K, Simbrunner J, et al. Cerebral arteriovenous malformations: diagnostic value of echo-enhanced transcranial Doppler sonography compared with angiography. AJNR Am J Neuroradiol 1999;20:101–6.

Uhlendorf V, Scholle FD, Reinhardt M. Acoustic behaviour of current ultrasound contrast agents. Ultrasonics 2000;38:81–6.

Umemura A, Yamada K, Masago A, Kanda Y, Matsumoto T, Shimazu N. Hemodynamic flow patterns evaluated by transcranial color-coded duplex sonography after STA-MCA bypass for internal carotid artery occlusion. Cerebrovasc Dis 2002;14:143–7.

Vanninen R, Manninen H, Koivisto K, Tulla H, Partanen K, Puranen M. Carotid stenosis by digital subtraction angiography: reproducibility of the European Carotid Surgery Trial and the North American Symptomatic Carotid Endarterectomy Trial measurement methods and visual interpretation. AJNR Am J Neuroradiol 1994;15:1635–41.

Venkatesh B, Shen Q, Lipman J. Continuous measurement of cerebral blood flow velocity using transcranial Doppler reveals significant moment-to-moment variability of data in healthy volunteers and in patients with subarachnoid hemorrhage. Crit Care Med 2002;30:563–9.

Vincent MA, Dawson D, Clark AD, Lindner JR, Rattigan S, Clark MG, et al. Skeletal muscle microvascular recruitment by physiological hyperinsulinemia precedes increases in total blood flow. Diabetes 2002;51:42–8.

Vingerhoets G, Luppens E. Cerebral blood flow velocity changes during dichotic listening with directed or divided attention: a transcranial Doppler ultrasonography study. Neuropsychologia 2001;39:1105–11.

van Zuilen EV, Moll FL, Vermeulen FE, Mauser HW, van Gijn J, Ackerstaff RG. Detection of cerebral microemboli by means of transcranial Doppler monitoring before and after carotid endarterectomy. Stroke 1995;26:210–3.

van Zuilen EV, Mess WH, Jansen C, Van der Tweel I, Van Gijn J, Ackerstaff GA. Automatic embolus detection compared with human experts: a Doppler ultrasound study. Stroke 1996;27:1840–3.

von Mitterwallner F. Variationsstatistische Untersuchungen an den basalen Hirngefässen. Acta Anat 1955;24:51–88.

von Reutern GM. Zerebraler Zirkulationsstillstand: Diagnostik mit der Dopplersongraphie. Dt Aerztebl 1991;88:B-2842–8.

von Reutern GM, Büdingen HJ. Ultrasound diagnosis of cerebrovascular disease: Doppler sonography of the extra- and intracranial arteries, duplex scanning. Stuttgart: Thieme, 1993.

Wardlaw JM, Cannon JC. Color transcranial "power" Doppler ultrasound of intracranial aneurysms. J Neurosurg 1996;84:459–61.

Wardlaw JM, Dennis MS, Merrick MV, Warlow CP. Relationship between absolute mean cerebral transit time and absolute mean flow velocity on transcranial Doppler ultrasound after ischemic stroke. J Neuroimaging 2002;12:104–11.

Wei K, Jayaweera AR, Firoozan S, Linka A, Skyba DM, Kaul S. Quantification of myocardial blood flow with ultrasound-induced destruction of microbubbles administered as a constant venous infusion. Circulation 1998;97:473–83.

Wei K, Le E, Bin JP, Coggins M, Thorpe J, Kaul S. Quantification of renal blood flow with contrast-enhanced ultrasound. J Am Coll Cardiol 2001;37:1135–40.

White PM, Wardlaw JM, Teasdale E, Sloss S, Cannon J, Easton V. Power transcranial Doppler ultrasound in the detection of intracranial aneurysms. Stroke 2001;32:1291–7.

Widder B. The Doppler CO_2 test to exclude patients not in need of extracranial/intracranial bypass surgery. J Neurol Neurosurg Psychiatry 1989;52:38–42.

Widder B, Paulat K, Hackspacher J, Mayr E. Transcranial Doppler CO_2 test for the detection of hemodynamically critical carotid artery stenoses and occlusions. Eur Arch Psychiatry Neurol Sci 1986;236:162–8.

Widder B, Kleiser B, Krapf H. Course of cerebrovascular reactivity in patients with carotid artery occlusions. Stroke 1994;25:1963–7.

Wiesmann M, Seidel G. Ultrasound perfusion imaging of the human brain. Stroke 2000;31:2421–5.

Withers CE, Gosink BB, Keightley AM, Casola G, Lee AA, van Sonnenberg E, et al. Duplex carotid sonography: peak systolic velocity in quantifying internal carotid artery stenosis. J Ultrasound Med 1990;9:345–9.

Wollschläger PB, Wollschläger G. Anterior cerebral–internal carotid artery and middle cerebral–internal carotid artery ratios. Acta Radiol Diagn (Stockh) 1966;5:615–20.

Wollschläger PB, Wollschläger G. The circle of Willis. In: Newton TH, Potts DG, editors. Radiology of the skull and brain, vol. 2: angiography. St. Louis: Mosby, 1974:324–368.

Yamamoto S, Nishizawa S, Tsukada H, Kakiuchi T, Yokoyama T, Ryu H, et al. Cerebral blood flow autoregulation following subarachnoid hemorrhage in rats: chronic vasospasm shifts the upper and lower limits of the autoregulatory range toward higher blood pressures. Brain Res 1998;782:194–201.

Zanette EM, Mancini G, De Castro S, Solaro M, Cartoni D, Chiarotti F. Patent foramen ovale and transcranial Doppler: comparison of different procedures. Stroke 1996;27:2251–5.

Zunker P, Wilms H, Brossmann J, Georgiadis D, Weber S, Deuschl G. Echo contrast-enhanced transcranial ultrasound: frequency of use, diagnostic benefit, and validity of results compared with MRA. Stroke 2002;33:2600–3.

4 Cerebral Veins

Examination

Special Equipment and Documentation

The equipment required and the procedure used to document the findings are more or less the same as those given above for the cerebral arteries (Chapters 2 and 3).

Examination Conditions

The examination conditions for both the patient and the examiner are largely the same as those described above for the cerebral arteries (Chapters 2 and 3). Since venous pressure relationships are low intracranially, changes in intra-abdominal pressure or in the position of the head relative to the body should be carefully noted. The PRF should be set to low levels in order to allow high-resolution imaging of low Doppler shift frequencies (slow velocities). The wall filter should be set to low values to avoid blocking slow flow components.

Patient and Examiner

The examinations should always be carried out in standardized conditions (e.g., always with the patient in a supine position), since the spontaneous flow relationships in the cerebral veins depend on the position of the body.

Conducting the Examination

The color signal image and correct adjustment of the pulsed-wave Doppler signal allow definite identification of the internal jugular vein by displaying venous flow signals that run in a caudal direction. The internal jugular vein can be imaged in longitudinal sections and cross-sections lateral to the carotid artery system.
Using transcranial duplex sonography, venous signals can be detected via a transtemporal access point as well as a transnuchal one (Hennerici 1990). The use of an occipital as well as a frontal access point has also been described (Aaslid et al. 1991, Stolz et al. 1999b). It has been suggested that the occipital approach can be used to allow investigation of the straight sinus by applying ultrasound at a depth of approximately 50 mm (Fig. 4.**1**). However, the use of the occipital as well as the frontal access point is not commonly accepted. The transtemporal approach is widely favored for routine examinations. Visualization of the venous structures often requires the administration of ultrasound contrast media (Ries et al. 1997). However, a combination of color flow transcranial duplex sonography with ultrasound contrast agents is the optimal approach for comprehensive imaging of the intracranial cerebral veins and allows identification of the cerebral basal veins on the basis of their anatomical relation to specific arteries (Andeweg 1996).

Functional Tests (Provoked Flow Signals)

A distinction is made between spontaneous and provoked flow signals.

Valsalva maneuver. The Valsalva maneuver modifies venous flow by increasing the abdominal pressure; flow in the jugular vein ceases, especially when data are recorded with the patient in a sitting position (Fig. 4.**2**). When the intracranial veins are being examined, even a very short reduction in flow velocity during a Valsalva test may sometimes help identify venous flow signals.

Deep inspiration. During deep inspiration, the jugular vein collapses, since the intrathoracic pressure is reduced, and simultaneously there is an initial increase in the caudally directed flow velocity. Doppler signals from the internal jugular vein can be located in the mid-neck region, between the sternocleidomastoid muscle and the carotid artery, and can be displayed both distally and proximally. It is important that the examination should be carried out with as little pressure exerted on the probe as possible.

Normal Findings

Anatomy and Findings

Cerebral drainage proceeds via superficial (external) and deep (internal) veins (Fig. 4.**1 a**, **b**) (for details, see Andeweg 1996). Both venous systems drain into the sinuses, and further drainage is provided by the internal jugular vein and the vertebral vein. The intracranial

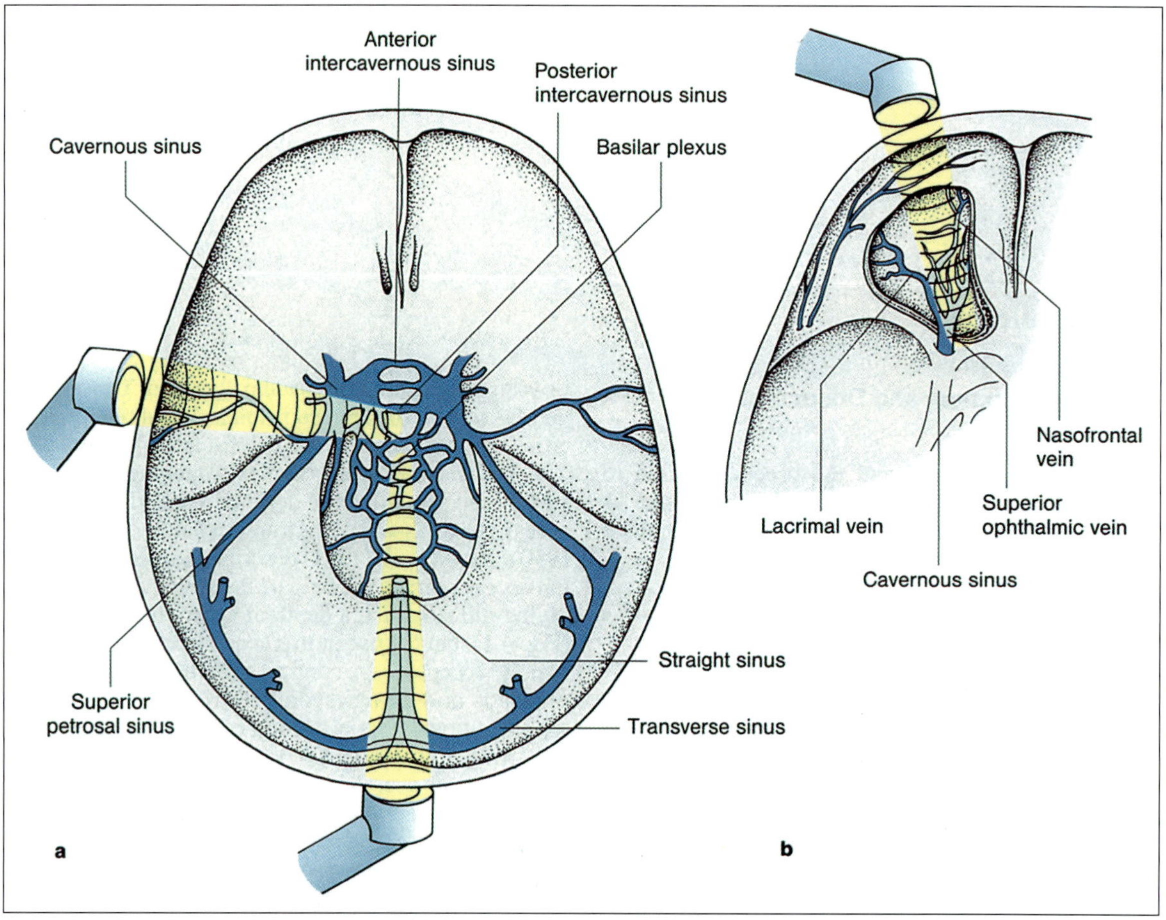

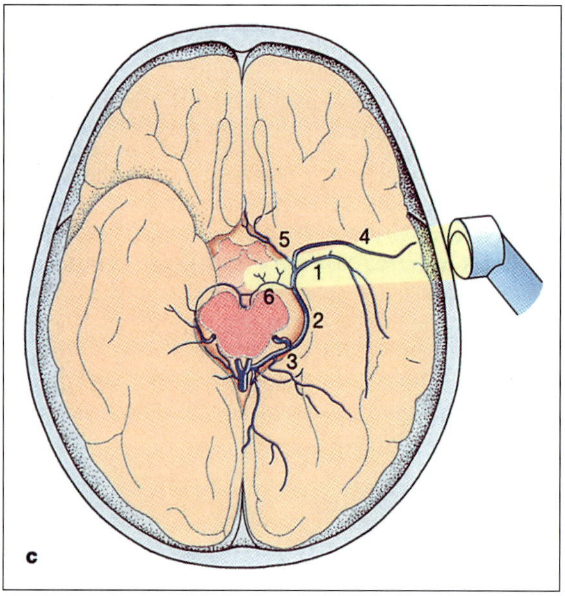

Fig. 4.**1** Schematic illustrations of the basal sinuses (**a**), periorbital veins (**b**), and basal cerebral veins (**c**). **c** Basal vein of Rosenthal: anterior segment (1), middle segment (2), posterior segment (3); deep middle cerebral vein (4), anterior cerebral vein (5), peduncular vein (6)

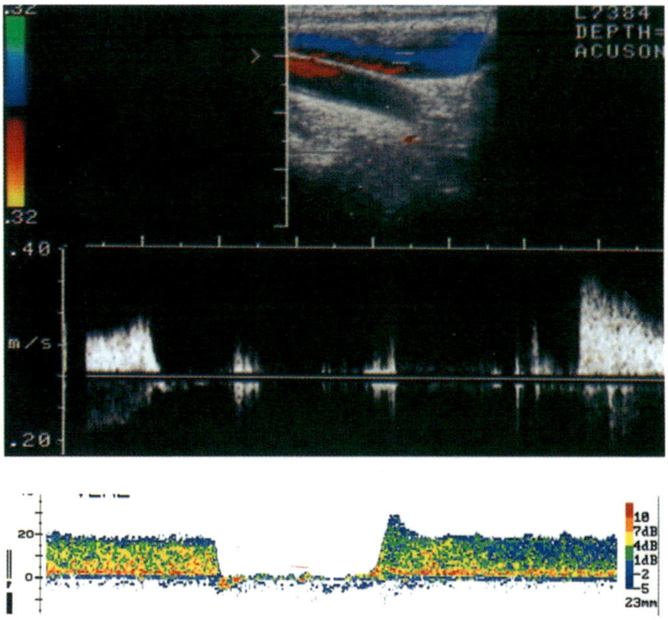

Fig. 4.**2** Display of Doppler spectra during the Valsalva maneuver. **a** Internal jugular vein. **b** Intracranial vein

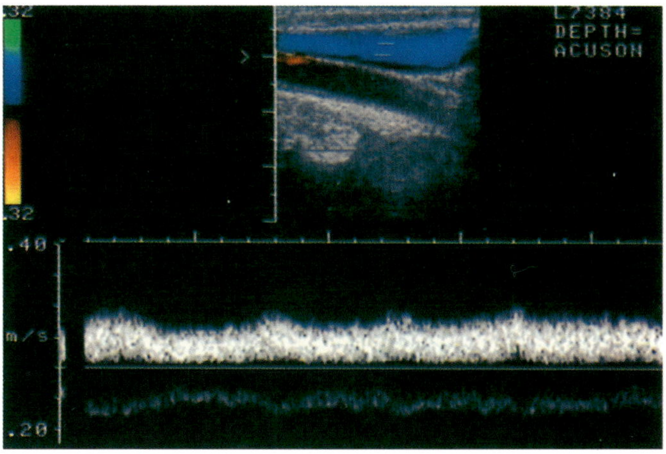

Fig. 4.**3** Display of the jugular vein using color flow duplex sonography (normal findings)

basal veins (anterior cerebral vein, deep middle cerebral vein, basal vein of Rosenthal, vein of Galen) form a circuit around the midbrain, comparable to the arterial circle of Willis. Additional, quantitatively smaller, venous drainage is provided by anastomoses between the sinuses, diploic veins, and emissary veins into the external jugular vein. There is further venous drainage via the cavernous sinus. Venous valves are rarely found in the neck and head region; the intracranial veins have no valves.

Extracranial Veins—Internal Jugular Vein

Color Flow Duplex Sonography

The B-mode allows the morphological relationships in the jugular veins to be evaluated; the pulsed-wave Doppler mode provides selective analysis of the venous flow signals without disturbance from arterial superimpositions. In color flow duplex imaging, it is possible to detect small veins that are otherwise hardly noticeable, such as the vertebral vein, in addition to the jugular venous system (Fig. 2.**12**, p. 34). In longitudinal section, the internal jugular vein is usually located lateral to the common carotid artery (Fig. 4.**3**). The vein is easily compressed. Depending on the direction of the pressure applied, a peaked venous image is produced below the sternocleidomastoid muscle (Steinke and Hennerici 1989). In some cases, venous valves can be depicted in the proximal venous segment. Sonographically, typical vascular wall pulsations can be detected. The maximum venous distension occurs during expiration, and the minimum venous diameter is reached at the end of inspiration (Fig. 4.**4**). Fluctuations in the caliber of the vessels due to respiration are superimposed by weak higher-frequency pulsations from the venous walls, which are caused by the heart. During the Valsalva maneuver, the vein dilates and the pulsations decrease.

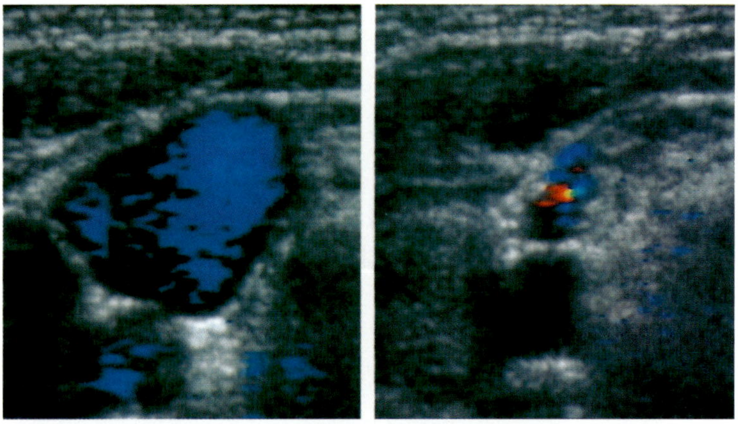

Fig. 4.**4** Varying dilation of the internal jugular vein during expiration (**a**) and inspiration (**b**)

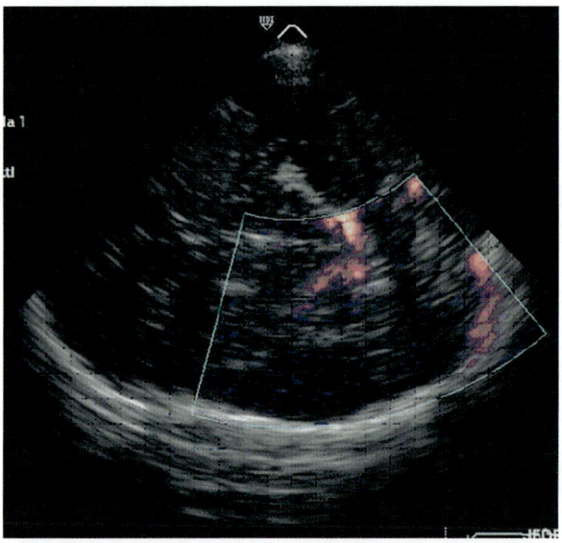

Fig. 4.**5** Display of the transverse sinus in an amplitude mode duplex image. The insonation route is from the *contralateral* temporal bone window

Often, streak-like internal echoes may be displayed, caused by slow-flowing blood. During deeper inspiration, the vascular lumen collapses, and retrograde blood flow simultaneously increases.

Color-coded flow signal imaging makes it easier to distinguish between veins and arteries. Even small-caliber veins such as the vertebral veins can usually be detected in the extracranial V_2 segment (Fig. 2.**18**). In color flow duplex sonography, imaging of the vertebral vein when the vertebral artery is missing is a criterion for vertebral artery occlusion.

Continuous-Wave Doppler Sonography

The internal jugular vein is located lateral to the common and internal carotid arteries. Usually, return flow from the upper venous system occurs passively, due to the inspiratory pressure decrease within the thorax. A biphasic progression can be observed in the Doppler sonogram, with dominant systolic and less developed diastolic flow (Baumgartner and Bollinger 1991). Acoustically, a predominantly unmodulated signal, which is affected by breathing, swallowing, and the Valsalva maneuver, is found (Fig. 4.**2**). The venous signals can easily be eliminated by applying light proximal or distal pressure with the hand.

Intracranial Veins

Color Flow Duplex Sonography

Although venous signals can often be identified from the way in which they produce a breath-like, band-shaped, and barely pulsatile signal during transcranial examination of the arterial system, systematic examination of the cerebral veins and sinuses with reproducible results is limited—even with color flow duplex sonography—to the basal vein of Rosenthal, the deep middle cerebral vein, the lateral transverse sinuses (Fig. 4.**5**), and, with a few reservations, to the rectus sinus, cavernous sinus, and inferior sagittal sinus (Hennerici 1990, Valdueza et al. 1996). As shown in Fig. 4.**1 c**, the basal vein of Rosenthal can be insonated via the transtemporal window adjacent to the P_2 segment of the posterior cerebral artery. Venous signals adjacent to the middle cerebral artery may derive from the deep middle cerebral vein. Pulsatile flow signals can also be recorded from the straight sinus and the transverse sinuses from ipsilateral and contralateral transtemporal windows, with a typical modulation effect following the Valsalva maneuver. Ultrasound contrast media that pass through the lungs are used with color flow transcranial duplex sonography (Ries et al. 1997). Although conclusive assessment of the usefulness of the procedure is not yet possible, this appears to be a suitable procedure for follow-up studies in patients with transverse sinus thrombosis and for differentiating between acute/chronic thrombosis and hypoplasia/aplasia, particularly if combined with magnetic resonance angiography (Figs. 4.**6**, 4.**10**) (Mattle et al. 1991).

Evaluation

The evaluation of findings in the cerebral veins is primarily qualitative. For transcranial recording of the basal vein of Rosenthal and the deep middle cerebral vein in adults, mean flow velocities of 4–17 cm/s (mean 10.1 ± 2.3 cm/s) have been reported. The scope of ultrasound diagnosis in the intracranial veins is still limited.

Pathological Findings

Principle

Since pathological findings here are rare, and since the accuracy of ultrasound examinations is limited, ultrasound diagnosis in the cerebral veins is less important than diagnosis in the veins of the extremities. Extracranially, the most frequent pathological finding is thrombosis of the jugular vein, often due to iatrogenic manipulation. Findings obtained from the intracranial cerebral veins provide information about direct and remote obstructions in the cerebral veins, sinus, and plexus.

Findings

Extracranial Veins

Jugular Vein Thrombosis

Color Flow Duplex Sonography

In thromboses of the internal jugular vein, the B-mode image shows a nonpulsatile, dilated blood vessel that has no detectable flow signals (Fig. 4.**7a**). Depending on the age of the thrombosis, irregular internal echoes that correspond to thrombotic material may be detected. The blood vessel cannot be compressed by exerting pressure with the probe. The Valsalva maneuver does not produce any further dilation of the vascular lumen. A flow signal is not detectable either with pulsed-wave Doppler sonography or in color flow imaging (Fig. 4.**7b**, **c**). It is important to ensure that very low flow velocities are also detected and represented in color-coded form.

If venous valves will not close, it may indicate long-standing tricuspid valve insufficiency or a prior jugular vein thrombosis (Brownlow and McKinney 1985).

Damaged venous valves, vascular wall thickening, and also floating thrombi have been detected in patients who have undergone frequent jugular venipuncture procedures.

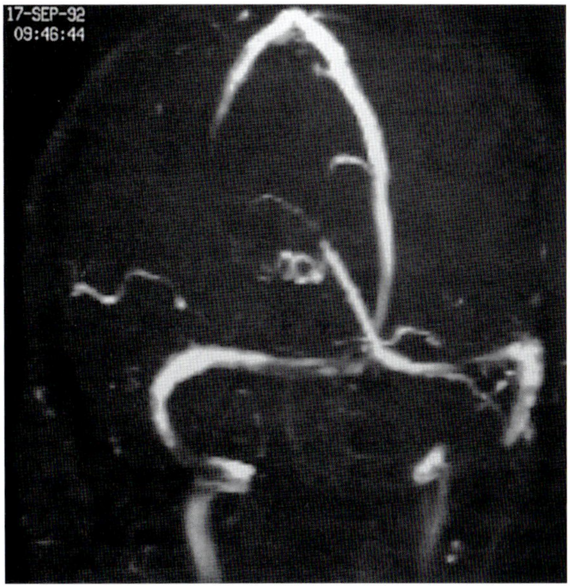

Fig. 4.**6** Noninvasive imaging of the intracranial venous system using magnetic resonance venography

Ectasias and Aneurysms

These can be detected without difficulty using *conventional* or *color flow duplex sonography.* In contrast to the findings with thrombosis, the Valsalva maneuver causes dilation, and the vein is compressible (Fig. 4.**8**).

Venous Congestion

When there is venous congestion, the disturbed venous drainage leads to decreased flow and venous dilatation (Steinke and Hennerici 1989).

Due to the close proximity of the cervical lymph nodes to the jugular vein, enlarged lymph nodes associated with metastatic processes may lead to displacement or compression of the jugular vein.

Arteriovenous Fistulas

Extracranial arteriovenous fistulas are very rare. Sonographically, the arterial and venous pathological findings involve increased flow velocities, and a volume increase in the afferent arteries and efferent veins (Fig. 4.**9**).

Intracranial Veins

Sinus Thrombosis

Noninvasive diagnosis of a sinus thrombosis using *transcranial Doppler sonography* is not reliable unless it is supported by concomitantly performed magnetic resonance venography or conventional cerebral angiography studies. When transcranial Doppler sonography is used in the acute phase, a flow delay through the

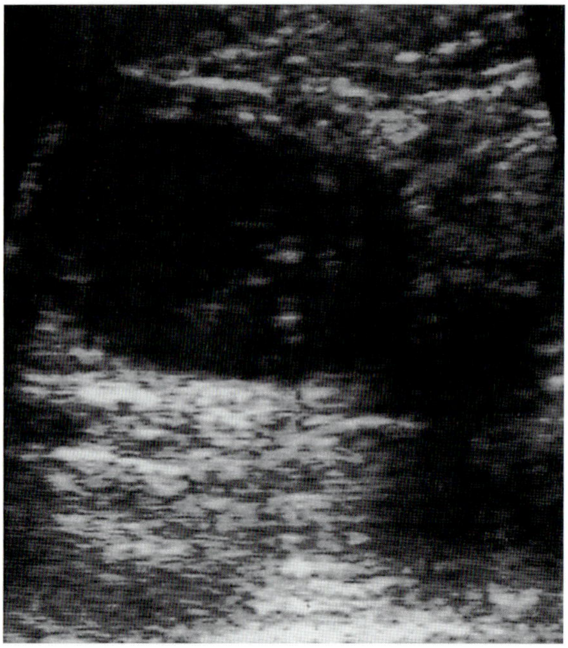

a

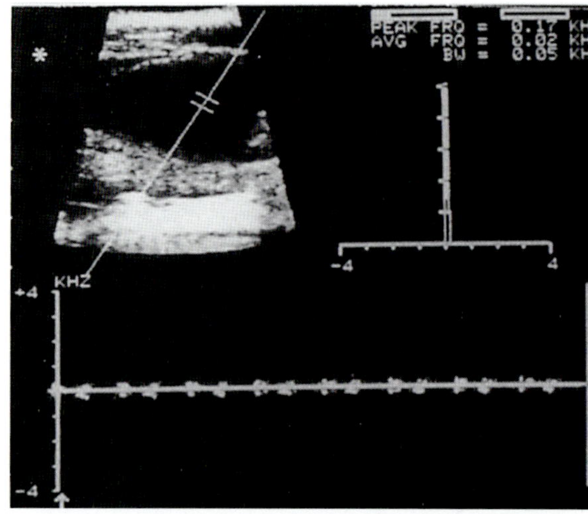

b

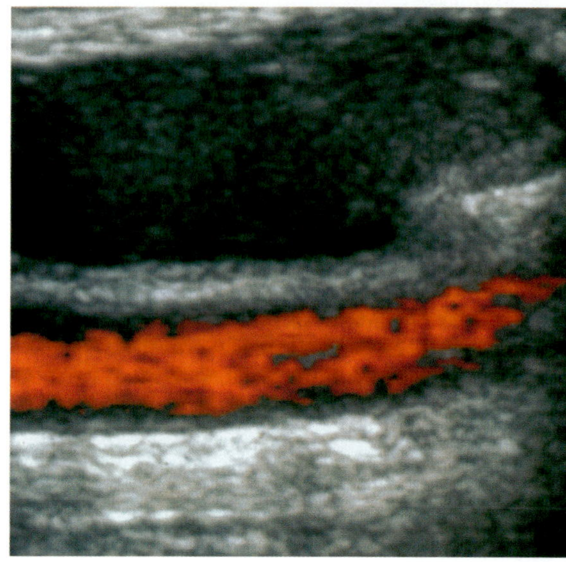

c

Fig. 4.**7** Internal jugular vein thromboses imaged using conventional duplex sonography (**a**, **b**) and with color flow duplex sonography (**c**). A maximally dilated blood vessel without pulsation and with partially hyperechoic material is seen (**a**, **c**). Inside the vein, pulsed-wave Doppler sonography (**b**) and color imaging (**c**) do not detect any flow signals. In the color flow image, the common carotid artery, with a normal flow signal, is seen dorsal to the vein

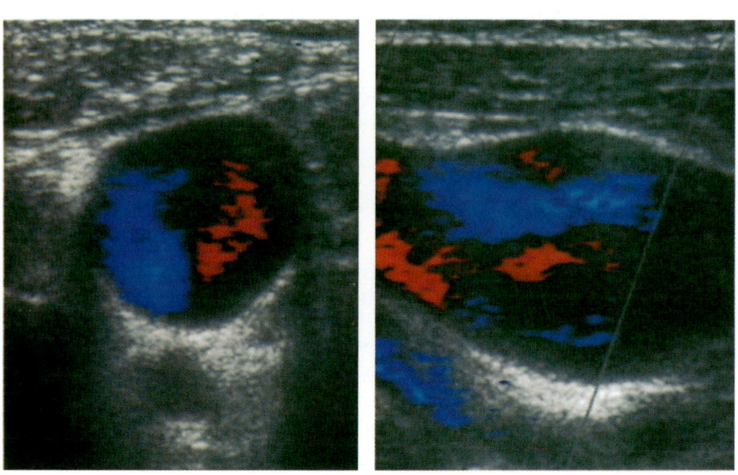

a b

Fig. 4.**8 a**, **b** A color flow duplex sonogram showing an ectatic internal jugular vein arterial branch, resulting from the obstructed venous drainage, may indicate a sinus thrombosis. In addition, flow velocity may be increased in the venous collateral vessels (Wardlaw et al. 1994, Valdueza et al. 1995, Ries et al. 1997). Normalization after recanalization of dural sinus thrombosis has also been monitored. *Transcranial color flow duplex sonography* using ultrasound contrast media can improve the diagnosis and can help identify otherwise undetectable low-flow signals hidden by the cerebral arteries. Systolic peak flow velocities with a symmetrical appearance of the blood flow spectra do not show any significant side-to-side differences. Velocities range from 15 to 20 cm/s (mean 17.5 ± 1.9 cm/s). In hypoplastic or partially blocked transverse sinuses, velocities were found to be significantly reduced (mean 9.4 ± 4.0 cm/s), but increased contralaterally (mean 28.4 ± 6.5 cm/s) (Ries et al. 1997). When combined with magnetic resonance venography, both noninvasive methods add complementary information and may help avoid conventional cerebral angiography (Fig. 4.**10**).

Monitoring of venous hemodynamics is an important issue and may help predict the outcome in patients with cerebral venous thrombosis (Stolz et al. 1999a, Valdueza et al. 1999). In a series of patients with acute sinus vein thrombosis, those with normal venous flow conditions or rapid recovery had a favorable outcome. The same features were observed in patients monitored after brain injury and subarachnoid hemorrhage (Mursch et al. 2001, 2002): those with high mean flow velocities, above the normal values in the basal veins, were reported to have a good outcome. In addition, patients with postischemic space-occupying brain edema who were monitored by Stolz et al. (2002) were found to have better outcomes if flow velocities in the straight sinus decreased and the midline shift was less than 1.5 cm. Similar studies of blood flow velocity monitoring have also shown a linear relationship between maximum flow velocity values and mean intracranial pressure (Schoser et al. 1999). Observations of microembolic signals detected in patients with superior sagittal sinus thrombosis are interesting, but have not yet been confirmed, and the prognostic significance of these has yet to be determined (Valdueza et al. 1997).

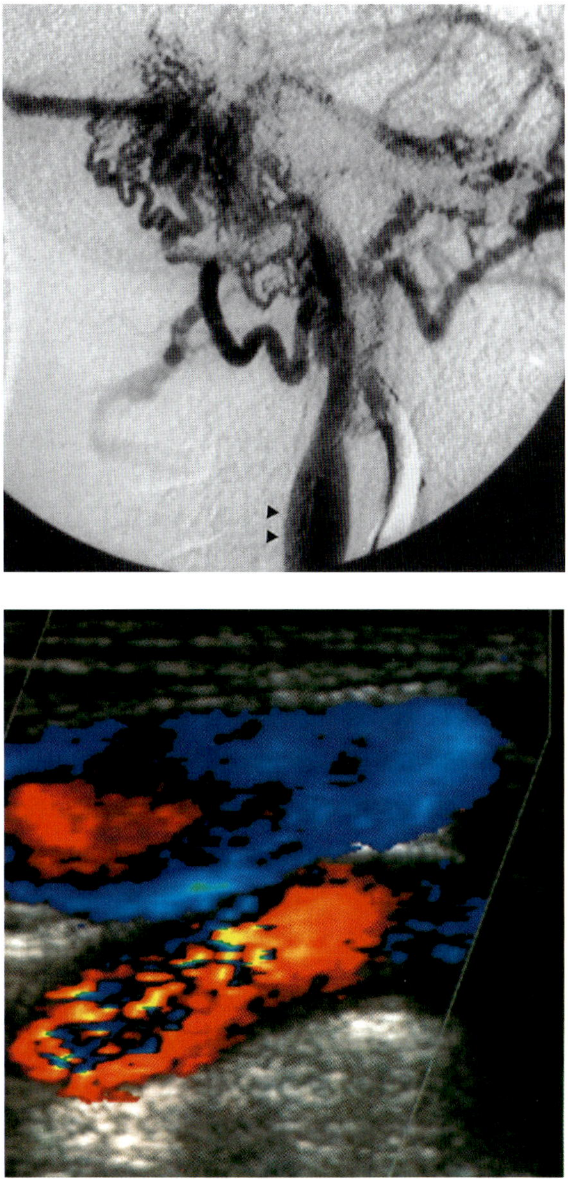

Fig. 4.**9** Angiogram (**a**) and color flow duplex sonogram (**b**) in a patient with a dural fistula. Significant areas of turbulence and areas of signal-void flow distortion are located distally within the internal jugular vein (**a** arrowheads, **b** upper blood vessel). There are abnormally high flow velocities in the common carotid artery as a result of the arteriovenous shunt

Cavernous Sinus Fistula

Carotid cavernous fistulas consist of a connection between the internal carotid artery and the cavernous sinus or dural branches of the external carotid artery. The typical finding using *continuous-wave or pulsed-wave Doppler sonography* at the medial canthus is a hissing, high-frequency, machine-like sound, with flow directed toward the probe. This signal is caused by drainage through the orbital veins, in which the blood flow is normally directed from exterior to interior. Fistulas between external carotid artery branches and the cavernous sinus often show normal findings in extracranial Doppler sonography. In these cases, *transcranial Doppler sonography* can produce diagnostically significant findings by displaying the fistula directly (Fig. 3.**26**, p. 122).

Arteriovenous Malformations

Although venous angiomas and cavernomas usually escape sonographic detection, the venous drainage of arterial angiomas can normally be imaged well using *transcranial Doppler sonography,* especially when there is a high shunt volume.

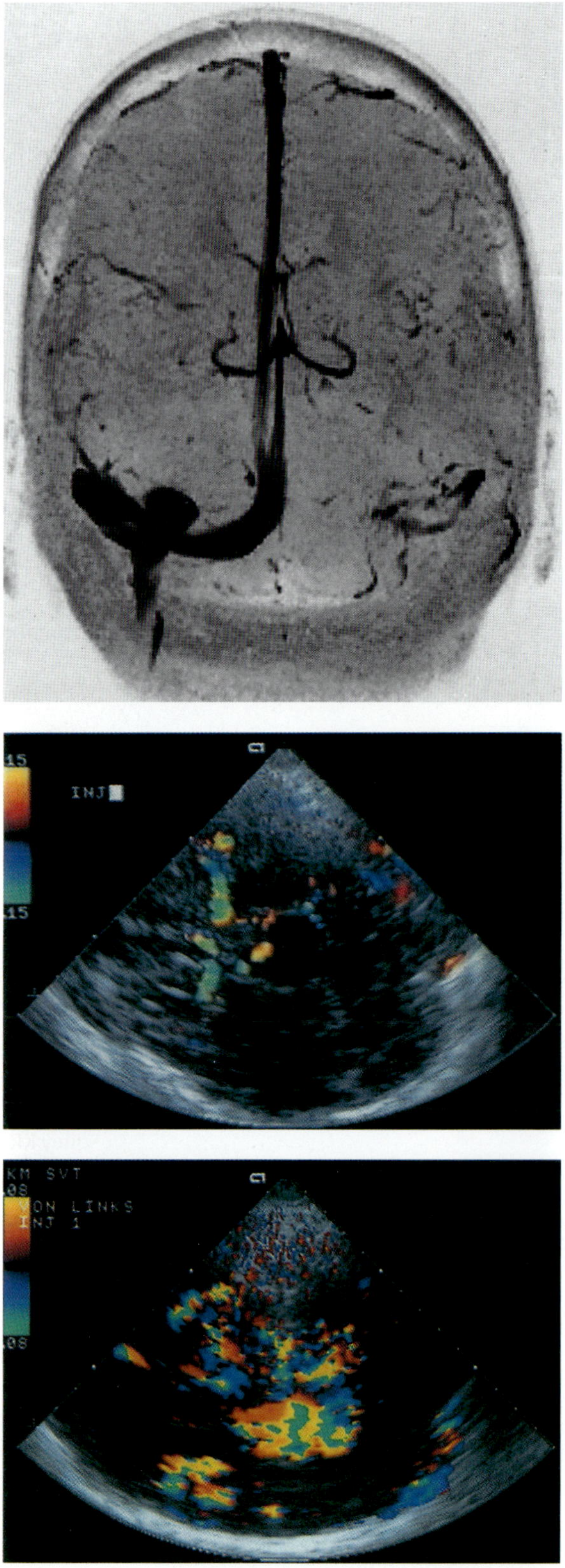

a b c

Fig. 4.**10 a** A thrombosis in the transverse sinus and ipsilateral jugular vein, imaged noninvasively in magnetic resonance angiography. The corresponding transcranial duplex analysis demonstrates normal flow in the contralateral transverse sinuses after administration of ultrasound contrast medium (**b** before injection, **c** after injection)

Venous Malformations

Although a small number of patients suffer from malformations of the intracranial veins and sinuses, the vein of Galen aneurysmal malformation is the most common type in neonatal and infantile age groups (Gerards et al. 2003). Management of children and adults (Fig. 4.**11**) with high-flow shunts, as well as prenatal diagnosis of these conditions, is challenging. There have been several promising case reports of successful interventional treatment after early diagnosis (Heling et al. 2000, Strowitzki et al. 2001).

Evaluation

The evaluation includes the spontaneous flow signals, the way in which they change after the Valsalva maneuver, and an assessment of vascular compressibility.

Sources of Error

Extracranial application. Exerting too much surface pressure with the probe often leads to errors when examining the extracranial venous system.

When there is an occlusion or high-grade stenosis of the innominate artery, marked hemodynamic changes occur distally in the distribution area of the carotid artery. Signals from the common carotid artery sometimes have a breath-like, venous quality, and confusing these with venous signals is a common mistake.

Intracranial application. The major disadvantages are difficulties in identifying and monitoring the cerebral veins and sinus in older patients, nearby high-flow signals (arteries, arteriovenous malformations, shunt volumes, dural fistulas, etc.) or total occlusion in hidden segments of the highly variable venous system.

Diagnostic Effectiveness

There are no data on the sensitivity and specificity of the method, since very few cases have been reported in the literature (Terwey et al. 1981, Albertyn and Alcock 1987, Bloching et al. 1989, Wardlaw et al. 1994, Valdueza et al. 1995, Ries et al. 1997). However, it can be assumed that using ultrasound to detect a jugular vein thrombosis is highly accurate. Duplex sonography is increasingly being used to clarify vascular relationships before a central venous catheter is placed. Ultrasound-guided puncture can reduce the number of incorrect punctures, and the technique is therefore likely to be particularly helpful in high-risk patients (Mallory et al. 1990).

Diagnosing intracranial sinus thromboses with ultrasound is more difficult and should always be supplemented by other imaging studies.

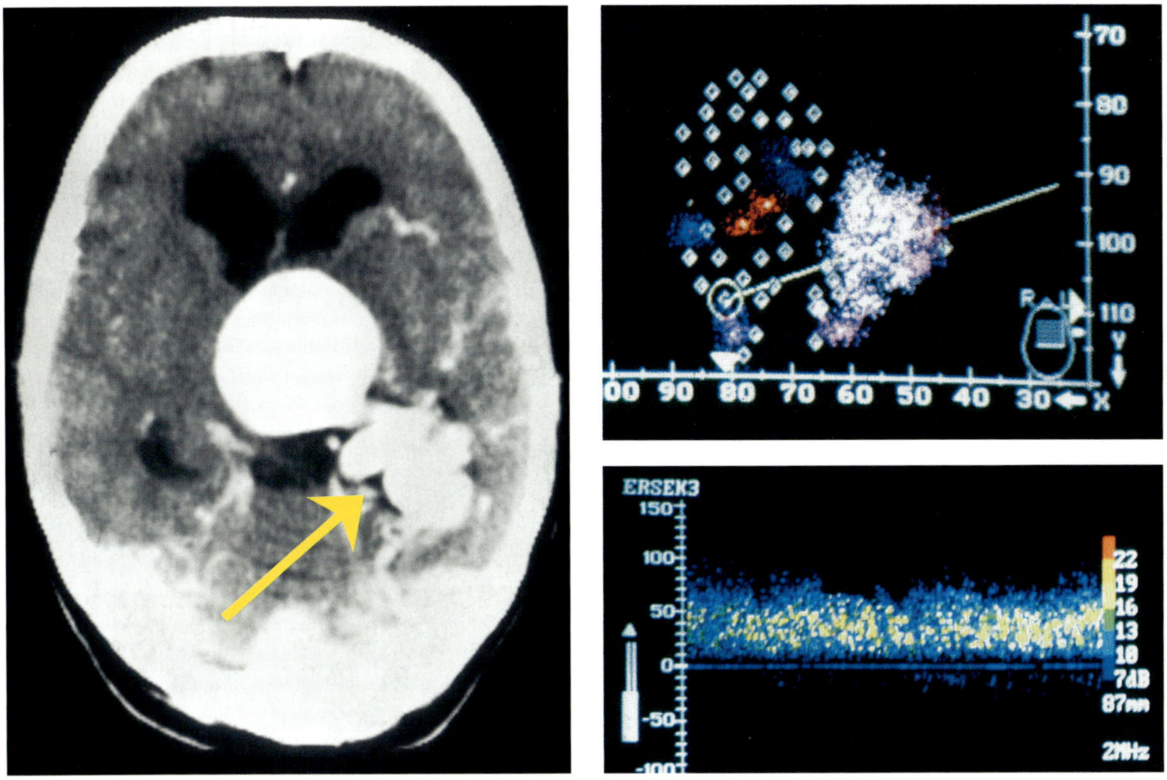

a b c

Fig. 4.**11** Hugh venous aneurysm (in the great vein of Galen) and efferent veins on computed tomography (CT) (**a**) and ultrasound imaging (**b**). The CT shows marked contrast filling of the aneurysm and efferent vein (arrow in **a**). **b** Ultrasound mapping (using a transtemporal approach, with the axial plane corresponding to the CT) shows areas without flow signals (open squares) or with barely detectable flow signals (blue and red dots). Conversely, high flow velocities (the white and pink spots can be identified and registered in the Doppler spectrum from the efferent vein (**c**).

References

Aaslid R, Newell DW, Stooss R, Sorteberg W, Lindegaard KF. Assessment of cerebral autoregulation dynamics from simultaneous arterial and venous transcranial Doppler recordings in humans. Stroke 1991;22:1148–54.

Albertyn LE, Alcock MK. Diagnosis of internal jugular vein thrombosis. Radiology 1987;162:505–8.

Andeweg J. The anatomy of collateral venous flow from the brain and its value in aetiological interpretation of intracranial pathology. Neuroradiology 1996;38: 621–8.

Baumgartner I, Bollinger A. [Diagnostic importance of the jugular veins; in German]. Vasa 1991;20:3–9.

Bloching H, Reuss JA, Seitz K, Rettenmaier G. [Thromboses of the subclavian vein and jugular vein and superior vena cava: sonographic diagnosis and control of treatment using tissue-type plasminogen activator therapy; in German.] Ultraschall Med 1989;10:314–7.

Brownlow RL Jr, McKinney W. Ultrasonic evaluation of jugular venous valve competence. J Ultrasound Med 1985;4:169–72.

Gerards FA, Engels MA, Barkhof F, van den Dungen FA, Vermeulen RJ, van Vugt JM. Prenatal diagnosis of aneurysms of the vein of Galen (vena magna cerebri) with conventional sonography, three-dimensional sonography, and magnetic resonance imaging: report of 2 cases. J Ultrasound Med 2003;22:1363–8.

Heling KS, Chaoui R, Bollmann R. Prenatal diagnosis of an aneurysm of the vein of Galen with three-dimensional color power angiography. Ultrasound Obstet Gynecol 2000;15:333–6.

Hennerici M. Ultrasound diagnosis of cerebrovenous flow disturbances. In: Einhäupl K, Kempski O, Baethmann A, eds. Cerebral sinus thrombosis: experimental and clinical aspects. New York: Plenum Press, 1990:201–209.

Mallory DL, McGee WT, Shawker TH, Brenner M, Bailey KR, Evans RG, et al. Ultrasound guidance improves the success rate of internal jugular vein cannulation: a prospective, randomized trial. Chest 1990;98:157–60.

Mattle HP, Wentz KU, Edelman RR, Wallner B, Finn JP, Barnes P, et al. Cerebral venography with MR. Radiology 1991;178:453–8.

Mursch K, Wachter A, Radke K, Buhre W, Al-Sufi S, Munzel U, et al. Blood flow velocities in the basal vein after subarachnoid haemorrhage: a prospective study using transcranial duplex sonography. Acta Neurochir (Wien) 2001;143:793–9 [discussion 799–800].

Mursch K, Muller CA, Buhre W, Lang JK, Vatter H, Behnke-Mursch J. Blood flow velocities in the basal cerebral vein after head trauma: a prospective study in 82 patients. J Neuroimaging 2002;12:325–9.

Ries S, Steinke W, Neff KW, Hennerici M. Echocontrast-enhanced transcranial color-coded sonography for the diagnosis of transverse sinus venous thrombosis. Stroke 1997;28:696–700.

Schoser BG, Riemenschneider N, Hansen HC. The impact of raised intracranial pressure on cerebral venous hemodynamics: a prospective venous transcranial Doppler ultrasonography study. J Neurosurg 1999;91:744–9.

Steinke W, Hennerici M. [Duplex sonography of the internal jugular vein; in German.] Ultraschall Med 1989;10:72–6.

Stolz E, Kaps M, Dorndorf W. Assessment of intracranial venous hemodynamics in normal individuals and patients with cerebral venous thrombosis. Stroke 1999a;30:70–5.

Stolz E, Kaps M, Kern A, Dorndorf W. Frontal bone windows for transcranial color-coded duplex sonography. Stroke 1999b;30:814–20.

Stolz E, Gerriets T, Babacan SS, Jauss M, Kraus J, Kaps M. Intracranial venous hemodynamics in patients with midline dislocation due to postischemic brain edema. Stroke 2002;33:479–85.

Strowitzki M, Fox T, Hamann G. Transcranial color-coded sonography of a vein of Galen arteriovenous malformation in an adult. J Neuroimaging 2001;11:205–8.

Terwey B, Krier C, Gerhardt P. [The demonstration of jugular venous thrombosis by high resolution ultrasound; in German.] Röfo 1981;134:557–9.

Valdueza JM, Schultz M, Harms L, Einhäupl KM. Venous transcranial Doppler ultrasound monitoring in acute dural sinus thrombosis: report of two cases. Stroke 1995;26:1196–9.

Valdueza JM, Schmierer K, Mehraein S, Einhäupl KM. Assessment of normal flow velocity in basal cerebral veins: a transcranial Doppler ultrasound study. Stroke 1996;27:1221–5.

Valdueza JM, Harms L, Doepp F, Koscielny J, Einhäupl KM. Venous microembolic signals detected in patients with cerebral sinus thrombosis. Stroke 1997;28:1607–9.

Valdueza JM, Hoffmann O, Weih M, Mehraein S, Einhäupl KM. Monitoring of venous hemodynamics in patients with cerebral venous thrombosis by transcranial Doppler ultrasound. Arch Neurol 1999;56:229–34.

Wardlaw JM, Vaughan GT, Steers AJ, Sellar RJ. Transcranial Doppler ultrasound findings in cerebral venous sinus thrombosis: case report. J Neurosurg 1994;80:332–5.

5 Peripheral Arteries

Systolic Blood-Pressure Measurement

Examination

Special Equipment and Documentation

Unidirectional Doppler equipment is usually used for this examination method. The equipment reproduces the Doppler signal acoustically (using what is known as the “pocket Doppler”), but does not allow any measurement of flow direction. However, bidirectional Doppler equipment, which does detect the flow direction, can also be used for the examination.

Measurements are made using standard blood-pressure equipment. The cuff width must be appropriate for the placement area (e.g., upper arm, distal lower leg 13 cm; thigh 15–18 cm).

A *transmission frequency* of 8–10 MHz should usually be used to localize the Doppler signal at the arteries of the foot (posterior tibial artery or dorsal artery of the foot, or both), and the radial artery. A transmission frequency of 4–5 MHz may be more effective for deeper arteries (popliteal artery, brachial artery).

It is obligatory to keep a written numerical record of the measured pressure values, describing the measurement area (cuff position) and the arteries measured (probe position).

Examination Conditions

Patient and Examiner

The examination is carried out with the patient *reclining*. A preliminary resting period, lasting about 5–10 minutes in healthy individuals, is important. In pathological cases, the resting period should be extended to 15–20 minutes, since the poststenotic pressure decreases with the slightest exertion, and only returns to its original value slowly. The examiner should be in a comfortable position, possibly sitting.

Conducting the Examination

The blood-pressure cuff is placed round the selected measurement area avoiding any external pressure. If possible, the pulse is first palpated at the artery to be measured, and contact gel is then applied. Using the Doppler ultrasound probe, the examiner searches for the optimal acoustic signal. Prior to inflating the cuff, the examiner should stabilize the probe by supporting it with his fingers, and/or hand. The patient must keep the leg or arm that is being measured still (Fig. 5.**1**).

The measurement procedure is the same as used in the conventional Riva-Rocci method. The cuff is inflated to above the systolic pressure until the Doppler signal disappears. The cuff is then *slowly* deflated (2–5 mmHg/s) until the signal is heard again. The pressure reading at this moment on the manometer scale is noted, and it corresponds to the systolic arterial pressure *under the cuff.*

It is important to distinguish between the cuff and probe positions. The pressure measured corresponds to the pressure in the artery or arteries underneath the cuff. Independently of the cuff location, the Doppler signal is usually recorded at the ankle (posterior tibial artery or dorsal artery of the foot, or both) (Fig. 5.**2**).

> **Note:**
> Measurement area—cuff position
> Measured artery—probe position

If the Doppler signal can no longer be detected at very low pressures, the patient’s upper body can be raised. The hydrostatically elevated pressure will then be within the measurable range, and the difference to the

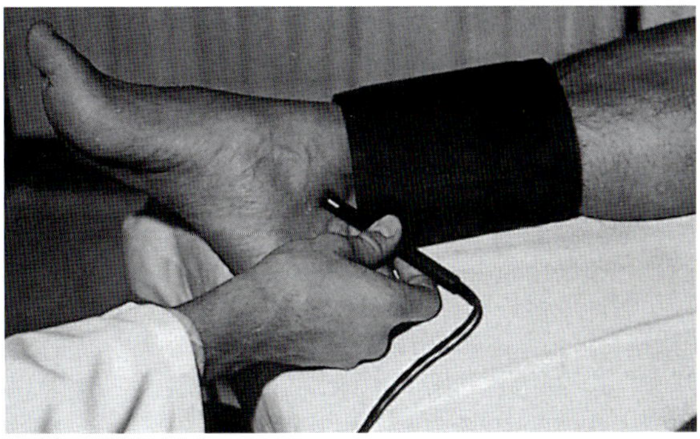

Fig. 5.**1** Measuring peripheral systolic pressure on the ankle using the Doppler ultrasound technique. The area being measured is the distal lower leg, and the artery being measured here is the posterior tibial artery

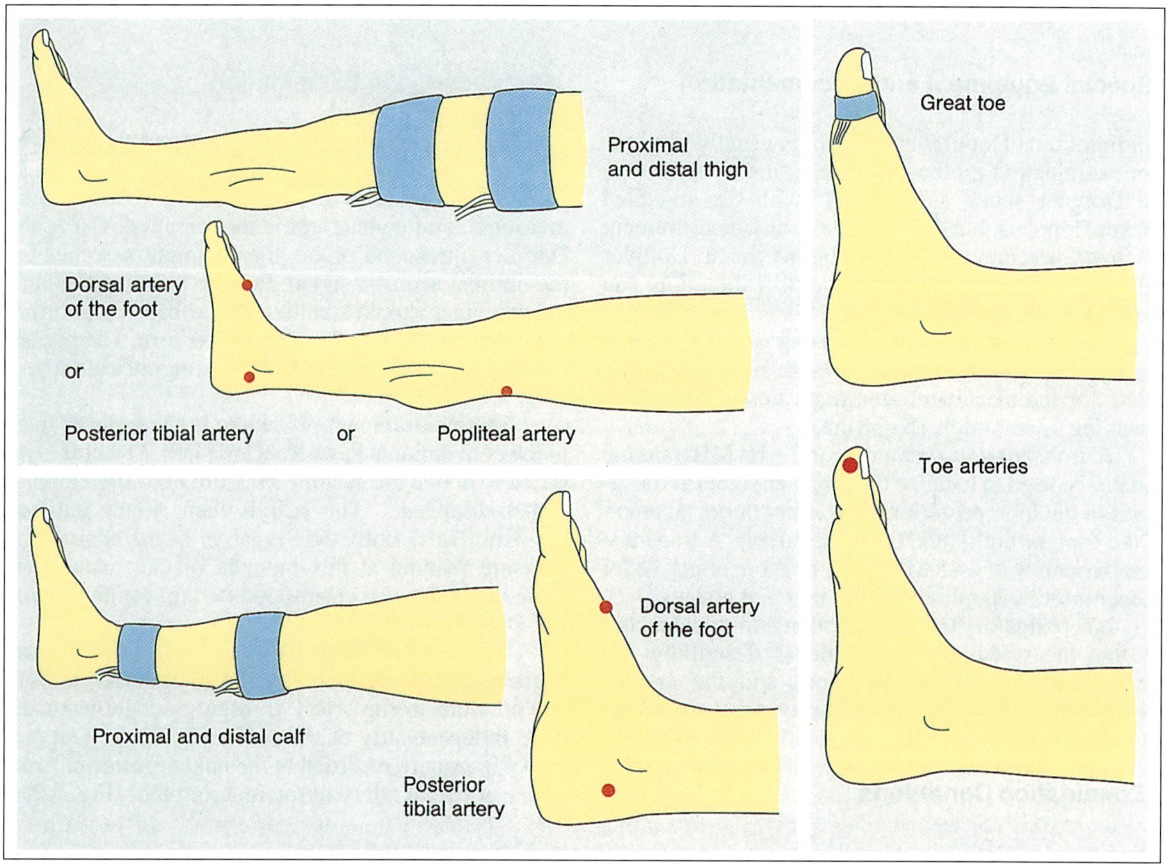

Fig. 5.2 Measurement areas and arteries for measurement in the lower extremity. The different cuff positions and arterial regions that are suitable for measuring with the continuous-wave Doppler probe are shown. If the Doppler signal can no longer be detected at very low pressures, the patient's upper body can be raised. The hydrostatically elevated pressure then lies within the measurable area, and the difference from the pressure in the ankle arteries while reclining can be deducted from the measured value (1.0 cm H_2O, 0.736 mmHg)

ankle pressure while reclining has to be deducted from the measured value (1.0 cm $H_2O \geq 0.736$ mmHg).

Examination Sequence

Pelvic and Leg Arteries

Measurement of Pressure at Rest

The systolic blood pressure in the lower extremities is always compared to the systemic pressure. The undisturbed arm pressure is used as the reference pressure. This value should be determined directly before or after measuring the leg pressure, using either the ultrasound Doppler method or the conventional stethoscope technique. The comparative studies carried out by Thulesius (1971) showed a very good correlation between the two measured values. A prior clinical examination (palpation, auscultation) and comparison of the blood-pressure values on each side must be carried out to exclude a vascular obstruction in the arm (subclavian artery, brachial artery).

The preferred area for leg measurement is the distal lower leg; the arteries measured are the posterior tibial artery or the dorsal artery of the foot, or both. All prior flow obstructions can be detected here, and it is also a good location for placing the blood-pressure cuff without too much interference from the musculature.

Measurement of Pressure Post Exercise

Defined active exercises that exert the leg muscles are used to differentiate borderline values. Simple exercise tests that have proved useful are tiptoeing and knee bending, which can be used without any difficulty (Köhler et al. 1972). More precisely defined exercise is possible using the treadmill ergometer (Cachovan and Maass 1984, Mahler 1990).

For patients in whom exercise testing is not possible, reactive hyperemia after arterial compression may be used (Fronek et al. 1973b, Mahler et al. 1975).

The following exercises are recommended as standards for the different exercise tests:

- Tiptoeing 40 times
- Knee bending 20 times
- Walking 200 m, 120 steps/minute
- Treadmill 100 m, 3.2 km/h, at a 12.5° gradient
- Hyperemia: 3 minutes of arterial compression above the knee with a wide blood-pressure cuff

Measurement of Segmental Pressure

Measuring the pressure at several locations in the thigh and lower leg is only necessary if the ankle pressure is pathological, and the obstruction level in the vascular system needs to be established. The usual *cuff placement locations* and corresponding cuff widths are:

- Arm 13 cm
- Proximal thigh 15–16 cm
- Distal thigh 15–16 cm
- Proximal lower leg 15–16 cm
- Distal lower leg 13 cm
- Great toe (proximal joint) 2–3 cm

Care should always be taken to ensure that the blood-pressure cuffs are of the appropriate width and length for the specific placement area (Fig. 5.**3**). The arteries of the ankle (posterior tibial artery, or dorsal artery of the foot) are usually the ones measured. However, some authors (Franzeck et al. 1981) do recommend placing the probe close to the cuff (e.g., popliteal artery for thigh measurements).

Measurement of Toe Pressure

Measurement of the systolic toe pressure is carried out using a special cuff (width approximately 25 mm), and the result is compared to the distal lower leg pressure. Sensing devices that can be used include the ultrasound Doppler probe, a strain gauge, or photoplethysmography. This important measurement is made more difficult by the fact that ordinary blood-pressure equipment does not allow pressure determination in small cuffs without special adaptation. In a practical clinical setting, it is preferable to use special equipment that fully automates the simultaneous blood-pressure measurements in the distal lower leg and toe, using a mercury strain gauge, and can also display the information graphically.

Upper Extremity Arteries

In the upper extremities, measurements of the systolic blood pressure using the ultrasound Doppler method are rarely made, since conventional blood-pressure measurement with the stethoscope is simpler and produces identical results (Thulesius 1971). Alternatively, a blood-pressure cuff can be placed round the upper arm or forearm, and the radial artery or ulnar artery can be used as the artery to be measured by the Doppler probe (Fig. 5.**4**).

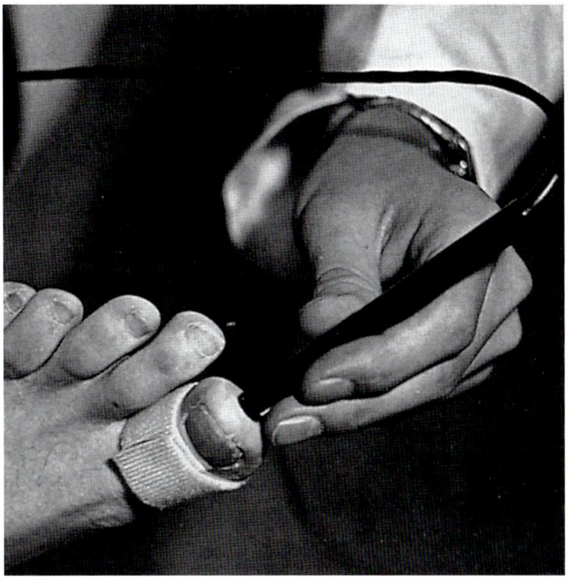

Fig. 5.**3** Toe pressure measurement, showing the special cuff at the basal joint of the toe and the position of the Doppler probe (8–10 MHz)

Measurement of Digit Pressure

Using special blood-pressure cuffs (width 20–25 mm), it is also possible to measure systolic blood pressure in the fingers. The cuff is placed around the distal or middle portion of the finger, and distal to this a high-frequency ultrasound Doppler probe is used to locate the arterial signal (Fig. 5.**5**).

However, it is more common to use plethysmographs, which provide suitably small cuffs, automatic inflation and deflation, with the aid of a strain gauge (Gundersen 1972, Nielsen et al. 1973).

Normal Findings

Principle

The most important and simplest parameter for demonstrating undisturbed arterial circulation in the extremities is obtained by measuring the blood pressure in the peripheral arteries. Using the ultrasound Doppler method, it has also become possible to detect the flow in blood vessels in which Korotkoff sounds can no longer be auscultated (Strandness et al. 1966, Schoop and Levy 1969).

The procedure used corresponds to the conventional Riva-Rocci arterial blood-pressure measurement, with the Doppler probe replacing the stethoscope.

However, only the systolic blood pressure can be measured, not the diastolic pressure. When the suprasystolic obstacle to blood flow is released, the ultrasound waves from the Doppler probe meet the peak

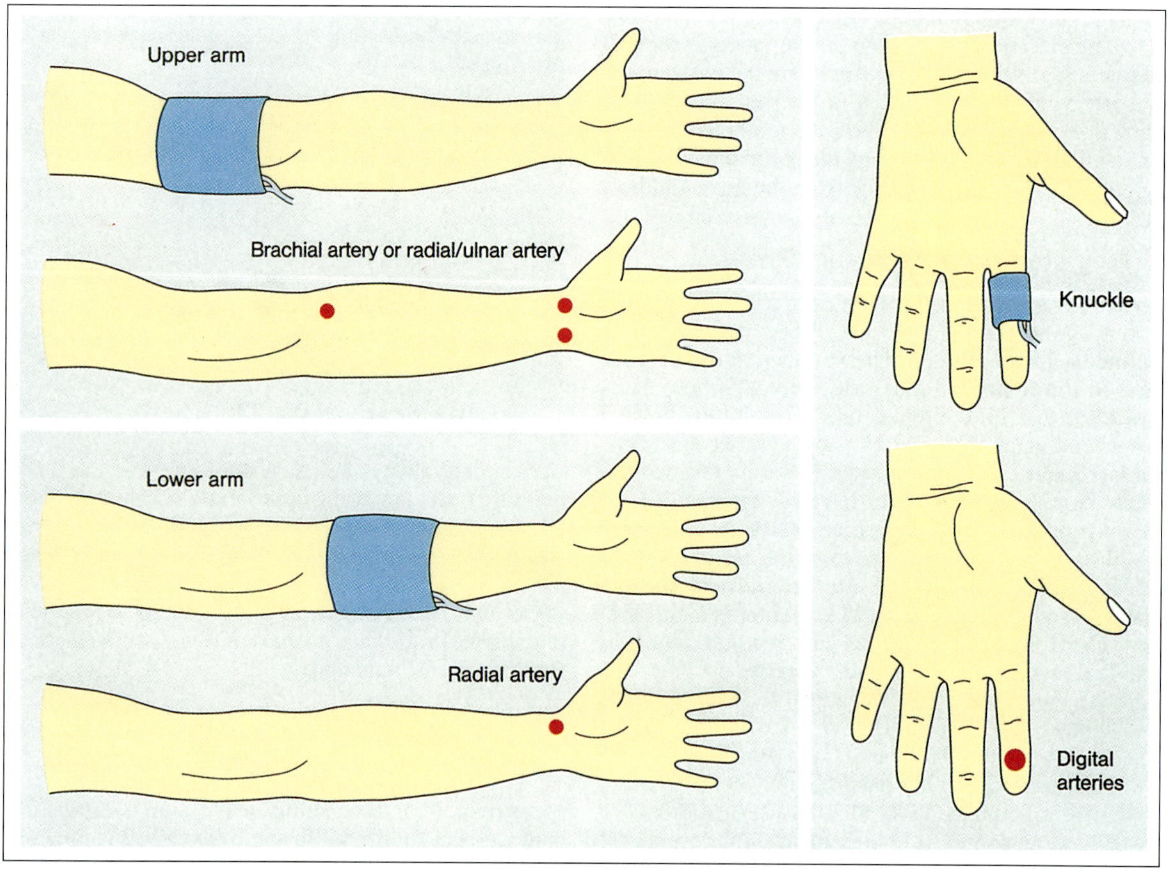

Fig. 5.**4** Measurement areas and arteries for measurement in the upper extremity. The placement points for the blood-pressure cuff and the most favorable positions for registering the Doppler signals are shown

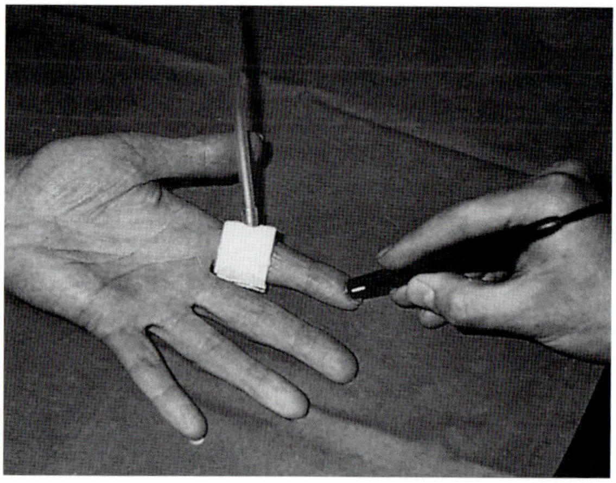

Fig. 5.**5** Measuring arterial blood pressure in the finger, showing the position of the cuff and the ultrasound Doppler probe (8–10 MHz)

flow of the blood that is running freely again. In contrast to the Korotkoff sounds, the signal does not cease during the diastolic phase. A special characteristic of pressure measurement in the distal lower leg, termed *ankle pressure measurement*, is that the normal peripheral systolic pressures are approximately 10–20 mmHg higher than the systolic arm pressure.

Wetterer et al. (1971) showed that “pressure pulses at a greater distance from the heart appear later than the central pulse, and differ from it both in their course and pressure amplitude.” The systolic pressure increases substantially in a peripheral direction, while the diastolic pressure, and also the mean pressure, decrease slightly (Fig. 5.**6**). This amplitude increase is called “amplification.” It is explained by the back-and-forth movement of the waves in the arterial system, which when superimposed on one another create what is known as a “standing wave” (Busse 1982). These interesting phenomena, which are important in the evaluation of blood-pressure measurements, are caused by wave reflection. Reflection occurs wherever wave resistance increases; in the arteries of the extremities, this happens peripherally. All of the longer arterial branches are individual elastic systems with

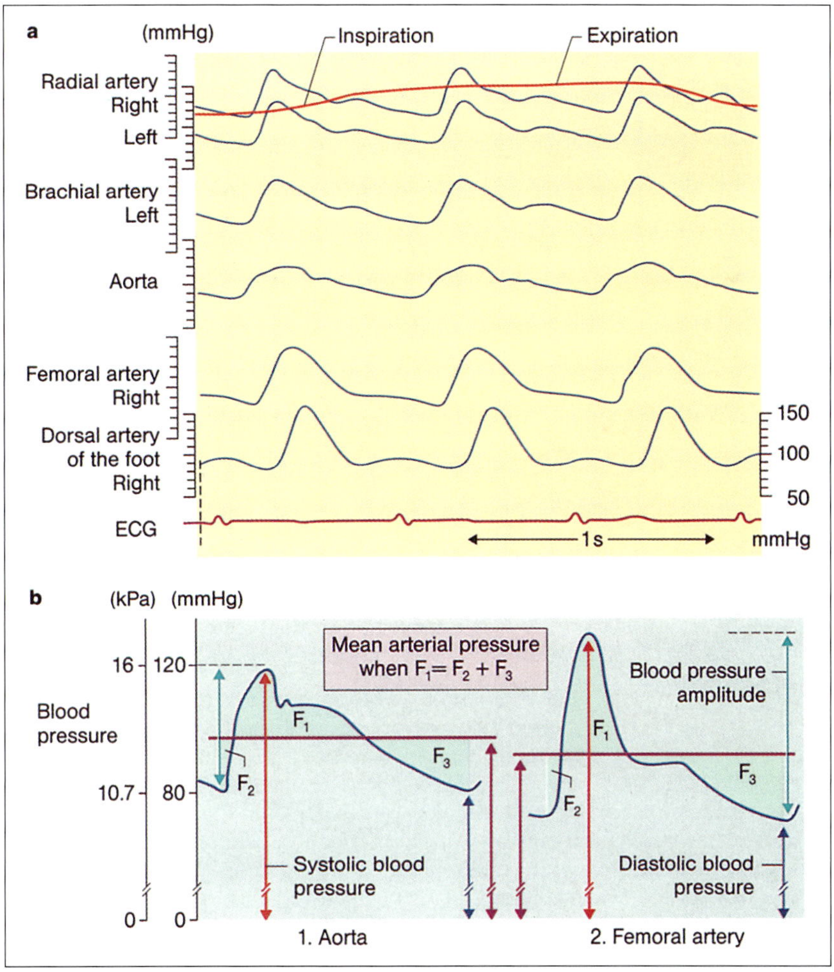

Fig. 5.**6a** Direct and simultaneously registered pressure pulses from the aorta and from the arteries of the upper and lower extremities (adapted from Wetterer). The pressure pulse amplitudes show increased systolic pressure and slightly reduced diastolic pressure—i.e., an increase in the pressure amplitude toward the periphery (aorta 121/84 mmHg, dorsal artery of the foot 158/83 mmHg).
b Although the average aortic pressure decreases when proceeding toward the femoral artery, systolic pressure in the peripheral arteries is higher than systolic pressure in the aorta. The diagram illustrates this principle (from S. Silbernagel and A. Despopoulos, *Pocket Atlas of Physiology*, 4th edition, Stuttgart: Thieme, 1991)

their own variable reflection factors, which can show wide differences, corresponding to the particular vascular tone.

Findings

Pelvic and Leg Arteries

Measurement of Pressure at Rest

When the vascular system in the leg is patent, the peak systolic ankle pressure, as mentioned above, is higher than the systolic arm pressure.

Schoop and Levy (1969) and Bollinger et al. (1970) thus demonstrated that, on average, peripheral pressures in the leg were 10 mmHg higher than the arm pressures.

Measurement of Pressure Post Exercise

After physical exertion (ergometry, tiptoeing, knee bending, treadmill), the peripheral systolic pressure increases in proportion to the elevated systemic pressure. A peripheral pressure decrease is usually not noted in healthy individuals. If it does occur, it is slight and brief (30–60 s) (Mahler et al. 1976, Ludwig et al. 1985) (Fig. 5.**7**).

Measurement of Segmental Pressure

If the distal lower leg pressure is higher than the arm pressure, then further pressure measurements with the blood-pressure cuff located proximal to the lower leg are not required, because the ankle pressure detects the sum of all preceding flow obstructions.

However, it is not possible to exclude isolated foot and toe obstructions with the usual ankle pressure measurement. In doubtful cases, a supplementary *toe pressure measurement* is recommended (p. 151) (Knoblich et al. 1986).

Upper Extremity Arteries

To exclude flow obstructions, which occur most frequently in the proximal subclavian artery, the blood-pressure cuff is usually placed round the upper arm to

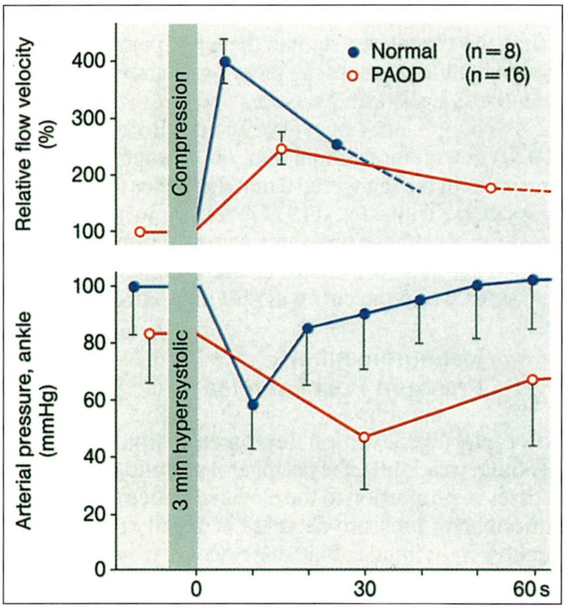

Fig. 5.**7** The flow velocity and the peripheral systolic pressure at rest and after 3 min of suprasystolic arterial compression, in normal individuals and in patients with arterial occlusive disease. In healthy individuals, the systolic ankle pressure shows a brief 40 % decrease after exercise, which normalizes within 60 s. In pathological cases, the maximum pressure decrease does not occur until 30 s have elapsed, and a return to normalcy has not yet taken place after 60 s. The relative flow velocity increases in a normal individual by more than 300 % and also returns to its normal value within 60 s. In pathological cases, the increase in the flow velocity is less dramatic, but persists for a longer time (adapted from Mahler 1990)

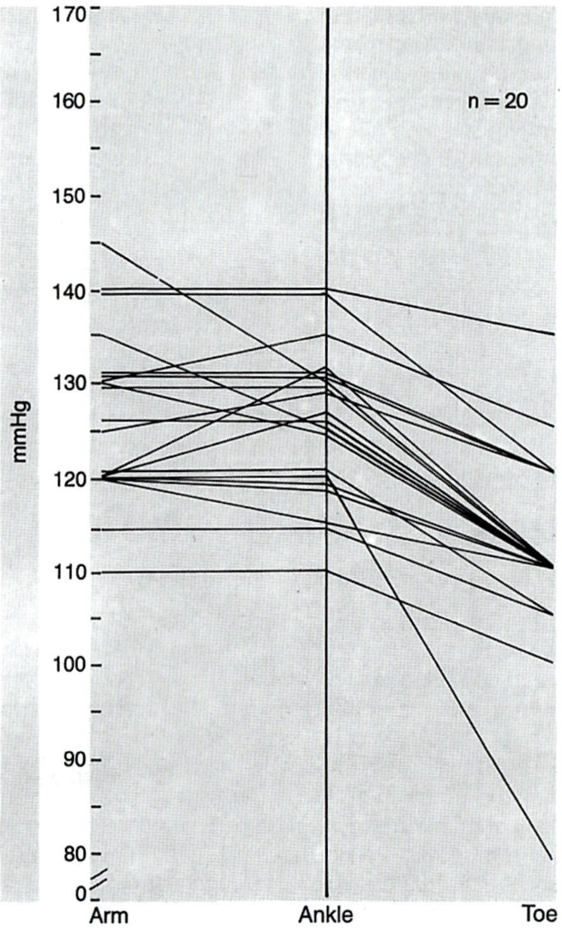

Fig. 5.**8** Measuring the pressure at the toes in 20 normal individuals. In a healthy individual, the arterial pressures within the toes are usually 10–15 mmHg below the pressure at the arm arteries. This is due to damping of the pressure amplitude in the small blood vessels

measure the radial artery, and the systemic pressure is measured and compared bilaterally. For special inquiries, the forearm and digit pressure can also be measured.

Evaluation

Measurement of Ankle Pressure

Statements about the *resting* peripheral pressures can be made in three forms:

- Absolute value
- Arm–ankle pressure difference (ΔP)
- Ankle/arm pressure index (ratio)

The varying significance of the individual measurements becomes clear in pathological findings (p. 158).

Analogously, the following borderline values for ankle pressure in healthy individuals can be given:

- Absolute value: ankle pressure usually 10–20 mmHg above the systemic pressure
- Pressure difference: negative, approximately 10–20 mmHg
- Pressure index: > 1

After exercise, the pressure in a healthy individual returns to its original value after 1 minute at the very latest (Mahler 1990). Since physical activity often causes the systemic pressure to rise, care must be taken to measure not only ankle pressure, but also arm pressure after exertion.

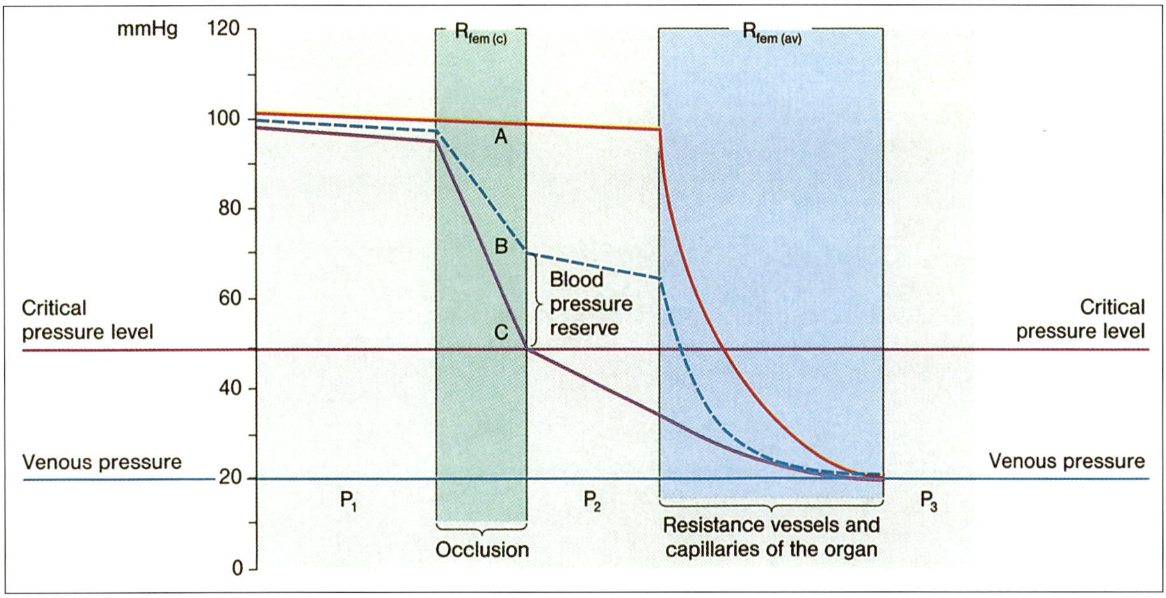

Fig. 5.**9** A diagram of the critical pressure level and blood-pressure reserve in normal and pathological conditions, relative to the average arterial pressure (adapted from Rieger). Pressure behavior in the lower extremities is shown:

A In a normal situation, with a retained blood-pressure reserve of AC
B In an arterial occlusion with reasonable compensation and a still adequate pressure reserve of BC
C In several occlusions with no pressure reserve
P_1 Preocclusive pressure
P_2 Postocclusive pressure
P_3 venous pressure
$R_{fem(c)}$ = Collateral resistance
$R_{fem(av)}$ = Organ resistance

Measurement of Toe Pressure

On average, the pressure is approximately 10–15 mmHg lower than the arm pressure (Fig. 5.**8**). Carter (1993) explained this as being due to the damping of the pressure amplitude in the small-caliber digital arteries of the toes, and also due to the peripheral resistance, which in turn is additionally affected by skin temperature. As an explanation, Gunderson (1972) emphasized that the location of the cuff at the toe is approximately 20 cm above the ankle examination area, so that the hydrostatic pressure is lower at the toe (20 cm hydrostatic pressure—14.6 mmHg).

Measurement of Digit Pressure

On average, the finger pressure is 5.0 mmHg above the systolic pressure. This is even more dependent on the vasomotor situation than in the toes (Knoblich et al. 1986).

Pathological Findings

Principle

Proximal to an occlusion, the arterial pressure does not increase, since excluding a small subsidiary circulatory system does not have any significant effect on the systemic blood pressure (Rieger 1993).

Distal to an occlusion, the *resting* pressure decreases, depending both on the number and length of the occlusions and on the extent of the collateral circulatory system (Schoop 1974). In contrast, the blood flow at rest is only detrimentally affected by extensive occlusive processes involving several levels. The *critical pressure level* is important here. This term refers to the postocclusive pressure that still allows undisturbed blood flow through the capillaries. It lies at 40–50 mmHg, sufficiently high above the venous pressure (20–25 mmHg) to maintain the arteriovenous blood flow (Rieger 1993) (Fig. 5.**9**).

All forms of *exercise* (walking, treadmill, tiptoeing, knee bending) result in vasodilation. The increased blood requirements of the musculature are met by decreased peripheral resistance and increased flow volume. This occurs because of a pressure decrease due to the increased friction. Reactive hyperemia following suprasystolic arterial compression also has the same effect: the systolic blood pressure decreases, and the

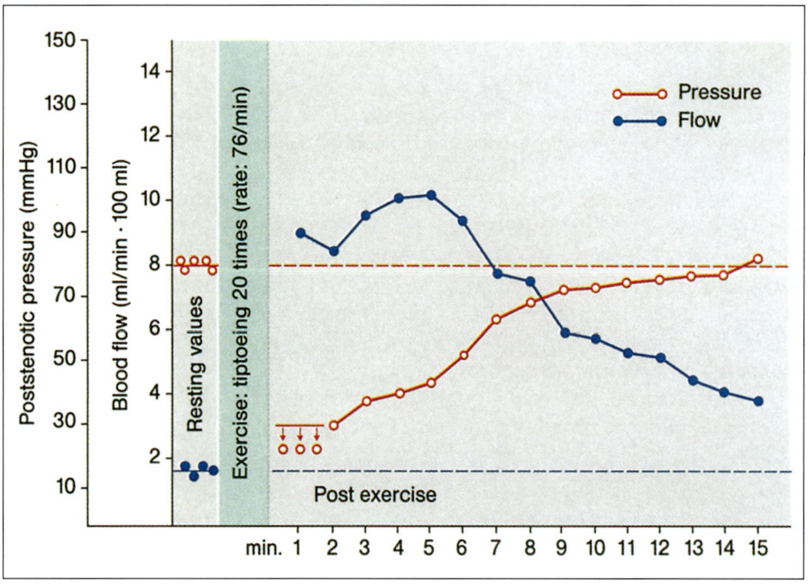

Fig. 5.**10** Pressure and flow following exercise (muscular work) (adapted from Köhler). In an individual suffering from vascular disease (occlusion of the femoral artery), postocclusive systolic pressure (measured using the ultrasound Doppler method) decreases after exercise and falls into an area that is not measurable, only returning to the initial value after approximately 15 min of rest. The length of the recuperation time indicates poor compensation. By contrast, the circulation flow clearly increases after exercise and has still not returned to the resting value after the same period

Table 5.**1** Ankle pressure values in relation to the clinically significant circulatory reserve in peripheral arterial disease

Absolute pressure in the distal lower leg (mmHg)	Circulatory reserve
> 100	Good
≈ 80–100	Satisfactory
≈ 50–80	Moderate
< 50 (critical pressure level)	Insufficient

Table 5.**2** The pressure gradient—i.e., the pressure difference between the systemic and peripheral systolic pressures—and its significance in indicating the level of hemodynamic compensation level in peripheral arterial disease

Pressure difference Δp in the arm/ankle (mmHg)	Hemodynamic compensation
< 20	Very good
≈ 40	Good
≈ 60	Satisfactory
≈ 80	Moderate
≈ 100	Poor

Table 5.**3** Although the ankle–arm index (quotient) is highly dependent on the systemic systolic blood pressure, using it makes it easier to follow the clinical course in peripheral artery disease

Systemic pressure (mmHg)	Peripheral pressure (mmHg)	Quotient
220	110	0.5
160	80	0.5
120	60	0.5

flow velocity increases. The ankle pressure is inversely proportional to the flow velocity (Sumner and Strandness 1969, Köhler et al. 1972, Bollinger et al. 1973, Mahler 1990) (Fig. 5.**10**).

Findings

Pelvic and Leg Arteries

Measurement of Pressure at Rest

When resting values are being estimated, the *absolute value* of the systolic blood pressure corresponds approximately to the *circulatory reserve*. This is the value that most reliably indicates whether an extremity is endangered. The benchmark values given in Table 5.**1** have proved significant.

The *pressure difference* between the systemic and peripheral pressures provides information about the hemodynamic *compensation* for occlusions or stenoses (Table 5.**2**). It indicates the pressure decrease in an individual case, and shows a marked dependence on the systemic pressure, so that this value cannot be used as an absolute indicator of the seriousness of disease (Rode and Schütz 1977). The pressure difference is often also called the pressure gradient. However, the terms "pressure gradient" and "pressure difference" are not synonymous. "Pressure gradient" refers to a pressure difference over a specified vascular length.

The *ankle/arm index*—i.e., the pressure ratio between the peripheral and systemic pressures—is preferred at many clinical centers. However, this theoretical number provides no information about the systemic pressure or the absolute peripheral pressure, so that the value is only of limited use as a single parameter. The same ratio can be based on very different peripheral values, for example (Table 5.**3**).

On the other hand, the ankle/arm index has good predictive value in distinguishing a healthy patient from a sick one (Ouriel and Zarins 1982). It is well suited to the evaluation of large patient groups, and for follow-up studies (Yao et al. 1969).

Measurement of Pressure Post Exercise

When the ankle pressure borders on pathological values (pressure 0–5 mmHg below the systemic pressure), exercise may reveal low-grade stenoses that do not become hemodynamically significant before muscular exertion (Yao 1970, Thulesius 1978).

Using the various exercises (tiptoeing 40 times, or knee bending 20 times), it can be established whether the pelvic or thigh flow obstruction is the hemodynamically more active one in a two-level occlusion (Neuerburg-Heusler et al. 1989). Looking at different types of exercise tests, Mahler (1990) showed that tiptoeing (30 times) is almost no different from treadmill ergometry, while hyperemia following arterial compression causes a slightly smaller pressure decrease (see Fig. 5.**14**).

Measurement of Segmental Pressure

If the ankle pressure is pathological—i.e., lower than the arterial systemic pressure—the occlusion or stenosis level can be located using segmental blood-pressure measurements (Fronek et al. 1978, Rutherford et al. 1979, Neuerburg-Heusler et al. 1989).

The usual cuff placement positions are described on p. 151. The pressure difference between the arm and the cuff placement area indicates an interposed vascular system obstruction. In addition, the pressure differences between the cuff locations on a leg are evaluated in longitudinal section.

The pressure difference between the measurement areas makes it possible to diagnose and locate obstructions that follow in tandem (Figs. 5.**11**, 5.**12**).

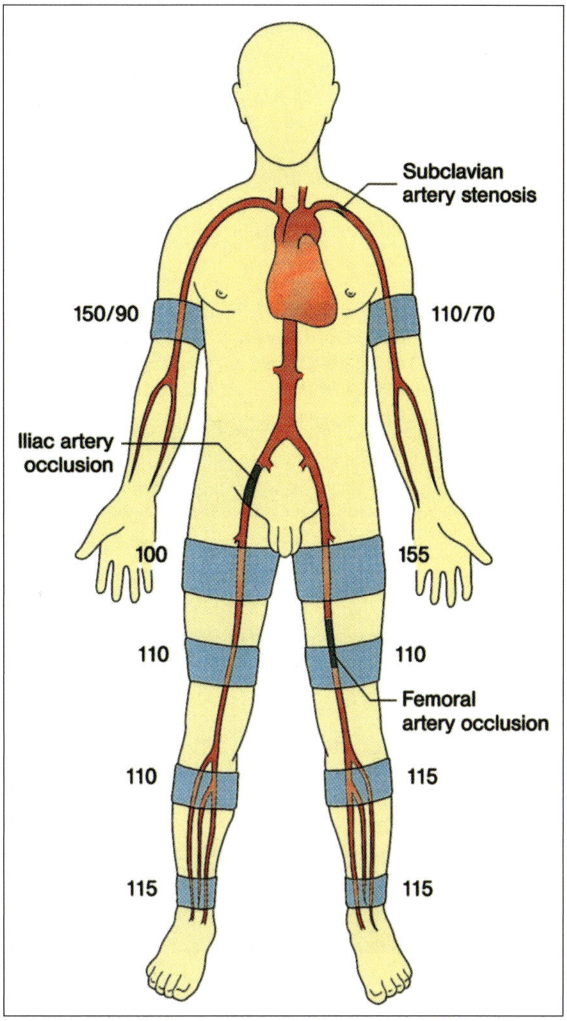

Fig. 5.**11** Segmental systolic pressure measurement, showing the pressure differences between the individual cuff placement areas in relation to various occlusion locations. The model shows:
- A left subclavian artery stenosis
- A right iliac artery occlusion
- A left-sided, distal occlusion of the femoral artery

Using pressure in the ankle arteries as the sole measurement method would not make it possible to identify the level of an occlusion

Fig. 5.**12** Segmental pressure measurement at various locations in the lower extremities. Cuff placement on the proximal thigh (A), distal thigh (B), proximal lower leg (C), the distal lower leg (D) relative to systemic pressure (S) ▷
Group 1 Normal results
Group 2 Isolated occlusions of the pelvic arteries
Group 3 Isolated occlusions of the thigh arteries
Group 4 Occlusions at several levels
When segmental pressure measurement is carried out, it can be seen that isolated occlusions at the pelvic level (group 2) are clearly distinct from isolated occlusions in the thigh arteries (group 3). The curves separate further due to an additional pressure decrease as one moves toward the periphery. In contrast, occlusions at several levels (group 4) can only be differentiated from isolated occlusions of the iliac artery through an additional decrease in peripheral pressure

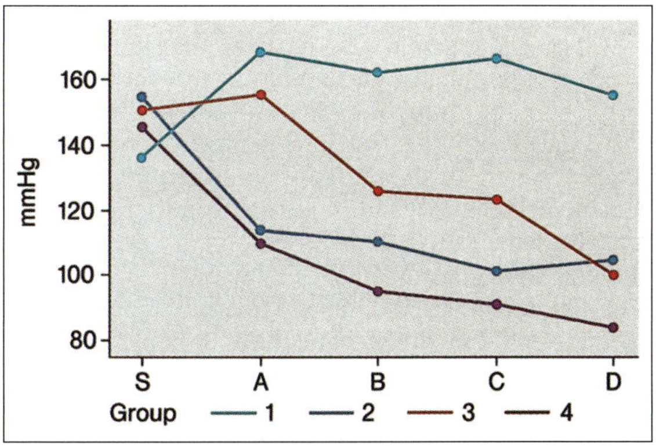

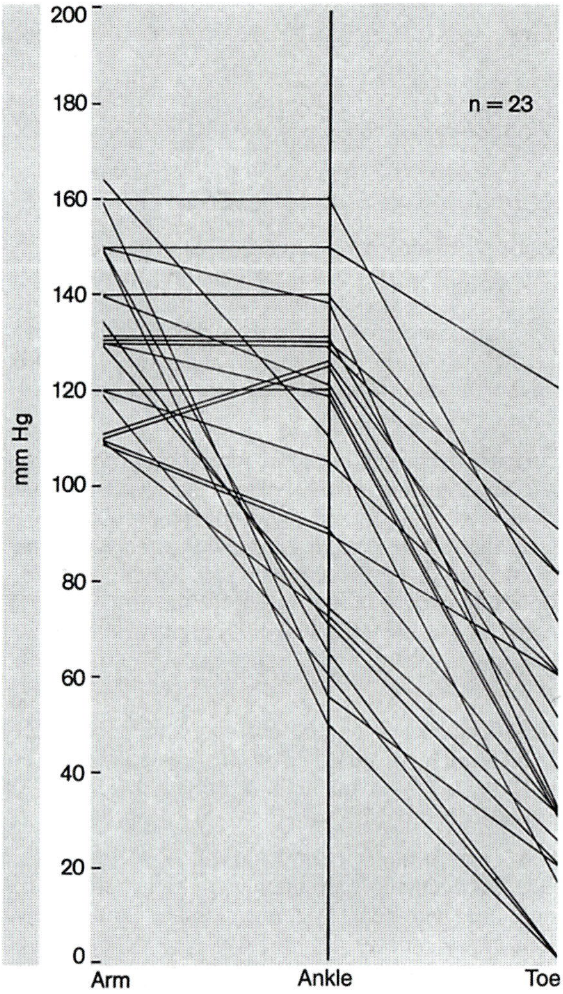

Fig. 5.**13** The pressure difference between the ankle artery and the great toe in patients with isolated or combined occlusions of the forefoot or great toe, or both. Preceding vascular system obstructions are present in some cases

Measurement of Toe Pressure

In patients suffering from endangiitis and diabetes, it is also important to examine the arteries of the foot, which are often affected on an isolated basis. Using toe pressure measurement (p. 151), the absolute toe pressure, as well as the pressure difference between the lower leg and the toe, are examined. This detects any additional occlusions in the forefoot region, which may be missed using ankle pressure measurement only. In a pathological case, the pressure difference is more than 20 mmHg. One of our own studies yielded a value of between 45 and 80 mmHg (Fig. 5.**13**) (Knoblich et al. 1986). Absolute toe pressures of less than 30 mmHg indicate an ischemic endangerment of the foot or the lower leg. Bowers et al. (1993) set a threshold value of 40 mmHg, and in the follow-up they found that 19 of 56 patients (34%) with stable claudication had a clear deterioration in their condition.

Upper Extremity Arteries

Measurement of Pressure at Rest

Bilateral comparison of the blood-pressure values should always be carried out in patients who are suspected of having arterial occlusive disease. Clinically unremarkable subclavian artery stenoses or occlusions can occasionally be detected; auscultating above the supraclavicular fossa can provide an additional indication of stenosis. Flow sounds detected above the supraclavicular fossa must always be clarified. Initially, it needs to be established whether or not the sound is actually coming from the heart. In addition, measuring the blood pressure can provide a general orientation regarding the hemodynamic significance of the finding. Blood-pressure differences are not classified as pathological until they exceed 20 mmHg. It is by no means rare to encounter blood-pressure differences in healthy individuals; these are due to nonsimultaneous blood-pressure measurements. In particular, there is a time difference between the measurements taken on each side.

Measurement of Pressure Post Exercise

Here, too, an exercise test can be used to help distinguish between slight side-to-side blood-pressure differences. The test is usually conducted using pushups or dumbbells. However, this type of study is carried out much less frequently here than in the lower extremities.

Measurement of Digit Pressure

Studies done by Hirai (1978), as well as our own studies, have shown that fingers with segmental arterial occlusions in only one of the two arteries have pressures in the normal range. When there are occlusions in both arteries, or their afferent arteries, the pressures are pathological. On average, pressures clearly decrease by approximately 60 mmHg.

Evaluation

Evaluating the measured *resting pressures* is dealt with in different ways. In the Anglo-American countries, the ratio of the systolic leg and arm pressures is usually calculated, while in the German-speaking countries, measuring the absolute values is preferred. As described in Table 5.**4**, the individual measurements can provide different information about the hemodynamic situation:

- *Absolute value:* circulatory reserve
- *Pressure difference:* hemodynamic compensation
- *Ratio* (ankle/arm index): degree of severity

Table 5.4 The importance of systolic pressure measurements in occlusive arterial disease

- Detecting or excluding obstructions
- Assessing circulatory reserve (residual peripheral pressure)
- Assessing hemodynamic compensation (pressure difference)
- Assessing the degree of severity (arm–ankle quotient)
- Localization (pressure measurement at different levels in the body)

Each of these evaluation modalities offers important information, and enables the examiner to classify the extent of arterial circulatory disturbance.

The values obtained can be entered on data sheets provided for the purpose, or can be plotted graphically.

Sources of Error

Resting period. An adequate resting period for the patient, lasting 10–20 minutes before measurements are taken, has to be observed (the patient can also rest in a sitting position). When there are poorly compensated flow obstructions, the resting period should be longer, in order to avoid measuring *erroneously low* values.

Cuff width. Sometimes, one may forget to adapt the cuff width and length to the placement area. If the same blood-pressure cuff used at the arm and the distal lower leg is applied to the thigh, *erroneously high* values may be measured. In the thigh, wider and longer cuffs should be used than those used for the arm and distal lower leg (p. 151). Conical cuffs improve the applied pressure. At the digital arteries, narrow cuffs have to be used.

Deflation. Releasing the cuff pressure too quickly (> 5.0 mmHg/s) causes *erroneously low* values. Care must be taken to maintain a deflation velocity of around 2.0 mmHg/s.

Arm–leg measurement interval. The time difference between the measurements should not be too long. Intra-individual systemic blood-pressure fluctuations can occur.

Distance between the measurement area and the probe position. If there are *extensive* occlusions located at several levels between the cuff position and the probe position—and only if this is the case—the blood flow through the collaterals may be slightly delayed. *Erroneously low* values result (Franzeck et al. 1981). In such cases, therefore, it is preferable to measure an artery lying near the cuff.

Subclavian obstruction. If subclavian stenoses or occlusions are missed, the systemic blood-pressure values measured are erroneously low; when peripheral pressures are also low, this may give a false impression of a patent vascular system.

Flow velocity in the arteries measured. If the blood flow velocity in the arteries that are being measured is too low, it is not possible to receive a Doppler signal. This phenomenon occurs at pressures below 30–40 mmHg (p. 149).

Media sclerosis. In individuals suffering from diabetes, media sclerosis, which occurs frequently and can lead to *erroneously high* values, should be considered. It should be suspected when the ankle pressure exceeds the arm pressure by more than 50 mmHg, or when the Doppler signal cannot be suppressed using a 300 mmHg cuff pressure. In such cases, one must use procedures that register the pulse instead (i.e., oscillography).

Edema. In a solid edema, especially a lipedema, adequate arterial compression may fail. *Erroneously high* values result.

Hypertension. When the systemic pressure is pathologically elevated, the absolute poststenotic values present a picture that is too favorable (*erroneously high* pressure). Since there is no linear relationship between a change in the systemic pressure and the peripheral pressure (Matsubara et al. 1980), the measurement should always be repeated after the systemic pressure has normalized.

Measuring the digital artery pressure. Measurement of the digital artery pressure is not capable of detecting segmental occlusions affecting individual finger arteries (Hirai 1978). The same is true of arterial toe pressure measurement. These pressures are clearly decreased when compared to the ankle pressure only if occlusions are also present in the afferent arteries of the forefoot.

Measurement of pressure post exercise. After physical exertion to evaluate both extremities, two examiners have to carry out the examination simultaneously, or there has to be an adequate rest period between the measurements of the right and left sides. The leg that has the lower resting pressure should be measured first, because the recovery time post exercise is otherwise too long in pathological cases.

Diagnostic Effectiveness

Peripheral systolic pressure measurement, which is relatively simple methodologically, provides a great deal of useful information.

Decreased ankle pressure confirms the *diagnosis* of occlusive arterial disease. This is extremely useful in differentiating a variety of leg complaints.

Table 5.**5** The accuracy of segmental systolic pressure measurements at rest using the ultrasound Doppler technique in stenoses greater than 50% and occlusions in arteries at the pelvic and thigh level

Vascular region	Rutherford et al. (1979) (n = 205)		Flanigan et al. (1981) (n = 95)		Lynch et al. (1984) (n = 345)		Neuerburg-Heusler et al. (1989) (n=144)		Moneta et al. (1993) (n = 151)	
	Sensitivity (%)	Specificity (%)	Sensitivity (%)	Specificity (%)	Sensitivity (%)	Specificity (%)	Sensitivity (%)	Specificity (%)	Sensitivity (%)	Specificity (%)
Isolated aortoiliac	75	97	79	56	97	50	60	96	59	86
Isolated femoropopliteal	60	97	–	–	89	68	88	96	85	53
Aortoiliac + femoropopliteal	78	97	–	–	55	–	80	96	–	–

The *absolute value* provides information about the available circulatory reserve, while the *pressure gradient* reflects the degree of hemodynamic compensation.

Finally, *pressure measurement at different levels of the body* allows the level of a flow obstruction to be identified (Table 5.**4**).

The value of systolic leg pressure measurement lies in the fact that it is simple to use, time-saving, and cost-effective, with easily reproducible results (Carter 1969, Grüntzig and Schlumpf 1974).

When a resting pressure is detected that is higher than the undisturbed arm pressure, this finding excludes any hemodynamically significant vascular system stenoses below and proximal to the blood-pressure cuff. With borderline results (arm and ankle pressure with the same magnitude), an exercise test should be completed to allow better discrimination (Yao et al. 1969, Carter 1972, Mahler 1990). In diabetics with media sclerosis, it is sometimes not possible to measure the peripheral systolic blood pressure, or only an erroneously high result is produced. This value can therefore not be used to exclude other diagnoses.

In contrast, it is not possible to differentiate between an *occlusion* and a *stenosis.* For this purpose, simple auscultation with a stethoscope at an appropriate location (pelvis, groin, adductor canal, popliteal region) is still used when duplex equipment is not available. However, recognizing and evaluating the sounds of stenosis requires some practice. In addition, *locating* an occlusion or a stenosis within a particular vascular level is not possible, nor is it possible to establish the length of the occlusion.

According to Rutherford et al. (1979), the *sensitivity* in detecting isolated lesions is 93% (n = 103). According to this study, occlusions at several levels are more difficult to locate, even using segmental measurements; the sensitivity amounts to only 86%. Ouriel and Zarins (1982) were able to demonstrate a 99% overall sensitivity for detecting occlusive arterial disease in the legs; further results are shown in Table 5.**5**. Wall changes such as plaques and ulcerations—the early forms of arteriosclerosis—are not detectable, since they do not cause any decreases in pressure, even during exertion.

The consensus of the reports in the literature is that the *specificity* of peripheral systolic pressure measurements in most studies is very high (Table 5.**5**). In a study involving 108 angiographically normal lower extremities, the peripheral systolic pressure was higher than the arm pressure in 83% of cases, the same in 15%, and a lower ankle pressure was detectable only in 2% (Köhler and Krüpe 1985).

Doppler pressure measurements are significant, because these noninvasively measured systolic pressures have the same value as pressure measurements obtained using more invasive procedures (Bollinger et al. 1976, Köhler and Lösse 1979). *Measured digital artery pressures* are a different matter. According to our own results and information from the literature, these show a somewhat larger scatter range, which can be explained by methodological problems (Gundersen 1972, Nielsen et al. 1973, Knoblich et al. 1986).

Differentiating between normal and pathological groups is markedly improved using *exercise tests* (Thulesius 1971, Ludwig et al. 1985). The same applies to the use of hyperemia following suprasystolic arterial compression (Fronek et al. 1973a, Mahler et al. 1975), although this has a somewhat smaller effect on the pressure decrease than physical exertion (Fig. 5.**14**).

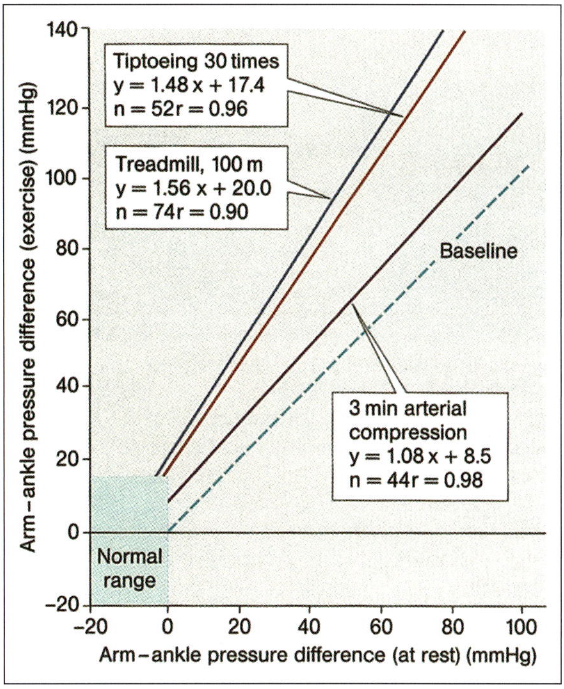

Fig. 5.**14** Various forms of exercise test that can confirm the diagnosis of arterial obstructions in cases of borderline pressure differences at rest (adapted from Mahler 1990). Following arterial compression (3 min), a resting pressure difference of 20 mmHg increases to 30 mmHg. After the patient has tiptoed 30 times, it increases to 46 mmHg, and after 100 m on the treadmill, it increases to 51 mmHg

Color Flow Duplex Sonography and Continuous-Wave Doppler Sonography

Examination

Special Equipment and Documentation

Color Flow Duplex Sonography

For the vasculature of the extremities, linear scanners with a 5–7.5 MHz B-mode transmission frequency are preferred (see Fig. 1.**14**, p. 12). These allow clear demonstration of arteries located relatively close to the surface.

In contrast, in the pelvic region, sector scanners (transmission frequency 3.5–5 MHz) may be easier to use, especially near the groin, as they have a smaller transducer, which also provides better depth resolution. It is essential to use pulsed-wave, continuous-wave, or color flow Doppler mode ultrasound (2.5–5 MHz), since evaluating the hemodynamics in the arteries of the extremities is of primary importance, while pure B-mode imaging provides information about vascular wall deposit morphology and plaque location, which is of secondary importance only. However, the imaging of true or false aneurysms, and determining their form and location, are best accomplished with B-mode imaging.

When documenting the findings, important regions should always be recorded, such as the femoral bifurcation, the origin of the superficial femoral artery, the deep femoral artery, and also the popliteal artery. Pathological findings, such as stenoses or occlusions, should be depicted separately. Color flow Doppler imaging is no substitute for precise measurement of the flow velocity using the pulsed-wave Doppler mode or formal analysis of the Doppler velocity waveform.

Continuous-Wave Doppler Sonography

Ultrasound is applied to the arteries in the same locations used to palpate the pulse. In the lower body, these are: the groin, the popliteal region, the lateral margin of the lateral malleolus, and the dorsum of the foot. In the upper body, the subclavian artery is mainly searched for via a supraclavicular access point, and the radial artery is detected at the pulse palpation point.

The flow velocity waveforms are obtained either by zero-crossing, or as frequency spectra. Due to the greater complexity of the equipment, exclusive use of pulsed-wave Doppler sonography has not become widespread for routine examinations. It is only used in specialized research situations.

Examination Conditions

Patient and Examiner

The patient is examined at the pelvis or leg while lying down. As in blood-pressure measurements, he or she should have rested for some 10–20 minutes beforehand. This is also important when measuring the flow

a
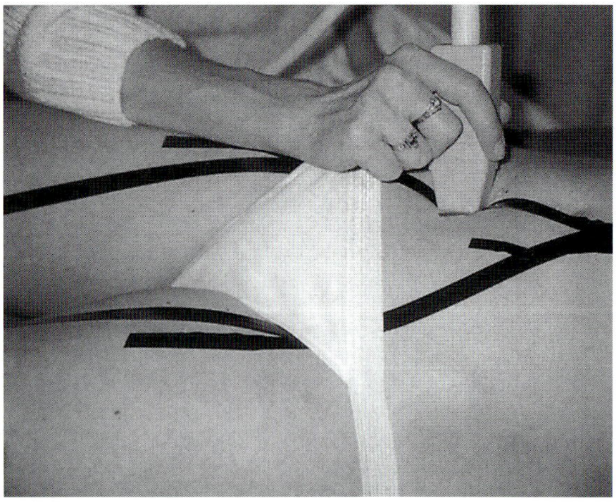

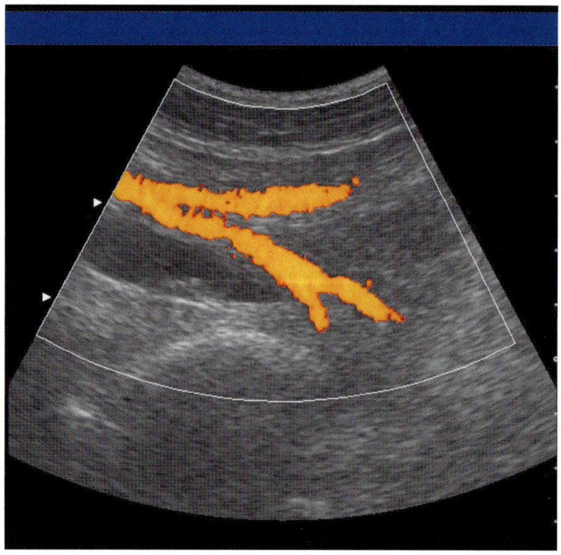
b

Fig. 5.**15a** Applying ultrasound to the pelvic vascular system using a sector scanner from a duplex system. The position of the transducer is appropriate for examining the external iliac artery at its origin. Depending on the artery being examined, the position of the transducer is displaced along the lines indicated, representing sectional planes, which are shown here as an example. **b** Corresponding to the position of the transducer in **a**, this shows the separation of the common iliac artery into the external and internal iliac artery on power mode imaging. The internal iliac artery is depicted longitudinally and shows branching along its course

velocity in healthy individuals, since incorrect pathological results can thereby be avoided.

Ultrasound is applied at the groin with the patient in a supine position, and at the popliteal region with the patient prone, or in a supine position with the knee bent.

The pelvis, thigh, and lower leg should also be examined in supine position, while the popliteal artery is easier to evaluate in prone position (Figs. 5.**15**, 5.**16**).

It is easier to assess the subclavian artery with the patient in a semi-sitting position; ultrasound is applied from a posterior—i.e., cranial—direction. The examiner should be seated during all procedures to facilitate stable positioning of the transducer or probe. The arm arteries are most effectively located in an abducted arm resting on a supporting surface (see Fig. 5.**19 a–c**).

a
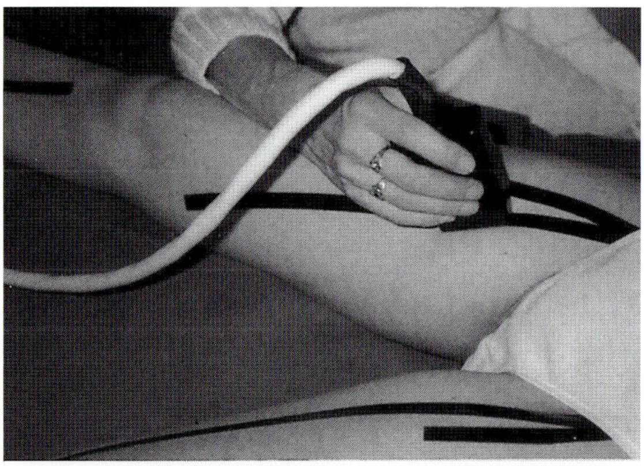

b
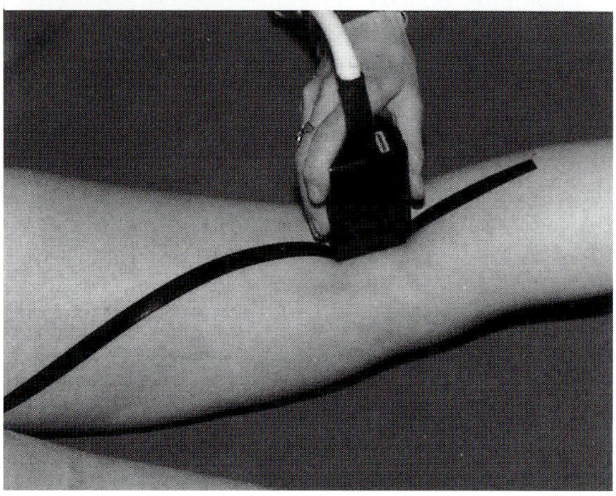

Fig. 5.**16a** Applying ultrasound to the superficial femoral artery with a linear scanner at the transition zone from the proximal to the middle segment. **b** Positioning of the linear scanner of a duplex system on a longitudinal plane above the popliteal region to evaluate the popliteal artery. A sectional plane is marked from the dorsal side to depict the middle section of the superficial femoral artery as far as the trifurcation. Particularly in the femoropopliteal transition zone, care needs to be taken to ensure that the ultrasound is applied continuously, even when the patient changes position

Conducting the Examination

Color Flow Duplex Sonography

B-mode. Imaging of the peripheral arterial vascular system in B-mode concentrates on the pulsating arterial wall, which is more hyperechoic than the venous wall. When the anatomical course is known, the arteries are usually displayed in longitudinal section. The cross-sectional view can sometimes be useful in locating the blood vessels. In a specialized investigation, only specific segments are usually examined. Since exact description of plaque morphology and location is not diagnostically decisive, the number of sectional planes does not have the same significance as in the carotid arteries. Instead, *locating* and *quantifying* vascular system obstructions in order to plan appropriate treatment is more important here.

Doppler mode. The Doppler sample volume serves to measure the flow velocity and detect all types of disturbed flow, from laminar to turbulent flow. Systematic examination of the total lengths of vessel (pelvis, thigh, lower leg) using the pulsed-wave Doppler mode requires considerable time. The specific area of interest can be identified using prior simple examination procedures (segmental pressure measurements, or directional continuous-wave Doppler sonography, or both). Within the area of interest, the hemodynamic effect of stenoses, or the occlusion length, can then be quantified through systematic examinations with the Doppler mode.

Color flow Doppler mode. The color-coded display of blood flow has greatly facilitated the examination of the arteries of the extremities. Arterial flow is usually represented in red, and venous flow in blue. It is necessary to note any regions with lighter coloring (increased flow velocity) and to identify gaps in the vasculature. These are additionally identified through outflow into thin collateral blood vessels. The color flow Doppler mode is important for imaging the lower leg arteries, which without this method would be extremely difficult to display and classify topographically. Even when color flow duplex sonography is used, however, exact measurement of the flow velocity in the regions of interest has to be carried out using the pulsed-wave Doppler mode.

Continuous-Wave Doppler Sonography

While this is being done, the probe is shifted with minimal movements, medially and laterally, proximally and distally. Acoustically, an attempt is made to obtain the signal with the loudest volume and highest frequency. It should be displayed on the monitor as a maximum amplitude curve, but without any venous superimposition.

The Doppler frequencies are documented as an frequency spectrum (pp. 12–15). With spectral analysis, automatic calculation of indices is usually possible. Quantitative measurement of the maximum systolic velocity with continuous-wave Doppler sonography is limited, since the incident angle of the ultrasound beam cannot be determined. The maximum frequency is therefore analyzed, and a qualitative formal analysis of the Doppler waveform is usually acceptable. In contrast to the carotid vascular system, stenoses are seldom directly located in the peripheral arteries. These arteries, which lie too deep anatomically, cannot be continuously evaluated with the probe.

Examination Sequence

Pelvic and Leg Arteries

Color Flow Duplex Sonography

To image the complete *pelvic and leg vascular systems* unilaterally or bilaterally, the examination begins at the distal aorta, with a search being made for aneurysmal dilations or plaques in cross-section and longitudinal section. The iliac vasculature is followed distally to the bifurcation of the femoral artery. Care should be taken to note wall deposits in B-mode imaging, as well as stenoses or occlusions using color flow or pulsed-wave Doppler (Fig. 5.**15 a, b**).

The ultrasound examination can also start at the common femoral artery, proceeding proximally to the aortic bifurcation. If the pulsed-wave Doppler sample volume is gradually included, this procedure has the advantage that it can identify any suspicion of preceding flow obstructions in the vascular system of the common femoral artery. However, due to the deep location of the pelvic vasculature, especially in patients with obesity or meteorism, continuous imaging of the vascular course may be difficult or impossible.

Subsequently, the examination of the *vascular system of the thigh* also starts at the groin. The transducer position, in a ventrodorsal sagittal plane, allows imaging of the common femoral artery and the bifurcation area initially, and in longitudinal section, the superficial femoral artery and the deep femoral artery can be depicted (see Fig. 5.**24**). Along its distal course, the femoral artery is increasingly examined in a sectional plane, moving from ventromedial to dorsolateral (Fig. 5.**16 a**).

Near the femoral bifurcation, the origin of the deep femoral artery must be displayed without fail; stenoses of the origin often cannot be seen in angiography. Indirect examination methods do not allow successful differentiation between the superficial femoral artery and the deep femoral artery. Following the deep femoral artery further distally is often difficult, since the artery changes course in a dorsal direction. A critical location for the superficial femoral artery is its transition into the popliteal artery in the adductor canal. Here, the patient's position has to be changed from supine to prone; the artery at this point is more deeply entrenched in the muscles.

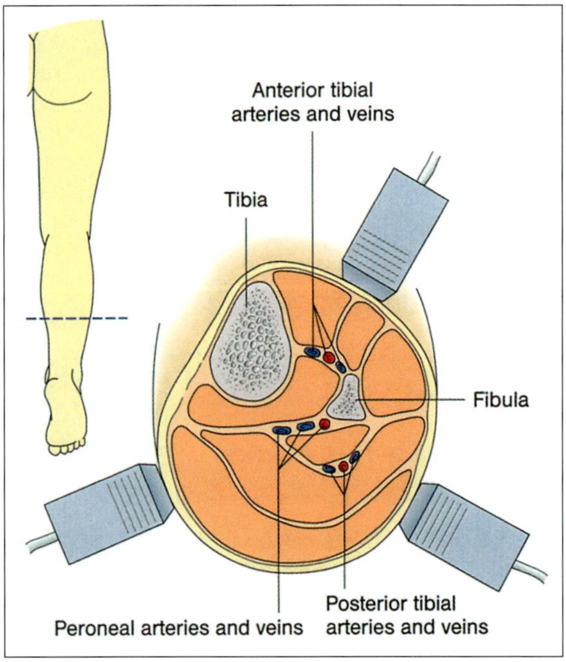

Fig. 5.**17** An anatomical cross-section through the lower leg from the proximal to the middle section, showing the three arteries of the lower leg: the anterior tibial artery, the fibular artery, and the posterior tibial artery, with the corresponding veins accompanying them. The transducer positions at which ultrasound should be applied to the vascular bundles are indicated

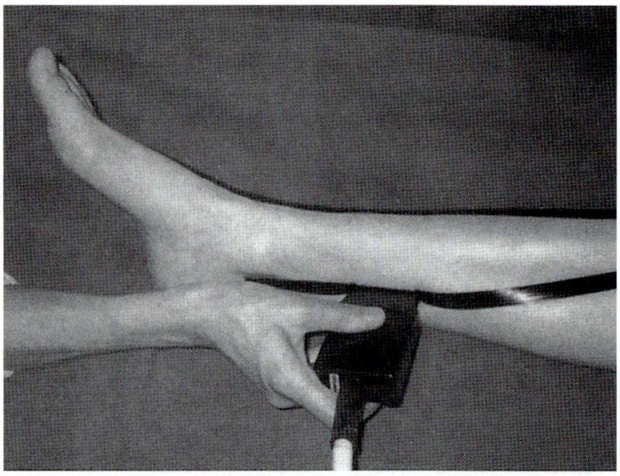

a

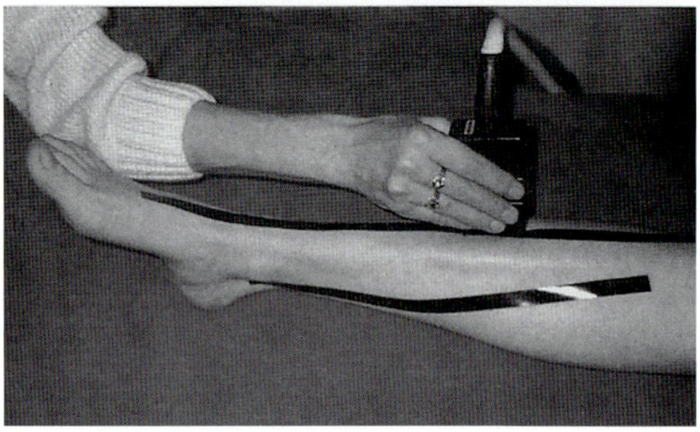

b

Fig. 5.**18** The lower leg arteries evaluated along their course using a linear scanner. **a** The posterior tibial artery at the distal lower leg. **b** The anterior tibial artery in the middle section of the lower leg

The distal femoral artery, or the proximal segment of the popliteal artery, is examined with ultrasound from a dorsal direction (Fig. 5.**16 b**). The *popliteal region* at the posterior aspect of the knee can be readily evaluated with ultrasound. Further distally, ramification into the three lower leg arteries (trifurcation) occurs. Usually, this can still easily be imaged. Imaging of the superficial femoral artery and the popliteal artery should be done in longitudinal section, so that flow velocity measurements can be made with angle correction.

The distal *lower leg arteries* are easier to follow starting from a distal direction (ankle level) and proceeding proximally. However, precise topographical knowledge (Fig. 5.**17**) is a necessary precondition for successfully distinguishing between the three arteries (posterior tibial artery, anterior tibial artery, and peroneal artery). Color flow Doppler mode makes it much easier to follow the arteries proximally (Fig. 5.**18**). Here, as in other arterial segments, vascular system stenoses and occlusions are easier to detect by the usual signs (color lightening, color interruption) (see Fig. 7.**33**, p. 324). Quantitative measurement of the flow velocity is possible using the adjusted Doppler sample volume.

In the groin and popliteal regions, attention should be given to the aneurysms that accompany dilative arteriopathy, which are not infrequent.

The groin region requires special attention, since iatrogenic vascular damage (hematomas, false aneurysms, and arteriovenous fistulas) occurs here most often after catheter interventions. Due to the wider-lumen access channel that is used, these complications are being observed with increasing frequency after coronary angioplasty (Steinkamp et al. 1992).

Continuous-Wave Doppler Sonography

The most important location to examine is the groin region, since cuff methods—i.e., pressure measuring procedures—cannot be used in the pelvic vascular system. The pulse of the common femoral artery is relatively easy to find. This is where the Doppler signal is registered.

Ultrasound is preferably applied to the popliteal artery, with the patient in a prone or lateral position. It is often difficult to achieve a favorable incident angle.

The Doppler flow curves from the posterior tibial artery and the dorsal artery of the foot—relatively

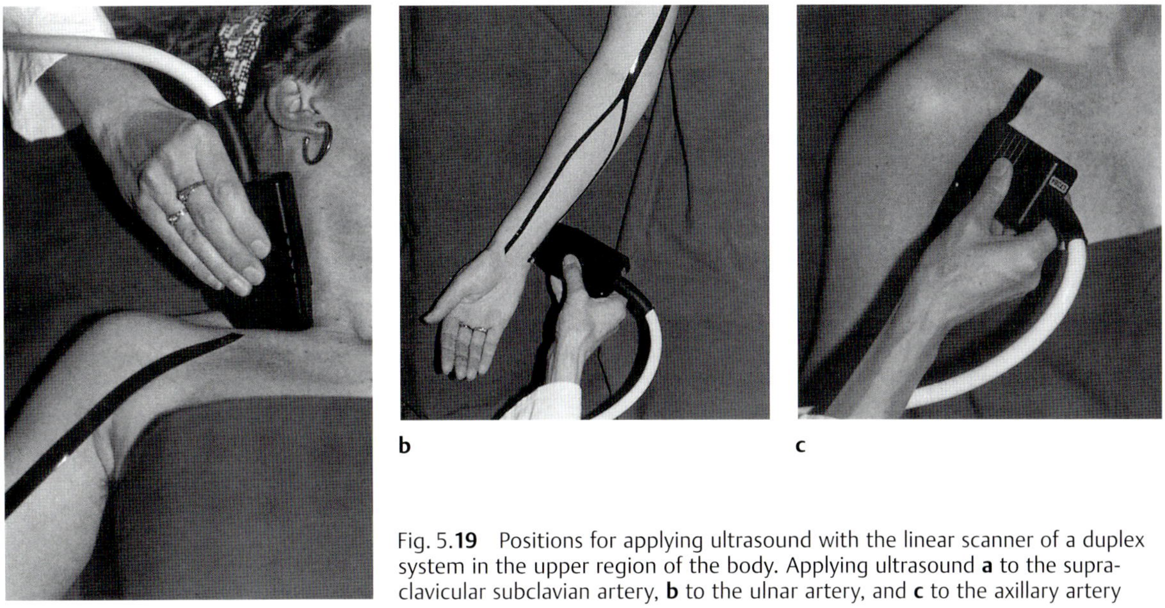

a b c

Fig. 5.**19** Positions for applying ultrasound with the linear scanner of a duplex system in the upper region of the body. Applying ultrasound **a** to the supraclavicular subclavian artery, **b** to the ulnar artery, and **c** to the axillary artery

small-caliber blood vessels located near the surface—are somewhat more difficult to detect, and are therefore recorded less often.

Upper Extremity Arteries

Color Flow Duplex Sonography

Using a side-to-side comparison of the measured pressure and continuous-wave Doppler sonography at the upper extremities, precisely quantified information concerning the patency of the vascular system or the presence of flow obstructions can be obtained. The most important region for applying duplex sonography is the origin of the subclavian artery, which can be imaged well with a little practice, and with appropriate patient positioning (Fig. 5.**19 a**). Since the proximal subclavian artery and the innominate artery can be classified as arteries supplying the brain—via the vertebral artery located near the origin—the corresponding section above should be referred to (pp. 44–46).

Another important region is the passage of the subclavian artery between the clavicle and the first rib. Occasionally, this is mechanically constricted, leading to functional or permanent stenoses (see Fig. 5.**39**). Due to the osseous structures, it is not easy to image this region, even with duplex sonography. The further courses of the subclavian artery, the axillary artery (Fig. 5.**19 c**), and the brachial artery, and the two forearm arteries, the radial artery and ulnar artery (Fig. 5.**19 b**), are continuously imaged only when there is a clinical suspicion of vascular stenoses and occlusions. The evaluation of hemodialysis fistulas is particularly important, however. These can be searched for with high-resolution transducers in the upper arm and forearm. Quantitative measurement of the flow rate using the adjusted Doppler mode is important, and this can be done with duplex sonography, since both the incident angle and the vascular diameter can be determined (p. 191).

Using modern high-resolution ultrasound probes (10–15 MHz), the flow curves in the *finger arteries* can also be obtained using duplex sonography. A strong dependence of the flow pattern on room and skin temperature is observed in the digital arteries (Voelckel et al. 1998), which leads to reduced perfusion in the fingers on exposure to cold. The examination is much easier to carry out after the fingers have been placed in a warm-water bath. On the one hand, this ensures that the examination temperature is kept constant and reproducible as far as possible, while on the other it means that functional stenoses in the vessels and vascular spasms caused by cold can be overcome. In addition, compression artefacts caused by excessive pressure from the ultrasound probe on the digital arteries, which lie very superficially, can be avoided.

Continuous-Wave Doppler Sonography

The subclavian artery can be defined as an artery supplying the extremities and also, via the origin of the vertebral artery, as an artery supplying the brain. The data registration area is described on p. 34.

The other arteries of the arm (axillary artery, brachial artery, radial artery, ulnar artery) are rarely directly examined with ultrasound; the Doppler flow curves are obtained at the usual palpation points.

The digital arteries in the fingers can also be located, and appropriate flow velocity curves can be reg-

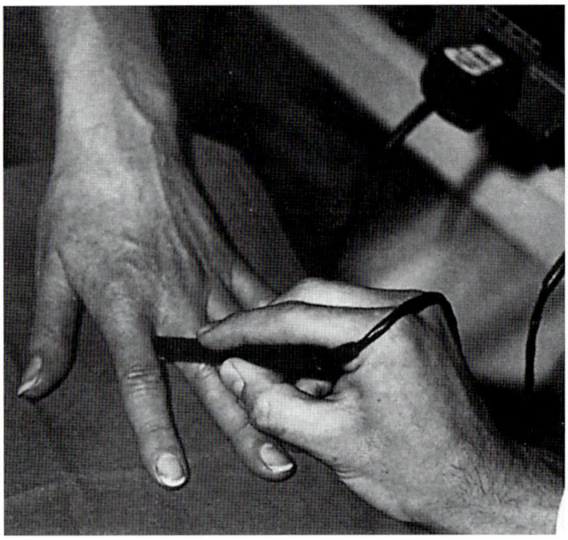

Fig. 5.**20** The way in which to hold the probe and the measurement position for registering curves when applying Doppler sonography at the proximal digital arteries

istered (Fig. 5.**20**). However, this cannot detect or exclude segmental occlusions with certainty. Measuring the pressure should be the preferred method (p. 151).

Normal Findings

Principle

At rest. The normal flow pulses in the extremities are triphasic, with a steep systolic ascent (acceleration), a narrow peak, a quick descent (deceleration), and a backflow component in the early diastole that is approximately one-fifth of the height of the systolic forward flow. In the late diastole, a short forward flow peak is also registered (Fig. 5.**21 a**).

The formation of this characteristic wave phenomenon is basically explained by a primary pulse wave which is reflected in the periphery, proceeds retrogradely through the arterial system, and is reflected again at the aortic valve, which has closed in the meantime (backflow component). The wave then again

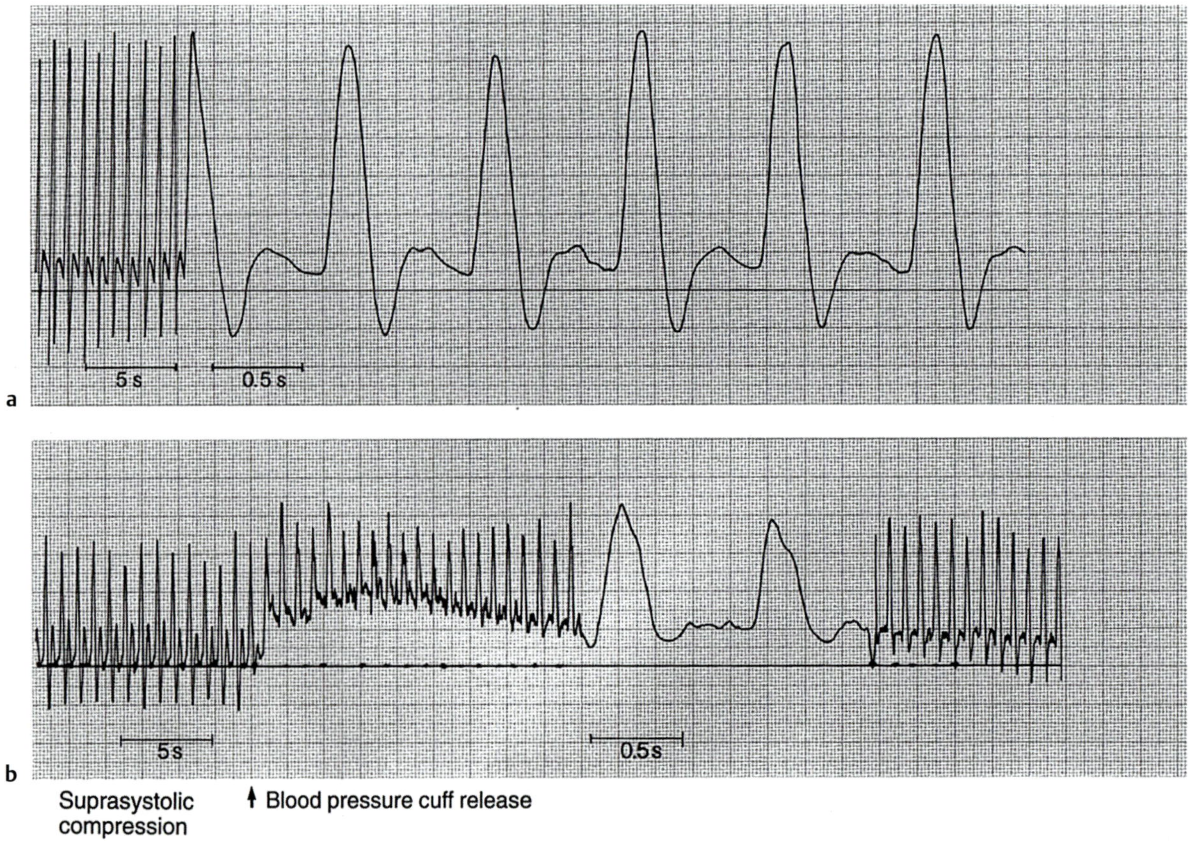

Fig. 5.**21** A flow velocity curve in a healthy individual, **a** at rest and **b** after suprasystolic arterial compression, registered from the common femoral artery above the groin. At rest, there is a characteristic tricyclic curve with a steep systolic increase, early diastolic backflow, and diastolic forward flow. During suprasystolic arterial compression, the backflow component deepens and a decrease in the diastolic flow while resting is seen, due to increased peripheral resistance. After the cuff is released, the hyperemic phase results in a clear increase in the diastolic flow velocity and a return to normal conditions due to a decrease in the peripheral resistance

proceeds peripherally, and there causes the second forward peak (dicrotism, "ping-pong phenomenon") (Busse et al. 1975).

The form of the flow pulse in the extremities is determined by the high peripheral resistance present in the terminal vascular regions (skin, muscles) under resting conditions. In normal circumstances, the flow velocity curves must show the same, typically triphasic, formal elements, at the usual registration points—groin, popliteal region, and ankle, as previously described. Occasionally, the amplitude height decreases distally, since the flow velocity decreases when passing parallel connecting blood vessels with decreasing diameter (Rieger 1993).

Post exercise. In a healthy individual, the systolic pressure briefly decreases after exercise, and the flow velocity increases significantly, especially in its diastolic component (decreased peripheral resistance), due to the higher blood supply required by the muscles. However, it returns to approximately normal values within 1 minute (Figs. 5.**7**, 5.**21 b**) (Sumner and Strandness 1969, Mahler et al. 1976).

Color Flow Duplex Sonography

B-mode. In physiological conditions, the B-mode serves to determine the vascular diameter and course. It is possible to exclude aneurysmal dilations and atherosclerotic wall deposits, or thrombotic occlusions.

Doppler mode. The criteria for evaluating flow velocity curves, which in duplex sonography are usually obtained with the pulsed-wave Doppler sample volume and are displayed as a frequency analysis, correspond to the formal analysis using continuous-wave Doppler sonography, mentioned above. However, quantitative measurement of the maximum velocity, mean velocity, and other parameters is also possible, since the angle between the Doppler beam and vascular axis, as well as the diameter of the vessel can be calculated.

Color flow Doppler mode. The color flow Doppler mode makes it easier to locate arteries, follow their course, and in normal cases to detect a patent arterial vascular system. The vascular course and diameter deviations, such as positional anomalies and hypoplasias, are more easily recognized with this method than with pure B-mode imaging.

Findings and Anatomy

Classifying peripheral arterial disease by anatomical region is based on the Ratschow location types (Heberer et al. 1974).

- Pelvic level (pelvic type)
- Thigh level (thigh type)
- Calf and foot arteries (peripheral type)

The occlusion level determines not only the clinical symptoms, but also requires differentiated therapy. We have retained this tried and tested classification system, even for normal findings. Of course, there is often what is known as multilevel disease.

Pelvic Level

Anatomy

The *distal abdominal aorta* belongs to the pelvic level. At the fourth lumbar vertebra, and at an angle of between 60° and 80°, the aorta branches into the common iliac arteries. After nearly 5–6 cm, these divide into the *external iliac artery* and the *internal iliac artery*. With its large ventral and dorsal main branches, the latter supplies the viscera in the small pelvis, and contributes to the collateral circulation in pathological cases. The external iliac artery follows a course through the small pelvis, without giving off any branches. Shortly before its entry into the lacuna of vessels, it gives off two smaller blood vessels: the *inferior epigastric artery* and the *deep circumflex iliac artery* (Fig. 5.**22**).

Color Flow Duplex Sonography

B-mode. If the preceding examinations using Doppler pressure measurement or continuous-wave Doppler sonography, or both, have excluded significant vascular system obstructions, only the pelvic vascular system is normally examined with ultrasound, in order not to miss any aneurysms or to acquire morphological information. According to measurements made by several authors, the sonographically measured diameter of the external iliac artery is 0.80 cm (Table 5.**6**). For the common iliac artery, radiological anatomical measurements on average give a figure of 0.83 cm (left side) and 0.89 cm (right side) (Luska 1990, p. 169).

Doppler mode. The Doppler mode is only used for specialized investigations. According to Jäger et al. (1985), the flow velocity in the external iliac artery is on average 119.3 cm/s (Table 5.**6**).

Color Flow Doppler Mode. It is easier to locate the arteries using color flow Doppler mode, especially at the origin of the internal iliac artery (Fig. 5.**23**). After femoral artery punctures, it is important to exclude a false aneurysm or a hematoma.

Continuous-Wave Doppler Sonography

Due to inadequate penetration depth and insufficient scope for arterial identification in the small pelvis, it is not possible to apply ultrasound to the pelvic vasculature using continuous-wave Doppler sonography.

A normal triphasic Doppler waveform obtained from the common femoral artery at the groin *indirectly* confirms the patency of the pelvic vascular system (Fig. 5.**21 a**).

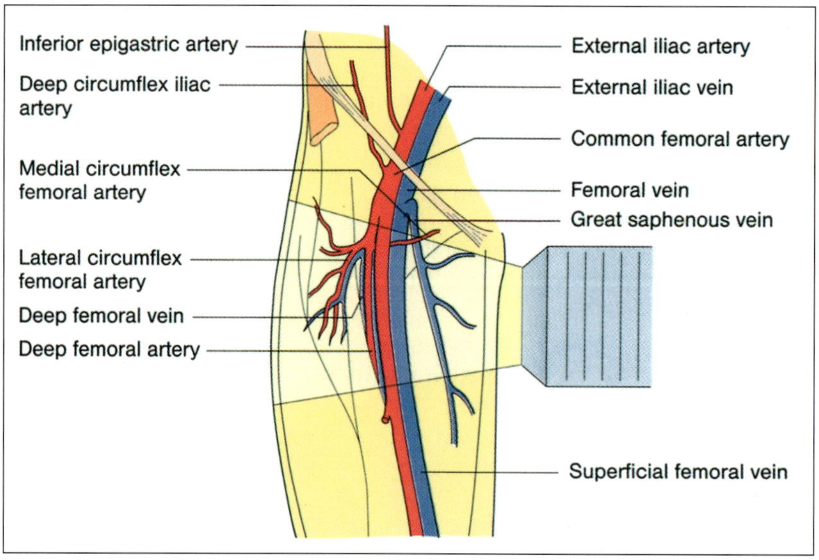

Fig. 5.**22** The anatomy of the common femoral artery and vein, and the branching of the common femoral artery into the superficial femoral artery and deep femoral artery with its side branches, the medial circumflex femoral artery and lateral circumflex femoral artery. Distal to the inguinal ligament, the femoral vein has a curved junction with the great saphenous vein

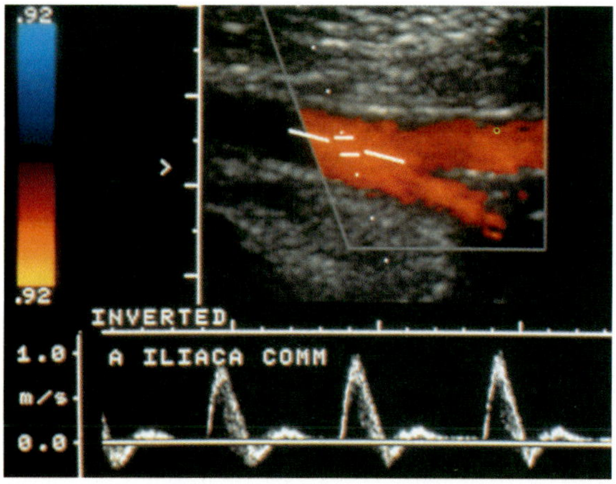

Fig. 5.**23** A color-coded image of the distal common iliac artery, showing its branching into the larger-caliber external iliac artery and the smaller-caliber internal iliac artery, and also displaying the corresponding tricyclic flow velocity waveform. The consistently intense color indicates a uniform mean flow velocity within the vascular bifurcation

Thigh Level

Anatomy

Below the inguinal ligament, the external iliac artery is called the *common femoral artery* (not according to anatomy textbooks, but among angiologists, radiologists, and vascular surgeons). This arterial segment is classified as belonging to the thigh level, and in its short, stretched course near the surface, it is particularly well accessible to all types of examination.

Approximately 4.0 cm below the inguinal ligament, the artery branches into the *superficial femoral artery* and the *deep femoral artery*. The deep femoral artery supplies the thigh muscles, and shortly after the origin gives off two larger branches, the *medial* and *lateral circumflex femoral arteries* (Fig. 5.**22**). At adductor canal level, the terminal branches of the deep femoral artery collateralize with smaller thigh branches from the superficial femoral artery, and form the most important collateral circulation for femoral artery occlusions proximal to the adductor canal. The superficial femoral artery continues as the *popliteal artery,* which begins with its proximal segment at the lower margin of the adductor canal, and reaches to the lower margin of the popliteal muscle. The popliteal artery belongs to the femoropopliteal segment, and in the knee region gives off several small branches—*genicular arteries* and *sural* arteries—which together form the *articular rete of the knee.*

Color Flow Duplex Sonography

B-mode. Using B-mode imaging alone to depict the femoral artery and the deep femoral artery is not recommended, since the exclusion of vascular system stenoses is not morphologically certain enough with this method, and it may be difficult to differentiate the vasculature. According to Jäger et al. (1985), in normal individuals the diameter of the common femoral artery is 0.82 cm, and that of the proximal superficial femoral artery is 0.60 cm, with its distal segment giving an average figure of 0.54 cm (Table 5.**6**).

However, the B-mode does allow the exclusion of wall changes, ectasias, aneurysms, and also hematomas after vascular punctures.

Doppler mode. Since the bifurcation of the deep femoral artery is a preferred location for arteriosclerotic stenoses, and can often not be assessed with sufficient certainty using angiography, the Doppler mode should be used for the identification and evaluation of this vascular segment. In addition, course variations in the femoral bifurcation can be detected, especially those affecting the origin of the deep femoral

Table 5.**6** Results from measuring the vascular diameter and the maximal systolic flow velocity (V_{syst}) including the standard deviation (SD) in normal individuals using duplex sonography in several studies

	Jäger et al. (1985)	Jäger et al. (1985)	Luska et al. (1990)	Sacks et al. (1992)	Hatsukami et al. (1992)	Strauss et al. (1989) Karasch et al. (1991b)
	Diameter ± SD (cm)	V_{syst} ± SD (cm/ s)	V_{syst} ± SD (cm/s)	V_{syst} ± SD (cm/s)	V_{syst} ± SD (cm/s)	V_{syst} ± SD (cm/s)
External iliac artery	0.79 ± 0.13	119.3 ± 21.7	102	112.0 ± 49.0	98 ± 17.5	-
Common femoral artery	0.82 ± 0.14	114.1 ± 24.9	104	90.0 ± 41.0	80 ± 16	-
Proximal superficial femoral artery	0.60 ± 0.12	90.8 ± 13.6	99	89.0 ± 23.0	73 ± 10	-
Distal superficial femoral artery	0.54 ± 0.11	93.6 ± 14.1	-	74.0 ± 21.0	56 ± 12	-
Deep femoral artery	-	-	-	-	64 ± 15	60.0 ± 14.0
Popliteal artery	0.52 ± 0.11	68.8 ± 13.5	62	59.0 ± 12.0	53 ± 17	89.7

V_{syst}: systolic maximal velocity; SD = standard deviation.

artery, the many branches of which may originate directly from the superficial femoral artery (Fig. 5.**22**). Evaluating the entire length of the vascular segment (20–30 cm) using conventional Doppler mode is very time-consuming, and can only be recommended when a specialized investigation requires it. By contrast, the popliteal artery, which lies near the surface, is easy to examine from a dorsal direction.

The flow velocity in normal individuals has been examined by several authors (Table 5.**6**). In the common femoral artery, Jäger et al. (1985) measured the peak systolic velocity as 114.1 cm/s; Sacks et al. (1992) arrived at a figure of 90.0 cm/s. For the proximal superficial femoral artery, the results obtained by Jäger and Sacks varied only minimally (90.8 cm/s versus 89.0 cm/s).

Strauss et al. (1989) measured the flow velocity of the deep femoral artery at its origin in healthy individuals as 60 cm/s ± 14 cm/s on average.

Table 5.**7** lists the normal values in cm/s for the individual speeds of maximum systolic flow (V_{max}), maximum diastolic flow (V_{diast}) and maximum backflow (V_{DIP}) in the pelvic and leg arteries, as given in a recent study. In a separate study using comparative measurements by two examiners using two separate duplex apparatuses, the mean flow velocity in the popliteal artery was measured as 89.7 cm/s (Karasch et al. 1991a). Jäger et al. (1985) and Sacks et al. (1992) obtained values of only 68.8 cm/s and 59.0 cm/s, respectively.

Color flow Doppler mode. Using color coding, the thigh level is especially easy to evaluate, due to the extended course of the femoral artery. The entire distance can be rapidly followed to the adductor canal. In this region, however, a change in position is usually necessary, and continuously displaying the vascular course is somewhat more difficult. With practice, it can be reliably evaluated (Fig. 5.**24**). In contrast, the popliteal artery, which is located near the surface, is easily displayed from a dorsal direction.

Table 5.**7** Normal values for maximum systolic velocity (V_{max}), maximum diastolic velocity (V_{diast}), and maximum backflow velocity (V_{DIP}) in the pelvic and leg arteries (adapted from Wuppermann et al. 2000)

	V_{max}	V_{DIP}	V_{diast}
Distal aorta	50–120	20–35	5–30
Common iliac artery	90–130	20–40	9–25
External iliac artery	100–140	30–50	11–26
Common femoral artery	90–140	30–50	8–25
Superficial femoral artery	80–110	25–45	8–21
Popliteal artery	55–82	18–38	4–16

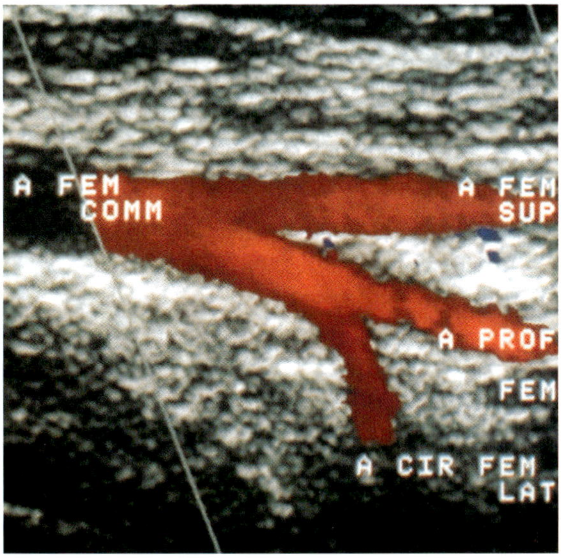

Fig. 5.**24** A color-coded image showing the femoral bifurcation in longitudinal section, including the origin of the superficial femoral artery and the deep femoral artery with its side branch, the lateral circumflex femoral artery

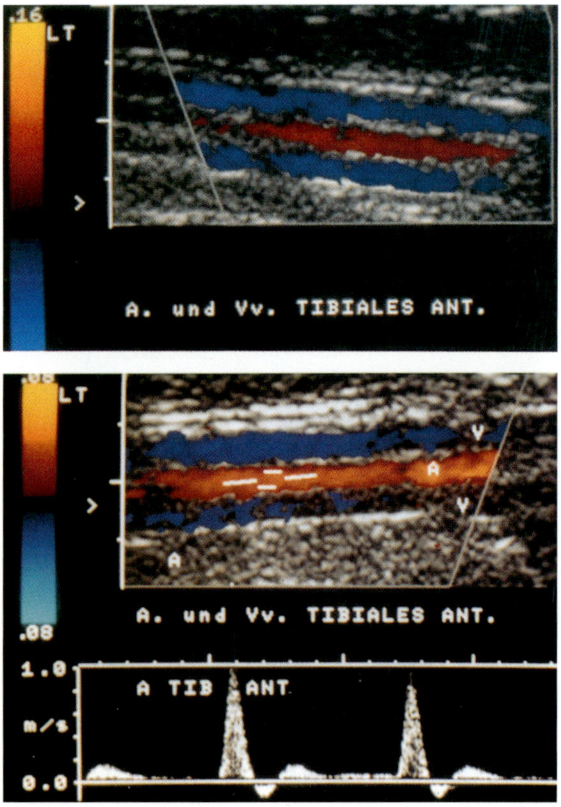

a

b

Fig. 5.**25a** A color-coded image showing the anterior tibial artery with the concomitant veins in longitudinal section. **b** Applying ultrasound with a linear scanner at the level of the distal lower leg and depicting the corresponding flow velocity curve

Continuous-Wave Doppler Sonography

Hemodynamically significant flow obstructions in the thigh region are again *indirectly* excluded by applying ultrasound to the popliteal artery. A triphasic Doppler curve obtained above the popliteal artery indicates that the subsequent vascular system is patent. In slender extremities, it is also possible to apply ultrasound continuously to the superficial femoral artery using the continuous-wave probe. However, due to methodological problems, this form of examination has not come into widespread use as a standard.

Calf and Foot Arteries

Anatomy

The first branch originating from the popliteal artery is the *anterior tibial artery,* which proceeds deeply downward between the lower leg extensors and, between the two malleoli, supplies the foot as the dorsal artery of the foot. One of its terminal branches makes a turn and forms the five dorsal metatarsal arteries. After the origin of the anterior tibial artery, the popliteal artery is called the *tibiofibular trunk* (see p. 260).

The *posterior tibial artery* is the largest terminal branch, which proceeds downward in stretched form between the soleus muscle and the deep flexors. Behind the medial malleolus, it supplies the sole of the foot with its terminal branches, the tibial plantar artery and the fibular plantar artery, and then forms the main afferent path to the deep plantar arch. The *peroneal artery* runs parallel to the posterior tibial artery, and in the ankle region joins with the malleolar rete and the fibular rete.

Color Flow Duplex Sonography

Conventional duplex sonography has rarely been used to examine these arteries, and has only been attempted in order to answer specific questions at highly specialized medical centers. Color flow duplex sonography has made it practical to image and evaluate this vascular region, but when the findings are normal, it does not form part of the routine program used to exclude certain diagnoses. Exact topographical knowledge of the way in which the three lower leg arteries are arranged within the muscles (Fig. 5.**17**) is definitely required for proper evaluation and differentiation of these arteries (Fig. 5.**25**).

Continuous-Wave Doppler Sonography

The lower leg arteries are clearly identified in the ankle region with the continuous-wave probe. Doppler waveforms can be successfully registered from both the posterior tibial artery and the dorsal artery of the foot. In this region, measuring the ankle pressure, however, is better than directly registering the pulse.

Upper Extremity Arteries

Anatomy

As arteries supplying the extremities, the *subclavian arteries* are classified as belonging to this level; in their proximal segment, they also count as arteries supplying the brain. The left subclavian artery originates directly from the aortic arch. The right subclavian artery originates from the *innominate artery,* which after 3–5 cm divides into the *common carotid artery* and the *subclavian artery*. Through its large side branch, the *vertebral artery*, the subclavian artery becomes both an artery supplying the extremities and also one supplying the brain (see p. 258d).

Distal to the lower margin of the clavicle, the artery is called the *axillary artery,* which can have varying locations, depending on the position of the arm. After giving off several branches supplying the muscles, the axillary artery, on entering the arm, is termed the *brachial artery* (see p. 259). The *deep brachial artery,* which is joined by the radial nerve, originates medially at the upper arm. Approximately 1.0 cm below the elbow, the brachial artery divides into the *radial artery* and the *ulnar artery*. On the radial side, the ventral side of the forearm, the radial artery becomes what is known as the pulse artery, and in the palm it forms the *deep palmar arch*. The main afferent path to the *superficial palmar arch* proceeds through the *ulnar artery*. The two palmar vascular arches are connected one below the other with the contralateral artery and give off vasculature to the fingers. Each finger has two blood supplies through the *proper palmar digital arteries* (Hafferl 1969, p. 784).

Color Flow Duplex Sonography

B-mode. The origin of the subclavian artery from the aortic arch is easy to display when the patient is suitably positioned (p. 165). Continuing to apply ultrasound to the artery is difficult beneath the clavicle, due to ultrasound signal loss. The transition zones into the axillary artery and the brachial artery, and also into the ulnar artery and the radial artery in the forearm, can be readily followed using B-mode sonography to exclude pathological structures (Fig. 5.**26**).

Doppler-mode. The flow waveform of the subclavian artery obtained with Doppler mode shows the same triphasic components as every blood vessel that supplies the extremities. A high systolic forward flow is seen, with a steep increase and a quick decrease. In the early diastole, a backflow component is detected, and in the late diastole, a short forward flow is seen.

Color flow Doppler mode. Displaying the entire arterial course is significantly quicker with the color flow Doppler mode. This can become important incidentally in intensive-care medicine, since the subclavian artery or the brachial artery can serve as a guide for easier identification of the accompanying veins. Before a hemodialysis shunt is placed, the arterial component of the vessel should be examined.

Continuous-Wave Doppler Sonography

Using Doppler sonography, the same waveforms can be registered at the brachial artery palpation points, shortly above the bend of the elbow, at the radial artery, and the ulnar artery. The same indices can be used as with Doppler waveforms from the lower extremities. However, these evaluations do not have the same significance as in the lower extremities, since level-oriented diagnosis does not play a significant role here and the quantitative value that is usually used is bilateral comparison of arm blood-pressure measurements.

Evaluation

Color Flow Duplex Sonography

Evaluating the pelvic and leg arteries over their long vascular distance is complicated, particularly because of the deep arterial location within the small pelvis.

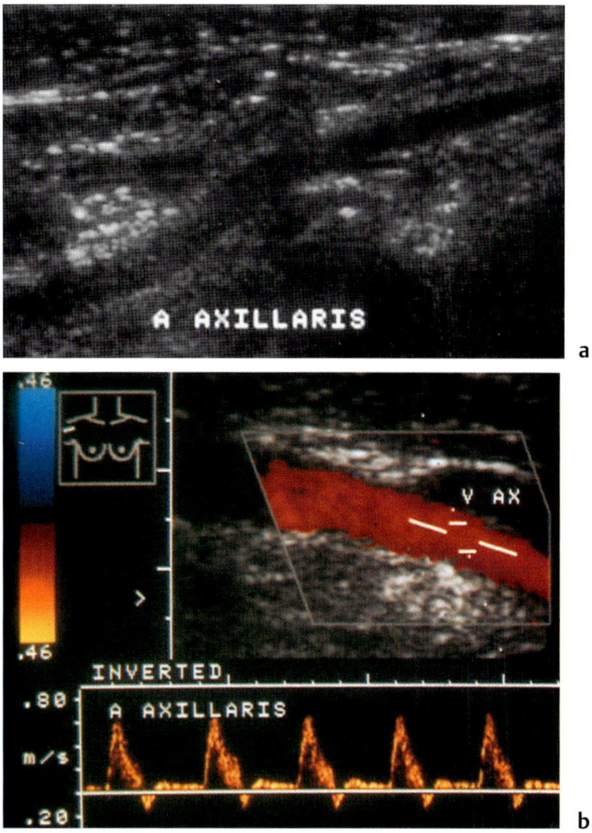

Fig. 5.**26** Applying ultrasound to the axillary artery. **a** A composite presentation of the axillary artery in longitudinal section using B-mode. **b** The frequency spectrum recorded from the axillary artery has a tricyclic form, typical of the arterial velocity waveform in an extremity. The positive diastolic component is less pronounced than in the lower extremities

The course of the external iliac artery, which disappears into the depths, is a particularly critical point. Several authors describe the imaging of the vascular system of the thigh in the adductor canal as being difficult; the positional change required at the transition into the proximal popliteal artery further complicates the continuity of the examination.

Displaying the lower leg arteries without using color provides little information.

B-mode. The vascular diameter can be determined in B-mode. In addition, the image is useful for excluding true aneurysms and, especially after femoral artery punctures, for excluding a false aneurysm or hematoma, or an arteriovenous fistula. In the popliteal region, a search can be made for popliteal cysts.

Doppler mode. With the integrated pulsed-wave Doppler mode, the same parameters can be evaluated as when using continuous-wave Doppler sonography. As in all vascular regions, it is important to note the angle to what is known as the vascular axis, which should not exceed 60°, and which should be kept at approximately the same value when following the course of an axis and

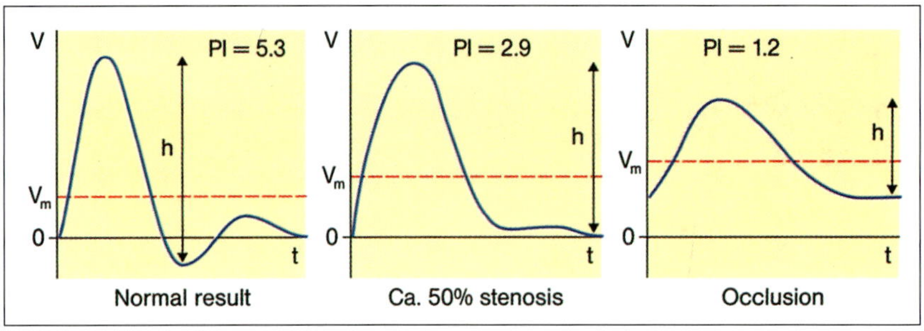

Fig. 5.**27** The pulsatility index (PI) in both normal and pathological flow pulses within the extremities (adapted from Gosling and King 1978). The PI is calculated from the height (h) of the maximum forward and backward flow, divided by the mean flow velocity (V_m)

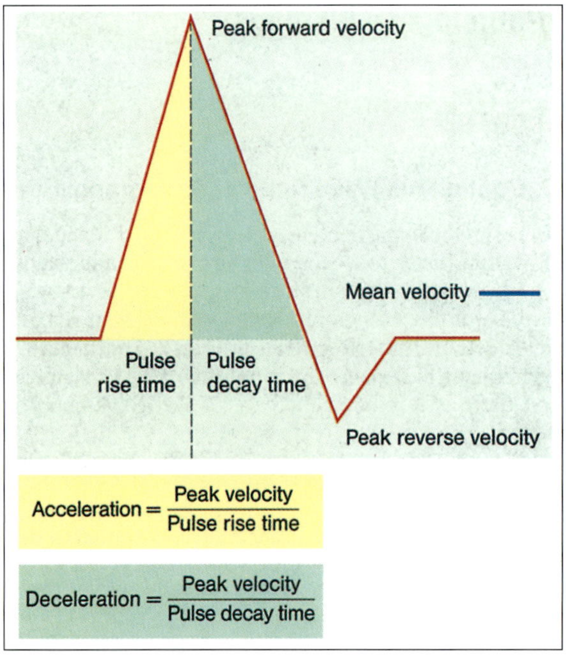

Fig. 5.**28** Criteria for measuring a normal flow pulse in the extremities (adapted from Fronek 1989). The parameters of acceleration and deceleration are calculated from the values measured for the systolic peak flow and the pulse increase and pulse decrease times. Modifications of these criteria are also used in other indices

making quantitative measurements. The advantage in doing this lies in the ability to convert the Doppler frequencies into velocities by a known ultrasound angle. This means that the comparison of the measured values is independent of the transducer transmission frequency selected, which often fluctuates due to the varying depth of the peripheral vasculature. In this procedure, the B-mode transmission frequency and the Doppler transmission frequency need not necessarily have the same high or low value. Evaluating the Doppler frequency or flow velocity and the advantages and disadvantages of each of these measurements are discussed in Chapter 1, pp. 4–5 (Phillips et al. 1989).

Color flow Doppler mode. The color flow Doppler mode makes it easier to detect the vasculature in all regions. Excluding vascular system stenoses or occlusions, pathological structures, or aneurysmal dilations, can be done quickly and reliably.

Continuous-Wave Doppler Sonography

The assessment of a normal flow waveform from the extremities is usually only qualitative, following the clear formal criteria given above (see Fig. 5.**29**).

Quantitative measurement of different parameters of the flow velocity curve increases the precision of the information, especially when pathological findings have to be excluded.

Different indices have been developed to measure the flow pulses. Among these, the *pulsatility index* (PI), in particular, has gained widespread acceptance, following Gosling and King (1974). For this calculation, the height between the positive systolic peak and the negative diastolic backflow is measured, and this quantity is divided by the average flow velocity (mean) (Fig. 5.**27**).

$$\text{Pulsatility index (PI)} = \frac{\text{Height of the maximum forward and reverse velocity}}{\text{Mean velocity}}$$

The advantage of the PI lies in the fact that it is relatively independent of the angle between the probe and the vascular axis. This is so because a potential error affects both the numerator and the denominator to the same degree. An index of around five indicates a normal value.

The *"damping factor"* (DF) is the ratio between the proximal pulsatility index and the distal pulsatility index:

$$\text{Damping factor} = \frac{PI_{proximal}}{PI_{distal}}$$

It is a sensitive parameter for detecting intervening intraluminal stenoses—even low-grade ones. Normally, a value of one is calculated (Fitzgerald and Carr 1977).

The *"transit time"* is sometimes measured. This expression refers to the pulse wave propagation from one registration point to the next. This value allows the stiffness of the arterial wall to be assessed. However, the Doppler waveform has to be simultaneously registered at two different points, and possible time delays in the pulse increase have to be calculated (Humphries et al. 1980).

Fronek et al. (1976) measured the pulse increase and decrease times, and divided them by the peak velocity (Fig. 5.**28**). The most favorable parameters for discrimination purposes were the peak velocity and the deceleration (Baker et al. 1986).

Cobet et al. (1986) and Scharf et al. (1988) examined, both in vitro and in vivo, the pulsatility index (PI), the resistance index (RI) (pp. 351–352), and the systolic half-life (SHL) in relation to different degrees of stenosis and varying peripheral resistances. They found that the SHL reacted with the greatest sensitivity to all proximal and distal stenotic changes.

Finally, mention should be made of the indices developed using the *Laplace transformation*, which were introduced by Skidmore and Woodcock (1980).

The practical use of such indices usually depends on the ability to evaluate the flow pulses digitally. Windeck et al. (1992) compared four indices—the pulsatility index, the damping factor, the systolic time delay index, and the "height/width index." The latter index showed the greatest sensitivity (92.5%) in diagnosing preceding vascular system obstructions, both before and after angioplasty. An index which they developed themselves, the *curve broadening index*, which uses a complicated measuring procedure to calculate both the pulse increase and decrease times at certain curve points, even reached a sensitivity of 100%.

Pathological Findings

Color Flow Duplex Sonography

B-mode. The opportunity to carry out morphological evaluation of vascular wall deposits, the vascular diameter, and the vascular course in B-mode imaging has made an outstanding contribution to diagnosis. Previously, aneurysmal dilations could only be suspected through palpation, and the total extent of them could not be recognized using angiography (see Fig. 7.**25**, pp. 308–309). Hematomas and cysts—e.g., Baker's cysts in the popliteal region—are now accessible noninvasively, and can be distinguished from aneurysms.

Doppler mode. It is essential to use pulsed-wave Doppler sonography when formally analyzing the Doppler waveform, qualitatively or quantitatively determining the stenosis grade, calculating indices, differentiating the vasculature, and evaluating the peripheral resistance.

Color flow Doppler mode. As mentioned above (p. 163), it was only with the introduction of the color flow Doppler mode that it first became practicable to examine the long vascular segment from the aorta to the distal lower leg, with targeted placement of the Doppler sample volume. Locating vascular system stenoses, differentiating occlusion and stenosis, and determining the length of an occlusion can all be carried out quickly and reliably. Arteriovenous fistulas and the inflow and outflow of blood to and from a pseudoaneurysm can also be diagnosed without difficulty (see Fig. 7.**26**, pp. 310–311).

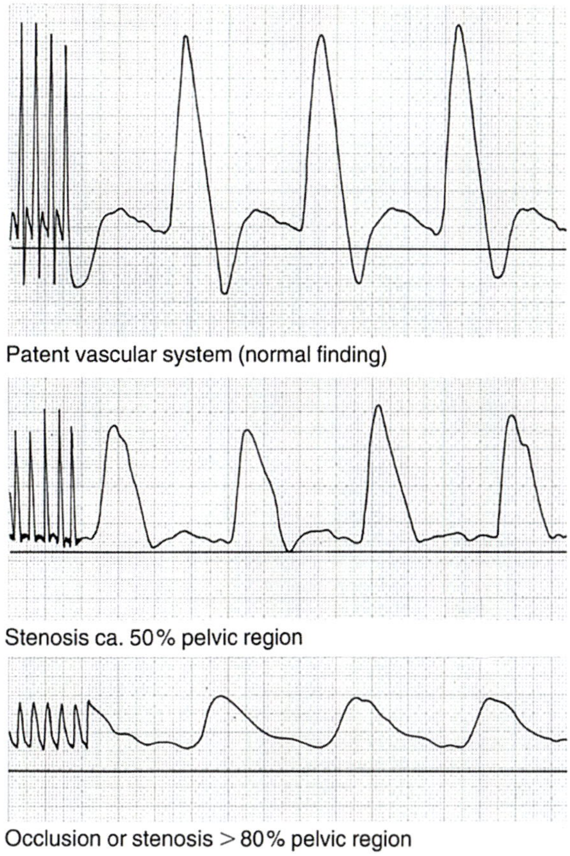

Fig. 5.**29** The Doppler waveform from the common femoral artery, obtained with continuous-wave Doppler sonography at the groin in a patent vascular system. The triphasic waveform that is typical of an artery in the extremities is seen. In medium-grade pelvic stenoses, the amplitude height decreases and the flow reversal disappears. Postocclusively, a monophasic waveform with a low amplitude height and clear diastolic flow is recorded (decreased peripheral resistance). The pulse rise and pulse decay times are clearly lengthened

Continuous-Wave Doppler Sonography

At rest. Due to the decreased peripheral resistance, the flow rate distal to *occlusions* largely remains stable. The flow pulses are similar in form to those found in regions of physiologically decreased organ resistance (e.g., cerebral arteries, renal arteries). The backflow component is absent, the amplitude height is lower, and there is a relative increase of the diastolic component (Busse 1982) (Fig. 5.**29**).

Post exercise. This examination is not of very great importance in pathological cases, since even in healthy individuals, a hyperemic reaction—i.e., a flow velocity increase—especially in the diastolic component, can be observed after exercise.

Stenoses can rarely be directly detected at the typical registration points (groin, popliteal region) in the arteries supplying the legs. Iliac artery stenoses are

Table 5.**8** Comparison of the arterial systems supplying the extremities and the brain

Parameter	Peripheral arteries	Carotid arteries
Peripheral resistance	High	Low
Region supplied	One transporting artery	Several transporting arteries
Vascular length	Long	Short
Exertion test	Muscular work Hyperemia	CO_2 test
Collaterals	Variable, depending on location	Preformed circle of Willis
No. of stenoses	Mainly affect several levels	Often isolated stenoses
Significance of stenoses	Severity Only higher-grade stenoses effective	Morphology Lower-grade stenoses also cause embolism

usually located deep in the pelvic region, which is not accessible to continuous-wave Doppler sonography. In addition, a superficial femoral artery stenosis, usually found in the adductor canal, often cannot be located with the usual continuous-wave ultrasound probes (4–5 MHz transmission frequency). However, when applying ultrasound to a superficial stenosis (e.g., in the common femoral artery, beneath the inguinal ligament), the noise phenomena and the flow pulse characteristics are identical to those obtained in the easily accessible distribution area of the carotid artery: the flow velocity increases, and what is termed "systolic peak reversal" occurs (see Fig. 7.**20**, pp. 298–300).

Grading of Stenoses

In the peripheral arteries, the graduation of the stenoses is subject to different hemodynamic criteria from those in the carotid stream, since the peripheral resistance is different, and above all because a series of consecutive stenoses, or occlusions and stenoses, complicates the hemodynamic assessment (see below) (Table 5.**8**).

Principle

Using frequency spectrum analysis, it is possible to carry out qualitative and quantitative evaluations.

Qualitative evaluation. Even in the peripheral arteries, it has proved valuable to carry out an approximate assessment of the severity of the constriction of the flow using a qualitative determination of the alteration of the waveform. The Doppler curve scheme proposed by Jäger et al. (1985) has gained widespread acceptance. The method uses the pulsed-wave Doppler mode of a conventional duplex apparatus (Fig. 5.**30**).

Quantitative evaluation. As in the carotid stream, grading a stenosis by measurements of the peak systolic velocity serves as a quick way of obtaining quantitative information concerning the peripheral arteries. The rule of thumb is that a doubling of the starting velocity indicates a stenosis greater than 50% (Jäger et al. 1985, Moneta et al. 1993). Other authors have determined an optimal threshold value of 200 cm/s (Cossman et al. 1989) or 180 cm/s (Ranke et al. 1992). Since the average velocity—e.g. in the femoral artery—is 90 cm/s, a stenosis of over 50% is assumed at a velocity of 180 cm/s. However, this is valid for isolated stenoses only (Allard et al. 1994). Additional parameters for frequency analysis such as end-diastolic velocity (EDV) or mean frequency are used less often.

Since measurements of the absolute velocity do not take into account distal or proximal stenoses or occlusions, the following physical laws (Neuerburg-Heusler and Karasch 1996) have proved their value alongside the familiar pulsatility index and resistance index.

Principle of the Continuity Equation

The continuity equation states that the volume flow in all consecutive sections of the vascular system remains constant. This means that the flow velocity in regions of reduced diameter (stenoses) must increase in order to maintain the volume flow, which is calculated according to the following formula (Fig. 5.**31**):

$$Q = A \times v$$

where Q = volume flow, A = vessel cross-sectional area, and v = mean flow velocity.

Since according to the continuity equation $Q_1 = Q_2$

are equal, then $A_1 \times v_1 = A_2 \times v_2$

or $$\frac{v_1}{v_2} = \frac{A_2}{A_1}$$

where 1 = prestenotic and 2 = intrastenotic.

The flow velocities thus behave in an inverse proportion to the vessel's cross-section.

When the flow velocity in a narrowing is compared with the prestenotic flow velocity according to the law of continuity, then in purely arithmetical terms it must be possible to use the above formula to calculate the reduction in the cross-sectional area at the point of narrowing. This velocity ratio has been used with considerable success as a way of grading stenoses, and it is known as the peak velocity ratio (PVR).

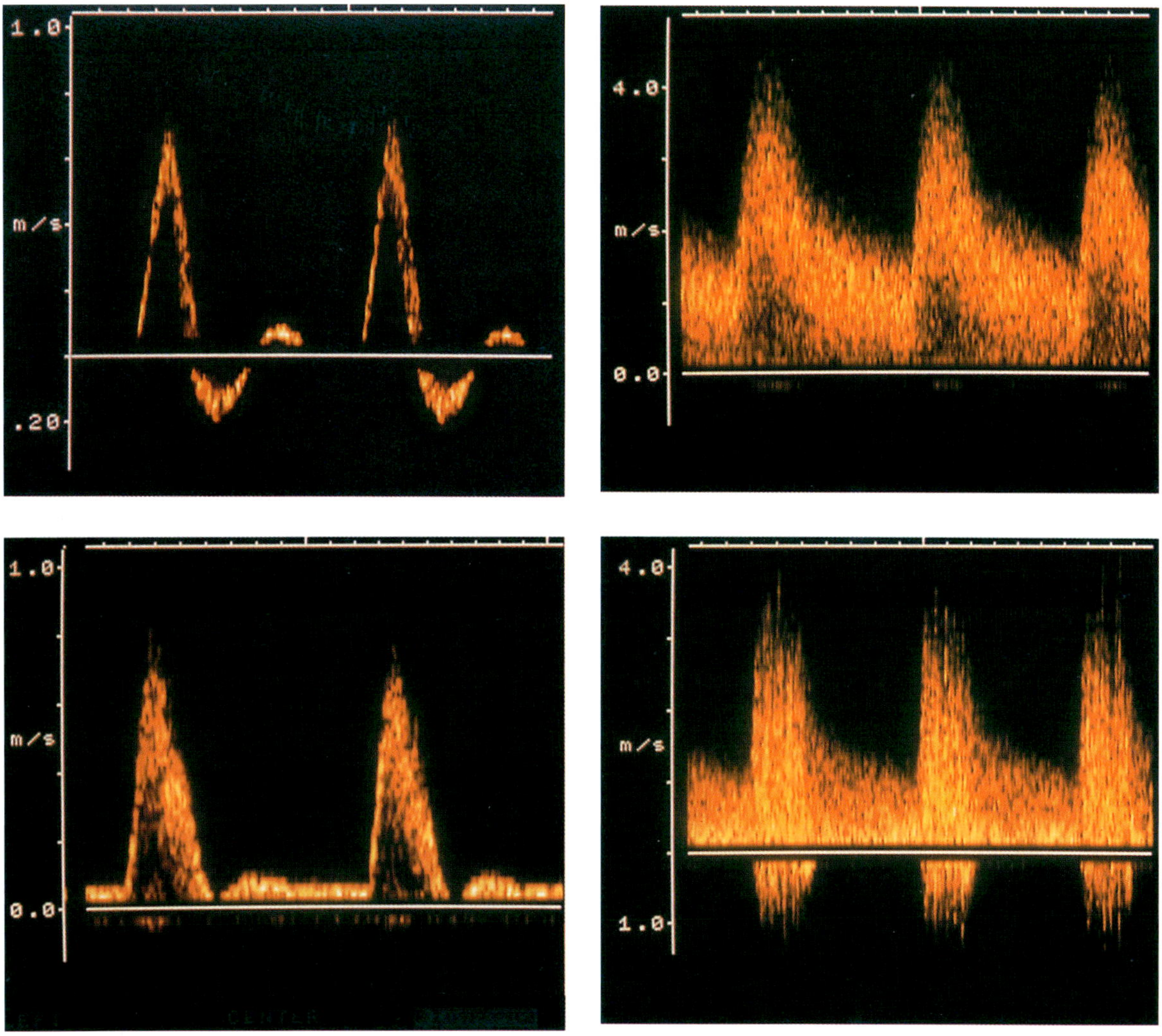

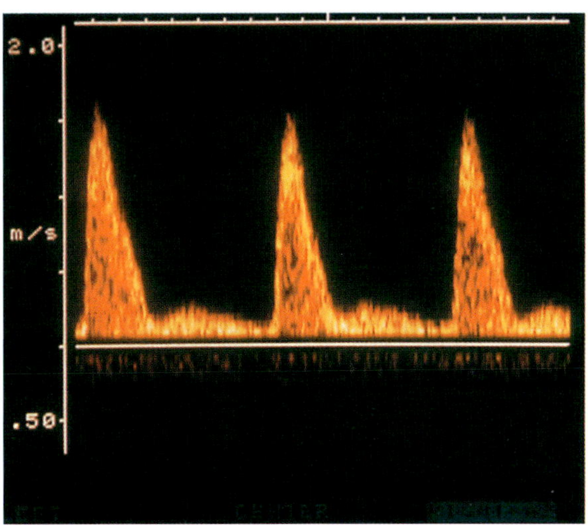

Fig. 5.**30** Qualitative grading of stenoses in the arteries of the extremities according to typical frequency and time spectral patterns

- **a** Normal findings, with a narrow frequency band and tricyclic curve structure
- **b** Flow disturbances in minimal wall deposits with a reduction in the diameter of less than 25 %. Slight spectral broadening is observed. However, the systolic window is open
- **c** Low-grade stenosis. The systolic window has dissolved, but there is not yet any increase in the flow velocity, or only a minimal one. The backflow component is missing
- **d** The image is the same as that for a high-grade stenosis. The maximum systolic velocity is clearly elevated (> 4 m/s). The diastolic peak velocity is also elevated and includes spectral broadening. The spectrum shown here is taken directly from the stenosis
- **e** High-grade stenosis. This Doppler frequency–time spectrum is taken from the same stenosis as in **d**, but now from just behind the stenosis. The peak velocity is close to 4 m/s, and there is a backflow component in the systole and a massive increase in the amplitude at low frequencies. Just behind the stenosis, the inverse flow components are visible underneath the zero line

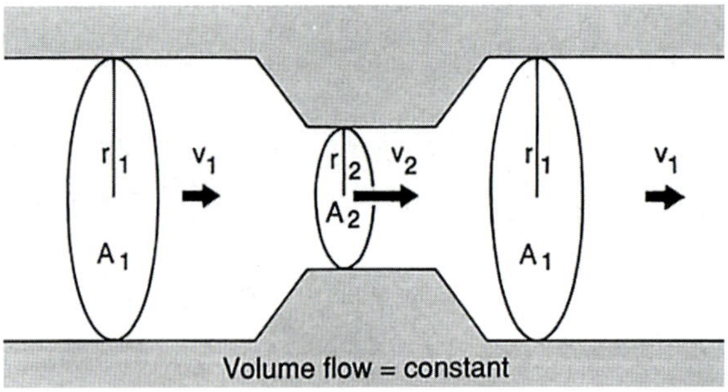

Fig. 5.**31** Principle of the law of continuity. Due to variation in the mean velocity of flow (v), the volume flow (Q) at each point in the vessel remains the same even when the area of the vessel's cross-section (A) changes. An increase in the flow velocity can therefore be used to calculate the extent of the reduction in the cross-section and the degree of stenosis (adapted from Wolf and Fobbe 1993)

The use of the *peak velocity* in the equation is based on the experimental studies carried out by Johnson et al. (1989) that demonstrate that the intrastenotic spectral peak velocity correlates well with cross-sectional area ($r = 0.93$), while the intensity-weighted mean velocities showed only a poor correlation. In cardiology, however, it is more usual to use the mean (or averaged) velocity to calculate the valvular cross-sectional area.

Principle of the Bernoulli Equation

In addition to measurement of the reduction in the cross-section in the stenotic area, measurement of the pressure loss caused by the stenosis is also important. This value is calculated according to the Bernoulli equation, known as the law of the maintenance of constant energy in the movement of fluids:

$$E = P + {}^1/_2\, p\, v^2 + p\, g\, h$$

where P	=	potential energy
$^1/_2\, p\, v^2$	=	kinetic energy
p g h	=	gravitational energy
p	=	density of the fluid
v	=	velocity
g	=	gravitational constant
h	=	height

In a *stationary liquid*, the static or hydrostatic pressure is constituted by the gravitational pressure (P) and the influence of gravitational forces (p g h). In a flowing liquid, part of the total energy is given off with the movement of the liquid, and is described as kinetic energy ($^1/_2\, p\, v^2$).

When flow velocities increase in the stenosis (see the continuity equation), kinetic energy is required, which is used up at the expense of the static pressure. Since the sum total of the energy remains constant, and since the gravitational energy does not change when the vessel region being measured remains at the same height, the gravitational pressure must fall so that the same total is derived from the three parameters.

The Bernoulli equation thus states that the energy level in front of the stenosis and at the stenosis is the same:

$$P_1 + {}^1/_2\, p\, v_1^2 + p\, g\, h = P_2 + {}^1/_2\, p\, v_2^2 + p\, g\, h$$

If the Bernoulli equation is then simplified by factors that stay constant or are negligible, the pressure gradient can be calculated as follows:

$$P_1 - P_2 = \Delta p = 4 \times (v_2^2 - v_1^2)$$

Assuming that the intrastenotic velocity (v_2) is high and that the prestenotic velocity (v_1) is low—i.e. below 1 m/s—the latter can be ignored as a subtrahend. With further simplification, the formula thus becomes:

$$\Delta p = 4 \times v_2^2$$

This formula has already been in use for some time to calculate the pressure loss in cardiac valvular stenoses. In the area of the peripheral arteries, the principle has already been tested in vivo and in vitro in several studies (p. 199), and has shown satisfactory correlation coefficients of $r = 0.8–0.9$. However, a recent study by Legemate et al. (1993), comparing the measurement of the pressure gradient according to the Bernoulli equation with the calculation of the stenosis grade according to the velocity difference ΔPSV (peak systolic velocity), showed that the use of the Bernoulli equation gave a markedly lower correlation quotient of $r = 0.62$ than the measurement of the velocity difference ΔPSV, at $r = 0.72$.

Findings

Pelvic Level

Prior to the introduction of duplex sonography, the diagnosis of pelvic occlusions and stenoses was questionable, since cuff methods were not applicable in this region and neither pressure measurements nor oscillographic examinations were possible. Even continuous-wave Doppler sonography was only able to provide indirect information about pelvic vascular system stenoses. Differentiating between the proximal and distal site of the lesion always presented difficulties. Clinically, only palpating the pulse at the groin and auscultation were practicable for orientation purposes.

Even if isolated pelvic occlusions occur in only 10.2% of the cases, two-level iliofemoral occlusions are still very frequent. According to Vollmar (1996), one-third of peripheral occlusive processes occur in the aortoiliac segment. However, 82% of these are combined with peripheral occlusions. Individual evaluation of the hemodynamic effect of an obstruction at each level is important, due to the availability of lumen recanalization techniques.

Color Flow Duplex Sonography

B-mode. Due to the deep location of the arteries and the often unfavorable ultrasound application angle, continuous imaging of the vascular wall structure is not always possible. It is more important to locate the plaques and then determine, using Doppler mode, whether or not a hemodynamically significant stenosis is present. Occlusions and stenoses are most often encountered near the aortic bifurcation, the common iliac artery, and at the origin of the external iliac artery.

Imaging of the internal iliac arteries is difficult and not always possible, and this is reserved for specialized investigations concerning vascular impotence, or collateralization of an iliac artery flow obstruction (Fig. 5.**32**).

Doppler mode. The Doppler mode displays the flow velocity curves with frequency analysis. Usually, the groin region is examined first, to see whether the pelvic vascular system is patent or obstructed. Qualitative evaluation of the form of the recording is sufficient (Fig. 5.**30**). When a definite or suspected indication of preceding pelvic obstruction is encountered, the stenosis grade must be determined, both *qualitatively,* using formal analysis of the Doppler waveform and also *quantitatively,* by measurements.

Color flow Doppler mode. Locating the deep pelvic arteries, with their markedly curved course, has become very much easier with the addition of color. Using color flow duplex sonography to display the pelvic vascular system has made angiographic controls largely superfluous following recanalization techniques such as angioplasty, stent implantation, and bypass operations.

Continuous-Wave Doppler Sonography

Registering Doppler waveforms from the groin is a practicable and fast method of indirect examination, although it only provides information about higher-grade stenoses of the pelvic vascular system.

It is possible to assess the extent of the aortoiliac obstruction approximately from the form of the flow pulse. In the subsequent vascular system, stenoses of less than 50% cause no change in the recording form. When the systolic peak is unchanged with regard to both the amplitude height and the steepness rate of the increase, only a reduced backflow component results. Qualitatively, this waveform image is no different from that obtained in a patent pelvic vascular system with subsequent femoral artery occlusions (Figs. 5.**33**, 5.**34**).

By contrast, preceding stenoses of more than 50%, and occlusions, both have a clearly decreased amplitude height and reduced steepness in the systolic rate of increase, along with a delayed diastolic decrease, without any backflow component.

A decreased Doppler waveform is predominantly caused by obstructions in the preceding vascular sys-

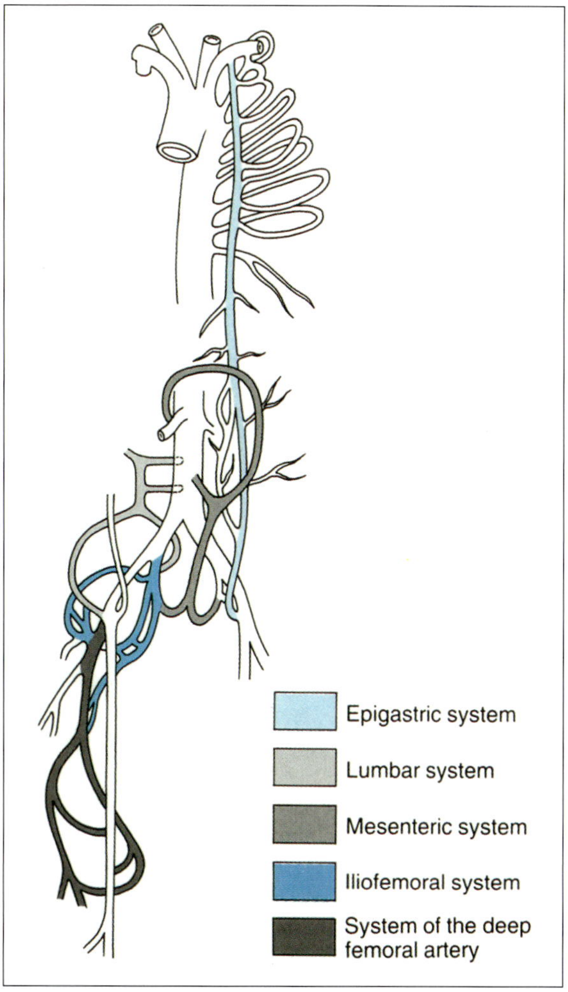

Fig. 5.**32** Potential collateral pathways for bypassing aortic, aortoiliac, and femoral occlusions.
Epigastric collateralization: blood flow via the superior epigastric artery to the inferior epigastric artery, with a connection to the common femoral artery
Lumbar collateralization: bridging of an aortoiliac flow obstruction via the lumbar arteries with a connection to the circumflex iliac artery and from there to the common femoral artery, or via the iliolumbar arteries with a connection to the internal iliac arteries
Mesenteric collateralization: collateralization via the superior mesenteric artery to the inferior mesenteric artery (Riolan anastomosis) and then via the superior rectal artery to the inferior rectal artery, with a connection to the internal iliac artery
Iliofemoral collateralization: branches of the internal iliac artery (superior gluteal artery, inferior gluteal artery, obturator artery) connect with the medial circumflex femoral artery or lateral circumflex femoral artery as branches of the deep artery of the thigh
Femoroprofundal collateralization: occlusions of the superficial femoral artery can be collateralized via the deep femoral artery and its side branches, which connect with the genicular arteries to reach the popliteal artery

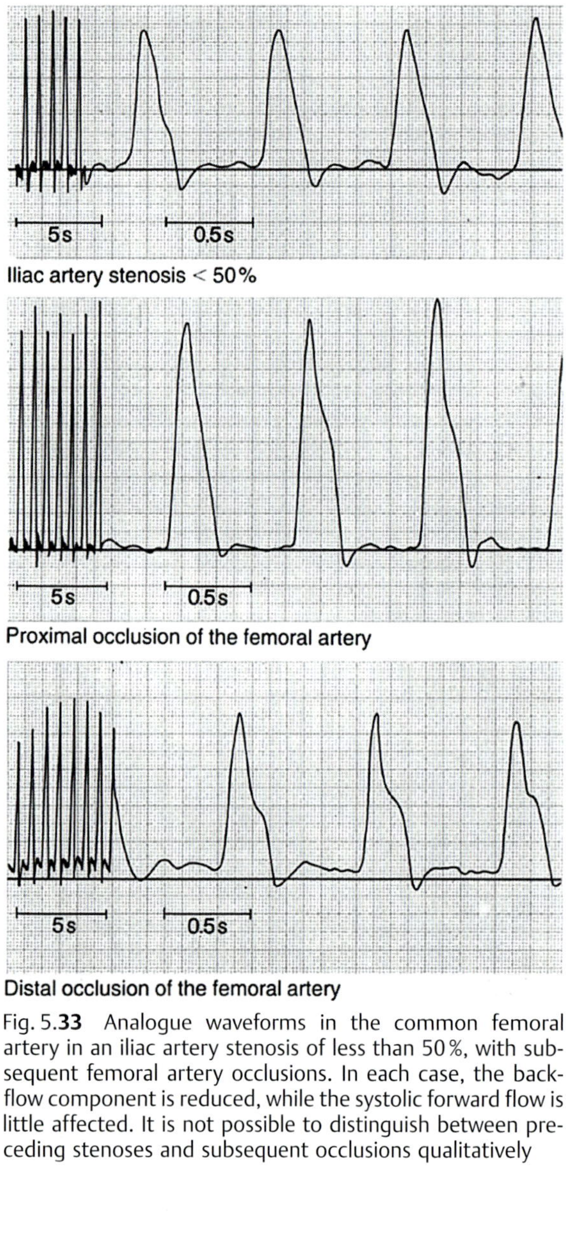

Fig. 5.**33** Analogue waveforms in the common femoral artery in an iliac artery stenosis of less than 50%, with subsequent femoral artery occlusions. In each case, the backflow component is reduced, while the systolic forward flow is little affected. It is not possible to distinguish between preceding stenoses and subsequent occlusions qualitatively

tem; subsequent occlusions only decrease the backflow component (Neuerburg-Heusler et al. 1981).

Thigh Level

Femoral artery occlusion is the most frequent manifestation of obstructive artérial disease in the arteries of the extremities (approximately 48%). It usually originates at the adductor canal, where a mechanical physiological arterial narrowing caused by the tendon of the great adductor muscle is thought to serve as the initiating mechanism for vascular wall lesions. In this region, stenoses also occur quite often. Occasionally, the occlusion extends to the origin of the deep femoral artery, which with its strong flow hinders appositional growth. Since femoral artery occlusions and stenoses are readily accessible to angioplastic techniques, noninvasive diagnosis in this region is particularly important.

Evaluating the common femoral artery before and after using the catheter in coronary or peripheral angioplasty is vital. *Beforehand*, plaques and stenoses are located, and unintentional plaque ablation is avoided; *after* these procedures, dissections, pseudoaneurysms (see Fig. 5.**42**), and hematomas can be detected.

Preferred locations for *stenoses* are the bifurcation, with the origins of the superficial femoral artery and deep femoral artery, and also the adductor canal (Figs. 5.**35**, 5.**36**). The most important collateral in this region is the deep femoral artery, which as a blood vessel capable of significant throughput can compensate well for occlusions above the adductor canal.

Color Flow Duplex Sonography

B-mode. The arterial images allow detection and structural evaluation of plaques and, most importantly, of aneurysms (pp. 186–187) the preferred locations of which in the extremities are the common femoral artery and the popliteal artery (see Figs. 7.**24**, 7.**25**, pp. 305–309).

Pseudoaneurysms are primarily located near the common femoral artery and in the proximal segments of both the superficial femoral artery and the deep femoral artery (see Figs. 7.**26**, 7.**27**, pp. 310–313; see Fig. 5.**44**).

Doppler mode. The qualitative and quantitative evaluation of stenoses using the Doppler mode is described above. Most importantly, at the femoral bifurcation, it becomes possible to differentiate stenoses as belonging to either the superficial femoral artery or the deep femoral artery, since the Doppler waveform with the B-mode image can be specifically registered from both of the originating blood vessels. By contrast, the origin of the deep femoral artery is often depicted in superimposed form when using angiography (Strauss et al. 1989).

Color flow Doppler mode. Continuous application of ultrasound to the femoral artery vascular system, a preferred location for atherosclerosis, has been very much simplified through color coding. Stenoses can be quickly recognized by their lighter color, so that, guided by the color, exact placement of the pulsed-wave sample volume becomes easier. Femoral artery occlusions can be located by the color interruption and by the originating collateral vasculature; the occlusion length can be reliably measured (Karasch et al. 1993) (Fig. 5.**37**; see Fig. 5.**53**). Even stenoses in tandem, or occlusions, are rarely missed with color flow duplex sonography. Difficulties can arise when the subsequent flow velocity is too small for adequate coding. Additional findings are given in the following sections (pp. 182 ff).

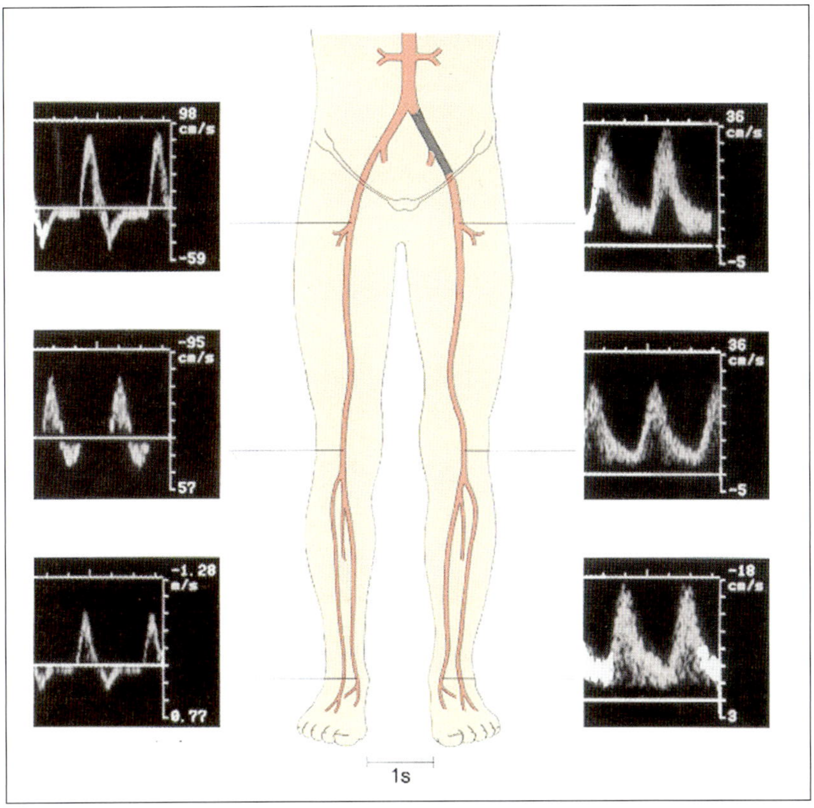

Fig. 5.**34** Doppler frequency–time spectra at various measurement points in side-to-side comparison, in a case of left-sided iliac occlusion

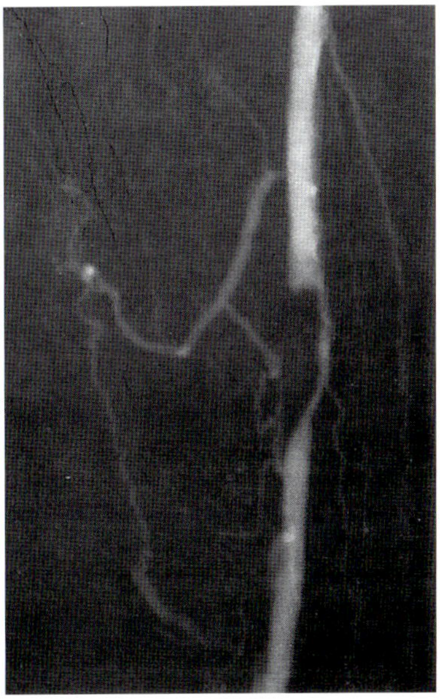

a

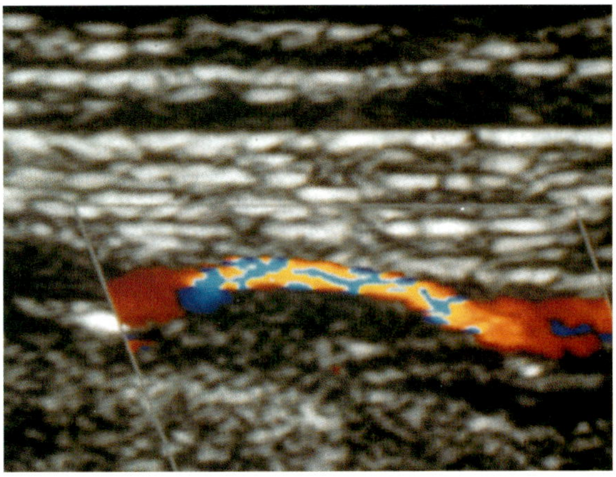

b

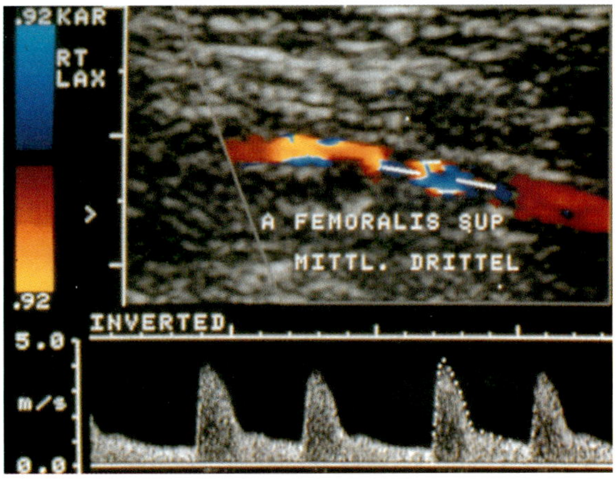

c

Fig. 5.**35** A threadlike stenosis in the middle third of the superficial femoral artery. **a** Depiction using selective angiography, with **b** the corresponding image in color flow mode using the duplex system. There is pronounced aliasing in the stenosis canal. **c** The Doppler flow waveform shows a systolic peak velocity of 4.36 m/s, with spectral broadening and a diastolic velocity of 1.03 m/s

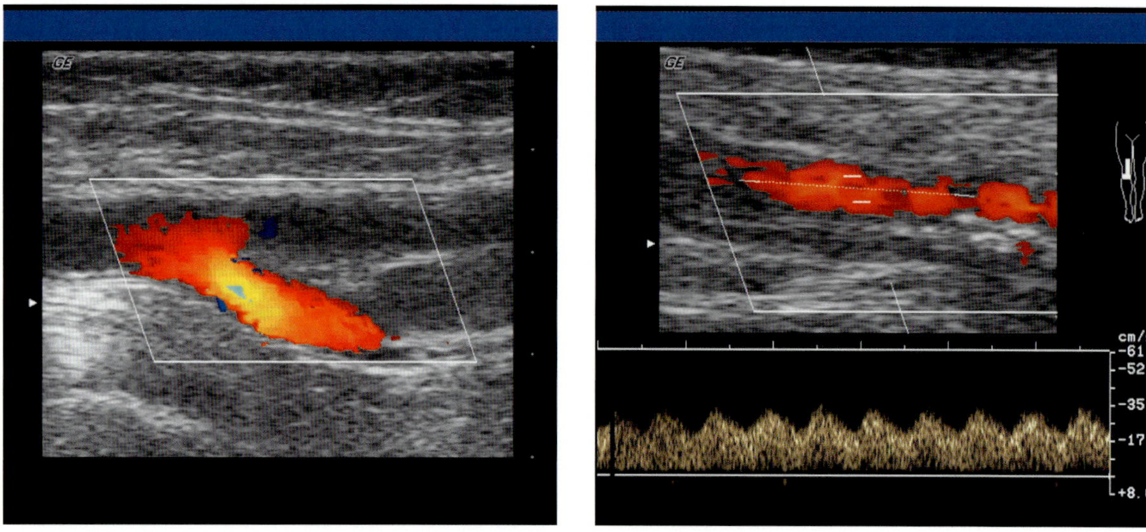

a b

Fig. 5.**36a** Outflow occlusion in the superficial femoral artery, with clear visualization of the deep femoral artery. In the vascular segment that is proximally occluded, flow is no longer evident on color duplex sonography in B-mode, despite good imaging conditions. **b** Doppler frequency–time spectrum from the superficial femoral artery distal to an occlusion. The spectrum shows a monophasic curve, with clearly reduced pulsatility. The systolic increase and decrease in the flow velocity is markedly delayed. The flow velocity is generally clearly reduced, reaching a systolic maximum of 33 cm/s. Collateralization of this occlusion must be assessed as poor

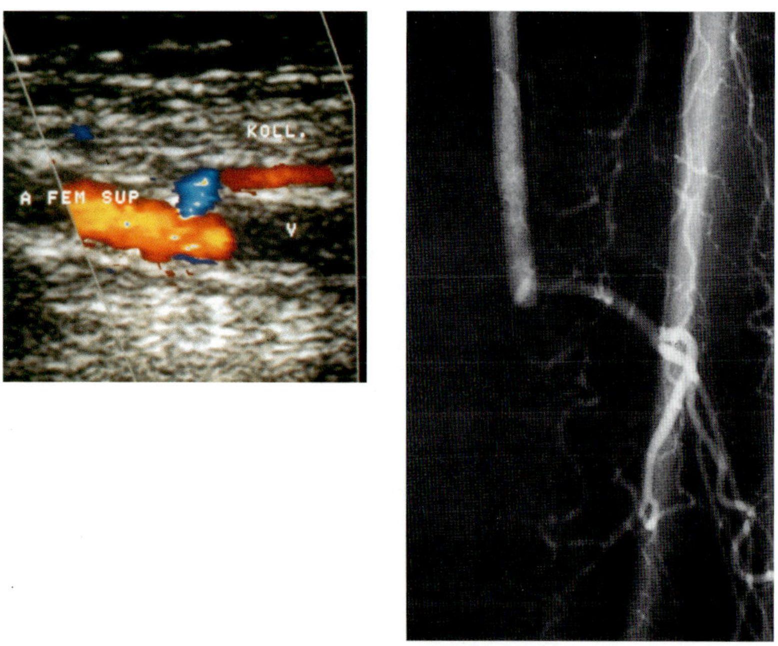

a b

Fig. 5.**37a** A color flow image of an occlusion in the distal third of the femoral artery. The branching collateral shows strong coloring. The blue color at the origin of the collateral reflects a change in the flow direction. **b** The corresponding angiogram shows the occlusion with the branching, vigorous collateral

Continuous-Wave Doppler Sonography

Detecting vascular system stenoses in the course of the superficial femoral artery is usually only done indirectly, by applying ultrasound to the popliteal artery. In the process, both qualitative and semiquantitative assessment of the magnitude of the hemodynamic effect are possible. It is not possible to examine the deep femoral artery either with clinical palpation or with continuous-wave Doppler sonography.

Calf and Foot Arteries

Arterial occlusions in this region are found in 18.4% of leg occlusions. The anterior tibial artery is affected most frequently, the fibular artery least frequently. Iso-

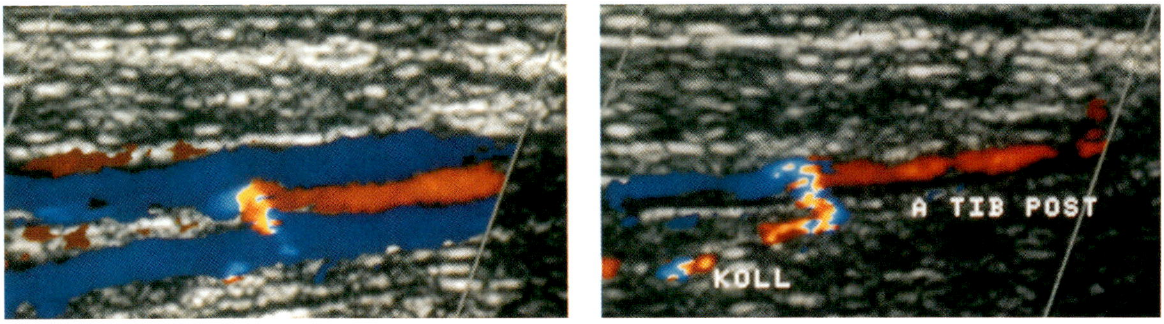

a b

Fig. 5.**38** An occlusion of the posterior tibial artery in the distal lower leg. **a** Refilling through a collateral. The artery follows a course between the accompanying veins, while proximal to this (on the left in the image), the absence of an arterial band can be observed. **b** Imaging in a different plane to verify the collateral, which shows a high flow velocity with aliasing

lated arterial occlusions of the foot are clearly less frequent, and occur predominantly in diabetics and in those suffering from endangiitis. Noninvasive diagnosis of these is only possible in summary form, by measuring the pressure at the toes. Magnification selective angiography has to be used to confirm the diagnosis. Combined with thigh occlusions, the distal arteries are often affected, forming a large group of the combined occlusive types (26%).

Color Flow Duplex Sonography

In this region, conventional duplex sonography plays a subordinate role, since finding the arteries without the benefit of color coding can be tedious and time-consuming.

Color flow Doppler mode. With color flow duplex sonography, it has become possible to image all three lower leg arteries. However, the examination is only carried out as part of a specialized investigation, since even with color coding, the time needed is quite considerable, and the topographical classification is not simple. The procedure requires some experience (Fig. 5.**38**). Practiced examiners, however, report short examination times and good results (Langholz et al. 1991, Larch et al. 1993). In the foot arteries, color coding allows rapid measurement of the flow velocity, so that from this value indirect indications of the collateral circulation in lower leg arterial occlusions can be obtained.

Continuous-Wave Doppler Sonography

It is not difficult to register the flow recording from the posterior tibial artery, the anterior tibial artery, and sometimes also from the peroneal artery, at the ankle. However, pressure measurements are used more often in this region. With these relatively narrow-caliber arteries, imaging of the Doppler waveform may be somewhat imprecise, since the optimal angle cannot always be easily adjusted.

Upper Extremity Arteries

Subclavian Artery and Arm Arteries

Atherosclerotic wall changes in the arteries of both upper extremities are very much less frequent than in the arteries supplying the legs. Stenoses or occlusions occur frequently only at the origin of the subclavian artery from the thoracic aorta. According to Vollmar (1996), 16% of supra-aortic arterial occlusions are subclavian artery occlusions. According to the joint study, the left subclavian artery is affected in 14.9% of cases and the right subclavian artery in only 9.2% of cases (Hass et al. 1968; Fields et al. 1970).

When there are hemodynamically significant stenoses in this region, the vertebral artery serves as a collateral supplying the arm. This flow reversal within the vertebral artery, which is also important for the cerebral circulation ("subclavian steal"), is discussed in Chapter 2.

Afferent collateral flow through the vertebral artery is usually more than sufficient to supply the axillary artery, the brachial artery, and the forearm arteries, so that symptoms such as arm claudication are not noticed until higher levels of exertion are applied (carrying heavy objects, or working above head level).

When there are embolic occlusions affecting the digital arteries of the hand, it is important to look for aneurysms in the subclavian artery. These can sometimes occur due to a neurovascular compression syndrome (p. 182, Fig. 5.**39**). Arteriosclerotic stenoses in the arteries of the arm are a rarity. Inflammatory causes such as Takayasu's arteritis (see p. 184), or forms of arteritis of varying pathogenesis, predominate (see Fig. 7.**34**).

The ultrasound examination is also important in detecting occlusions after intra-arterial punctures or catheter manipulations, and in testing hemodialysis shunt output.

Color Flow Duplex Sonography

B-mode. Imaging of the subclavian artery, which originates on the left from the aortic arch and on the right from the innominate artery, is adequate with proper patient positioning and practice, although direct evaluation of this region is not simple. It is important to detect aneurysms distal to the clavicle, which may be due to a mechanical stenosis, and sometimes appear as a poststenotic dilation.

Doppler mode. Under visual B-mode control or using color, the sample volume is used to quantify the stenoses. The velocity curves are evaluated using the criteria applied in the arteries of the extremities (p. 195). It is often sufficient to use indirect criteria in the poststenotic or postocclusive area, since stenoses of the *subclavian artery* vascular system are quantitatively assessed by measuring the arm pressure. Imaging of the subclavian steal phenomenon is discussed in Chapter 2.

Color flow Doppler mode. Again, color flow Doppler mode simplifies the imaging of the arteries and allows faster diagnosis in specialized investigations.

Arteries of the Hand and Fingers

Checking perfusion in the superficial and deep palmar arch can be important in vascular injuries, after puncture-related vascular spasms, dissections, and occlusions, and before placement of a dialysis shunt. The functional clinical *Allen test* has an error rate of 13% and cannot be used in patients who are in post-traumatic vascular shock (Fronek 1989). The *palmar test* described by Ruland et al. (1988), using the Doppler probe, showed better results. The safest criterion in the latter study was *flow reversal* in the distal radial artery during proximal obturating compression of the radial artery on the edge of the brachial muscle. Another useful method was to test the direction of flow in the radial artery of the thumb during compression of the radial artery at the typical pulse palpation point. By contrast, an increase in flow velocity in the ulnar or radial artery during obturating compression of the other artery showed too wide a range of variation.

In cases of disturbances in the *acral* arterial circulation, functional Raynaud's syndromes need to be distinguished from acral vascular occlusions, which have previously tended to be diagnosed using pulse-recording procedures or pressure measurements in the digital arteries. However, reliable imaging of pathological findings in the palmar digital arteries can be successfully achieved using high-resolution linear ultrasound probes at around 10–13 MHz.

The following color duplex sonography criteria can be used to identify *digital artery stenoses* and *occlusions:*

- Local segmental increases in flow velocity of more than 50% in comparison with the preceding vascular segment
- Aliasing phenomena on color Doppler despite a correctly adjusted pulse repetition rate and color enhancement
- Segmental vascular discontinuities in color Doppler mode
- Visible collateral vessels (color Doppler)
- Circumscribed arterial dilation
- Increased tortuosity of the vessels (in Buerger's disease)
- Reduction in the maximum systolic flow velocity to less than 15 cm/s

Continuous-Wave Doppler Sonography

The *axillary artery and the brachial artery* are rarely examined with continuous-wave Doppler sonography, because atherosclerotic occlusions and stenoses are a rarity in this region. Obstructions are more commonly caused by vasculitis, or are due to catheter placements.

In principle, the velocity recording from the *digital arteries of the hand* can be registered using Doppler sonography. However, continuous tracing of the course of the palmar digital arteries is not reliably possible.

Special Sets of Findings

Neurovascular Compression Syndrome of the Shoulder Girdle

In the shoulder girdle, the subclavian artery, along with the subclavian vein and the brachial plexus, as a vascular nerve bundle, have to pass through three naturally narrow points, and this can lead to functional arterial stenoses.

When there is continuous compression of the subclavian artery, organic changes with poststenotic dilation, culminating in an aneurysm, may also occur. Peripheral emboli in the digital arteries of the hand often indirectly indicate the pathogenesis (Fig. 5.**39**).

Figure 5.**39** shows the subclavian artery passing through osseous and muscular structures, and indicates the narrow points. Clinically, the costoclavicular syndrome causes functional or organic damage in 80% of cases.

Examination and Findings

Indirect clinical criteria for vascular stenosis can already be obtained from bilateral comparative blood-pressure measurements, unilateral pulse weakening, and auscultation of a stenosis bruit in the provocation position.

A narrow point itself can rarely be displayed in the B-mode image, since the bone structures obstruct ultrasound penetration. Imaging aimed at detecting aneurysms or poststenotic dilations is very important, as is the location of thrombotic components in an aneurysm, which is best done using color coding. Since ablation of small emboli occurs occasionally, occlusions affecting the digital arteries of the hand that are unclear in their origin always require that a search should be made for subclavian aneurysms.

Hypothenar Hammer Syndrome

Another rare cause of acral circulatory disturbances in fingers DIII–V is trauma, usually recurrent, to the ulnar edge of the hand (the hypothenar eminence). The *hypothenar hammer syndrome* arises in patients who regularly use the hypothenar as a striking instrument, and in handball and volleyball players—with similar forms of trauma—due to a traumatic occlusion of the ulnar artery and several digital arteries (Birrer and Baumgartner 2002). The pathogenesis can be assumed to involve vascular spasm with subsequent arterial thrombosis, or—more rarely—traumatic development of an ulnar aneurysm with thrombosis and embolization (Taute et al. 1998).

Endothelial Dysfunction

Recent discoveries have raised the suspicion that reduced endothelial vasoactivity might be able to provide an early indicator of arteriosclerosis. In addition to plethysmographic procedures, endothelial dysfunction can also be assessed using duplex-sonographic measurement of the diameter of the brachial artery. Changes in the vascular caliber following specific stimuli (such as nicotine, nitrate, a cold-pressure test, and intra-arterially administered vasoactive agents) are particularly important here (Corretti et al. 1998). However, disturbed endothelial function may also be seen in endocrine disturbances such as those seen in diabetes mellitus (Avena et al. 1998) and primary hyperparathyroidism (Kosch et al. 2000a, 2000b). The changes evident with these disease pictures are always associated with reduced endothelial reactiveness. Research on the influence of nutritional disturbances, with hypertriglyceridemia or hypercholesterolemia, on endothelial function is still ongoing (Schnell et al. 1999).

Inflammatory Vascular Diseases

Inflammatory vascular diseases in the area of the arm arteries are, in absolute terms, only a rare cause of peripheral circulatory disturbances, but from a differential-diagnostic and therapeutic point of view it is very important to be able to recognize them. Table 5.**9** shows the frequency of the different forms of arteritis in the large arteries in western Europe and America.

In terms of diagnosis, duplex sonography is particularly advantageous with inflammatory vascular changes that have hemodynamic effects, since in addition to functional information it also provides guidance on the quality of the vessel walls and perivascular tissue. A typical finding in most inflammatory vascular diseases is segmental involvement, with no evidence of arteriosclerotic wall lesions, alternating with morphologically and functionally completely healthy arterial segments. A few characteristic patterns of findings in various inflammatory vascular diseases are described below.

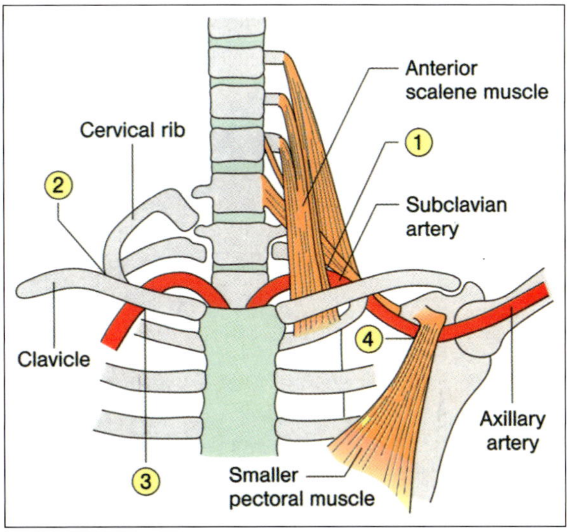

Fig. 5.**39** Muscular and osseous points of narrowing that the subclavian artery has to pass in its course. The compression syndromes are defined by the location of their origin (adapted from Kappert)
1 Anterior scalene syndrome
2 Cervical rib syndrome
3 Costoclavicular syndrome
4 Hyperabduction syndrome

Table 5.**9** Frequency of arteritis of the large arteries in western Europe and the USA per million inhabitants per year (adapted from Stiegler et al. 2002)

Arteritis	Rate per million/year
Takayasu arteritis	1–3*
Temporal arteritis	30–90
Polymyalgia rheumatica	290–540

* 10–30 per million population per year in Asia.

Thromboangiitis Obliterans (Winiwarter–Buerger Disease)

In Buerger's disease, it is usually the acral vessels in the upper and lower extremities that are affected (Diehm 1998). The disease picture is very rare in Europe, representing approximately 0.5% of all patients with peripheral arterial occlusive disease; in India, the proportion is 45–63%, in Korea and Japan it is 16–66%, and in some populations of Ashkenazi Jews it is 80% (Olin 2000).

Table 5.**10** B-mode and color Doppler criteria in Buerger's disease (thromboangiitis obliterans)

Vascular cross-section	Smoothly delimited
Occlusion material	Hypoechoic
Wall	Hypoechoic
Collaterals	Intraluminal corkscrews
Digital arteries	Disseminated occlusions
Luminal boundary	Smooth
Veins	Accompanying phlebitis, thrombosis

Table 5.**11** Ultrasound findings in giant-cell arteritis and arteriosclerotic vascular changes (adapted from Stiegler et al. 2002)

	Giant-cell arteritis	Arteriosclerosis
Vascular diameter	Normal to enlarged; reduced in the chronic stage	Normal/enlarged/reduced
Marginal zone reflex	Normal to absent	Enlarged or absent
Wall thickness	Segmentally enlarged	Disseminated enlargement or reduction
Luminal boundary	Smooth	Irregular
Occlusion material	Homogeneous, tending to be hypoechoic	Inhomogeneous, more hyperechoic

The typical duplex-sonographic signs found are corkscrew collateral vessels seen as screw-shaped winding blood courses *inside* closed vascular trunks (Amendt 1998). These typical collateral vessels, which may be connected with distal vascular segments at their origin, can also be found in the upper extremity both in the lower arm and in the palmar and digital arteries.

The precise origin of the phenomenon known as Martorell's sign angiographically is as yet still unclear. Possible explanations that have been discussed include intraluminal collaterals originating from a directly preceding vascular segment that is still patent (direct collaterals) or via collateral shoots sent off from parallel arteries (indirect collaterals). Alternatively, it might involve large-lumen vasa vasorum from the occluded vessels.

In thromboangiitis obliterans, the vascular walls in the occluded arteries do not show any morphological ultrasound abnormalities such as wall thickening, interrupted contours, or plaque deposition. The diameter of the artery, which is usually filled with hypoechoic thrombotic material, is markedly reduced. Due to the segmental pattern of the disease, the subsequent refilled vascular segment may have a completely normal appearance. A nonspecific finding that is nevertheless helpful for differential diagnosis is simultaneous evidence of epifascial or subfascial veins showing thrombotic or thrombophlebitic changes (Amendt 1998). Table 5.**10** provides a summary of typical color duplex-sonographic findings in patients with thromboangiitis obliterans.

Takayasu Arteritis

In accordance with the 1992 Chapel Hill Consensus Conference, Takayasu arteritis is regarded, along with Horton arteritis, as belonging to the group of vasculitides featuring involvement of the large arteries. This granulomatous form of arteritis, with autoimmune causes, mainly affects young women, and it is rare in Europe (Amendt 1998). The arteries of the upper extremities are a frequent site of manifestation, although involvement of the pelvic and leg arteries, aorta, visceral arteries, and coronary arteries, as well as aortic valve insufficiency, may also occur.

Typical findings include diffusely thickened vascular walls that increasingly narrow the lumen. The homogeneous inflammatory wall thickening is juxtaposed to the marginal zone reflex, which can usually still be demarcated, distant from the lumen. The vascular diameter is usually unchanged, although aneurysms may often develop in this form of vascular disease, as well as postoperatively.

Although Takayasu arteritis mainly affects the arteries of the upper extremities, involvement of the pelvic and leg arteries, aorta, and visceral arteries may also occur occasionally. Typical findings include diffusely thickened vascular walls that increasingly narrow the lumen. The homogeneous inflammatory wall thickening is juxtaposed to the marginal zone reflex, which can usually still be demarcated, distant from the lumen. The vascular diameter is usually unchanged, although aneurysms may often develop in this form of vascular disease, as well as postoperatively.

Horton-type giant-cell arteritis is barely distinguishable from Takayasu arteritis on duplex sonography. It can be differentiated from arteriosclerosis with color duplex sonography, using the criteria listed in Table 5.**11**.

Entrapment Syndrome

Entrapment syndrome is the term used for a compression syndrome in the popliteal artery in which the artery is intermittently compressed from outside, leading to intermittent ischemic syndromes.

Cause

The cause is a malformation of the popliteal artery and possibly also of the popliteal vein, or of the muscular structures in the popliteal fossa, particularly in the medial head of the gastrocnemius muscle during sporting activities. Compression of the vessels can occur during muscle contraction, particularly during plantar flexion of the foot and standing on the toes.

Classification

According to Insua et al. (1970), four types of compression syndrome can be distinguished and are easily recognized on duplex sonography:

- Type I: malformation of the popliteal artery, which courses round the medial head of the gastrocnemius muscle on the medial side.
- Type Ia: malformation of the medial head of the gastrocnemius muscle, which is attached more laterally and cranially than normal on the femur and consequently shifts the popliteal artery, which in turn courses again on the medial side of the medial head of the gastrocnemius.
- Type II: compression of the normally positioned popliteal artery and possibly also popliteal vein in the popliteal fossa, due to an abnormally attached lateral extension of the medial head of the gastrocnemius.
- Type IIa: compression of the normally positioned popliteal artery and possibly also popliteal vein by a plantar muscle with a variant course.

Findings

Types I and Ia (in Insua's classification) can be recognized on duplex sonography by the way in which the arteries and veins do not lie close together in the popliteal fossa; instead, a muscular structure appears in the popliteal fossa between the popliteal artery and vein.

In addition, excessive hypertrophy of the gastrocnemius muscle of the type seen in entrapment syndrome can also lead to intermittent claudication if the hypertrophied muscle head causes compression of the popliteal artery (and possibly popliteal vein) during contraction.

Effects of Compression

Due to the compression phenomena described above, structural damage to the vascular wall can occur during the course of the disease, which can in turn lead to secondary popliteal occlusion, poststenotic dilation, or aneurysm. Sacculations of the vascular wall of this type also carry a risk of arterio-arterial embolism in the lower leg arteries, from a thrombotic margin located on the vascular wall. In rare cases, a sequence of stenoses caused by different muscular compressions may be found, which require angiographic documentation to allow precise surgical planning.

Provocation Maneuvers

Various provocation maneuvers have become established for guidance when there is a suspected diagnosis of entrapment syndrome. During the examination, an attempt should be made to provoke compression of the artery and to document transient stenosis or occlusion of the vessel accordingly. This may be possible with the patient in the prone position during plantar flexion of the foot against resistance, or better—due to the stronger effort required—with the patient in a standing position on tiptoes.

Cystic Adventitial Degeneration

Cystic adventitial degeneration is a rare disease picture in which solitary or multiple unilocular or multilocular cystic structures in the adventitia of arteries near the joints may cause narrowing of the vessels, depending on how full the cysts are (Leu et al. 1977). The popliteal artery is affected in over 90% of cases (Dunant et al. 1973, Flanigan et al. 1979, Flückiger et al. 1991, Schäberle and Eisele 1996). The fullness of the cysts typically varies, leading to characteristically fluctuating symptoms with claudication and asymptomatic intervals, and marked restriction in walking distances. This disease picture should be considered particularly in younger patients with no cardiovascular risk factors who have the above symptoms.

Findings

On angiography, an hourglass-shaped impression in the artery is typically seen. However, if the cyst is not full and also if there is complete occlusion, angiographic evidence is difficult to obtain.

On ultrasonography, demonstration of mucous degeneration of the adventitia as a more hypoechoic area in the vascular wall is possible, largely independently of the state of fullness of the cysts. Ultrasound-guided fine-needle aspiration of the cyst contents can be offered to patients as an initial treatment option (Schäberle and Eisele 1996, Stiegler et al. 2002). If this treatment fails, surgical reconstruction of the diseased vascular segment is still possible.

Aneurysms

Arterial aneurysms are differentiated according to etiological, morphological, and clinical criteria (Table 5.**12**). They play a significant role not only in the abdomen, but also in the peripheral arteries. Both arteriosclerotic true aneurysms and iatrogenic false aneurysms are found (Fig. 5.**40**).

Clinically, suspicion of an aneurysm in the lower extremities is mainly raised by the acute appearance of claudication or ischemia, while subclavian aneurysms are mostly recognized by digital atheroembolisms. In a

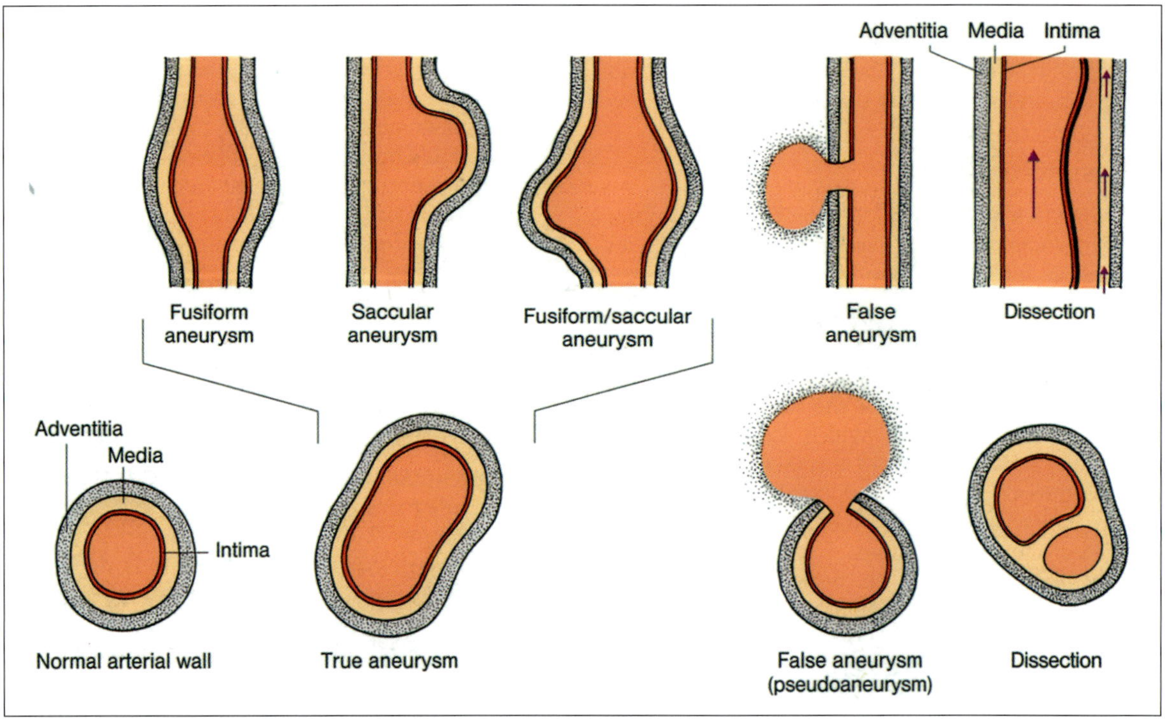

Fig. 5.**40** The various types of arterial aneurysm in longitudinal section and cross-section. A true aneurysm signifies arterial wall sacs in which all of the wall layers are retained, and which have various configurations. A false aneurysm results from iatrogenic interruption of all of the wall components, forming an undefined varying lumen in the perivascular tissue. Dissection refers to iatrogenic or spontaneous splitting of the media, creating a double lumen that can also cause aneurysmal arterial expansion

Table 5.**12** Classification of aneurysms

Etiological	Morphological	Clinical
Congenital	True	Ruptured
Arteriosclerotic	Saccular	Closed
Syphilitic	Fusiform	–
Traumatic	Saccular/fusiform	Open lumen
Mycotic	Dissecting	Thrombosed lumen
Poststenotic	False (spurious)	Partially thrombosed lumen
Iatrogenic	pseudoaneurysm	

multicenter study of popliteal aneurysms, Varga et al. (1994) found in 125 symptomatic popliteal aneurysms that there was claudication in 58 % of cases, ischemia in 56 %, and atheroembolism of the limbs in 11 %. According to a recent study by Mönig et al. (1996), infrapopliteal aneurysms have been reported in the world literature in only 15 cases so far.

True Aneurysm

The arteriosclerotic *true aneurysm* in the peripheral arteries is preferentially located in the common femoral artery and in the popliteal segment. It usually develops from a preceding dilative arteriopathy (Hepp and Pallua 1991, Loeprecht and Bruijnen 1991) (Figs. 5.**41**, 5.**42**). The low incidence of aneurysms in these locations (0.003–0.007 %) was analyzed in detail in a study by Lawrence et al. (1995), but no connection with atherosclerotic risk factors was found. However, the late appearance of the condition in patients aged over 65 is notable. By contrast, the subclavian artery aneurysm usually forms from a poststenotic dilation subsequent to a proximal mechanical stenosis, which is caused by the compression of narrow points in the shoulder girdle area, as shown above.

Vascular graft aneurysms are due to a wall insufficiency in venous grafts or in synthetic prostheses.

Findings

Pathological arterial dilation is recognizable in *sonographic B-mode imaging*, and the size and form of the aneurysm can be measured (Fig. 5.**43**). The distinction between thrombosed lumen and patent lumen is important. The patent lumen can be most effectively displayed and evaluated with color flow.

a 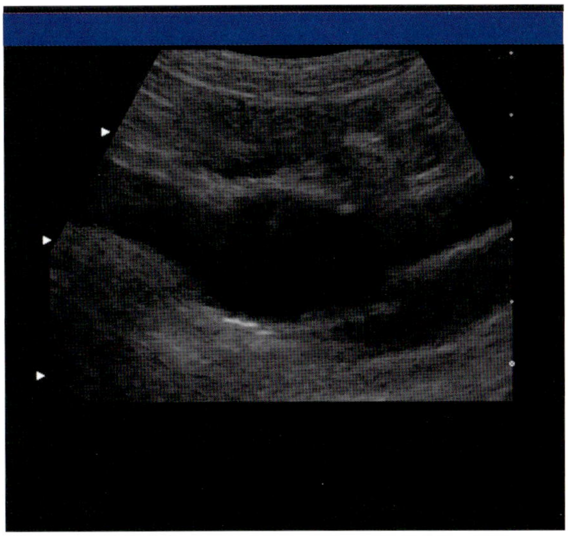

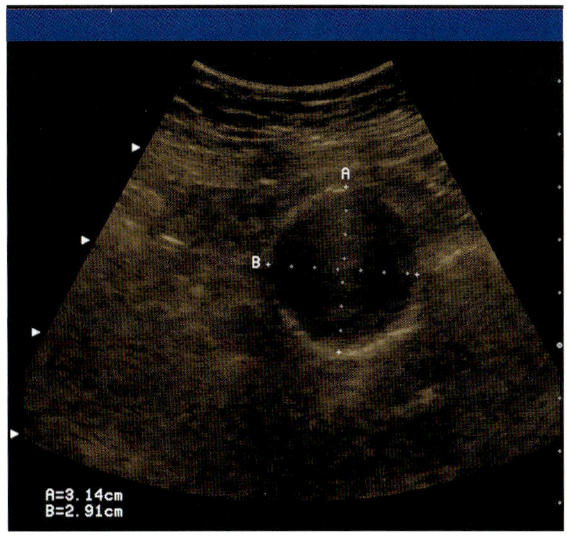 b

Fig. 5.**41a** A true aneurysm in the common iliac artery, seen in longitudinal section. The segmental dilation of the artery is clearly seen. **b** In cross-section, the complete extent of the aneurysm is evident. Brown-colored imaging was used in B-mode here (B-color).

a 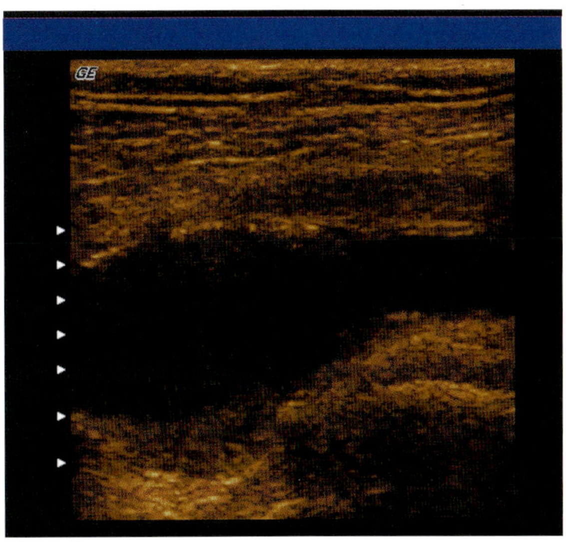

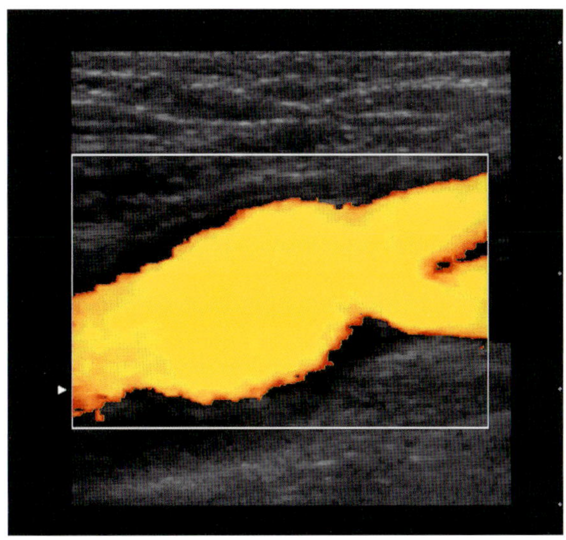 b

Fig. 5.**42a** A true aneurysm in the popliteal artery, seen in longitudinal section. The popliteal artery is a site of predilection for arterial aneurysms and should always be examined bilaterally when an aneurysm is present in another vascular region (e.g., in the abdominal aorta or iliac arteries), to ensure that the frequently occurring aneurysms in this area are not overlooked. **b** A true aneurysm in the insertion area of an aortofemoral bypass prosthesis, with the superficial femoral artery and deep femoral artery branching off, seen in power mode imaging.

In the periphery, the greatest risk is represented by aneurysmal thrombosis and the acute arterial occlusion associated with it, while the danger of rupture in this arterial region is minimal. The thrombotic deposits serve as an source of embolization into the acral regions of the upper and lower extremities. The subclavian aneurysm is a particularly frequent source of emboli.

False Aneurysm

While traumatic aneurysms predominated earlier, the injury pattern has now shifted toward iatrogenic vascular lesions. Due to the wide variety of diagnostic and therapeutic procedures that involve introducing catheters into the common femoral artery, the number of pseudoaneurysms in this region has significantly increased. An additional cause of false aneurysms is wound dehiscence, which is seen not infrequently after procedures involving prosthesis placement.

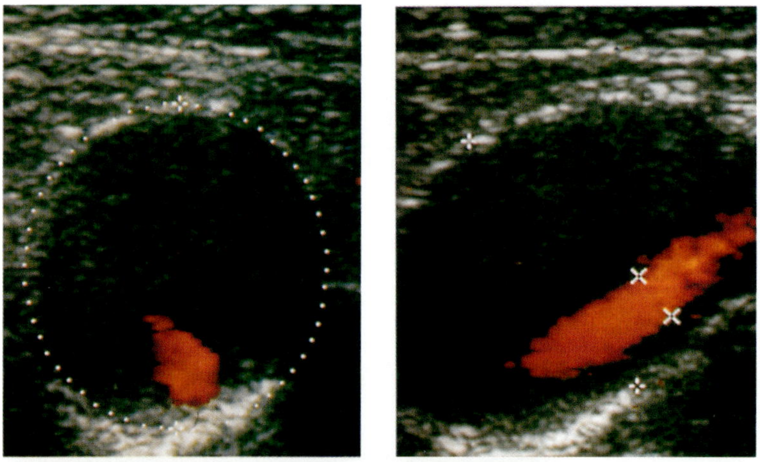

Fig. 5.**43** A large popliteal artery aneurysm, **a** in cross-section, **b** in longitudinal section, and **c** in longitudinal section with the corresponding frequency spectrum. The popliteal artery is a site of predilection for peripheral aneurysms. In this case, the lumen can be seen in both cross-section and longitudinal section, with minimal blood flow. The form of the flow waveform is relatively normal. The flow velocity at 40 cm/s is very low. **d** Corresponding angiograms

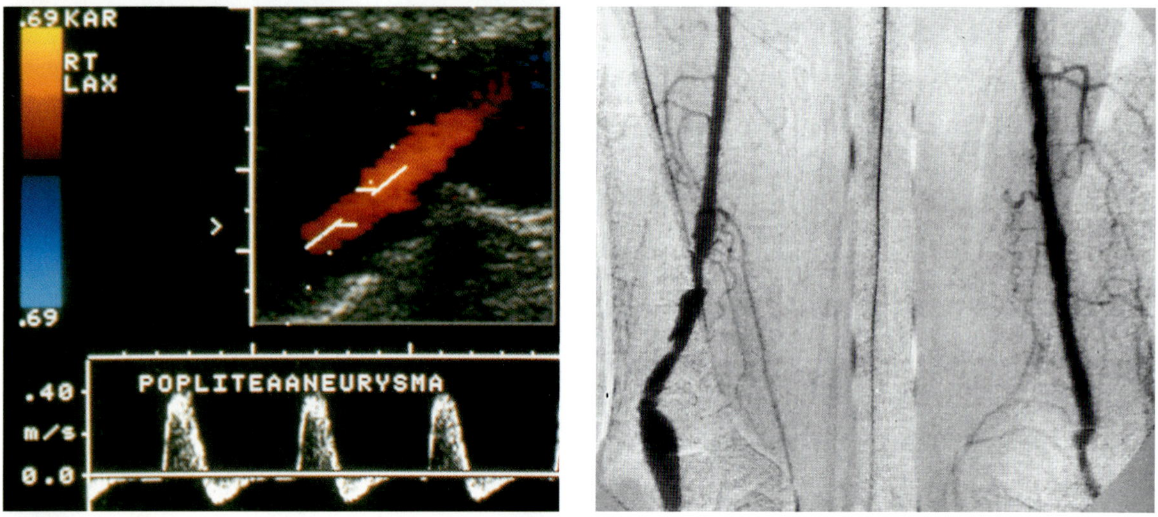

Findings

While the morphological diameter and size of an aneurysm can be determined with the *B-mode image,* and thrombotic components can be recognized, the *Doppler mode* registers a characteristic Doppler waveform, which is marked by the "to-and-fro" sign (Abu-Yousef et al. 1988) (Fig. 5.**44**). During systole, the arterial blood is pressed like a jet (forward flow) into the aneurysmal cavity, and in the diastole then flows back again into the originating artery (reverse flow) (Mitchell et al. 1987) (Fig. 5.**44a**). The Doppler waveform shows a pronounced pendular flow. The reverse flow component, which is broader and is present throughout the entire diastole, is clearly distinct from the brief backflow (dip) in normal arteries.

In the *color flow Doppler mode,* the aneurysmal neck or stalk is easy to identify, and the phenomenon of forward and backward flow can be impressively demonstrated in real time (Figs. 5.**45**).

When there is wound dehiscence, the original aneurysmal defect is often broader, and can lead to large, pulsating hematomas.

Compression Therapy

Although pseudoaneurysms, especially those with small volume, thrombose partially and spontaneously (Paulson et al. 1992), the cautious waiting approach, even with regular follow-up examinations, involves an inherent risk of aneurysmal rupture (Buchholz et al. 1991). Using color-guided compression of the aneurysmal stalk by the transducer, Fellmeth et al. (1991), were able to detect a thrombosis in 27 of 29 cases. In the meantime, these results have been confirmed by Sorrell et al. (1993), Do et al. (1993), and also in our own patients. While compression is being applied, regular assessment during the interval using color flow Doppler mode proves helpful in determining whether or not the femoral artery is being compromised. Other-

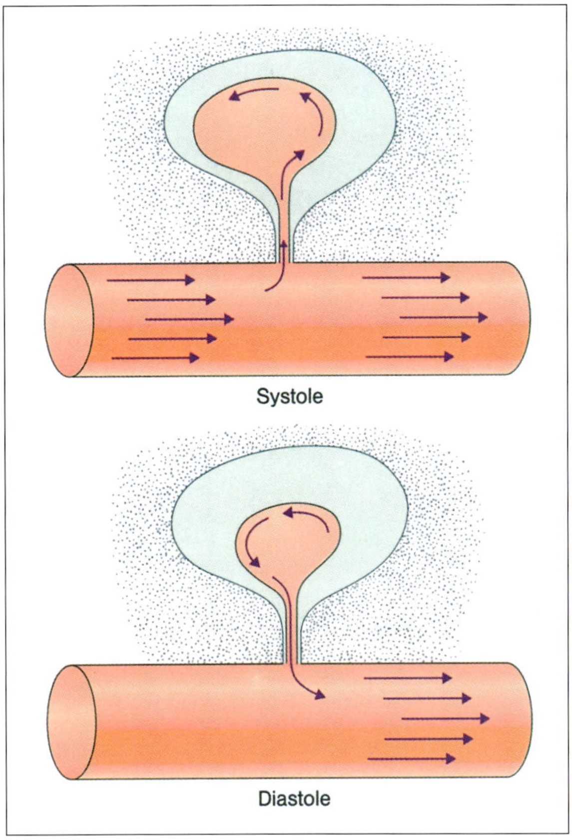

a

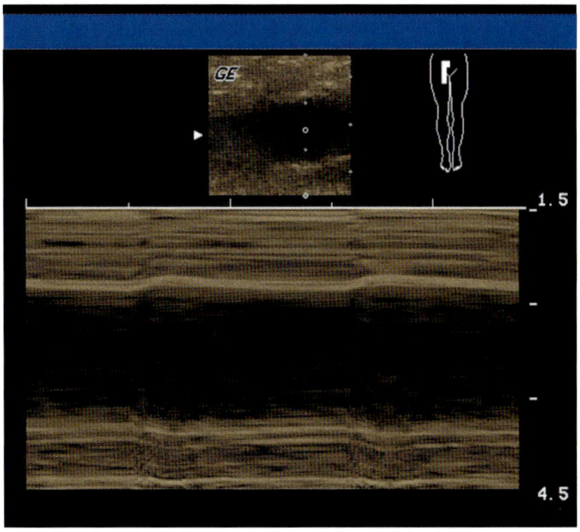

b

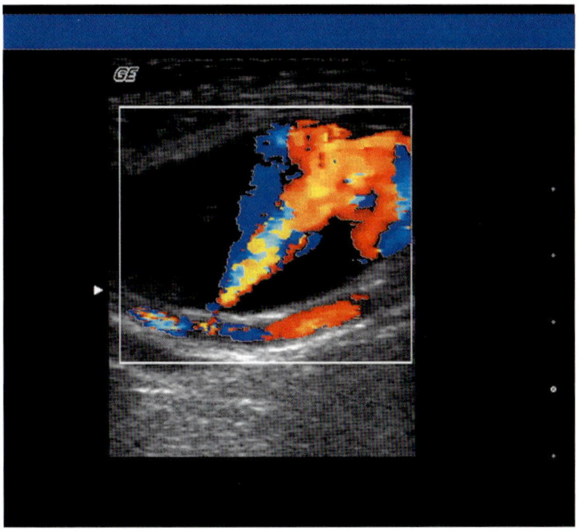

c

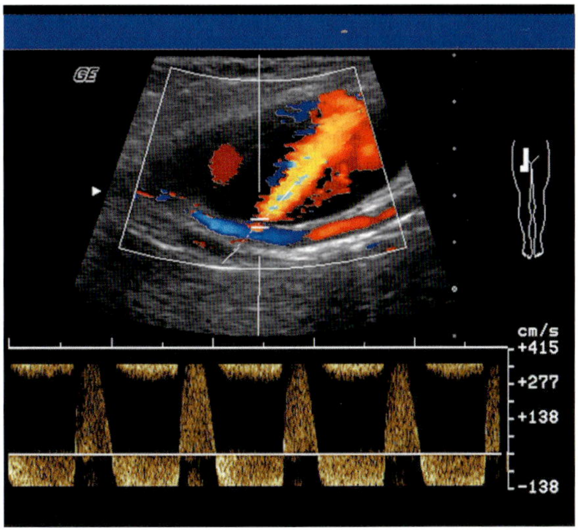

d

Fig. 5.**44a** Pseudoaneurysm. During systole, there is flow into the aneurysm, caused by damage done to the arterial wall, which burrows into the perivascular tissue. During diastole, blood is again pressed out of the aneurysmal sac. In the neck of the aneurysm, flow in the opposite direction results. **b** A pseudoaneurysm in the common femoral artery, seen in M-mode imaging. M-mode provides good visualization of pulsations in the aneurysmal cavity. In contrast to a pseudoaneurysm, a thrombosed hematoma shows no pulsations in B-mode or M-mode. **c**, **d** Imaging of a pseudoaneurysm in color Doppler mode (**c**) and with a Doppler frequency–time spectrum (**d**). Inflow into the pseudoaneurysm is immediately recognizable on color Doppler sonography. The pathognomonic back-and-forth flow in the aneurysmal neck is seen in the Doppler frequency–time spectrum

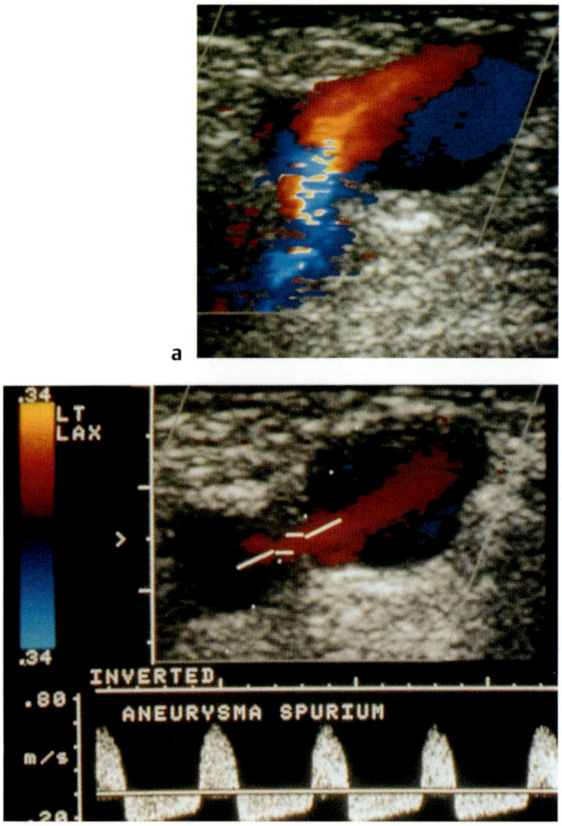

Fig. 5.**45** A pseudoaneurysm seen in color flow mode, in longitudinal section, near the left groin. **a** A summation image, with aliasing in the aneurysmal stalk. **b** During systole, there is forward flow into the aneurysm

Table 5.**13** Classification of arteriovenous fistulas according to etiological, morphological, and hemodynamic criteria

Etiology	
Congenital (10–15 %)	Mainly in the brain
	Hemangiomas
Acquired (80–90 %)	
– Traumatic	Penetrative vascular injuries
– Iatrogenic	Postinterventional
	Postoperative
– Spontaneous	Neoplasms
	Collapsed aneurysms
– Therapeutic	Dialysis shunt
	After thrombectomy in thrombosis
Morphology	
Single/Multiple	Direct fistula (80 %)
	Indirect fistula
	– With pseudoaneurysms
	– Without pseudoaneurysms
Hemodynamic effects	
Without cardiovascular reaction	
With cardiovascular reaction	Increased heart rate
	Increase in cardiac volume
	Cardiac insufficiency
	Venectasias
	Arterial aneurysms
	Longitudinal growth

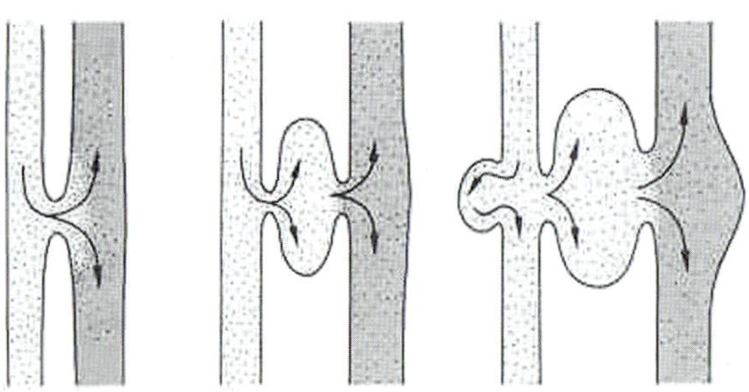

Fig. 5.**46** Morphological variants in arteriovenous fistulas caused by trauma. **a** In approximately 80 % of cases, the injuries result in direct short-circuit connections between an artery and a vein, also known as a "direct arteriovenous fistula." **b** In the remaining cases, an aneurysmal sac develops between the vessels, resulting in an "indirect arteriovenous fistula." **c** If the fistula persists for a longer period, additional aneurysmal wall sacculations develop in the artery and vein

wise, there is a danger of a thrombosis in the main blood vessel.

As an alternative to compression therapy, an injection of saline alongside the neck of the aneurysm or an ultrasound-guided thrombin injection into the pseudoaneurysm may be considered (Kang et al. 1998, Gehling et al. 1999, Sievert et al. 2000, Hughes et al. 2000, Pezullo et al. 2000). In a study of thrombin injection, it was shown that thrombosis was achieved in 15 of 20 pseudoaneurysms in a time of less than 20 s; a second injection was needed in four aneurysms, and failure was only observed in one case (Kang et al. 1998). Complications and recurrent occlusions were not observed in this group of patients, as the fast coagulation prevents the thrombus from being released through the narrow neck of the aneurysm and any thrombin that diffuses into the circulating blood is quickly neutralized by dilution and anticoagulation factors.

Arteriovenous Fistulas

An arteriovenous fistula is defined as a combination of arterial and venous injury at corresponding sites, creating a short-circuit connection between two vessels without an intervening capillary bed (Karasch et al. 1991c). Figure 5.**46** illustrates some important morphological variants of traumatic arteriovenous fistulas.

Arteriovenous fistulas can be distinguished according to their etiology, morphology, and hemodynamic effects. Table 5.**13** shows the various classification principles.

Principle and Etiology

A distinction is made between congenital and acquired fistulas (Table 5.**13**).

The acquired forms are especially significant for ultrasound diagnosis in the extremities. *Traumatic* fistulas appear after gunshot or knife wounds, while *iatrogenic* fistulas occur after surgery, mainly following orthopedic operations (Karasch et al. 1991b, Seifert et al. 1987) and after *diagnostic or therapeutic catheter manipulations.*

In a consecutive series of observations following cardiac catheterization, Steinkamp et al. (1992) reported an 18–20% occurrence of arteriovenous fistulas.

Pathophysiologically, the short-circuited circulatory system leads to a significant decrease in the flow resistance and to a volume loss from the arterial system into the venous low-pressure system. A pulse frequency increase and vasoconstriction in the peripheral circulation cause increased cardiac output, decreased blood pressure, and cardiac hypertrophy, which can lead to cardiac decompensation. These potential consequences are to be taken into account particularly in hemodialysis shunts.

Findings

Localizing the fistula canal can begin clinically. A thrill can be observed on palpation, and during auscultation, a pronounced, continuous systolic–diastolic sound (machinery murmur) is detected.

In the *B-mode image*, the canal between the artery and the vein can be seen. Often, this canal undergoes aneurysmal dilation and additional pseudoaneurysmal dilations and windings in the afferent artery and the afferent vein, which contains arterial blood.

Using *Doppler mode*, a turbulent, elevated flow velocity is often found in the fistula canal (shunt). In the arterial segment proximal to the fistula, there is an increase in systolic velocity and an appearance of diastolic blood flow (Strano et al. 1995). In the venous segment, there is arterial blood flow directed toward the heart (flow reversal). When the ultrasound application angle and the vascular diameter are known, the flow rate can be calculated.

The *color flow Doppler mode* makes it easier to locate the fistula canal, and the arterial backflow into the vein becomes dynamically visible. Color coding of course allows the examiner to recognize a change in the flow direction toward the heart, opposite to the arterial inflow (Fig. 5.**47**).

Hemodialysis Shunt

A distinction is made between internal arteriovenous shunts, which are formed with the body's own tissue, and prosthetic shunts with interposed vascular grafts (Wolf and Fobbe 1993). The oldest and best-known fistula between the radial artery and the cephalic vein is the Brescia–Cimino internal shunt. This is initially placed as far distally as possible near the hand, and is not moved proximally until it has become occluded. After the internal shunt has failed, or when there are unsuitable venous relationships, a synthetic implant is interposed.

Qualitative detection of an increase in the flow velocity in the shunt to the point of turbulence is possible even with *continuous-wave Doppler sonography.*

Conventional and color flow duplex sonography has expanded the diagnostic process, both morphologically and functionally. The shunt itself, the origin of stenoses in the sutured area of the venous or arterial connection, and also aneurysm formation or thromboses, can be morphologically detected (Middleton et al. 1989b).

Shunt volumes can be measured using the Doppler sample volume. This quantity is determined by electronically measuring the length of the vascular radius (r) and the mean flow velocity (V_m).

$$\text{Flow rate (mL/min)} = V_{mean}\ \text{(cm/min)} \times r^2\ (\text{cm}^2) \times \pi$$

Wittenberg et al. (1993) were able to show that several investigators repeatedly measuring the flow rates achieved an acceptable degree of reproducibility, with an 11% error rate. The following threshold values have become accepted as representing the norm (Kathrein et al. 1989):

- PTFE fistulas 614 ± 242 mL/min
- Brescia–Cimino fistulas 464 ± 199 mL/min
- Mean shunt volumes 514 mL/min

Since flow rates that are too low (less than 200–300 mL/min) can be corrected by widening the shunt, and those that are too high can be surgically reduced to eliminate the cardiac consequences, measuring the output is clearly important.

Temporary Shunts

Postoperatively, temporary shunts are often created in order to increase the flow into the arteries or veins operated on. This is often done after an operation for a fresh venous thrombosis, to prevent the high tendency toward re-thrombosis (Schmitz-Rixen et al. 1986).

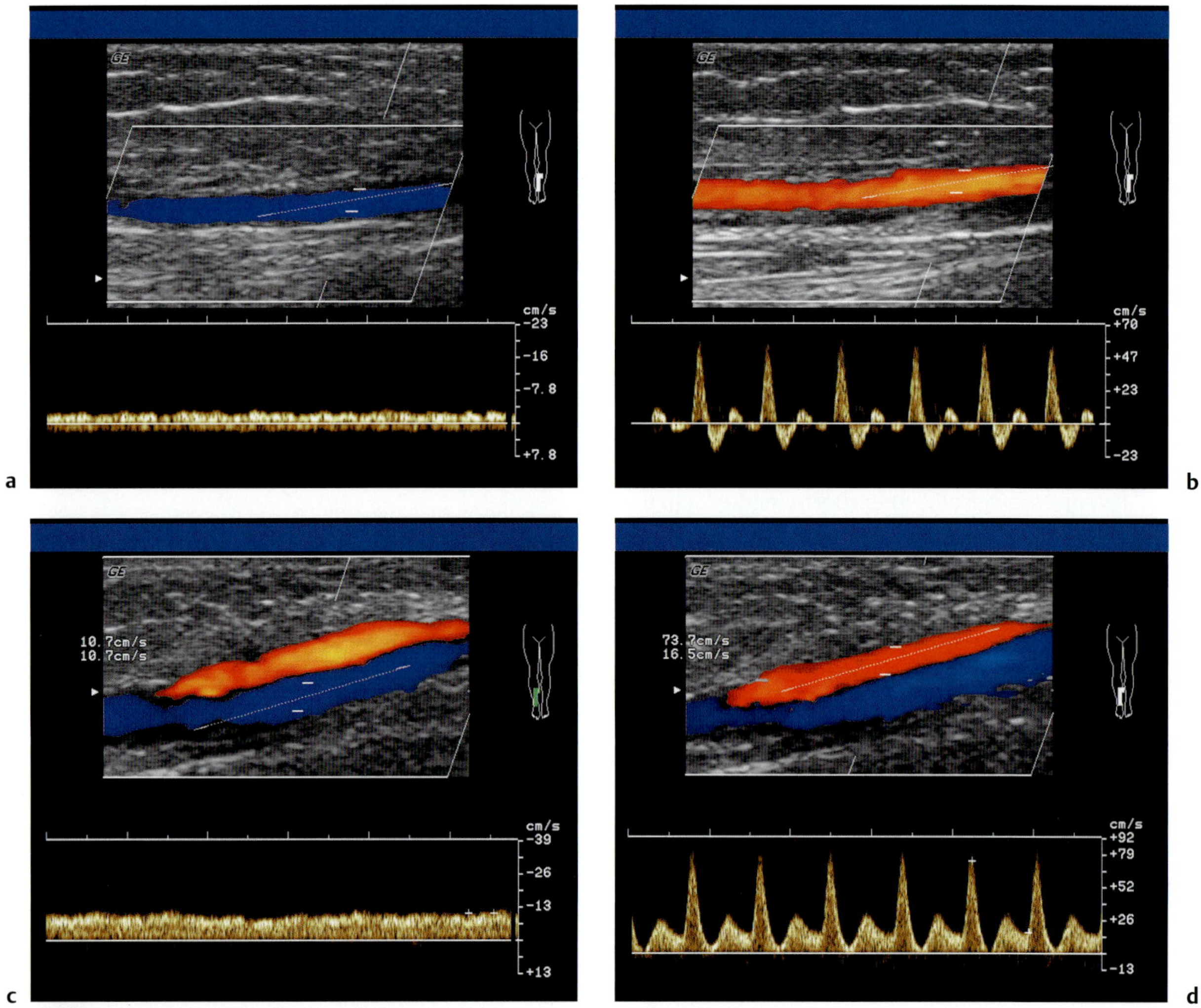

Fig. 5.**47** Indirect criteria for an arteriovenous fistula peripheral to the transducer site. This example shows congenital arteriovenous fistulas in the area of the right heel. **a** On the contralateral healthy side on the left, there is a normal low flow velocity of approximately 4 cm/s in the posterior tibial vein. **b** There is also a normal flow profile in the ipsilateral posterior tibial artery. **c** On the ipsilateral side with the subsequent arteriovenous fistula, an increased flow velocity of approximately 11 cm/s is seen in the efferent vein. **d** There is reduced resistance in the afferent posterior tibial artery, evident from the increased diastolic flow velocity

Monitoring Lumen Recanalization Techniques

Monitoring lumen recanalization measures after various angioplasty procedures, and evaluating stent implantations is as important as the follow-up examination after vascular surgery. Morphological monitoring, and measurements of the flow velocity in the patent arterial segment, in the interposed prostheses, and also in the bypasses, make it possible to recognize complications and potential recurrent occlusions at an early stage.

Angioplasty and Stenting

After percutaneous transluminal angioplasty, residual wall deposits and partial dissections can be detected with *B-mode imaging*.

Using *Doppler mode*, it is possible to evaluate vascular system stenoses and flow disturbances both qualitatively and quantitatively. Regular noninvasive follow-up examinations are possible (p. 202).

Implantation of a stent after angioplasty can be depicted with excellent quality in B-mode imaging, with or without color coding. Adequate expansion of the stent relative to the regular vascular diameter can be well documented in B-mode imaging, and undisturbed flow or stenotic narrowing can be quantitatively assessed using Doppler mode (Fig. 5.**48**).

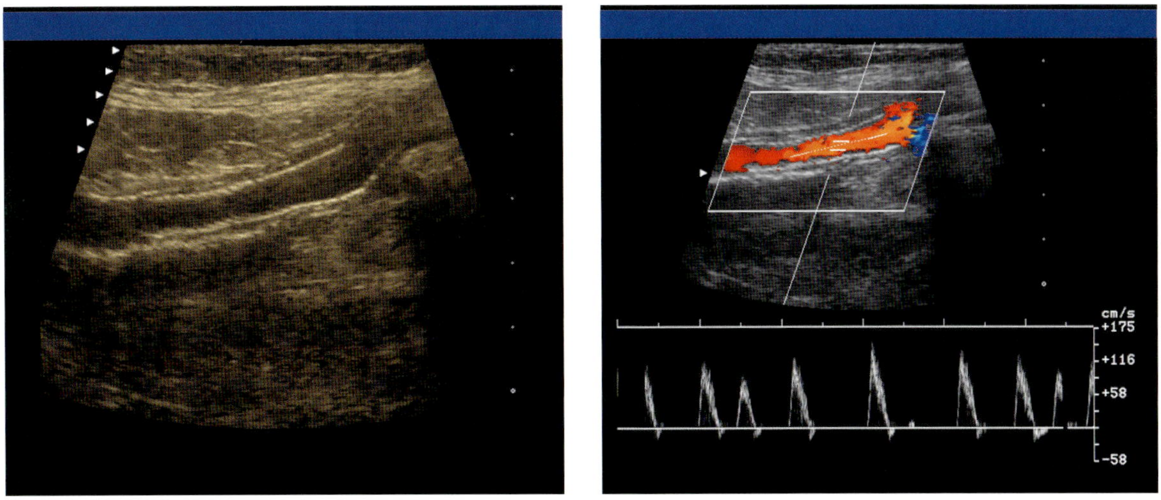

Fig. 5.**48** A stent in the area of the external iliac artery. **a** The B-mode longitudinal ultrasound section makes it possible to distinguish the stent from the vascular wall, due to its hyperechoic structure. **b** The color Doppler mode image shows that the stented vascular segment is freely patent. In the Doppler frequency–time spectrum, a normal flow velocity with absolute arrhythmia is obtained

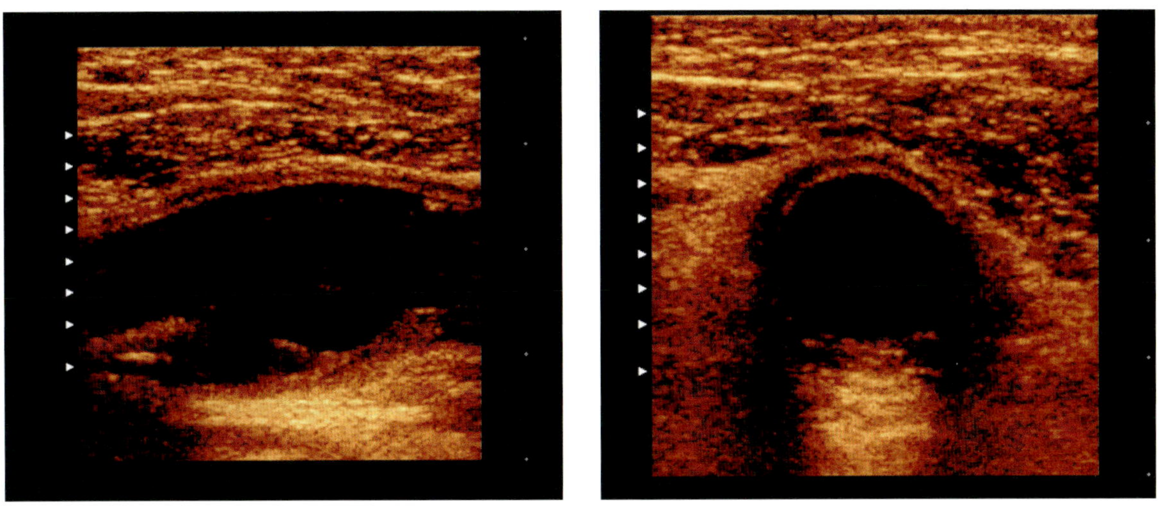

Fig. 5.**49** The initial part of the distal right bypass limb of an aortobifemoral Y prosthesis in longitudinal section (**a**) and cross-section (**b**). The prosthesis material appears clearly as a hyperechoic double contour, with an intervening hypoechoic layer, and is thus distinguishable from the original artery

Alternatively, the stent can be checked with *endovascular* ultrasound guidance. Arko et al. (1998) argue in favor of this method, although it requires additional equipment and costs. However, the rate of recurrent stenosis was markedly reduced. Katzenschlager et al. (2000) also tested ultrasound-guided local lysis of occlusions in the femoropopliteal region. They succeeded in achieving partial recanalization in this difficult region in 11 of 13 patients, with complete recanalization after subsequent angioplasty.

Duplex sonography is important not only for immediate and later checking of the results of angioplasty. The lumen can also be opened with ultrasound guidance under visual control. The catheter is incorporated into the duplex scanner for this procedure (Cluley et al. 1993). Katzenschlager et al. (1996) showed that percutaneous transluminal angioplasty can also be carried out without an integrated catheter if color-coded duplex guidance is used.

Vascular Prostheses and Bypasses

Follow-up observations of vascular prostheses and bypasses have already been carried out intensively for a long time using various ultrasound procedures. A bypass occlusion rate of approximately 30% within 5 years is considered normal (Bandyk 1993). Quantitative measurement of the flow velocity and qualitative formal analysis of the Doppler waveforms indicate possible bypass stenoses or aneurysms, as well as any reduced output that may reflect an impending reocclusion (Fig. 5.**49**).

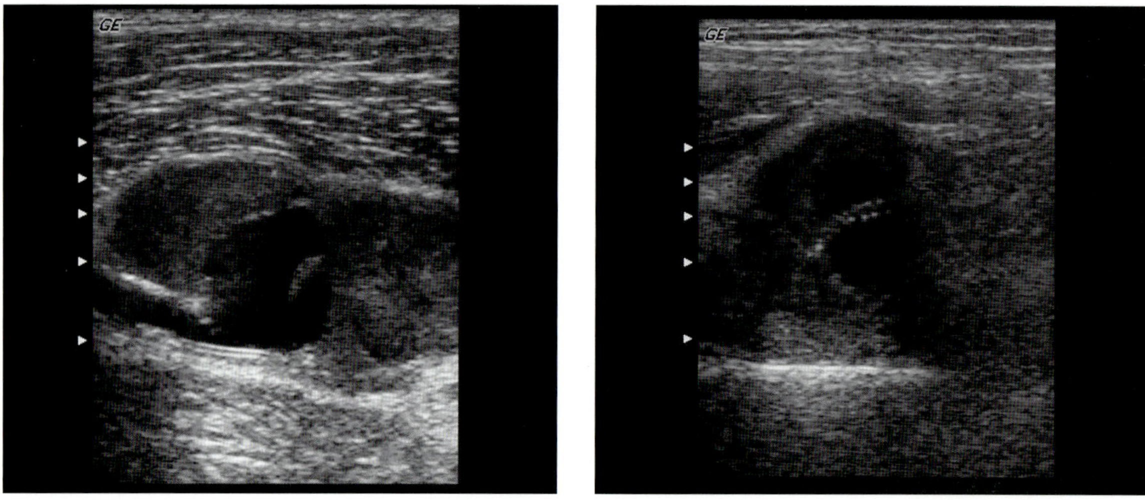

Fig. 5.**50** Rupture of a femoropopliteal bypass in the area of the distal anastomosis in the region of the start of the popliteal artery. **a** The prosthesis can be recognized on ultrasound by its hyperechoic structure in the longitudinal section. An extensive hypoechoic hematoma is seen around the prosthesis. **b** In the transverse section, the homogeneous hyperechoic structure of the bypass prosthesis and the hypoechoic hematoma surrounding it can be recognized

Various studies have provided values that can be used to predict pathological changes.

As early as 1985, Bandyk et al. were able to demonstrate using *continuous-wave Doppler sonography* that bypasses have differing flow velocities depending on their location (Bandyk et al. 1985). The femoropopliteal bypass showed a flow velocity of 90 cm/s; the femorotibial bypass yielded a flow velocity of 68 cm/s on average. A decrease in the peak systolic velocity to less than 45 cm/s was interpreted as being a certain sign of impending reocclusion. In evaluating 379 venous bypasses, Mills et al. (1990) were also able to demonstrate that a velocity of less than 45 cm/s indicated a preceding stenosis with a lumen obstruction of over 50 %.

In a later study using *conventional duplex sonography*, Bandyk et al. (1988) were able to show that formal analysis of the flow velocity recordings can indicate preceding or following stenoses. A monophasic configuration indicates a preceding or following stenosis with a lumen obstruction of 50–75 %. Pronounced pendular blood flow indicates a distal stenosis.

In a study carried out by Sladen et al. (1989), in which both decreased flow velocity (less than 45 cm/s) and elevated velocity (higher than 300 cm/s) were used as stenosis indicators, it was shown that of 15 stenoses with a lumen obstruction of over 80 %, only six showed a poststenotic flow velocity of less than 45 cm/s.

In a larger, prospective study by Gooding et al. (1991), the predicted value, less than 40 cm/s *reduced* peak systolic velocity, could only be confirmed with a 33 % sensitivity. However, the *elevated* peak systolic velocity was a more informative parameter.

In stenoses, Polak et al. (1990a) observed a peak velocity of between 117 cm/s and 225 cm/s. Absent diastolic backflow could also be used to indicate a subsequent stenosis. In a larger study by Buth et al. (1991), color flow duplex sonography was used in follow-up observations to examine five continuous-wave Doppler waveform parameters in 116 grafts. It was shown that a velocity index formed from the velocity ratio in a normal bypass and a stenotic bypass, with a threshold value of 0.65, identified all stenoses with a lumen obstruction of over 40 %. Taylor et al. (1992) also found that the velocity quotient had the greatest predictive value. However, they set their threshold value at over 2.0.

In addition to diagnosing stenoses and predicting impending bypass occlusions, it is important to search for pseudoaneurysms at the bypass insertion point or ruptures at the bypass site (Fig. 5.**50**).

Iatrogenic Arterial Dissection

In the arteries of the leg, including the pelvic region, iatrogenic dissection is not infrequently observed after angioplasty, and can be clearly documented using color flow duplex sonography. Iatrogenic dissections can cause occlusions, or can spontaneously regress without obstructing the lumen; both of these phenomena can be documented in their course using duplex sonography (Murphy et al. 1991). Spontaneous dissections, as encountered in the aorta and also in the carotid arteries, are rare in the arteries of the extremities.

Arterial Locks

After some interventional catheter procedures, arterial locks are left in the common femoral artery remain for a certain time to allow speedy access to the arterial vascular system. Figure 5.**51** shows the ultrasound appearance of an arterial lock in the common femoral artery.

Evaluation

Color Flow Duplex Sonography

B-mode. The interpretation of wall deposits and plaques follows the same criteria as those discussed above for the carotid artery vascular system. Plaques are characterized according to their location, size, and form, as well as their surface, echo structure, and echogenicity. However, as mentioned previously, due to the wide variety of possible wall deposits, the characterization of plaques in the long, stretched vascular bed of the extremities is not as significant as it is in the carotid arteries. The possible peripheral microembolization consequences are less important here than they are in the sensitive cerebral parenchyma.

The B-mode is especially useful in detecting aneurysms. They can be precisely located and also classified according to their form and size. Even the thrombotic component of aneurysms can usually be readily detected in the B-mode image. Vascular dilations, hematomas, and cysts can also be evaluated in B-mode.

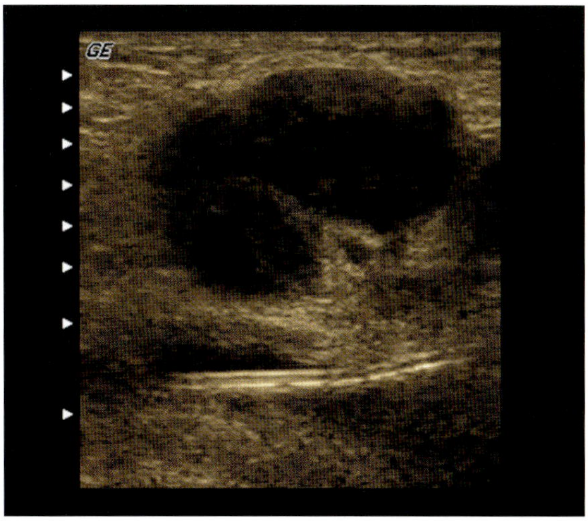

Fig. 5.**51** A longitudinal section through the common femoral artery. An arterial lock can be seen in the lumen of the artery, recognizable from its hyperechoic double contour, with a hypoechoic interior lumen. An encapsulated hematoma with a hypoechoic appearance, which developed after puncture of the artery, can be seen ventral to the lock

Doppler mode. Until now, predominantly *qualitative* Doppler waveform evaluations from the peripheral arteries have been used to estimate the degree of stenosis (Fig. 5.**30**). They make it possible to classify approximately the extent of vascular system stenoses. The following assessment criteria, adapted from Jäger and Landmann (1994), are recommended:

- *Normal findings:* triphasic Doppler curve with narrow frequency band.
- *1–25% stenosis:* triphasic curve shape, discrete widening of the frequency spectrum, flow velocity not increased.
- *25–50% stenosis:* triphasic curve shape, broadening of the frequency spectrum, flow velocity increased > 30%.
- *50–75% stenosis:* loss of reverse flow component, systolic window filled, flow velocity increased some 100–200%.
- *75–99% stenosis:* High diastolic flow, strong-amplitude signals around the zero line under the systolic peak. Usually an increase in flow velocity > 300% (only filiform stenoses produce a reduction in the velocity again).

Of course, it is also possible to calculate the pulsatility index with pulsed-wave Doppler mode. However, this does not have the same value as in continuous-wave Doppler sonography, since imaging techniques make it possible to measure the stenotic area directly.

Quantitatively, the peak systolic velocity values and the ratio of the maximum velocity within and 2 cm proximal (or distal) to the stenosis are usually used, the peak velocity ratio (PVR). Using receiver operating curves, Sacks et al. (1992) determined the following threshold values in stenoses of more than 50%:

Stenosis	V (cm/s)	Quotient
> 50%	120	1.4
> 70%	160	2.0
> 90%	180	2.9

These curves are constructed by plotting the sensitivity of the test against its specificity at various thresholds. Usually, the value of the optimal relationship between test sensitivity and specificity is the one chosen (Sumner et al. 1993).

However, these values are lower than in most other studies, in which stenoses of more than 50% indicate a threshold value of 180 cm/s for peak systolic velocity, and a ratio of 2.5. This agrees with in vivo results obtained by Whyman et al. (1993), who obtained a figure of 2.67 for the peak velocity ratio (PVR) for stenoses of more than 50%.

Color flow Doppler mode. As mentioned above, the color flow Doppler mode has become an indispensable aid in locating stenoses and determining the length of an occlusion. Color flow Doppler sonography can also be used to detect a false aneurysm following femoral artery punctures. Pendular flow in the leak can be impressively depicted using color, and compression therapy can be applied specifically under color control (pp. 187).

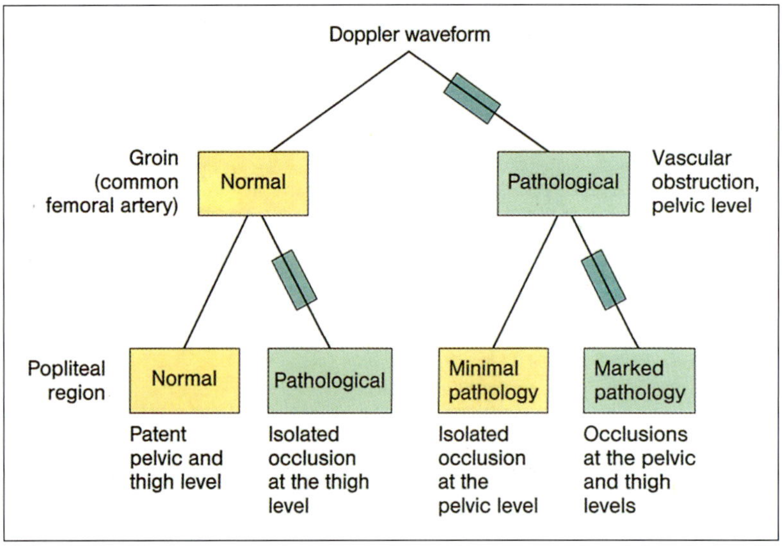

Fig. 5.**52** Determining the level of hemodynamically active vascular system obstructions using directional continuous-wave Doppler sonography. Formal analysis of the flow velocity curves at the usual registration points allows qualitative assessment of the presence of a high-grade stenosis or occlusion proximal to the registration point. Comparative evaluation of the Doppler curves at two typical registration points allows a topographical diagnosis

Continuous-Wave Doppler Sonography

As mentioned earlier (pp. 171–173), many indices have been developed that use quantitative values obtained from the measurement recording segments to determine a correlation with the extent of the stenosis.

The threshold value of the *pulsatility index* (PI) lies at around five, and often shows much higher values (up to 10 or 12). In a comprehensive study (of 175 aortoiliac segments) comparing the results to those from intra-arterial blood-pressure measurements, Johnston et al. (1983) demonstrated that 5.5 as a threshold value provided the smallest error range in detecting aortoiliac obstructions. In poorly compensated occlusions, the index can sink to a value of one, proximal to the registration point. In our own study (Neuerburg-Heusler et al. 1991b), the pulsatility index values obtained applying ultrasound to the common femoral artery were:

- In normal findings: PI of 8.5 ± 3.5
- Isolated iliac occlusions: PI of 2.8 ± 1.6
- Combined pelvic femoral artery occlusions: PI of 2.3 ± 1
- Isolated subsequent femoral artery occlusions still retained normal values: PI of 6.3 ± 2.6

Only preceding vascular system obstructions are detected with this index; subsequent occlusions are not evaluated.

In practical clinical use, qualitative evaluation of the waveform is sufficient (Fig. 5.**29**), since many results that complement one another are available.

In contrast to the pressure, which is not detrimentally affected by subsequent vascular system obstructions, the Doppler waveform is already altered in the prestenotic segment. The reverse flow component lessens, due to decreased peripheral resistance in the collateral or peripheral vasculature, or both. This recording is not distinguishable from poststenotic recordings, with minimal vascular system stenoses (less than 50%) proximal to the registration point (Fig. 5.**33**).

When occlusions and stenoses of more than 50% occur in isolated form, registering the flow pulses segmentally makes it possible to assign them to specific vascular levels. Occlusions covering several levels are difficult to classify topographically (Fig. 5.**52**).

Examining the circulation to the superficial and deep palmar arterial arches can be important in vascular injuries, and especially when placing a dialysis shunt. The functional clinical *Allen test* has a 13% error rate, and cannot be used in patients suffering from posttraumatic circulatory shock (Fronek 1989). The *palmar test* described by Ruland et al. (1988), which uses the Doppler probe, provides better results.

In this study, the most reliable criterion was *flow reversal* in the distal radial artery after proximal occlusive compression of the radial artery at the margin of the brachial muscle. Equally practicable was testing the flow direction in the radial artery of the thumb after compressing the radial artery at the typical pulse palpation point.

Sources of Error

Color Flow Duplex Sonography

Methodological Sources of Error

Transducer/transmission frequency. Choosing an unsuitable transducer or an inadequate transmission frequency can have a significantly adverse effect on the examination quality.

The transmission frequency, adjusted for the depth of the ultrasound application, has to be observed without fail (see the section on examination, p. 161).

Sectional plane. When an unfavorable section is used, plaques may be missed, or an incorrect diameter value may be measured.

Echogenicity. Thrombotic material in the vasculature can have the same acoustic impedance as the blood. In the B-mode image, the blood vessel or the aneurysm appears to be patent. It is only when using the Doppler mode (with or without color coding) that stenoses, occlusions, or thrombotic components can be detected. It is also important to note vascular wall pulsations.

Doppler mode. Positioning the Doppler mode at the necessary angle of less than 60° can present problems, especially in the pelvic region. The vasculature does not have a stretched course here. Guiding the probe is more difficult, and not as easily varied as it is in the extremities.

Color flow Doppler mode. The following fundamental problems associated with this procedure should be noted.

Flow velocity. A mean flow velocity is color-coded, and this only provides qualitative and semiquantitative information.

Angle dependence. When using color, the angle of the many pulsed-wave Doppler sample volumes used cannot be adequately adjusted for changes in the vascular course. This means that different color intensities do not always indicate stenoses, but may also be due to an alteration in the incident angle of the ultrasound in relation to the vascular axis.

Color change. When changing the true flow direction in relation to the transducer, the sudden color change proceeds through black. In the aliasing phenomenon, the color change proceeds through white, since it takes place at high frequencies.

Perivascular color artifacts. This phenomenon entails color clouds consisting of multicolored pixels, which can be projected onto the tissue surrounding stenotic vessels. Widder (1993) describes this as a "confetti effect." Middleton et al. (1989a) discovered that these artifacts vary with the heartbeat, and are most prominent during systole (less noticeable during diastole). The authors assume that turbulent flow conducts the vibrations into the perivascular tissue, and there causes Doppler shifts that are reproduced in color. They mainly observed these phenomena in arteriovenous fistulas, hemodialysis shunt stenoses, and also in renal artery stenoses, pelvic stenoses, and stenoses in the leg arteries.

Continuous-Wave Doppler Sonography

Incident angle of ultrasound. Independent of the evaluation method used (analogue registration or spectral analysis), the most important error source in continuous-wave Doppler sonography is the fact that the ultrasound application angle is not known. At an acute incident angle of less than 30° a flow acceleration is simulated; at an angle of more than 60°, the amplitude height of the Doppler waveform is reduced; at a 90° angle, no signal is registered.

Identifying the arteries. Properly assigning the artery signal can pose difficulties; not every signal that is located belongs to the artery sought. The anatomical course, the neighboring arteries, the side branches, and possibly also the collaterals, have to be known.

In regions in which peripheral compression maneuvers are possible (oscillating vibration, suprasystolic arterial compression), identification is easier (e.g., subclavian artery, brachial artery). In the thigh, for example, it is not possible to apply ultrasound separately to the superficial femoral artery and the deep femoral artery.

Vascular diameter. The vascular diameter is not known, so that the flow velocity can be slightly disturbed or somewhat slower in a wide or pathologically enlarged blood vessel. In a narrow blood vessel, the flow velocity increases.

Hematoma, obesity. Particularly in the groin region, applying ultrasound can be more difficult due to a hematoma (following angiography), or in cases of massive obesity.

Superimposed veins (popliteal region) can also complicate the identification of the arteries.

Other Sources of Error

Rest period. An adequate rest period is necessary, both in healthy patients and in pathological cases. After physical exertion, the flow velocity increases while the peripheral resistance decreases, and the resulting monophasic curve simulates a preceding vascular system obstruction.

Noting positional anomalies. Positional anomalies, which are especially frequent at the origin of the deep femoral artery and its branches, must be known. A distal deep femoral artery occlusion may be missed. When there is a superficial femoral artery occlusion, the deep femoral artery, which then assumes a steep position, is often interpreted as being the superficial femoral artery, and its side branches are viewed as the deep femoral artery. Loop formations affecting the pelvic arteries should not be misinterpreted as an occlusion. The points of origin of the lower leg arteries also have variants. Specifically, the anterior tibial artery can originate from the popliteal artery at varying levels.

Classifying space-occupying lesions. When using *B-mode imaging* alone, it is not easy to differentiate between a hematoma, a false aneurysm, and an arteriovenous fistula. Complications occurring after procedures in the groin region, which are being carried out more and more often, can only be verified using differential diagnosis with the Doppler mode, or even better using the *color flow Doppler mode*. Depicting the fistula canal (arteriovenous fistula, false aneurysm), in particular, is much easier when the inflow into the narrow channel can be color-coded.

Recognizing and distinguishing occlusion and stenosis. A stenosis can be falsely deduced:
- When a collateral is stenotic at its origin, and is characterized as being the main blood vessel.
- When the grading of a stenosis proceeds according to the morphological image, which almost always proves to be misleading.

Grading the stenosis is made more difficult by preceding or subsequent vascular system obstructions. If a stenosis is apparent in the area of the bifurcation, the peak velocity ratio (PVR) can only be calculated using the peak flow velocity distal to the stenosis.

Measuring occlusion length. The occlusion length can be *underestimated* when proximal or distal collaterals in the vascular course simulate a patent vascular system.

The occlusion length can be *overestimated:*
- When plaques in the proximal or distal occlusive segment extinguish the color through ultrasound shadows; and/or
- When a significantly reduced postocclusive flow velocity can no longer be color-coded.

Diagnostic Effectiveness

Color Flow Duplex Sonography

In contrast to the arteries supplying the brain, many noninvasive procedures for diagnosing vascular system obstructions in the arteries supplying the extremities have already been available for a long time. These allow both qualitative and quantitative evaluation of the hemodynamic effects of vascular system stenoses. Locating occlusions and stenoses with respect to the vascular level is possible both with clinical methods and with equipment. Segmental pressure measurements using the ultrasound Doppler method (p. 150) and directional Doppler sonography (see above) have increasingly come to predominate.

A gap existed with regard to conclusive *differentiation* between *occlusion* and *stenosis*, *localization* of occlusions and stenoses within the vascular levels, and *measuring the length* of an occlusion. An additional weak point shared by all previous procedures was differentiation between isolated pelvic occlusions and *combined* pelvic and thigh occlusions.

Conventional duplex sonography, and particularly the color flow method, have now closed this gap. Although *conventional duplex sonography* with careful arterial scanning already made it possible to differentiate between occlusion and stenosis, confirmed the location of vascular system obstructions, and allowed semiquantitative stenosis grading (Jäger et al. 1985, Kohler et al. 1987a), this method required considerable time to carry out, and this was an obstacle to its widespread use.

Using *color flow duplex sonography* has made following the arteries easier in varying degrees. Color-guided sample volume placement in the stenotic areas has simplified the identification of the highest-velocity region within the stenosis. In addition, the color interruption has made it possible to recognize occlusions and determine their extent more quickly.

Various studies have confirmed the validity of examinations using duplex sonography, both with and without color (Table 5.**14**). Comparisons between conventional and color flow duplex sonography provide recognizable, but not marked, improvements in the findings, due to the color flow Doppler mode. Jäger et al. (1985) and Kohler et al. (1987a) were able to obtain very good results using conventional duplex sonography (sensitivity 77% and 98%, respectively, specificity 81% and 92%, respectively). They used a simplified grading process with spectral analysis patterns (Fig. 5.**30**) that is still very useful for a rapid diagnosis.

More comprehensive studies using color flow duplex sonography, such as those carried out by Cossman et al. (1989), Koennecke et al. (1989), Landwehr et al. (1990), and Polak et al. (1990b), produced sensitivity figures of between 87% and 97%, and had an outstanding specificity lying between 95% and 99%. Increasingly, the validity of the method for individual vascular segments when using duplex sonography is also being assessed (Table 5.**15**). A limiting factor is that almost all studies evaluate stenoses and occlusions that are over 50%. In differentiating stenoses of less than 50% from those of more than 50%, there are certain methodological inaccuracies involved, both when using angiography and also when applying duplex sonography.

The diagnostic validity of Duplex sonography increasingly has to be measured against another noninvasive examination procedure—magnetic resonance arteriography (MRA). However, investigations completed by Mulligan et al. (1991) and Baumgartner et al. (1993) show that duplex sonography is clearly superior. Correlation with the angiographic findings produced the following results:

	Mulligan et al. (1991)	Baumgartner et al. (1993)
MRA	71%	81%
Duplex sonogram	93%	88%

Continuous-Wave Doppler Sonography

With directional continuous-wave Doppler sonography, it is possible to register flow velocity curves—the visual, *qualitative evaluation* of which makes it possible

Table 5.**14** The accuracy of angiographically controlled studies using conventional and color flow duplex sonography in diagnosing stenoses of diverse grades and occlusions of the pelvic and leg arteries

Author, year	Patients (n)	Segments (n)	Sensitivity (%)	Specificity (%)	Threshold value for the measurement procedure
Conventional duplex sonography					
Jäger et al. (1985)[1]	30	338	77	98	Qualitative spectrum, PSV > 100 %
Kohler et al. (1987a)[1]	32	393	82	92	Qualitative spectrum, PSV > 100 %
Color flow duplex sonography					
Stenoses > 50 % + occlusions					
Cossman et al. (1989)[2]	61	629	87	99	PSV > 200 cm/s
Koennecke et al. (1989)[3]	53	344	97	96	PSV 30–100 %
Landwehr and Lackner (1990)[3]	52	132	92	99	Qualitative
Polak et al. (1990a)[1]	17	238	88	95	PSV > 100 %
Legemate et al. (1991)[4]	61	918	84	96	PVR 2.5
Ranke et al. (1992)[1]	62	121	87	94	PVR > 2.4
	62	121	66	80	PSV > 180 cm/s
Strauss et al. (1993b)[1]	598	1460	88	86	PSV >100 %
Occlusions					
Cossman et al. (1989)[5]	61	560	81	99	–
Legemate et al. (1991)[5]	61	105	91	99	–
Karasch et al. (1993)[5]	94	150	98	–	–

1. Obstructions of the common femoral artery up to and including the distal popliteal artery > 50 %.
2. Obstructions of the iliac artery up to and including the trifurcation > 30 %.
3. Obstructions of the iliac artery up to and including two arteries of the lower leg > 20 %.
4. Aorta up to and including the popliteal artery > 50 %.
5. Femoropopliteal occlusions.

PSV: peak systolic velocity; PVR: peak velocity ratio.

Table 5.**15** The validity of angiographically controlled studies using conventional and color flow duplex sonography in diagnosing stenoses > 50 % and occlusions pertaining to vascular segments of the pelvic and leg arteries

Arteries (segments)	Conventional duplex				Color flow duplex							
	Jäger et al. (1985) (n = 338)		Kohler et al. (1987a) (n = 393)		Cossman et al. (1989) (n = 629)		Koennecke et al. (1989) (n = 344)*		Legemate et al. (1991) (n = 918)		Moneta et al. (1992) (n = 286)	
	Sensitivity (%)	Specificity (%)	Sensitivity (%)	Specificity (%)	Sensitivity (%)	Specificity (%)	Sensitivity (%)	Specificity (%)	Sensitivity (%)	Specificity (%)	Sensitivity (%)	Specificity (%)
Iliac	81	100	89	90	81	98	96	98	79/92[2]	94/96[2]	89	99
Common femoral	56	46	67	98	80	100	–	–	57	98	76	99
Superficial femoral	76	97	84	93	91	96	98	100	73/100/92[3]	96/98/95[3]	87	98
Deep femoral	86	100	67	81	80	100	–	–	71	94	83	97
Popliteal	80	100	75	97	90	100	88	98	75	93	67	99
Lower leg	–	–	–	–	–	–	95/94[1]	92	–	–	90	93/92[1]
All segments	77	98	82	92	87	99	97	96	84	96	–	–

1. Anterior tibial artery/posterior tibial artery.
2. Common iliac artery/external.
3. Proximal superficial femoral artery, middle, distal.

* Obstructions of all grades.

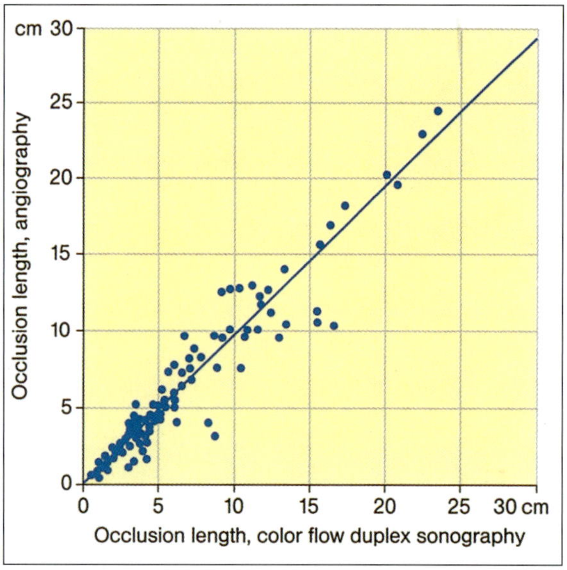

Fig. 5.**53** The correlation between the length (in centimeters) of an occlusion in the superficial femoral artery or popliteal artery, measured angiographically and sonographically (using color flow duplex mode), based on data from 98 lower extremities; $y = 0.96\,x + 0.27$; correlation coefficient $r = 0.95$; significance level $P = 0.001$

to assess the presence of preceding or subsequent vascular system obstructions.

Registering what is known as the *groin pulse curve* is particularly important, since cuff methods cannot be used in the pelvic region, and quantitative pressure measurements are therefore not possible.

Segmental flow velocity curves are valuable in *diagnosing the level* of vascular system stenoses. The examination can be carried out more quickly than segmental pressure measurement.

A normally configured triphasic flow pulse in the groin (common femoral artery), or at other registration points, excludes the presence of a hemodynamically active flow obstruction in the preceding afferent arteries.

The grade of preceding vascular system obstructions can be estimated by formal analysis (Fig. 5.**38**). Using visual interpretation in a prospective study, Walton et al. (1984) demonstrated an 87% sensitivity in occlusions and stenoses with a lumen obstruction of more than 50% in the aortoiliac region. By contrast, stenoses of less than 50% cannot be distinguished from subsequent femoral artery occlusions using formal analysis (Neuerburg-Heusler et al. 1981).

According to comparison of several indices completed by Humphries et al. (1980), all individually evaluated *quantitative* measurements that were employed—i.e., parameters such as the pulsatility index, the "damping factor," the pulse wave duration time, and the pulse increase time—failed to distinguish between normal and low-grade pathological findings. By contrast, Fitzgerald and Carr (1977), combining several indices, were able to obtain a sensitivity of 93% in detecting obstructions in 187 arterial segments.

When examining the aortoiliac level alone, Nicolaides et al. (1976) arrived at a sensitivity of 86%. For the pulsatility index, Johnston et al. (1984) measured a sensitivity of 96% when no subsequent vascular system obstruction was present. When the femoropopliteal area was also affected, the sensitivity dropped to 83%. According to this study, in which several indices were compared, the pulsatility index is the value that has the best hit rate and a high specificity of 98%.

Our own results evaluating the PI (pulsatility index), obtained from different sets of findings, showed that for isolated stenoses in the pelvic arteries, a sensitivity of 80% is possible, while for occlusions at two levels a sensitivity of 100% could be obtained. Subsequent femoral artery occlusions in a patent pelvic vascular system could only be suspected with a sensitivity of 41%. The total overall specificity was 88%. In this study, the pulsatility index of 5.0 was used as the threshold value between normal and pathological findings (Neuerburg et al. 1989).

Stenosis Grading

Currently, there are still no unanimously accepted criteria for grading a stenosis; this contrasts with the extremely exact stenosis grade determination in the cerebral arteries (Widder et al. 1986). In general, a doubling of the maximum systolic velocity is interpreted as indicating a stenosis of more than 50% (Moneta et al. 1993). By contrast, Leng et al. (1993) believed that one can assume a stenosis of more than 50% only when the peak velocity has undergone an increase of over 300%. Allard et al. (1991), who evaluated many velocity waveform parameters in 379 segments, arrived at a sensitivity of only 50%, with a specificity of 98%.

In a paper by Sacks et al. (1992), an attempt was made to distinguish between 50%, 70%, and 90% stenoses by evaluating 558 segments using a quotient formed from the flow velocity 2.0 cm proximal to the stenosis and the flow velocity within the stenosis itself, known as the *peak velocity ratio* (PVR). The results obtained (stenosis over 50%: 71% sensitivity, stenosis over 70%: 80% sensitivity, stenosis over 90%: 85% sensitivity, with a specificity of between 93% and 97%) are not yet comparable to those from the carotid artery vascular system. Ranke et al. (1992) evaluated the sensitivity and specificity of stenosis grading in 10% increments between 20% and 90%, and found sensitivities between 84% and 96%, with specificities ranging from 75% to 97%. They examined the peak systolic velocity within the stenosis and also the peak velocity ratio in relation to the velocity both in the proximal and the distal segment (Table 5.**14**). However, as mentioned above, angiographic evaluation of the degree of a stenosis as a gold standard is problematic, since imaging of the pelvic and leg regions is usually only practicable in a single plane.

In a study by de Smet et al. (1996), various parameters for the Doppler curve (PSV, PSV ratio, PSV difference, EDV) were investigated in 112 aortoiliac segments in comparison with arteriographic diameter reduction and invasive blood-pressure measurement. The best result was achieved with the PSV ratio in detecting aortoiliac stenoses of more than 50%, with a threshold value of 2.8, giving a sensitivity of 86% and a specificity of 84%. Arteriographic stenoses of more than 75% were measured by a PSV ratio of 5, although with a lower sensitivity of 65% and a higher specificity of 91%.

A different method of measuring the stenosis grade is to *calculate the pressure gradient* according to the Bernoulli equation (analogous to measuring the pressure gradient at the aortic valve). However, this equation can only be used in larger arteries, such as the iliac vascular system.

Langsfeld et al. (1988) were able to attain a correlation coefficient of $r = 0.9$ in 11 patients. They believe that this method provides the best correlation, especially for higher-grade stenoses with a peak velocity of between 2 m/s and 4.5 m/s. In 28 patients, Strauss et al. (1993a) obtained a correlation coefficient of $r = 0.8$ when comparing the mean pressure decrease through stenoses. Legemate et al. (1993) evaluated the sensitivity and specificity of this procedure with 81% and 88% values, respectively, for hemodynamically significant stenoses. In addition to measuring the pressure gradient according to Bernoulli, they also compared the difference between the maximum systolic velocity in the stenosis and the lowest maximum velocity proximal or distal to the stenosis. For this parameter, they obtained the same sensitivity and specificity rates—81% and 88%, respectively. In this study, the correlation with invasively measured pressures was $r = 0.62$ for the Bernoulli measurement and $r = 0.72$ for the maximum velocity measurement.

Determining Occlusion Length

In some studies, separate evaluation of the *differentiation between stenosis and occlusion* is not carried out. However, particularly in relation to therapeutic lumen recanalization techniques such as angioplasty, information regarding the exact location and length of an occlusion is very significant. In some studies, detecting occlusions was successful with a sensitivity of 81–98% and a specificity of 99% (Table 5.**14**).

A more recent study by our group, which evaluated 100 extremities in 94 patients, showed that determining the *femoral artery occlusion* length with color flow duplex sonography gave a correlation coefficient of $r = 0.95$, fully comparable with the results obtained using angiography (Karasch et al. 1993) (Fig. 5.**53**). Segmental classification of the location of the occlusion was also successful in 95% of the cases. It is expected that purely diagnostic angiographic procedures will increasingly lose their importance, or that they will only be selectively used in combination with angioplasty.

With regard to the validity of magnetic resonance angiography (MRA) in comparison with conventional intra-arterial angiography, data are available from studies published in the mid-1990s. In comparison with intraoperative angiography (as the gold standard), the sensitivity of MRA and preoperative conventional angiography for distinguishing between patent and occluded vascular segments was 83% and 85%, respectively (Baum et al. 1995). Both procedures had a specificity of 81%. For recognizing vascular segments suitable for bypasses, the sensitivity was 82% and 77%, respectively. The specificity of MRA in this setting, at 84%, was much lower than that of preoperative angiography, at 92%.

In two early direct comparisons of magnetic resonance angiography with duplex-sonographic findings, using conventional angiography as the control, duplex sonography was found to be superior. There was agreement with the angiographic findings in 71% and 81% of cases, respectively, and for duplex sonography in 93% and 88% of cases (Mulligan et al. 1991, Baumgartner et al. 1993). In view of the marked technical improvements that have been made in the field of magnetic resonance angiography in recent years, as well as in color duplex sonography, magnetic resonance contrast agents, and high-field devices, it can be expected that the accuracy of the two procedures will continue to increase (Ho et al. 1998). Limiting factors on routine and generalized use of MRA are currently still its relatively poor availability and the high costs of the examination, so that noninvasive color duplex-sonographic examinations are likely to maintain their outstanding diagnostic value for some considerable time to come.

Aneurysms

A further area of application for duplex sonography is in recognizing and differentiating *aneurysms*; differentiating a true aneurysm from a false aneurysm is clearly successful with color coding (pp. 186–188). The incidence of *false aneurysms* after diagnostic and therapeutic catheterization procedures has been examined in various larger studies using color flow duplex sonography. In 1120 patients following cardiac catheterization examinations, Moll et al. (1991) found a false aneurysm in 4% of the cases. After diagnostic catheterization procedures, Habscheid and Landwehr (1989) discovered pseudoaneurysms in 5% of the cases. After 144 coronary angioplasties, eight aneurysms (6%) were diagnosed by Kresowik et al. (1991). By contrast, when symptomatic patients were examined in whom femoral artery lesions were suspected following catheter procedures, the rate of pseudoaneurysms was I the range of 35–43% (Gross-Fengels et al. 1987, Steinkamp et al. 1992).

Ultrasound-guided compression closure of pseudoaneurysms, first introduced by Fellmeth et al. (1991), is now in widespread use, although in some studies (Kresowik et al. 1991, Johns et al. 1991), spontaneous thromboses without any subsequent complications were observed. The success rate of compression therapy was found by Sorrell et al. (1993) to be 91%.

In our own hospital, 50 pseudoaneurysms were diagnosed using color duplex sonography in the year

2000, in patients with postinterventional symptoms after 2710 coronary angiography examinations, including 1623 catheter interventions (percutaneous transluminal coronary angioplasty, stent implantation, intravascular ultrasound, aortic valvuloplasty). This represents a postinterventional complication rate of 1.85% for pseudoaneurysms.

In a prospective long-term study, Hajarizadeh et al. (1995) achieved an overall success rate of 95%. During an 18-month follow-up period, there were no late recurrences or significant changes in ankle brachial pressure. Even with uninterrupted anticoagulation, Dean et al. (1996) achieved a satisfactory success rate of 73%.

Angioplasty and Stenting

Changes in vascular diameter and in the vessel walls are increasingly being examined post angioplasty directly and in long-term studies using the duplex procedure. In the study by Mewissen et al. (1992), the open lumen rate after 6–12 months in patients with normal findings after angioplasty was 84%, while only 15% with a velocity ratio ≥ 2.0 after percutaneous transluminal angioplasty (PTA) showed no occlusions or residual stenoses. In the study by Sacks et al. (1994), the open lumen rates with a velocity ratio of less than 2.0 and with a velocity ratio of 2.0 or more were 54% and 74%, respectively. Henderson et al. (1994) showed that in 98% of 49 lesions, a significant increase in lumen diameter was seen immediately after angioplasty. The overall diameter of the vessel, including the vessel wall, was increased in 87.7%. The mean change in the lumen diameter was 2.1 mm, and the mean change in the overall vessel diameter was 1.6 mm; after 6 months, these figures were 1.1 mm and 1.2 mm, respectively.

In a study by Spijkerboer et al. (1996) on iliac stenoses, no significant difference was found between PTA results with "residual stenoses" (≥ 2.5 PVR) and those of 2.5 PVR or less, although the boundary value was set fairly high. A large proportion of the residual stenoses showed significant improvement within one year of PTA—an observation that had already been made in angiographic and clinical check-ups before the duplex period. Baumgartner et al. (1996) also confirmed that a residual stenosis of less than 50% after endovascular treatment was not associated with an increased risk of recurrent stenosis during the follow-up period, but they found that a hypoechogenic endoluminal wall thickening immediately after endovascular procedures appears to be predictor of recurrent stenosis.

Vascular Prostheses and Bypasses

The patency of bypasses and vascular prostheses, and also the prediction of *bypass reocclusions,* have been examined in several studies using simple continuous-wave Doppler sonography. Bandyk et al. discovered in 1985 that a reduction of the maximum systolic velocity to less than 40 cm/s indicates an impending bypass occlusion. In addition, they found in a later study (1988) that a transformation of the triphasic Doppler waveform into a monophasic configuration, when related to a decreased velocity, indicates proximal or distal segmental stenoses between 50% and 75%. The staccato configuration of the Doppler waveform with low physiological flow and pronounced reverse flow, was viewed as being an additional criterion associated with distal bypass stenoses.

Grigg et al. (1988) viewed a very low velocity ratio of 1.5 as being the threshold value for detecting stenoses of more than 50% in bypasses. In 14 patients with 92 graft segments, Polak et al. (1990a) documented a 95% sensitivity and a 100% specificity for detecting and locating stenoses using color flow duplex sonography; a peak velocity ratio of more than 2.0 served as the threshold value.

In a prospective study involving 54 patients with 62 bypasses, Trattnig et al. (1992) were able to detect a bypass stenosis by purely qualitative optical means, using the lightening of the color coding. These angiographically controlled cases produced a sensitivity of 92% and a specificity of 100%, but without determining the stenosis grade.

Follow-up studies conducted by Idu et al. (1993) and Mattos et al. (1993), using color flow duplex sonography, also showed that the increased flow velocity criterion within a stenosis is a more important parameter for indicating imminent bypass occlusions than decreased velocity. Mattos et al. (1993) were able to demonstrate that flow velocity doubling, or a peak velocity ratio of more than 2.0, requires a bypass revision. They also showed that the patency rates of revised stenoses in comparison with stenoses that had not undergone revision were 88% and 63%, respectively.

A similar observation was made by Idu et al. (1993) in a follow-up study over five years. In evaluating stenoses of more than 50% with color flow duplex sonography, it was shown that bypass occlusions occurred in 57% of the unrevised stenoses, and only in 9% of the revised grafts. With B-mode image and color coding, other postoperative bypass complications, such as aneurysms, hematomas, and seromas can also be quickly diagnosed.

The studies mentioned show that color flow duplex sonography has clearly improved the potential for noninvasive diagnosis of bypass stenoses and impending reocclusions, and that a threshold value for the peak velocity ratio at 2.0 currently appears to be the most reliable parameter for indicating stenoses of more than 50%. Since early bypass revision clearly produces an increased patency rate (Mattos et al. 1993), systematic follow-up bypass evaluations using duplex sonography appear to be indicated postoperatively.

Upper Body Arterial Occlusive Disease

This is clearly less important, since those suffering from peripheral vascular disorders show an affected upper body region in only 5–10% of cases (Edwards and Porter 1993). In the *large arteries*, subclavian artery stenoses and subclavian artery occlusions occur most frequently. The significance of these is discussed along with the cerebral arteries in Chapter 2. Takayasu's syndrome or other inflammatory vascular diseases can sometimes cause extensive brachial artery stenoses or other arterial stenoses in the arm.

When there are disturbances in the circulation of the *small arteries of the hands*, a distinction needs to be made between a functional Raynaud syndrome and acrovascular occlusions. The latter have so far been diagnosed more often using pulse-registering procedures, or by measuring the pressure in the digital arteries of the hand. However, high-resolution transducers now allow the imaging of individual digital arteries.

In a study by Langholz et al. (1997) including 450 digital arteries (45 hands), the sensitivity and specificity of color duplex sonography for diagnosing digital artery stenoses, in comparison with selective angiography of the hand, were 86.3% and 93.1%, respectively. In a later study, the same research group (Ladleif et al. 1998) also achieved a sensitivity of 86.9% (95% CI, 81–91%) and a specificity of 93.8% (95% CI, 91–96%) in a later total of 500 digital arteries examined (50 hands, 46 patients; digital artery occlusions in 34%). The positive predictive value and negative predictive value were 88.4% and 93%, respectively. Flow velocities of less than 15 cm/s indicate preceding stenoses or occlusions (Steins et al. 1998). In rare cases, arteriovenous malformations and hemangiomas can be identified in the digital arteries (Hübsch and Trattnig, in Frühwald and Blackwell 1992).

References

Abu-Yousef MM, Wiese JA, Shamma AR. The “to-and-fro” sign: duplex Doppler evidence of femoral artery pseudoaneurysm. AJR Am J Roentgenol 1988;150:632–4.

Ahn SS, Rutherford RB, Becker GJ, Comerota AJ, Johnston KW, McClean GK, et al. Reporting standards for lower extremity arterial endovascular procedures. Society for Vascular Surgery/International Society for Cardiovascular Surgery. J Vasc Surg 1993;17:1103–7.

Allard L, Langlois YE, Durand LG, Roederer GO, Beaudoin M, Cloutier G, et al. Computer analysis and pattern recognition of Doppler blood flow spectra for disease classification in the lower limb arteries. Ultrasound Med Biol 1991;17:211–23.

Allard L, Cloutier G, Durand LG, Roederer GO, Langlois YE. Limitations of ultrasonic duplex scanning for diagnosing lower limb arterial stenoses in the presence of adjacent segment disease. J Vasc Surg 1994;19:650–7.

Amendt K. Takayasu-Arteriitis. In: Amendt K, Diehm C, editors. Handbuch akrale Durchblutungsstörungen. Epidemiologie, Pathogenese, Diagnostik und Therapie. Heidelberg: Johann Ambrosius Barth Verlag, 1998: 153–84.

Arko F, Mettauer M, McCollough R, Patterson D, Manning L, Lee S, et al. Use of intravascular ultrasound improves long-term clinical outcome in the endovascular management of atherosclerotic aortoiliac occlusive disease. J Vasc Surg 1998;27:614–23.

Avena R, Curry KM, Sidawy AN, Simpkins JF, Neville RF, Mitchell ME, et al. The effect of occult diabetic status and oral glucose intake on brachial artery vasoactivity in patients with peripheral vascular disease. Cardiovasc Surg 1998;6:584–9.

Baker AR, Prytherch DR, Evans DH, Bell PR. Doppler ultrasound assessment of the femoro-popliteal segment: comparison of different methods using ROC curve analysis. Ultrasound Med Biol 1986;12:473–82.

Baker JD. Poststress Doppler ankle pressures: a comparison of treadmill exercise with two other methods of induced hyperemia. Arch Surg 1978;113:1171–3.

Bandyk DF. Monitoring during and after distal arterial reconstruction. In: Bernstein EF, editor. Vascular diagnosis. St. Louis: Mosby, 1993: 579–87.

Bandyk DF, Cato RF, Towne JB. A low flow velocity predicts failure of femoropopliteal and femorotibial bypass grafts. Surgery 1985;98:799–809.

Bandyk DF, Seabrook GR, Moldenhauer P, Lavin J, Edwards J, Cato R, et al. Hemodynamics of vein graft stenosis. J Vasc Surg 1988;8:688–95.

Bandyk DF, Bergamini TM, Towne JB, Schmitt DD, Seabrook GR. Durability of vein graft revision: the outcome of secondary procedures. J Vasc Surg 1991;13:200–8; discussion 209–10.

Bandyk DF, Mills JL, Gahtan V, Esses GE. Intraoperative duplex scanning of arterial reconstructions: fate of repaired and unrepaired defects. J Vasc Surg 1994;20:426–32; discussion 432–3.

Barras JP, Bollinger A. Korrelation zwischen transkutaner (Ultraschall-Doppler-Technik) und intraarterieller (Mikromanometrie) Messung des Fussarteriendruckes. In: Kriessmann A, Bollinger A, editors. Ultraschall-Doppler-Diagnostik in der Angiologie. Stuttgart: Thieme, 1978

Baum RA, Rutter CM, Sunshine JH, Blebea JS, Blebea J, Carpenter JP, et al. Multicenter trial to evaluate vascular magnetic resonance angiography of the lower extremity. American College of Radiology Rapid Technology Assessment Group. JAMA 1995;274:875–80.

Baumgartner I, Maier SE, Koch M, Schneider E, von Schulthess GK, Bollinger A. [Magnetic resonance arteriography, duplex sonography and conventional arteriography for the evaluation of peripheral arterial occlusive disease; in German.] RöFo 1993;159:167–73.

Baumgartner I, Zwahlen I, Do DD, Redha F, Mahler F. Color-coded duplex sonography for evaluation of femoro-popliteal restenosis after percutaneous catheter atherectomy and subsequent transluminal balloon angioplasty. J Vasc Invest 1996;2:125–30.

Belkin M, Schwartz LB, Donaldson MC, Mannick JA, Whittemore AD. Hemodynamic impact of vein graft stenoses and their prediction in the vascular laboratory. J Vasc Surg 1997;25:1016–21; discussion 1022.

Bernstein EF. Vascular diagnosis. St. Louis: Mosby, 1993.

Birrer M, Baumgartner I. Work-related vascular injuries of the hand: hypothenar hammer syndrome. N Engl J Med 2002;347:339.

Bollinger A, Mahler F, Zehender O. [Combined pressure and flow measurements in the evaluation of arterial circulatory disorders; in German.] Dtsch Med Wochenschr 1970;95:1039–43.

Bollinger A, Schlumph M, Butti P, Gruntzig A. Measurement of systolic ankle blood pressure with Doppler ultrasound at rest and after exercise in patients with leg artery occlusions. Scand J Clin Lab Invest Suppl 1973;128:123–8.

Bollinger A, Barras JP, Mahler F. Measurement of foot artery blood pressure by micromanometry in normal subjects and in patients with arterial occlusive disease. Circulation 1976;53:506–12.

Bowers BL, Valentine RJ, Myers SI, Chervu A, Clagett GP. The natural history of patients with claudication with toe pressures of 40 mmHg or less. J Vasc Surg 1993;18:506–11.

Buchholz J, Scherf F, Bucker-Nott HJ, Hilgenberg J. [Iatrogenic femoral pseudoaneurysm–an urgent surgical indication; in German.] Vasa 1991;20:261–6.

Busse R, editor. Kreislaufphysiologie. Stuttgart: Thieme, 1982.

Busse R, Wetterer E, Bauer RD, Pasch T, Summa Y. The genesis of the pulse contours of the distal leg arteries in man. Pflügers Arch 1975;360:63–79.

Buth J, Disselhoff B, Sommeling C, Stam L. Color-flow duplex criteria for grading stenosis in infrainguinal vein grafts. J Vasc Surg 1991;14:716–26; discussion 726–8.

Cachovan M, Maass U. Vergleichende Untersuchungen zur Beurteilung der Gehstrecke bei Claudicatio intermittens. In: Mahler F, Nachbur B, editors. Zerebrale Ischämie. Berne: Huber, 1984.

Calligaro KD, Syrek JR, Dougherty MJ, Rua I, McAffee-Bennett S, Doerr KJ, et al. Selective use of duplex ultrasound to replace preoperative arteriography for failing arterial vein grafts. J Vasc Surg 1998;27:89–94; discussion 94–5.

Carter SA. Clinical measurement of systolic pressures in limbs with arterial occlusive disease. JAMA 1969;207:1869–74.

Carter SA. Response of ankle systolic pressure to leg exercise in mild or questionable arterial disease. N Engl J Med 1972;287:578–82.

Carter SA. Role of pressure measurement. In: Bernstein EF, editor. Vascular diagnosis. St. Louis: Mosby, 1993: 486–512.

Cluley SR, Brener BJ, Hollier L, Schoenfeld R, Novick A, Vilkomerson D, et al. Transcutaneous ultrasonography can be used to guide and monitor balloon angioplasty. J Vasc Surg 1993;17:23–30; discussion 30–1.

Cobet U, Scharf R, Klemenz A, Millner R, Blumenstein G, Wiegand E. Rechnergestütztes, richtungssensitives Ultraschall-Dopplersystem zur quantitativen Blutflusscharakterisierung in Arterien. Z Klin Med 1986;41:523–7.

Coleman SS, Anson BJ. Arterial patterns in the hand based upon a study of 650 specimens. Surg Gynecol Obstet 1961;113:409–24.

Corretti MC, Plotnick GD, Vogel RA. Smoking correlates with flow-mediated brachial artery vasoactivity but not cold pressor vasoactivity in men with coronary artery disease. Int J Card Imaging 1998;14:11–7.

Cossman DV, Ellison JE, Wagner WH, Carroll RM, Treiman RL, Foran RF, et al. Comparison of contrast arteriography to arterial mapping with color-flow duplex imaging in the lower extremities. J Vasc Surg 1989;10:522–8; discussion 528–9.

Coughlin BF, Paushter DM. Peripheral pseudoaneurysms: evaluation with duplex US. Radiology 1988;168:339–42.

Crawford ES, Hess KR. Abdominal aortic aneurysm. N Engl J Med 1989;321:1040–2.

de Smet AA, Visser K, Kitslaar PJ. Duplex scanning for grading aortoiliac obstructive disease and guiding treatment. Eur J Vasc Surg 1994;8:711–5.

de Smet AA, Ermers EJ, Kitslaar PJ. Duplex velocity characteristics of aortoiliac stenoses. J Vasc Surg 1996;23:628–36.

Dean SM, Olin JW, Piedmonte M, Grubb M, Young JR. Ultrasound-guided compression closure of postcatheterization pseudoaneurysms during concurrent anticoagulation: a review of seventy-seven patients. J Vasc Surg 1996;23:28–34, discussion 34–5.

Diehm C. [Early education of peripheral arterial occlusive disease patients regarding athero-thrombotic risks; in German.] Fortschr Med 1998;116:40–1.

Do DD, Zehnder T, Mahler F. [Color-coded Doppler sonography in iatrogenic spurious aneurysms in the groin; in German.] Dtsch Med Wochenschr 1993;118:656–60.

Dousset V, Grenier N, Douws C, Senuita P, Sassouste G, Ada L, et al. Hemodialysis grafts: color Doppler flow imaging correlated with digital subtraction angiography and functional status. Radiology 1991;181:89–94.

Dunant JH, Eugenidis N. Cystic degeneration of the popliteal artery. Vasa 1973;2:156–9.

Edwards JM, Porter JM. Evaluation of upper extremity ischemia. In: Bernstein EF, editor. Vascular diagnosis. St. Louis: Mosby, 1993: 630–40.

Eichlisberger R, Frauchiger B, Schmitt H, Jäger K. [Pseudoaneurysm following arterial catheterization: diagnosis and follow-up; in German.] Ultraschall Med 1992;13:54–8.

Fellmeth BD, Roberts AC, Bookstein JJ, Freischlag JA, Forsythe JR, Buckner NK, et al. Postangiographic femoral artery injuries: nonsurgical repair with US-guided compression. Radiology 1991;178:671–5.

Fields WS, Maslenikov V, Meyer JS, Hass WK, Remington RD, Macdonald M. Joint study of extracranial arterial occlusion, V: progress report of prognosis following surgery or nonsurgical treatment for transient cerebral ischemic attacks and cervical carotid artery lesions. JAMA 1970;211:1993–2003.

Fitzgerald DE, Carr J. Peripheral arterial disease: assessment by arteriography and alternative noninvasive measurements. AJR Am J Roentgenol 1977;128:385–8.

Flanigan DP, Burnham SJ, Goodreau JJ, Bergan JJ. Summary of cases of adventitial cystic disease of the popliteal artery. Ann Surg 1979;189:165–75.

Flanigan DP, Gray B, Schuler JJ, Schwartz JA, Post KW. Correlation of Doppler-derived high thigh pressure and intra-arterial pressure in the assessment of aorto-iliac occlusive disease. Br J Surg 1981;68:423–5.

Flückiger F, Steiner H, Rabl H, Waltner F. [Cystic adventitia degeneration of the popliteal artery: sonographic confirmation of the diagnosis; in German.] Ultraschall Med 1991;12:84–6.

Franzeck UK, Bernstein EF, Fronek A. The effect of sensing site on the limb segmental blood pressure determination. Arch Surg 1981;116:912–6.

Fronek A. Noninvasive diagnostics in vascular disease. New York: McGraw-Hill, 1989.

Fronek A, Johansen KH, Dilley RB, Bernstein EF. Noninvasive physiologic tests in the diagnosis and characterization of peripheral arterial occlusive disease. Am J Surg 1973a;126:205–14.

Fronek A, Johansen K, Dilley RB, Bernstein EF. Ultrasonographically monitored postocclusive reactive hyperemia in the diagnosis of peripheral arterial occlusive disease. Circulation 1973b;48:149–52.

Fronek A, Coel M, Bernstein EF. Quantitative ultrasonographic studies of lower extremity flow velocities in health and disease. Circulation 1976;53:957–60.

Fronek A, Coel M, Bernstein EF. The importance of combined multisegmental pressure and Doppler flow velocity studies in the diagnosis of peripheral arterial occlusive disease. Surgery 1978;84:840–7.

Frühwald F, Blackwell DE. Atlas of color-coded Doppler sonography: vascular and soft tissue structures of the upper extremity, thoracic outlet, and neck. Vienna: Springer, 1992.

Gale SS, Scissons RP, Salles-Cunha SX, Dosick SM, Whalen RC, Pigott JP, et al. Lower extremity arterial evaluation: are segmental arterial blood pressures worthwhile? J Vasc Surg 1998;27:831–8; discussion 838–9.

Gehling G, Werner D, Schmidt A, Ludwig J, Daniel WG. POP – Perkutane Okklusion von Pseudoaneurysmen der A. femoralis: ein neues minimalinvasives Verfahren. Vasa 1999;28(Suppl 55):96.

Gooding GA, Perez S, Rapp JH, Krupski WC. Lower-extremity vascular grafts placed for peripheral vascular disease: prospective evaluation with duplex Doppler sonography. Radiology 1991;180:379–86.

Gosling RG, King DH. Continuous wave ultrasound as an alternative and complement to X-rays in vascular examinations. In: Reneman R, editor. Cardiovascular applications of ultrasound. Amsterdam: North-Holland, 1974.

Gosling RG, King DH. Processing arterial Doppler signals for clinical data. In: de Vlieger M, et al. Handbook of clinical ultrasound. New York: Wiley, 1978.

Grigg MJ, Nicolaides AN, Wolfe JH. Detection and grading of femorodistal vein graft stenoses: duplex velocity measurements compared with angiography. J Vasc Surg 1988;8:661–6.

Gross-Fengels W, Beyer D, Lorenz R, Kristen R. [Imaging of iatrogenic aneurysms and AV fistulas of the lower extremity with IV-DSA and sonography; in German.] Röntgenblätter 1987;40:131–6.

Grüntzig A, Schlumpf M. The validity and reliability of post-stenotic blood pressure measurement by Doppler ultrasonic sphygmomanometry. Vasa 1974;3:65–71.

Gundersen J. Segmental measurements of systolic blood pressure in the extremities including the thumb and the great toe. Acta Chir Scand Suppl 1972;426:1–90.

Habscheid W, Landwehr P. [Pseudoaneurysm of the femoral artery following heart catheterization: a prospective ultrasound study; in German.] Z Kardiol 1989;78:573–7.

Hafferl A. Lehrbuch der topographischen Anatomie. Berlin: Springer, 1969.

Hajarizadeh H, LaRosa CR, Cardullo P, Rohrer MJ, Cutler BS. Ultrasound-guided compression of iatrogenic femoral pseudoaneurysm failure, recurrence, and long-term results. J Vasc Surg 1995;22:425–30; discussion 430–3.

Hartmann I. [dissertation]. Aachen, Germany: University of Aachen, 2002.

Hass WK, Fields WS, North RR, Kircheff II, Chase NE, Bauer RB. Joint study of extracranial arterial occlusion, 2: arteriography, techniques, sites, and complications. JAMA 1968;203:961–8.

Hatsukami TS, Primozich J, Zierler RE, Strandness DE Jr. Color Doppler characteristics in normal lower extremity arteries. Ultrasound Med Biol 1992;18:167–71.

Heberer G, Rau G, Schoop W, editors. Angiologie. Grundlagen, Klinik und Praxis. Stuttgart: Thieme, 1974.

Helvie MA, Rubin JM, Silver TM, Kresowik TF. The distinction between femoral artery pseudoaneurysms and other causes of groin masses: value of duplex Doppler sonography. AJR Am J Roentgenol 1988;150:1177–80.

Henderson J, Chambers J, Jeddy TA, Chamberlain J, Whittingham TA. Serial investigation of balloon angioplasty induced changes in the superficial femoral artery using colour duplex ultrasonography. Br J Radiol 1994;67:546–51.

Hepp W, Pallua N. Das Anastomosenaneurysma der Leiste: Chirurgische Therapie und Ergebnisse. In: Sandmann W, Kniemeyer HW, Kolvenbach R, editors. Aneurysmen der grossen Arterien. Diagnostik und Therapie. Berne: Huber, 1991.

Hirai M. Arterial insufficiency of the hand evaluated by digital blood pressure and arteriographic findings. Circulation 1978;58:902–8.

Ho KY, Leiner T, de Haan MW, Kessels AG, Kitslaar PJ, van Engelshoven JM. Peripheral vascular tree stenoses: evaluation with moving-bed infusion-tracking MR angiography. Radiology 1998;206:683–92.

Hosten N, Puls R, Sahimbas O, Balzer J, Urbank A, Felix R. [Color Doppler ultrasonography in peripheral artery occlusive disease: continuous application of a signal enhancer; in German.] RöFo 1998;169:495–8.

Hughes MJ, McCall JM, Nott DM, Padley SP. Treatment of iatrogenic femoral artery pseudoaneurysms using ultrasound-guided injection of thrombin. Clin Radiol 2000;55:749–51.

Humphries KN, Hames TK, Smith SW, Cannon VA, Chant AD. Quantitative assessment of the common femoral to popliteal arterial segment using continuous wave Doppler ultrasound. Ultrasound Med Biol 1980;6:99–105.

Idu MM, Blankenstein JD, de Gier P, Truyen E, Buth J. Impact of a color-flow duplex surveillance program on infrainguinal vein graft patency: a five-year experience. J Vasc Surg 1993;17:42–52; discussion 52–3.

Idu MM, Buth J, Hop WC, Cuypers P, van de Pavoordt ED, Tordoir JM. Vein graft surveillance: is graft revision without angiography justified and what criteria should be used? J Vasc Surg 1998;27:399–411; discussion 412–3.

Insua JA, Young JR, Humphries AW. Popliteal artery entrapment syndrome. Arch Surg 1970;101:771–5.

Jäger KA, Landmann J, editors. Praxis der angiologischen Diagnostik. Stufendiagnostik und rationelles Vorgehen bei arterieller und venöser Durchblutungsstörung. Berlin: Springer, 1994.

Jäger KA, Phillips DJ, Martin RL, Hanson C, Roederer GO, Langlois YE, et al. Noninvasive mapping of lower limb arterial lesions. Ultrasound Med Biol 1985;11:515–21.

Johns JP, Pupa LE Jr, Bailey SR. Spontaneous thrombosis of iatrogenic femoral artery pseudoaneurysms: documentation with color Doppler and two-dimensional ultrasonography. J Vasc Surg 1991;14:24–9.

Johnson EL, Yock PG, Hargrave VK, Srebro JP, Manubens SM, Seitz W, et al. Assessment of severity of coronary stenoses using a Doppler catheter: validation of a method based on the continuity equation. Circulation 1989;80:625–35.

Johnston KW, Kassam M, Cobbold RS. Relationship between Doppler pulsatility index and direct femoral pressure measurements in the diagnosis of aortoiliac occlusive disease. Ultrasound Med Biol 1983;9:271–81.

Johnston KW, Kassam M, Koers J, Cobbold RS, MacHattie D. Comparative study of four methods for quantifying Doppler ultrasound waveforms from the femoral artery. Ultrasound Med Biol 1984;10:1–12.

Kang SS, Labropoulos N, Mansour MA, Baker WH. Percutaneous ultrasound guided thrombin injection: a new method for treating postcatheterization femoral pseudoaneurysms. J Vasc Surg 1998;27:1032–8.

Karasch T. Duplexsonographie bei peripherer arterieller Verschlusskrankheit. Vasa 1995;24(Suppl 45):27.

Karasch T. Ergometrie bei peripherer arterieller Verschlusskrankheit. In: Löllgen H, Erdman E, editors. Ergometrie. Belastungsuntersuchungen in Klinik und Praxis. Berlin: Springer, 2000a: 301–8.

Karasch T. Neue Entwicklungen in der farbkodierten Duplexsonographie. Ultraschall Med 2000b;21 (Suppl 1):25.

Karasch T. Strömungsdarstellung im B-Bild. Die B-flow-Technologie. Med Klin 2001;96(abstracts vol. I):92.

Karasch T, Rieger H. Zur Problematik des Therapiekonzeptes beim Poplitea-Aneurysma. In: Imig H, Schröder A, editors. Varizen, Poplitea-Aneurysmen. Darmstadt, Germany: Steinkopff, 1995: 97–102.

Karasch T, Schmidt R, Neuerburg-Heusler D. Aneurysma spurium mit arteriovenöser Fistel nach operativer Patella-alta-Korrektur. Ultraschall Klin Prax 1990a;5:198.

Karasch T, Veit J, Hermann F, Neuerburg-Heusler D. Duplexsonographische Messungen der maximalen und mittleren Strömungsgeschwindigkeit im interapparativen und interindividuellen Vergleich am Modell der Arteria poplitea. Vasa 1990b;19(Suppl 32):167–71.

Karasch T, Rieser R, Neuerburg-Heusler D. Validität der farbduplexsonographischen Bestimmung der Verschlusslänge und -lokalisation in der A. femoralis superficialis. Ultraschall Klin Prax 1991a;6:130.

Karasch T, Rieser R, Neuerburg-Heusler D. Bestimmung der Verschlusslänge und -lokalisation in Extremitätenarterien: Farbduplexsonographie versus Angiographie. Vasa 1991b;20(Suppl 33):295–6.

Karasch T, Rieser R, Neuerburg-Heusler D, Roth FJ. [Varicosis of the great saphenous vein as a main symptom of iatrogenic arteriovenous fistula; in German.] Dtsch Med Wochenschr 1991c;116:1871–4.

Karasch T, Rieser R, Grun B, Strauss AL, Neuerburg-Heusler D, Roth FJ, et al. [Determination of the length of the occlusion in extremity arteries: color duplex ultrasound versus angiography; in German.] Ultraschall Med 1993;14:247–54.

Karasch T, Valdes KG, Dohmann-Scheurle C, Strauss A, Roth FJ, Rieger H. Farbkodierte Duplexsonographie arterieller Strecker-Stents. Vasa 1995;24(Suppl 45):100.

Karasch T, Roth FJ, Rieger H. [Coarctation of the abdominal aorta; in German.] Dtsch Med Wochenschr 1996;121:159–64.

Kathrein H, Konig P, Weimann S, Judmaier G, Dittrich P. [Non-invasive morphologic and functional assessment of arteriovenous fistula in dialysis patients with duplex sonography; in German.] Ultraschall Med 1989;10:33–40.

Katzenschlager R, Ahmadi A, Minar E, Koppensteiner R, Maca T, Pikesch K, et al. Femoropopliteal artery: initial and 6-month results of color duplex US-guided percutaneous transluminal angioplasty. Radiology 1996;199:331–4.

Katzenschlager R, Ahmadi A, Atteneder M, Ugurluoglu A, Koppensteiner R, Minar E, et al. Colour duplex sonography-guided local lysis of occlusions in the femoro-popliteal region. Int Angiol 2000;19:250–4.

Knoblich S, Krings W, Schulte M, Neuerburg-Heusler D. Messung des systolischen Blutdruckes an den Digital-Arterien mit Hilfe der Strain-Gauge-Technik. Vasa 1986;15(Suppl 15):13.

Koennecke HC, Fobbe G, Hamed MM, Wolf KJ. [Diagnosis of arterial vascular diseases of the lower extremities with color-coded duplex sonography; in German.] RöFo 1989;151:42–6.

Köhler M, Krüpe M. [Specificity and normality of peripheral systolic pressure measurement by the ultrasound Doppler technic in healthy angiography-imaged extremities; in German.] Z Kardiol 1985;74:39–45.

Köhler M, Lösse B. [Simultaneous measurement of systolic blood pressure with the ultrasound Doppler technique and blood method in the human radial artery; in German.] Z Kardiol 1979;68:551–6.

Köhler M, Hinger HU, Zahnow W. [Evaluation of compensation in chronic arterial obstruction using the ultrasonic Doppler method and venous occlusion plethysmography; in German.] Z Kreislaufforsch 1972;61:401–12.

Kohler TR, Nance DR, Cramer MM, Vandenburghe N, Strandness DE Jr. Duplex scanning for diagnosis of aortoiliac and femoropopliteal disease: a prospective study. Circulation 1987a;76:1074–80.

Kohler TR, Nicholls SC, Zierler RE, Beach KW, Schubart PJ, Strandness DE Jr. Assessment of pressure gradient by Doppler ultrasound: experimental and clinical observations. J Vasc Surg 1987b;6:460–9.

Kosch M, Hausberg M, Kisters K, Barenbrock M. [Alterations of arterial vessel wall properties in hyperparathyroidism; in German.] Med Klin (Munich) 2000a;95:267–72.

Kosch M, Hausberg M, Vormbrock K, Kisters K, Rahn KH, Barenbrock M. Studies on flow-mediated vasodilation and intima-media thickness of the brachial artery in patients with primary hyperparathyroidism. Am J Hypertens 2000b;13:759–64.

Kresowik TF, Khoury MD, Miller BV, Winniford MD, Shamma AR, Sharp WJ, et al. A prospective study of the incidence and natural history of femoral vascular complications after percutaneous transluminal coronary angioplasty. J Vasc Surg 1991;13:328–33; discussion 333–5.

Kröger K, Buss C, Renzing-Kohler K, Santosa F, Rudofsky G. Segmental manifestation of peripheral atherosclerosis and its association to risk factors. Vasa 2000;29:199–203.

Ladleif et al. 1998

Landwehr P, Lackner K. [Color-coded duplex sonography before and after PTA of the arteries of the lower extremities; in German.] RöFo 1990;152:35–41.

Landwehr P, Tschammler A, Schaefer RM, Lackner K. [The value of color-coded duplex sonography of a dialysis shunt; in German.] RöFo 1990;153:185–91.

Langholz J, Stolke O, Heidrich H, Behrendt C, Blank B, Feáler B. Farbkodierte Duplexsonographie von Unterschenkelarterien Darstellbarkeit in Zuordnung zu Fontainestadien. Vasa 1991;20(Suppl 33):209.

Langholz J, Ladleif M, Blank B, Heidrich H, Behrendt C. Colour coded duplex sonography in ischemic finger artery disease: a comparison with hand arteriography. Vasa 1997;26:85–90.

Langsfeld M, Nepute J, Hershey FB, Thorpe L, Auer AI, Binnington HB, et al. The use of deep duplex scanning to predict hemodynamically significant aortoiliac stenoses. J Vasc Surg 1988;7:395–9.

Larch E, Ahmadi R, Koppensteiner R, Minar E, Schnürer G, Stümpflen A, et al. Farbkodierte Duplexsonographie (FD) zur Beurteilung der Unterschenkelarterien (US) bei peripherer arterieller Verschlusskrankheit. Vasa 1993;22(Suppl 41):14.

Larch E, Minar E, Ahmadi R, Schnurer G, Schneider B, Stümpflen A, et al. Value of color duplex sonography for evaluation of tibioperoneal arteries in patients with femoropopliteal obstruction: a prospective comparison with anterograde intraarterial digital subtraction angiography. J Vasc Surg 1997;25:629–36.

Lawrence PF, Lorenzo-Rivero S, Lyon JL. The incidence of iliac, femoral, and popliteal artery aneurysms in hospitalized patients. J Vasc Surg 1995;22:409–15; discussion 415–6.

Legemate DA, Teeuwen C, Hoeneveld H, Ackerstaff RG, Eikelboom BC. Spectral analysis criteria in duplex scan-

ning of aortoiliac and femoropopliteal arterial disease. Ultrasound Med Biol 1991;17:769–76.
Legemate DA, Teeuwen C, Hoeneveld H, Eikelboom BC. How can the assessment of the hemodynamic significance of aortoiliac arterial stenosis by duplex scanning be improved? A comparative study with intraarterial pressure measurement. J Vasc Surg 1993;17:676–84.
Leng GC, Whyman MR, Donnan PT, Ruckley CV, Gillespie I, Fowkes FG, et al. Accuracy and reproducibility of duplex ultrasonography in grading femoropopliteal stenoses. J Vasc Surg 1993;17:510–7.
Leu HJ, Bollinger A, Pouliadis G, Brunner U, Soyka P. [Pathology, clinical correlates, radiology and surgery of cystic adventitial degeneration of the peripheral blood vessels, 1: pathogenesis and histology of cystic adventitial degeneration of the peripheral blood vessels; in German.] Vasa 1977;6:94–9.
Loeprecht H, Bruijnen H. Femoro-popliteale Aneurysmen. In: Sandmann W, Kniemeyer HW, Kolvenbach R, editors. Aneurysmen der grossen Arterien. Diagnostik und Therapie. Berne: Huber, 1991.
Ludwig M, Trübestein G, Leipner N. Die Bedeutung der Dopplerdruckmessung unter Belastung bei vorliegender arterieller Verschlusskrankheit im Bereich der unteren Extremität. Herz Kreislauf 1985;4:189–91.
Ludwig M, Strauss A, Stein H, Horz R, Koop H, Arning C, et al. Leitlinien zur Qualitätssicherung in der Ultraschalldiagnostik der Gefässe. Ultraschall Med 1998;19:M96–101.
Lundell A, Lindblad B, Bergqvist D, Hansen F. Femoropopliteal-crural graft patency is improved by an intensive surveillance program: a prospective randomized study. J Vasc Surg 1995;21:26–33; discussion 33–4.
Luska G, Risch U, Pellengahr M, von Boetticher H. [Color-coded Doppler sonographic studies of the morphology and hemodynamics of the pelvic and leg arteries in normal subjects; in German.] RöFo 1990;153:246–51.
Luzsa G. Röntgenanatomie des Gefässsystems. Budapest: Akadémiai Kiadó, 1972.
Lynch TG, Hobson RW, Wright CB, Garcia G, Lind R, Heintz S, et al. Interpretation of Doppler segmental pressures in peripheral vascular occlusive disease. Arch Surg 1984;119:465–7.
Mahler F. Systolische Druckmessung nach Belastung. In: Kriessmann A, Bollinger A, Keller H, editors. Praxis der Doppler-Sonographie. Stuttgart: Thieme, 1990.
Mahler F, Brunner HH, Fronek A, Bollinger A. [The ankle arterial pressure during reactive hyperemia in healthy patients and patients with arterial occlusive disease; in German.] Schweiz Med Wochenschr 1975;105:1786–8.
Mahler F, Koen L, Johansen KH, Bernstein EF, Fronek A. Postocclusion and postexercise flow velocity and ankle pressures in normals and marathon runners. Angiology 1976;27:721–9.
Matsubara J, Neuerburg-Heusler D, Schoop W. Über das Verhalten des poststenotischen systolischen Blutdruckes bei Änderungen des arteriellen Systemdruckes. In: Müller-Wiefel H, editor. Gefässersatz. Baden-Baden, Germany: Witzstrock, 1980.
Mattos MA, van Bemmelen PS, Hodgson KJ, Ramsey DE, Barkmeier LD, Sumner DS. Does correction of stenoses identified with color duplex scanning improve infrainguinal graft patency? J Vasc Surg 1993;17:54–64; discussion 64–6.
Mewissen MW, Kinney EV, Bandyk DF, Reifsnyder T, Seabrook GR, Lipchik EO, et al. The role of duplex scanning versus angiography in predicting outcome after balloon angioplasty in the femoropopliteal artery. J Vasc Surg 1992;15:860–5; discussion 865–6.
Middleton WD, Erickson S, Melson GL. Perivascular color artifact: pathologic significance and appearance on color Doppler US images. Radiology 1989a;171:647–52.
Middleton WD, Picus DD, Marx MV, Melson GL. Color Doppler sonography of hemodialysis vascular access: comparison with angiography. AJR Am J Roentgenol 1989b;152:633–9.
Mills JL, Harris EJ, Taylor LM Jr, Beckett WC, Porter JM. The importance of routine surveillance of distal bypass grafts with duplex scanning: a study of 379 reversed vein grafts. J Vasc Surg 1990;12:379–86; discussion 387–9.
Mitchell DG, Needleman L, Bezzi M, Goldberg BB, Kurtz AB, Pennell RG, et al. Femoral artery pseudoaneurysm: diagnosis with conventional duplex and color Doppler US. Radiology 1987;165:687–90.
Moll R, Habscheid W, Landwehr P. [The frequency of false aneurysms of the femoral artery following heart catheterization and PTA; in German.] RöFo 1991;154:23–7.
Moneta GL, Yeager RA, Antonovic R, Hall LD, Caster JD, Cummings CA, et al. Accuracy of lower extremity arterial duplex mapping. J Vasc Surg 1992;15:275–83; discussion 283–4.
Moneta GL, Yeager RA, Lee RW, Porter JM. Noninvasive localization of arterial occlusive disease: a comparison of segmental Doppler pressures and arterial duplex mapping. J Vasc Surg 1993;17:578–82.
Mönig SP, Walter M, Sorgatz S, Erasmi H. True infrapopliteal artery aneurysms: report of two cases and literature review. J Vasc Surg 1996;24:276–8.
Mulligan SA, Matsuda T, Lanzer P, Gross GM, Routh WD, Keller FS, et al. Peripheral arterial occlusive disease: prospective comparison of MR angiography and color duplex US with conventional angiography. Radiology 1991;178:695–700.
Murphy TP, Dorfman GS, Segall M, Carney WI Jr. Iatrogenic arterial dissection: treatment by percutaneous transluminal angioplasty. Cardiovasc Intervent Radiol 1991;14:302–6.
Neuerburg-Heusler D, Karasch T. Farbkodierte Duplexsonographie – Erweiterung der gefässdiagnostischen Möglichkeiten? In: Maurer PC, Dörrler J, von Sommoggy S, editors. Gefässchirurgie im Fortschritt. Neuentwicklungen, Kontroversen, Grenzen, Perspektiven. Stuttgart: Thieme, 1991a: 2–13.
Neuerburg-Heusler D, Karasch T. Farbkodierte Duplexsonographie der Extremitätenarterien. Möglichkeiten in Diagnostik und Therapiekontrolle. Vasa 1991b;20(Suppl 32):113–21.
Neuerburg-Heusler D, Karasch T. [Determining the degree of stenosis of peripheral arteries: hemodynamic and ultrasound principles; in German.] Vasa 1996;25:109–13.
Neuerburg-Heusler D, Karasch T. Aktuelle Bedeutung der Ultraschalldiagnostik. Vasa 1999;28(Suppl 55):5.
Neuerburg-Heusler D, Schoop W. [Ultrasonic Doppler diagnosis in angiology; in Germany.] Hippokrates 1978;49:231–46.
Neuerburg-Heusler D, Voigt T, Roth FJ. Doppler-Sonographie der Leistenarterie. In: Breddin K, editor. Thrombose und Atherogenese. Pathophysiologie und Therapie der arteriellen Verschlusskrankheit. Bein-Beckenvenen-Thrombose. Baden-Baden, Germany: Witzstrock, 1981.
Neuerburg-Heusler D, Schlieszus A, Schulte M. Topographische Diagnose der peripheren arteriellen Verschlusskrankheit (PAVK). Vergleich von segmentaler systolische Druckmessung (SSP) und Oszillographie (OSZ) in Ruhe und nach zwei differenzierten Belastungen. Vasa 1989;18(Suppl 27):369.

Nicolaides AN, Gordon-Smith IC, Dayandas J, Eastcott HH. The value of Doppler blood velocity tracings in the detection of aortoiliac disease in patients with intermittent claudication. Surgery 1976;80:774–8.

Nielsen PE, Bell G, Lassen NA. Strain gauge studies of distal blood pressure in normal subjects and in patients with peripheral arterial disease: analysis of normal variation and reproducibility and comparison to intraarterial measurements. Scand J Clin Lab Invest Suppl 1973;128:103–9.

Olin JW. Thromboangiitis obliterans (Buerger's disease). N Engl J Med 2000;343:864–9.

Ouriel K, Zarins CK. Doppler ankle pressure: an evaluation of three methods of expression. Arch Surg 1982;117:1297–300.

Paulson EK, Hertzberg BS, Paine SS, Carroll BA. Femoral artery pseudoaneurysms: value of color Doppler sonography in predicting which ones will thrombose without treatment. AJR Am J Roentgenol 1992;159:1077–81.

Pezzullo JA, Dupuy DE, Cronan JJ. Percutaneous injection of thrombin for the treatment of pseudoaneurysms after catheterization: an alternative to sonographically guided compression. AJR Am J Roentgenol 2000;175:1035–40.

Phillips DJ, Powers JE, Eyer MK, Blackshear WM Jr, Bodily KC, Strandness DE Jr, et al. Detection of peripheral vascular disease using the Duplex Scanner III. Ultrasound Med Biol 1980;6:205–18.

Phillips DJ, Beach KW, Primozich J, Strandness DE Jr. Should results of ultrasound Doppler studies be reported in units of frequency or velocity? Ultrasound Med Biol 1989;15:205–12.

Polak JF, Donaldson MC, Dobkin GR, Mannick JA, O'Leary DH. Early detection of saphenous vein arterial bypass graft stenosis by color-assisted duplex sonography: a prospective study. AJR Am J Roentgenol 1990a;154:857–61.

Polak JF, Karmel MI, Mannick JA, O'Leary DH, Donaldson MC, Whittemore AD. Determination of the extent of lower-extremity peripheral arterial disease with color-assisted duplex sonography: comparison with angiography. AJR Am J Roentgenol 1990b;155:1085–9.

Ranke C, Creutzig A, Alexander K. Duplex scanning of the peripheral arteries: correlation of the peak velocity ratio with angiographic diameter reduction. Ultrasound Med Biol 1992;18:433–40.

Rieger H. Pathophysiologische Aspekte der arteriellen Verschlusskrankheit. Klinikarzt 1984;13:323–34.

Rieger H. Pathophysiologie des akuten und chronischen Arterienverschlusses. In: Alexander K, editor. Gefässkrankheiten. Munich: Urban & Schwarzenberg, 1993.

Rode V, Schütz RM. Doppler-Ultraschall-Messungen: Relevanz ihrer Masszahlen für die klinische und die Funktionsdiagnostik. Herz Kreislauf 1977;9:422–6.

Ruland O, Borkenhagen N, Prien T. [The Doppler palm test; in German.] Ultraschall Med 1988;9:63–6.

Rutherford RB, Lowenstein DH, Klein MF. Combining segmental systolic pressures and plethysmography to diagnose arterial occlusive disease of the legs. Am J Surg 1979;138:211–8.

Sacks D, Robinson ML, Marinelli DL, Perlmutter GS. Peripheral arterial Doppler ultrasonography: diagnostic criteria. J Ultrasound Med 1992;11:95–103.

Sacks D, Robinson ML, Summers TA, Marinelli DL. The value of duplex sonography after peripheral artery angioplasty in predicting subacute restenosis. AJR Am J Roentgenol 1994;162:179–83.

Schäberle W, Eisele R. [Ultrasound diagnosis, follow-up and therapy of cystic degeneration of the adventitia: 2 case reports and review of the literature; in German.] Ultraschall Med 1996;17:131–7.

Scharf R, Cobet U, Millner R. [The value of ultrasound Doppler pulse curve parameters in the assessment of stenotic arterial diseases; in German.] Ultraschall Med 1988;9:67–71.

Schlieszus A, Herrmann F, Schulte M, Neuerburg-Heusler D. Topographische Diagnose der peripheren arteriellen Verschlusskrankheit. Vergleich von segmentaler systolischer Druckmessung und Oszillographie in Ruhe und nach zwei differenzierten Belastungsarten. Klin Wochenschr 1988;66(Suppl 13):219.

Schmitz-Rixen T, Horsch S, Erasmi H, Schmidt R, Pichlmaier H. Die arterio-venöse Fistel als Hilfsmittel zur Revaskularisation der Unterschenkelarterien. Akt Chir 1986;21:46–51.

Schnell GB, Robertson A, Houston D, Malley L, Anderson TJ. Impaired brachial artery endothelial function is not predicted by elevated triglycerides. J Am Coll Cardiol 1999;33:2038–43.

Schoop W. Pathophysiologie der Arterien und der arteriellen Durchblutung. In: Heberer G, Rau G, Schoop W, editors. Angiologie. Stuttgart: Thieme, 1974.

Schoop W, Levy H. Messung des systolischen Blutdrucks distal eines Extremitätenarterienverschlusses mit Hilfe der Ultraschall-Doppler-Technik. Verh Dtsch Ges Kreislaufforsch 1969;35:455–61.

Seifert H, Jager K, Mona D, Segantini P, Bollinger A. [Arteriovenous fistula and pseudoaneurysm following arthroscopic meniscectomy; in German.] Vasa 1987;16:389–92.

Sievert H, Baser A, Pfeil W, Fach A, Scherer D, Spies H, et al. [The treatment of iatrogenic spurious aneurysm of the femoral artery by direct thrombin injection; in German.] Dtsch Med Wochenschr 2000;125:822–5.

Silbernagel S, Despopoulos A. Pocket atlas of physiology, 4th edition. Stuttgart: Thieme, 1991.

Skidmore R, Woodcock JP. Physiological interpretation of Doppler-shift waveforms, 2: validation of the Laplace transform method for characterization of the common femoral blood-velocity/time waveform. Ultrasound Med Biol 1980;6:219–25.

Sladen JG, Reid JD, Cooperberg PL, Harrison PB, Maxwell TM, Riggs MO, et al. Color flow duplex screening of infrainguinal grafts combining low- and high-velocity criteria. Am J Surg 1989;158:107–12.

Sorrell KA, Feinberg RL, Wheeler JR, Gregory RT, Snyder SO, Gayle RG, et al. Color-flow duplex-directed manual occlusion of femoral false aneurysms. J Vasc Surg 1993;17:571–7.

Spijkerboer AM, Nass PC, de Valois JC, Eikelboom BC, Overtoom TT, Beek FJ, et al. Iliac artery stenoses after percutaneous transluminal angioplasty: follow-up with duplex ultrasonography. J Vasc Surg 1996;23:691–7.

Steinkamp HJ, Jochens R, Zendel W, Zwicker C, Hepp W, Felix R. [Catheter-induced femoral artery lesions: diagnosis with B-mode ultrasound, Doppler ultrasound and color Doppler ultrasound; in German.] Ultraschall Med 1992;13:221–7.

Steins A, Hahn M, Volkert B, Duda S, Schott U, Junger M. [Color duplex ultrasound of the finger arteries: an alternative to angiography? In German.] Hautarzt 1998;49:646–50.

Stiegler H, Mendler G, Baumann G. Prospective study of 36 patients with 46 popliteal artery aneurysms with nonsurgical treatment. Vasa 2002;31:43–6.

Strandness DE Jr. Duplex scanning for diagnosis of peripheral arterial disease. Herz 1988;13:372–7.
Strandness DE Jr. The gold standard in the diagnosis of vascular disease. In: Labs KH, Jäger KA, Fitzgerald DE, Woodcock JP, Neuerburg-Heusler D, editors. Diagnostic vascular ultrasound. London: Arnold,1992.
Strandness DE Jr, McCutcheon EP, Rushmer RF. Application of a transcutaneous Doppler flowmeter in evaluation of occlusive arterial disease. Surg Gynecol Obstet 1966;122:1039–45.
Strano A, Pinto A, Alletto G. Congenital arteriovenous fistulas of the limbs: pathophysiological and diagnostic aspects. In: Blov S, Loose DA, Weber J, editors. Vascular malformations. Reinbek, Germany: Einhorn-Presse, 1989: 66.
Strauss AL, Schaberle W, Rieger H, Neuerburg-Heusler D, Roth FJ, Schoop W. [Duplex ultrasound studies of the deep femoral artery; in German.] Z Kardiol 1989;78:567–72.
Strauss AL, Scheffler A, Rieger H. [Doppler ultrasound determination of pressure decrease in model peripheral arterial stenoses; in German.] Vasa 1990;19:207–11.
Strauss AL, Schäberle W, Rieger H, Roth FJ. Use of duplex scanning in the diagnosis of arteria profunda femoris stenosis. J Vasc Surg 1991;13:698–704.
Strauss AL, Roth FJ, Rieger H. Noninvasive assessment of pressure gradients across iliac artery stenoses: duplex and catheter correlative study. J Ultrasound Med 1993a;12:17–22.
Strauss AL, Sandor D, Karasch T, Roth FJ, Brocai DRC, Neuerburg-Heusler D, et al. Wertigkeit der Farbduplexsonographie in der arteriellen Gefässdiagnostik. Vasa 1993b;22(Suppl 41):15.
Strauss AL, Karasch T, Sandor D, Brocai DRC, Neuerburg-Heusler D, Roth FJ. Normwerte und Grenzwerte zur farbduplexsonographischen Diagnostik von Becken- und Beinarterienstenosen. Ultraschall Klin Prax 1993c;8:195.
Strauss AL, Karasch T, Roth FJ. Accuracy of duplex scanning in the prediction of pressure gradients across peripheral artery stenoses. Bildgebung 1993d;60:294–6.
Strauss AL, Sandor D, Karasch T, Roth FJ, Brocai DRC. Vergleich der Farbduplexsonographie mit der arteriellen Angiographie. Bildgebung/Imaging 1994;61(Suppl 2):35.
Strauss AL, Roth FJ, Lutter EEM, Dohmann-Scheurle C, Karasch T, Rieger, H. Determinanten der Langzeitergebnisse der iliakalen und femoropoplitealen Ballon-Angioplastie. Vasa 1995a;24(Suppl 45):59.
Strauss AL, Weber G, Karasch T, Roth FJ, Rieger H. Quantifizierung hämodynamisch wirksamer Arterienstenosen mit der Farbduplex-Sonographie. In: Ludwig M, Straub H, Arning C, editors. Doppler-/Duplex-Sonographie in der Intensivmedizin. Bonn: Kagerer, 1995b: 18–24.
Sukkar I, Pappadakis K, Reilly D, Grigg M, Christopolous D, Salmasi AM, et al. Quantitative assessment of aortoiliac disease by ultrasound Doppler spectral analysis. J Cardiovasc Tech 1990;9:213–26.
Sumner DS. Evaluation of noninvasive testing procedures: data analysis and interpretation. In: Bernstein EF, editor. Vascular diagnosis. St. Louis: Mosby, 1993: 39–63.
Sumner DS, Strandness DE Jr. The relationship between calf blood flow and ankle blood pressure in patients with intermittent claudication. Surgery 1969;65:763–71.
Taute BM, Podhaisky H. [Quantitative sonography of peripheral arteriovenous malformations; in German.] Ultraschall Med 1998;19:275–9.
Taylor PR, Tyrrell MR, Crofton M, Bassan B, Grigg M, Wolfe JH, et al. Colour flow imaging in the detection of femorodistal graft and native artery stenosis: improved criteria. Eur J Vasc Surg 1992;6:232–6.
Thiele BL, Bandyk DF, Zierler RE, Strandness DE Jr. A systematic approach to the assessment of aortoiliac disease. Arch Surg 1983;118:477–81.
Thompson MM, Smith J, Naylor AR, Nasim A, Sayers RD, Boyle JR, et al. Microembolization during endovascular and conventional aneurysm repair. J Vasc Surg 1997;25:179–86.
Thulesius O. Beurteilung des Schweregrades arterieller Durchblutungsstörungen mit dem Doppler-Ultraschall-Gerät. In: Bollinger A, Brunner U, editors. Messmethoden bei arteriellen Durchblutungsstörungen. Berne: Huber, 1971.
Thulesius O. Systemic and ankle blood pressure before and after exercise in patients with arterial insufficiency. Angiology 1978;29:374–8.
Tordoir JH, de Bruin HG, Hoeneveld H, Eikelboom BC, Kitslaar PJ. Duplex ultrasound scanning in the assessment of arteriovenous fistulas created for hemodialysis access: comparison with digital subtraction angiography. J Vasc Surg 1989;10:122–8.
Trattnig S, Maier A, Sautner T, Schwaighofer B, Breitenseher M, Karnel F. [Saphenous vein bypass stenoses, early diagnosis with color-coded Doppler sonography; in German.] Ultraschall Med 1992;13:67–70.
Treiman GS, Lawrence PF, Galt SW, Kraiss LW. Revision of reversed infrainguinal bypass grafts without preoperative arteriography. J Vasc Surg 1997;26:1020–8.
Tullis MJ, Primozich J, Strandness DE Jr. Detection of "functional" valves in reversed saphenous vein bypass grafts: identification with duplex ultrasonography. J Vasc Surg 1997;25:522–7.
Varga ZA, Locke-Edmunds JC, Baird RN. A multicenter study of popliteal aneurysms. Joint Vascular Research Group. J Vasc Surg 1994;20:171–7.
Voelckel S, Bodner G, Voelckel W, Stadlwieser C, De Koekkoek P, Springer P. [Doppler ultrasound determination of vascular resistance in arteriovenous shunts of the finger tip; in German.] Ultraschall Med 1998;19:181–6.
Vogelberg KH, Helbig G, Stork W. Doppler sonographic examination of reactive hyperemia in the diagnosis of peripheral vascular disease. Klin Wochenschr 1988;66:970–5.
Vollmar J. [Traumatic arteriovenous fistulas: report of 190 cases; in German.] Zentralbl Chir 1964;89:1930–9.
Vollmar J. Rekonstruktive Chirurgie der Arterien, 4th ed. Stuttgart: Thieme, 1996.
Walton L, Martin TR, Collins M. Prospective assessment of the aorto-iliac segment by visual interpretation of frequency analysed Doppler waveforms: a comparison with arteriography. Ultrasound Med Biol 1984;10:27–32.
Wells PNT. Instrumentation including color flow mapping. In: Taylor KJW, Burns PN, Wells PNT, editors. Clinical application of Doppler ultrasound. New York: Raven Press, 1988: 26–45.
Wetterer E, Bauer RD, Pasch T. Arteriensystem. In: Schütz E, editor. Physiologie des Kreislaufs, vol. 1. Berlin: Springer, 1971.
Wenz W, Beduhn D. Extremitätenangiographie. Berlin: Springer, 1976.
Weskott HP. [B-flow: a new method for detecting blood flow; in German.] Ultraschall Med 2000;21:59–65.
Whyman MR, Hoskins PR, Leng GC, Allan PL, Donnan PT, Ruckley CV, et al. Accuracy and reproducibility of duplex

ultrasound imaging in a phantom model of femoral artery stenosis. J Vasc Surg 1993;17:524–30.

Widder B. [Significance of technical parameters in color-coded duplex ultrasound for vascular studies; in German.] Ultraschall Med 1993;14:231–9.

Widder B, von Reutern GM, Neuerburg-Heusler D. [Morphologic and Doppler sonographic criteria for determining the degree of stenosis of the internal carotid artery; in German.] Ultraschall Med 1986;7:70–5.

Windeck P, Labs KH, Jaeger KA. How useful are acceleration- and deceleration-based Doppler indices? A trial on patients with percutaneous transluminal angioplasty. Ultrasound Med Biol 1992;18:525–34.

Wittenberg G, Landwehr P, Moll R, Tschammler A, Buschmann B, Krahe T. [Interobserver variability of dialysis shunt flow measurements using color coded duplex sonography; in German.] RöFo 1993;159:375–8.

Wolf KJ, Fobbe F. Farbkodierte Duplexsonographie. Stuttgart: Thieme, 1993.

Wuppermann T, Capell F, Dittrich O, Evers EJ, Müller W, Naumann E. Becken- und Beinarterien. In: Wuppermann T, editor. Ultraschallkurs Gefässe. Munich: Urban & Fischer, 2000:211–34.

Yao ST. Haemodynamic studies in peripheral arterial disease. Br J Surg 1970;57:761–6.

Yao ST, Hobbs JT, Irvine WT. Ankle systolic pressure measurements in arterial disease affecting the lower extremities. Br J Surg 1969;56:676–9.

6 Peripheral Veins

Examination

Special Equipment and Documentation

Color Flow Duplex Systems

A 5-MHz linear scanner is preferable for displaying the deep veins of the extremities. For the vena cava and the iliac veins, a 3.5–5.0 MHz sector scanner provides more effective depth resolution. When the color flow Doppler mode is being used, care should be taken to ensure that the flow is coded blue and that the color scale is selected to detect low flow velocities ("slow flow technique"). Otherwise the color display of slow venous flow will not be adequate.

There is a limiting factor in the fact that the color-coding setting in some types of color flow duplex equipment simultaneously determines the deflection of the Doppler spectrum direction, so that separate directional change is not possible. In these cases, the preferred documentation type has to be chosen—either the blue venous flow coding, usual in angiology, or the recommended recording format, using the polarity that results in signals directed toward the probe having an upward deflection.

Continuous-Wave Doppler Equipment

Unidirectional ultrasound Doppler equipment can be used to provide general diagnostic orientation. Bidirectional Doppler equipment is preferable, however, as it can indicate the flow direction and can graphically register and document the findings. Pulsed Doppler systems do not have a significant place in routine clinical practice.

The transmission frequency varies according to the region being examined: 4–5 MHz at the groin and the popliteal artery, 8–10 MHz at the ankle. For documentation purposes, it is recommended that flow directed toward the heart should be adjusted to indicate an upward deflection, and that a slow paper speed should be selected.

Examination Conditions

Patient and Examiner

The patient's position needs to change in accordance with the region being examined and the clinical objectives. The examiner should be in a sitting position if possible, in order to ensure stable holding of the transducer or the probe with low contact pressure. The procedure described below, which was developed for continuous-wave Doppler sonography, is also generally valid for duplex sonography (Table 6.**1**).

Table 6.**1** External examination conditions for applying ultrasound to the venous system in the lower extremity using continuous-wave Doppler sonography

Venous system	Patient's position	Region to apply ultrasound in	Provocation maneuvers
Deep veins			**Compression**
Iliac veins	Reclining/supine	Groin	Thigh
Femoral vein	Reclining/supine	Groin/thigh	Thigh
Popliteal vein	Reclining/prone (with the talocalcanean joint supported from below)	Back of the knee	Calf
Lower leg veins	Standing or seated	Ankle	Sole of the foot, lower leg if appropriate
Superficial veins			**Valsalva maneuver**
Great and small saphenous veins	Standing	Along their anatomical course	
Perforating veins			**Compression**
Dodd—thigh Boyd—proximal lower leg Cockett—distal lower leg	Seated	Above the anatomical region or above the "blow-out"	Distal/proximal, using tourniquets

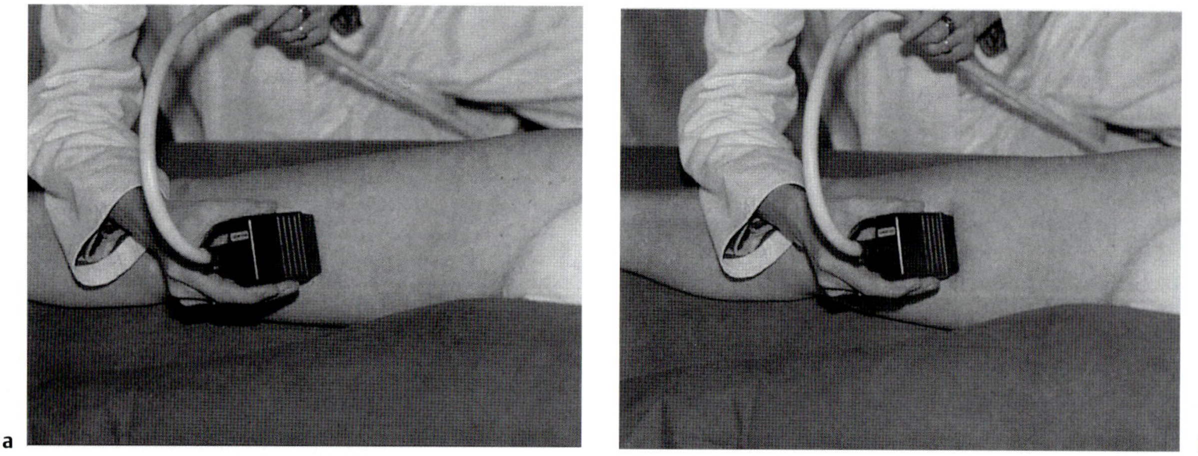

Fig. 6.**1** Examination of the superficial femoral vein in transverse section with the transducer of a duplex apparatus, with the patient supine. Compression maneuvers for excluding or detecting a thrombosis, without compression (**a**) and with compression (**b**)

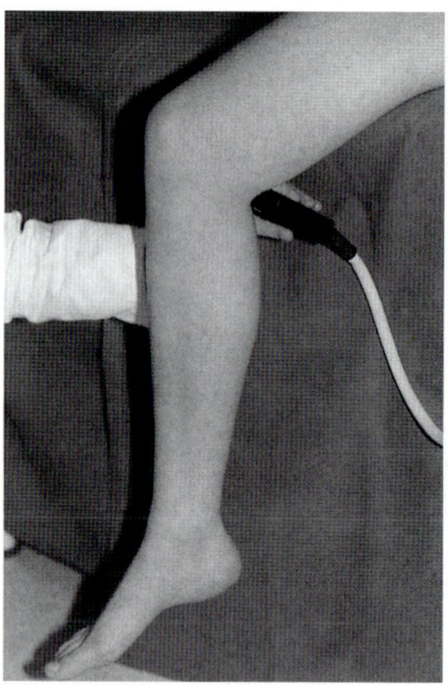

Fig. 6.**2** Applying the duplex transducer to the popliteal vein at the back of the knee, in transverse section, with the patient sitting

Deep Pelvic and Leg Veins

For ultrasound examinations of the *iliac, common femoral,* and *superficial femoral veins,* the patient should be in a supine position. Having the upper body slightly elevated is helpful, as it causes better venous filling in the extremities (Fig. 6.**1**).

The prone position is preferable for examining the *popliteal vein*; the ankles should be slightly elevated using a suitable support. The patient's position can be varied depending on the objectives of the examination and the clinical findings. Examining the popliteal vein is also possible with the patient in the supine or lateral position, or sitting (Fig. 6.**2**). The supine position is recommended mainly for bedridden and elderly patients, for whom the prone position is difficult. Due to the improved display, the *lower leg veins* should be examined while the patient is seated, or with the leg held lower.

Superficial Leg Veins

For examination of the *great saphenous vein* with its branches, and the *small saphenous vein*, the patient should be standing (Fig. 6.**3**). If necessary, the examination can also be carried out with the patient supine. The leg being examined should not be carrying any body weight. In contrast, the *perforating veins* are better examined with the patient seated (see Fig. 6.**6**).

Upper Extremity Veins

The patient should be in the supine position. In this position, the subclavian vein, axillary vein, and brachial vein can be located; the examiner sits next to or behind the patient. The deep and superficial veins of the forearm can also be examined with the patient seated (Talbot 1986).

Conducting the Examination

Pelvic and Leg Veins

Color Flow Duplex Sonography

B-mode. In a transverse section, the veins are displayed in a round or oval cross-section, which increases by about 50–100% when the Valsalva maneuver is applied.

In longitudinal section, the venous lumen is anechoic, and the wall cannot be distinguished consistently. There is no arterial pulsation. However, there is visible modulation due to respiration. A patent vein can be completely compressed when pressure is applied with the transducer, while the accompanying pulsating artery cannot be compressed by the same pressure. In the presence of an acute thrombosis, the diameter of the vein is usually clearly larger than that of the artery, and compression is not effective, or only partially so. Sometimes the thrombus can be recognized by its greater echo intensity. In longitudinal section, the head of a thrombus may be clearly distinguishable from the patent vascular lumen (see Fig. 6.**20**).

Doppler mode. With the adjusted Doppler mode, reactions to respiratory maneuvers, to deliberate pressure exertion, and also signal augmentation are examined in the same manner as will be described for continuous-wave Doppler sonography. In addition to the morphological image, functional parameters describing the venous flow can be obtained (see Fig. 6.**25**).

Color flow Doppler mode. When diagnosing a thrombus, the location and extent of the thrombotic material are more easily recognized with this method than in a pure B-mode image, since thrombus, especially if it is fresh, has the same echogenicity as blood. In addition, partially thrombosed thrombi, surrounded by flowing blood, or partially recanalized venous segments, can be clearly detected (see Fig. 6.**21**).

Respiratory maneuvers, such as cessation of inspiratory flow, and reactions during the Valsalva maneuver and compression maneuvers, can be detected with certainty using the color reproduction dynamics. Venous reflux during the Valsalva maneuver can be recognized as a reversal flow with a color coding change. Turbulent flow patterns are sometimes also present (Fig. 7.**41** p. 339). Diagnosing insufficient valves in this way is more certain and faster than when using the Doppler mode. As in the arterial system, locating the lower leg veins is significantly easier with color coding. The direction of the blood flow in the perforating veins can be detected very easily.

Continuous-Wave Doppler Sonography

The spontaneous venous signal is characterized by respiration-dependent modulations, which sound like the howling of wind, or roaring waves. Usually, the vein is found directly medial to the artery, at the typical palpation points of the arteries of the extremities. In the popliteal region, the vein lies dorsal to the artery, which means it is closer to the transducer when the patient is in the prone position.

The various respiration and provocation tests are summarized in Table 6.**2**.

Spontaneous Respiration (S-sounds)

This term describes the observation and recording of the spontaneous venous flow. The phasic increase and decrease of the venous flow during quiet respiration is detected acoustically, and an appropriately modulated flow directed toward the heart is registered.

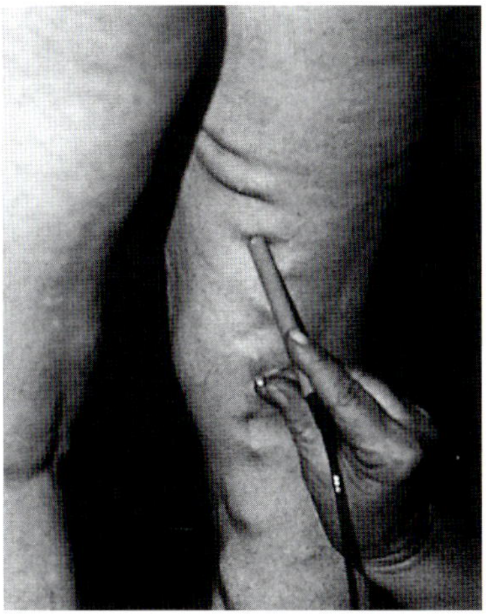

Fig. 6.**3** Using continuous-wave Doppler sonography to examine the great saphenous vein in the distal thigh, with the patient standing. The clear varicose dilation of the superficial vein is easily recognizable

Table 6.**2** Functional tests of venous flow during recording of Doppler curves

Spontaneous signals
S-sounds (spontaneous sounds)
- *Respiration at rest*

Provoked signals
A-sounds (augmented sounds)
- *Maneuvers during respiration*
  - Deepened (forced) respiration
  - Valsalva maneuver
- *Manual signal amplification*
  - Distal compression/decompression
  - Proximal compression/decompression
  - Modulation
  - Tourniquets

During inspiration, the blood flow in the veins of the lower extremities is minimal; on expiration, blood flows more intensely toward the heart. In the upper extremities, the opposite sequence of events occurs. During inspiration, blood flow into the thorax increases, and on expiration it decreases (see Fig. 6.**13**).

Augmented Sounds (A-sounds)

Increased Respiration

During deepened but not forced respiration, an almost complete cessation of blood flow occurs in the late inspiratory phase. Expiration causes an increase in venous flow directed toward the heart. In the presence of venous thrombosis, continuous flow is registered, due to changed pressure-flow relationships (p. 230, Fig. 6.**18**).

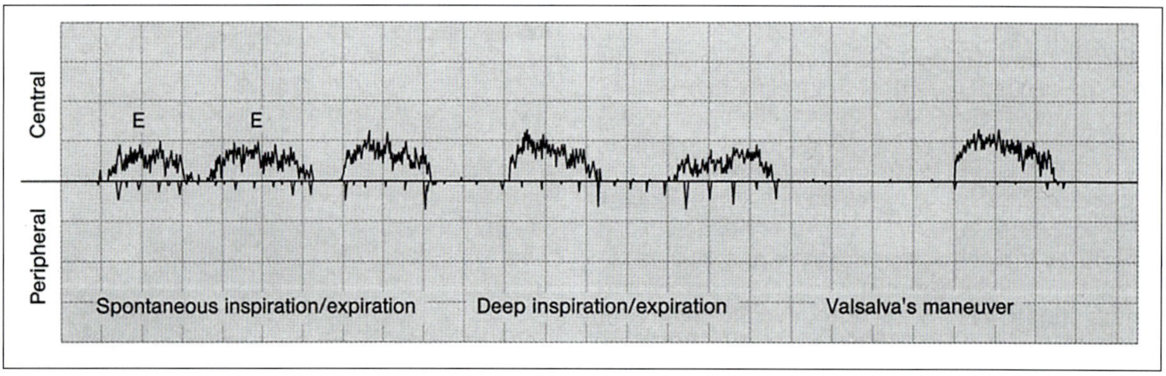

Fig. 6.**4** A recording of the continuous-wave Doppler waveform in a healthy individual at the popliteal vein during spontaneous respiration, deep (forced) respiration, and the Valsalva maneuver. Venous flow cessation is seen, due to elevated intra-abdominal pressure following forced respiration into the abdominal cavity

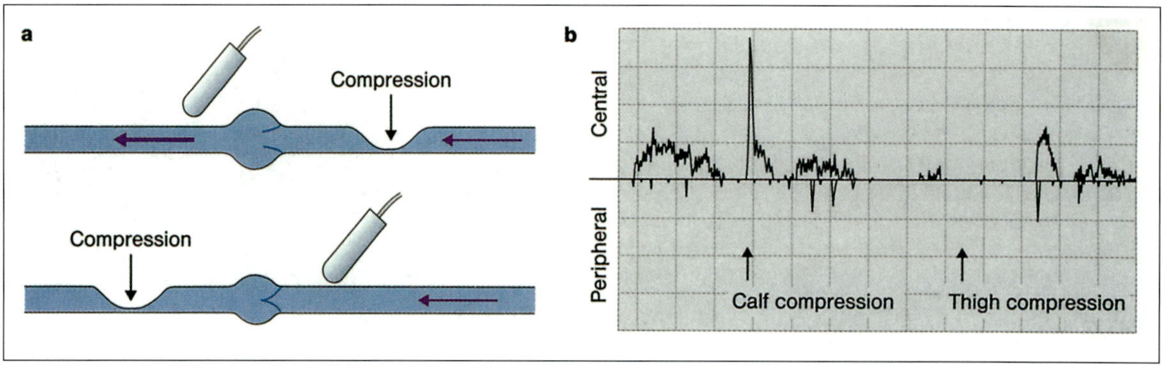

Fig. 6.**5** Distal and proximal compression maneuvers (A-sounds) in a normal patient. **a** The venous valve is open during distal compression, and closes during proximal compression to prevent retrograde flow into the periphery. **b** The Doppler waveform of the popliteal vein shows peak flow during distal calf compression and flow cessation during proximal thigh compression, with a subsequent, brief peak flow toward the heart

Valsalva Maneuver

The Valsalva maneuver involves exerting strong pressure while the patient's mouth is closed after deep inspiration. This causes increased thoracic and abdominal pressure and inflow congestion into the right heart. The maneuver causes the centripetal venous flow to cease (Fig. 6.**4**). When the venous valves, especially those in the groin, are insufficient, centrifugal reflux of varying duration occurs in the extremities, due to the marked abdominal pressure elevation. Immediately after the maneuver has been completed, elevated reactive flow toward the heart is registered (see Fig. 6.**18 b**).

Valsalva maneuver should be practiced with the patient first, since patients often do not exert the required sudden lower abdominal pressure in the optimal way.

Manual Signal Amplification

Signal augmentation, known as the A-sound (augmented sound), refers to rapid manual compression executed distal or proximal to the registration area, or both.

Veins peripheral to the popliteal artery are sometimes difficult to locate. When compression is applied *distal* to the registration point, the venous signal is amplified, making it easier to locate the venous flow. Normally, brief flow peaks toward the heart are registered, due to distal compression. If the valves are insufficient, peripheral reflux results when the compression is released.

Proximal compression is used, as is the Valsalva maneuver, to test valvular function. Sufficient valves produce a cessation of blood flow during compression, and subsequently, during decompression, cause increased flow toward the heart (Fig. 6.**5**). Insufficient valves produce distal reflux.

Venous modulation is a variation of signal augmentation. It plays a role in locating the *superficial* veins when using the continuous-wave Doppler probe. Within the anatomical course of a superficial vein, testing is usually carried out distal to a registration point to see whether the signals are propagated with quick, repetitive modulation (Straub and Ludwig 1990). Using modulation, which usually occurs at an anatomically

a
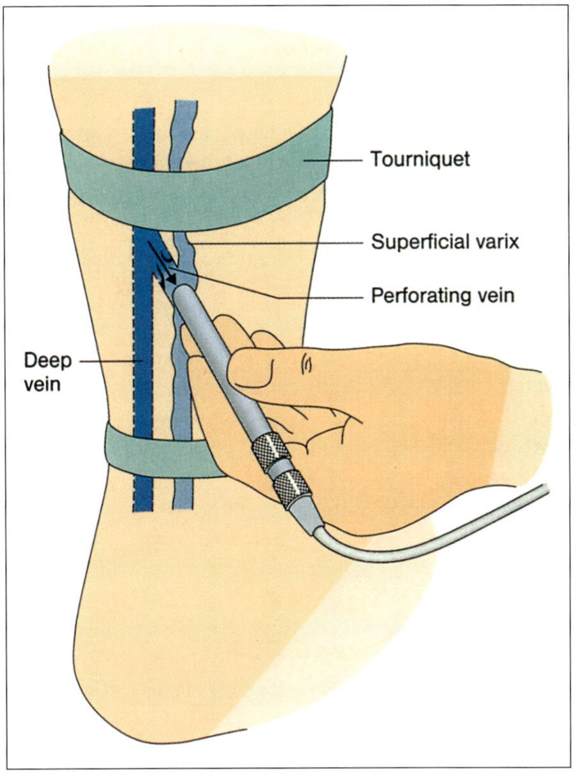

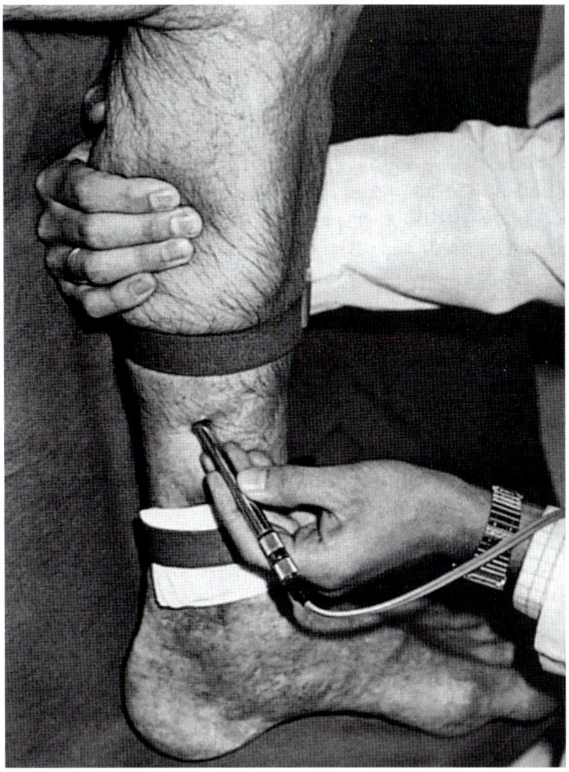
b

Fig. 6.**6** Testing for insufficiency of the perforating veins. **a** The probe is placed in the direction of the perforating vein. During distal or proximal compression, venous flow is registered through the insufficient valve of the perforating vein toward the surface and approaching the probe. The superficial varices have to be compressed with tourniquets so that epifascial venous flow is not incorrectly registered. **b** Conducting the examination with the patient sitting. Proximal compression is used here as the patient had a post-thrombotic syndrome, with insufficient deep venous valves. If the deep venous system is intact, distal compression should be employed. However, this usually results in weaker signals, since only the foot is available for compression

known location, the course of a vein or varix can be easily followed, and its patency can be evaluated.

Testing for Insufficiency of Perforating Veins

With the patient seated, the dangling leg is examined. Proximal and distal to the suspected area (a blow-out, or a known anatomical region), a tourniquet is placed to compress the superficial varices. Subsequently, the tissue distal or proximal to the tourniquet, or both, is compressed. Insufficient perforating vein valves cause reflux that flows out through the deep veins via the perforating vein (flow toward the probe), subsequently producing internally directed, retrograde suction (flow away from the probe) (Bjordal 1981). Proximal compression is only effective when there are no valves or insufficient valves between the compression area and the perforating vein. The tourniquet obstructs the propagation of venous flow through the epifascial varices. If the perforating veins are intact, then neither distal nor proximal compression will subsequently result in flow toward the probe (Fig. 6.**6**).

Upper Extremity Veins

Color Flow Duplex Sonography

B-mode. As in the lower extremities, the most important criterion in pure B-mode sonography is the *compressibility* of the deep veins (axillary vein, brachial vein), which allows their patency to be evaluated. A patent vein is very easily compressed. Care must therefore be taken to use an externally supported light touch when guiding the transducer; this is particularly true when the epifascial veins are being imaged. Noncompressible segments, which usually have a dilated lumen in comparison with the artery, indicate thrombosis or thrombophlebitis. In addition, the latter is often recognizable through increased intraluminal echoes.

Doppler mode. The Doppler mode can be used to examine venous flow specifically, with its phasic modulation and its reactions to provocation tests. The veins should be imaged in longitudinal section, so that a favorable incident ultrasound angle can be adjusted for the Doppler sample volume. Particularly in regions in which venous compressibility is uncertain or cannot be assessed at all, as in the subclavian artery below the clavicle, it is indispensable to register Doppler waveforms to confirm the diagnosis (Longley et al. 1993).

Color flow Doppler mode. The color-coded display of the venous system is also helpful in this region, especially for veins in the forearm that have a narrow lumen.

All reactions to the provocation maneuvers described above can of course also be recognized in color Doppler mode. However, assessment in this case is more qualitative in nature, while alterations in venous flow can be captured better—and in particular can also be documented—using integrated pulsed Doppler.

Continuous-Wave Doppler Sonography

Spontaneous Respiration (S-sounds)

Respiratory modulation is observed and registered in the same way as in the lower extremities. However, the phenomena observed are the opposite of those in the legs: during inspiration venous flow increases, and on expiration it decreases (Strandness and Sumner 1972) (see Fig. 6.**12 b**).

Augmented Sounds (A-sounds)

Increased Respiration

Venous flow modulation cannot be increased as significantly as it can in the lower extremities. Near-cessation of blood flow does not occur during expiration.

Valsalva Maneuver

As in the lower body, the test using forced respiration causes a cessation of blood flow, followed by a typical acceleration of flow. If an incomplete thrombotic occlusion is present, only a minimally increased flow velocity can be obtained after the Valsalva maneuver.

Manual Signal Amplification

As in the lower extremities, increased flow velocity (peak flow) can be obtained by quick compression distal to the probe position. Proximal compression tests the valve function.

Examination Sequence

The recording of values from the examination points should be carried out proceeding from proximal to distal, beginning on the asymptomatic side, and side-to-side comparisons should be made. Pelvic veins and lower leg veins are often evaluated with ultrasound proceeding in the opposite direction.

Pelvic and Leg Veins

Color Flow Duplex Sonography

The examination usually starts in the *groin region,* where the femoral vein is easy to find directly medial to the femoral artery. Initially, the course of the vein is followed distally, and after only a few centimeters (approximately 5–9 cm), the junction of the great saphenous vein with the common femoral vein can be identified (see Fig. 6.**14**). Further distally and dorsally lies the junction of the deep femoral vein. The superficial femoral vein is examined with ultrasound proceeding from an anterior direction, both parallel and medial to the superficial femoral artery, all the way to the adductor canal, until the vein passes dorsally in front of the superficial femoral artery and the popliteal artery.

Proceeding from a dorsal direction, the *popliteal vein* is examined with ultrasound at the posterior aspect of the knee, where it is easily located and lies close to the transducer (Fig. 6.**2**). The division into the three lower leg veins is usually located easily distal to the interarticular space of the knee (48 %). However, it should be borne in mind that the lower leg veins are found in pairs accompanying the corresponding artery, and that the division can also be encountered as a variant located near the popliteal vein, but above the interarticular space (40 %) (May and Nissl 1966).

Experienced examiners can follow the *lower leg veins* in transverse or longitudinal section, or both (Fig. 6.**7**). Sometimes, in normal findings, the veins are not displayed in a completely continuous form. In contrast, a thrombosed, dilated vein is easily recognized, and its lack of compressibility confirms the diagnosis of lower leg venous thrombosis (Habscheid and Wilhelm 1988).

The epifascial veins (great and small saphenous veins) are continuously followed: the search usually proceeds in a proximal to distal direction.

Examining the *pelvic veins,* which can be done either initially or at the end of the examination sequence, depending on the clinical objectives, also starts at the groin, where the femoral vein becomes the external iliac vein above the inguinal ligament. In its distal segment, the vessel can be displayed well and easily compressed. However, when it bends, proceeding deeper into the body, its compressibility is no longer guaranteed. Although the ease with which the veins can be displayed clearly depends on the extent of a patient's obesity, it is always difficult to follow the external iliac vein along its further course, or to follow the common iliac vein to the vena cava within the deeper regions of the body. The vena cava is searched for on the right of the umbilical region.

Continuous-Wave Doppler Sonography

Pelvic veins cannot be assessed with the continuous-wave Doppler probe. The examination starts at the common femoral artery in the *groin.* The inguinal pulse is felt first and then, after contact gel is applied, the

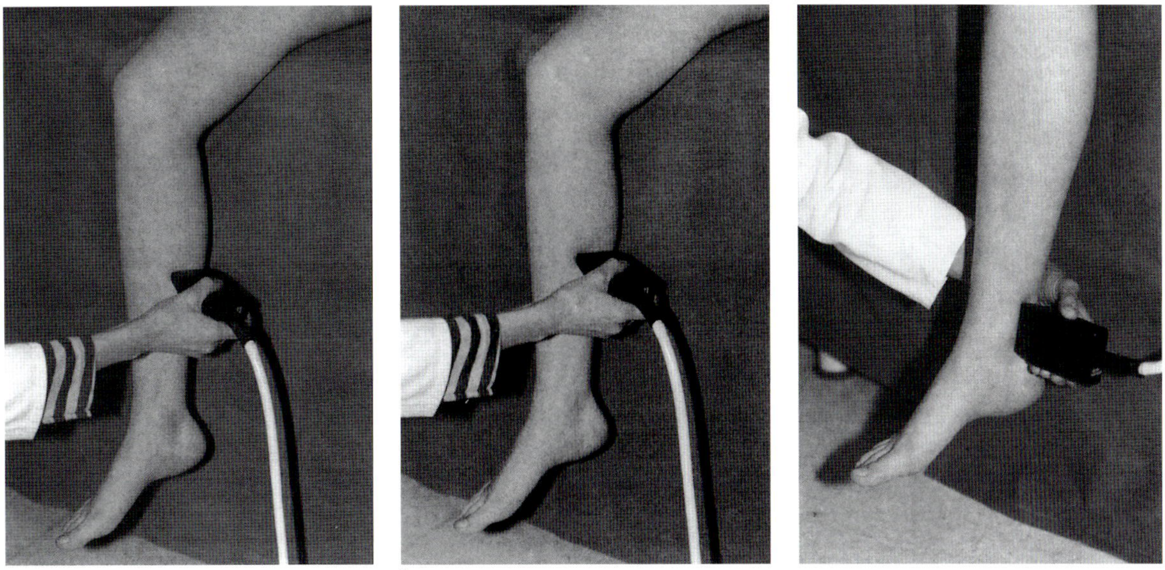

a b c

Fig. 6.**7** Applying the duplex transducer to the posterior tibial veins in the loosely hanging lower leg. Transverse sections without compression (**a**), with compression (**b**), and in longitudinal section posterior to the medial malleolus (**c**)

probe makes contact to search for the typical, whiplike arterial signal. The probe is then displaced slightly in a medial direction until the low noise of the femoral vein, modulated by respiration, is heard. Using the various maneuvers described previously, tests for patency and valve function are carried out. If the vein is completely open, the venous signal is affected by respiratory fluctuations. Distal to thromboses, there is continuous flow (Fig. 7.**37**, pp. 332–333).

When the venous valves are intact, the Valsalva maneuver causes a complete cessation of blood flow, or at the most a very brief moment of reflux, which means that the blood in the femoral vein does not come to a standstill until a point slightly distal to the groin, due to a valve ("early incompetence"). If the groin valves are insufficient, a continuous low noise and distal flow occurs during pressure exertion.

Differentiation between reflux in the great saphenous vein and the femoral vein is achieved by placing a light tourniquet. This compresses the great saphenous vein, so that the return flow either ceases or can only proceed through the femoral vein (Fronek 1989).

Similarly, the *popliteal vein* is located and evaluated at the posterior aspect of the knee. The venous signal here is more difficult to hear spontaneously than in the groin, since the angle of the probe to the vessel is often too steep. The signal can be amplified by compressing the calf.

In routine practice, one can often dispense with evaluating the posterior tibial vein and the dorsal vein of the foot, since the signals can usually only be detected by foot compression (A-sounds).

The superficial veins are easy to follow using Doppler sonography. The junction of the main collecting vein, the *great saphenous vein*, can often be located medial to the femoral vein, approximately 9 cm below the inguinal ligament. Clinical assessment follows the vein along its anatomical course (Fig. 6.**3**), and the segment that shows reflux is evaluated with successive Valsalva maneuvers or proximal compression.

In valvular insufficiency, the course of the *small saphenous vein* can also be identified when the flow is modulated by repeated compression of the distal segments.

Insufficient *perforating veins* are usually only examined in their typical anatomical location (Fig. 6.**6**) when a clinically suspicious region is noticed (hyperthermia, redness, blow-out) or chronic venous insufficiency of unclear etiology is present.

Upper Extremity Veins

Color Flow Duplex Sonography

Usually, the veins of the upper extremity are followed in an ascending manner and in cross-section, starting from the bend of the elbow. This allows imaging of both the *brachial vein,* which is paired, and also the superficial veins, namely the *basilic vein* and the *cephalic vein.* For the superficial, delicate veins, the contact pressure applied with the transducer must be very light. Imaging these veins and testing their patency is very important in connection with the placement of permanent venous catheters (in intensive-care medicine). Imaging the *axillary vein* as far as the clavicle can sometimes be difficult, while the *subclavian vein* can be displayed both below and above the clavicle. Below the clavicle, the vein lies above the artery, and can be more easily depicted here than in its supraclavicular location,

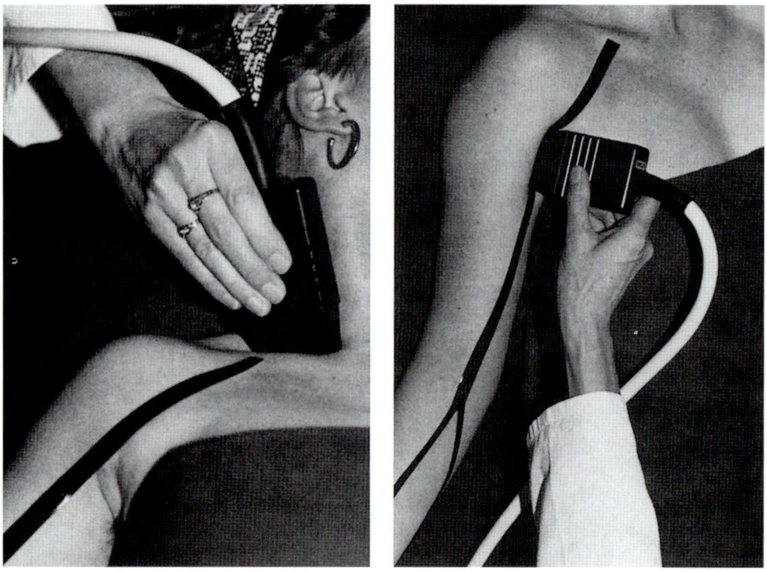

a b

Fig. 6.**8** Applying the duplex transducer to the subclavian vein (**a**) and the axillary vein (**b**) in the shoulder region, using the linear scanner of a duplex apparatus. The course of the vein is marked in black on the skin, providing a guide for moving the transducer

where it lies underneath the artery. The clavicle itself extinguishes the ultrasound signal, so that it is not possible to follow the course of the subclavian vein continuously (Fig. 6.**8**).

Continuous-Wave Doppler Sonography

It has proved useful to start by carrying out side-to-side ultrasound comparison of the subclavian vein, and conducting the various maneuvers to test the patency of the vessels as described above. The probe is placed in a supraclavicular location, or below the clavicle if the artery covers the vein at that position.

Depending on the nature of the clinical situation—e.g., when a thrombosis is suspected—the axillary vein in the armpit and the brachial vein slightly above the bend of the elbow can be examined with Doppler sonography.

Evaluating patency after the placement of indwelling venous catheters sometimes requires assessment of the cephalic vein. The anatomical course of this vessel can be followed using the modulation test. Examinations of arteriovenous dialysis shunts in the upper arm or forearm can also be conducted with continuous-wave Doppler sonography, which provides general orientation.

Normal Findings

Principle

In a reclining position, the return flow of venous blood is predominantly modulated by respiration. The pressure difference between the periphery and the right atrium amounts to approximately 15 mmHg. Flowing from the legs to the heart, the blood passes through two "closed containers"—the abdominal and the thoracic cavities (Sumner 1984).

During *inspiration*, the diaphragm moves caudally, and the abdominal pressure increases. It exceeds the venous pressure in the legs, and the venous flow toward the heart stops. Simultaneously, thoracic pressure decreases, so that blood from the upper body region is moved or sucked into the thorax (Bollinger et al. 1970) (Fig. 6.**9a**).

On *expiration*, the process is reversed. The pressure within the abdominal "container" sinks, and the blood flows from the lower extremities, in which the pressure is now higher than in the abdomen, into the abdominal cavity. Flow into the thorax decreases. Bollinger et al. (1968) termed this process the "abdominothoracic two-phase pump."

During standing and walking, venous pressure increases hydrostatically by approximately 60–80 mmHg. However, since an identical pressure increase occurs in the arterial system, the arteriovenous pressure gradient remains the same. In addition, during walking, the pump-like action of the muscles comes into play, and this can lead to significantly increased blood flow toward the heart. In this process, the muscles are the energy source, and the veins are the bellows. The blood, which has been propelled toward the heart by the muscles, is prevented from returning through the venous valves (Sumner 1984) (Fig. 6.**9b**). During rest, however, the venous valves are constantly open, both when the body is reclining and when standing. They only close following an acute increase in intra-abdominal pressure (coughing, straining) and if there is a sudden change in body position (Schoop 1988).

Special provocation maneuvers and compression tests can be used to gather diagnostically useful information concerning venous function.

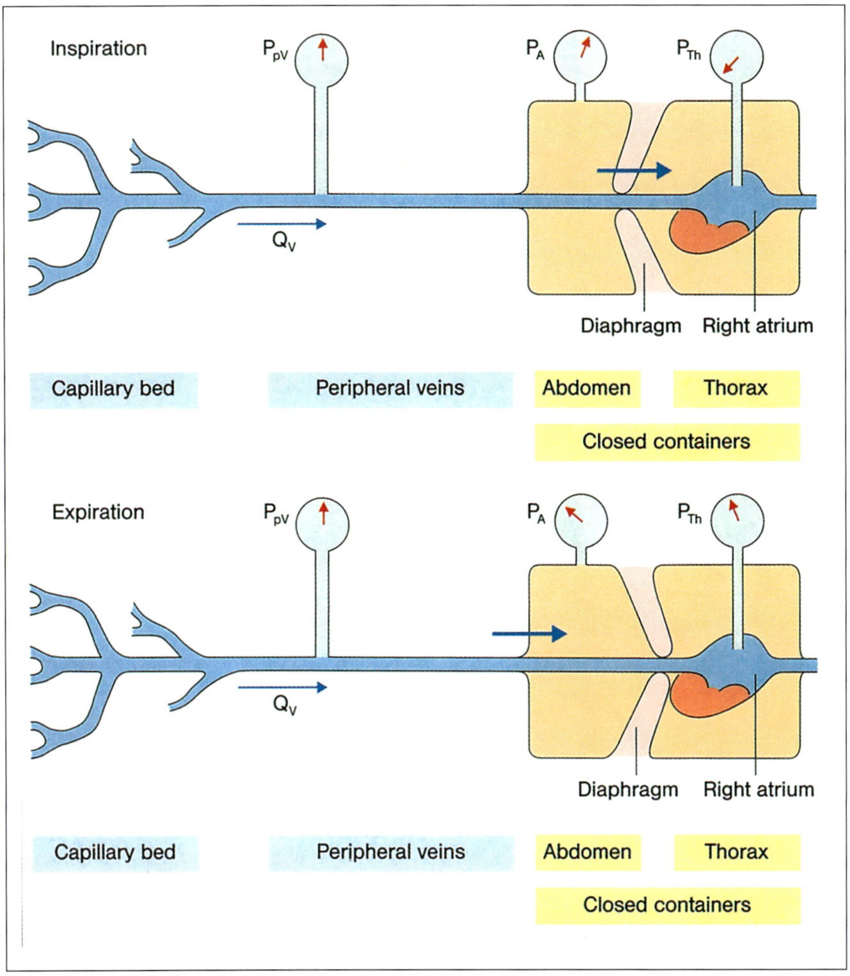

a

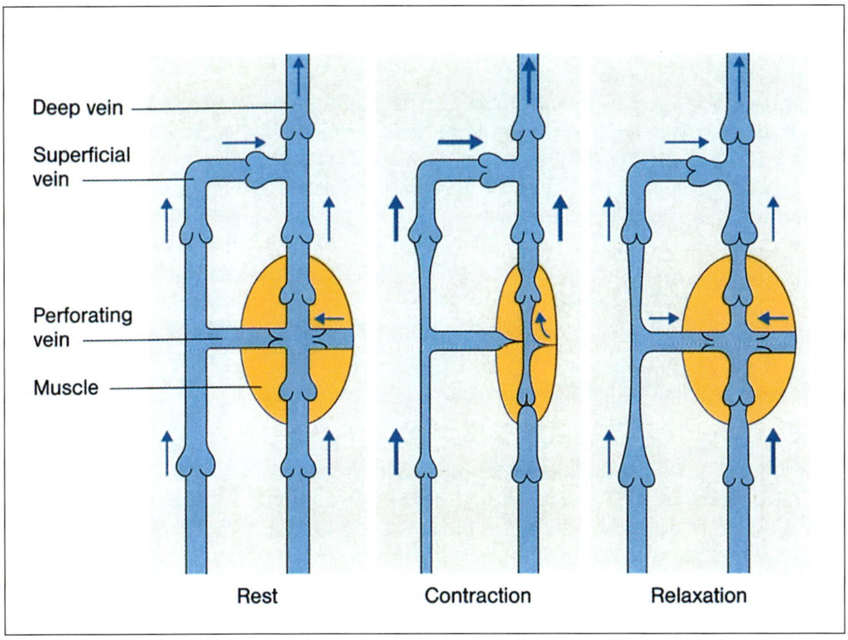

b

Fig. 6.**9 a** The physiology of venous flow toward the heart when the patient is lying down, illustrating the principle of the abdominal and thoracic "containers" (following Sumner), with changing pressure relationships that are respiration-dependent. **b** Physiology of the pump function provided by the muscles, which actively propel blood back to the heart when the body is in a standing position (adapted from Sumner). *Resting:* at rest, the deep and superficial venous valves are open. The valves of the perforating veins are closed. *Contraction:* during muscular contraction, the deep and superficial veins increasingly empty in a central direction. *Relaxation:* during muscular relaxation, blood is sucked from the superficial veins through the open valves of the perforating veins into the deep venous system, and is guided toward the center

P_{pV} Peripheral venous pressure
P_A Abdominal cavity pressure
P_{Th} Thoracic cavity pressure
Q_V Venous outflow

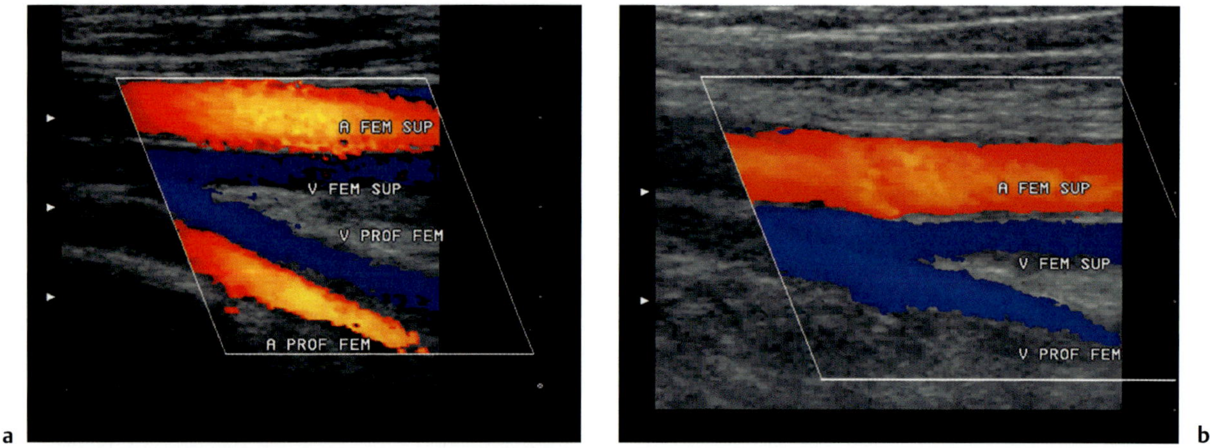

Fig. 6.**10** In a longitudinal sonographic section, the same anatomic region shown in Fig. 5.**23** (p. 168) is visualized with color flow duplex. The venous bifurcation of the superficial femoral vein (V FEM SUP) and deep femoral vein(V PROF FEM) lies 4–5 cm distal to the arterial vascular bifurcation of the superficial femoral artery (A FEM SUP) and deep femoral artery (A PROF FEM). **a** The course of the veins and arteries is parallel. **b** On the patient's left side, beneath the venous bifurcation, only the superficial femoral artery is visualized, as the deep femoral artery follows a different course not on this one plane

Color Flow Duplex sonography

Imaging procedures have the advantage of allowing the venous course, which is often very variable, to be followed, and pulsed-wave Doppler signals from the vascular lumen being examined can be registered specifically under B-mode imaging control. Precise knowledge of the vascular anatomy, including the many variations in the courses of the veins that are often seen, is an important and necessary precondition for correct classification of the findings (Fig. 6.**10**).

The *B-mode* provides almost continuous display of the venous vascular system, including both the deep and superficial veins of the upper and lower extremities, as well as the pelvic and abdominal regions. An evaluation of the venous lumen, excluding the presence of thrombotic components, and venous dilation following the Valsalva maneuver, can be used to exclude thromboses. The most important test for assessing the patency of a vein is still mechanical compression with the transducer, which is successful in a genuinely patent vein when slight downward pressure is applied (Fig. 6.**11**).

With the integrated *Doppler mode,* spontaneous venous flow and its phasic course can be examined. Venous valve sufficiency, both in the deep and in the superficial venous system, can be confirmed using the Doppler mode with the Valsalva maneuver or proximal compression.

Color flow Doppler mode makes it easier to locate the veins, with color intensity and flow direction coding. It is particularly useful in anatomically difficult regions such as the lower legs, where three pairs of veins have to be identified. In the pelvic region as well, however, color coding is extremely helpful in locating the deep common iliac veins (see Fig. 6.**19**).

Continuous-Wave Doppler Sonography

Continuous-wave Doppler sonography has become an inexpensive noninvasive procedure for venous diagnosis, and it can provide information on functional venous flow parameters. *Pulsed-wave Doppler sonography* is not generally used as the only examination method, since determining venous depth is not the primary aim. With increased use of duplex sonography equipment, examinations using the pulsed-wave Doppler mode are often carried out as part of duplex sonography.

Anatomy and Findings

Pelvic and Leg Veins

When dealing with the veins of the extremities, it is useful to divide the venous system into deep and superficial systems as well as a system of perforating veins. This differentiation is reflected in the discussion of the normal and pathological findings that follows.

Deep Venous System

Anatomy

The deep veins of the lower extremity and the pelvis follow a course within a vascular sheath parallel to the arteries, and have the same names as the arteries.

The deep veins of the foot—i.e., the dorsal veins of the foot and the medial and lateral plantar veins, which are all paired—are connected to the superficial rete at the dorsum and the sole of the foot through the perforating veins. They drain the blood into the veins of

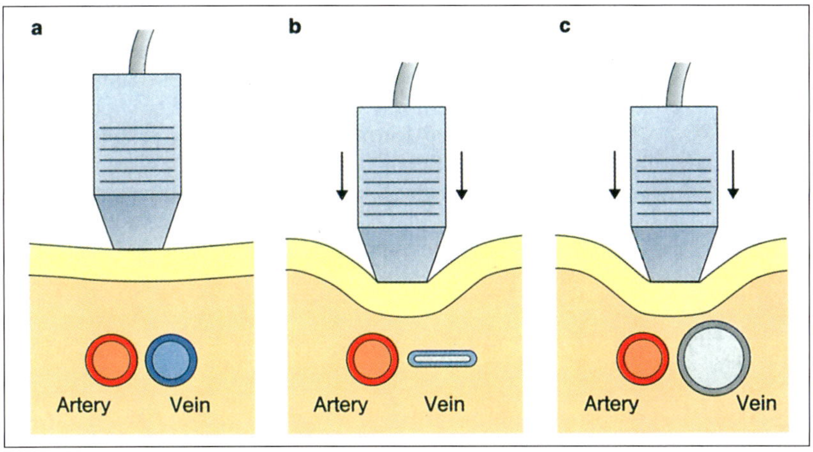

d
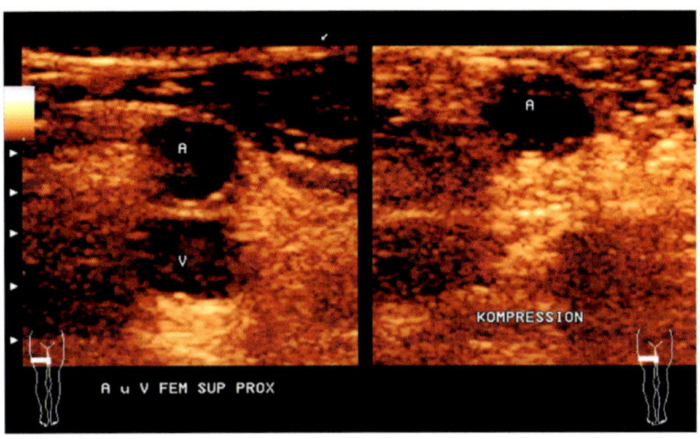

Fig. 6.**11** Compression testing to examine the patency of the deep veins (adapted from Habscheid). **a** With no external pressure applied, the veins normally appear to be approximately the same size. The arterial wall can be identified by its more intense reflections. **b** During compression, the vein collapses, while the size of arterial lumen remains unchanged. **c** In a pathological case, the freshly thrombosed vein, without any compression, is larger than the artery. During compression, it does not change its diameter at all, or only very slightly. **d** Compression sonography of the proximal superficial femoral vein. On the left-hand side, the lumen of the superficial femoral artery (A) and vein (V) is shown in a transverse plane. On the right-hand side, the lumen of the vein is completely collapsed under compression with the transducer

the lower leg, specifically the *posterior and anterior tibial veins* and the *fibular veins*, which also accompany the three corresponding arteries in pairs. They sometimes divide into three or four blood vessels. In the proximal segment, they fuse; their junction with the popliteal vein occurs either distal to the knee joint (46%), at the joint level (9%), or proximal to it (40%) (Weber and May 1990, p. 25).

The *popliteal vein* follows a dorsal course at the posterior aspect of the knee above the corresponding artery, and has a diameter of about 0.8 cm. After it enters the adductor canal, the popliteal vein becomes the superficial femoral vein and joins the deep femoral vein 2–7 cm below the inguinal ligament. From this point onward, it is termed the *common femoral vein* in both the angiological and radiological nomenclature. Even the superficial femoral vein shows variations from the normal unpaired form (62%). In 21% of cases, there is a division in the distal segment; in 13%, multiple veins are found. A completely paired superficial femoral vein only occurs in 3%.

Above the inguinal ligament, the femoral vein becomes the *external iliac vein* (diameter 1.2–1.4 cm), which follows a course parallel to the corresponding arteries, and unites with the internal iliac vein. It then continues as the *common iliac vein* (diameter 1.6–1.8 cm) into the inferior vena cava.

The junction of the common iliac veins with the inferior vena cava occurs to the right of the spine. For this reason, the left common iliac vein, with a length of 7.5 cm, is longer than the right common iliac vein, which is approximately 5.5 cm long. The right common iliac vein is located dorsal to the corresponding artery, while the left common iliac vein lies medial to the artery. Shortly before its junction, it is crossed by the right common iliac artery and pressed against the body of the fifth lumbar vertebra. At this point, the so-called *venous spur* can form (May 1974, p. 168)—a membrane that projects into the lumen and obstructs it. The spur is important, since it represents a thrombogenic obstruction to blood flow.

The number of venous valves varies. The most important are the subinguinal valve of the common femoral vein, located just below the inguinal ligament, and the valve of the popliteal vein, located at the level of the interarticular space of the knee. The superficial femoral vein and the common femoral vein have between one and six valves. Fairly consistently, a valve is found directly below the junction of the deep femoral vein, as well as in the adductor canal. Many valves are found in the lower leg veins, at intervals of approximately 2.2 cm.

A special group of veins located in the calf muscles are termed *muscle veins*. These are particularly impor-

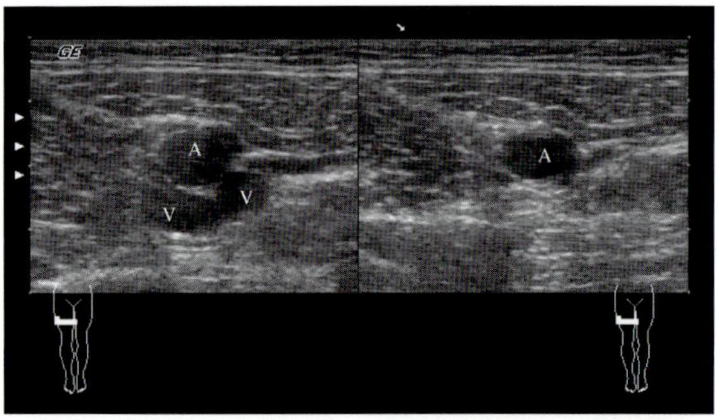

a

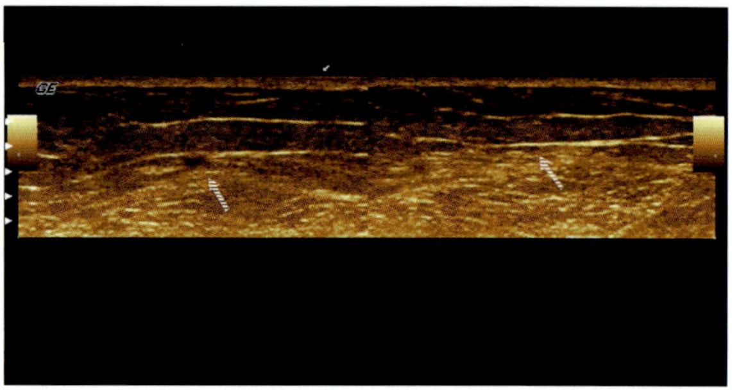

b

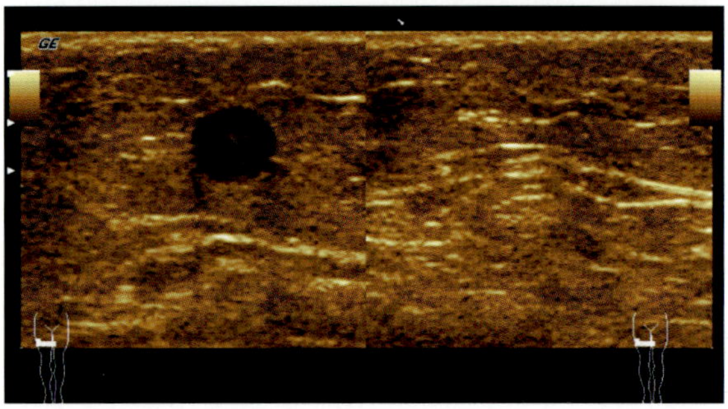

c

Fig. 6.**12** **a** Transverse section at the proximal thigh without (left side) and with compression (right side). The superficial femoral artery (A) is accompanied by two duplicated superficial femoral veins. During compression both lumina of the veins are compressed. **b**, **c** Transverse section of the great saphenous vein at the proximal thigh of two different patients. **b** Normal compression sonography (split-screen: left side without compression and right side with compression) of a vein with a very narrow lumen (arrows). **c** The same maneuver with a larger vessel caliber

tant due to their tendency to form localized thrombi. They can be regarded as indirect perforating veins to some extent, but are usually classified as belonging to the deep veins.

Two groups are distinguished: the *gastrocnemial veins* and the *soleal veins*. The soleal veins, which form three intramuscular longitudinal branches, are considered by Dodd and Cockett to be a broad, valveless venous sinus, which, like the gastrocnemial veins, flows into the popliteal vein.

Color Flow Duplex Sonography

B-mode. Imaging of the vein allows evaluation of the venous lumen, the diameter of the vessel, and its compressibility, and some authors therefore consider functional flow examinations to be unnecessary (Sullivan 1984, Habscheid and Wilhelm 1988).

However, this only applies to testing the patency in regions that can be directly compressed by the transducer (Fig. 6.**12**).

This is not necessarily the case in the pelvic region. In addition, examining valve function with the B-mode alone is usually not possible. Depicting the venous valves and their functioning is only feasible with very good imaging quality (see Fig. 6.**15**). However, in most regions, a general overview excluding a deep venous thrombosis can be confirmed by an experienced examiner using the method known as compression sonography in transverse section (see Table 6.**6**).

Doppler mode. As described below for continuous-wave Doppler sonography, the Doppler mode is used to register spontaneous flow modulations, as well as the provocation maneuvers that assess valvular function, both under visual control.

Color flow Doppler mode. The color flow Doppler mode makes it possible to locate the veins more quickly, in the lower leg and pelvic regions as well, and can therefore exclude venous thromboses with a greater degree of certainty. However, the distinction between respiration-modulated flow and continuous flow, as seen in thromboses, can only be made reliably with the Doppler mode. Continuous color representation in the venous segment being examined can exclude a thrombosis as well. Valsalva maneuvers can also be recognized using the color flow Doppler mode. Reflux is observed when there is a color change. With color flow Doppler sonography, it may often still be possible to detect venous flow in the lower legs without using provocation maneuvers when a venous signal can no longer be detected with continuous-wave Doppler sonography.

Continuous-Wave Doppler Sonography

The Doppler-sonographic examination of the deep veins proceeds from a proximal to a distal orientation, and uses the favorable areas for applying ultrasound described in Table 6.**1**. Functional evaluation of venous flow provides information about the venous segment located caudal to the probe. Two important aspects need to be evaluated: the *patency* and the *venous valve functioning* of the undisturbed vein. When the venous flow is genuinely free, the blood flow usually is phasic, decreases during inspiration, and increases again toward the heart during expiration (pp. 218–219). If the

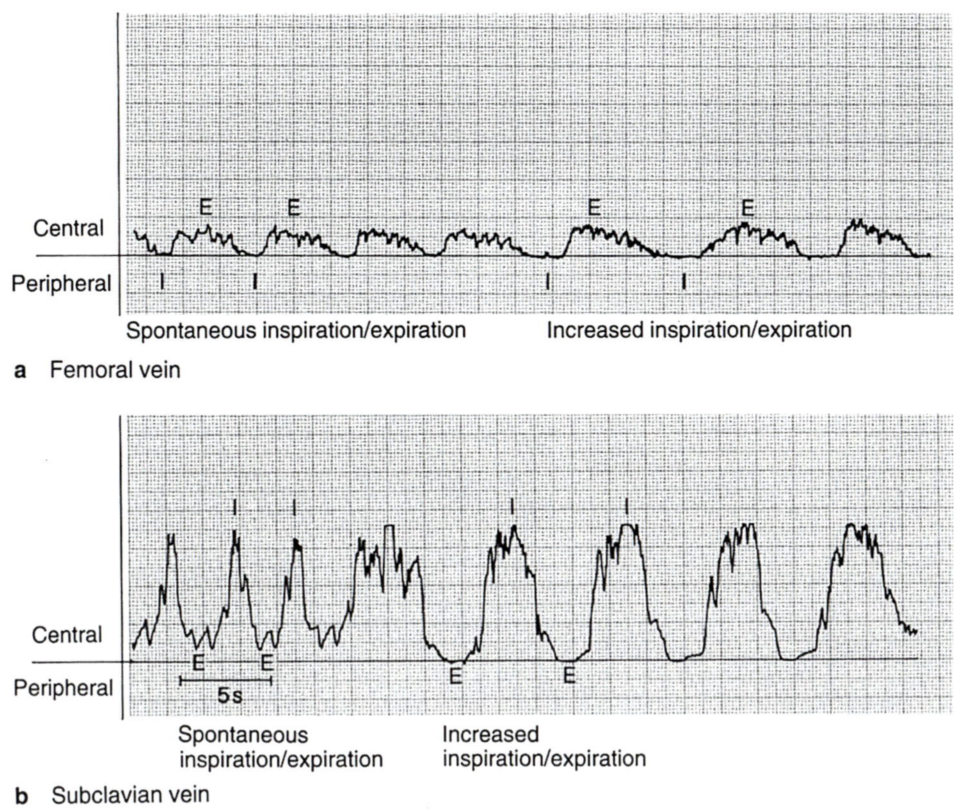

Fig. 6.**13** The graph shows physiological venous blood flow in the lower body, femoral vein (**a**), and in the upper body, subclavian vein (**b**) using continuous-wave Doppler sonography. **a** In the femoral vein, the spontaneous venous flow almost stops during inspiration, and flows centrally only during expiration. During forced inspiration and expiration, the modulation of the venous curve becomes even more pronounced. **b** The flow behavior in the subclavian vein is precisely the opposite. During spontaneous respiration, the marked, centrally directed flow peaks can be registered in inspiration. During expiration, the venous flow decreases. While forced inspiration and expiration is going on, the flow ceases briefly during expiration

findings are unclear, the spontaneous flow is tested during deep respiratory excursions (Fig. 6.**13**).

When no spontaneous flow can be registered, as may happen at distal registration points, especially at the ankle (posterior tibial veins), an A-sound, namely a flow peak (Fig. 6.**5**) directed toward the heart, is elicited by rapidly compressing the forefoot distally. Increased blood flow can be produced indirectly even during compression proximal to the registration point. On compression, the flow initially ceases if the valves are intact. However, after decompression, an overshooting flow directed toward the heart follows. In addition, A-sound propagation indirectly indicates the patency of the intervening venous segment.

Usually, the Valsalva maneuver is used to *assess valvular function*. An analysis of the curve can be used to ascertain whether the flow comes to a standstill, because outflow from the extremities ceases due to the elevated proximal pressure, and reflux does not occur when the valves are sufficient. After the maneuver, an elevated flow velocity toward the heart is registered for a short time (overshoot) (Köhler and Neuerburg 1978).

Superficial Venous System

Anatomy

The superficial and epifascial veins lie above the fascia within the subcutaneous fatty tissue and, in contrast to the deep veins, have no accompanying arteries along their course.

The most important superficial vein is the long *great saphenous vein,* which originates from the superficial veins at the medial margin of the dorsum of the foot, and continues to the lower leg in front of the medial malleolus. Here it follows a course along the medial side and ascends in to the thigh dorsolateral to the medial condyle of the femur. It proceeds through the saphenous opening into the deeper regions, and approximately at the level of the junction of the deep

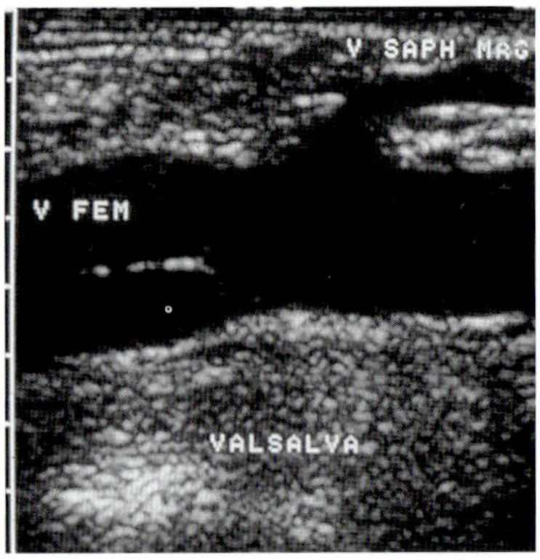

Fig. 6.**14** A B-mode image showing the femoral vein with the origin of the great saphenous vein during a Valsalva maneuver (normal results). There is a closed valve in the femoral vein during the Valsalva maneuver, and a chiasmatic shape at the junction with the great saphenous vein, in which return flow ceases during the Valsalva maneuver, as does that in the femoral vein

femoral vein—often somewhat proximal to this—curving from medioventral, joins the common (or superficial) femoral vein. The confluence has a characteristically arched, chiasmatic form (Fig. 6.**14**). The distal diameter of the great saphenous vein is 0.4–0.5 cm, and at the proximal junction point, which is often dilated into a funnel shape, it can range up to 2.0 cm. Approximately ten valves are encountered along its course. In 73 % of cases, a single trunk of the great saphenous vein is present; in 27 %, it is paired. Near its junction, there is a confluence of several veins: the superficial gastric vein, the external pudendal vein, the circumflex vein, the superficial medial femoral vein, the superficial lateral circumflex femoral vein, and the superficial circumflex iliac vein. This region has been described as a "venous star" due to the star-shaped form of the venous junction. The most important branches of the great saphenous vein are listed in Table 6.**3**.

Table 6.**3** Branches of the great saphenous vein, with location information relative to the course

Branch	Location
Lateral accessory vein	Anterior and exterior side of the thigh, possibly over the knee to the distal side
Medial accessory vein	Anterior and interior side of the thigh
Anterior arcuate crural vein (anterior arcuate vein)	Anterior and exterior side of the thigh
Posterior arcuate crural vein (posterior arcuate vein)	Interior side of the thigh

The second important collecting vein is the *small saphenous vein,* which originates from the lateral margin of the dorsum of the foot, and proceeds dorsal to the lateral malleolus to the dorsal lower leg. It continues onward and, approximately at mid-calf, runs through the fascia between the heads of the gastrocnemius muscle into the deeper regions, flowing into the popliteal vein usually at the level of the knee joint. The junction point is very variable: in 79 % of cases, it lies 3–5 cm above the interarticular space of the knee joint; in 14 %, it lies at interarticular space level or up to 3 cm above it; in 1 %, it is located distal to the interarticular space. In 50 % of cases, a single, unpaired vein is found, and in approximately one-third, the vein is paired (Weber and May 1990, p. 37). The number of valves is between six and twelve.

Color Flow Duplex Sonography

In general, the superficial veins can be displayed with high-resolution, high-frequency transducers. However, examining them is not normally part of routine diagnostic procedures.

The superficial veins in the lower body are often used as veins for bypass procedures. Testing the unobstructed patency of a vein and excluding paired segments and varicose abnormalities is important in these cases, especially when the great saphenous vein is used as an in-situ bypass. Since they are removed preoperatively, the valves do not have to be sufficient.

Continuous-Wave Doppler Sonography

Examining the *great* and *small saphenous veins* is only done when valvular insufficiency or thrombophlebitis is suspected.

If the valve in the region of the confluence of the great saphenous vein is found to be sufficient, no further examinations are undertaken unless a trunk varicosis, which can be caused by insufficient perforating veins, indicates distal insufficiency. The *small saphenous vein* is also only examined in varicosis, phlebitis, and chronic venous insufficiency.

Perforating Veins

Anatomy

The blood vessels that are termed *perforating veins* connect the superficial, epifascial veins and the deep, subfascial veins, while connections *within* the region of the superficial or the deep veins are made by the *communicating veins* (Weber and May 1990, p. 40). The perforating veins, which transport blood from the superficial venous system into the deeper regions, are of clinical significance. Reverse flow is normally prevented by valves.

The perforating veins are often referred to by the proper names of the anatomists who first described them (Table 6.**4**). They proceed through all the vascular

Table 6.**4** Perforating veins that have been given specific names (often proper names) due to the relative frequency of cases of insufficiency in these vessels

Perforating vein	Afferent and efferent areas	Location
Cockett's veins	Arcuate crural vein to the posterior tibial veins	Approx. 7, 13, or 18 cm above the sole of the foot
Bassi perforated vein	Small saphenous vein to the fibular veins	Approx. 5–7 cm above the sole of the foot, dorsolateral calf
Lateral perforating veins	Lateral branch of the great saphenous vein to the anterior tibial veins or veins of the gastrocnemius muscle	Distal third of the dorsolateral lower leg
May's veins (gastrocnemial point)	Small saphenous vein to the gastrocnemius muscle	About half-way up the thigh
Sherman's veins (24 cm perforating)	Arcuate crural vein to the posterior tibial veins	Center of the lower leg, about 24 cm above the sole of the foot
Boyd's veins	Great saphenous vein to the posterior tibial veins	A hand's breadth below the knee
Midcrural veins (lower, middle, upper group)	Anterior arcuate vein to the posterior tibial veins	Anterolateral side of the lower leg
Perforating popliteal veins	Popliteal vein to the surface vessels	In the centre of the popliteal rhomb
Dodd's veins	Superficial femoral vein to the great saphenous vein	A hand's breadth above the knee
Deep perforating veins	Extrafascial vessels to the deep femoral vein	Exterior side of the thigh
Hunter's veins	Great saphenous vein to the superficial femoral vein	Center of the thigh

regions of the extremities. Of the 95 groups described by van Limborgh (1963), only 18 groups, the majority of which are located in the lower leg, have clinical significance (Weber and May 1990, p. 349).

In the *lower leg*, eight groups are distinguished, including the medial, lateral, and posterior perforating veins, with Cockett veins I, II, and III having the greatest clinical significance.

The *Cockett* perforating veins lie in a defined region above the sole of the foot, and connect the great saphenous vein with the posterior tibial veins distally:

- Cockett I: 7 ± 1 cm
- Cockett II: 13.5 ± 1 cm
- Cockett III: 18.5 ± 1 cm

The group termed the *Boyd* veins is also well known. It lies approximately a hand's breadth below and medial to the knee joint, and is important because of its potential to supply the great saphenous vein in case of insufficiency.

The perforating veins of the *thigh* are also divided into three main groups: the medial, lateral, and posterior perforating femoral veins. The clinically most important veins here are the *Dodd perforating veins*, which are located at the medial distal thigh, and connect the great saphenous vein to the femoral vein.

Color Flow Doppler Sonography

The perforating veins can be located in B-mode images, so that functional tests can be carried out using the *Doppler mode* with definite sample volume placement. The *color flow Doppler mode* helps locate the connecting veins more quickly, and is useful in evaluating the flow direction (Fig. 7.**43**, p. 342).

Continuous-Wave Doppler Sonography

The method used to evaluate the perforating veins is described above in the section on examination (p. 215). Functional testing of these veins is not usually important. The perforating veins do not have any clinical significance until valvular insufficiency appears.

Upper Extremity Veins

Deep Venous System

Anatomy

The ulnar and radial veins collect blood from the deep palmar arch, and course in pairs alongside the corresponding arteries. They unite slightly above the elbow joint, and flow into the *brachial veins*, which may be either paired or singular.

The confluence of the brachial veins with the *axillary vein*, which is only paired in 1% of cases, is at the level of the inferior margin of the greater pectoral muscle. The superficial *cephalic vein* joins 2–3 cm distal to this point. The axillary vein contains a valve, and is approximately 3–5 cm long, with a diameter of 1.3 cm (0.8–1.9 cm). After passing below the clavicle, the axillary vein is called the subclavian vein.

The *subclavian vein* is separated from the artery of the same name by the anterior scalene muscle, and, on the right, flows into the innominate artery. On the left side, the vein, in contrast to the artery, is termed the left innominate vein after the junction with the internal jugular vein. The terminal valve is located in the venous angle; its functioning is influenced by the intrathoracic pressure.

Color Flow Duplex Sonography

As in the lower extremities, the veins of the arm can be followed from distal to proximal using *B-mode*, predominantly in transverse section, and their compressibility can be evaluated at the same time. The subclavian vein is examined proximal and distal to the clavicle. Since compression is not always possible at this location, the normal collapse of the subclavian vein after quick, deep inspiration is tested (Gooding et al. 1986). In B-mode imaging, definite intravascular identification of a central venous catheter can be tested (Möllmann et al. 1987). Subclavian vein punctures, and particularly axillary vein punctures, are much easier to do under visual control using B-mode imaging (Taylor and Yellowlees 1990).

Using the *Doppler mode*, phasic flow during normal respiration can be observed, registered and, if necessary, provocation maneuvers can be used (see Fig. 6.**27 a**).

Color flow duplex sonography allows faster documentation of the six important upper arm veins (the paired brachial veins, cephalic veins, and basilic veins) and the four forearm veins—i.e., the paired radial veins and ulnar veins. Continuous color intensity makes it possible to exclude thromboses.

Continuous-Wave Doppler Sonography

In the upper extremities, including the subclavian vein, the behavior of venous flow during respiration is the opposite of what it is in the lower body region. During inspiration, pressure in the thorax decreases, and the venous blood flows toward the heart. During expiration, the thoracic pressure increases, and the flow decreases, or even ceases during deep respiration (Fig. 6.**13 b**).

Forced respiration can also be used to assess valvular sufficiency in the upper extremities. Examinations using proximal compression are not usually done, and are not clinically relevant. However, quick distal compression, which produces A-sounds to help locate the veins, can be useful in this region.

Superficial Venous System

Anatomy

The most important collecting veins of the superficial veins of the arm are the cephalic vein and the basilic vein.

The *cephalic vein* proceeds from the radial margin of the dorsal venous rete of the hand, and follows a course diagonally above the radial margin of the forearm to the cubital fossa. From there, it continues in a stretched manner medially along the upper arm, 2–3 cm below the clavicle and into the axillary vein. It has six to ten valves.

The *basilic vein* starts from the ulnar margin of the dorsal venous plexus, and travels onward to the bend of the elbow, where it is joined by the cephalic vein. From there it continues into the medial bicipital sulcus and, approximately in the middle section of the upper arm, flows into the medial brachial vein at some depth. It contains four to eight valves, and may be paired, mainly in the forearm. There are two additional small branches that pass superficially on the volar side of the forearm—the median antebrachial vein and the median cubital vein—which flow into the cephalic vein and the basilic vein somewhat above the bend of the elbow.

Color Flow Duplex Sonography

As in the lower extremity, the superficial veins can be located using high-resolution, high-frequency transducers in *B-mode* imaging, with a water buffer if necessary. Examining these veins can be important prior to the placement of a hemodialysis shunt to confirm unobstructed vascular patency.

In *Doppler mode,* quantitative flow volume measurements can be made. These are important in measuring hemodialysis fistula output, and are discussed in Chapter 5 (p. 190).

The *color flow Doppler mode* is helpful in these investigations, since it makes it easier to follow the branching, superficial venous network, and facilitates the imaging of dialysis fistulas.

Continuous-Wave Doppler Sonography

Doppler sonography is normally only used to examine the larger collecting veins of the upper arm, the cephalic vein and the basilic vein, if there is any doubt about their patency prior to placing venous catheters. Assessing valvular insufficiency has no clinical role in the upper body.

The modulation test (p. 216) is recommended when the course of superficial veins is being followed.

Perforating Veins

Although the *upper extremities* also have many communicating veins, they have no clinical significance, and are therefore not specifically named and identified.

a b c d

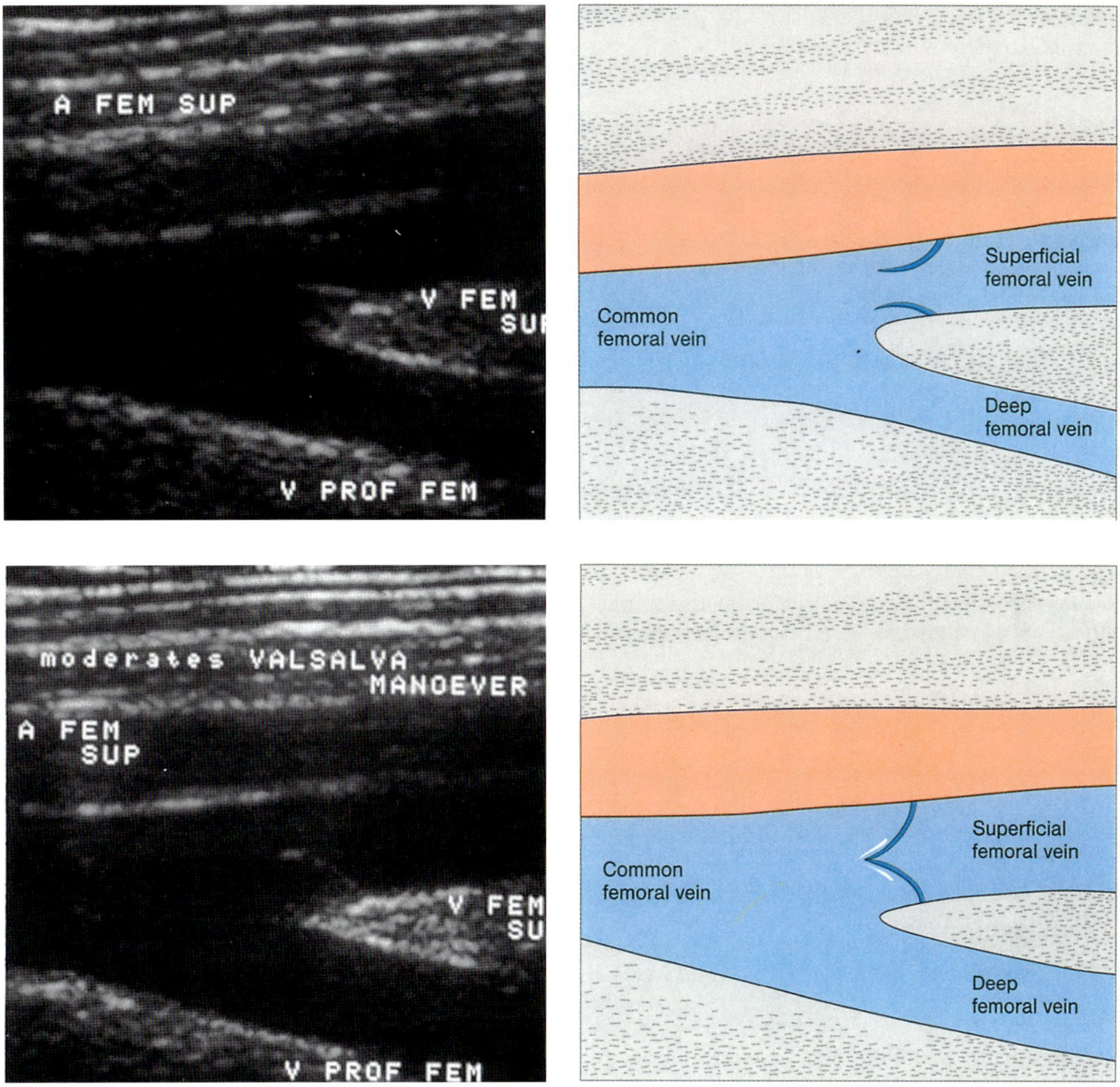

Fig. 6.**15** A longitudinal sonographic section through the bifurcation of the femoral vein. A few centimeters distal to the arterial vascular bifurcation (**a**, **c**), the confluence of the superficial femoral vein and the deep femoral vein is seen, forming the common femoral vein; the corresponding diagrams are shown below (**b**, **d**). The venous lumen is almost anechoic. In the accompanying superficial femoral artery, which is located ventrally, sparse reflections are recognizable. During expiration (**a**, **b**), the venous lumina have a normal diameter. Shortly below the confluence of the deep femoral vein, a venous valve in the proximal segment of the superficial femoral vein is open; the cusps of the valve lie flat on the venous wall. A moderate Valsalva maneuver causes an increased venous caliber. The cusps of the femoral vein valve have noticeably closed (**c**, **d**)

Evaluation

Patency Examination

Color Flow Duplex Sonography

In *"compression sonography,"* which involves compressing veins with the B-mode or duplex transducer, venous compressibility is preferably tested in transverse section. If the vein, in contrast to the accompanying artery, can be fully compressed, then venous thrombosis in the region being examined can be excluded with great certainty.

After the vein has been located in *B-mode imaging*, respiration-dependent signal modulation at rest and during deepened inspiration and expiration is observed using the *Doppler mode*.

In *color flow Doppler mode*, unobstructed venous patency is already detectable, without any compression, from the continuous color intensity both in cross-section and longitudinal section.

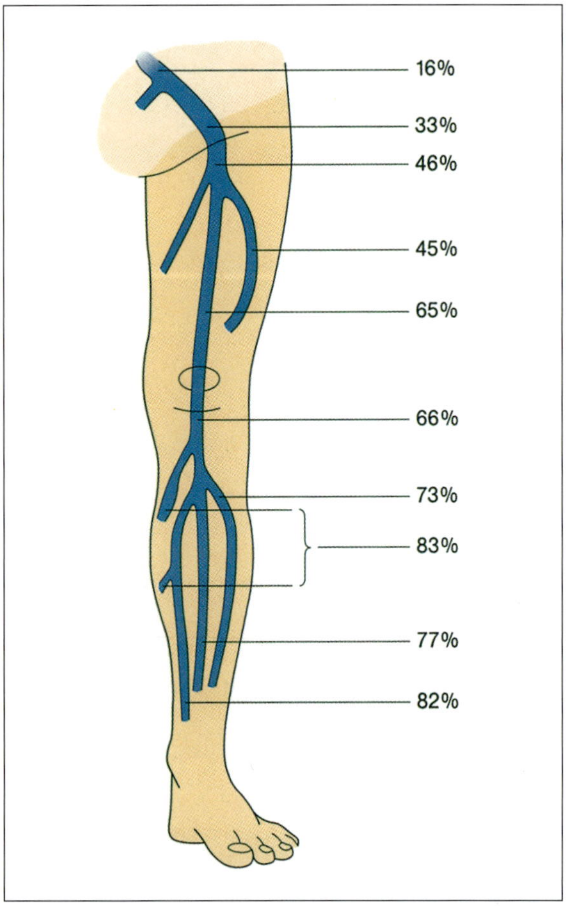

Fig. 6.**16** The location of thromboses and the frequency with which different venous segments are affected, expressed as percentage distributions (adapted from H.E. Schmitt, in Weber and May 1990)

These tests mainly apply to the deep venous system, but they can also be used for the superficial veins.

Continuous-Wave Doppler Sonography

Initially, the behavior of the venous blood flow is evaluated acoustically. As it is affected by respiration, the signal modulation is tested at rest and during deep inspiration and expiration. Flow cessation during the Valsalva maneuver is also examined. For overall orientation purposes, acoustic evaluation of the Doppler signals is sufficient.

It is advisable to register the venous flow as a Doppler frequency spectrum, in order to provide more precise documentation of the individual examination maneuvers (p. 211), and especially for follow-up examinations. Quantitative information is not usually provided in venous Doppler sonography. It is usually only the presence or absence of a reaction to respiratory or compression maneuvers that is of interest.

Valvular Function

Color Flow Duplex Sonography

In *B-mode*, the valves and their function can often be displayed morphologically. However, *Doppler mode* is usually also used to ensure a definite assessment of flow cessation. Assessment with *color flow Doppler mode* is even easier. During the Valsalva maneuver or proximal compression, no color change should take place (Fig. 6.**15**).

Continuous-Wave Doppler Sonography

In normal findings, an examination is carried out at the usual registration points to assess whether or not flow cessation occurs during the Valsalva maneuver. This confirms that definite valve closure has taken place. Alternatively, valve closure can also be provoked by proximal compression of the lower abdominal veins, or of the veins along the course of the extremity.

Pathological Findings

Principle

The division of the venous system into deep and superficial parts, as well as a system of communicating veins, is reflected in the different clinical pictures seen in pathological cases.

The most frequent disease of the deep venous system is acute *venous thrombosis*, with shorter or longer deep venous obstruction. Deep venous thrombosis, and its sequela, pulmonary embolism, is the leading cause of preventable in-hospital mortality in the USA and other developed countries. After spontaneous or drug-induced lysis, this usually leaves behind damaged venous valves. Subsequent clinical symptoms such as congestion, hyperpigmentation, secondary varices, and ulcers may appear, and this characteristic clinical picture is termed the post-thrombotic syndrome.

According to Kerr et al. (1990a), deep and superficial venous system thromboses occur above the knee in 51 % of cases, in 32 % below the knee, and in 17 % within the superficial veins. This retrospective data summary was carried out on the basis of 1084 consecutively studied extremities with acute venous thrombosis, which were examined using duplex sonography. In order of decreasing frequency, of the 3169 venous segments, 509 were popliteal vein thromboses (above the knee), 475 were located in the superficial femoral vein, 425 in the posterior tibial vein, 418 in the common femoral vein, and 314 thromboses were located in the great saphenous vein. A phlebography study by Schmitt (1977) provided percentage figures for thrombosis frequencies in a group of 383 patients (Fig. 6.**16**).

In superficial veins, *valvular insufficiencies* with their accompanying sequelae, varicose venous dilations, are important. In addition, local inflammatory re-

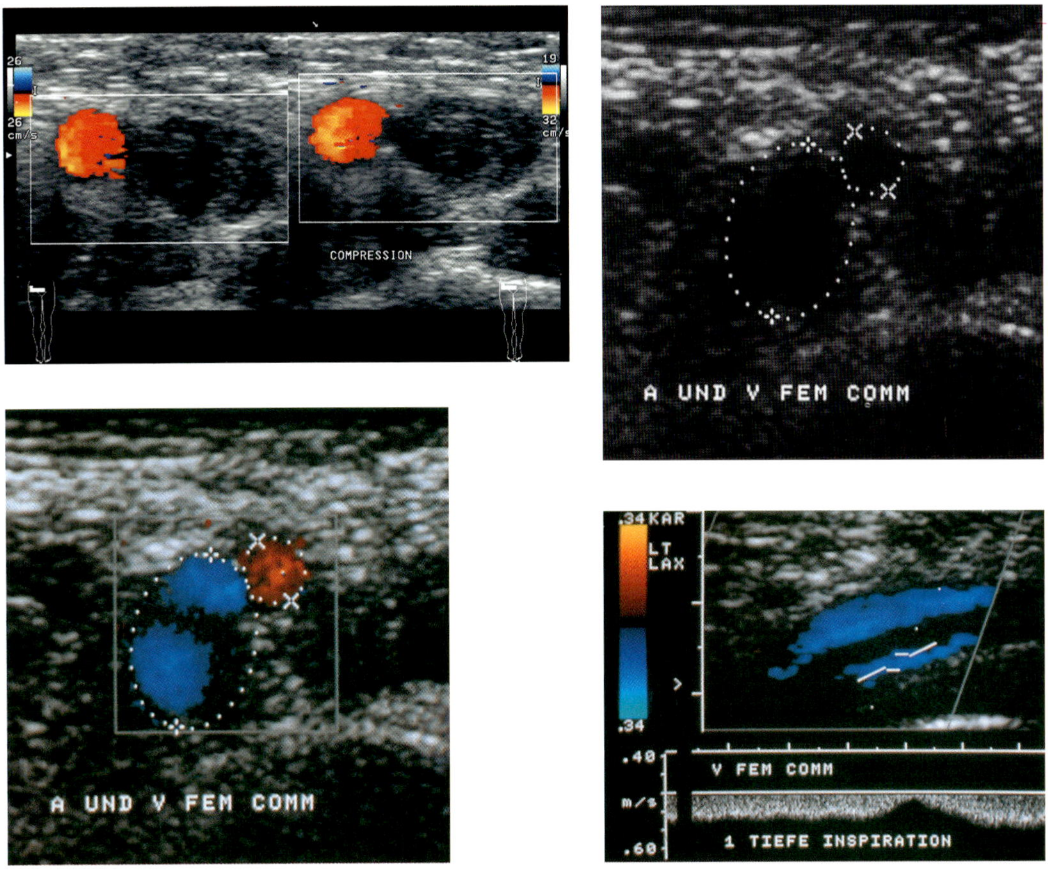

Fig. 6.**17 a** Complete thrombosis of the right common femoral vein. A transverse section, with the accompanying common femoral artery coded in red. During compression with the transducer (right side of the split screen), the shape of the vein becomes more oval, but the caliber is not reduced. **b–d** A partially recanalized common femoral vein thrombosis. **b** A transverse section at the left groin. The common femoral artery and the common femoral vein are shown, without compression. The vein is clearly dilated. **c** The same image after adding color flow Doppler mode. Only with color coding does the partial venous recanalization become clear. **d** The common femoral vein in longitudinal section, clearly displaying two recanalized canals at the margins of the vascular lumen. In the pulsed-wave Doppler mode spectrum, discrete respiratory modulation can be seen. During deep inspiration, the continuous venous flow while resting decreases. For technical reasons, the spectrum representing physiological venous flow was, exceptionally, registered below the baseline

actions such as *thrombophlebitis* are not uncommon in normal veins, or veins dilated due to varicosis.

Valvular insufficiency of the communicating veins, termed *perforating vein insufficiency*, also leads to drainage disturbances and secondary varicose dilations affecting the superficial veins.

All of these symptoms are caused by characteristic *morphological* and *functional* changes, which can be detected with ultrasound imaging procedures and diagnosed in their hemodynamic effects using Doppler devices.

Color Flow Duplex Sonography

It is possible to detect *deep venous thrombosis* using the B-mode component of duplex equipment, as well as with dedicated *B-mode equipment*. This is at its most reliable in regions that are readily accessible to compression with the transducer. The thrombosed vein shows no reaction on compression, or only a minimal one, and when compared to the normal vein it often contains scattered echo patterns (Fig. 6.**17 a**, **b**). Further size increase during a Valsalva maneuver is not possible.

Dilation, wall thickening, and internal structures within the venous lumen that are seen in the B-mode image of the superficial veins also suggest *thrombophlebitis* (see Fig. 6.**24**).

In *Doppler mode,* reflux phenomena accompanying valvular insufficiency can be clearly classified under visual control as belonging to the superficial or deep veins, and can also be followed along their course (Markel et al. 1992).

During the post-thrombotic phase, the spontaneous flow can again become phasic. Since the valves are damaged following thrombosis, deep venous *valvu-*

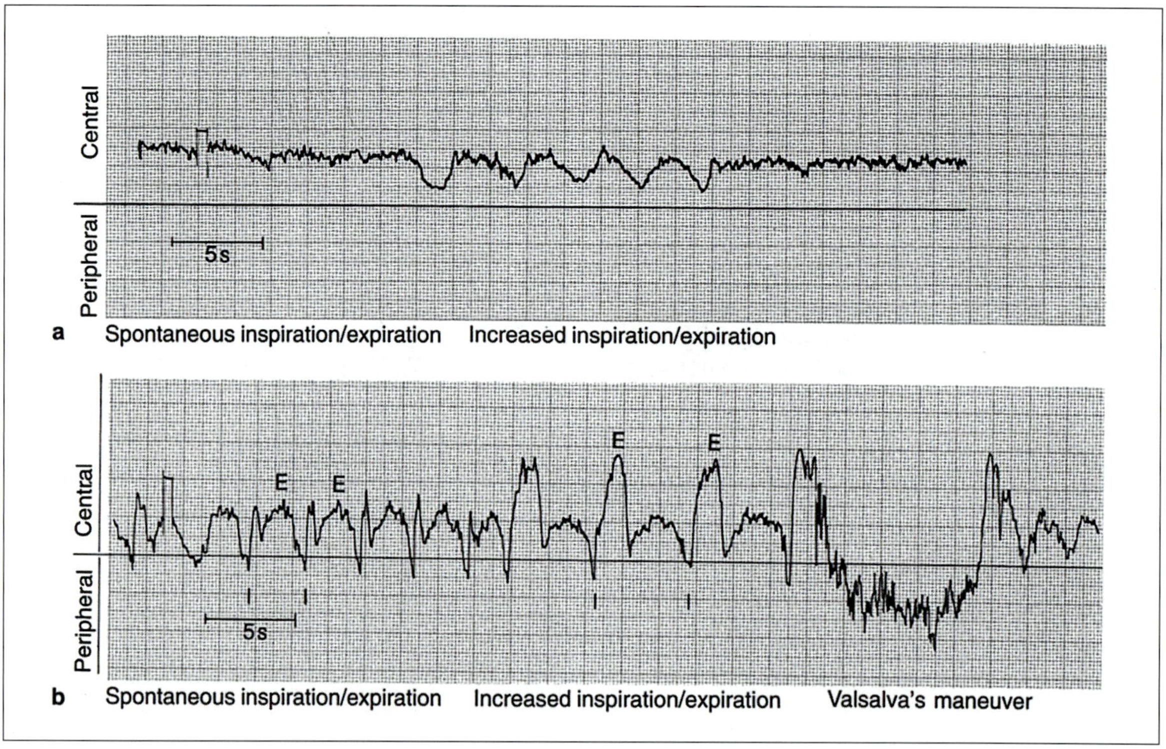

Fig. 6.**18** Pathological Doppler flow recordings registered from the femoral vein at the groin. **a** Venous flow during a pelvic venous thrombosis. During spontaneous respiration, there is a continuous, relatively high rate of flow without any modulation due to respiration. During forced inspiration and expiration, respiratory fluctuations are recognizable, although they are clearly less than those obtained in normal findings. During inspiration, centrally directed flow persists. **b** Reflux during a Valsalva maneuver. In patients with insufficient groin valves, retrograde flow into the deep or epifascial thigh veins occurs when pressure is exerted. This retrograde flow lasts until the reflux is stopped by a functioning valve, or until the maneuver ends. With continuous-wave Doppler sonography, it is not possible to differentiate, on the basis of the waveform, whether the reflux is in the superficial femoral vein or in the great saphenous vein. When duplex apparatus is not available, this has to be decided by compressing the great saphenous vein with a tourniquet

lar insufficiency, which can be detected with pulsed-wave Doppler, occurs subsequently. This means that during the Valsalva maneuver, *reflux* can be heard and documented peripherally (Fig. 6.**18 b**).

In *color flow Doppler mode,* thrombosis can be recognized by a color gap. Marginal flow in partial thromboses or during recanalization can be recognized with certainty (Fig. 6.**17 c**), and valvular insufficiencies can be detected by a color change during the Valsalva maneuver (change of flow direction).

Continuous-Wave Doppler Sonography

In the case of acute *venous thrombosis,* it is not possible to register a signal above the thrombotic occlusion itself. Distal to the occlusion, the venous pressure increases, and is higher than the intra-abdominal pressure during inspiration as well. Flow cessation or flow reduction during inspiration no longer occurs. Instead, depending on the quality and extent of the collateral circulatory system, a continuous trickle or elevated flow toward the heart is seen (Fig. 6.**18 a**) (Partsch and Lofferer 1971, Thulesius 1978).

In the *superficial venous system,* continuous-wave Doppler sonography can be used easily to detect valvular insufficiencies, which are recognized by a distinct reflux toward the periphery during the Valsalva maneuver. This occurs on proximal compression, as well as during the deflation phase following distal compression.

Findings

Pelvic and Leg Veins

Deep Venous System

As mentioned above, deep venous thrombosis in the leg represents the most common and serious disorder of the deep venous system. It may give rise to pulmonary embolism and chronic sequelae—particularly chronic venous insufficiency as a result of a post-thrombotic syndrome.

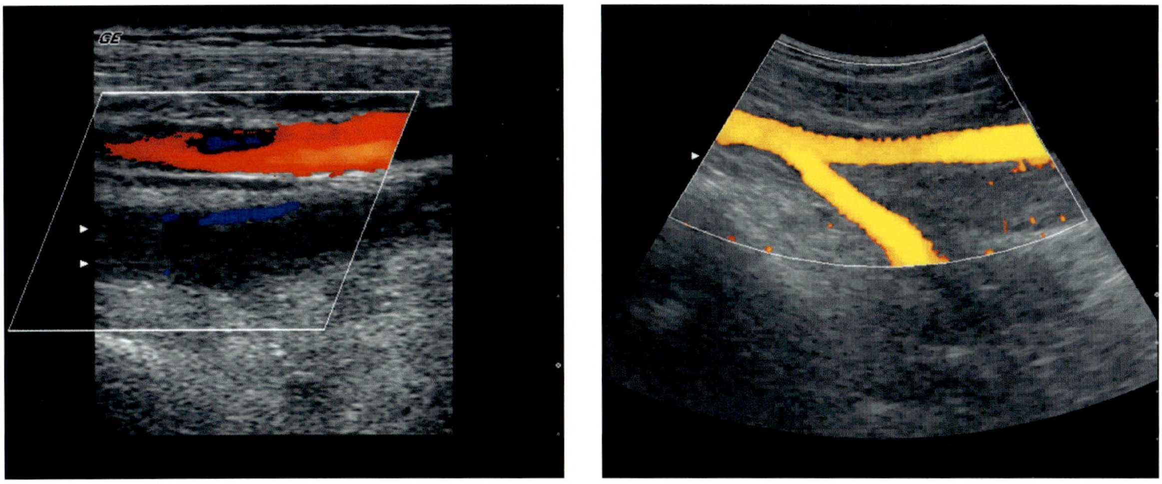

a b

Fig. 6.**19** Thrombosis in the common and external iliac vein. **a** Left external iliac artery and common and external iliac vein in longitudinal section. A nearly completely thrombosed common and external iliac vein with marginal ventral flow (blue) in the external iliac vein is recognizable. The diameter of the vessel lumen is dilated. Ventral to the vein, the arterial flow in the iliac artery is coded in red. **b** On the contralateral side in the same patient, the normal venous bifurcation with internal, external and common iliac vein is visualized in power mode

Thrombotic Processes

Color Flow Duplex Sonography

B-mode. Diagnosing a deep venous thrombosis in the femoral and popliteal veins has become the domain of pure B-mode imaging diagnosis, and is termed *compression sonography*.

This procedure was first described by Talbot (1982), who formulated the most important criteria for B-mode imaging diagnosis (Table 6.**5**). These observations have proved to be reliable, and have been confirmed by many studies (see Table 6.**7**) (Fig. 7.**41**, p. 339).

Compression testing is not possible in the pelvic region, where there is no resistance to compression. In addition, evaluating the lower leg veins solely with B-mode sonography requires good anatomical and topographical knowledge, practical experience, and commitment on the part of the examiner. It has been shown that thrombosed and dilated venous segments, which accompany the lower leg arteries as soft, tissue-like, noncompressible cords, are a good criterion for diagnosing thromboses, and are easily located (Figs. 6.**19**, 6.**20**). It is more difficult to follow the three patent paired veins continuously, however, and many authors believe this is not necessary (Habscheid and Landwehr 1990, Krings et al. 1990, Herzog et al. 1991, Yucel et al. 1991). The variable quality of B-mode imaging (depending on older or newer equipment) has a significant effect on whether these smaller-caliber veins can be located.

Doppler mode. Using Doppler mode, altered flow criteria, as described above, can be detected, and reactions to compression under visual control can be

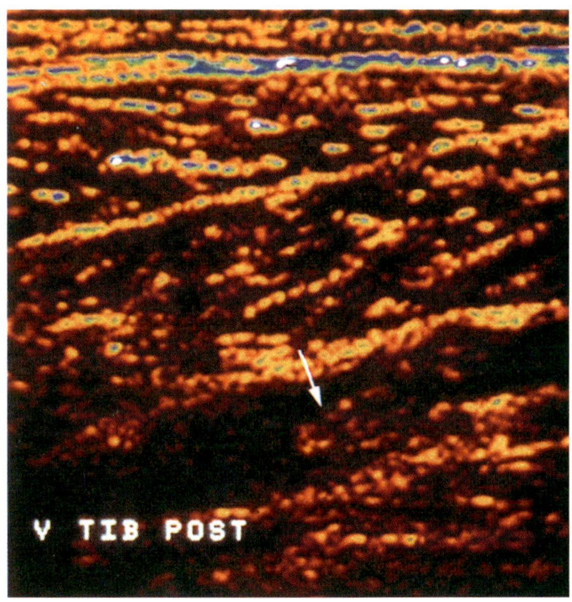

Fig. 6.**20** Posterior tibial vein thrombosis. The color-coded B-mode image shows the thrombus, the head of which is easily recognizable in the patent lumen (arrow). Close to the transducer, the muscles are seen

Table 6.**5** Criteria for detecting thrombosis using compression sonography

- Dilation of the thrombosed venous lumen
- Absent or clearly reduced compressibility
- Stationary internal echoes within the thrombosed lumen
- Absence of additional dilation during the Valsalva maneuver

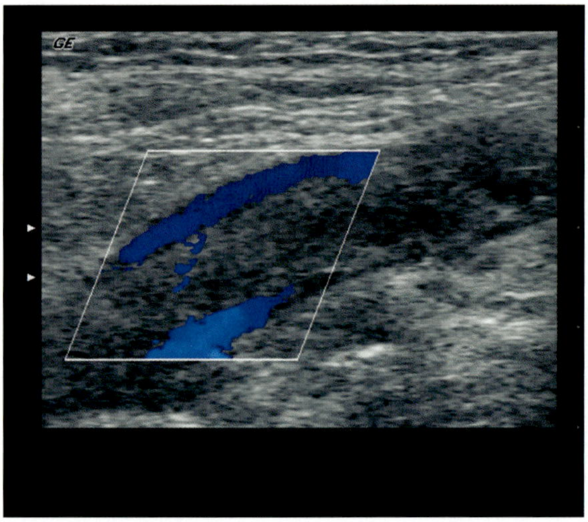

Fig. 6.**21** A parietal thrombus in the popliteal vein. With color coding, bilateral marginal flow can be seen in this longitudinal section through the popliteal vein, allowing assessment of a parietal thrombus (attached to the wall)

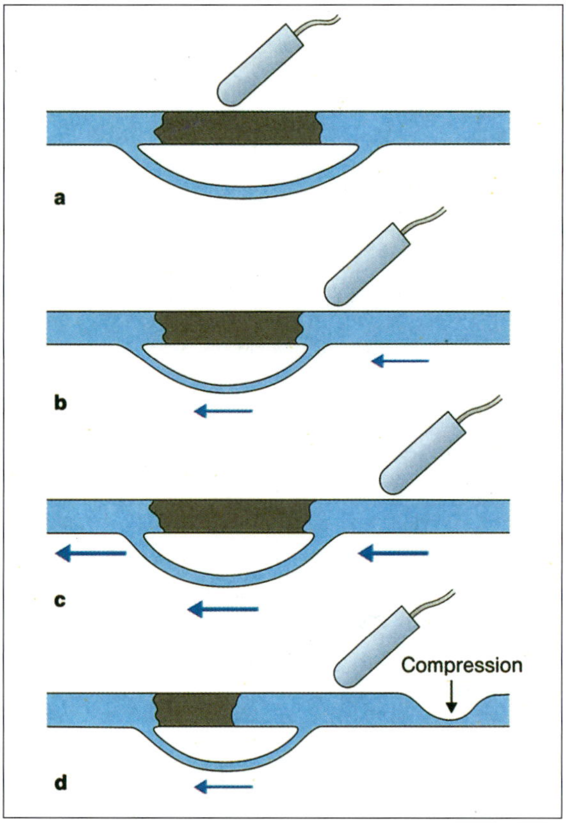

Fig. 6.**22** Doppler signals by a deep venous thrombosis and their dependence on the location of the Doppler probe, the quality of the collateral blood supply, and also the effect of distal compression (adapted from Strandness and Thiele). **a** Probe above the thrombosis: no signal. **b** Probe distal to the thrombosis, poor collateralization: continuous flow with a low velocity. **c** Probe distal to the thrombosis, good collateralization: continuous flow with a high flow velocity. **d** Doppler probe distal to the thrombosis, compression peripheral to the probe: low flow peaks directed toward the heart

specifically elicited in the deep and superficial venous segments. Comparison with the contralateral side is essential here.

Color flow Doppler mode. The color flow Doppler mode is particularly helpful in detecting marginal flow phenomena in partially thrombosed veins (Fig. 6.**21**). Reliable detection of thromboses is possible even when it is difficult to locate the veins (pelvic veins and lower leg veins) and compression sonography cannot be used, or only inadequately.

Continuous-Wave Doppler Sonography

It is decisive here to observe whether the examination is carried out at the typical registration points in the groin and popliteal region, proximal or distal to the thrombotic occlusion, or whether the probe is lying above the thrombotically occluded vein. In the latter case, no signal is registered, or minimal or increased continuous flow, depending on the vascular diameter, is detected from the collaterals running parallel to the occluded region. Peripheral to the occlusion, continuous flow is the clearest criterion for flow obstruction. Proximal to the occlusion, the flow is still predominantly modulated by respiration (Fig. 6.**22 a–c**) (Strandness 1990, p. 169).

There is a characteristic reaction to *distal compression,* which causes an abrupt flow cessation or a very low flow peak following compression when the probe lies distal to the occlusion (Fig. 6.**22 d**). The compression maneuver therefore causes a reaction opposite to that seen in free flow conditions, in which compression results in increased flow (A-sounds) being registered toward the center. When the probe is placed proximally, no flow reaction to *distal* compression is detectable, or only a very limited one (Fig. 7.**37**, pp. 332–333).

The clearest appearance of these phenomena results when ultrasound is applied above the groin, and they become somewhat less noticeable toward the periphery. In the Anglo-American literature, proximal and distal compression has been reported to be very useful (Sigel et al. 1972, Barnes et al. 1976, Sumner and Lambeth 1979, Strandness and Thiele 1981). However, these phenomena should not be overvalued. In the German medical literature, Partsch (1976) and Wuppermann (1986) found no convincing diagnostic improvement when additionally using signals elicited by compression.

Muscle Vein Thromboses

Like deep vein thromboses, muscle vein thromboses may also cause pain in the vessel region involved. In addition, the risk lies in the potential ascent or descent, with thromboid growth in the large deep conducting veins. For this reason, every Doppler-sonographic examination—particularly if the deep veins are free—

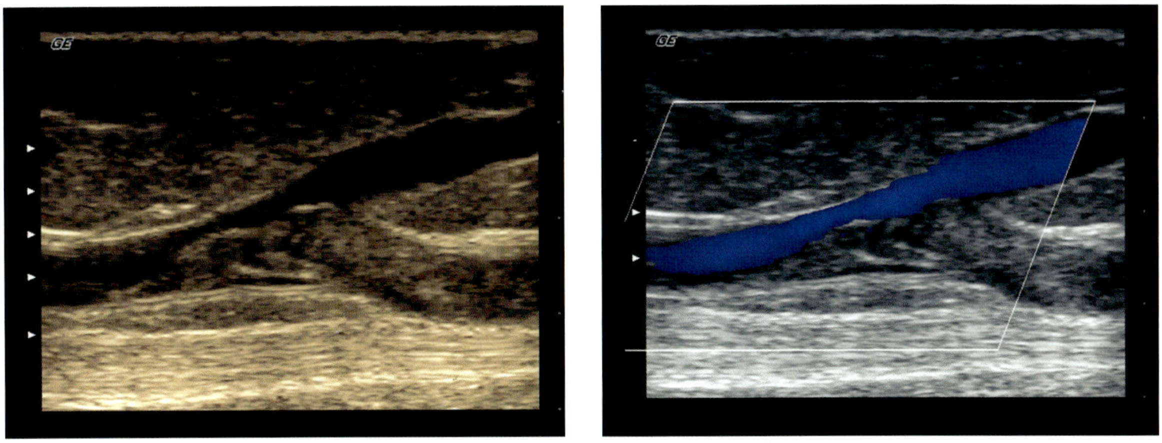

a b

Fig. 6.**23** Thrombosis of a muscle vein in the left proximal thigh in a longitudinal plane. The thrombotic material in the deeper vein can be detected due to its higher echogenicity (**a**). The peak of the thrombosis reaches the venous junction and juts into the incompletely thrombosed branch. With the color mode (**b**), normal flow in the unaffected vein is clearly visualized

should focus on the presence of muscle vein thromboses. They are often located in the lower leg, but may also occur in the thigh, as can be seen in Fig. 6.**23**. During the sonographic examination, the region in which the patient reports the strongest pain should be examined very carefully for muscle vein thromboses.

Color Flow Duplex Sonography

B-mode. The same criteria apply in the diagnosis of muscle vein thromboses as in the detection of thrombotic venous occlusions in the deep veins or epifascial surface veins. Apart from the recognition of the thrombotic material, which usually has a higher echogenicity in comparison with that of blood, the absence of compressibility of the muscle vein by the linear transducer has decisive significance.

Doppler mode. The pulsed Doppler mode is only of very minor importance, particularly since the speed of the blood flow in the muscle veins at rest is very slow.

Color flow Doppler mode. Basically, the points mentioned above on thromboses in the deep veins of the leg also apply to the use of color Doppler or power modes.

Valvular Insufficiency

Deep Venous System

Various studies (Strandness et al. 1983, Markel et al. 1992) have shown that valvular insufficiency develops after deep venous thrombosis at varying time intervals, depending on the recanalization process (p. 249). It does not usually appear until five or six months later, and signs of insufficiency are primarily seen in the valves below the knee, documented by reflux during the Valsalva maneuver or proximal compression.

Color Flow Duplex Sonography

B-mode. With good local resolution in B-mode, larger venous valves can be recognized and their failure to close during the Valsalva maneuver can be observed (Talbot 1986). However, in most cases it is not possible to depict all of the valves, and in particular it is not normally possible to demonstrate their insufficiency.

Doppler mode. Like continuous-wave Doppler sonography, the Doppler mode allows evaluation of the reflux reaction, its duration, and—with systematic scanning—the length of the reflux area.

Color flow Doppler mode. In color flow Doppler mode, reflux can be recognized by a change in the color coding. At first, the changed flow direction can also lead to turbulences, which can be detected as differently colored pixels. When venous flow toward the heart is color-coded as blue, the reflux is coded red, for example. The examination can be completed in different specific regions of the deep venous system, and the end point of the reflux area can be identified visually.

Continuous-Wave Doppler Sonography

Valvular insufficiency in the deep conducting veins is already detectable with continuous-wave Doppler sonography. Whether the refluxes occur in the deep or superficial veins has to be differentiated.

A reflux that is registered at the *groin* can flow into the superficial femoral vein or great saphenous vein. When the great saphenous vein is compressed with a

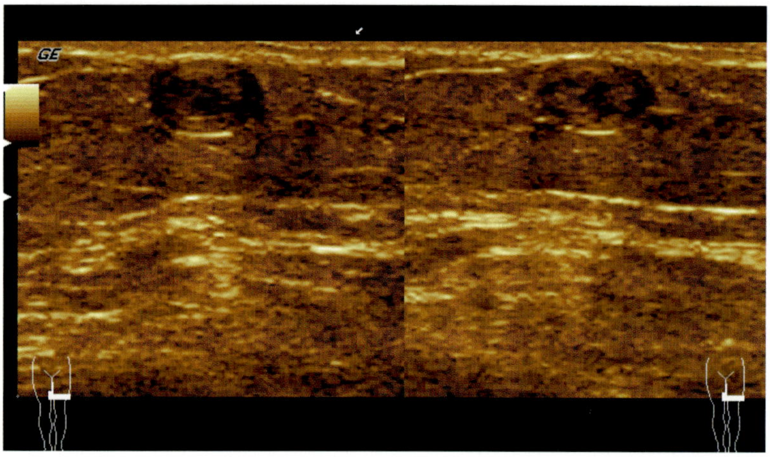

Fig. 6.**24** Great saphenous vein thrombophlebitis in the distal third of the thigh. The dilated superficial vein is visible, with a thickened wall and "frozen" valvular cusps. The lumen does not decrease under compression (left side in the split-screen mode)

tourniquet, reverse flow into the epifascial veins is avoided, and assignment to the deep venous system becomes easier.

Similarly, when Doppler sonography is being used to examine the popliteal region, reflux can be differentiated by obstructing blood flow to the great saphenous vein and the small saphenous vein in the distal lower leg. If the reflux persists, then it involves the deep veins. If the reflux ceases, then it was located in the superficial, epifascial venous system.

In *fresh thrombosis*, valve closure does not occur during the Valsalva maneuver, but the centripetal blood flow is generally preserved. Despite the elevated abdominal pressure (20–80 mmHg), venous pressure in the leg is still higher than the abdominal venous pressure.

When the venous thrombosis is fresh, the Valsalva maneuver should only be used if absolutely necessary, since there is a danger that the increased pressure may cause thrombi to be released, with subsequent pulmonary embolism (Kriessmann et al. 1990, p. 67).

Superficial Venous System

Detecting insufficient valves in the great saphenous vein and the small saphenous vein is clinically very important, and they can be reliably diagnosed even with continuous-wave Doppler sonography. Usually, the *valvular insufficiency* can be suspected due to a superficial varicosis. Clinically, taking the case history clarifies whether primary or secondary varicosis is involved. The latter often occurs in the post-thrombotic syndrome, which can be confirmed with duplex sonography. The deep venous system should therefore always be examined, and valve closure in the deep conducting veins should be tested. *Thrombophlebitis*—i.e., epifascial vein inflammation—can also be verified with duplex sonography using high-resolution transducers (Fig. 6.**24**).

Color Flow Duplex Sonography

B-mode. In the B-mode image, the dilated confluence regions of the great saphenous vein and the small saphenous vein can be measured, and their diameters can be compared with those of the main veins—i.e., the femoral vein and the popliteal vein. Bork-Wölwer and Wuppermann (1991) showed that these values correlate closely with measurements of vascular diameters obtained from phlebography. B-mode imaging can also evaluate the suitability of the great saphenous vein for use as a bypass vein by excluding any residual inflammatory changes, paired veins, or aneurysmal dilations (Ruoff et al. 1987).

Doppler mode. Functional evaluation of the reflux area is completed in the same way as with continuous-wave Doppler sonography. Careful placement of the pulsed-wave sample volume allows clear differentiation between the superficial and deep veins (Fig. 6.**25**).

Color flow Doppler mode. It is much easier to follow the course of a reflux phenomenon using color flow Doppler sonography. The exact spot can be observed at which the reflux color change can no longer be detected (Fig. 7.**42**, pp. 340–341).

Continuous-Wave Doppler Sonography

In valvular insufficiency, *spontaneous flow* in the superficial veins is slightly pendular, and contains centrifugal and centripetal components during increased respiration. The response to the Valsalva maneuver is a long *reflux*, with a flow peak that is directed toward the periphery and followed by overshooting flow directed toward the heart (Fig. 6.**18 b**). The length of time of the reflux depends on the number of insufficient valves and whether the Valsalva maneuver is applied effectively. An examination is conducted throughout the anatomical course of the main epifascial veins to assess whether partial or complete valvular insufficiency is

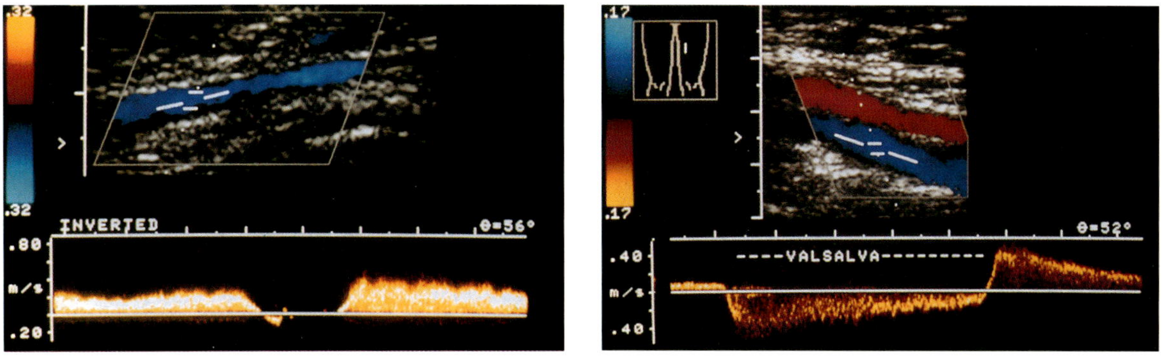

a b

Fig. 6.**25** Valsalva maneuver in the proximal third of the superficial femoral vein. **a** An example with sufficient valves. In the Doppler frequency spectrum, valve closure during the Valsalva maneuver is recognized by a flow cessation. **b** Pathological finding during a Valsalva maneuver. A long reflux is registered here. When the Valsalva maneuver is completed, the spectral analysis shows reactively increased flow toward the heart

involved, and particularly to determine how much of the vein the reflux affects.

According to Hach et al. (1977), varicosis of the great saphenous vein consists of four stages, which differ from one another in the length of the reflux that is observed:

- *Stage I:* valvular insufficiency in the confluence region and in the two venous valves located distal to it (proximal thigh).
- *Stage II:* valvular incompetence in the confluence region and in the femoral region of the great saphenous vein (distal thigh).
- *Stage III:* during the Valsalva maneuver, venous reflux to and beyond the knee (proximal lower leg).
- *Stage IV:* valvular insufficiency all the way to the ankle.

Assessing valvular insufficiency in the *small saphenous vein* using Doppler sonography has not become as important as it is in the great saphenous vein, because the confluence region has significant anatomical variants, and the course of the small saphenous vein in the deeper regions is not easy to follow. Without using B-mode, it is also difficult to distinguish whether a reflux in the popliteal region is coming from the popliteal vein or the small saphenous vein. Compressing or modulating the small saphenous vein behind the lateral malleolus is helpful in the identification process.

Collateral Veins

In the case of thrombotic deep venous occlusions, drainage of blood from the region takes place via collateral veins, with elevated pressure distal to the occlusion, in the direction of the heart.

Following May (1974, p. 141), drainage through the great saphenous vein is termed first-order collateral circulation; drainage through the small saphenous vein and the femoropopliteal vein into the veins of the buttocks is termed second-order collateral circulation; and drainage through the small saphenous vein and the superficial veins at the medial thigh that flow into the pudendal plexus is termed third-order collateral circulation. The continuous, usually elevated, drainage through these collateral veins can be clearly identified with a duplex scanner, preferably a color flow Doppler.

A spontaneous form of Palma's operation is also known. In a unilateral pelvic venous occlusion, this leads the venous flow suprapubically through the external pudendal veins to the contralateral side (Fig. 7.**38**, p. 334).

The patency of a spontaneous Palma or of Palma's operation, and the flow direction in each case, can be confirmed with continuous-wave Doppler sonography. However, it is better to evaluate patency and valvular insufficiency using conventional or color flow duplex sonography.

Regular examination of valvular sufficiency in the collateral veins is recommended. At the beginning of a thrombotic occlusion, the valves are usually still intact, but in chronic occlusion, valvular insufficiency results and secondary varicosis develops (Kriessmann et al. 1990, p. 76).

Thrombophlebitis

Initially, thrombophlebitis of the superficial veins is diagnosed on the basis of clinical symptoms such as reddening, hyperthermia, swelling, and pain in the affected epifascial venous segment. It has been shown that high-resolution transducers can determine the extent of the thrombosis and can occasionally locate thrombus growth in the patent lumen of the deep draining veins (femoral vein or popliteal vein). Lutter et al. (1991) were able to confirm that thrombophlebitis is associated with complications in 35% of cases, especially when it affects the great saphenous vein. The

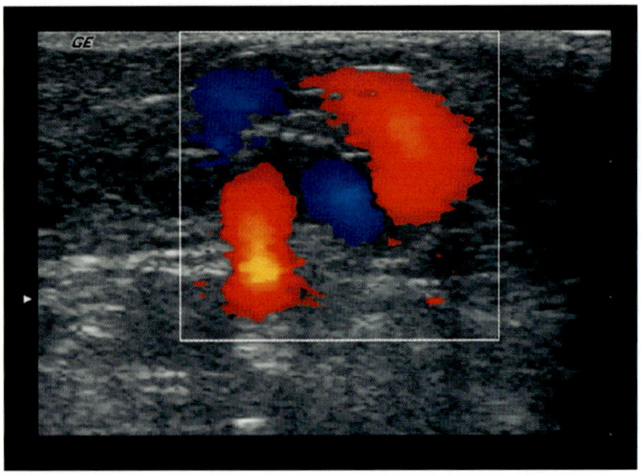

Fig. 6.**26** Color flow Doppler, showing an insufficient perforating vein in the lower leg, with enormous elongation and dilation of the vein. The insufficiency is identifiable due to the retrograde flow toward the transducer (red area in the middle of the picture)

complications entail either pulmonary embolism or deep venous involvement.

Color Flow Duplex Sonography

In *B-mode*, the vein sometimes appears to be irregularly configured, with a thickened wall. As in the deep veins, the lumen cannot be compressed. Depending on the echogenicity of the thrombus, a stronger reflection intensity in the lumen is noticeable even without compression (Fig. 6.**24**).

Using *Doppler mode*, the absent venous flow can be documented (Ruoff et al. 1987). With *color flow Doppler sonography*, Schönhofer et al. (1992) were able to demonstrate that thrombi partially adhering to the walls could be superimposed on phlebitis in the great saphenous vein. In individual cases, these thrombi can lead to pulmonary embolisms.

Perforating Veins

Valvular *insufficiency of the perforating veins* is clinically significant. It can lead to drainage disturbances and to the development of varicose malformations in the epifascial veins. Whenever the causes of venous congestion and ulcers are unclear, the examination of these veins should not be omitted.

Color Flow Duplex Sonography

The usually winding perforating veins can be depicted in *B-mode imaging*. The *Doppler mode* documents an unphysiological flow direction in the epifascial veins during compression, and retrograde suction during decompression. If a *color flow* duplex system is used, tourniquets are not necessary. Outward blood flow into the epifascial veins can be recognized by the same color coding as is seen in deep venous flow toward the center (Fig. 7.**43**, p. 342 and Fig. 6.**26**).

Continuous-Wave Doppler Sonography

A search is made for areas in which there is a gap in the fascia at anatomically typical locations. These are often recognized by redness or a "blow-out" (a protruding, limited, livid varix loop from a fascial opening). After sealing off the varices with tourniquets (see the examination section, p. 215), distal compression is applied, preferably at the foot or just above the examination area, causing flow from the interior to the exterior, toward the probe. When the compression is released, suction toward the interior can be recognized acoustically, and is also registered.

Upper Extremity Veins

Deep Venous System

Thrombotic Processes

Spontaneous venous thromboses in the arm, involving the axillary vein and the subclavian vein, occur relatively rarely (approximately 2% of all thromboses). However, they can arise acutely following physical exertion, and are usually related to a neurovascular compression syndrome (p. 182). They are therefore described as *"exertion-induced thromboses."* Deep venous thrombosis in the upper extremities is also known as *Paget–von Schroetter* syndrome (Hübsch et al. 1988). Central venous catheters are a not uncommon source of thrombosis, and this type of thrombosis is being diagnosed with increasing frequency in patients in intensive-care units (Weissleder et al. 1987, Horattas et al. 1988).

Color Flow Duplex Sonography

The criteria for diagnosing a thrombosis are the same as those in the lower extremities (p. 231).

B-mode. Compression testing using a B-mode or duplex transducer can be applied to the veins of the upper extremities in the same way as in the veins of the leg. In areas where this is not possible—e.g., in the subclavian vein below the clavicle and at its junction with the superior vena cava—the thrombosed vein can be recognized by its dilated lumen, with stationary texture patterns. The Valsalva maneuver does not cause an additionally increased lumen; also, the "sniff test" (rapid panting) does not cause venous collapse (Gooding et al. 1986) (Fig. 7.**44**, pp. 344–345).

Doppler mode. As with continuous-wave Doppler sonography, examinations with the Doppler mode also assess venous flow under visual control (Fig. 6.**27**).

Color flow Doppler mode. The color flow Doppler mode allows veins to be identified quickly, can detect occlusions or partial thromboses that are surrounded by flowing blood, and allows follow-up evaluations of

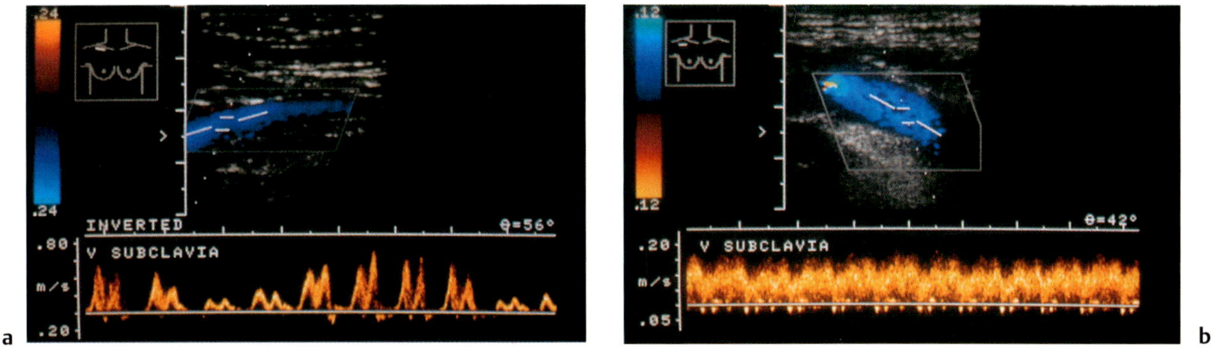

Fig. 6.**27** A color-coded B-mode image, showing the subclavian vein (blue) and the corresponding frequency spectrum of the venous flow. **a** Normal findings. Fluctuations during respiration can be seen. They decrease during deep expiration, while the flow velocity increases on inspiration. The double pulse of the venous flow due to tricuspid valve closure is easily recognized in the analytical spectral depiction. **b** Subclavian vein thrombosis. Continuous venous flow distal to a deep thrombosis of the subclavian vein. Respiratory fluctuations are not transmitted; the flow velocity is relatively high, indicating good collateral drainage

the progress of recanalization after deep venous thromboses (Fig. 6.**17**).

Continuous-Wave Doppler Sonography

As in the lower extremities, venous flow distal to the occlusion shows an absence of respiratory modulation and continuous flow toward the heart, while no venous flow can be registered directly above the thrombotic segment itself.

Due to good collateral blood flow, compression maneuvers that are applied proximal and distal to the thrombotic occlusion are less effective for differentiation purposes. The Valsalva maneuver usually produces an extensive flow cessation, but only results in a slight subsequent increase in flow in comparison with the healthy side (Kriessmann et al. 1990, p. 81).

Superficial Venous System

In the superficial venous system of the upper extremities, the most important blood vessels are the cephalic vein and the basilic vein. Primary or secondary varicosis is extremely rare. Welch and Villavicencio (1994) report three cases of unilateral segmental varicosis in the arm not caused by a malformation and without accompanying varicoses in the lower extremities. Superficial varicosis is one of the typical signs of congenital venous dysplasias (Klippel–Trénaunay syndrome; p. 238).

Superficial veins have become more important in intensive-care medicine with the introduction of central venous catheters and venous indwelling cannulas. In addition, they are used to place hemodialysis fistulas (p. 191). After longer-term placement in the body, venous catheters cause phlebitis, which is readily clinically recognizable by redness and pain, and does not always require additional clarification using sonography.

Color Flow Duplex Sonography

B-mode. Using B-mode sonography, the superficial veins can be examined with a high-resolution transducer, with a water buffer if necessary, and their patency can be confirmed, or inflammatory changes and thrombotic displacements can be recognized. B-mode is important for measuring the vascular diameter, which is necessary to calculate the flow volumes. B-mode imaging has become indispensable for monitoring the location of a central venous catheter.

Doppler mode. Doppler sonography registers a flow velocity waveform, and the computer equipment calculates a value for the average flow velocity. Secondary reticular collateral veins often appear in the upper arm and in the shoulder region after subclavian and axillary artery thromboses. Varicose malformation, however, is rare. With this value and the size measured in B-mode, the flow rate of the hemodialysis shunt can be determined.

Color flow Doppler mode. Imaging of the superficial veins has been made very much easier with color coding. Fast orientation is provided by using the color flow Doppler mode to diagnose the functional capacity of a hemodialysis shunt, and the presence of venous stenoses or venous thrombotic displacements. For quantitative measurements, additional use of the Doppler mode is necessary.

Continuous-Wave Doppler Sonography

The superficial veins can be located with the pen-shaped probe from continuous-wave Doppler equipment, and their patency can be evaluated. This can sometimes be important prior to venous punctures, or prior to the introduction of venous catheters. Assessment of valvular sufficiency has not yet formed part of these examinations.

Special Sets of Findings

Congenital Venous Dysplasias

Congenital vascular malformations are more important in the venous system than they are in the arterial component of the circulatory system. There are a number of characteristic clinical syndromes, and the symptoms of these should be mentioned here briefly. Since combined forms occasionally occur, Bollinger (1979, p. 244) recommends that specific naming of the syndromes should be avoided. Malan (1974) recommended a different classification. Only the clinical manifestations that are commonly recognized will be listed here, focusing on the atypical venous changes that are connected with them and can be confirmed using ultrasound diagnosis (Rutherford 1989). The extent of these changes is sometimes only recognized using imaging procedures. For additional information, the relevant textbooks can be referred to (Bollinger 1979, Vollmar et al. 1989) and the specialist literature can be consulted (Langer and Langer 1982, Voss 1984, Vollmar et al. 1989, Belov et al. 1989).

Klippel–Trénaunay Syndrome

(Diepgen et al. 1988)

- Unilateral, atypical varicosis
- Soft-tissue and bone hypertrophy affecting one extremity
- Nevus flammeus
- Deep venous system anomalies:
  - Valve agenesis
  - Aneurysmal venous transformations

F.P. Weber Syndrome

(Leipner et al. 1982)

- Secondary varicosis
- Soft-issue and bone hypertrophy affecting one extremity
- Arteriovenous fistulas
- Capillary hemangiomas

Servelle–Martorell Syndrome

(Martorell and Monserrat 1962, Servelle 1949)

- Atypical varicosis
- Skeletal hypoplasia
- Cavernous hemangiomatosis affecting one extremity
- Deep venous system anomalies:
  - Valve agenesis
  - Aneurysmal venous transformations

In more detail, the following sets of findings should be noted and, as far as possible, clarified using ultrasound diagnosis.

Atypical Varicosis

Unilateral, atypical varicosis of the extremities must be examined for its extent and its blood supply source. The examiner should check whether an arteriovenous fistula is supplying the varicosis, whether valvular agenesis involving the deep venous system is present, and whether apparent monomelic gigantism suggests the above clinical syndromes (Karasch et al. 1991).

Pelvic Venous Syndrome

Pelvic venous syndromes consist of a group of poorly understood disorders of the pelvic and gonadal venous circulation, with mild to severe symptoms of pelvic pain, dysuria, dysmenorrhea, dyspareunia, and sometimes the presence of vulval and pelvic varices (Scultetus et al. 2002). These symptoms can be caused by hypogastric vein tributary reflux, combined gonadal and hypogastric vein reflux, or incompetent tributaries of the internal iliac vein, which can be recognized with duplex sonography.

Hemangiomas

Capillary hemangiomas, such as nevus flammeus, should be distinguished from *cavernous* hemangiomas.

Nevus flammeus forms part of the classic clinical picture of Klippel–Trénaunay syndrome. The nevus is a capillary dysplasia located below the epidermis, with no tendency to grow further. It is usually congenital in origin, and can often regress. However, regression is rarely observed in the Klippel–Trénaunay syndrome. Since the clinical diagnosis is clear, Doppler or duplex sonography is not currently used to evaluate this condition. The high-resolution transducers (20 MHz) recently introduced in dermatology may be able to provide new morphological details.

Cavernous hemangioma consists of soft, venous cavernous structures that are filled with blood and lined with endothelium, in which chronic, intravascular clotting occurs. They are classified as tumors, and duplex sonography can be used to determine their depth and extent before any radiotherapy that may be required, and also in subsequent follow-up examinations (Handl-Zeller et al. 1989).

Arteriovenous Fistulas

The presence of arteriovenous fistulas in angiodysplasias suggests a possible F.P. Weber syndrome if the resulting secondary varices or aneurysmal structural disturbances appear unilaterally, and are related to increased growth of one extremity during childhood. Ultrasound diagnosis of arteriovenous fistulas is described on p. 190 (Fig. 6.**28**) (Partsch et al. 1975).

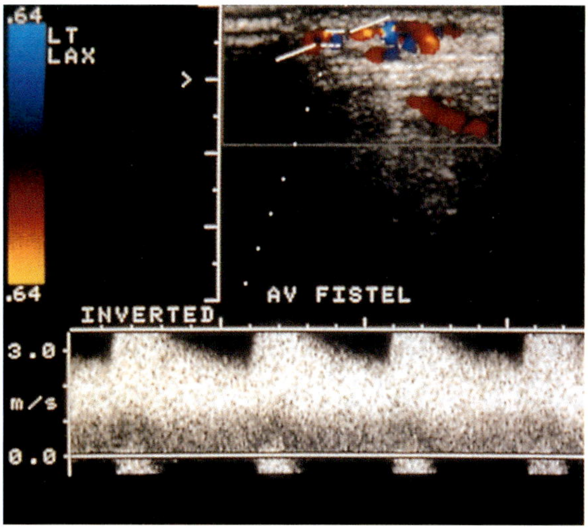

a

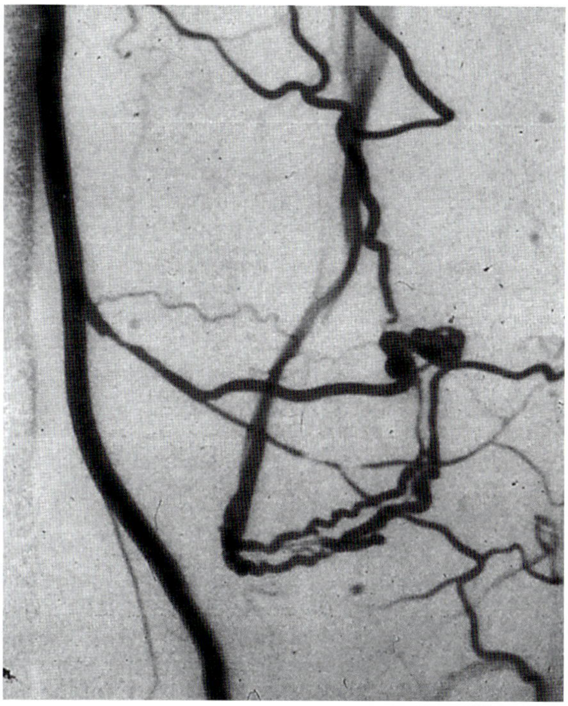

b

Fig. 6.**28 a** An in-situ arteriovenous fistula in a bypass of the great saphenous vein. The Doppler spectrum shows the fistula's high flow velocity of more than 3 m/s, similar to a high-grade stenosis with turbulence. **b** In the radiographic image of the same phase, the venous bypass, containing arterial flow, and a simultaneously filled vein are seen

Venous Hypoplasia and Aplasia

Vollmar et al. (1989) report finding a shorter or longer hypoplastic or aplastic segment of the deep venous system, with an accompanying increased resistance to drainage, in 50% of cases involving the clinical syndromes listed above.

So far, it has only been possible to detect these malformations using phlebography, but it should now also be feasible using color flow duplex sonography. In superficial varices that have an atypical course, the deep veins should always be examined to determine their course and the diameter of their lumen.

Venous Valve Agenesis and Hypoplasia

Doppler and duplex procedures have made it much easier to diagnose congenital valvular insufficiency. It has been shown that the syndromes listed above are associated with valvular agenesis, more often than was previously suspected. This means that the developing varicosis should not just be interpreted as manifesting congenital venectasias, but should also be viewed as reflecting secondary varicosis due to venous insufficiency. Partial and complete agenesis must be differentiated from each other. Very often, the valves of the deep main branches and the connecting veins are absent (Bollinger et al. 1971).

Venous Aneurysms

More extensive imaging diagnosis (Vollmar et al. 1989) has been able to show that Klippel–Trénaunay syndrome and Servelle–Martorell syndrome are often accompanied by venous aneurysms, and that they also show deep venous anomalies. The anomalies are almost always seen in these syndromes (Klippel–Trénaunay 96%, Servelle–Martorell 100%). According to the above study, venous leg angiodysplasias were seen in 80% of the cases, and arm angiodysplasias in only 20% of the cases. Aneurysmal transformations from normal venous structure appeared in half of the cases. Percentage figures for so-called cylindrical ectasias or aneurysmal transformations are not given. In venous graft aneurysms, Karkow et al. (1986) indicate a 50% venous dilation as a criterion for detecting the presence of aneurysms.

Secondary Arteriovenous Fistulas

Secondary arteriovenous fistulas are discussed on pp. 190–191 (arteries of the extremities). It is important to examine the venous component contributing to the fistula using Doppler, or—even better—color flow Duplex sonography. For example, stenoses in the vein forming the fistula are often encountered in hemodialysis shunts, and thrombi in the fistula may be found, or a combination of the two (Nonnast-Daniel et al. 1992).

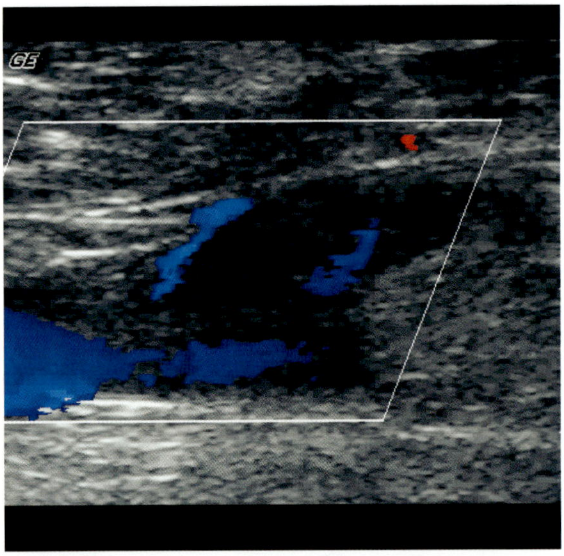

Fig. 6.**29** A recanalized thrombosis of the right great saphenous vein, showing vascular dilation and hypoechoic, intraluminal structures. Marginal flow is visible in the color flow Doppler mode

Acquired Venous Aneurysms

In contrast to the arterial sector, acquired venous aneurysms are rarely observed. They appear to be related to congenital wall weakness in primary varicosis, elevated volume and pressure loading, and also due to acquired arteriovenous fistulas (Vollmar et al. 1989). In addition, Friedman et al. (1990) suggest trauma and inflammatory processes as further factors causing venous aneurysms in the upper extremities. The observations made by Loose and Drewes (1983), based on 600 cases, showed that approximately 30% of the aneurysms occur in the lower extremities, and only 4.2% in the upper extremities. It can be expected that noninvasive investigation of the veins using B-mode imaging will in the future be able to identify clinically silent aneurysms more often. It is also important to detect aneurysmal dilation in bypass veins (Karkow et al. 1986).

Evaluation

The criteria described for normal findings should be considered when pathology is encountered. Beyond that, the following aspects are important.

Thrombotic Processes

Color Flow Duplex Sonography

Compression sonography is important in diagnosing thromboses, and can be documented with the *B-mode imaging* component of duplex equipment, or using dedicated B-mode imaging equipment without a Doppler component. It is very important to determine the location and the length of the thrombotic segments. The following criteria should be evaluated and documented:

- Venous display with the accompanying artery in transverse section
- Display without and with compression
- The distal and proximal ends of the thrombus, if its echogenicity is different from that of flowing blood

In a *post-thrombotic syndrome*, or if there is a suspicion that venous thromboses have occurred in the past, the irregular and thickened wall structures should be clearly recognizable in longitudinal section after recanalization. Compressibility is decreased, but not absent.

Uncertain B-mode findings, and a need to establish whether the margins of the thrombus are still surrounded by flowing blood, are important reasons for additional use of the *Doppler mode*. An absence of spontaneous flow, or an absence of modulation by respiration, confirms the diagnosis of fresh thrombosis, especially in regions that are usually not accessible to compression sonography (e.g., the pelvic region).

In *color flow duplex sonography*, an interruption of the color and imaging of the collateral veins clearly indicates the presence of a *thrombosis*. Results obtained from Doppler sonography are declining in importance due to the easy observation of the dynamic flow processes this method provides. Additional use of compression sonography to confirm the diagnosis is recommended. It should be noted in particular whether the thrombi are still surrounded by flowing blood (color), or have adhered to a vessel wall (Fig. 6.**29**).

Thromboses of the superficial veins, known as *thrombophlebitis* or varicophlebitis, can be examined with high-resolution transducers using the same criteria as those for deep venous thromboses (Fig. 6.**24**).

Continuous-Wave Doppler Sonography

Continuous flow through the collateral veins should be documented; the speed of the flow reflects the effectiveness of the collateralization. In uncertain findings, compression and decompression maneuvers should be carried out, and the results should be registered (p. 230). Note that respiration-independent, characteristic continuous flow, or flow that is only very slightly respiration-modulated, can only be detected peripheral to a thrombotic occlusion, or parallel to it.

Valvular Insufficiency

Color Flow Duplex Sonography

Testing for valvular insufficiency is largely a functional examination, in which morphological assessment of the venous lumen plays a lesser role. However, even in *B-mode imaging*, so-called frozen valves can be detected if the resolution is good. These show no reaction to proximal compression, and in particular do not close (Zwiebel 1992) (Fig. 6.**24**).

Valvular insufficiency in the *deep conducting veins* is examined using the Valsalva maneuver. A short reflux to the first valve is not pathological. The duration of the reflux reflects the length of the valvular insufficiency, but it also depends on the intensity of the Valsalva maneuver.

In the *superficial veins*, the distal end point of the reflux can be established by repeated testing along the course of the great saphenous vein. Wuppermann et al. (1981) were able to show that the measured length of the reflux area obtained with Doppler sonography and the length of the reflux measured by phlebography correlate exactly. Assigning reflux phenomena to the individual venous regions (deep venous system, superficial venous system, connecting veins) is simpler and more reliable than with continuous-wave Doppler sonography.

The fastest and most impressive way of detecting valvular insufficiency is using the *color flow Doppler mode*. The color change that occurs during the Valsalva maneuver clearly indicates the changed flow direction (Fig. 7.**42**, pp. 340–341). In a purely visual manner, the length of the color change along the course of the veins shows the extent of the valvular insufficiency in all three venous regions (deep, superficial, and connecting veins). Since this is a dynamic process, the static documentation provided by color flow duplex images can only reflect this flow reversal inadequately. Theoretically, one could code the color differently in the same location. Videotape recording is a suitable form of documentation.

Recent research has sought to identify the anatomic distribution of venous reflux and to quantify venous reflux time (Weingarten et al. 1993, van Ramshorst et al. 1994). A total reflux time of 9.66 s was predictive of ulceration.

Welch et al. (1996) considered segmental reflux to be present if the valve closure time was greater than 0.5 s, and system reflux was considered to be present if the sum of the segments was greater than 1.5 s. Van Bemmelen et al. (1989) report that the examination was conducted with the aid of distal cuff deflation, with the patient in the supine position, with pressures of 80 mmHg in the thigh and 100 mmHg in the calf.

Continuous-Wave Doppler Sonography

When evaluating *insufficiency of the perforating veins*, it is important to locate the characteristic regions. The Cockett perforating veins, located on the medial lower leg, are preferentially affected, especially those at a defined region above the sole of the foot (p. 225). During compression, forward flow toward the probe and return flow on decompression can be recognized acoustically and registered (Straub and Ludwig 1990, Wuppermann 1986).

Sources of Error

In addition to the many *objective* findings that can adversely affect diagnostic certainty, there are also *subjective* limitations, as in all ultrasound investigations, due to insufficient knowledge and practice on the part of the examiner.

The distinction between false-positive and false-negative results presented below is based on the fact that false-positive results restrict the specificity of the method—i.e., the certainty that it will be capable of recognizing normal findings as being normal; while false-negative results affect its sensitivity—i.e., its ability to confirm pathological findings with a high degree of probability.

False-Positive Results

Thoracic respiration is often found in young women and asthenic individuals. The respiratory pressure fluctuations are minimal in this type of respiration, and venous return flow is almost continuous. However, during deep inspiration, or when the patient is practicing abdominal respiration, it is possible to provoke a flow cessation, and this does not succeed if a thrombosis is present (Bollinger et al. 1970).

Congenital aplasia or hypoplasia of the venous valves. This finding can lead to misinterpretation of a venous valvular insufficiency when using continuous-wave Doppler sonography. Depending on the quality of the B-mode image, the venous valves cannot always be recognized with certainty.

Absent venous signal. Due to minimal flow velocities, which are often present in the venous system, venous signals can sometimes not be detected with Doppler sonography. In addition, color coding requires the use of a special mode of the equipment, known as the slow flow technique.

Edema, painful swelling. Extreme edema may clinically simulate a thrombosis, and complicate the examination as a result of mechanically more difficult compression and also as a result of the painful swelling itself. Occasionally, local edemas conceal a hematoma (Fig. 6.**30**), a muscle tear, or a Baker cyst, which can

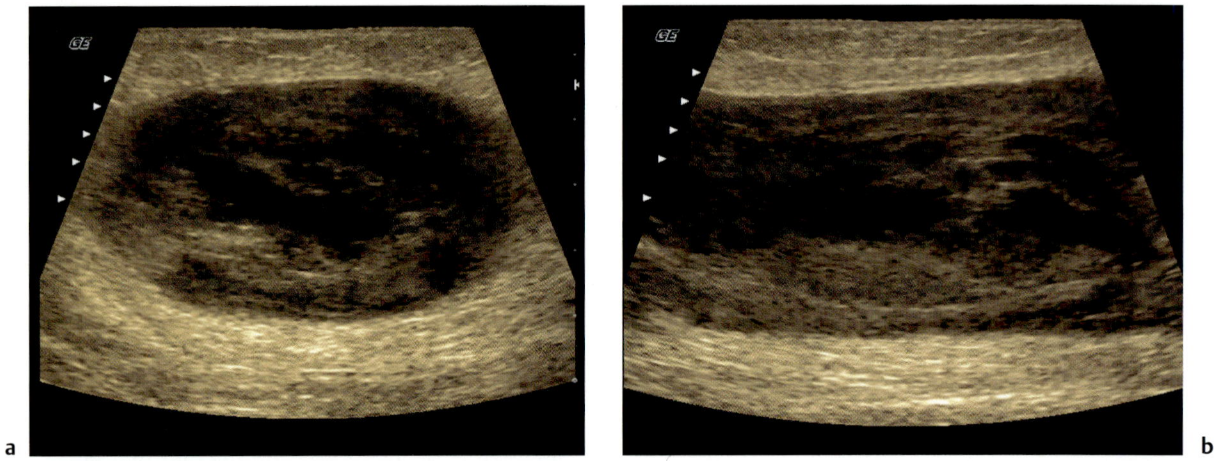

Fig. 6.**30** An extensive hematoma in the lower leg in transverse (**a**) and longitudinal planes (**b**). In this patient, a small spontaneous hemorrhage in the calf during oral anticoagulation therapy was initially misinterpreted as a venous thrombosis. Due to intravenous heparin therapy, the hematoma expanded to an extent approaching a compartment syndrome

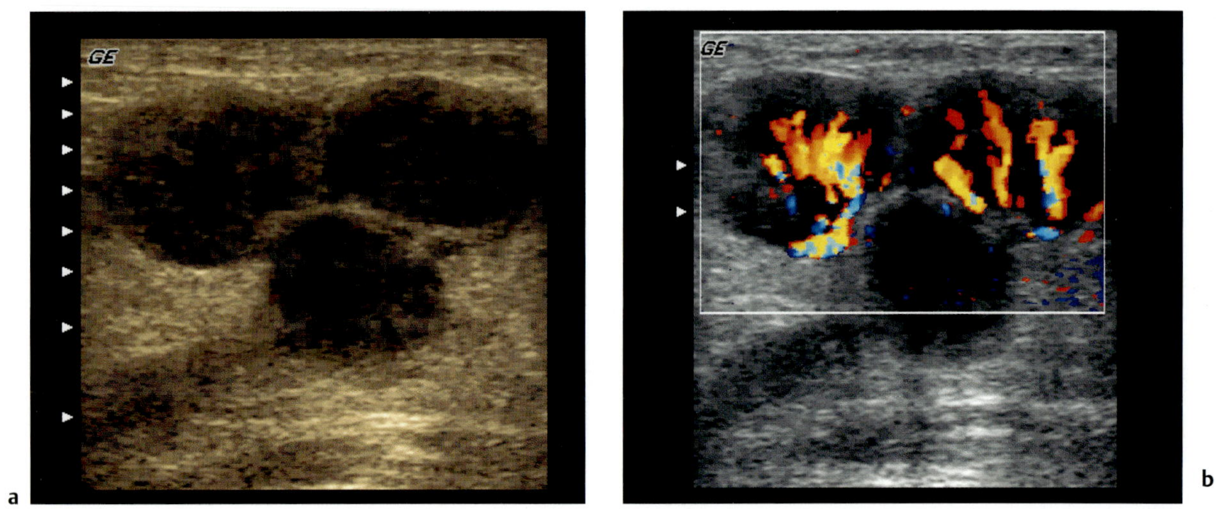

Fig. 6.**31** B-mode (**a**) and color flow Doppler (**b**) in a transverse plane of the groin. Three hypoechogenic structures in a projection of a typical vessel region are seen in the B-mode image. A thrombosis of a duplicated common femoral vein was suspected. Color flow Doppler imaging revealed the vasculature of two enlarged lymphatic nodes in this patient with Hodgkin's disease

only be differentiated using B-mode imaging (Cronan et al. 1988, Borgstede and Clagett 1992).

External venous compression. When it is not possible to use B-mode imaging to confirm the diagnosis, external compression from a tumor or a cyst may be incorrectly interpreted as being due to a thrombosis (Fig. 6.**31**).

Excessive contact pressure exerted by the transducer may result in the compressed vein not being detected).

False-Negative Results

Nonocclusive thrombi. In nonocclusive thrombi, respiratory modulation may remain unchanged, so that these thrombi may escape detection if B-mode imaging is not used, and especially if the color flow Doppler mode is not used.

Good collateralization of a thrombotic occlusion. If there is good collateralization, the respiratory flow modulation may appear to be completely normal when functional tests are conducted in the Doppler mode.

Parietal thrombi, or thrombi covering a short distance. These can be missed using B-mode sonography

Table 6.**6** The accuracy of phlebographically controlled studies in the diagnosis of deep venous thrombosis in the leg using **B-mode sonography**

First author	Year	Patients/ phlebographies (n)	Thromboses (n)	Sensitivity (%)	Specificity (%)
Sullivan	1984	23/23	12	100	92
Raghavendra	1986	20/20	14	100	100
Aitken	1987	46/42	16	94	100
Cronan	1987	51/51	28	89	100
Gaitini	1988	45/45	23	87	91
Habscheid	1988	146/146	127	94	97
O'Leary	1988	53/50	25	88	96
Rollins	1988	63/46	35	87	98
Lensing*	1989	220/220	77	91	99
Habscheid*	1990	238/301	153	96	100
Herzog*	1991	113/101	57	88	98
Langholz	1991	64/64	25	76	88*

Including calf vein thromboses.

if an additional functional examination with Doppler ultrasound is not used to assess the venous flow.

Unsuccessful provocation maneuvers. If the vein is caught at a tangent, compression with the transducer can fail, especially in a longitudinal section.

The *Valsalva maneuver*, which tests the capacity for dilation, may be carried out incorrectly. Patients often exhale deeply instead of breathing into the abdomen, or they may be afraid of urination or flatulence.

The distal or proximal *compression maneuver* may not be executed above the afferent vein, or may be more difficult due to a painful edema.

Multiple veins. If the clinical presentation indicates a suspected thrombosis, a search should always be made for possibly paired veins, which are mainly encountered in the popliteal and the distal femoral region. However, the clinical manifestation may often be very obscure, since the second parallel vein functions as a collateral (Quinn and Vandeman 1990).

Perforating veins. Failing to detect insufficient perforating veins when a "blow-out" or fascial gap is not recognizable. Inadequately excluding epifascial veins when testing for insufficiency of the perforating veins.

Unfavorable patient position. The information yield of the findings can be improved if care is taken to note the patient's changing positions when ultrasound is applied to the veins (Table 6.**1**). Proper positioning can enhance sufficient venous filling, and can also make it easier to move the transducer.

Diagnostic Effectiveness

Thrombotic Processes

In 1982, Talbot listed criteria for detecting acute deep venous thrombosis using duplex sonography:

- Absent compressibility
- Increased cross-sectional diameter
- Increased intraluminal reflection intensity
- No vascular caliber variation during respiration maneuvers
- No Doppler signal

These criteria were confirmed in a small number of cases (11 and 23 patients, respectively) in the earliest studies, carried out by Raghavendra et al. (1984) and Sullivan et al. (1984), and are still valid today. These two research groups only used real-time B-mode equipment, without a Doppler mode. Raghavendra et al. (1984) considered that the echo pattern and the absence of caliber variation were more important than the absent compressibility. They achieved a sensitivity of 100%, and a similar specificity. In the same year, Sullivan et al. (1984) published a study including 215 extremities. In 23 phlebographically controlled results, the sensitivity and specificity values in the lower body region were 92% and 100%, respectively (Table 6.**6**).

In a large study conducted by Dauzat et al. (1986), which evaluated 145 patients with 100 thromboses, B-mode sonography as well as continuous-wave Doppler sonography were used (sensitivity 94%, specificity 100%). Further studies by Aitken and Godden (1987) and by Cronan et al. (1987) using B-mode sonography followed. Cronan et al. (1987) were able to show that assessing *venous compressibility as the sole diagnostic parameter* was sufficient.

Table 6.7 The accuracy of phlebographically controlled studies in diagnosing deep venous thrombosis in the leg using **duplex sonography**

First author	Year	Patients/ phlebographies (n)	Thromboses (n)	Sensitivity (%)	Specificity (%)
Dauzat [1]	1986	145/145	100	94	100
Appelman	1987	112/112	52	96	97
Elias [2]	1987	430/854	268	98	95
Langsfeld	1987	58/65	24	100	78
Vogel	1987	54/54	25	91–94	100
Barnes [2]	1989	78/309[3]	14	86	97
Killewich*	1989b	47/50	38	92	92
Stapff* [1]	1989	49/50	26	96	71
Krings	1990	–/235	–	93–100	96–99
de Valois	1990	180/101	61	92	90
Wright	1990	84/71	34	91	95
Comerota	1990	103/72	44	96	93
van Ramshorst	1991	117/120	64	91	95*

* Including venous thromboses in the lower leg.
1 B-mode sonography + continuous-wave Doppler sonography.
2 Additional continuous-wave Doppler sonography.
3 Preoperative and postoperative.

Table 6.8 The accuracy of phlebographically controlled studies in diagnosing deep venous thrombosis in the leg with **color flow duplex sonography**

First author	Year	Patients/ phlebographies (n)	Thromboses (n)	Sensitivity (%)	Specificity (%)
Persson*	1989	264/24	16	100	100
Fobbe	1989	103/129	58	96	97
Foley	1989	47/47	19	89	100
Fürst	1990	75/102	39	95	99
Grosser*	1990	180/180	154	94	99
Rose*	1990	69/75	32	79	88
Langholz	1991	116/116	65	100	94
van Gemmeren	1991	114/141	74	96	97
Mattos	1992	75/77	32	100	98
Schweizer [1]	1993	78/70	70	98	100*

* Including venous thromboses in the lower leg.
1 Intensified by using contrast medium.

In the same year—5 years after Talbot's pioneering research in 1982—the first additional studies using *duplex sonography* appeared: Appelman et al. (1987), Elias et al. (1987), Langsfeld et al. (1987) and Vogel et al. (1987). In a study including 430 patients, Elias et al. achieved a high sensitivity of 98 % and a specificity of 95 %. Despite the use of duplex sonography, continuous-wave Doppler sonography was initially applied in this study, so that—in contrast to pure B-mode studies—it was also possible to evaluate the iliac veins (n = 57) and inferior vena cava (n = 47). In 1987, Elias et al. were the first to describe the examination of the *lower leg veins* using what is known compression sonography, in controlled conditions. They detected thromboses in this area with a sensitivity of 91 % (84 of 92 cases). Habscheid and Wilhelm (1988) achieved a sensitivity of 89 % and a specificity of 100 % in 77 extremities that had undergone phlebography and contained 29 lower leg thromboses (see Table 6.**10**).

Many more studies using the various ultrasound procedures followed (Tables 6.**6**, 6.**7**). Some of these only used B-mode compression sonography; some used duplex sonography or B-mode combined with continuous-wave Doppler sonography (Krings et al. 1990). Parallel to this, the first studies carried out using *color flow duplex sonography* also produced good, although not significantly better, results (Table 6.**8**).

Today, the specificity of duplex sonography is very high, and it is therefore clinically safe to rely on a single negative color duplex examination including the calf veins in patients who have no progressive symptoms from the affected limb. Noren et al. (2002) reported on a consecutive series of patients with suspected deep venous thrombosis who were referred for color duplex examination over a 1-year period. A total of 341 patients with negative duplex findings were followed up for 3 months clinically or by reviewing hospital charts, or both, and by consulting the official registry of health care and causes of death. Deep vein thrombosis following a negative duplex examination was only diagnosed in one of the patients. The patient concerned had accentuated symptoms, and a thrombus in the peroneal vein was detected at a subsequent phlebography. None of the other patients with negative duplex findings developed signs of, or had treatment initiated for, deep venous thrombosis or pulmonary embolism during the 3-month period after the duplex investigation. Five patients died, but none of the deaths was related to thromboembolism. The results indicate that the method is clinically safe.

Overall, the indicated sensitivity values for conventional and color flow duplex sonography usually lie between 90% and 95%; the specificity lies between 95% and 100%. The results obtained depend not only on methodological variations in access and evaluation criteria, but are primarily influenced by whether or not findings obtained from the pelvic and lower leg veins are included. This is shown in Tables 6.**6** and 6.**8**. Since the disease prevalence and clear field tables were not provided in all of the published studies, results for the predictive value are not listed.

In a study published by Heijboer et al. in 1993, a very high positive predictive value of 94% was obtained for pathological findings using compression sonography (84 of 89 pathological results with B-mode examinations were confirmed using phlebography), and the importance of repeat examinations (on the second and eighth days) was demonstrated.

Becker et al. (1989) published a critical analysis summarizing 15 studies conducted between 1984 and 1988, which diagnosed deep venous thromboses using B-mode sonography alone (eight reports) and B-mode sonography including Doppler sonography (seven reports). The analysis showed the many notable sources of methodological error that can appear when designing and conducting such studies.

For *lower leg thromboses*, more studies (Table 6.**9**) indicate lower sensitivity and specificity values than for the thigh and pelvic regions. However, with increasing experience, satisfactory diagnostic information can also be obtained here, especially with the application of color flow Doppler mode. In some studies, the accuracy of identification of the different lower leg arteries is listed separately (Grosser et al. 1991). In a prospective study including 75 patients with isolated calf vein thrombi, Lohr et al. (1991) reported that the thrombi propagated in 24 patients (32%), and into the popliteal or large veins of the thigh in 46% of these cases.

In a retrospective study, Mattos et al. (1996) showed that imaging of three paired calf veins is feasible in 94% of cases and that calf vein thrombosis is present in two-thirds of limbs with documented acute symptomatic thrombosis of the deep veins.

More attention still needs to be given to the study of *floating thrombi* in the deep and superficial veins. According to a study by Voet and Afschrift (1991), these cause *pulmonary embolisms* with an incidence of 9%.

Several research teams have examined the application of duplex scanning in patients with suspected pulmonary embolism. Killewich et al. (1993) showed that, in the case of angiographically confirmed pulmonary emboli, color flow duplex sonography registered a positive finding in only 44% of cases. In 664 patients with suspected pulmonary embolism, Matteson et al. (1996) detected a confirmed deep venous thrombosis in only 13% of cases; the authors consider that ventilation/perfusion scanning is the more important diagnostic measure.

Nicholls et al. (1996) studied venous thromboemboli in a new way. Transcranial Doppler monitoring is an established means of identifying emboli in the arterial circulation. Nicholls et al. extended this technique to identify emboli in the venous circulation. Of 60 patients with deep vein thrombosis, embolism was demonstrated in 43%. In patients taking heparin, the embolism disappeared within 72 h.

If sonography is nondiagnostic, venography should be considered. Magnetic resonance venography can differentiate an acute occlusion from chronic thrombus, but because of its high cost and limited availability, it is not yet used for the routine diagnosis of lower extremity venous occlusion only. Venography has long been considered the gold standard for identifying proximal venous occlusion. Both computed tomography scanning and magnetic resonance imaging (MRI), however, can diagnose acute pelvic vein or inferior vena cava occlusion even more accurately. MRI is preferred because it is noninvasive, does not require contrast administration, involves no exposure to ionizing radiation, and is highly accurate and reproducible (Haage et al. 2002).

Although continuous-wave Doppler sonography has been replaced in some areas by B-mode sonography (for diagnosing thrombosis) and by conventional and color flow duplex sonography, it is still used quite appropriately for preliminary diagnosis and for specific investigations.

In areas such as the pelvic region, where compression testing using real-time B-mode is difficult or impossible, functional evaluation of venous flow using simple continuous-wave Doppler sonography, or the pulsed-wave Doppler mode of a duplex apparatus, is still important. Validation of this method of detecting or excluding *venous thromboses in the pelvis and thigh* was carried out by many authors in the era before duplex sonography (Table 6.**10**). Overall, good results were obtained, especially with regard to the method's specificity; the sensitivity is subject to wider fluctuations, and depends on both the vascular level and the grade of the lumen obstruction. For the *lower leg* veins, Barnes et al. (1976) and Sumner and Lambeth (1979) obtained a sensitivity of 94% and 91%. Flanigan et al. (1978) did not report comparable figures (sensitivity 36%).

Testing *valvular function* is another role for Doppler sonography. However, assignment to specific regions, the deep or the superficial veins, is only possible

Table 6.**9** The validity of phlebographically controlled studies using **B-mode Duplex and color flow duplex sonography** in diagnosing the level of deep venous thrombosis

First author	Year	Patients/ phlebographies (n)	Iliac vein		Common femoral vein		Superficial femoral vein		Popliteal vein		Lower leg	
			Sensitivity (%)	Specificity (%)	Sensitivity (%)	Specificity (%)	Sensitivity (%)	Specificity (%)	Sensitivity (%)	Specificity (%)	Sensitivity (%)	Specificity (%)
B-mode sonography												
Habscheid	1988	104/146	–	–	100	99	95	99	95	99	89	100
Habscheid	1990	238/301	–	–	100	100	97	99	97	100	94	99
Herzog	1991	13/101	78	78	–	–	100	100	98	98	60	97
Duplex sonography												
Dauzat[1]	1986	145/145			100				94		62	–
Vogel	1987	54/54			91				94		–	–
Elias	1987	430/854			100				98		91	96
Stapff	1989	49/50	100	100	–	–	100	95	100	89	92	50
Krings[1]	1990	235/235	93	99	100	96	96	99	97	98	93	96
Color flow duplex sonography												
Fürst	1990	75/102	–	–	95	99	93	97	90	99	72	100
Rose	1990	69/75	100	100			92	100			73	86
Langholz	1991	116/65	100	100			99	94	100	96	89	91
Yucel	1991	45/45	–	–	–	–	–	–	–	–	88	96
Grosser	1991	325/325			95–100	99–100			94–100	99–100	93–100	99–100
Mattos	1992	75/77	–	–	94	100	100	98	85	92	94	81
Schweizer[2]	1993	78/70	100	100	–	–	96	100	100	95	95	100
Söldner	1993	84/103			95–100	98	95	98	97	100	93	97

[1] Or B-mode sonography + continuous-wave sonography.

[t] Intensified with contrast medium.

Table 6.**10** The accuracy of phlebographically controlled studies in the diagnosis of deep venous thrombosis of the leg using **continuous-wave Doppler sonography**. Most of the studies were concerned with thromboses in the pelvis or thigh, or both. Individual evaluation of the vascular level was not undertaken in most of the studies

First author	Year	Phlebographies (n)	Sensitivity (%)	Specificity (%)
Sigel	1968	121	86	88
Grüntzig	1971	42	92	93
Sigel	1972	248	85	91
Strandness	1972	57	79	95
Yao	1972	50	87	94
Holmes	1973	71	100	85
Barnes	1976	122	96	92
Dosick	1978	160	99	92
Sumner	1979	75	95	89
Hanel	1981	183	91	92
Bendick	1983	140	54	90

with aids (tourniquets). Excluding insufficient valves at the origin in the groin region was successfully achieved by Hach et al. (1977), with a specificity of 91%. In the same study, the sensitivity was 93.6%, while Sigel et al. (1972), for example, only reported a figure of 76%. In 1981, Wuppermann et al. were able to demonstrate that the length of the reflux area and the phlebographically determined reflux length correlate precisely. This is important in classifying the stage of a superficial varicosis, according to Hach et al. (1977).

Examining the perforating veins using continuous-wave Doppler sonography is a difficult procedure. There have been no studies to validate this procedure, to our knowledge.

Thrombosis of the Shoulder and Arm Veins

Deep venous thrombosis affecting the *shoulder and arm veins* is a rare occurrence, and is described in the literature as having a frequency of between 1% and 2% of all thromboses. If more recent investigations, which primarily focus on catheter-related thrombosis in the superior thoracic aperture (Habscheid and Landwehr 1992), are taken into account, a higher incidence of upper body venous thromboses may be suspected. The validity of (color flow) duplex sonography in diagnosing venous thromboses in the shoulder–arm venous region is very satisfactory. A more comprehensive study involving 99 phlebographies completed by Knudson et al. (1990) reported a sensitivity of 78% and a specificity of 92%. The sensitivity may decrease if proximal subclavian thromboses cannot be directly visualized, and only the absence of a color signal and not the Doppler spectra are used to make the diagnosis.

With regard to the causes of thromboses in the upper extremities, subclavian vein, and axillary vein regions, the data provided by Kerr et al. (1990b) and Hingorani et al. (1997) are listed in Table 6.**11**.

Table 6.**11** Etiology of thromboses in the subclavian and axillary vein regions

Thrombosis in the upper extremities	Kerr (1990b) (n = 345)	Hingorani (1997) (n = 170)
Catheter thrombosis	61%	65%
Carcinoma	31%	38%
Compression syndrome	6%	0.6%
Other causes	2%	–

To assess the accuracy of duplex ultrasonography in the diagnosis of deep venous thrombosis of the upper extremities, Baarslag et al. (2002) designed a prospective study of duplex ultrasonography in comparison with venography. In 126 consecutive inpatients and outpatients with suspected thrombosis of the upper extremities contrast venography after duplex ultrasonography was judged independently. A three-step protocol, involving compression ultrasonography, color ultrasonography, and color Doppler ultrasonography, was used. The sensitivity, specificity, and probability ratios for ultrasonography as a whole were calculated. Venography and ultrasonography were not feasible in 23 of 126 patients (18%) and one of 126 patients (0.8%), respectively. The results of ultrasonography were inconclusive in three patients. Venography demonstrated a thrombosis in 44 of 99 patients (44%); in 36 patients (36%), the thrombosis was related to intravenous catheters or malignant disease. Venous incompressibility correlated well with thrombosis, whereas only 50% of isolated flow abnormalities proved to be thrombosis-related.

Post-Thrombotic Venous System

Post-thrombotic syndrome develops in 40–60% of patients with deep venous thrombosis. Factors that are important in the development of post-thrombotic syndrome include venous reflux, deep vein obstruction, and calf muscle pump dysfunction. Duplex sonography has introduced a new dimension into this process, in that it allows measurement of the extent of the thrombus and can assess both length and width. In addition, subsequent valvular insufficiency can be assigned chronologically and an indication of the age of the thrombus can be derived from its echogenicity and caliber decreases during the subsequent stages. Also, echographic criteria for distinguishing between a fresh venous thrombosis and a re-thrombosis in post-thrombotic changes have been compiled.

Reflux and calf muscle pump dysfunction in relationship to the severity of post-thrombotic syndrome were evaluated by Haenen et al. (2002) in a 2-year follow-up study of patients with acute deep venous thrombosis. The study included 86 legs, and the 2-year follow-up period was completed for 70 legs. Significantly more reflux was found in previously thrombosed vein segments, with odds ratios of 1.8 after 3 months, 2.1 after 6 months, 2.5 after 12 months, and 3.2 after 24 months. Multiple regression analysis showed that the most important risk factor for early clinical signs of post-thrombotic syndrome was superficial reflux in months 3, 6, and 12 ($P \leq 0.02$). Deep reflux did not have a synergistic relationship with superficial reflux in correlation with the clinical signs of post-thrombotic syndrome. The supine venous pump function test was not able to predict the development of post-thrombotic syndrome. More reflux develops in previously thrombosed vein segments. As early as after the third month, patients with superficial reflux have an increased risk of developing the first clinical signs of post-thrombotic syndrome.

During the long-term follow-up, deep venous thrombosis in the calf may lead to significant post-thrombotic disease. Reflux in the primarily uninvolved popliteal vein is frequent and may be associated with more severe disease (Saarinen et al. 2002).

The advent of the widespread availability of color-coded duplex sonography has made it possible to diagnose isolated venous thromboses in the lower leg safely and to check clinical and hemodynamic changes during the course. McLafferty et al. (1998) carried out a follow-up study (mean follow-up 3.4 years) which showed that 62% of the patients were completely asymptomatic; only 26% had an abnormal valvular closure time and 23% had a pathological vein recovery time. Surprisingly, a prolongation of the valvular closure time was found primarily in segments that had not been thrombosed at the first examination.

In a recent study in a homogeneous Mediterranean population, 694 patients with unilateral or bilateral symptoms and signs of chronic venous insufficiency in the lower limbs were investigated using color duplex scanning (Giannoukas et al. 2002). On the basis of the patients' history and ultrasound findings, the limbs were classified into those with post-thrombotic and those with primary chronic venous insufficiency. The clinical presentation, based on the clinical, etiological, anatomical, and pathophysiological (CEAP) classification, was correlated with the anatomic distribution of venous reflux.

Most of the symptomatic limbs (537 of 656; 81.5%) with primary chronic venous insufficiency belonged to classes 1 to 3. In these limbs, reflux confined to the superficial veins was very common (424 of 656; 64.5%), whereas the prevalence of deep and perforator vein reflux was 8% and 25.5%, respectively. In contrast, most of the limbs (69.5%) with post-thrombotic chronic venous insufficiency belonged to classes 4 to 6, had a complex pattern of reflux, and involvement of deep and perforator veins was common (86.5% and 48%, respectively). No reflux was found in either limb on duplex scanning in approximately one-quarter of the patients (24%) with suspected primary chronic venous insufficiency. Most of the patients (48%) had telangiectasia. Bilateral reflux was found in 71% of the patients with primary chronic venous insufficiency.

The clinical presentation was poorer in limbs with post-thrombotic chronic venous insufficiency than in those with a primary disease. Post-thrombotic insufficiency was associated with a complex pattern of reflux, affecting mostly the deep and perforator veins, whereas superficial reflux was the most common pattern in limbs with primary chronic venous insufficiency. Surgery aimed at eliminating superficial reflux would therefore confer only a minimal benefit in limbs with post-thrombotic insufficiency, but would treat the majority of limbs with primary chronic venous insufficiency. The high prevalence of bilateral reflux found in patients with primary insufficiency suggests a bilateral predisposition, which supports the hypothesis of the existence of a generalized venous disease (Giannoukas et al. 2002).

Age of Venous Thrombosis

Richter et al. (1992) quantitatively analyzed thrombus echogenicity in 45 patients with 65 venous thromboses; the authors were not able to detect a relationship between echogenicity and the age of the thrombus. Similar results were obtained by Appelman et al. (1987) and Murphy and Cronan (1990). The latter authors also concluded that a decreasing venous diameter in the different vascular segments correlated better with the age of the thrombus than the echo structure did. For example, the diameter of the superficial femoral vein during the first week amounted to 8 mm, and after 5 weeks, it averaged only 6 mm. O'Shaughnessy and FitzGerald (1996) investigated the aging process of a fresh thrombosis in 100 patients in a sequence of five investigations over 1 year and established that calf vein thrombi resolve quickly, with few long-term effects. The organization of thrombi above the knee was found to be influenced by the patient's age and physical condition.

Methods of differentiating between a fresh initial thrombosis and an acute *re-thrombosis* in a previously damaged venous system have only been investigated in a few studies. Rollins et al. (1988) found an acute venous thrombosis in 63 legs from a total of 76 extremities examined. An initial thrombosis was present in 15 cases, and an acute re-thrombosis was detectable in 48 instances. The authors successfully detected a chronic venous thrombosis, as distinct from an acute venous thrombosis, with a sensitivity of 94% and a specificity of 95%. Since the clinical indications of completed thromboses are often imprecise, a great deal of experience is required to make this differentiation successfully solely on the basis of the B-mode image. In a long-term study of 43 patients with thromboses, Gaitini et al. (1990) also discussed this problem, and recommended that ultrasound findings should be obtained about 6–12 months after venous thromboses, in order to be able to differentiate later acute re-thromboses from post-thrombotic ultrasound results.

A study by the Strandness group (Meissner et al. 1995) discusses methods of distinguishing between re-thrombosis, propagation, and new contralateral thrombi. In a prospective examination of 204 extremities, re-thrombosis occurred in 31%, propagation in 30%, and new thrombi in 6% within 1 year. Unfortunately, the study does not indicate whether clinical symptoms were involved in these findings.

From these studies, it can be concluded that involutional developments in the thrombus—caused by retraction and fragmentation, fibroblast ingrowth, and intrathrombotic lysis—are often already complete within 3 months. They are certainly complete by the time 6 months have elapsed. Veins that are not patent at this time usually remain chronically occluded, or only partially recanalized. Data on the percentage of veins that still show residual effects 180 days after a thrombosis vary between 14% (Killewich 1989a) and 48% (Murphy and Cronan 1990).

A recent study used ultrasound-based reconstructive elasticity imaging to determine the age of a thrombosis. Blood clots consisting of fibrin harden as they develop and organize. Imaging of clot elasticity may make it possible both to detect and differentiate between clots, thus providing an urgently needed noninvasive method of thrombosis staging (Emelianov et al. 2002).

Valvular Insufficiency

Locating and quantifying venous reflux during the Valsalva maneuver or a compression maneuver has taken on a new dimension with duplex sonography. Retrograde venous flow can be reliably classified as belonging to the deep or superficial veins. Problems arise when *quantifying* the causal respiratory or compression maneuvers. Van Bemmelen et al. (1989) examined the reflux resulting during distal cuff deflation; the compression intensity and its duration were predetermined. Nicolaides's group quantified reflux automatically according to its velocity and quantity (Vasdekis et al. 1989).

Special attention has been given to *measuring reflux in the post-thrombotic state*. In a comprehensive follow-up study conducted by Markel et al. (1992), including 123 patients who had had deep venous thrombosis, 17% of 106 patients with sufficient valves during the initial examination showed valvular insufficiency even during the first week. After 1 month, this figure had risen to 40%, and after 1 year, 69% had demonstrable valvular insufficiency, which developed in the recanalized veins in 40–50% of cases.

The venous function parameters available—such as diameter, area measurements, and flow velocities—were found to have a wide range of standard deviation in the study by Hirschl and Bernt (1990). No correlation was found with the levels of seriousness of the manifest venous insufficiency.

Labropoulos et al. (1994) found that the severity of chronic venous insufficiency in 217 legs increased in limbs with combined superficial and deep venous distal reflux. Reflux confined to the deep veins alone is less harmful. Shami et al. (1993) recorded similar findings in a smaller study.

The best method of quantifying venous reflux is still a matter of debate. Weingarten et al. (1996) found that the total limb reflux time correlated significantly with a specific air–plethysmographic variable. A total reflux time greater than 9.66 s was found to be predictive of venous ulceration in this study. By contrast, Rodrigues et al. (1996) found that duplex-derived valve closure times do not correlate with the magnitude of reflux and should not be used to quantify the degree of reflux.

Evers and Wuppermann (1997) consider that a reflux speed of > 10 cm/s is a safe criterion for post-thrombotic reflux. They compared manual reflux provocation with provocation using the Valsalva maneuver and a cuff. They concluded that provocation using the cuff test was the most effective method, with good reproducibility.

Despite the methodological problems associated with the quantitative measurement of reflux, purely *qualitative* evaluation of venous valvular insufficiency and also evaluation of the location and extent of reflux has proved to be important both prognostically and in relation to the etiology of chronic venous insufficiency. The *pathogenesis of chronic venous insufficiency* should be identified as being due to one of the following:

- Deep venous thrombosis
- Conducting vein insufficiency
- Perforating vein insufficiency
- Primary varicosis

Insufficient Perforating Veins

The diagnosis of *insufficient perforating veins* has been revived with the use of color flow duplex sonography. Insufficiency can be clearly demonstrated through the change in the direction of flow within the perforating vein on distal compression. Franzeck et al. (1993) and Stiegler et al. (1994) produced comprehensive validating studies. However, Franzeck et al. (1993) confirmed that there was still a significant discrepancy between clinically suspected cases of venous insufficiency (n = 62), those ascertained using continuous-wave Doppler sonography (n = 37), and those detected using color flow duplex sonography (n = 20). Using color flow duplex sonography, Stiegler et al. (1994) compared preoperatively diagnosed cases of insufficient perforating veins (n = 334) in 94 patients with primary varicosis with the intraoperative findings. The success rate with which the previously diagnosed and marked insufficient perforating veins were surgically confirmed was 96%. By contrast, the perforating veins detected in 31 phlebograms were only confirmed surgically in 65% of cases. This study shows that preoperative localization of insufficient perforating veins using color flow duplex sonography can produce a significant improvement in the rate of targeted surgical removal.

A study by Pierik et al. (1997) did not fully confirm these good results. In a smaller number of patients (n = 20), they only found a sensitivity of 79.2% for the detection of insufficient perforating veins, although with a specificity of 100%. The authors also compared their preliminary color duplex sonographic findings with the intraoperative findings.

Thrombophlebitis

The imaging of superficial venous thrombotic inflammations using B-mode sonography has only begun to attract attention, since the clinical diagnosis of thrombophlebitis is usually quite clear (Skillman et al. 1990). However, the first large studies, published by Pulliam et al. (1991) and Lutter et al. (1991), showed that the inflammatory thrombotic processes in the superficial veins are occasionally combined with complications and thrombus growth into the deep venous system. In a large retrospective investigation including 12 856 venous examinations using duplex sonography, Lutter et al. (1991) found 186 patients with primary superficial phlebitis. Of these, 57 patients (31%) had a complication. A pulmonary embolism occurred in eight instances (4%); an expansion of the thrombosis into the deep venous system was observed in 28% (n = 53). In 34% of cases, great saphenous vein phlebitis was associated with complications; small saphenous vein phlebitis yielded complications in 16% of the cases. Isolated local phlebitis led to complications in only 8% of the occurrences.

Krause et al. (1998) provided a retrospective report on the approach in 398 patients suffering from thrombophlebitis of the great or small saphenous vein. They carried out surgery in 56 patients after a color duplex sonographic diagnosis of thrombus ascent, floating thrombus, or thrombosis in the opening of the great saphenous vein. The duplex findings frequently differed from the intraoperative findings. The rate of deep vein thromboses was 5% higher in surgical patients than in the overall group (0.75%); the same applied to the rate of pulmonary embolisms, at 3.5 compared with 0.5%.

In general, however, the following predisposing factors apply for complications during a thrombophlebitis:

- Age > 60 years
- Female sex
- Previous deep vein thrombosis
- Confinement to bed
- Systemic infections

In 1996, Chengelis et al. reviewed the records of 263 patients with superficial venous thrombosis. Thirty patients (11%) showed progression, the most common form of which was progression of the disease from the greater saphenous vein into the common femoral vein (70%). The authors recommend anticoagulation therapy in patients with proximal great saphenous vein thrombosis.

Special Sets of Findings

Congenital venous dysplasias are not observed sufficiently often for studies assessing the accuracy of duplex sonography examination to be available. However, there is no disputing the immense diagnostic benefit which the imaging procedures have provided, together with functional Doppler-sonographic examinations of venous flow. Color coding of the Doppler shift has made the diagnosis of unusual anatomical formations—e.g., arteriovenous fistulas and aneurysms—much easier. The same is true of *secondary arteriovenous fistulas* and *acquired venous aneurysms*. Ritter et al. (1993) published an overview examining the frequency and distribution of venous aneurysms in the literature and in their own patients (152 of 1000 phlebograms). The authors report that the aneurysms were found in the following locations: 65% in the lower extremity, 14% in the upper extremity, and the remainder in the head, neck, thorax, and abdomen.

In 1995, Calligaro et al. published a literature review covering 20 years (1973–1993), reporting 52 neck and face aneurysms, 19 aneurysms of the upper vena cava, 32 abdominal aneurysms, and 44 aneurysms of the extremities. The authors consider that prophylactic surgical correction is required in the deep veins of the extremities, whereas with other locations they recommend a duplex-sonographic observation unless surgical removal is indicated for cosmetic reasons.

References

Aitken AG, Godden DJ. Real-time ultrasound diagnosis of deep vein thrombosis: a comparison with venography. Clin Radiol 1987;38:309–13.

Appelman PT, De Jong TE, Lampmann LE. Deep venous thrombosis of the leg: US findings. Radiology 1987;163:743–6.

Baarslag HJ, van Beek EJ, Koopman MM, Reekers JA. Prospective study of color duplex ultrasonography compared with contrast venography in patients suspected of having deep venous thrombosis of the upper extremities. Ann Intern Med 2002;136:865–72.

Barnes RW, Russell HE, Wu KK, Hoak JC. Accuracy of Doppler ultrasound in clinically suspected venous thrombosis of the calf. Surg Gynecol Obstet 1976;143:425–8.

Barnes RW, Nix ML, Barnes CL, Lavender RC, Golden WE, Harmon BH, et al. Perioperative asymptomatic venous thrombosis: role of duplex scanning versus venography. J Vasc Surg 1989;9:251–60.

Becker DM, Philbrick JT, Abbitt PL. Real-time ultrasonography for the diagnosis of lower extremity deep venous thrombosis. The wave of the future? Arch Intern Med 1989;149:1731–4.

Belov S, Loose DA, Weber J, editors. Vascular malformations. Reinbek (Germany): Einhorn-Presse, 1989.

Bendick PJ, Glover JL, Holden RW, Dilley RS. Pitfalls of the Doppler examination for venous thrombosis. Am Surg 1983;49:320–3.

Bjordal RJ. Die Zirkulation in insuffizienten Vv. perforantes der Wade bei venösen Störungen. In: May R, Pansch H, Staubesand J, editors. Venae perforantes. Munich: Urban & Schwarzenberg 1981.

Bollinger A. Funktionelle Angiologie. Stuttgart: Thieme, 1979.

Bollinger A, Mahler F, de Sepibus G. Diagnostik peripherer Venenerkrankungen mit Doppler-Strömungsdetektoren. Dtsch Med Wochenschr 1968;46:2197–2201

Bollinger A, Rutishauser W, Mahler F, Grüntzig A. [Dynamics of return flow from the femoral vein; in German.] Z Kreislaufforsch 1970;59:963–71.

Bollinger A, Wirth W, Brunner U. [Valve agenesis and dysplasia of leg veins: morphological and functional studies; in German.] Schweiz Med Wochenschr 1971;101:1348–53.

Borgstede JP, Clagett GE. Types, frequency, and significance of alternative diagnoses found during duplex Doppler venous examinations of the lower extremities. J Ultrasound Med 1992;11:85–9.

Bork-Wolwer L, Wuppermann T. [Improvement in noninvasive diagnosis of the greater and lesser saphenous vein insufficiency with duplex sonography; in German.] Vasa 1991;20:343–7.

Calligaro KD, Ahmad S, Dandora R, Dougherty MJ, Savarese RP, Doerr KJ, et al. Venous aneurysms: surgical indications and review of the literature. Surgery 1995; 117:1–6.

Chengelis DL, Bendick PJ, Glover JL, Brown OW, Ranval TJ. Progression of superficial venous thrombosis to deep vein thrombosis. J Vasc Surg 1996;24:745–9.

Comerota AJ, Katz ML, Greenwald LL, Leefmans E, Czeredarczuk M, White JV. Venous duplex imaging: should it replace hemodynamic tests for deep venous thrombosis? J Vasc Surg 1990;11:53–9 [discussion 59–61].

Coulier B. Hyperechogenicity of medial gastrocnemial veins during ultrasound scanning of the calf in sitting patients: a normal variant. Eur Radiol 2002;12:1843–8.

Cronan JJ, Dorfman GS, Scola FH, Schepps B, Alexander J. Deep venous thrombosis: US assessment using vein compression. Radiology 1987;162(1 Pt 1):191–4.

Cronan JJ, Dorfman GS, Grusmark J. Lower-extremity deep venous thrombosis: further experience with and refinements of US assessment. Radiology 1988;168:101–7.

Dauzat MM, Laroche JP, Charras C, Blin B, Domingo-Faye MM, Sainte-Luce P, et al. Real-time B-mode ultrasonography for better specificity in the noninvasive diagnosis of deep venous thrombosis. J Ultrasound Med 1986;5:625–31.

de Valois JC, van Schaik CC, Verzijlbergen F, van Ramshorst B, Eikelboom BC, Meuwissen OJ. Contrast venography: from gold standard to "golden backup" in clinically suspected deep vein thrombosis. Eur J Radiol 1990;11:131–7.

Diepgen TL, Bassukas ID, Hornstein OP. Atypisches Klippel–Trenaunay–Weber Syndrom mit Osteohypotrophie. Akt Dermatol 1988;14:299–301

Dosick SM, Blakemore WS. The role of Doppler ultrasound in acute deep vein thrombosis. Am J Surg 1978;136(2):265–8.

Elias A, Le Corff G, Bouvier JL, Benichou M, Serradimigni A. Value of real time B mode ultrasound imaging in the diagnosis of deep vein thrombosis of the lower limbs. Int Angiol 1987;6:175–82.

Emelianov SY, Chen X, O'Donnell M, Knipp B, Myers D, Wakefield TW, et al. Triplex ultrasound: elasticity imaging to age deep venous thrombosis. Ultrasound Med Biol 2002;28:757–67.

Evers EJ, Wuppermann T. [Characterization of post-thrombotic reflux with color-coded duplex ultrasound diagnosis; in German.] Vasa 1997;26:190–3.

Flanigan DP, Goodreau JJ, Burnham SJ, Bergan JJ, Yao JS. Vascular-laboratory diagnosis of clinically suspected acute deep-vein thrombosis. Lancet 1978;ii:331–4.

Fobbe F, Koennecke HC, el Bedewi M, Heidt P, Boese-Landgraf J, Wolf KJ. [Diagnosis of deep venous thrombosis of the leg using color-coded duplex sonography; in German.] RöFo 1989;151:569–73.

Foley WD, Middleton WD, Lawson TL, Erickson S, Quiroz FA, Macrander S. Color Doppler ultrasound imaging of lower-extremity venous disease. AJR Am J Roentgenol 1989;152:371–6.

Franzeck UK, Billeter M, Schultheiss R, Bollinger A. Vergleich von klinischer Untersuchung, Doppler-Ultraschall und Farb-Duplex-Sonographie bei Patienten mit insuffizienten Perforansvenen. Vasa 1993;(Suppl 41):46.

Friedman SG, Krishnasastry KV, Doscher W, Deckoff SL. Primary venous aneurysms. Surgery 1990;108:92–5.

Fronek A. Noninvasive diagnostics in vascular disease. New York: McGraw-Hill, 1989.

Fürst G, Kuhn FP, Trappe RP, Mödder U. [The diagnosis of deep venous thromboses of the leg: color Doppler sonography versus phlebography; in German.] RöFo 1990;152:151–8.

Gaitini D, Kaftori JK, Pery M, Weich YL, Markel A. High-resolution real-time ultrasonography in the diagnosis of deep vein thrombosis. RöFo 1988;149:26–30.

Gaitini D, Kaftori JK, Pery M, Markel A. Late changes in veins after deep venous thrombosis: ultrasonographic findings. RöFo 1990;153:68–72.

Giannoukas AD, Tsetis D, Ioannou C, Kostas T, Kafetzakis A, Petinarakis I, et al. Clinical presentation and anatomic distribution of chronic venous insufficiency of the lower limb in a typical Mediterranean population. Int Angiol 2002;21:187–92.

Gooding GA, Hightower DR, Moore EH, Dillon WP, Lipton MJ. Obstruction of the superior vena cava or subclavian veins: sonographic diagnosis. Radiology 1986;159:663–5.

Grosser S, Kreymann G, Guthoff A, Taube C, Raedler A, Tilsner V, et al. [Color-coded Doppler sonography for phlebothrombosis; in German.] Dtsch Med Wochenschr 1990;115:1939–44.

Grosser S, Kreymann G, Kuhns A. [Value of color-coded duplex sonography in diagnosis of acute and chronic venous diseases of the lower extremity; in German.] Ultraschall Med 1991;12:222–7.

Grüntzig A, Bollinger A, Zehender O. [Possibilities and limits of qualitative vein diagnosis using Doppler ultrasound (results of a blind study); in German.] Klin Wochenschr 1971;49:245–51.

Haage P, Krings T, Schmitz-Rode T. Nontraumatic vascular emergencies: imaging and intervention in acute venous occlusion. Eur Radiol 2002;12:2627–43.
Habscheid W, Landwehr P. [Diagnosis of acute deep leg vein thrombosis with compression ultrasonography; in German.] Ultraschall Med 1990;11:268–73.
Habscheid W, Landwehr P. Duplexsonographie als Diagnoseverfahren bei Venenthrombosen der oberen Thoraxapertur. Med Welt 1992;43:137–41.
Habscheid W, Wilhelm T. [Diagnosis of deep leg vein thromboses by real-time sonography; in German.] Dtsch Med Wochenschr 1988;113:586–91.
Hach W, Girth E, Lechner W. Einteilung der Stammvarikose der V. saphena magna in 4 Stadien. Phlebol Proktol 1977;6:116–23
Haenen JH, Janssen MC, Wollersheim H, Van't Hof MA, de Rooij MJ, van Langen H, et al. The development of postthrombotic syndrome in relationship to venous reflux and calf muscle pump dysfunction at 2 years after the onset of deep venous thrombosis. J Vasc Surg 2002;35:1184–9.
Handl-Zeller L, Hübsch P, Hohenberg G, Schratter A, Wickenhauser J. [B-mode and color-coded Doppler sonography in irradiation planning in Klippel–Trenaunay syndrome with Kasabach–Merritt symptoms; in German.] Ultraschall Med 1989;10:41–3.
Hanel KC, Abbott WM, Reidy NC, Fulchino D, Miller A, Brewster DC, et al. The role of two noninvasive tests in deep venous thrombosis. Ann Surg 1981;194:725–30.
Heijboer H, Buller HR, Lensing AW, Turpie AG, Colly LP, ten Cate JW. A comparison of real-time compression ultrasonography with impedance plethysmography for the diagnosis of deep-vein thrombosis in symptomatic outpatients. N Engl J Med 1993;329:1365–9.
Herzog P, Anastasiu M, Wollbrink W, Herrmann W, Holtermuller KH. [Real-time sonography in deep thrombosis of the pelvic and leg veins: a prospective comparison with phlebography; in German.] Med Klin (Munich) 1991;86:132–7.
Hingorani A, Ascher E, Lorenson E, DePippo P, Salles-Cunha S, Scheinman M, et al. Upper extremity deep venous thrombosis and its impact on morbidity and mortality rates in a hospital-based population. J Vasc Surg 1997;26:853–60.
Hirschl M, Bernt R. Normalwerte, Reproduzierbarkeit und Aussagekraft duplexsonographischer Kriterien in der Venenfunktionsdiagnostik. Ultraschall Klin Prax 1990;5:81–4.
Holmes MC. Deep venous thrombosis of the lower limbs diagnosed by ultrasound. Med J Aust 1973;1:427–30.
Horattas MC, Wright DJ, Fenton AH, Evans DM, Oddi MA, Kamienski RW, et al. Changing concepts of deep venous thrombosis of the upper extremity: report of a series and review of the literature. Surgery 1988;104:561–7.
Hübsch PJ, Stiglbauer RL, Schwaighofer BW, Kainberger FM, Barton PP. Internal jugular and subclavian vein thrombosis caused by central venous catheters: evaluation using Doppler blood flow imaging. J Ultrasound Med 1988;7:629–36.
Karasch T, Rieser R, Neuerburg-Heusler D, Roth FJ. [Varicosis of the great saphenous vein as a main symptom of iatrogenic arteriovenous fistula; in German.] Dtsch Med Wochenschr 1991;116:1871–4.
Karkow WS, Cranley JJ, Cranley RD, Hafner CD, Ruoff BA. Extended study of aneurysm formation in umbilical vein grafts. J Vasc Surg 1986;4:486–92.
Kerr TM, Cranley JJ, Johnson JR, Lutter KS, Riechmann GC, Cranley RD, et al. Analysis of 1084 consecutive lower extremities involved with acute venous thrombosis diagnosed by duplex scanning. Surgery 1990a;108:520–7.
Kerr TM, Lutter KS, Moeller DM, Hasselfeld KA, Roedersheimer LR, McKenna PJ, et al. Upper extremity venous thrombosis diagnosed by duplex scanning. Am J Surg 1990b;160:202–6.
Killewich LA, Bedford GR, Beach KW, Strandness DE Jr. Diagnosis of deep venous thrombosis: a prospective study comparing duplex scanning to contrast venography. Circulation 1989a;79:810–4.
Killewich LA, Bedford GR, Beach KW, Strandness DE Jr. Spontaneous lysis of deep venous thrombi: rate and outcome. J Vasc Surg 1989b;9:89–97.
Killewich LA, Nunnelee JD, Auer AI. Value of lower extremity venous duplex examination in the diagnosis of pulmonary embolism. J Vasc Surg 1993;17:934–8 [discussion 938–9].
Knudson GJ, Wiedmeyer DA, Erickson SJ, Foley WD, Lawson TL, Mewissen MW, et al. Color Doppler sonographic imaging in the assessment of upper-extremity deep venous thrombosis. AJR Am J Roentgenol 1990;154:399–403.
Köhler M, Neuerburg D. [Determination of valvular function using an ultrasonic technic and auscultation during Valsalva's maneuver; in French.] Phlébologie 1978;31:269–72.
Krause U, Kock HJ, Kröger K, Albrecht K, Rudofsky G. Prevention of deep venous thrombosis associated with superficial thrombophlebitis of the leg by early saphenous vein ligation. Vasa 1998;27:34–8.
Kriessmann A, Bollinger A, Keller HM. Praxis der Doppler-Sonographie. Stuttgart: Thieme, 1990.
Krings W, Adolph J, Diederich S, Urhahne S, Vassallo P, Peters PE. [The diagnosis of deep venous thrombosis of the lower extremities using high-resolution real-time and CW-Doppler sonography: accuracy and limitations; in German.] Radiologe 1990;30:525–31.
Labropoulos N, Leon M, Nicolaides AN, Sowade O, Volteas N, Ortega F, et al. Venous reflux in patients with previous deep venous thrombosis: correlation with ulceration and other symptoms. J Vasc Surg 1994;20:20–6.
Langer M, Langer R. [Radiologic aspects of the congenital arteriovenous malformations, Klippel–Trenaunay type, and Servelle–Martorell type; in German.] RöFo 1982;136:577–82.
Langholz J, Heidrich H. [Sonographic diagnosis of deep pelvic/leg venous thrombosis: is color-coded Duplex sonography "superfluous"? In German.] Ultraschall Med 1991;12:176–81.
Langsfeld M, Hershey FB, Thorpe L, Auer AI, Binnington HB, Hurley JJ, et al. Duplex B-mode imaging for the diagnosis of deep venous thrombosis. Arch Surg 1987;122:587–91.
Leipner N, Janson R, Kuhr J. [Angiomatous dysplasia (Weber type); in German.] RöFo 1982;137:73–7.
Lensing AW, Prandoni P, Brandjes D, Huisman PM, Vigo M, Tomasella G, et al. Detection of deep-vein thrombosis by real-time B-mode ultrasonography. N Engl J Med 1989;320:342–5.
Lohr JM, Kerr TM, Lutter KS, Cranley RD, Spirtoff K, Cranley JJ. Lower extremity calf thrombosis: to treat or not to treat? J Vasc Surg 1991;14:618–23.
Longley DG, Finlay DE, Letourneau JG. Sonography of the upper extremity and jugular veins. AJR Am J Roentgenol 1993;160:957–62.

Loose DA, Drewes J. Venous aneurysm. Union Internationale de Phlébologie. 8eme Congrès mondiale. Abstract no. 149. Brussels: Union Internationale de Phlébologie, 1983.

Lutter KS, Kerr TM, Roedersheimer LR, Lohr JM, Sampson MG, Cranley JJ. Superficial thrombophlebitis diagnosed by duplex scanning. Surgery 1991;110:42–6.

McLafferty RB, Moneta GL, Passman MA, Brant BM, Taylor LM Jr, Porter JM. Late clinical and hemodynamic sequelae of isolated calf vein thrombosis. J Vasc Surg 1998;27:50–6 [discussion 56–7].

Malan E. Vascular malformations (angiodysplasias). Milan: Carlo Erba Foundation, 1974.

Markel A, Manzo RA, Bergelin RO, Strandness DE Jr. Valvular reflux after deep vein thrombosis: incidence and time of occurrence. J Vasc Surg 1992;15:377–82 [discussion 383–4].

Martorell F, Monserrat J. Atresic iliac vein and Klippel-Trenaunay syndrome. Angiology 1962;13:265–7.

Matteson B, Langsfeld M, Schermer C, Johnson W, Weinstein E. Role of venous duplex scanning in patients with suspected pulmonary embolism. J Vasc Surg 1996;24:768–73.

Mattos MA, Londrey GL, Leutz DW, Hodgson KJ, Ramsey DE, Barkmeier LD, et al. Color-flow duplex scanning for the surveillance and diagnosis of acute deep venous thrombosis. J Vasc Surg 1992;15:366–75 [discussion 375–6].

Mattos MA, Melendres G, Sumner DS, Hood DB, Barkmeier LD, Hodgson KJ, et al. Prevalence and distribution of calf vein thrombosis in patients with symptomatic deep venous thrombosis: a color-flow duplex study. J Vasc Surg 1996;24:738–44.

May R. Chirurgie der Bein- und Beckenvenen. Stuttgart: Thieme, 1974.

May R, Nissl R. [Phlebographic studies of the anatomy of leg veins; in German.] Fortschr Geb Röntgenstr Nuklearmed 1966;104:171–83.

Meissner MH, Caps MT, Bergelin RO, Manzo RA, Strandness DE Jr. Propagation, rethrombosis and new thrombus formation after acute deep venous thrombosis. J Vasc Surg 1995;22:558–67.

Möllmann M, Wagner W, Böttcher HD, Lawin P. Lagekontrolle zentralvenöser Katheter mittels Ultraschall. Ultraschall 1987;8:215–7.

Murphy TP, Cronan JJ. Evolution of deep venous thrombosis: a prospective evaluation with US. Radiology 1990;177:543–8.

Nicholls SC, O'Brian JK, Sutton MG. Venous thromboembolism: detection by duplex scanning. J Vasc Surg 1996;23:511–6.

Nonnast-Daniel B, Martin RP, Lindert O, Mugge A, Schaeffer J, vd Lieth H, et al. Colour Doppler ultrasound assessment of arteriovenous haemodialysis fistulas. Lancet 1992;339:143–5.

Noren A, Ottosson E, Rosfors S. Is it safe to withhold anticoagulation based on a single negative color duplex examination in patients with suspected deep venous thrombosis? A prospective 3-month follow-up study. Angiology 2002;53:521–7.

O'Leary DH, Kane RA, Chase BM. A prospective study of the efficacy of B-scan sonography in the detection of deep venous thrombosis in the lower extremities. J Clin Ultrasound 1988;16:1–8.

O'Shaughnessy AM, FitzGerald DE. Organization patterns of venous thrombus over time as demonstrated by duplex ultrasound. J Vasc Investig 1996;2:75–81.

Partsch H. "A-sounds" or "S-sounds" for Doppler ultrasonic evaluation of pelvic vein thrombosis. Vasa 1976;5:16–9.

Partsch H, Lofferer O. [Studies of the venous reflux from the lower limb using a directional ultrasonic detector; in German.] Wien Klin Wochenschr 1971;83:781–9.

Partsch H, Lofferer O, Mostbeck A. [The diagnosis of arteriovenous fistulas in angiodysplasias of the extremities: criteria for the indication to surgery; in German.] Vasa 1975;4:288–95.

Persson AV, Jones C, Zide R, Jewell ER. Use of the triplex scanner in diagnosis of deep venous thrombosis. Arch Surg 1989;124:593–6.

Pierik EG, Toonder IM, van Urk H, Wittens CH. Validation of duplex ultrasonography in detecting competent and incompetent perforating veins in patients with venous ulceration of the lower leg. J Vasc Surg 1997;26:49–52.

Pulliam CW, Barr SL, Ewing AB. Venous duplex scanning in the diagnosis and treatment of progressive superficial thrombophlebitis. Ann Vasc Surg 1991;5:190–5.

Quinn KL, Vandeman FN. Thrombosis of a duplicated superficial femoral vein: potential error in compression ultrasound diagnosis of lower extremity deep venous thrombosis. J Ultrasound Med 1990;9:235–8.

Raghavendra BN, Rosen RJ, Lam S, Riles T, Horii SC. Deep venous thrombosis: detection by high-resolution real-time ultrasonography. Radiology 1984;152:789–93.

Raghavendra BN, Horii SC, Hilton S, Subramanyam BR, Rosen RJ, Lam S. Deep venous thrombosis: detection by probe compression of veins. J Ultrasound Med 1986;5:89–95.

Richter G, Böhm S, Görg C, Schwerk WB. Verlaufsuntersuchungen zur Echogenität venöser Gerinnungsthromben. Ultraschall Klin Prax 1992;7:69–73.

Ritter H, Weber J, Loose DA. [Venous aneurysms; in German.] Vasa 1993;22:105–12.

Rodriguez AA, Whitehead CM, McLaughlin RL, Umphrey SE, Welch HJ, O'Donnell TF. Duplex-derived valve closure times fail to correlate with reflux flow volumes in patients with chronic venous insufficiency. J Vasc Surg 1996;23:606–10.

Rollins DL, Semrow CM, Friedell ML, Calligaro KD, Buchbinder D. Progress in the diagnosis of deep venous thrombosis: the efficacy of real-time B-mode ultrasonic imaging. J Vasc Surg 1988;7:638–41.

Rose SC, Zwiebel WJ, Nelson BD, Priest DL, Knighton RA, Brown JW, et al. Symptomatic lower extremity deep venous thrombosis: accuracy, limitations, and role of color duplex flow imaging in diagnosis. Radiology 1990;175:639–44.

Ruoff BA, Cranley JJ, Hannan LA, Aseffa N, Karkow WS, Stedje KG, et al. Real-time duplex ultrasound mapping of the greater saphenous vein before in situ infrainguinal revascularization. J Vasc Surg 1987;6:107–13.

Rutherford RB. New approaches to the diagnosis of congenital vascular malformations. In: Belov S, Loose DA, Weber J, editors. Vascular malformations. Reinbek (Germany): Einhorn-Presse, 1989.

Saarinen JP, Domonyi K, Zeitlin R, Salenius JP. Postthrombotic syndrome after isolated calf deep venous thrombosis: the role of popliteal reflux. J Vasc Surg 2002;36:959–64.

Schmitt HE. Aszendierende Phlebographie bei tiefer Venenthrombose. Stuttgart: Huber, 1977.

Schönhofer B, Bechtold H, Renner R, Bundschu HD. [Ultrasonographic findings in varicose phlebitis of the great saphenous vein: evidence of thrombus growth and detachment with asymptomatic lung embolism; in German.] Dtsch Med Wochenschr 1992;117:51–5.

Schoop W. Praktische Angiologie. 4th ed. Stuttgart: Thieme, 1988.

Schweizer J, Oehmichen F, Brandl HG, Altmann E. [Color-coded duplex ultrasound and contrast medium enhanced duplex ultrasound in deep venous thrombosis; in German.] Vasa 1993;22:22–5.
Scultetus AH, Villavicencio JL, Gillespie DL, Kao TC, Rich NM. The pelvic venous syndromes: analysis of our experience with 57 patients. J Vasc Surg 2002;36:881–8.
Servelle M. Les malformations congénitales des veines. Rev Chir 1949;68:88.
Shami SK, Sarin S, Cheatle TR, Scurr JH, Smith PD. Venous ulcers and the superficial venous system. J Vasc Surg 1993;17:487–90.
Sigel B, Popky GL, Wagner DK, Boland JP, Mapp EM, Feigl P. Comparison of clinical and Doppler ultrasound evaluation of confirmed lower extremity venous disease. Surgery 1968;64:332–8.
Sigel B, Felix WR Jr, Popky GL, Ipsen J. Diagnosis of lower limb venous thrombosis by Doppler ultrasound technique. Arch Surg 1972;104:174–9.
Skillman JJ, Kent KC, Porter DH, Kim D. Simultaneous occurrence of superficial and deep thrombophlebitis in the lower extremity. J Vasc Surg 1990;11:818–23 [discussion 823–4].
Söldner J, Lösch W, Richter EI, Zeitler E. Radiologische Diagnostik der Venenerkrankungen bei geriatrischen Patienten. Phlebol Proktol 1993;22:253–8.
Stapff M. Betzl G, Küffer GV, Hahn D, Spengel FA. Stellenwert der Duplex-Sonographie in der Diagnostik der tiefen Bein- und Beckenvenenthrombose. Bildgebung/Imaging 1989;56:52–6.
Stiegler H, Rotter G, Standl R, Mosavi S, von Kooten HJ, Weichenhain B, et al. [Value of color duplex ultrasound in diagnosis of insufficiency of perforant veins: a prospective study of 94 patients; in German.] Vasa 1994;23:109–13.
Strandness DE Jr. Duplex scanning in vascular disorders. New York: Raven Press, 1990.
Strandness DE Jr, Sumner DS. Ultrasonic velocity detector in the diagnosis of thrombophlebitis. Arch Surg 1972;104:180–3.
Strandness DE Jr, Thiele BL. Selected topics in venous disorders: pathophysiology, diagnosis, and treatment. Mount Kisco (New York): Futura, 1981.
Strandness DE Jr, Langlois Y, Cramer M, Randlett A, Thiele BL. Long-term sequelae of acute venous thrombosis. JAMA 1983;250:1289–92.
Straub H, Ludwig M. Der Doppler-Kurs. Doppler-Sonographie der peripheren Arterien und Venen. Munich: Zuckschwerdt, 1990.
Sullivan ED, Peter DJ, Cranley JJ. Real-time B-mode venous ultrasound. J Vasc Surg 1984;1:465–71.
Sumner DS. Venous anatomy and pathophysiology. In: Hershey FB, Barnes RW, Sumner DS, editors. Noninvasive diagnosis of vascular disease. Pasadena: Appleton Davies, 1984.
Sumner DS, Lambeth A. Reliability of Doppler ultrasound in the diagnosis of acute venous thrombosis both above and below the knee. Am J Surg 1979;138:205–10.
Talbot SR. Use of real-time imaging in identifying deep venous obstruction: a primary report. Bruit 1982;65:41.
Talbot SR. B-mode evaluation of peripheral arteries and veins. In: Zwiebel WJ, editor. Introduction to vascular ultrasonography. 2nd ed. Orlando, FL: Grune and Stratton, 1986: 351.
Taylor BL, Yellowlees I. Central venous cannulation using the infraclavicular axillary vein. Anesthesiology 1990;72:55–8.
Thulesius O. Pathophysiologische Gesichtspunkte über den venösen Rückstrom. In: Kriessmann A, Bollinger A, editors. Ultraschall-Doppler-Diagnostik in der Angiologie. Stuttgart: Thieme, 1978.
van Bemmelen PS, Bedford G, Beach K, Strandness DE. Quantitative segmental evaluation of venous valvular reflux with duplex ultrasound scanning. J Vasc Surg 1989;10:425–31.
van Gemmeren D, Fobbe F, Ruhnke-Trautmann M, Hartmann CA, Gotzen R, Wolf KJ, Distler A, Schulte KL. [Diagnosis of deep leg vein thrombosis with color-coded duplex sonography and sonographic determination of the duration of the thrombosis; in German.] Z Kardiol 1991;80:523–8.
van Limborgh J. La nomenclature des veines communicantes des 1'extremité inférieure. Rapport du Comité de Nomenclature de la Société Beneluxienne de Phlébologie. Amsterdam: Institute of Anatomy, 1963.
van Ramshorst B, Legemate DA, Verzijlbergen JF, Hoeneveld H, Eikelboom BC, de Valois JC, et al. Duplex scanning in the diagnosis of acute deep vein thrombosis of the lower extremity. Eur J Vasc Surg 1991;5:255–60.
van Ramshorst B, van Bemmelen PS, Hoeneveld H, Eikelboom BC. The development of valvular incompetence after deep vein thrombosis: a follow-up study with duplex scanning. J Vasc Surg 1994;19:1059–66.
Voet D, Afschrift M. Floating thrombi: diagnosis and follow-up by duplex ultrasound. Br J Radiol 1991;64:1010–4.
Vogel P, Laing FC, Jeffrey RB Jr, Wing VW. Deep venous thrombosis of the lower extremity: US evaluation. Radiology 1987;163:747–51.
Vollmar JF, Paes E, Irion B, Friedrich JM, Heymer B. [Aneurysmic transformation of the venous system in venous angiodysplasias of the limbs; in German.] Vasa 1989;18:96–111.
Voss EU. Angiodysplasia of extremities. In: Nobbe F, Hammann H, editors. Gefässchirurgie. Möglichkeiten and Grenzen. Bad Oeynhausen, Germany: TM Verlag, 1984.
Weber J, May R. Funktionelle Phlebologie: Phlebographie, Funktionstests, interventionelle Radiologie. Stuttgart: Thieme, 1990.
Weingarten MS, Branas CC, Czeredarczuk M, Schmidt JD, Wolferth CC Jr. Distribution and quantification of venous reflux in lower extremity chronic venous stasis disease with duplex scanning. J Vasc Surg 1993;18:753–9.
Weingarten MS, Czeredarczuk M, Scovell S, Branas CC, Mignogna GM, Wolferth CC Jr. A correlation of air plethysmography and color-flow-assisted duplex scanning in the quantification of chronic venous insufficiency. J Vasc Surg 1996;24:750–4.
Weissleder R, Elizondo G, Stark DD. Sonographic diagnosis of subclavian and internal jugular vein thrombosis. J Ultrasound Med 1987;6:577–87.
Welch HJ, Villavicencio JL. Primary varicose veins of the upper extremity: a report of three cases. J Vasc Surg 1994;20:839–43.
Welch HJ, Young CM, Semegran AB, Iafrati MD, Mackey WC, O'Donnell TF Jr. Duplex assessment of venous reflux and chronic venous insufficiency: the significance of deep venous reflux. J Vasc Surg 1996;24:755–62.
Wright DJ, Shepard AD, McPharlin M, Ernst CB. Pitfalls in lower extremity venous duplex scanning. J Vasc Surg 1990;11:675–9.
Wuppermann T, Exler U, Mellmann J, Kestilä M. Non-invasive quantitative measurement of regurgitation in insufficiency of the superior saphenous vein by Doppler-

ultrasound: a comparison with clinical examination and phlebography. Vasa 1981;10:24–7.
Yao ST, Gourmos C, Hobbs JT. Detection of proximal-vein thrombosis by Doppler ultrasound flow-detection method. Lancet 1972;1:1–4.
Yucel EK, Fisher JS, Egglin TK, Geller SC, Waltman AC. Isolated calf venous thrombosis: diagnosis with compression US. Radiology 1991;179:443–6.
Zwiebel WJ. Introduction to vascular ultrasonography. Philadelphia: Saunders, 1992.

7 Case Histories

Anatomical Overview

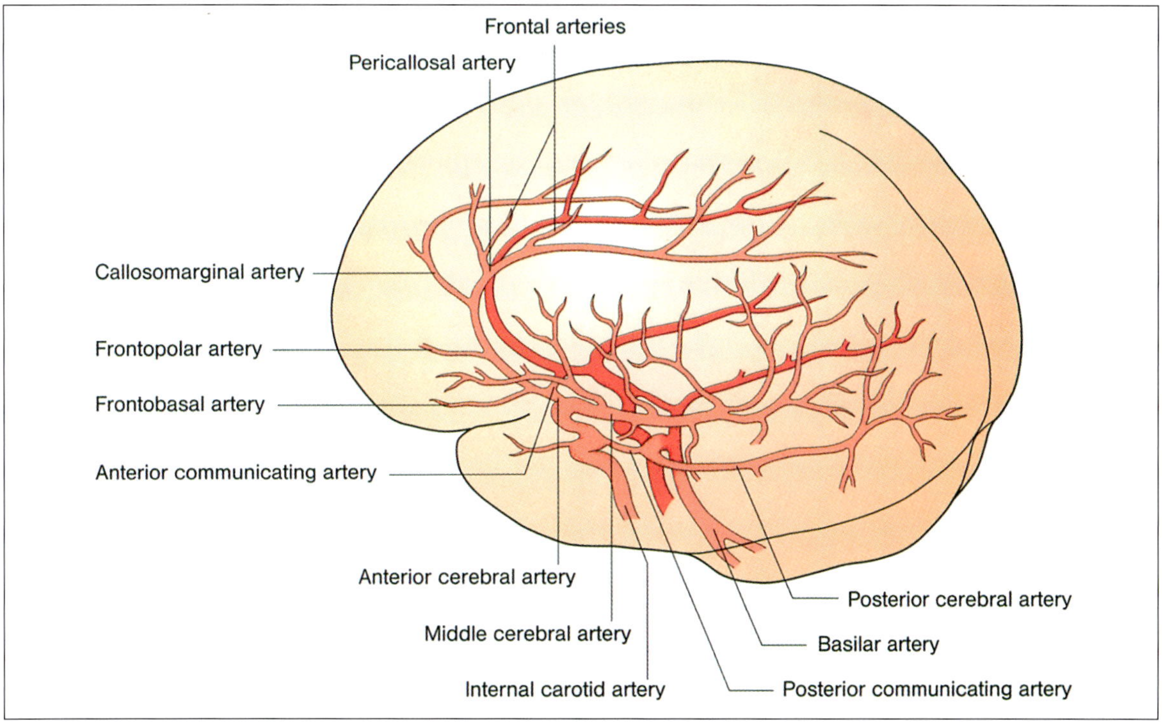

a

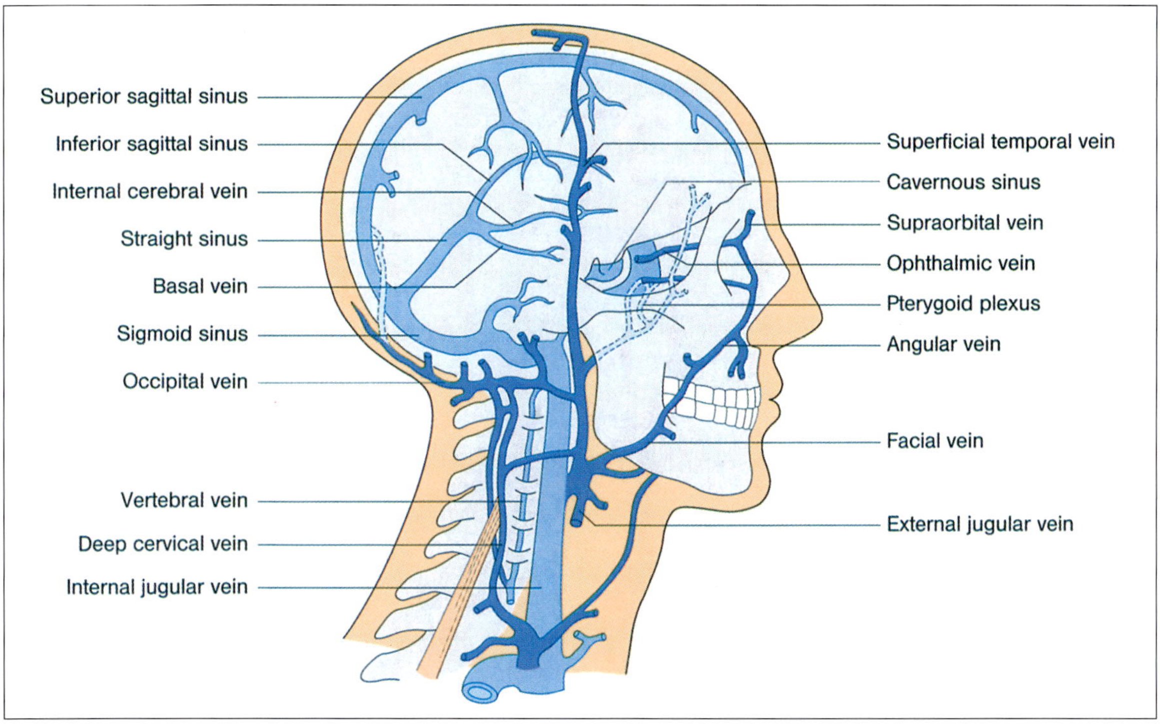

b

Fig. c–f ▷

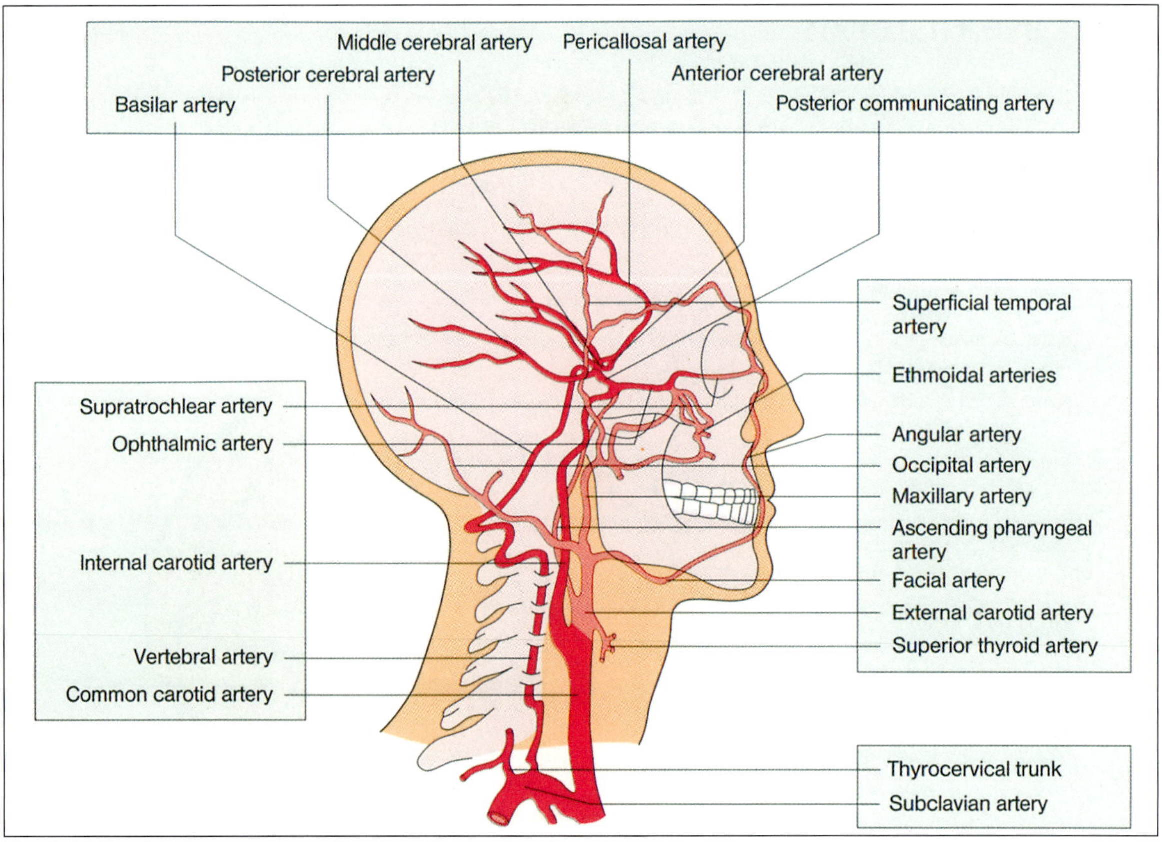
Middle cerebral artery
Pericallosal artery
Posterior cerebral artery
Anterior cerebral artery
Basilar artery
Posterior communicating artery
Superficial temporal artery
Ethmoidal arteries
Supratrochlear artery
Ophthalmic artery
Angular artery
Occipital artery
Maxillary artery
Ascending pharyngeal artery
Internal carotid artery
Facial artery
External carotid artery
Superior thyroid artery
Vertebral artery
Common carotid artery
Thyrocervical trunk
Subclavian artery

c

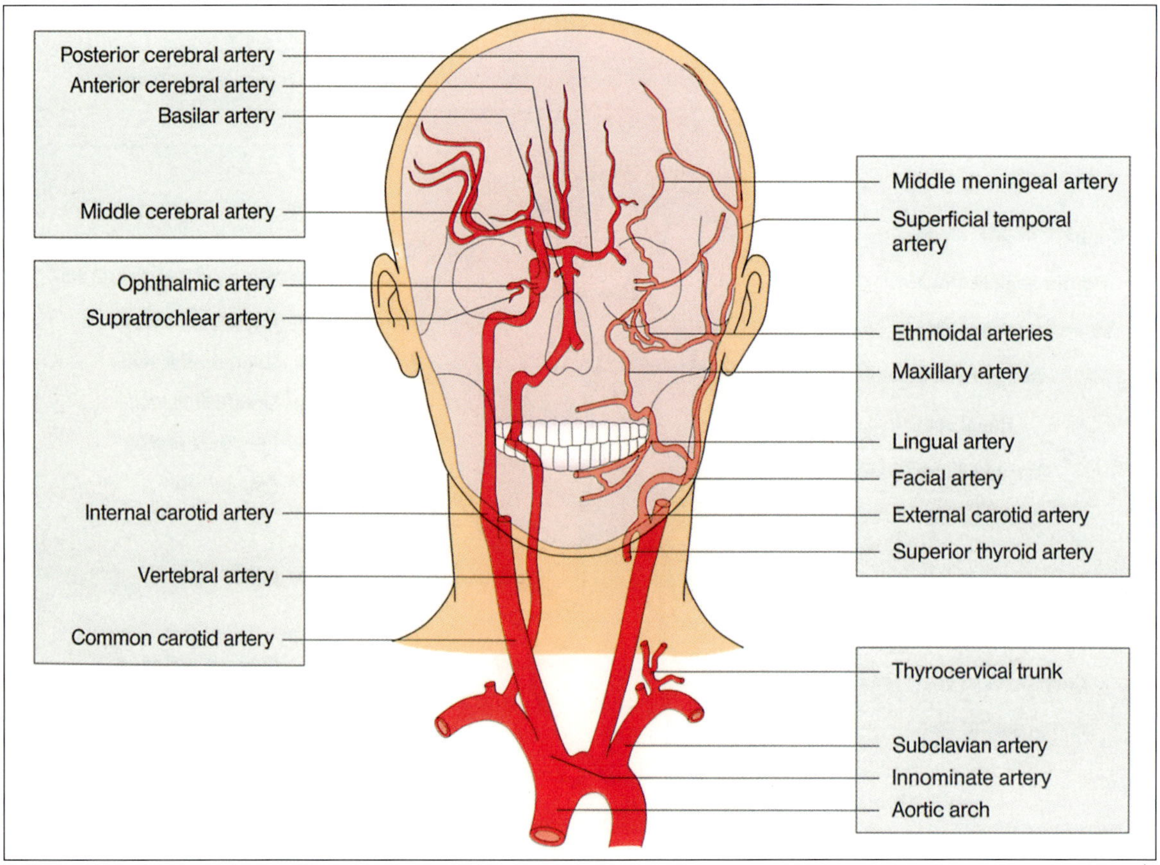
Posterior cerebral artery
Anterior cerebral artery
Basilar artery
Middle cerebral artery
Middle meningeal artery
Superficial temporal artery
Ophthalmic artery
Supratrochlear artery
Ethmoidal arteries
Maxillary artery
Lingual artery
Facial artery
Internal carotid artery
External carotid artery
Superior thyroid artery
Vertebral artery
Common carotid artery
Thyrocervical trunk
Subclavian artery
Innominate artery
Aortic arch

d

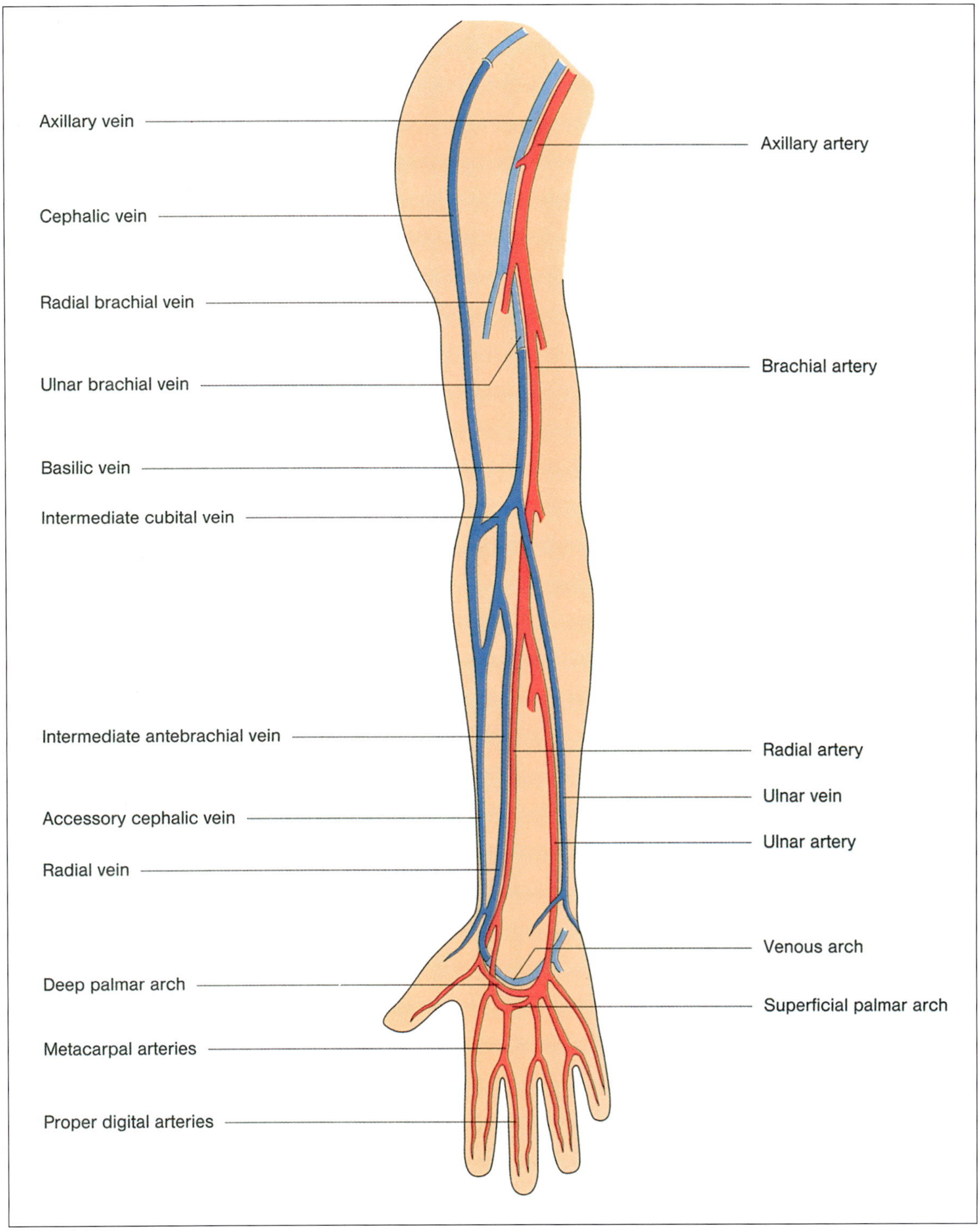
Axillary vein
Cephalic vein
Radial brachial vein
Ulnar brachial vein
Basilic vein
Intermediate cubital vein
Intermediate antebrachial vein
Accessory cephalic vein
Radial vein
Deep palmar arch
Metacarpal arteries
Proper digital arteries
Axillary artery
Brachial artery
Radial artery
Ulnar vein
Ulnar artery
Venous arch
Superficial palmar arch

e

Fig. f ▷

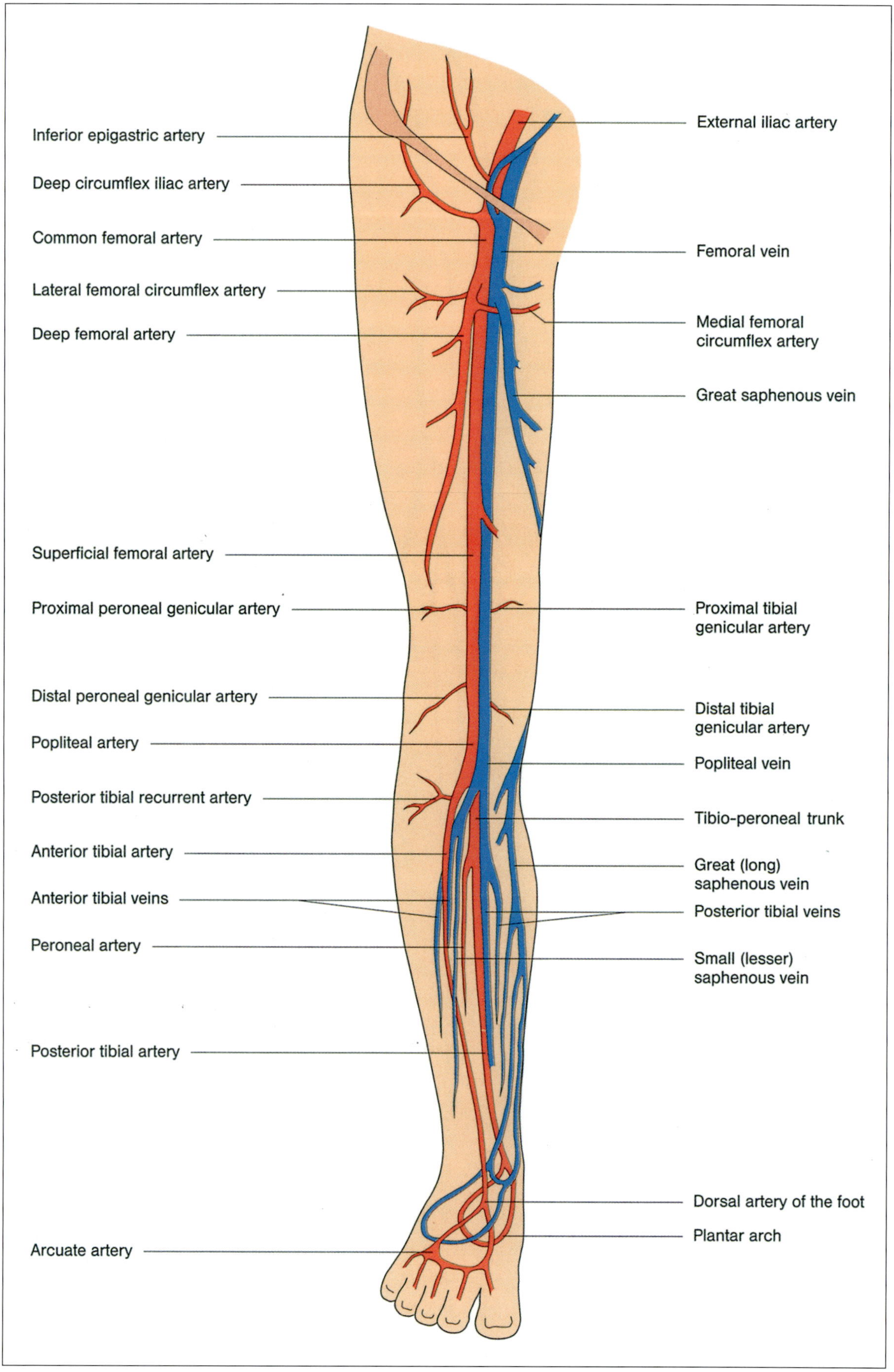
Inferior epigastric artery
Deep circumflex iliac artery
Common femoral artery
Lateral femoral circumflex artery
Deep femoral artery
Superficial femoral artery
Proximal peroneal genicular artery
Distal peroneal genicular artery
Popliteal artery
Posterior tibial recurrent artery
Anterior tibial artery
Anterior tibial veins
Peroneal artery
Posterior tibial artery
Arcuate artery
External iliac artery
Femoral vein
Medial femoral circumflex artery
Great saphenous vein
Proximal tibial genicular artery
Distal tibial genicular artery
Popliteal vein
Tibio-peroneal trunk
Great (long) saphenous vein
Posterior tibial veins
Small (lesser) saphenous vein
Dorsal artery of the foot
Plantar arch

f

Documentation and Examination Report

Image documentation should be straightforward and self-explanatory. This is particularly important when the detection or exclusion of pathological vascular wall processes or intraluminal structures forms a significant portion of the diagnosis. In this case, anatomical structures and individual Doppler signals must be clearly labeled and named.

Various kinds of data storage media can be directly connected to the ultrasound equipment and used to register image and spectral data. Color printers, video systems, or foil printers that make images available during the investigation are used. Additional image data processing syestems are also employed – e.g., magnetic or optic disks, and/or radiographs. It is highly advantageous to record representative examination segments on videotape/digital videofiles (mpeg) along with the acoustic information regarding the Doppler shift. The section orientation chosen must be evident from the documentation, and patient identification must be unmistakable.

The written report of the examination and its results should summarize the clinical data, the problem to be investigated, the ultrasound findings obtained, their interpretation, and if desired, recommendations for further diagnosis and/or therapy.

A description of the findings can either be in written from, or can be presented graphically in a diagram. Table 7.**1** lists the *minimal* requirements for documenting findings in a clinical investigation using duplex sonography. An anatomical illustration of the vascular regions examined can incorporate pathological findings (plaques, stenoses, occlusions) and is especially useful in follow-up examinations for purposes of comparison. For clinical trials specific documentation is mandatory and has to be controlled by ultrasound core labs and advising review panels (Touboul et al. 2004, see Chapter 2).

Table 7.**1** Formal requirements for proper image documentation in duplex sonography

Patient identification
Examination date
Indication leading to the examination
Formulation of the question
Blood vessel identification
B-mode image parameters
Section
– sagittal
– transverse
– frontal
– longitudinal
Doppler parameters
– Transmission frequency
– Pulse repetition frequency
– Filter adjustment
– Sample volume size
– Flow direction
– Baseline position

A comprehensive report of the findings obtained with duplex sonography describes the sonographically detected structures as well as the frequency spectra or blood flow velocities derived from individual vascular segments. This initially purely descriptive account uses the criteria oulined in Table 7.**2**, followed by a summary and evaluation of the findings, closing with a diagnosis based on all studies and functional tests performed. Examination findings that include pathological changes should be recorded in more detail.

Table 7.**2** Content guidelines to describe findings from a duplex sonography study

B-mode image component	Doppler component
Depiction Good Average Poor *Anatomy* Regular Variation/anomaly Kinking/coiling Aneurysm, dilatation *Vascular caliber* Interior wall diameter Exterior wall diameter (remaining lumen) Intima-media thickness *Pulsations* Transverse Axial Systolic/diastolic *Vascular wall structures* Localization Form Size Surface contour – Continuous, smooth – Interrupted, rough Echo pattern – Strength (strong/weak) – Size (progressive/regressive) – Echogenicity (anechoic/echogenic) – Texture (homogeneous/heterogeneous) – Ultrasound shadow (present/absent) Plaque motion (uniform/discrepant) Plaque volume Embolic activity Compressibility of veins	*Flow detection* *Flow direction* Physiological Retrograde Alternating forward and backward (pendular) *Flow velocity* Flow curve amplitude – Decreased – Elevated Frequency – High (kHz) – Low (kHz) Spectral broadening *Flow pattern* Parallel, laminar Disturbed Secondary vortex Absence Limited acceleration Disturbances in heart rhythm *Others* Valsalva maneuver Echocontrast studies Functional studies Breathing/breathholding tests

Extracranial Cerebral Arteries

1 Plaques (Right) and Mild Degree of Stenosis (Left) in the Internal Carotid Artery

Patient: R.M., female, age 63.
Clinical diagnosis: neurologically asymptomatic.

CAD:	–	Vascular surgery:	–	Diabetes mellitus:	–
PAOD:	+	Heredity:	–	Hyperlipidemia:	–
TIA, stroke:	–	Hypertension:	–	Smoking:	+

Doppler sonogram *(spectral analysis)* (**c**). The supratrochlear artery bilaterally presents with physiological blood flow, and the waveforms in both ex-tracranial carotid systems are normal. On the left immediately after the bifurcation, an increase in the sys-tolic and diastolic flow velocity is seen, accompanied by acoustic turbulence phenomena without a significant change in the Doppler spectrum. On the right side at the same location, only a broadening of the systolic window can be seen. Note the different scales of the ordinate!

Color flow duplex sonogram (**b**). *Left:* From the distal segment of the CCA into the proximal ICA, there is inhomogeneous plaque formation, without any significant echo-shadow. The flow separation zone usually seen in the carotid sinus is absent. Instead, as far as the end of the plaque formation, a color reversal is seen as a sign of incipient flow turbulence. It is possible to interpret the color change in the distal segment of the ICA as a niche formation. Also noticeable with color bleaching is a flow acceleration, which serves as an initial sign of a stenosis in the ICA. Estimated degree of stenosis: mild, not hemodynamically significant (≤70–80%).

Right: Proceeding from the terminal segment of the CCA and extending into the ICA, there is inhomogeneous plaque formation without any significant echo-shadow. Color bleaching is observed as a sign of flow acceleration. The often detectable flow separation boundary in the carotid sinus is absent.

Angiogram (**a**). *Left:* From a lateral viewpoint, extensive plaque formation is seen, with a relatively smooth surface, proceeding from the distal segment of the CCA to the upper proximal section of the ICA. In its distal segment, the plaque formation appears to have ill-defined boundaries, which may be a possible expression of a slight niche formation. The estimate of the degree of stenosis is affected by whether the local or distal degree of stenosis is used as the basis for determining this value. The local stenosis grade is approximately 70%, and the distal approximately 60%.

Right: From the distal end of the CCA up to and beyond the proximal ICA, there is a relatively smooth-surfaced and not significantly lumen-constricting plaque formation at the posterior wall of the blood vessel, without any detectable ulceration and without any significant degree of stenosis whether the grade of the stenosis is calculated locally or distally (<50%).

Comment. Localized vascular wall changes that do not significantly obstruct the lumen can be imaged in several sections in the echotomogram. When compared to the angiogram, ultrasound produces a higher detection rate, and surface structures can be better imaged. With color flow sonography, it is also often possible to image ulcerative changes on the surface of the plaques, and the image is probably better than with angiography. Spectral analysis of Doppler signals on its own can at best provide indirect indications of a plaque formation that is not yet hemodynamically significant, e.g., abnormal audiosignals, a reduction in the width of the systolic window, or an increased appearance of low-frequency signal components. Carotid plaque formation at the origin of the ICA is often encountered with a loss of flow separation in the carotid sinus. However, this is not always the case, and separation phenomena may also be missed in the anatomical variant of an absent carotid sinus.

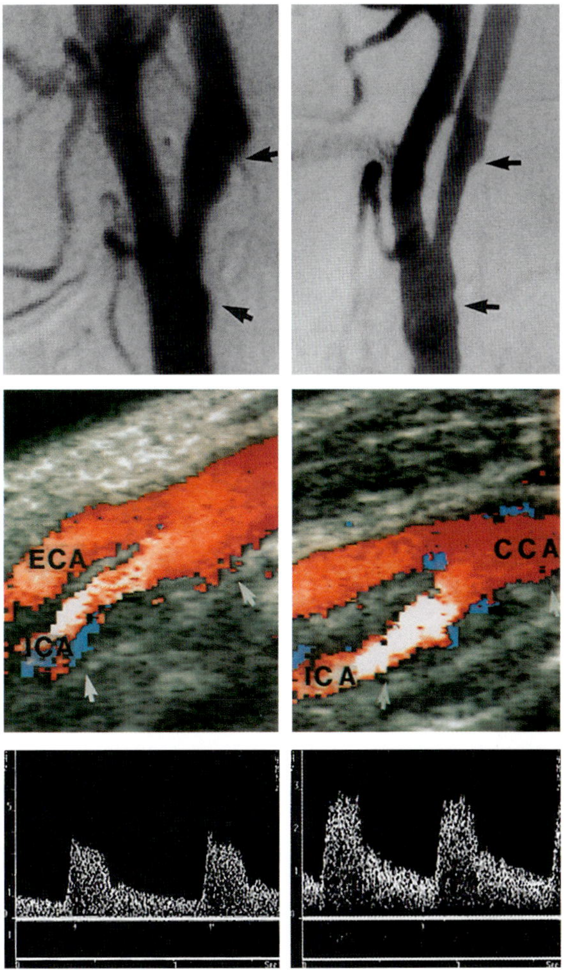

a b c

Left Right

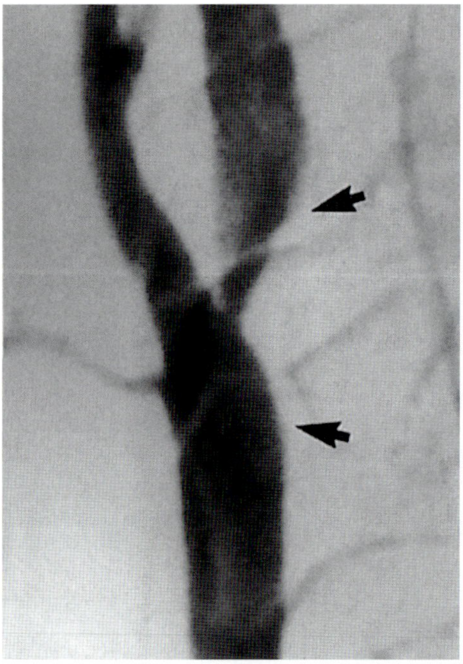

a

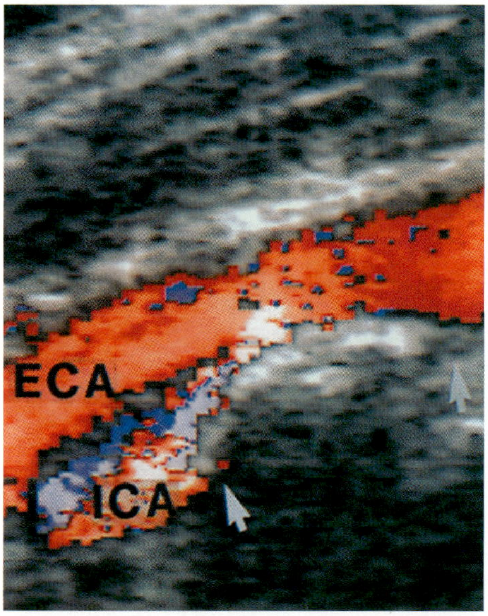

b

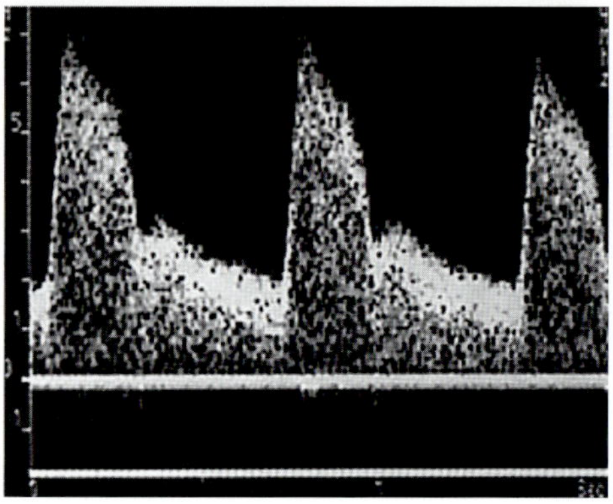

c

2 Moderate Degree of Stenosis of the Right Internal Carotid Artery

Patient: S.F., male, age 58.
Clinical diagnosis: stroke in the left hemisphere.

CAD:	–	Vascular surgery:	–	Diabetes mellitus:	–
PAOD:	–	Heredity:	–	Hyperlipidemia:	+
TIA, stroke:	+	Hypertension:	–	Smoking:	+

Doppler sonogram *(spectral analysis)* (**c**). The supratrochlear artery bilaterally has physiological flow. The carotid artery system on the left is normal. Carotid system on the right: CCA and ECA are normal. In the ICA, there is flow acceleration with systolic deceleration in the proximal ICA (stenosis region). There are poststenotic irregularities in the audiosignal, with preserved, but reduced flow waveforms.

At the site of stenosis, there is a broadening of the spectrum and an increase of the systolic peak frequency to 7 kHz.

Color flow duplex sonogram (**b**). At the transition of the distal CCA into the proximal ICA and extending beyond the carotid sinus, echo-intense plaque formation is seen, with a smooth surface clearly constricting the lumen in the longitudinal section. There is a clear bleaching of color at the origin of the ICA and a loss of the separation zone in the carotid sinus, with an irregular color mixture distal to the stenosis and superior to the bifurcation in the ICA.

Angiogram (**a**). Moderate degree of stenosis of the ICA after the bifurcation, with a smooth surface.

Comment. The stenosis can be well depicted using the echo -impulse imaging B-mode and can be graded as a moderate stenosis according to the flow parameters from either the Doppler sonogram or the color flow duplex analysis. In the angiogram as well, a moderate degree of stenosis is seen, particularly when the distal degree of stenosis is measured (70%). So far the patient has remained free of symptoms of ischemia in its territory.

3 Dissection of the Left Internal Carotid Artery

Patient: A.T., male, age 39.
Clinical diagnosis: Horner's syndrome on the left, TIA with sensorimotor hemiparesis on the right.

CAD:	–	Vascular surgery:	–	Diabetes mellitus:	–
PAOD:	–	Heredity:	–	Hyperlipidemia:	–
TIA, stroke:	+	Hypertension:	–	Smoking:	–

Doppler sonogram *(spectral analysis)* (**a, b**). The supratrochlear artery shows asymmetric, orthograde flow with a reduced flow amplitude on the left, but no change following compression of the branches of the ECA.

The entire extracranial carotid system on the right is normal. In a side-to-side comparison, the flow velocity in the left CCA is reduced. From the bifurcation up to the submandibular region in the ICA, there is a detectable slosh phenomenon with synchronous forward-and-backward components in the systole but an absent signal during diastole (upper spectrum).

The vertebral and the subclavian arteries are normal on both sides.

In a follow-up examination after three weeks, the left ICA has completely recanalized (lower spectrum).

Color flow duplex sonogram (**b**). In the longitudinal section, forward-and-backward signal components in blue–red color coding are seen next to one another in the proxima ICA. Distally, an area free of flow signals marks the proximal end of the dissection. Corresponding Doppler signals characterize partial recanalization with systolic forward-and-backward signal components, but with diastolic forward flow preservation.

Angiogram (**c**). Proximally, there is a thread-like occlusion/subtotal stenosis of the ICA without a connection to the intracranial vasculature (thick arrow). The carotid siphon is supplied into the preclinoid segment through the anastomosis of the maxillary artery from the ophthalmic artery, in which blood flows in a retrograde direction. By way of this anastomosis, good filling of the middle cerebral artery also results. After recanalization, the high-grade submandibular stenosis is still seen in the follow up examination as the point of origin for the dissection (thin arrow). It has irregularities in the intraosseous canal, but the perfusion of the carotid sinus is once again normal and physiological.

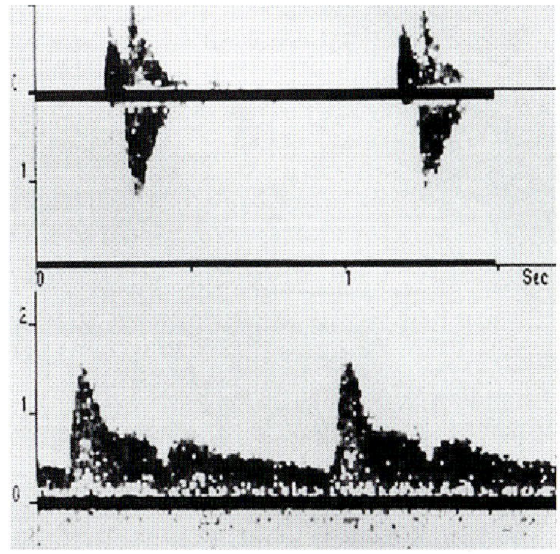

a

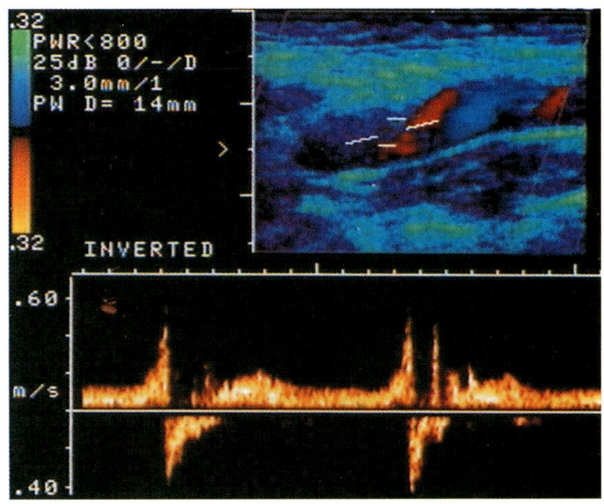

b

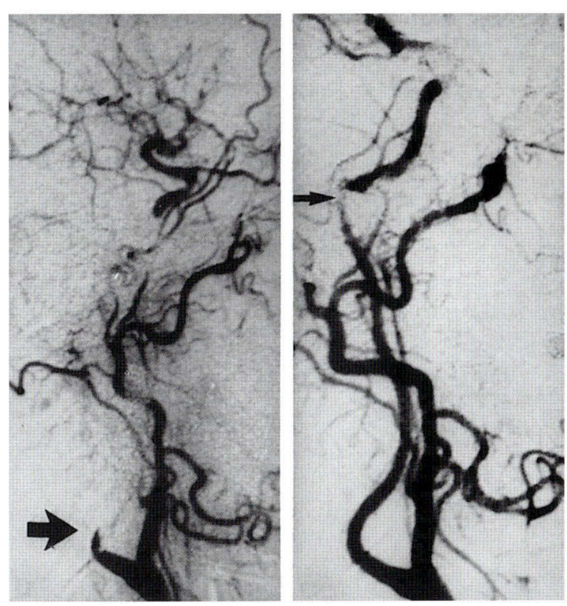
c

Comment. Much more often than previously suspected, the highly characteristic Doppler-sonographic correlate of a dissection is seen in young patients with TIA/stroke. The dissection can appear spontaneously, or may follow traumatic events accompanied by the fully developed picture of focal ischemia with facial and neck pain and Horner's syndrome. It can also appear with very few symptoms, or may even be completely asymptomatic. In combination with the color flow duplex sonogram, it is almost always possible to make the diagnosis when the flow signal is carefully followed *over the entire neck region.* Monthly follow-up assessments are important, since in the majority of the cases spontaneous recanalization occurs.

4 Findings after Stent Placement

Patient: K.B., male, age 63.
Clinical diagnosis: TIA in the left hemisphere.

CAD:	+	Vascular surgery:	–	Diabetes mellitus:	–
PAOD:	+	Heredity:	+	Hyperlipidemia:	+
TIA, stroke:	+	Hypertension:	+	Smoking:	–

Color flow duplex sonogram. *Left:* From the distal segment of the common carotid artery (CCA) to the distal internal carotid artery (ICA), the stent can be visualized as a hyperechoic signal in longitudinal as well as cross-sectional slices (**h–m**). Flow velocities within the stent are slightly increased (maximum peak velocity 150 cm/s) (**h–j**), and B-mode imaging shows mild residual stenosis, with a 46% reduction in the area (**k–l**). (Most interventional specialists tend not to completely normalize or even to dilate the arterial lumen, but stop the angioplasty when the artery still has mild stenosis). *Right:* The blood flow velocity is normal throughout the extracranial carotid system (proximal ICA 120 cm/s) (**d–g**).

Angiogram. Before stent placement, the left bifurcation shows a high-grade 95% stenosis of the left ICA (proximal, over a length of approximately 1 cm) (**a**). After angioplasty, the stent is visible in place (**b**). At the final control examination, with the stent in place, there is a persistent mild stenosis after the angioplasty and stenting (**c**).

Magnetic resonance angiogram (center image). After stenting, there is some loss of the artificial flow signal at the entrance of the stent (arrow), which does not indicate new stenosis. The mild residual stenosis in the middle of the stent is also visible.

Comment. It is possible to display the stent with ultrasound in almost all cases in the early period after placement (the first few months). During the long-term follow-up, the ultrasound display of the stent deteriorates, probably due to endothelialization. Ultrasound is the method of choice for evaluating the flow pattern in the stented segment.

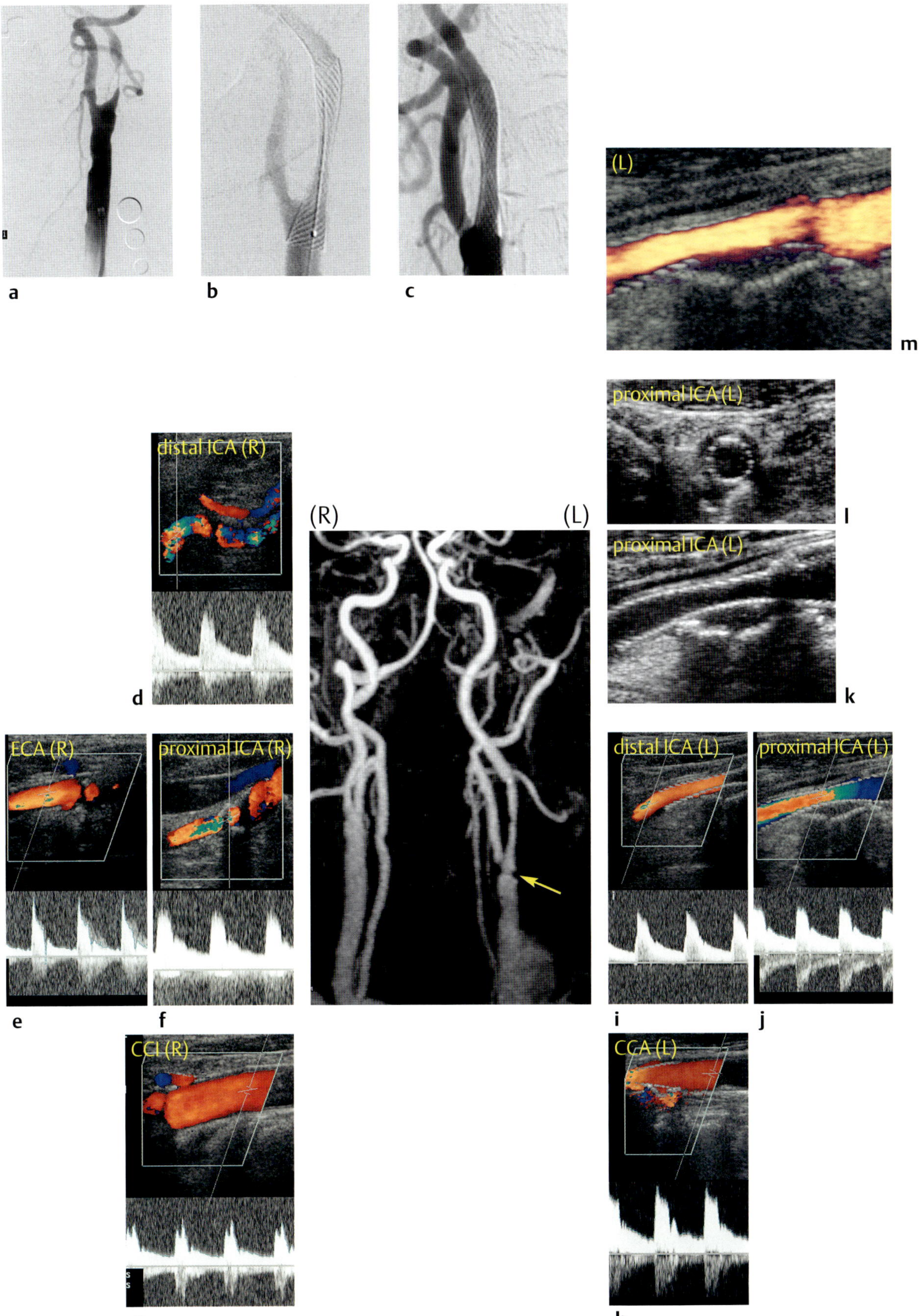
a
b
c
(L)
m
proximal ICA (L)
l
proximal ICA (L)
k
distal ICA (R)
d
(R)
(L)
ECA (R)
e
proximal ICA (R)
f
distal ICA (L)
i
proximal ICA (L)
j
CCI (R)
g
CCA (L)
h

5 Multivessel Disease

Patient: G.S., male, age 66.
Clinical diagnosis: Acute right hemispheric transient ischemic attacks, with left-sided hemiparesis lasting several minutes. Residual right-sided sensorimotor hemiparesis (for 2 years).

CAD:	+	Vascular surgery:	–	Diabetes mellitus:	+
PAOD:	+	Heredity:	+	Hyperlipidemia:	–
TIA, stroke	+	Hypertension:	+	Smoking:	+

Initial Examination

Color flow duplex sonogram. *Left*: The entire lumen of the common carotid artery (CCA) is filled with echodense material (**k**). Some color signals only show very short and subtle oscillating Doppler signals (wall pulsations). The lumina of the external carotid artery (ECA) and internal carotid artery (ICA) are patent, but significant arteriosclerotic wall changes are seen in the B-mode image. Oscillating color changes (red/blue alterations) are visible in both carotid branches and Doppler analysis reveals a pathological flow pattern in the ECA and ICA, with oscillating flow in both segments (**l**, **m**). Net flow (during diastole) in the ECA (**l**), however, is retrograde, while net flow in the ICA (**m**) is orthograde, indicating collateralization of the ICA via the ECA.

The vertebral artery (VA) (diameter 3.5 mm) shows low intermittent flow, with a resistance pattern (**j**), due to a high-grade stenosis of the subclavian artery with transient subclavian steal phenomenon.

Right: Proceeding from the terminal segment of the CCA (**h**) and extending into the ICA, inhomogeneous plaque formation is seen, with some echo shadow and an irregular surface. Color bleaching ("aliasing") is observed as a sign of flow acceleration. Doppler registrations show normal maximum peak velocities in the CCA (110 cm/s) (**h**) but a pathological spectral pattern, with some increased pulsatility. Registrations of the ECA (**g**) and particularly the ICA (**f**) show maximum peak velocities > 5 m/s, with low-frequency components and distal flow separations in the ICA in particular. There is increased flow velocity throughout the entire course of the ICA up to the mandible (**e**), with distal maximum peak velocity values of > 3 m/s.

The VA (diameter 1.2 mm) is hypoplastic, with barely measurable low orthograde flow (20 cm/s) (**i**).

Transcranial color flow duplex sonogram (not shown). Stenoses of the right middle cerebral artery (MCA; 240 cm/s at an insonation depth of 66 mm, 90 cm/s at an insonation depth of 50 mm) and the right posterior cerebral artery (PCA; 180 cm/s at an insonation depth of 69 mm). Normal flow in the left MCA (71 cm/s at an insonation depth of 57 mm) and PCA (35 cm/s at an insonation depth of 66 mm).

Magnetic resonance imaging (MRI) and magnetic resonance angiography (MRA). MRI shows chronic left hemispheric infarction and subcortical small ischemic lesions on the right as well (**b**). On the right side, intracranial MRA shows that the carotid artery and vertebral artery are the arteries with the greatest flow (**a**). However, there is residual flow in the left ICA, underlining the fact that the ICA is patent despite occlusion of the CCA. Right MCA and PCA stenosis are hardly visible.

Angiogram. Display of the right carotid bifurcation shows a massive calcified arteriosclerotic plaque, barely allowing numerical quantification of the degree of stenosis but confirming a high-grade, hemodynamically relevant stenosis of the carotid bulb (**c**). The results after right angioplasty and stent placement are shown in (**d**).

Follow-Up after 6 Months

Color flow duplex sonogram. *Left*: The CCA findings are unchanged (**v**). Flow in the ECA (**w**) and ICA (**x**, **y**) is significantly altered, with notable retrograde flow in the ECA and significantly higher flow in the ICA. The Doppler spectra of the flow show reduced pulsatility.

The VA (**u**; diameter 3.5 mm) shows low permanent retrograde flow, with a resistance pattern. There is no change in the findings in the subclavian artery, but the subclavian steal is now permanent.

Right: Proceeding from the terminal segment of the CCA (**s**) and extending into the ICA (**q**, **p**), there is visualization of the stent and also of inhomogeneous plaque formation, with echo shadowing. Color bleaching ("aliasing") is observed as a sign of flow acceleration. Doppler registrations of the CCA reveal normal maximum peak velocities (96 cm/s) but an almost pathological spectral pattern, with increased pulsatility. Despite angioplasty and stenting, registrations of the ECA (**r**) and particularly of the ICA (**g**) show maximum peak velocities > 5 m/s, with low-frequency components and distal flow separations in the ICA in particular. At this point, however, the distal flow in the ICA (**p**) is lower than at the initial examination (110 cm/s) with lower pulsatility.

The VA findings are unchanged (**t**).

Angiogram. Imaging of the right carotid bulb shows the stent with suspected recurrent stenosis (**n**, **o**).

Comments: The flow velocities in the right-sided carotid stenosis are influenced by the occlusion of CCA on the left, causing an increase in velocity on the right side in order to compensate. As neither vertebral artery is capable of making a significant contribution to the compensation, the velocities in the right carotid system are directly reflecting the collateral capacity. Very high maximum peak velocities therefore need to be corrected for this factor, suggesting a degree of stenosis of around 80% initially. The stenosis is hemodynamically

Initial Examination

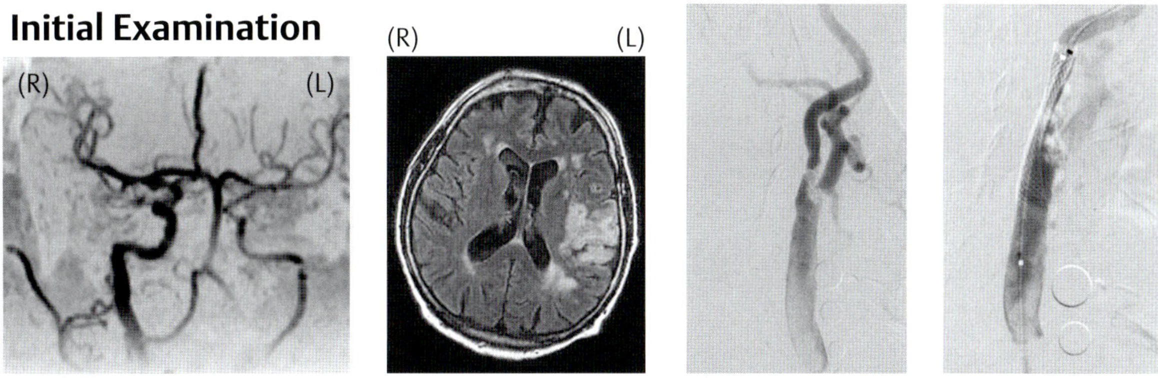

a b c d

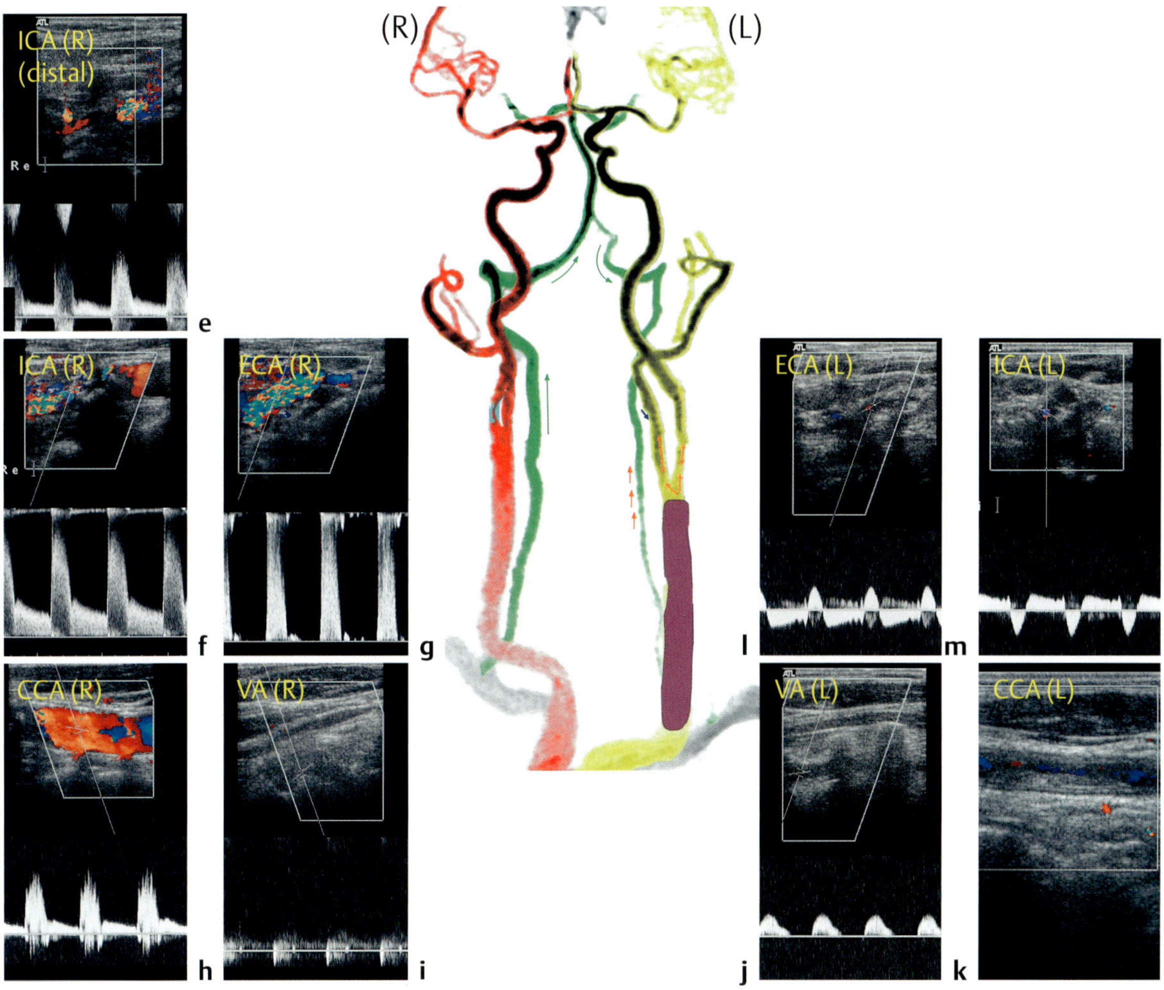

e f g l m h i j k

Follow-Up after 6 Months

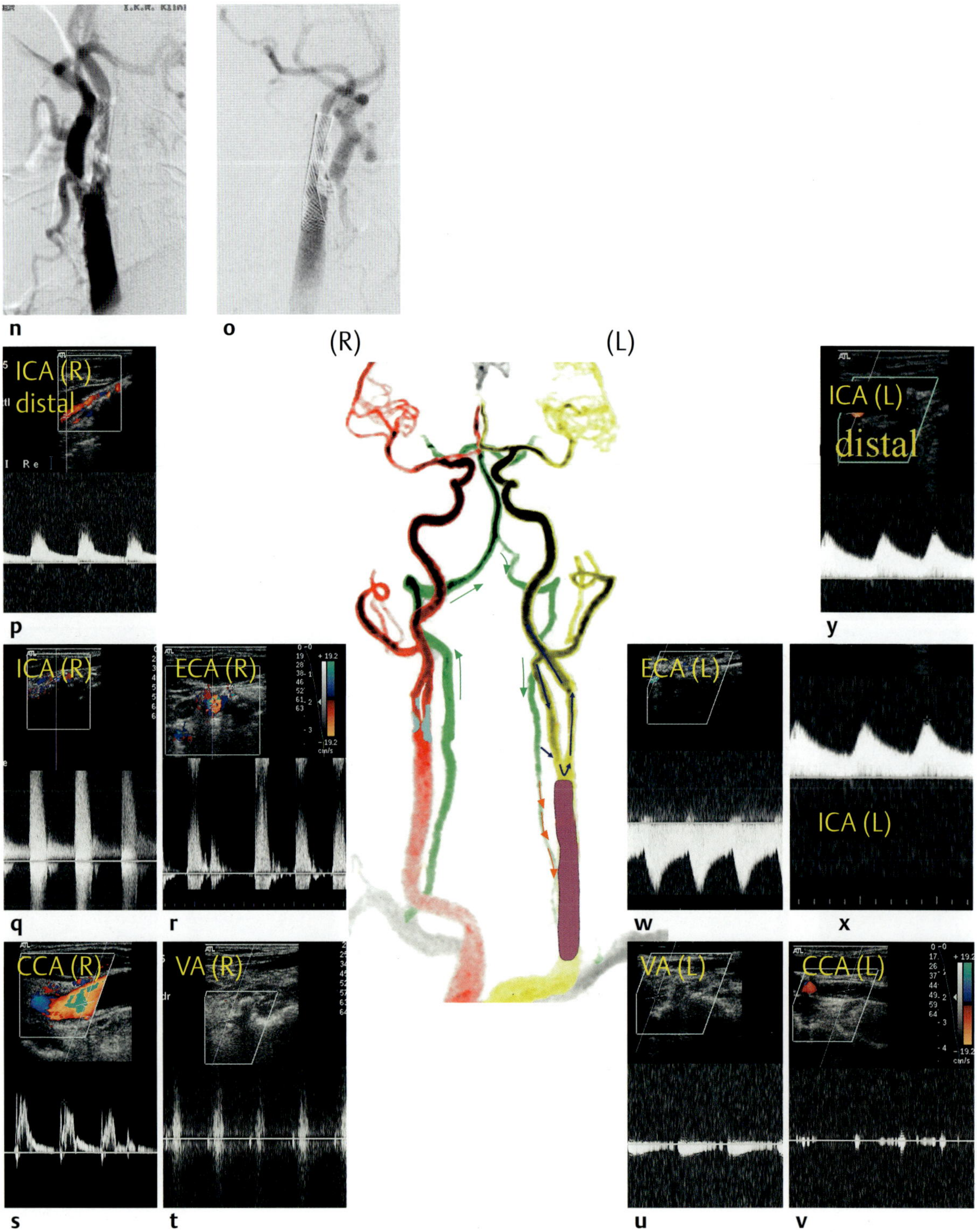

significant due to its clear influence on both the proximal and distal flow profiles. The follow-up examination reveals a mild stenosis after angioplasty and stenting though hemodynamically more pronounced than structural for collateral demands although the left carotid system shows collateral improvement. This case demonstrates that angiography is incapable of providing a "true" measurement of stenoses but contributes structural aspects to hemodynamic ones from ultrasound studies.

6 Severe Stenosis of the Innominate Artery

Patient: L.K., male, age 58.
Clinical diagnosis: Asymptomatic. Blood pressure is 170/90 mmHg on the left and 105/80 mmHg on the right.

CAD	–	Vascular surgery:	–	Diabetes mellitus:	–
PAOD:	–	Heredity:	–	Hyperlipidemia:	+
TIA, stroke:	–	Hypertension:	+	Smoking:	+

Color flow duplex sonogram. There is a clear difference between the left and the right sides. The left common carotid artery (CCA) (**f**), internal carotid artery (ICA) (**g**), and external carotid artery (ECA) (**h**) show a regular spectrum pattern, with flow velocities within the upper normal range. On the right side, there is oscillating flow in the CCA (**d**), pathological retrograde flow with systolic deceleration in the ICA (**a, b**), and reduced flow in the ECA oscillating (**c**). The vertebral artery (VA) on the right (**e**) shows retrograde flow, indicating filling of the right subclavian artery via the right VA.

Comment. The principal features of hemodynamically relevant stenosis of the innominate artery are bizarre and spontaneously fluctuating pathological flow patterns on the pathological side (CCA, ICA, ECA, and VA) and normal or accelerated flow profiles on the contralateral side. The patterns can change spontaneously according to systematic blood pressure changes.

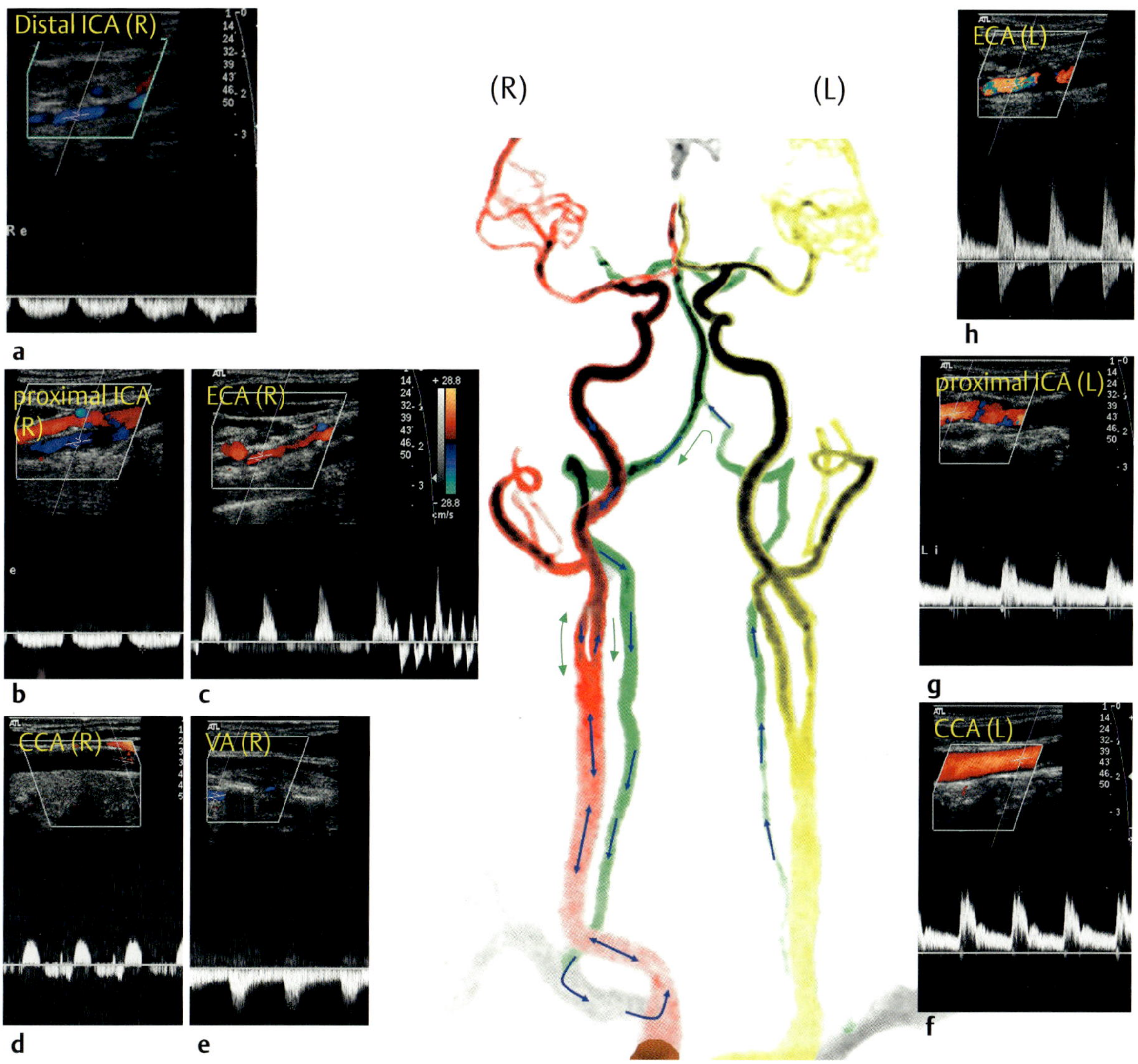

7 Ipsilateral Multilevel Carotid Disease: Stenosis at the Siphon and the Bifurcation of the Left Internal Carotid Artery in the Neck

Patient: H.S., male, age 64.
Clinical diagnosis: stroke in the left hemisphere.

CAD:	–	Vascular surgery:	–	Diabetes mellitus:	–
PAOD:	–	Heredity:	–	Hyperlipidemia:	+
TIA, stroke:	+	Hypertension:	+	Smoking:	+

Doppler sonogram *(analogue registration)* (**c**) and *(spectral analysis)* (**d**). The systolic flow velocity in the left supratrochlear artery is lower than on the right. Flow is bilaterally orthograde. When compared to the right side, the left CCA, with a slight change in the waveform, clearly has a reduced diastolic flow velocity. Above the bifurcation, there is a localized diastolic flow acceleration with an increase in the systolic amplitude modulation. Further cranially, the flow signals are again reduced.

In the spectrum at the site of the extracranial stenosis (**d**), a clear broadening of the window can be seen, with an accumulation of low-frequency signals, especially during the diastole, and slightly elevated frequency values.

The right carotid system is normal. The vertebral and subclavian arteries are bilaterally normal.

Transorbital and extracranial pulsed wave Doppler sonograms (**e**). Using the transorbital access on the right, symmetrical spectra from the carotid segment can be registered at a depth of 70 mm, before (C_4 segment, curve deflection upward) and after (C_2 segment, curve deflection downward) the origin of the ophthalmic artery (average flow velocity 26 cm/s or 666 Hz). On the left, the proximal stenosis of the siphon can be recognized from the flow acceleration, with an increase in the average flow velocity (46 cm/s or 1177 Hz) in the presence of an extremely disturbed audio-signal, extending into the poststenotic segment. With a slight tilting of the probe at the same location, the distal carotid siphon can also be evaluated. A stenosis is not detectable here.

Angiogram (**a**). A moderate degree of stenosis distal to the bifurcation of the ICA and additional mild-grade stenoses at the origin of the ophthalmic artery in the intracranial carotid segment (carotid siphon) are demonstrated.

Color flow duplex sonogram (**b**). At the origin of the ICA, there is a small, smooth, homogeneous plaque formation, with moderate constriction of the lumen. Slight flow acceleration is demonstrated by the color bleaching in the color-coded image. At the carotid sinus, the usual flow separation zone, with blue-coded color components, is missing.

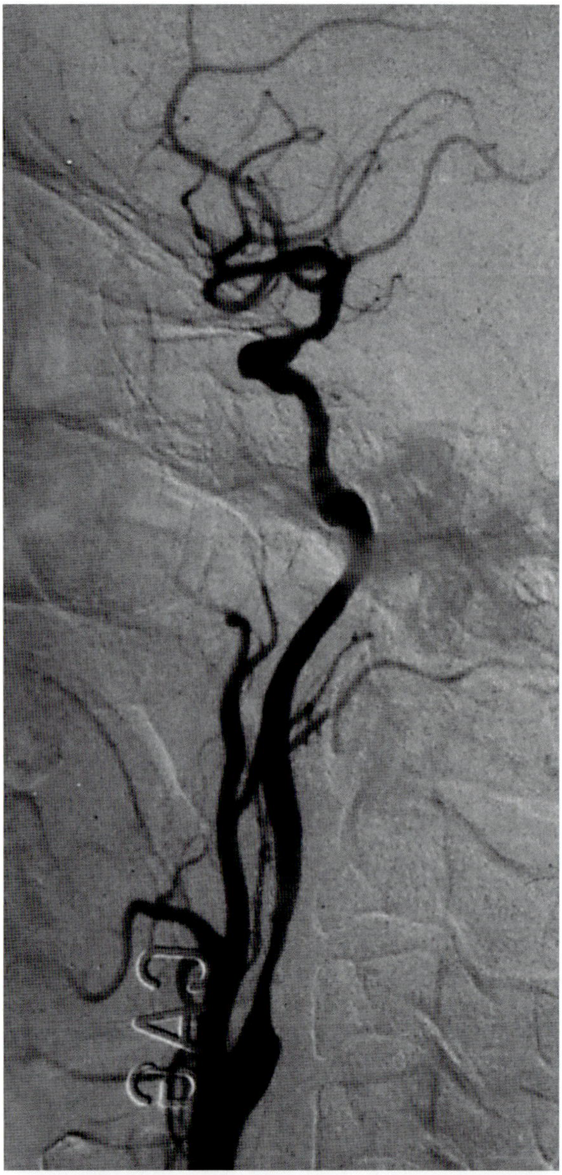

a

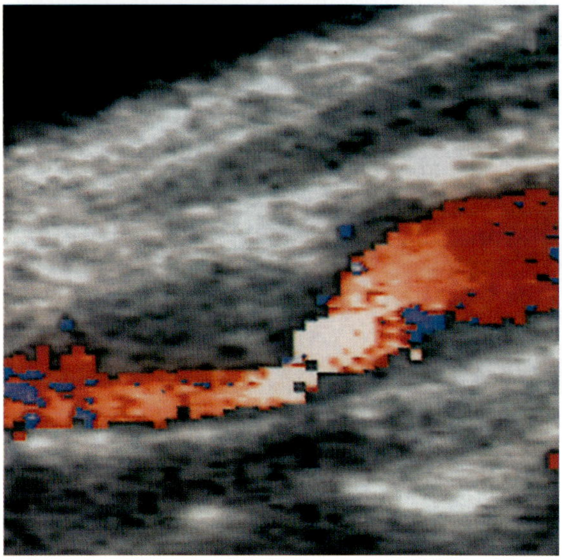

b

Comment. The mild flow acceleration at the origin of the ICA and the plaque formation in the B-mode image do not explain the clear asymmetry between the two diastolic flow velocities in the CCAs. From this finding, it has to be concluded that a second flow obstruction introducing a hemodynamically significant obstruction should, exist beyond the immediate area of direct ultrasound application. A comparison of the fronto-orbital terminal branches of the ophthalmic artery, which show greater flow on the right, suggests that the location of the stenosis should be sought on the left before the origin of this blood vessel. Alternatively, a stenosis at the origin of the left CCA might be possible. Pulsed wave Doppler sonography with an examination of both regions (immediately supra-aortal from the supraclavicular fossa, and in the carotid siphon from a transorbital approach) allows further differentiation. However, as in the present example, flow obstructions in the proximal siphon segment of the carotid artery are easier to detect than distal plaque formations.

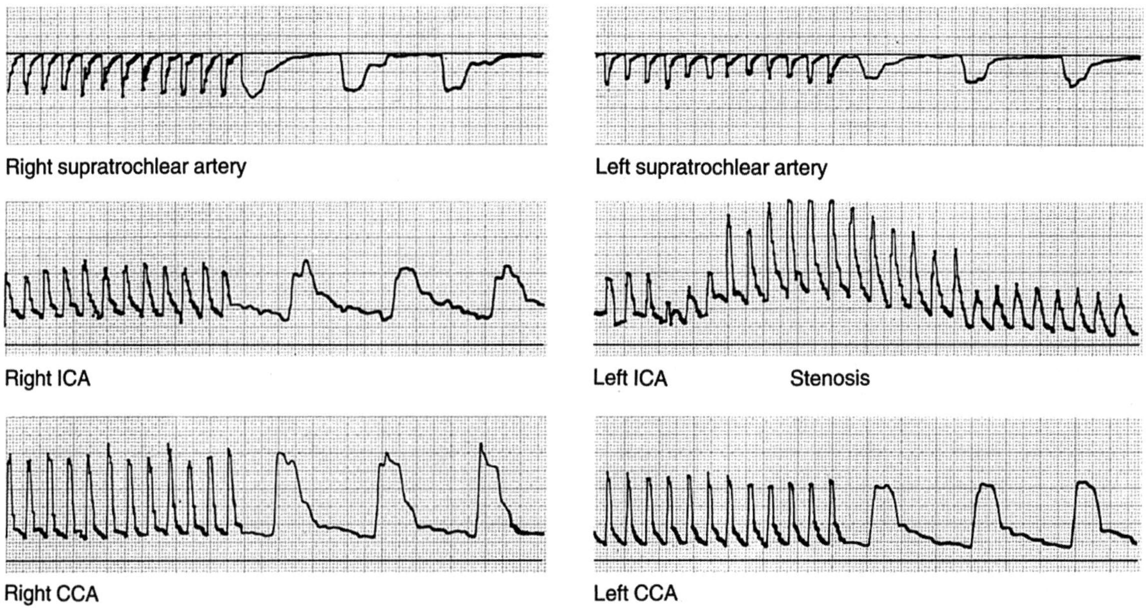

c

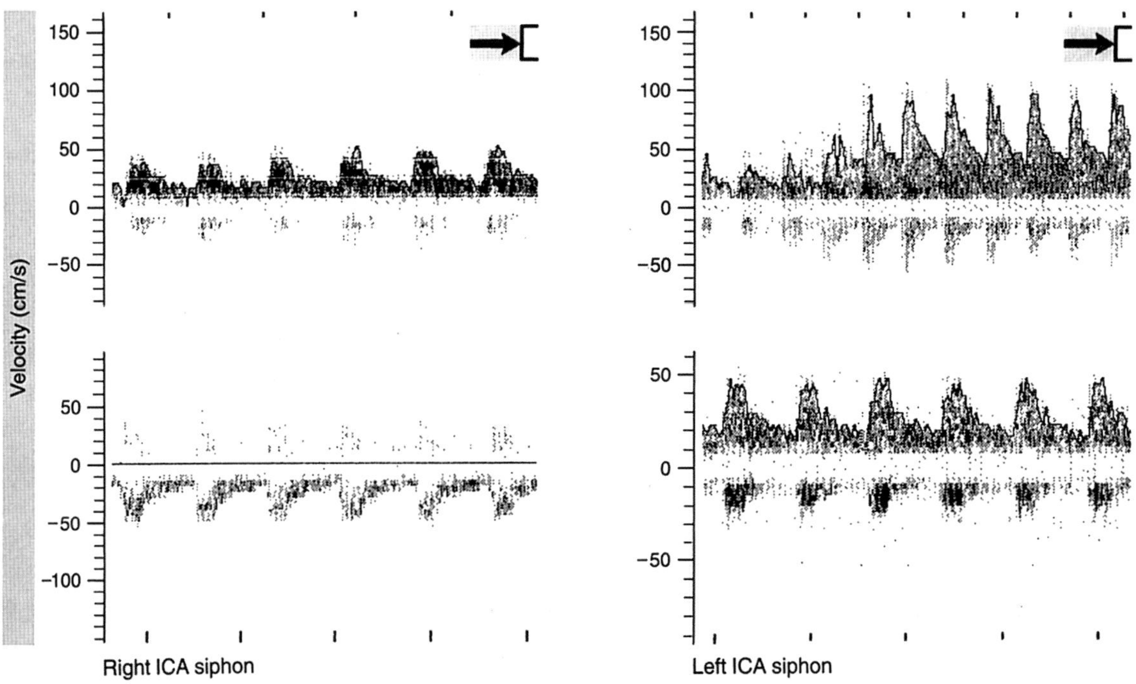

d

8 Permanent Subclavian Steal Phenomenon on the Left

Patient: E.E., male, age 58.
Clinical diagnosis: asymptomatic.

CAD:	+	Vascular surgery:	–	Diabetes mellitus:	+
PAOD:	–	Heredity:	–	Hyperlipidemia:	–
TIA, stroke:	–	Hypertension:	–	Smoking:	–

Extracranial continuous wave Doppler sonogram *(analogue registration)* (**b**). The left vertebral artery has a pathological waveform and a positive compression effect: diastolic residual flow is absent during vascular compression of the left upper arm, and after reopening there is a transient diastolic flow acceleration for several cardiac cycles (reactive hyperemia). The left subclavian artery shows a similar waveform and compression effect (not shown). The right vertebral artery has a normal flow signal but also shows a positive compression effect during *left-sided* upper arm compression (vertebrovertebral shunt). The right subclavian artery is normal.

Extracranial pulsed wave Doppler sonogram *(spectral analysis)* (**c**). At the origin of the left subclavian artery close to the aorta, no signal can be registered. Distally, as in the vertebral artery, a pathological waveform is seen. In comparison, the right subclavian artery appears to be normally depicted at its origin from the innominate artery.

Angiogram (**a**). During retrograde angiography, the right vertebral artery, basilar artery, also the left vertebral artery are filled from the brachial artery. In later images, proximal occlusion of the subclavian artery is seen, with the blood supply of the distal subclavian artery proceeding through the ipsilateral retrograde vertebral artery. (The images shown are 1.0 s, 1.7 s, and 2.3 s after contrast medium injection).

Comment. A subclavian steal phenomenon that can only be demonstrated by either Doppler sonography, angiography, or both, but does not cause any neurological or peripheral deficits in the affected upper extremity should not be described as *subclavian steal syndrome and needs no further treatment.*

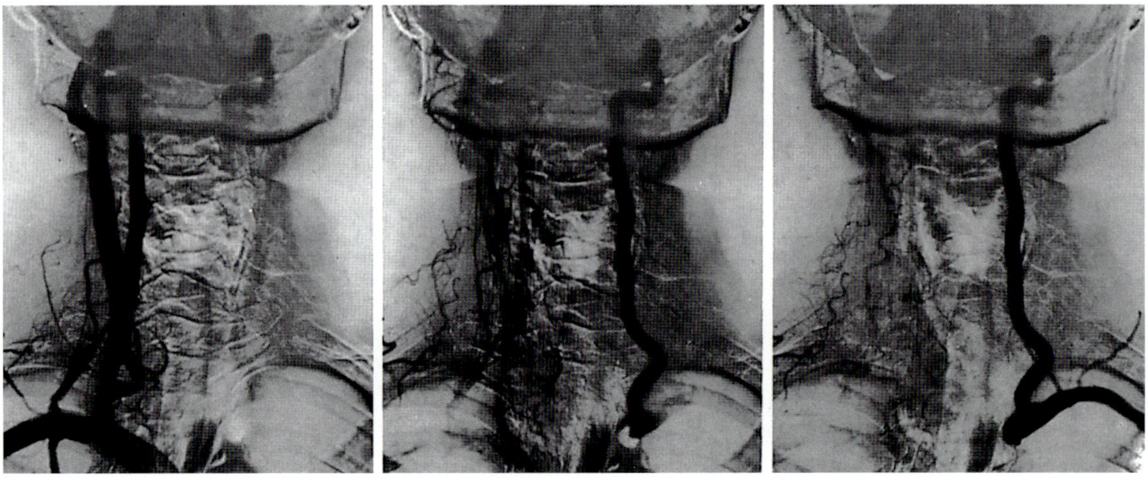

a

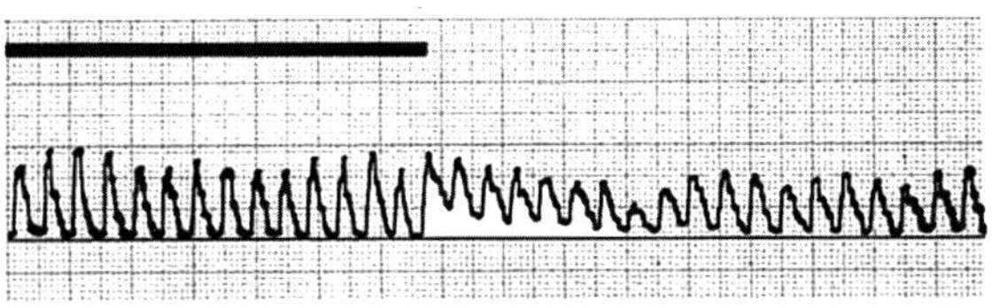

Right vertebral artery at the mastoid process

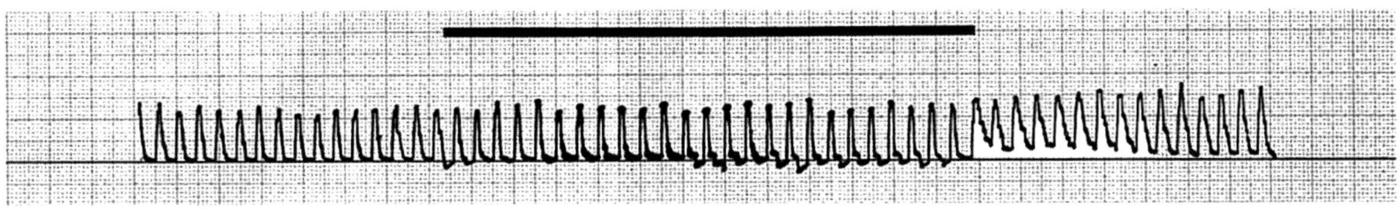

Left vertebral artery at the mastoid process

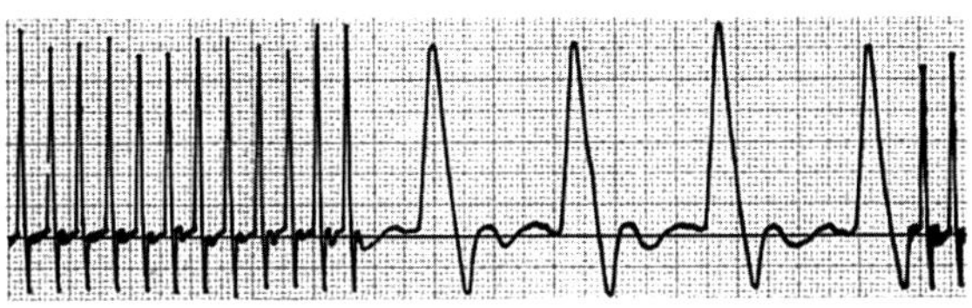

Distal right subclavian artery

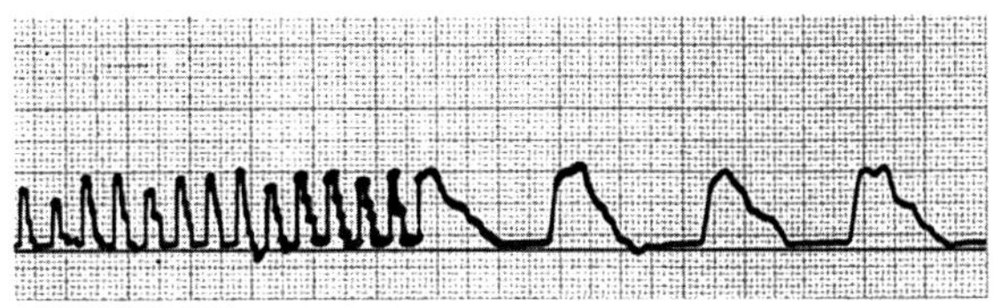

Distal left subclavian artery

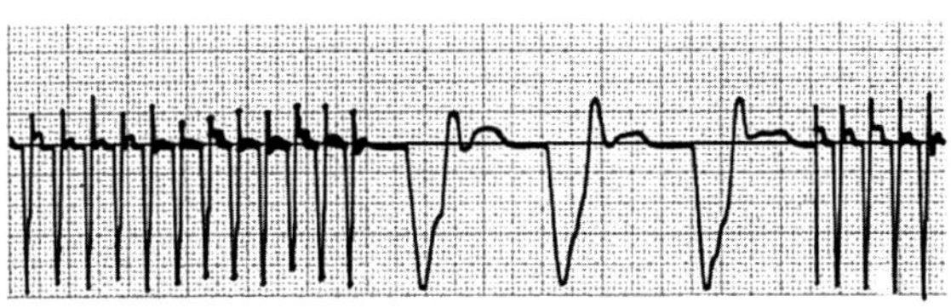

Proximal right subclavian artery

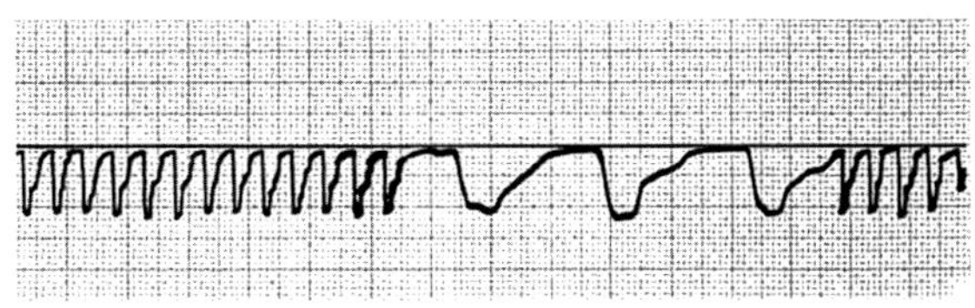

Proximal left subclavian artery

b

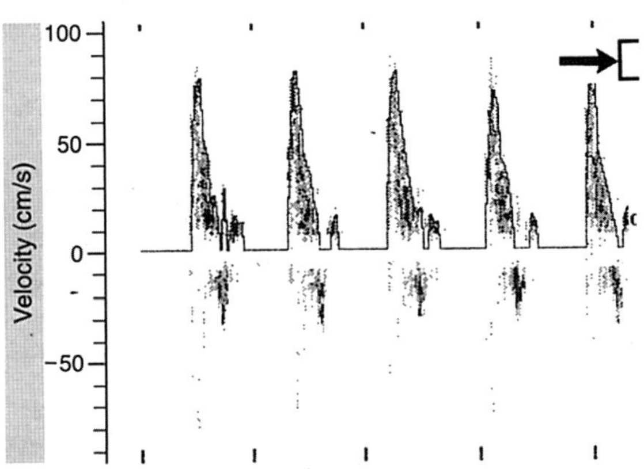

Right subclavian artery (normal, depth: 55 mm)

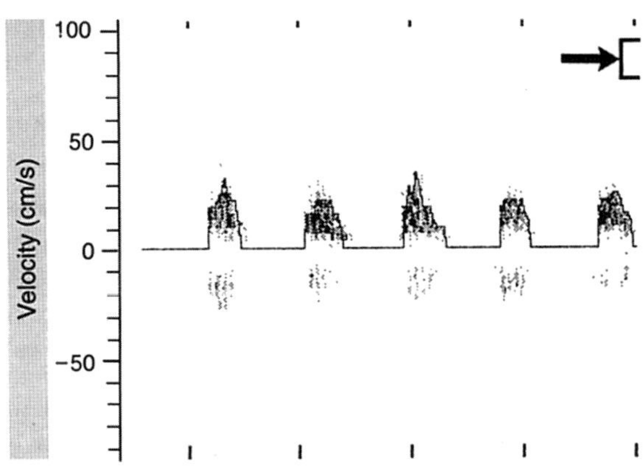

Distal left subclavian artery (occlusion, depth: 50 mm)

c

9 Takayasu's Aortic Arch Arteritis

Patient: C.P., female, age 36.
Clinical diagnosis: Acute left hemispheric stroke as the initial manifestation of Takayasu's arteritis. Blood pressure is 90/60 mmHg on the left and 105/80 mmHg on the right.

CAD	–	Vascular surgery:	–	Diabetes mellitus:	–
PAOD:	–	Heredity:	–	Hyperlipidemia:	–
TIA, stroke:	+	Hypertension:	–	Smoking:	–

Extracranial color duplex sonogram. The lumen of both common carotid arteries (CCAs) is significantly reduced over their entire course (**g**, **i**). No circumscribed increase in the flow velocities is detected. The wall changes show echo characteristics lying between those of the wall echo and the low echo of the lumen. The area reduction is 90% on the left and 75% on the right. The Doppler spectrum distal to the obstruction in the internal carotid artery (ICA) on both sides (the left side being more abnormal than the right side) shows a pathological pattern, with systolic deceleration (**j**). The vertebral arteries are normal, with orthograde flow on both sides (**f**, **h**). There is loss of early diastolic backflow in both subclavian arteries, indicating relevant proximal obstruction. The cross-sectional display of the aortic arch (sector probe 6 MHz) shows luminal narrowing and low-echo stenotic wall changes (**k**), as in the CCAs.

Transcranial duplex sonogram. The basilar artery (**e**) shows normal flow. There is a slight right–left side difference between the middle cerebral arteries (MCAs) (**b**, **d**).

Magnetic resonance imaging (a), including MR angiography (time-of-flight) (**c**). Left hemispheric stroke shows an embolic (territorial) pattern (**a**). Both MCAs are patent, and peripheral branching shows no side differences, indicating complete recanalization (**c**). Flow signal in the basilar artery is normal. Extracranial magnetic resonance angiography (right center image) and of the abdominal aorta (**l**) confirm Takayasu's disease.

Comment. Obstructive processes near the aorta can be evaluated optimally with duplex sonography. An abnormal spectrum pattern in both CCAs or even both vertebral arteries, as seen in this case, is a clear indication of proximal disease.

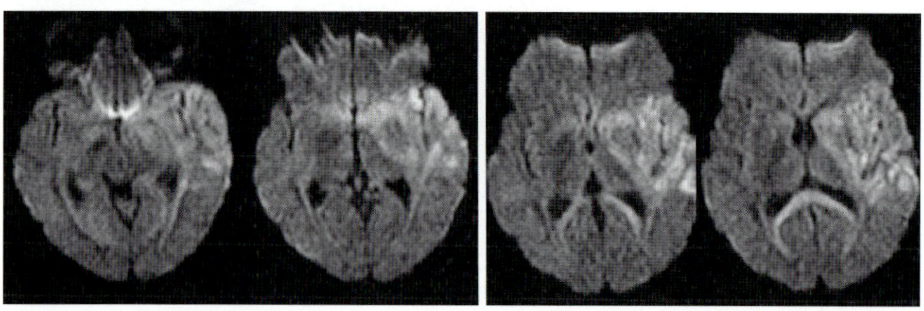

a

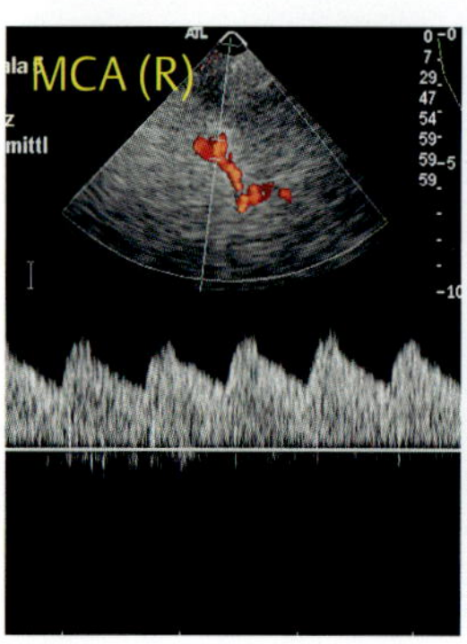

b

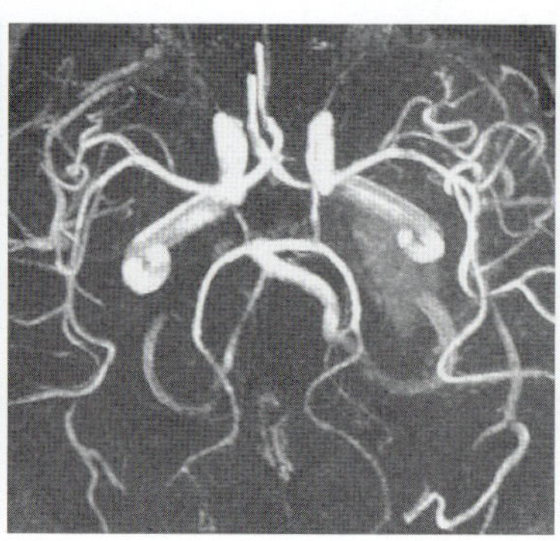

c

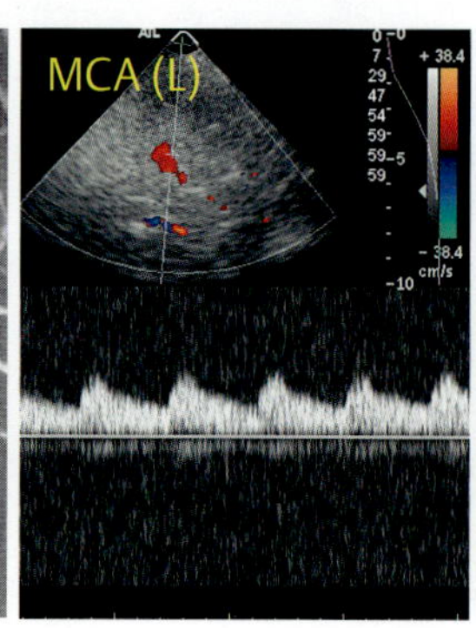

d

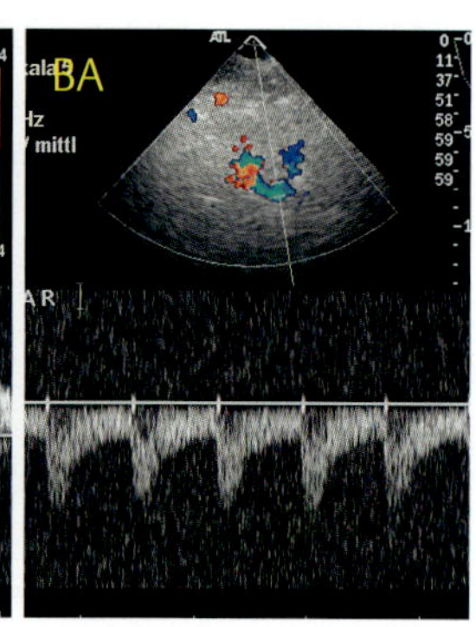

e

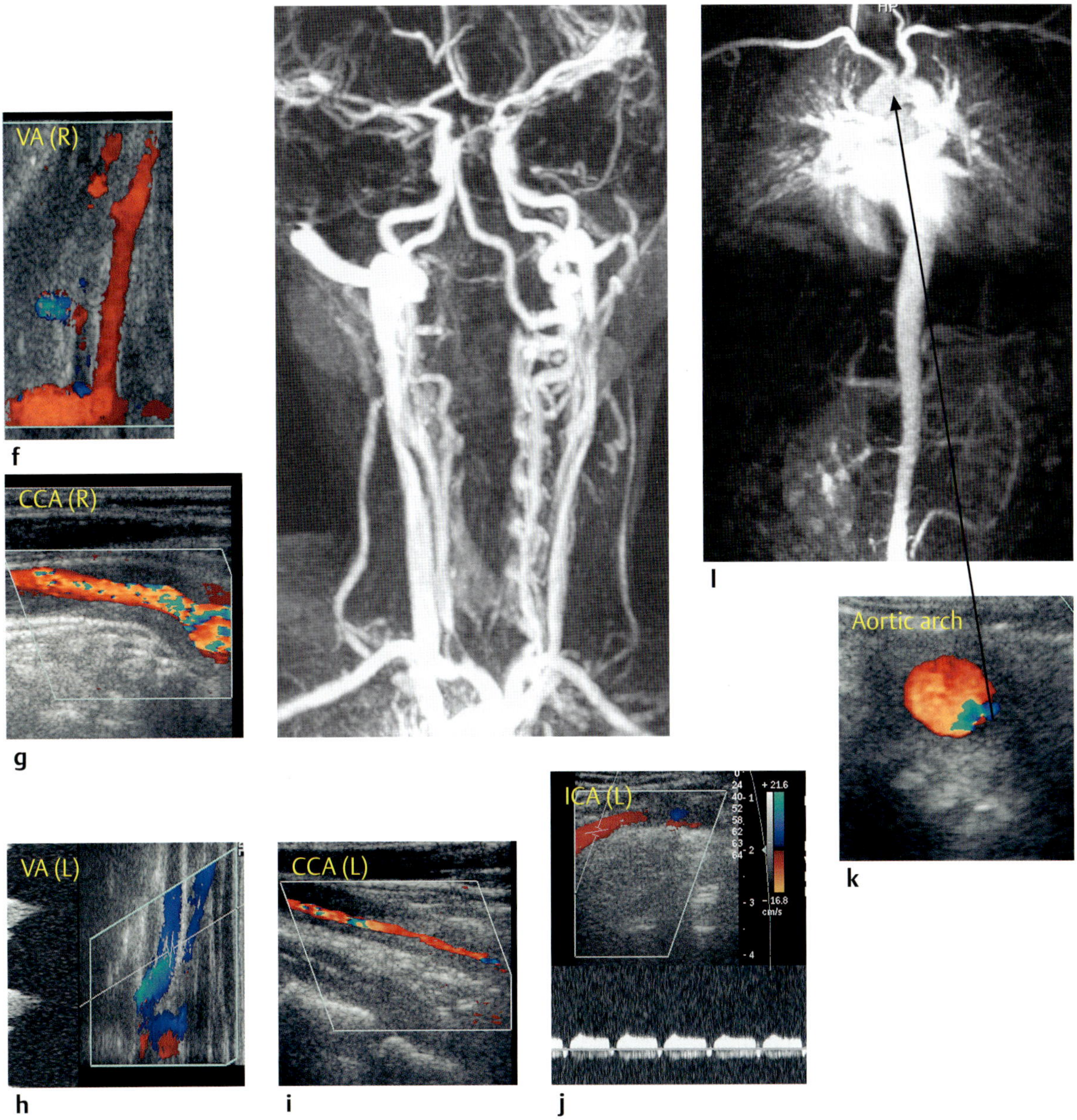
VA (R)
f
CCA (R)
g
l
Aortic arch
k
ICA (L)
+ 21.6
– 16.8
cm/s
VA (L)
h
CCA (L)
i
j

10 Dissection of Both Vertebral Arteries

Patient: A.S., female, age 36.
Clinical diagnosis: exacerbation of neck pain after a chiropractic procedure. Two days later, double vision, ataxia, dysarthria, and brainstem infarction developed.

CAD:	–	Vascular surgery:	–	Diabetes mellitus:	–
PAOD:	–	Heredity:	–	Hyperlipidemia:	–
TIA, stroke:	+	Hypertension	–	Smoking	+

Color flow duplex sonogram. Examination of the carotid system reveals normal measurements on both sides. Display of the vertebral artery (VA) on the right shows reduced flow and a high resistance pattern (**d**), with transition to a typical "to-and-fro" signal. Color mode reveals red–blue changes within the cardiac cycle. B-mode imaging (**c**) shows a normal lumen diameter (2.8 mm), with hypoechoic material obstructing the arterial lumen (implying dissection). Color duplex sonography of the left VA (**e**) shows forward-and-backward signal components in blue–red color coding only in some segments of the VA, while other segments do not show any color signals. Doppler registration (**f**) reveals very low systolic signals that can either be interpreted as reflecting only wall oscillations or very low flow (the resolution threshold of the ultrasound method indicating at least functional occlusion). In the B-mode display, the VA diameter is normal, but the same hypoechoic material as on the right side is seen occluding the artery.

Angiogram (**a**). Irregular lumen narrowing over the entire extracranial segment of the VA is seen when injecting the right VA. No passage of contrast can be seen when injecting the left side, and there is contrast backflow from the VA junction to the left side when injecting into the right VA. These findings support the diagnosis of total occlusion in the presence of VA hypoplasia.

Magnetic resonance imaging (MRI) and magnetic resonance angiography (MRA). MRA did not show any significant or at least valid flow signals in projections to any area of the VAs (not shown). MRI axial T2-weighted and fluid-attenuated inversion recovery (FLAIR) imaging shows arterial hematomas on both sides (A, B), with a very subtle flow void in the right VA but no residual flow in the left VA (**b**).

Comment. Ultrasound is the only method that routinely allows display of the external arterial wall, even in the presence of obstruction. All other angiographic techniques only show the patent lumen. However, MRI provides the most specific display of arterial hematomas. Conventional angiography shows intraarterial flow but fails to diagnose the etiology of an occlusion.

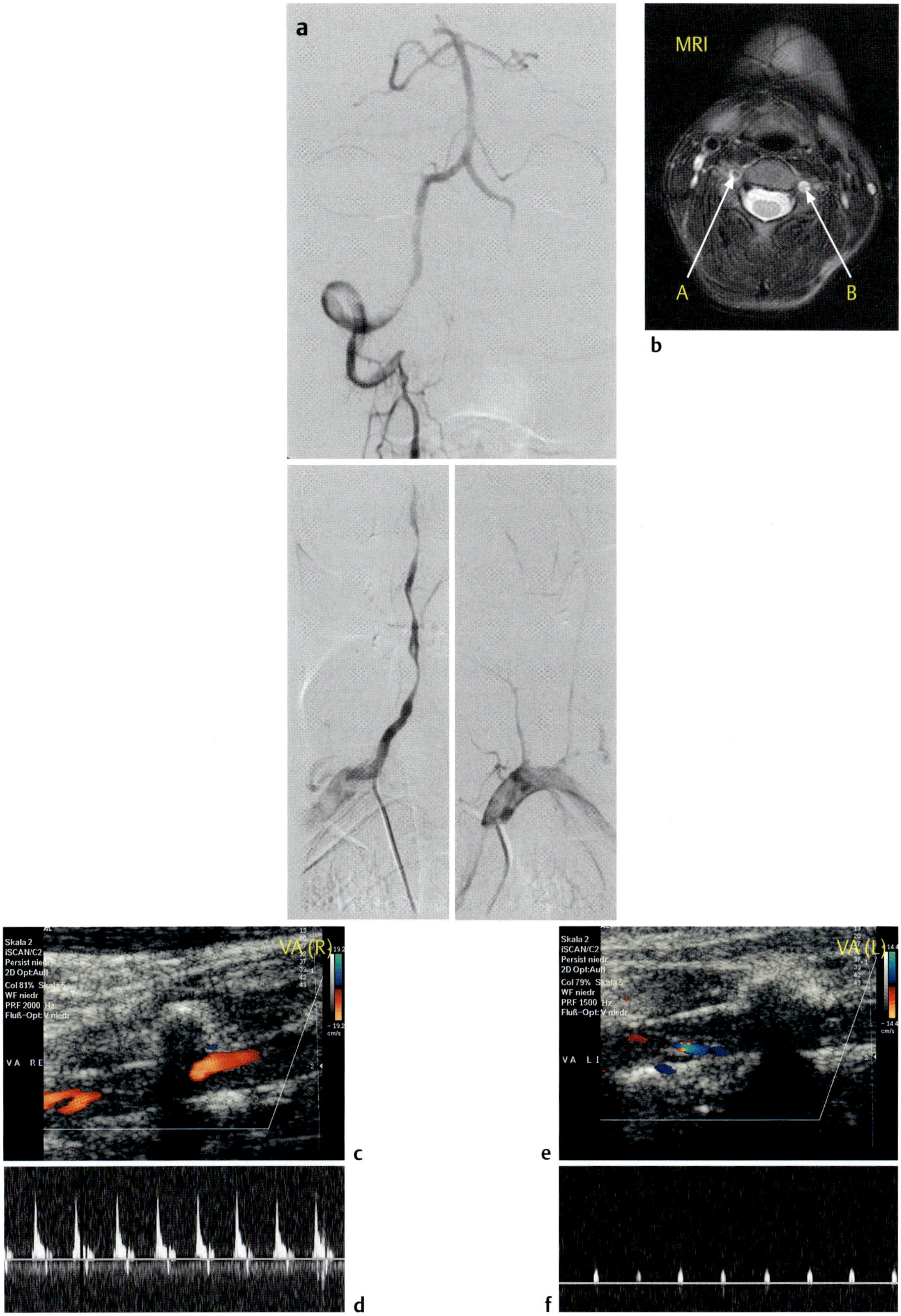
a
MRI
A
B
b
Skala 2
iSCAN/C2
Persist niedr
2D Opt:Aufl
Col 81% Skala 5
WF niedr
PRF 2000 Hz
Fluß-Opt: V niedr
VA (R)
VA R
c
d
Skala 2
iSCAN/C2
Persist niedr
2D Opt:Aufl
Col 79% Skala 5
WF niedr
PRF 1500 Hz
Fluß-Opt: V niedr
VA (L)
VA L
e
f

11 Stenosis of the Vertebrobasilar Arteries

Patient: A.W. male, age 64.
Clinical diagnosis: Brainstem ischemia (right-sided hemiparesis and gait ataxia).

CAD	–	Vascular surgery:	–	Diabetes mellitus:	–
PAOD:	–	Heredity:	–	Hyperlipidemia:	–
TIA, stroke:	+	Hypertension:	+	Smoking:	–

Color flow duplex sonogram. The fronto-orbital branches of the ophthalmic arteries and both extracranial carotid systems are normal. The vertebral arteries present clear asymmetries, with a small systolic and low diastolic flow signal in the left vertebral artery, compared to a normal signal in the right vertebral artery.

At the transition of the vertebral/basilar arteries a high velocity signal can be localized (**c**).

Transnuchal Doppler sonograms (multigate registration—spectral analysis) (**a**). A clear flow acceleration (with maximum velocities of approximately 190 cm/s) is seen in the intracranial transition zone between the two vertebral arteries and the basilar arteries (78–83 mm depths). Distally, there is a markedly reduced Doppler signal (93–98 mm depths).

Angiogram. In conventional angiography (**b**) a high-grade stenosis at the transition from the vertebral artery to the basilar artery at the origin of the PICA is seen, with subsequent filling of the basilar artery and its branches up to the top of the basilar artery.

Comment. It is often not possible to locate the stenosis between the confluence of the two vertebral arteries and the basilar artery using the Doppler-sonographic findings alone; even with color flow duplex sonograms display of the correct anatomy may be difficult (**c**). Sometimes MRA provides sufficient noninvasive confirmation of the Doppler findings, which in turn compensate for potential MRA artifacts, in some cases like in ours invasive angiography is still necessary.

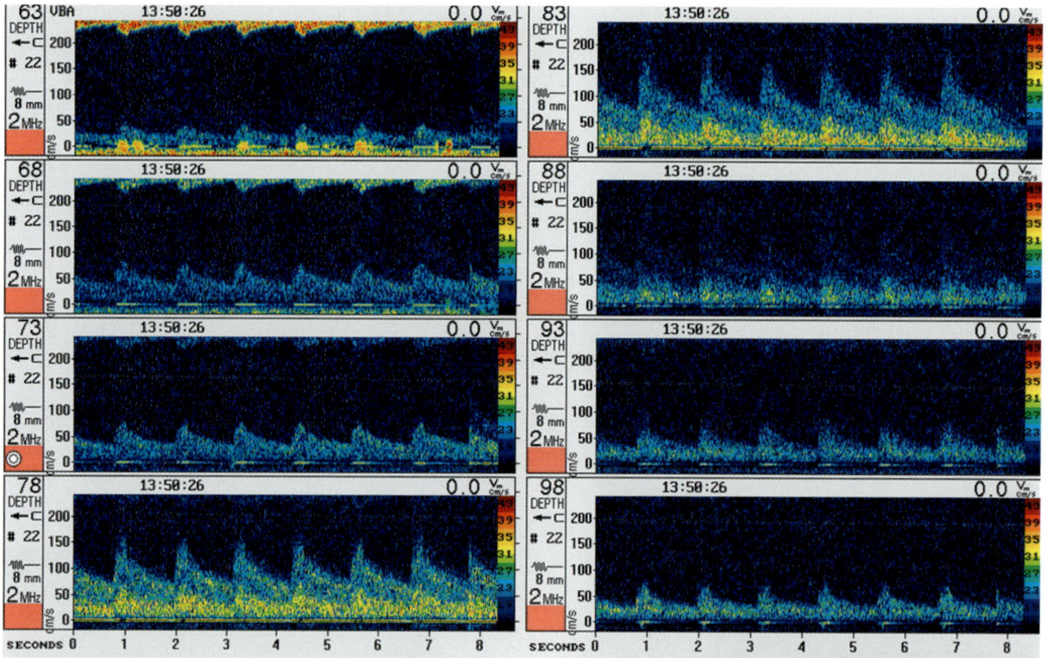

a

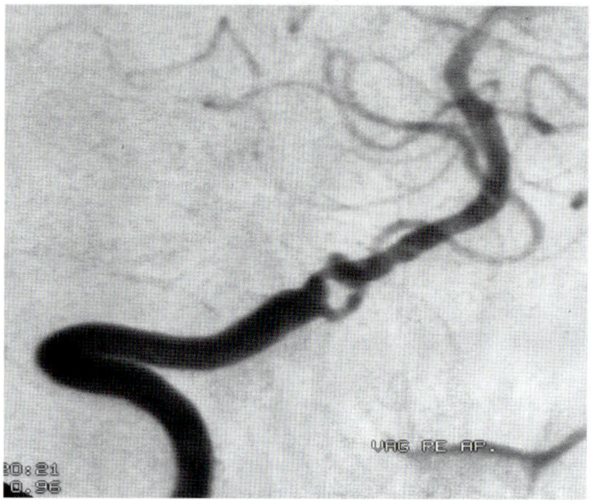

b

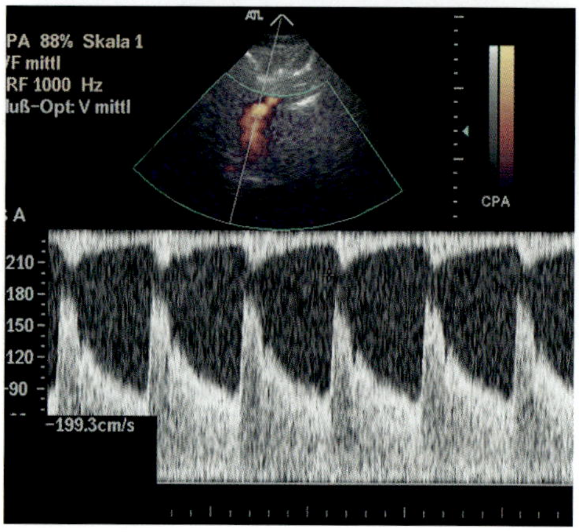

c

Intracranial Arteries

12 Stenosis of the Middle Cerebral Artery (Right M_1 Segment)

Patient: A.M., female, age 63.
Clinical diagnosis: stroke in the right middle cerebral artery territory.

CAD	–	Vascular surgery:	–	Diabetes mellitus:	–
PAOD:	+	Heredity:	+	Hyperlipidemia:	–
TIA, stroke:	+	Hypertension:	+	Smoking:	+

Extracranial Doppler sonogram (analogue registration). The supratrochlear arteries show an orthograde blood flow direction bilaterally, but there is slight asymmetry in the diastolic signal favoring the left side. When compared side-to-side, flow curves in the internal carotid artery (ICA) are almost the same with regard to their amplitude modulation, while the diastolic flow in both blood vessels is slightly reduced on the right. Amplitude modulation and flow velocity are clearly reduced in the right common carotid artery (CCA).

The vertebral and the subclavian arteries are normal bilaterally.

Transcranial Doppler sonogram (spectral analysis) (**b**). There is a maximum increase in peak systolic cerebral blood flow velocities at an insonation depth of 57 mm (cerebral blood flow volume > 250 cm/s), with low-frequency components indicating significant stenosis. Poststenotically, the flow signal is attenuated (52, 48 and 44 mm depths). However, the flow velocities are not reduced but show normal systolic peak flow velocities (approximately 100 cm/s). The transcranial color duplex image (**a**) shows aliasing at the site of stenosis. Doppler analysis confirms the high flow velocities already shown with multigate transcranial Doppler imaging.

Magnetic resonance angiogram (**c**). Moderate stenosis of the M_1 segment in the right middle cerebral artery.

Comment. In the present case, the intracranial and extracranial findings complement one another to the extent that there are extracranial flow changes that can be classified more precisely with transcranial Doppler sonography.

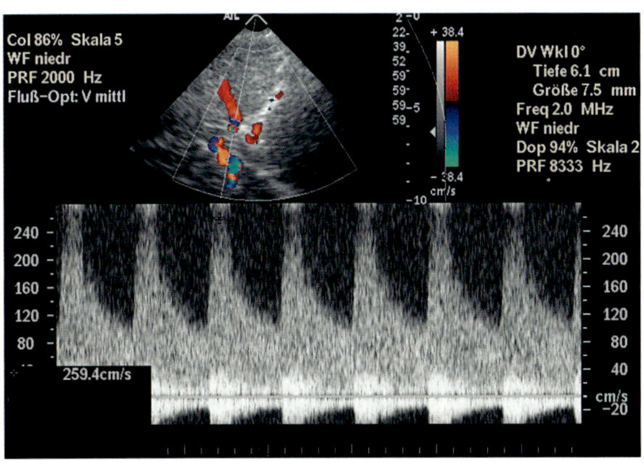

a

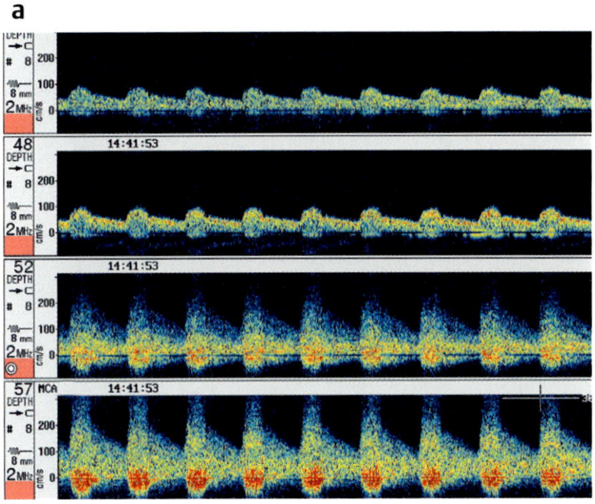

b

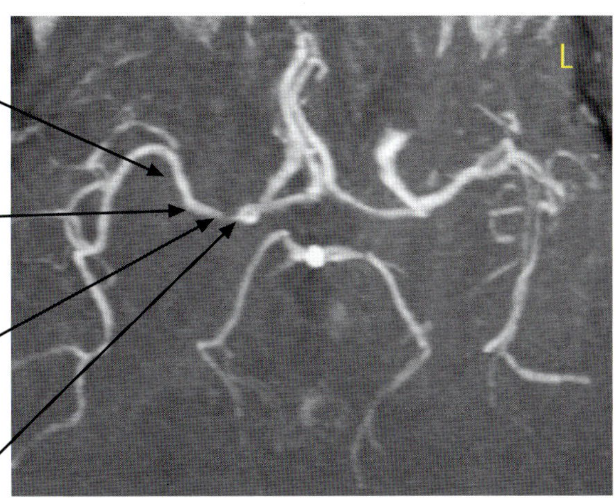

c

13 MCA Occlusion due to Carotid Stenosis

Patient: R.S., male, age 59.
Clinical diagnosis: stroke in the left middle cerebral artery territory.

CAD	+	Vascular surgery:	–	Diabetes mellitus:	–
PAOD:	–	Heredity:	–	Hyperlipidemia:	+
TIA, stroke:	+	Hypertension:	+	Smoking:	–

Initial Examination 2 h after Onset of Stroke

Transcranial duplex sonogram (**b**). *Left MCA:* there is a subtle color signal in projection to the middle cerebral artery (MCA) from the proximal origin toward an insonation depth of about 45 mm. Doppler flow registration shows oscillating flow around the zero line (to-and-fro), indicating occlusion of the distal MCA main stem or the majority of the M_2 branches. *Right:* normal color signals and Doppler spectra (flow velocity at an insonation depth of 50 mm = 105 cm/s.—not shown).

Bilateral transcranial Doppler monitoring shows the significant side-to-side difference (**a**) (left: yellow, right: red).

Extracranial Color duplex sonogram (**c**). There is the same oscillating signal in the internal carotid artery (ICA) as in the MCA, indicating the hemodynamic relevance of the distal MCA obstruction and evidence of functional occlusion. The common carotid artery (CCA) shows a resistance pattern, and the CCA Doppler spectrum is adapted to that of the external carotid artery (ECA; externalization). The ECA is normal. The B-mode and color display, as well as Doppler flow velocities and the spectral pattern, are normal on the right side.

Magnetic resonance imaging (MRI) and magnetic resonance angiography (MRA). Severe hypoperfusion (**e**) is seen in the entire MCA territory, with a peak in the temporal area. There is no signal on a projection of the left MCA (**d**) and left ICA, indicating occlusion or very low oscillating flow (functional occlusion).

Follow-Up Examination after Thrombolysis 6 h after the Onset of Stroke

Transcranial duplex sonogram. Restoration of normal color signals in the left MCA and a normal Doppler flow profile (**g**) in duplex mode. Bilateral registrations reveal normal cerebral blood flow velocities on both sides (left MCA: flow velocity at an insonation depth of 50 mm = 92 cm/s; right MCA: flow velocity at an insonation depth of 52 mm = 101 cm/s) (**f**).

Extracranial color duplex sonogram (**h**). The left proximal ICA shows increased flow velocities with maximum systolic peak velocities > 190 cm/s, but it is difficult to detect the peak velocities precisely. In addition, there are low-frequency components that indicate the hemodynamic relevance of the stenosis. Other indicators of high-grade stenosis are reduced flow with increased pulsatility in the CCA and reduced flow in the distal ICA. Examination of the right carotid system constantly show normal results.

MRI and MRA. There is significant reperfusion (**j**), with normalization of the temporal and frontal aspects of the MCA territory, but still some hypoperfusion in the parietal region. The intracranial MRA (**i**) shows recanalization of the ICA and MCA.

Follow-Up Examination after Carotid Endarterectomy 12 h after the Onset of Stroke

Transcranial duplex sonogram. There are normal flow registrations in all of the basal brain supplying arterial segments.

Extracranial color duplex sonogram. The left proximal ICA (**k**) shows a patent lumen after carotid surgery with normal Doppler spectrum and normal flow velocities in the entire ICA (60/28 cm/s systolic/diastolic peak velocity).

MRI. Complete reperfusion with normal perfusion (**l**) on the left side.

Comment. Adequate ultrasound diagnosis of vascular pathology in the acute stroke setting is sometimes difficult (e.g., due to the lack of a temporal bone window). However, ultrasound can add important information in individual cases. Despite the lack of any signal from the ICA and MCA on MRA in the present case, ultrasound reveals circumscribed MCA occlusion, with at least some oscillating flow in the proximal MCA, in the T-junction, and particularly in the ICA. This finding is confirmed by the favorable result after thrombolysis—complete occlusive thrombosis of the ICA and MCA is unlikely to reopen after standard intravenous thrombolysis. The stroke is the result of MCA occlusion, most likely due to an embolus that had its origin in a high-grade carotid stenosis.

2 h after Stroke

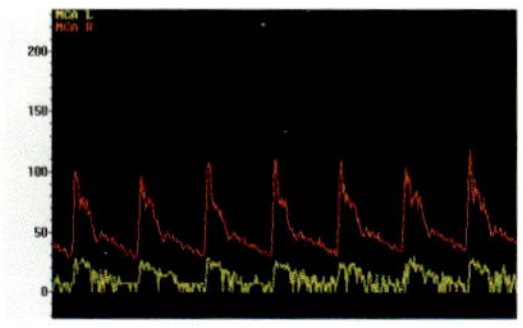

a

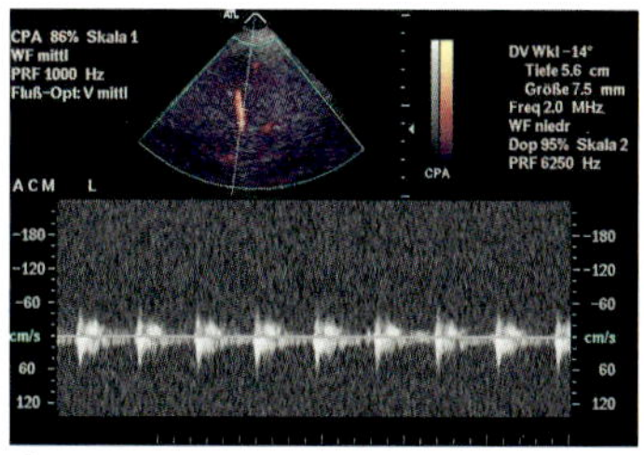

b

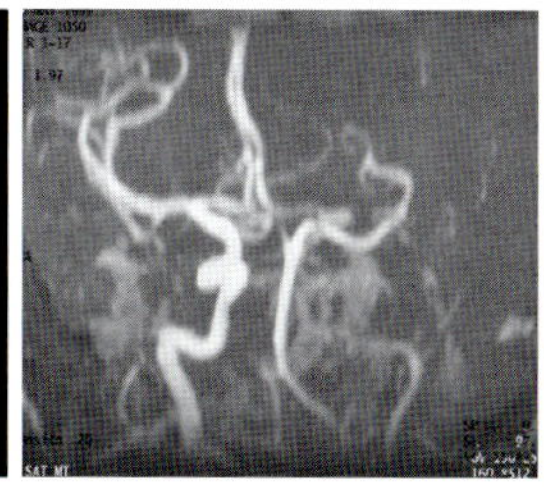

d

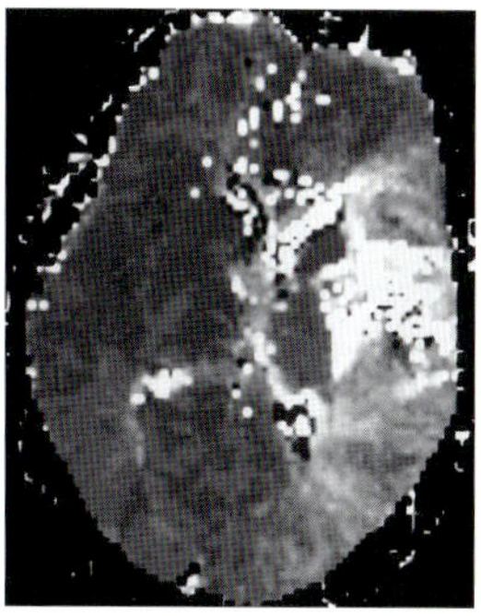

e

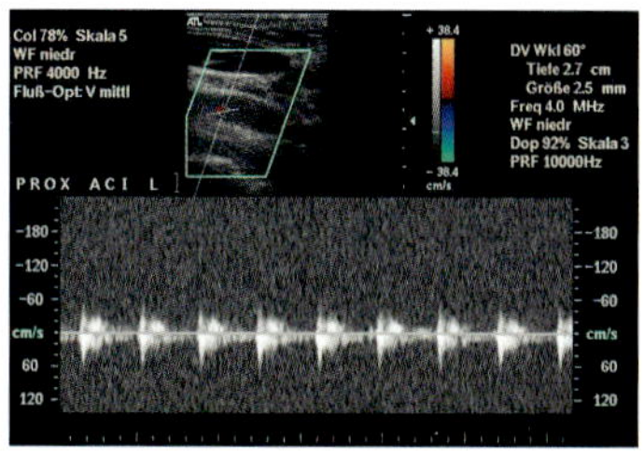

c

6 h after Stroke

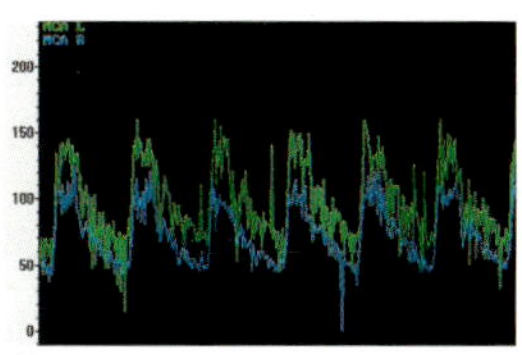

f

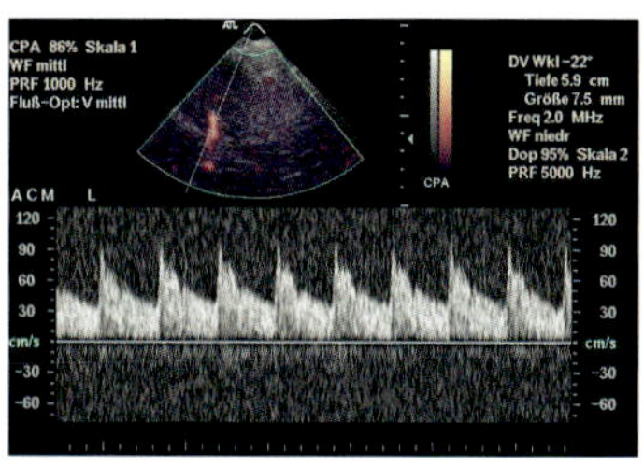

g

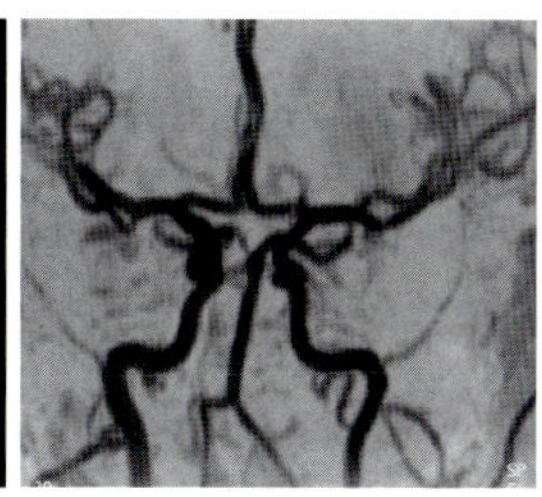

i

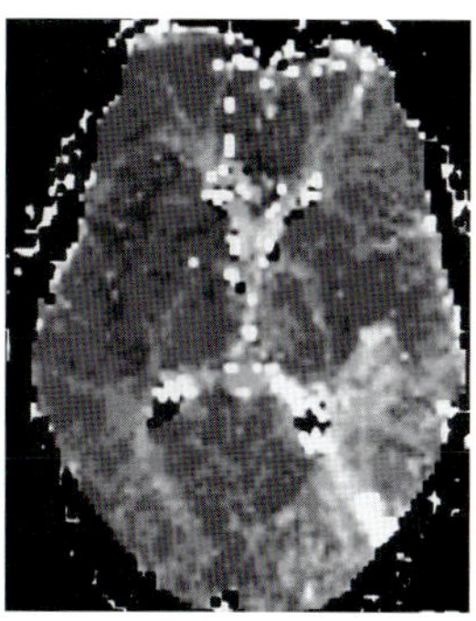

j

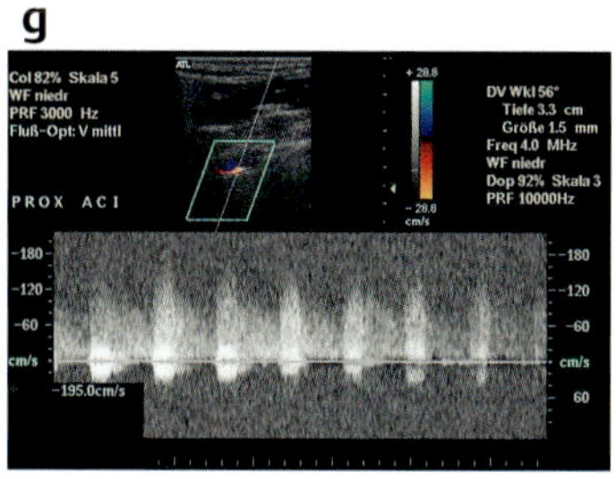

h

12 h after Stroke

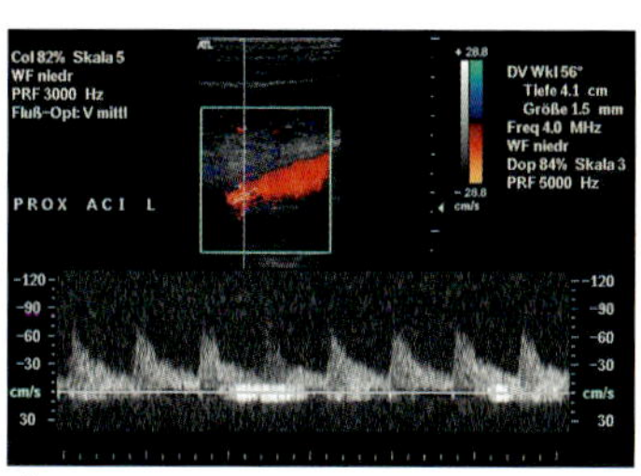

k

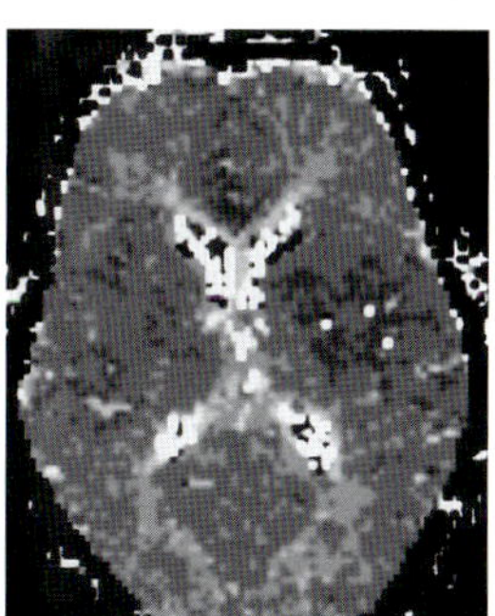

l

14 Basilar Aneurysm due to Dolichoectasia

Patient: L.Z., male, age 78.
Clinical diagnosis: brainstem infarction; double vision, ataxia, dysarthria, facial paralysis on the left, paralysis during swallowing.

CAD	+	Vascular surgery:	–	Diabetes mellitus:	+
PAOD:	–	Heredity:	+	Hyperlipidemia:	–
TIA, stroke:	+	Hypertension:	+	Smoking:	+

Extracranial color duplex sonogram. There are regular flow velocities in both carotid systems, an increase in the intima–media thickness (IMT; left 1.4 mm, right 1.9 mm), and mild plaque formation in both carotid bifurcations, without significant luminal narrowing. Both vertebral arteries (**g**, **k**) (VAs) show reduced blood flow velocities (52/24 cm/s right V_2 segment; 56/24 cm/s) and discretely increased pulsatility, although their diameters are normal (right 2.8 mm, left 2.7 mm) throughout the extracranial course.

Transorbital and transcranial Doppler and duplex sonograms. Intracranial insonation also shows reduced flow in both VAs. Flow in the basilar artery (BA) (**h–j**) is significantly reduced (28/16 cm/s at an insonation depth of 87 mm), with oscillation flow at an insonation depth of between 95 and 105 mm.

Magnetic resonance imaging (MRI) (a, b) and magnetic resonance angiography (MRA) (c). There is a hyperintense lesion covering about 80 % of the entire area of the brainstem, with a discrete eccentric flow void. MRA fails to show the BA completely, because of the very low flow velocities.

Computed-tomographic angiography (CTA). Single slices (**d**, **e**) show a patent lumen with very low flow. Three-dimensional reconstruction (**f**) shows the extent and volume of the aneurysm.

Comment. Careful examination of the extracranial vertebral arteries can often indicate intracranial vertebral or even basilar pathology. Duplex sonograms are needed for detailed interpretation of low flow conditions.

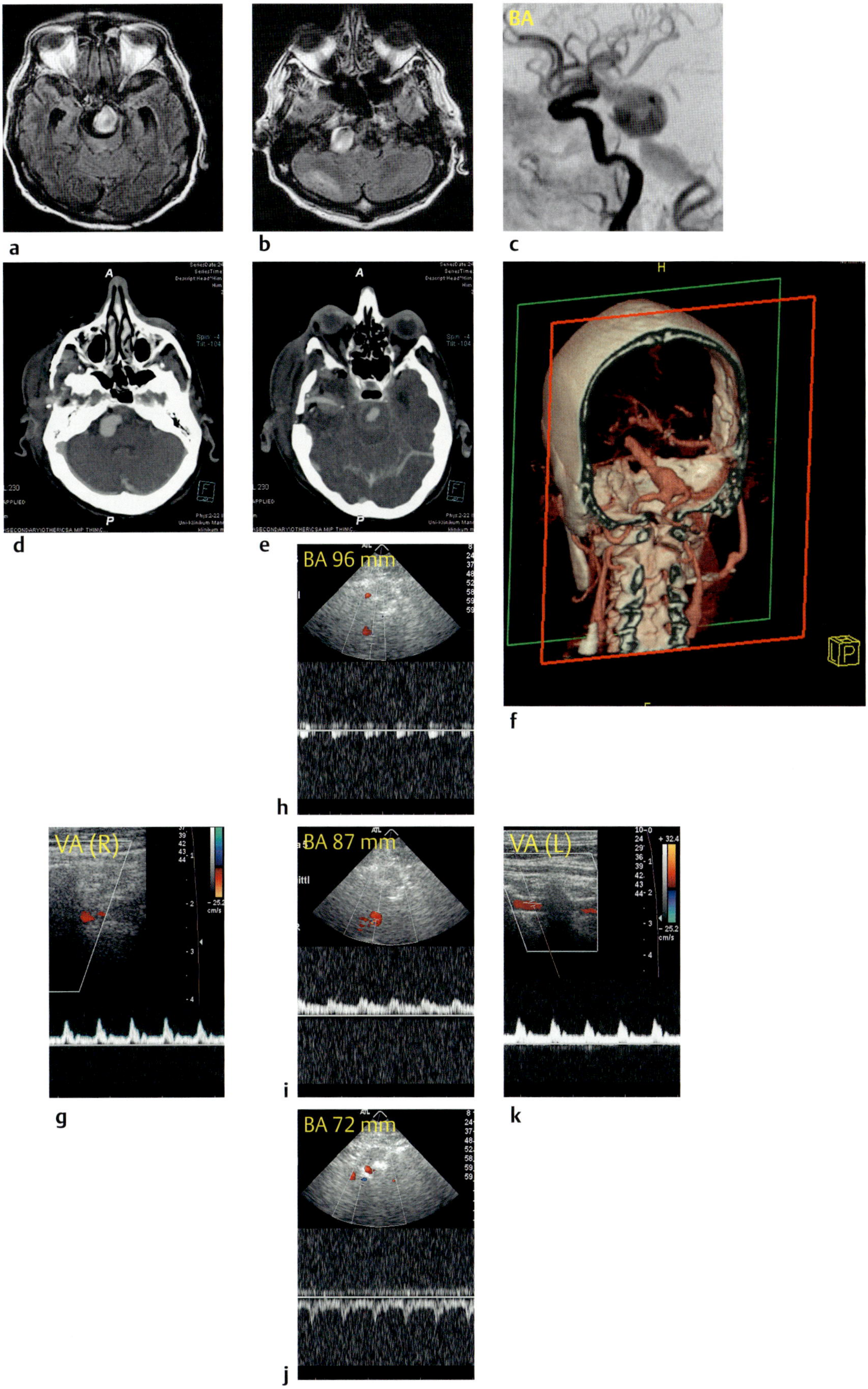
a
b
BA
c
A
P
d
A
P
e
H
f
BA 96 mm
h
VA (R)
g
BA 87 mm
i
VA (L)
k
BA 72 mm
j

15 Transient Internal Carotid Artery Fistula on the Left

Patient: L.S., male, age 50.
Clinical diagnosis: double vision (abducens paresis), visual loss in the left eye, tinnitus.

CAD	–	Vascular surgery:	–	Diabetes mellitus:	–
PAOD:	–	Heredity:	–	Hyperlipidemia:	–
TIA, stroke:	–	Hypertension:	–	Smoking:	–

On Admission

Color duplex. The right extracranial carotid system (**c**, common carotid artery: 50/100 cm/s; **b**, internal carotid artery: 58/92 cm/s) is normal. In the left common carotid artery (CCA) (**d**, 60/120 cm/s) and more markedly in the left internal carotid artery (ICA) (**e**, 138/178 cm/s), there are elevated diastolic flow velocities larger than the systolic velocities, leading to reduced amplitude modulation and decreased pulsatility.

Transcranial duplex sonograms (spectral analysis). There is a clear flow acceleration in the intracranial carotid siphon on the left (**f**), while the middle cerebral artery (MCA) (**g**) is normal. With transtemporal duplex sonography, elevated peaks (right 68 cm/s, left 148 cm/s) are seen in the siphon, as well as bidirectional high signal amplitudes of slow flow velocities near the baseline, which is a sign of functional stenosis.

Magnetic resonance angiogram (**a**). Flow signal in projection to the siphon and the cavernous sinus indicates arteriovenous malformation (internal carotid artery fistula).

Ultrasound Follow-Up 6 Days Later

Color duplex. The right extracranial carotid system (**j**, CCA: 23/74 cm/s; **i**, ICA: 26/62 cm/s; **i**) is normal. In the left CCA (**k**, 32/90 cm/s) and more markedly in the left ICA (**l**, 30/67 cm/s), there are elevated diastolic flow velocities larger than the systolic velocities, leading to reduced amplitude modulation and decreased pulsatility.

Transcranial duplex sonograms. There is normalization of flow in the siphon (**m**). With transtemporal duplex sonography, normal systolic peak velocities are seen (right 67 cm/s, left 65 cm/s) in the siphon. The spectral pattern is normal.

Magnetic resonance angiography (MRA) (h) (carried out after the ultrasound examination, but on the same day). MRA verifieds the ultrasound findings and also shows normalization of flow, with normal filling of the brain-supplying arteries. Persistent small aneurysm of the ICA.

Comment. This is a rare case of spontaneous occlusion of aninternal carotid artery-cavernous sinus fistula.

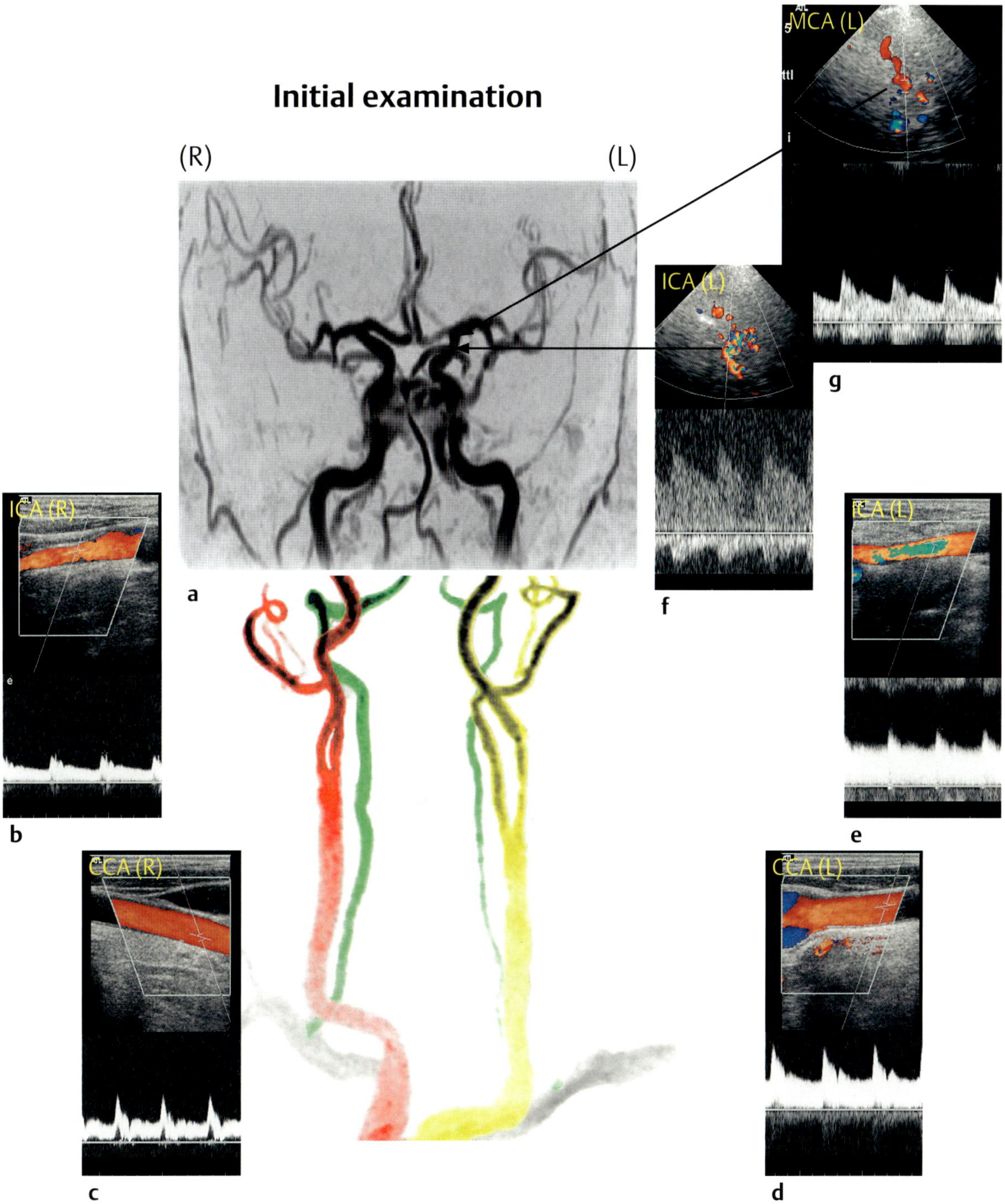
Initial examination
(R)
(L)
MCA (L)
ICA (L)
ICA (R)
ICA (L)
CCA (R)
CCA (L)
a
b
c
d
e
f
g

Follow-up examination (6 days)

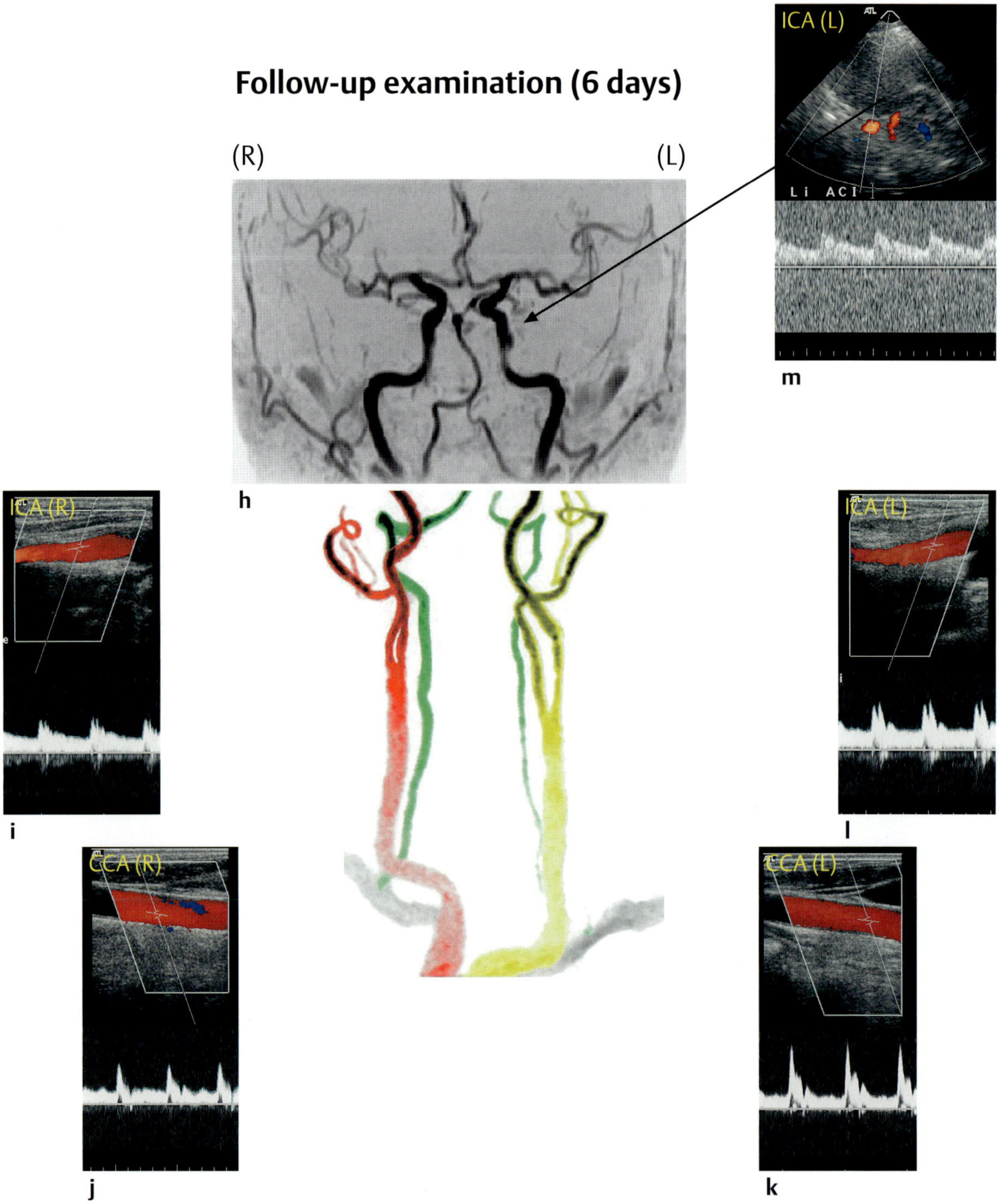

16 Acute Stroke due to Middle Cerebral Artery Occlusion with Spontaneous Recanalization

Patient: M.W., female, age 30.
Clinical diagnosis: Left hemispheric acute stroke with hemiplegia on the right, global aphasia, disturbance in consciousness, patent foramen ovale.

CAD:	–	Vascular surgery:	–	Diabetes mellitus:	–
PAOD:	–	Heredity:	–	Hyperlipidemia:	–
TIA, stroke	+	Hypertension:	–	Smoking	+

Initial Examination 4 h after the Onset of Stroke

Color duplex sonogram. The extracranial examination shows a normal Doppler spectrum, color pattern, and arterial lumen in B-mode in the right carotid system (**c**, **d**). There is a high resistance pattern, with diastolic zero flow in the left common carotid artery (CCA) (**e**) and left internal carotid artery (ICA) (**f**). The transcranial examination shows that, despite a patent temporal bone window with normal color signals in projection to the posterior cerebral artery (PCA) (**g**) and with reduced flow in projection to the ICA (siphon) (**i**), there is no signal from the anterior cerebral artery (ACA) or middle cerebral artery (MCA) even after contrast administration (**h**), indicating occlusion of the MCA.

Magnetic resonance imaging (MRI) and magnetic resonance angiography (MRA; time-of-flight). There is a large area of MCA ischemia (on diffusion-weighted imaging) (**a**). The MRA shows complete signal loss from the ICA, MCA and ACA (**b**), indicating occlusion of the T-junction.

Follow-Up Examination 3 Days after the Initial Examination

Color duplex sonogram. The extracranial examination shows a normal Doppler spectrum, color pattern, and arterial lumen in B-mode in the right carotid system (**m**, **n**). Doppler spectra from the left side still show some lower velocities and higher pulsatility than on the right side, but there is significant normalization (**o**, **p**). The transcranial examination shows a patent MCA, with color signals in color duplex mode (**l**) in projection to the MCA. Doppler registrations reveal normal flow velocities in the PCA (**q**) and stenotic signals in the proximal MCA (**r**) and ACA (**s**), indicating high-grade residual proximal stenosis of the T-junction.

MRI and MRA (time-of-flight). There is a large MCA infarction (fluid-attenuated inversion recovery, FLAIR), with some swelling on the left (**j**). MRA shows recanalization of the MCA (**k**).

Comment. Unilateral occurrence of a resistance pattern in Doppler spectra from the extracranial carotid, without display of locally circumscribed extracranial lumen narrowing, indicates distal severe obstruction of the intracranial ICA segment, the T-junction, or the very proximal MCA. The slight side-to-side difference in the extracranial carotid Doppler flow velocities seen at the follow-up examination is probably due to the severe residual stenosis on the left.

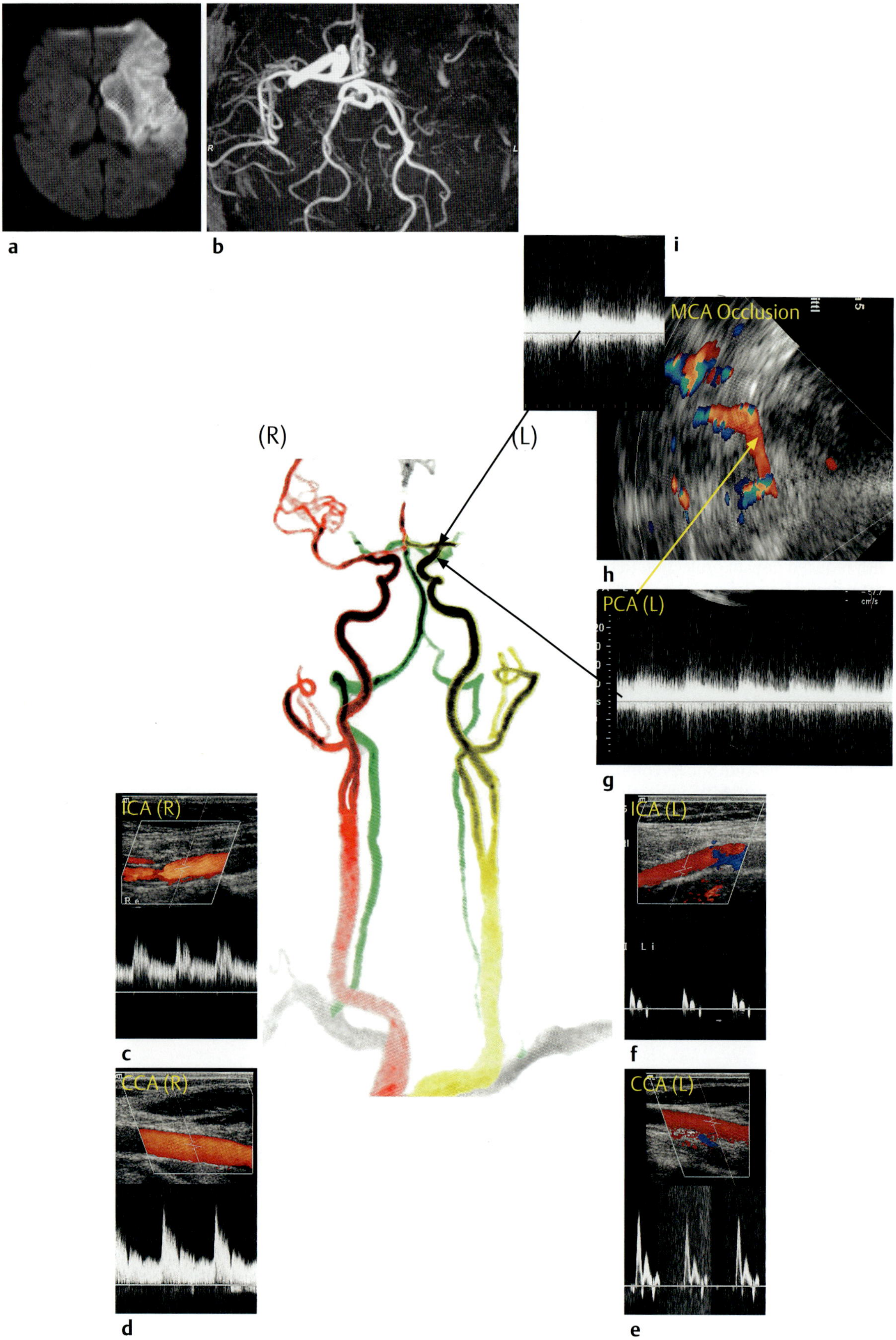
a
b
i
MCA Occlusion
(R)
(L)
h
PCA (L)
g
ICA (R)
ICA (L)
c
f
CCA (R)
CCA (L)
d
e

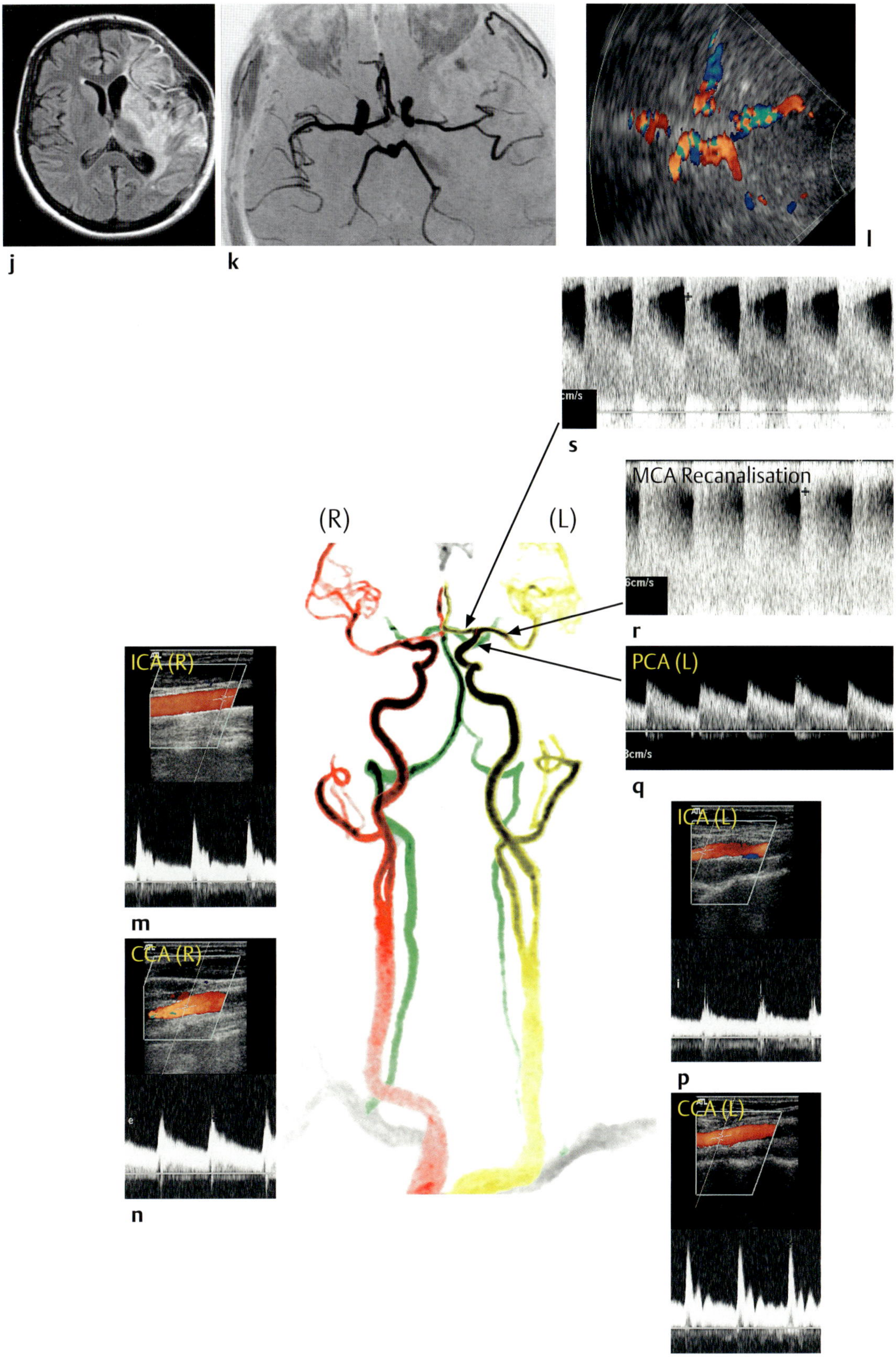
j
k
l
s
MCA Recanalisation
r
(R)
(L)
PCA (L)
q
ICA (R)
m
CCA (R)
n
ICA (L)
p
CCA (L)
o

Peripheral Arteries

Pelvic Level

17 Stenosis of the Iliac Artery ≤ 50%

Patient: E.R., male, age 44.
Clinical diagnosis: peripheral arterial occlusive disease **(PAOD), stenosis of the right iliac artery,** stenosis of the left iliac artery.

CAD:	–	Vascular surgery:	–	Diabetes mellitus:	–
PAOD:	+	Heredity:	–	Hyperlipidemia:	–
TIA, stroke:	–	Hypertension:	–	Smoking:	+

Case report. For the previous two years the patients had suffered hip and thigh claudication on the left side, and for the previous six months thigh claudication on the right. Systemic lysis with streptokinase showed a clear improvement. However, bilateral residual stenoses still remained, more pronounced on the left than on the right. Subsequently, treatment with a balloon catheter on the left was undertaken.

Examination results:

	Pulse status*		Auscultation**	
	Right	Left	Right	Left
Abdomen			+	+
Common femoral artery	(+)	+	+	(+)
Superficial femoral artery			–	–
Popliteal artery	(+)	(+)	–	–
Posterior tibial artery	(+)	((+))		
Dorsal artery of the foot	–	((+))		

* Pulse status: + = normal pulse, (+) = slightly attenuated pulse, ((+)) = significantly attenuated pulse, – = pulse not palpable.
** Auscultation: – = no flow sounds, (+) = short, soft flow sound (mainly systolic), + = clear flow sound (usually extending into the diastole).
This key also applies to the subsequent tables. (The omission of a bracket on one side of the plus sign implies an intermediate grade.)

Doppler sonogram *(analogue registration)* (**b**). *Right:* The common femoral artery has a slightly reduced amplitude and a missing late-diastolic reverse flow component. The right popliteal artery also shows a reduced amplitude height, a delayed increase in its steepness, and a missing reverse flow component. The same applies to the right posterior tibial artery. In the dorsal artery of the foot, a moderate registration is seen, with an unfavorable ultrasound angle, but with a minimal reverse flow component.

Angiogram (**a**). Central venous digital subtraction angiography (DSA). *Right:* low-grade stenosis of the common iliac artery. *Left:* After completion of angio-plasty, wall changes in the dilated region persist, along with a minimal residual pressure gradient.

Comment. It was possible to open up the high-grade obstructions in the pelvic region in this still relatively young patient to a large extent without surgery, using systemic lysis and angioplasty. The residual pressure gradient extending to the ankle amounted to only 10 mmHg on the right and 30 mmHg on the left. On the right, the pressure at the distal lower leg was higher than at the proximal thigh. This indicates that there is no additional intervening flow obstruction and that the collateral blood supply is very good. On the left, the pressures at the proximal thigh and the distal lower leg are nearly the same (slight fluctuations can be due to intraindividual changes in blood pressure). The pain-free walking distance increased from 200 m to 1000 m in this patient.

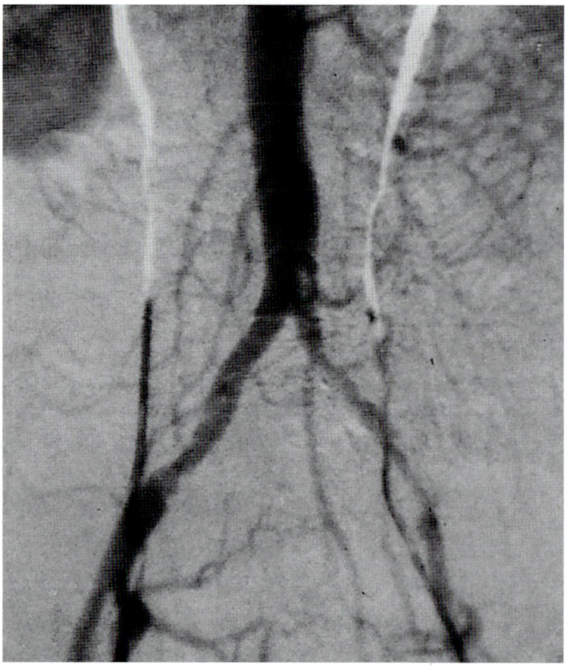

a

Case Figs. **17 b–d** ▷

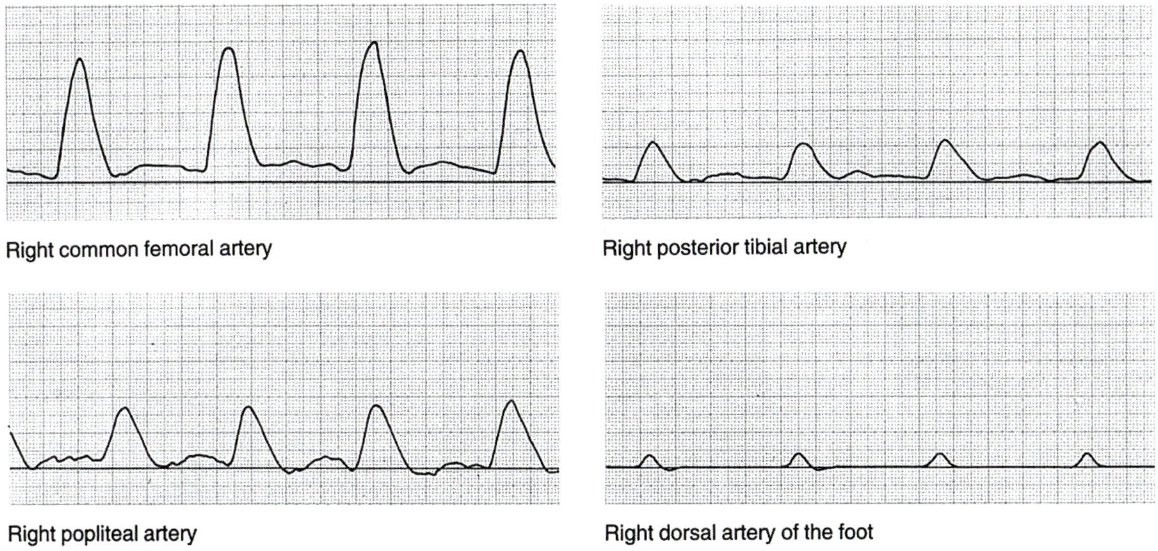

b CW Doppler sonograms

Cuff position	Right registration point		Left registration point	
	Posterior tibial artery	Dorsal artery of the foot	Posterior tibial artery	Dorsal artery of the foot
Proximal thigh	105	95	100	95
Distal lower leg	115	120	95	90

Systolic pressure measurement (Doppler ultrasound technique; mmHg)

Example using color flow duplex sonography. In a *different* patient (H.I., male, age 63), imaging showed a hemodynamically effective low-grade stenosis at the origin of the left external iliac artery obtained with color flow duplex sonography. In the color flow Doppler mode, the arterial lumen constriction can be recognized by a plaque protruding into the lumen from a dorsal direction (**c**). Aliasing here indicates an elevated average flow velocity. As a result of vascular wall vibrations, discrete color artifacts are formed in a projection on the perivascular tissue.

In a different sonographic sectional plane, the slightly broadened *Doppler frequency spectrum* documents the elevated maximal systolic flow velocity (270 cm/s), with longer diastolic forward flow and still retained triphasic wave form (**d**).

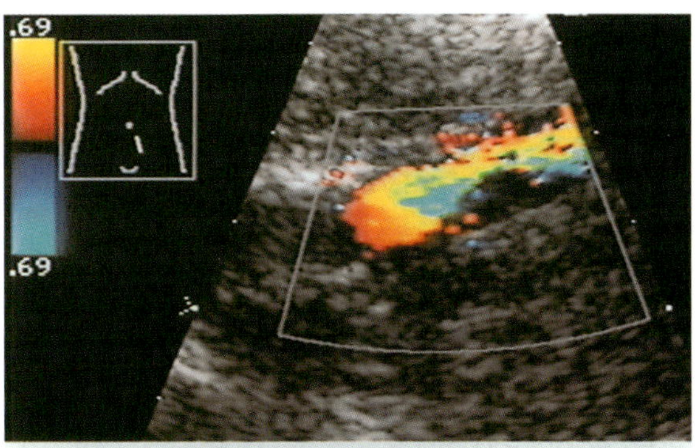

c

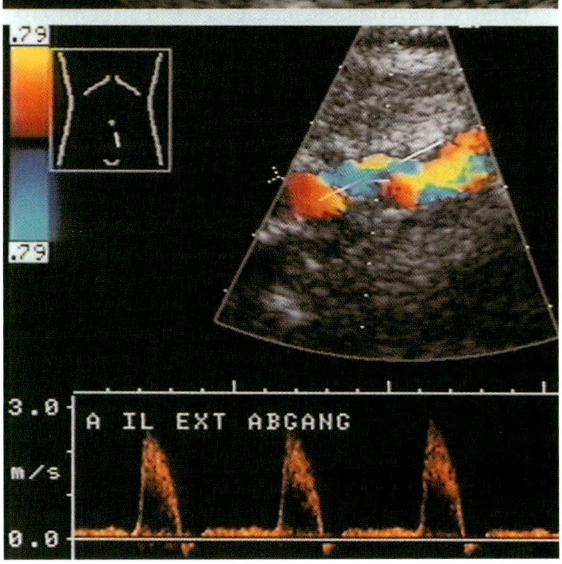

d

Case Figs. **17 b–d**

18 Stenosis of the Iliac Artery > 50%

Patient: M.E., male, age 56.

Clinical diagnosis: **PAOD, high-grade stenosis of the right iliac artery;** differential diagnosis: occlusion of the left iliac artery.

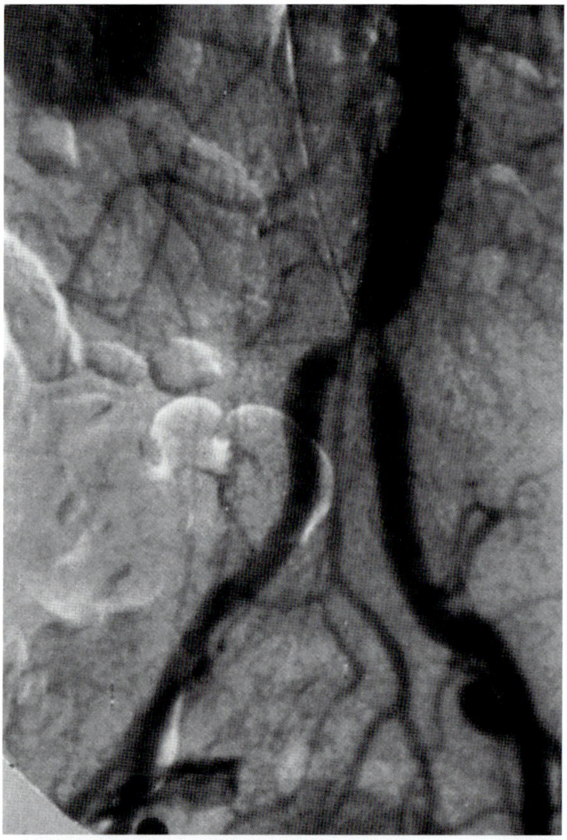

a

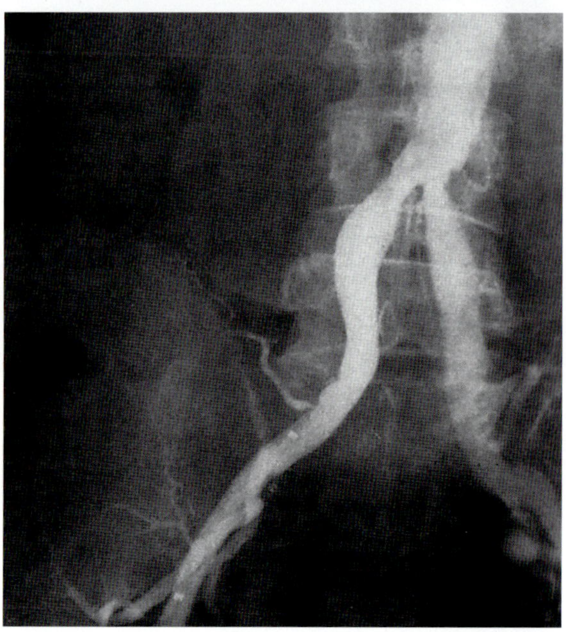

b

Case Figs. **18 c, d** ▷

CAD:	–	Vascular surgery:	–	Diabetes mellitus:	–
PAOD:	+	Heredity:	–	Hyperlipidemia:	–
TIA, stroke:	–	Hypertension:	–	Smoking:	+

Case report. For the previous nine months, there had been claudication in the region of the right buttock, mainly when climbing stairs. Level walking distance was 500–600 m (120 steps/min). The sole risk factor is a high consumption of nicotine (40 cigarettes per day, since teenage years).

Examination results:

	Pulse status		Auscultation	
	Right	Left	Right	Left
Abdomen			–	–
Common femoral artery	((+))	+	–	–
Superficial femoral artery			–	–
Popliteal artery	((+))	+	–	–
Posterior tibial artery	–	+		
Dorsal artery of the foot	–	+		

Doppler sonogram *(analogue registration)* (**c**). *Right:* Over the common femoral artery, a delayed systolic increase in steepness, decreased amplitude height, and absent late diastolic reverse flow are seen. The popliteal artery also presents a clearly reduced amplitude height and absent reverse flow component. Similar results are found over the posterior tibial artery and the dorsal artery of the foot. *Left:* Normal flow pulses are present.

Angiogram. *Right:* There is a high-grade stenosis at the origin of the common iliac artery and an additional occlusion at the origin of the internal iliac artery (**a**). Following angioplasty, one notes only a minimal constriction of the lumen (**b**). *Left:* A minimal stenosis is detected at the origin of the common iliac artery. The drainage relationships in both legs appear to be normal.

Comment. Usually, an isolated stenosis of the iliac artery is so well-compensated for through collaterals that the arm–ankle pressure gradient on the right is not very large (35 mmHg). In this case, it is seen that the pressure still shows a tendency to increase between the proximal thigh and the distal lower leg. This is a sign of good collateral filling without a subsequent flow obstruction being present.

The stenosis of the left iliac artery has hardly any hemodynamic effect. Since the patient insisted on recanalization therapy despite being able to walk a good distance, angioplasty was successfully conducted on the right side, and a complete absence of complaints was achieved.

Clinically, it was not possible to distinguish between a stenosis of the highest possible grade and an occlusion because no flow sound could be auscultated.

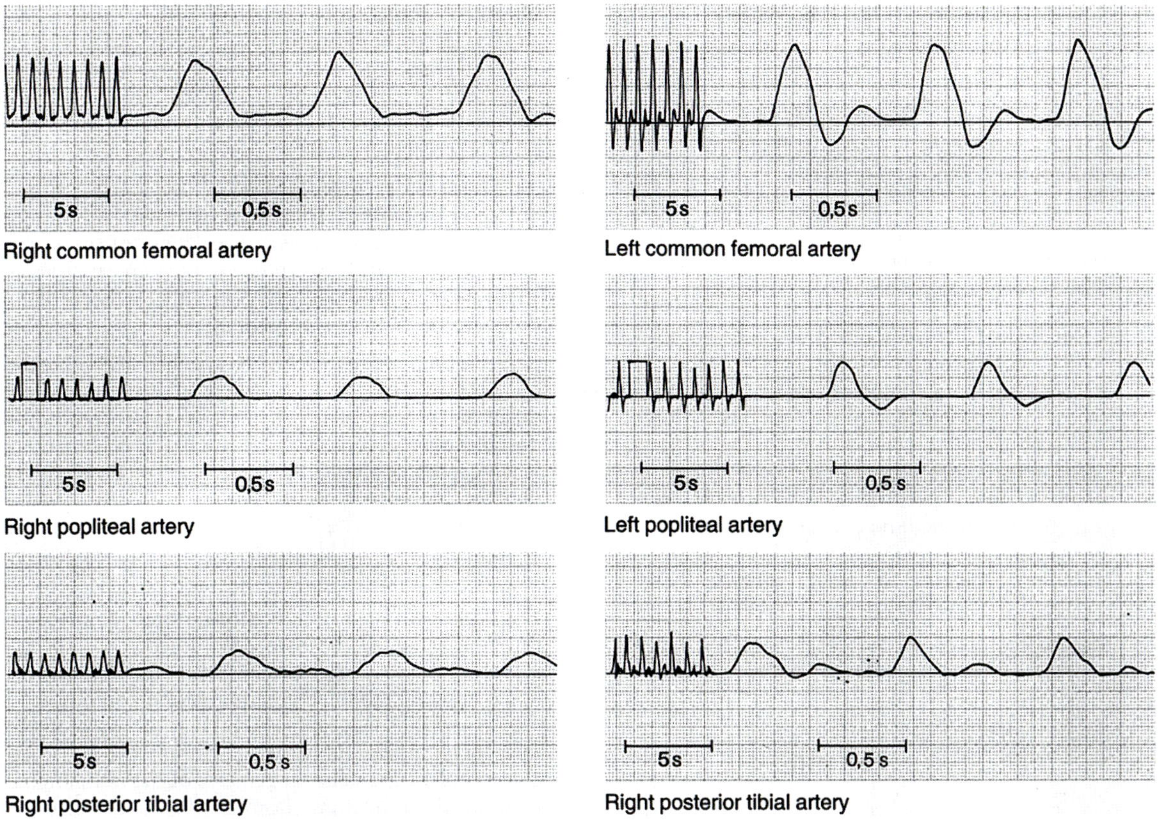

c CW Doppler sonograms

Cuff position	Right registration point		Left registration point	
	Posterior tibial artery	Dorsal artery of the foot	Posterior tibial artery	Dorsal artery of the foot
Proximal thigh	95	90	130	135
Distal lower leg	105	90	145	125

Systolic pressure measurement (Doppler ultrasound technique; mmHg)

Example with color flow duplex sonography. A different patient (A.S., male, age 82) presented with a high-grade stenosis of the iliac artery, which already had a stent positioned in a vascular segment (**d**). With the lighter colors in the stenosis (aliasing, with a color change from light yellow over white to light blue), the color flow Doppler reflects a locally elevated average flow velocity. Dorsal to the artery, sections of the external iliac vein can be recognized from the blue color coding.

The *Doppler frequency spectrum* from the stenotic vascular segment is monophasic. The maximal systolic, end-diastolic and average velocity has increased to 396, 40 and 186 cm/s respectively. The pulsatility index is 1.9.

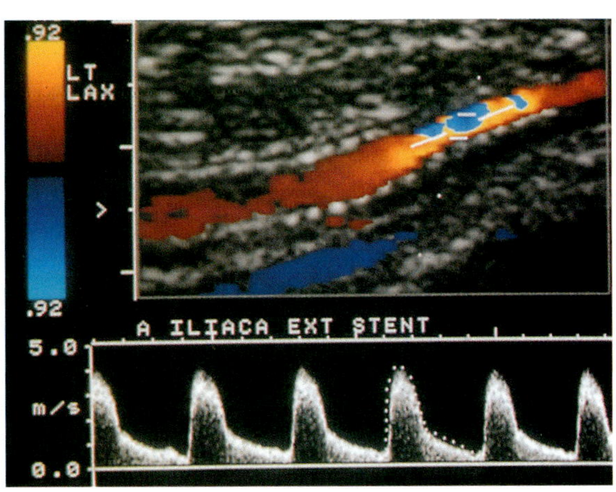

d

Case Figs. **18 c, d**

19 Occlusion of the Iliac Artery

Patient: T.M., male, age 51.
Clinical diagnosis: **PAOD, occlusion of the right iliac artery,** low-grade stenosis of the left iliac artery; differential diagnosis: high-grade stenosis of the right iliac artery.

CAD:	+	Vascular surgery:	–	Diabetes mellitus:	–
PAOD:	+	Heredity:	+	Hyperlipidemia:	–
TIA, stroke:	–	Hypertension:	–	Smoking:	+

Case report. During the previous two years, the patient had suffered from hip and leg claudication on the right side, predominantly on inclined walking surfaces. Fifteen years before the first myocardial infarc-tion had occurred. One year previously, a further in-farction had taken place. There were no serious complaints of angina pectoris. A continuing risk factor was a high level of nicotine consumption, which had not been significantly reduced. In addition, there was a familial predisposition to vascular disease.

Examination results:

	Pulse status Right	Pulse status Left	Auscultation Right	Auscultation Left
Abdomen			+⁺	⁺+
Common femoral artery	(+)	+	–	+
Superficial femoral artery			–	–
Popliteal artery	(+)	+	–	–
Posterior tibial artery	(+)	+		
Dorsal artery of the foot	–	+		

Doppler sonogram *(analogue registration)* (**a**). *Right:* The flow pulse in the groin is extremely attenuated, and has no diastolic flow while at rest. Through the posterior tibial artery, a relatively good flow pulse is detected. *Left:* The flow pulses are almost normal, and have a reverse flow component.

Angiogram (**b**). *Right:* There is an occlusion of the common iliac artery from its origin, with refilling in the distal vascular segment. The length of the occlusion is approximately 2 cm without any additional, subsequent occlusions. *Left:* A minimal stenosis of the common iliac artery is present.

Comment. Despite the clear postocclusive attenuation of the flow velocity and an extremely pathological groin pulse waveform indicating a serious preceding vascular system obstruction in the pelvic region, there is good compensation in the absence of subsequent occlusion. The patient has hardly any difficulty in walking (1000 m walking distance) with a pressure of 80 mmHg at the ankle arteries. Treatment to open the lumen is not necessary.

Example using color flow duplex sonography. In a different patient (R.T., male, age 49) with an *occlusion of the left external iliac artery,* the blood vessel refills at the transition to the common femoral artery (**c**).

B-mode sonographic imaging allows definite identification of the external iliac artery (arrows). Because of the minimal echogenicity of the intraluminal material the reflections from the vascular walls can still be clearly differentiated.

Using the *color flow Doppler mode,* the refilled arterial segment can be quickly identified in the region of postocclusive, peripheral filling, due to the visually displayable flow. No color signals can be registered from the occluded region. The light yellow color distal to the occlusion, in comparison to the subsequent vascular system, corresponds to the relatively higher flow velocity of confluent collateral blood vessels.

Cuff position	Right registration point: Posterior tibial artery	Right registration point: Dorsal artery of the foot	Left registration point: Posterior tibial artery	Left registration point: Dorsal artery of the foot
Proximal thigh	80	75	110	105
Distal lower leg	80	80	135	135

Systolic pressure measurement (Doppler ultrasound technique; mmHg)

Case Figs. **19 a–c** ▷

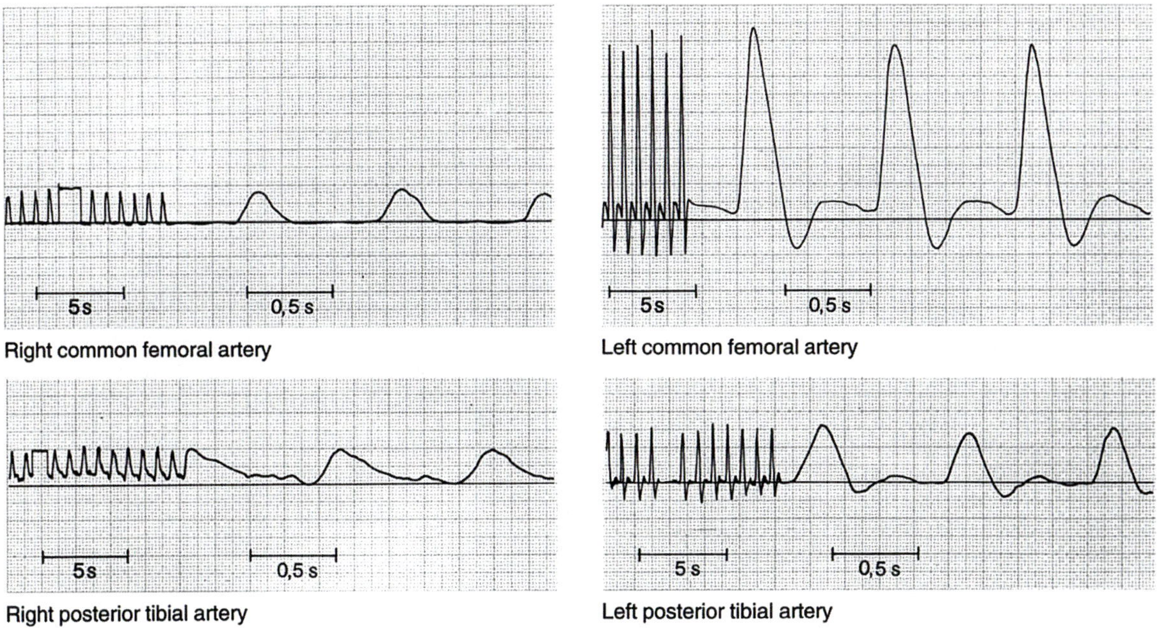

a CW Doppler sonograms

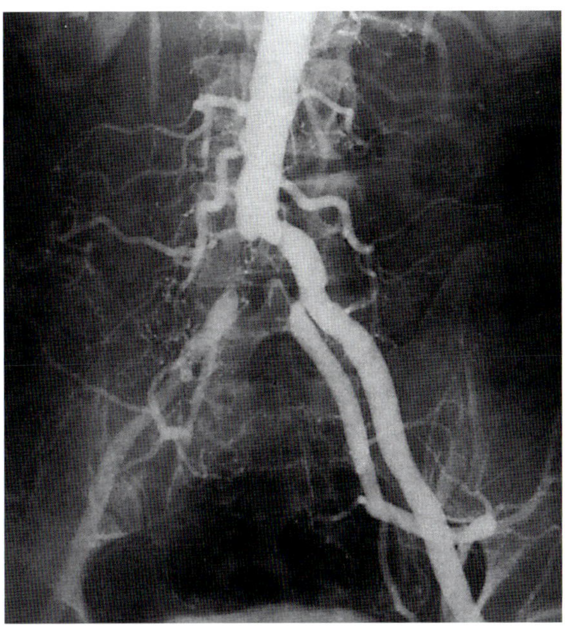

b

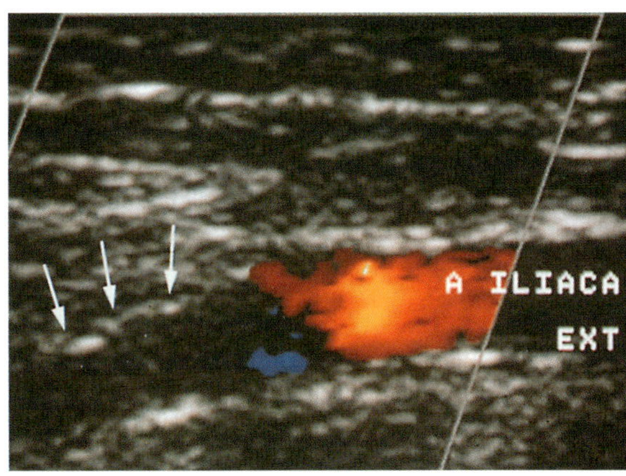

c

Case Figs. **19 a–c**

Thigh Level

20 Stenosis of the Femoral Artery

Patient: K.K., male, age 66.
Clinical diagnosis: amputation of the right thigh following trauma, **PAOD, stenosis of the left superficial femoral artery.**

CAD:	+	Vascular surgery:	–	Diabetes mellitus:	–
PAOD:	+	Heredity:	–	Hyperlipidemia:	+
TIA, stroke:	–	Hypertension:	–	Smoking:	+

Case report. On the right side, the thigh had been amputated following a war injury in 1945. In 1971, a bypass was completed using a Y prosthesis. During the previous year, there had been incidents of calf claudication on the left side, with the walking distance limited to approximately 100 m. Clinically, there was a suspicion of stenosis of the left femoral artery, to be reopened using angioplasty.

Examination results:

	Pulse status		Auscultation	
	Right	Left	Right	Left
Abdomen			–	–
Common femoral artery	+	+	+	+
Superficial femoral artery			–	+
Popliteal artery		+		–
Posterior tibial artery		(+)		
Dorsal artery of the foot		(+)		

Doppler sonogram *(analogue registration)* (**c**). Both *before* and *after* angioplasty, the *common femoral artery* shows a more or less unchanged, normal wave form.

Intrastenotically, a systolic peak reversal in the *superficial femoral artery* can be registered.

In the *popliteal artery* and the *posterior tibial artery,* reduced amplitudes of the flow velocity are noted prior to the catheter intervention. The wave form in the popliteal artery is still triphasic, and in the posterior tibial artery it is monophasic.

After dilatation of the femoral artery stenosis, normalization of the flow waveforms in the following vascular system confirms the success of the procedure.

Doppler frequency spectrum. Registering the Doppler frequency shift from the stenotic region reveals a *massive flow disturbance,* with vortex formations and *high systolic maximal frequencies* up to 9 kHz (**a**). The wave form is no longer triphasic, and in the systole flow above and below the baseline is registered.

After angioplasty, the Doppler frequency spectrum from the dilated region again shows a narrow frequency band, with a triphasic waveform and a systolic maximal frequency of approximately 2.7 kHz (**b**).

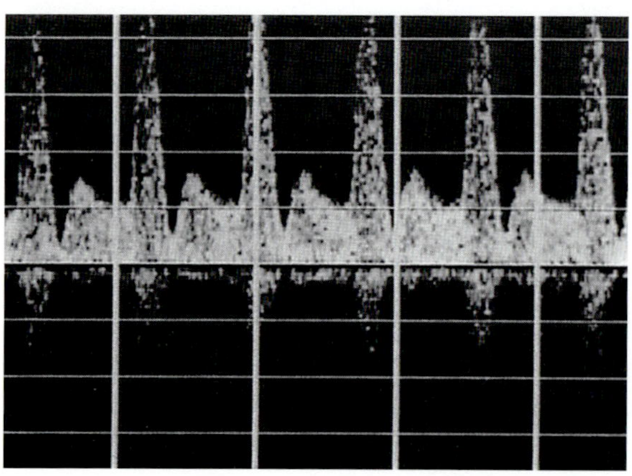

a Left superficial femoral artery intrastenotically **prior** to catheter intervention (ordinates: 1 graduation = 2 kHz)

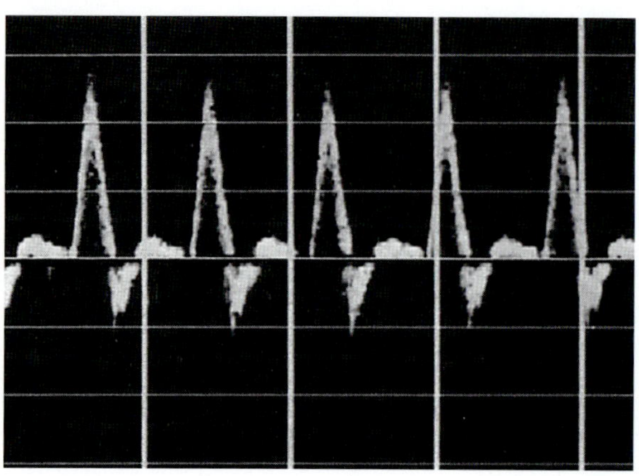

b Left superficial femoral artery intrastenotically **after** catheter intervention (ordinates: 1 graduation = 1 kHz)

Case Figs. **20 c–f** ▷

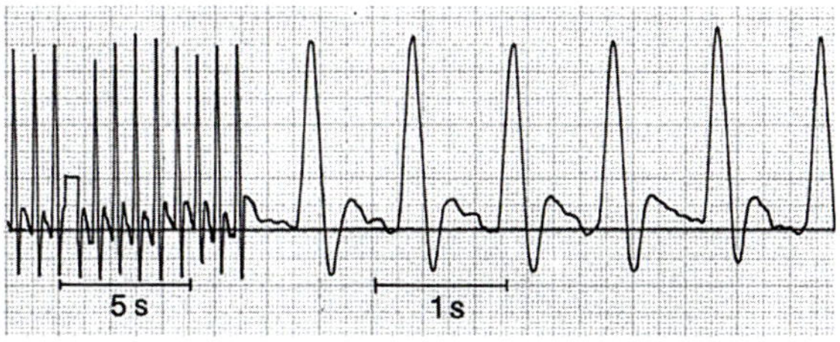

Left common femoral artery **prior** to catheter treatment

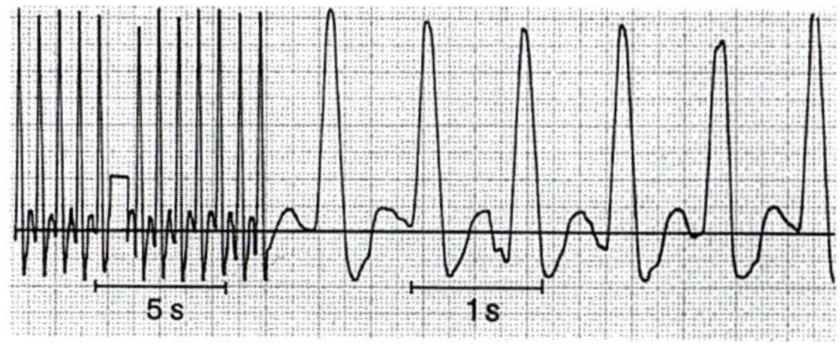

Left common femoral artery **after** catheter treatment

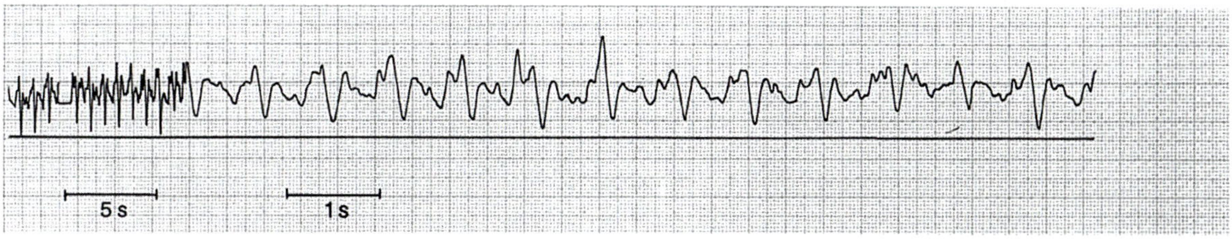

Left superficial femoral artery, sample volume in the stenosis with peak reversal, **prior** to catheter treatment

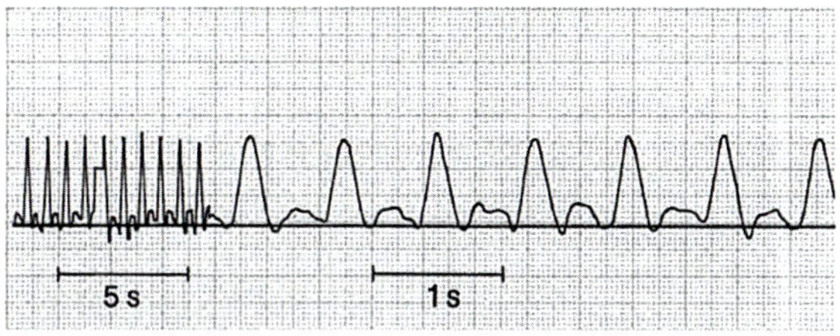

Left popliteal artery **prior** to catheter treatment

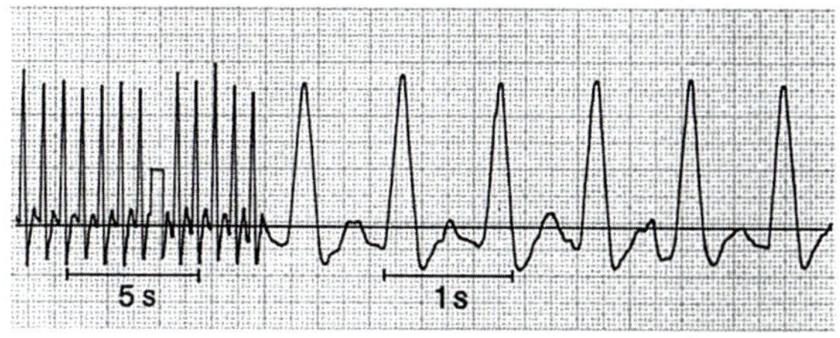

Left popliteal artery **after** catheter treatment

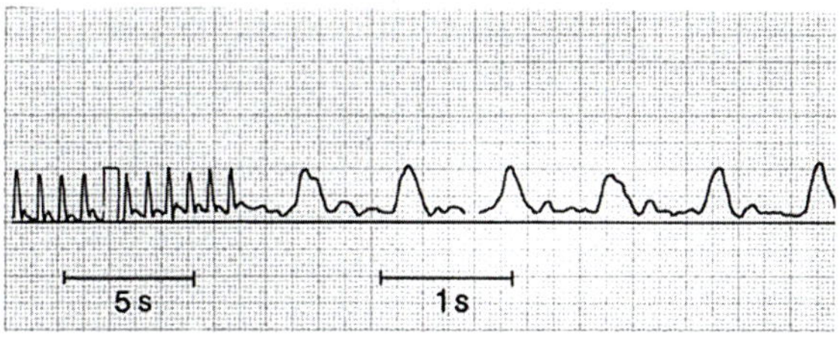

Left posterior tibial artery **prior** to catheter treatment

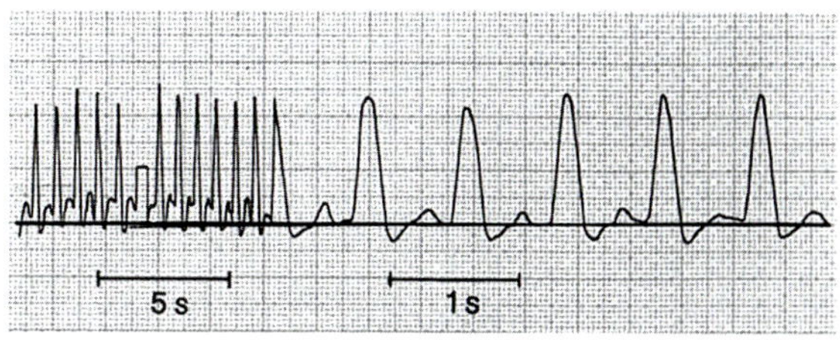

Left posterior tibial artery **after** catheter treatment

c CW Doppler sonograms

Case Figs. **20 c–f**

Systolic pressure measurement (Doppler ultrasound technique; mmHg) before catheter treatment

Cuff position	Right registration point		Left registration point	
	Posterior tibial artery	Dorsal artery of the foot	Posterior tibial artery	Dorsal artery of the foot
Proximal thigh			175	165
Distal lower leg			90	100

Systolic pressure measurement (Doppler ultrasound technique; mmHg) after catheter treatment

Cuff position	Right registration point		Left registration point	
	Posterior tibial artery	Dorsal artery of the foot	Posterior tibial artery	Dorsal artery of the foot
Proximal thigh			155	155
Distal lower leg			110	110

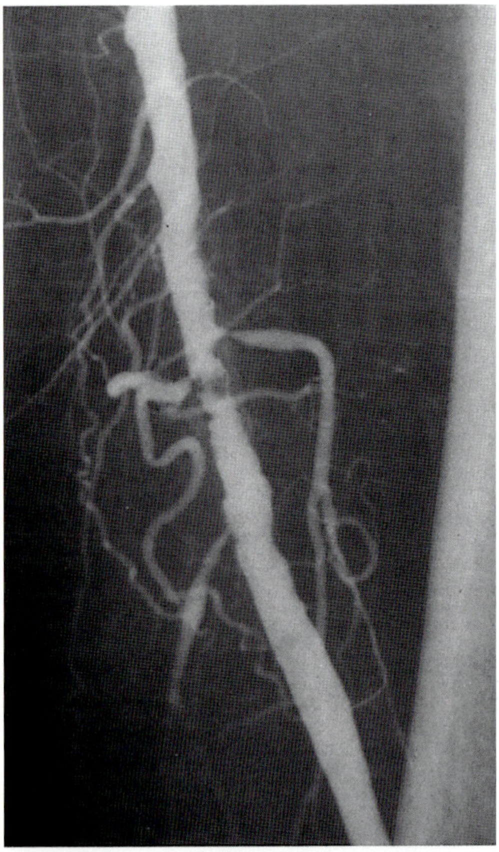
d

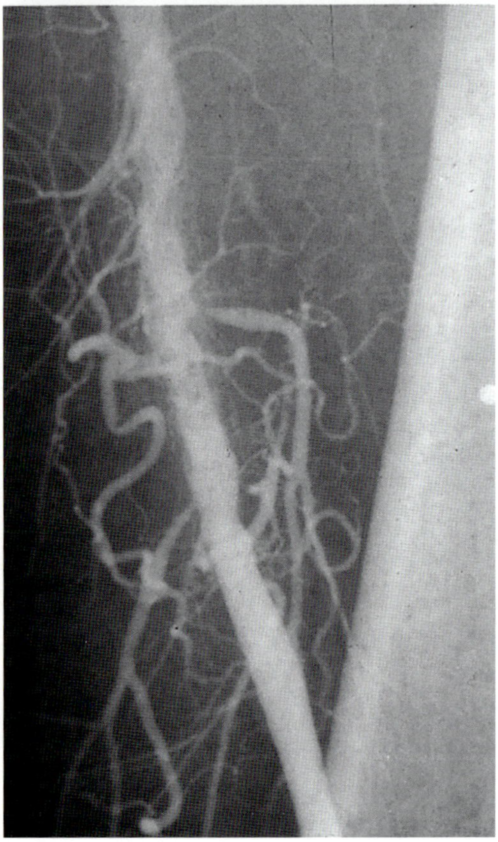
e

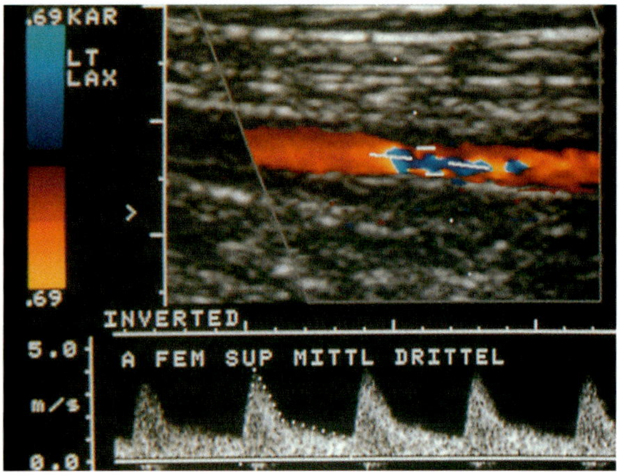

f

Angiogram. *Left:* (**d**) There is a significant stenosis of the superficial femoral artery in the distal third. An occlusion of the anterior tibial artery is seen in the distal segment. Stenoses of the posterior tibial artery are present. (**e**) After angioplasty, the stenosis of the femoral artery is extensively dilated.

Comment. All methods of measurement were able to locate the stenosis of the femoral artery in these patients. The stenosis was already clinically suspected due to a bruit in the adductor canal. The jump in pressure between the proximal thigh (175 mmHg) and the distal lower leg (90 mmHg) amounted to 85 mmHg prior to the lumen-opening procedure, and 45 mmHg after percutaneous transluminal angioplasty. The hemodynamic effect of the treatment was shown using functional measurements, and also in the angiogram. The walking distance increased to 300 m.

Example using color flow duplex sonography, in a different patient with a stenosis over a short distance in the left femoral artery in the middle third of the blood vessel (**f**). With the help of the *color flow Doppler mode,* the flow velocity can be measured in a more targeted manner, due to the lighter colors in the stenoses. The color change due to aliasing (light yellow to white to light blue) indicates the region of maximum flow acceleration.

The *Doppler frequency spectrum* from this vascular segment is monophasic, and high maximum sys-tolic, end-diastolic, and average velocities of 401 cm/s, 105 cm/s, and 178 cm/s respectively can be measured. The pulsatility index lies in the pathologic range (1.66).

21 Occlusion of the Femoral Artery (Short)

Patient: I.G., male, age 50.
Clinical diagnosis: **PAOD stage IIa on the left,** status after implantation of a **bifurcation prosthesis, occlusion of the left femoral artery.**

CAD:	–	Vascular surgery:	+	Diabetes mellitus:	–
PAOD:	+	Heredity:	–	Hyperlipidemia:	+
TIA, stroke:	–	Hypertension:	–	Smoking:	+

Case report. For approximately the previous five years, the patient had suffered calf claudication, initially on the left side, and later also on the right. Ten months previously, the pain-free walking distance decreased to below 100 m, and a bifurcation prosthesis was implanted due to bilateral occlusions of the external iliac arteries.

Following an initially unlimited walking ability, four months later a calf claudication on the left appeared after the patient had walked a distance of approximately 150 m. Fibrinolytic therapy for an occlusion of the femoral artery with an ultrahigh dose of streptokinase followed over a period of two days. This treatment was initially successful. After a few days, a reocclusion of the superficial femoral artery occurred, with a retained increase in the pain-free walking distance up to 600 m.

On admission, there was renewed claudication in the left thigh after a distance of 200 m.

Examination results. Well-healed scars following the implantation of a Y prothesis were seen.

	Pulse status		Auscultation	
	Right	Left	Right	Left
Abdomen			–	–
Common femoral artery	+	+	–	+
Superficial femoral artery			–	–
Popliteal artery	+	–	–	–
Posterior tibial artery	+	–		
Dorsal artery of the foot	+	–		

Color flow duplex sonogram. Using a longitudinal section almost parallel to the axis of the left superficial femoral artery, the short arterial occlusion (18.8 mm) can be seen at the transition from the middle to the distal third of the blood vessel, due to the absent color coding of the lumen (**a**).

Proximal to the occlusion, only a narrow color band with a slow flow velocity is depicted. *Distal* to the occlusion, there is a typical finding of a confluent collateral blood vessel with a high flow velocity (aliasing), recognizable by the light colors, in the region of arterial *refilling* (*). In this blood vessel, the flow direction is oriented toward the transducer. Dor-solateral to the femoral artery, the femoral vein is seen as an accompanying structure.

Angiogram. In the selective image during percutaneous transluminal angioplasty, the approximately 20-mm long occlusion of the superficial femoral artery is shown at the transition from the middle to distal third of the blood vessel (**b**). There are three patent arteries in the lower leg.

Comment. Recanalization of the approximately 2-cm long occlusion of the superficial femoral artery succeeded without leaving a residual stenosis. However, four months after the balloon dilatation, a renewed angioplasty was necessary to remove a filiform stenosis of the femoral artery due to a myointimal proliferation in the region of the previous balloon dilatation. This procedure was also successful.

With regard to the examination technique itself, it should be added that, particularly when dealing with longer vascular occlusions that involve the superficial femoral artery, using duplex sonography for orientation purposes through the accompanying femoral vein is very helpful in anatomically classifying the different structures in the thigh.

The color flow duplex sonogram (**a**) shows the close spatial relationship between these two vessels, so that in addition to B-mode imaging even of occluded vessels, based on their vascular walls and internal luminal structures, it is also possible to differentiate successfully between the original artery and collateral blood vessels.

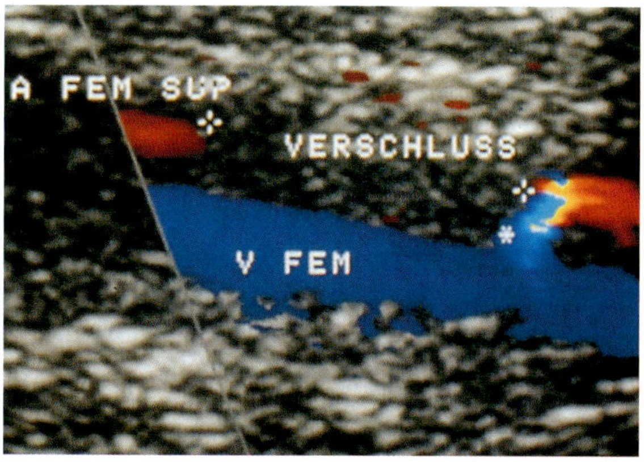

a

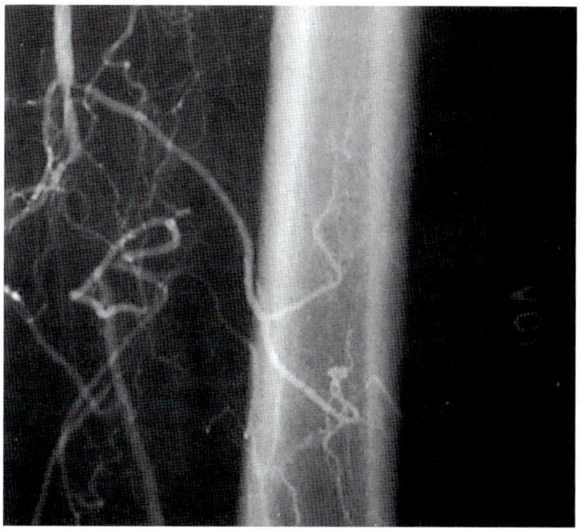
b

22 Occlusion of the Femoral Artery (Longer)

Patient: F.B., male, age 52.
Clinical diagnosis: **PAOD stage IIa on the right, reocclusion of the right superficial femoral artery.**

CAD:	+	Vascular surgery:	–	Diabetes mellitus:	+
PAOD:	+	Heredity:	–	Hyperlipidemia:	+
TIA, stroke:	–	Hypertension:	+	Smoking:	+

Case report. Approximately 18 months previously, symptoms appeared for the first time in the form of a calf claudication on the right side. This was treated by an angioplasty of an occlusion of the right superficial afemoral artery, with good results. For approximately three months, a clear decrease in the pain-free walking distance occurred, down to 50–80 m. The patient had occasional nocturnal pain in the right lower leg.

Examination results:

	Pulse status		Auscultation	
	Right	Left	Right	Left
Abdomen			–	–
Common femoral artery	+	+	–	+
Superficial femoral artery			–	–
Popliteal artery	–	+	–	–
Posterior tibial artery	–	+		
Dorsal artery of the foot	–	–		

Systolic pressure measurement (Doppler ultrasound technique; mmHg):

	Right	Left
Brachial artery	190	
Posterior tibial artery	–	170
Anterior tibial artery	80	150

Color flow duplex sonogram. A ventromedial sectional plane proceeding from the medial side of the thigh on the right shows the bifurcation of the femoral artery, with the proximal segment of the superficial femoral artery (coded red) (**c**).

The *beginning of the occlusion* can be localized at the gap in the color coding. Proximal to the occlusion, a stenosis is recognized in a collateral given off laterally, due to the visible lumen constriction and the aliasing, which indicates an elevated flow velocity.

Over a 16-cm stretch in the superficial femoral artery, no blood flow can be registered using color flow Doppler sonography, or in the pulsed wave Doppler mode. This is consistent with vascular occlusion. In the region of the distal third of the blood vessel, peripheral refilling is detected, with two collaterals joining (light blue) (**d**).

Angiogram. Catheter angiography during percutaneous transluminal angioplasty shows the proximal beginning of the occlusion of the right superficial femoral artery, several centimeters distal to the bifurcation of the femoral artery (**a**). Directly proximal to the occlusion, a collateral with a stenosis at its origin proceeds laterally. The trunk of the deep femoral artery is also occluded a few centimeters after its origin.

Angiography demonstrates the peripheral refilling of the superficial femoral artery via collateral blood vessels in its distal third (**b**).

Comment. For this patient, a noninvasive individual treatment concept was developed, consisting of intra-arterial arteriography in preparation for a simultaneous catheter intervention.

After an initially successful catheter dilatation, there was still persistent occlusion of the deep femoral artery, and a further reocclusion of the superficial femoral artery occurred five months later. This too was successfully recanalized.

Following catheter dilatation of newly appearing residual stenoses of the ipsilateral femoral artery four and nine months later, only wall deposits without any hemodynamic effects were detectable with color flow duplex sonography in the right superficial femoral artery, 24 months after the final balloon dilatation. In retrospect, the concept of repeated angioplasty has proved an effective one.

Case Figs. **22 a–d** ▷

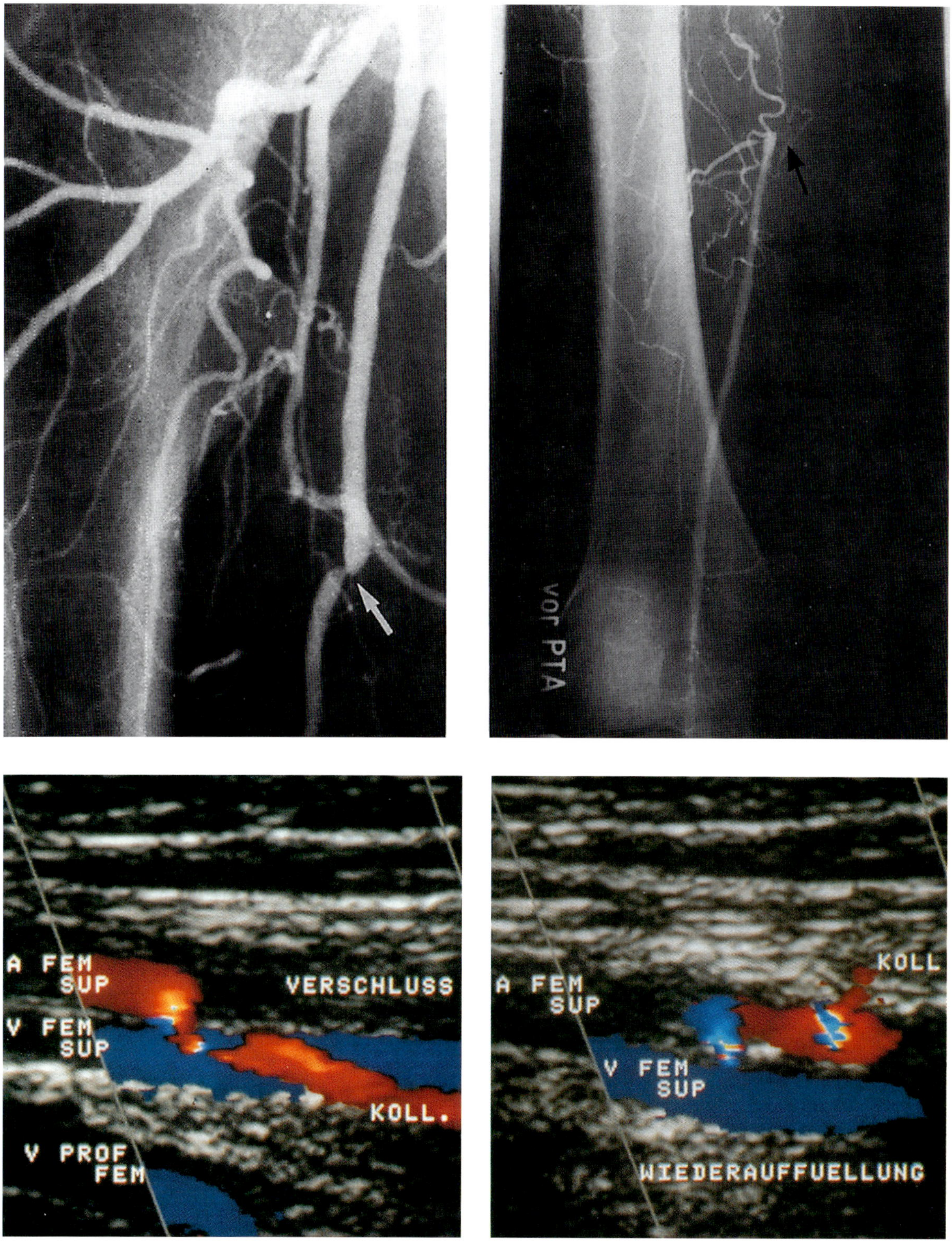

Case Figs. **22 a–d**

23 Occlusion of the Popliteal Artery

Patient: G.K., male, age 59.
Clinical diagnosis: **PAOD stage IIa on the right,** suspected **stenosis of the right popliteal artery,** and **stenosis of the left femoral artery.**

CAD:	–	Vascular surgery:	+	Diabetes mellitus:	+
PAOD:	+	Heredity:	–	Hyperlipidemia:	+
TIA, stroke:	–	Hypertension:	+	Smoking:	+

Case report. Bilateral calf claudication had persisted for about one year, on the right more than on the left. Angioplasty of a stenosis of the right popliteal artery had been carried out ten months previously. Suspicion of restenosis.

Examination results. A 59-year-old patient in a good general state of health and with good nutritional status, with the following angiological findings:

	Pulse status		Auscultation	
	Right	Left	Right	Left
Abdomen			–	–
Common femoral artery	+	+	–	–
Superficial femoral artery			–	+
Popliteal artery	+	(+	+	–
Posterior tibial artery	–	(+		
Dorsal artery of the foot	(+)	(+)		

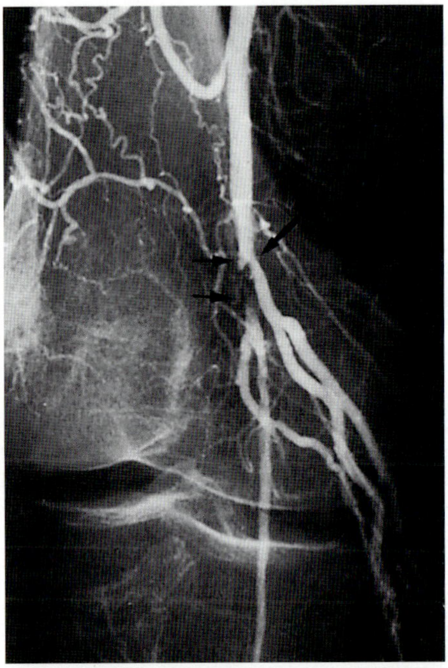
a

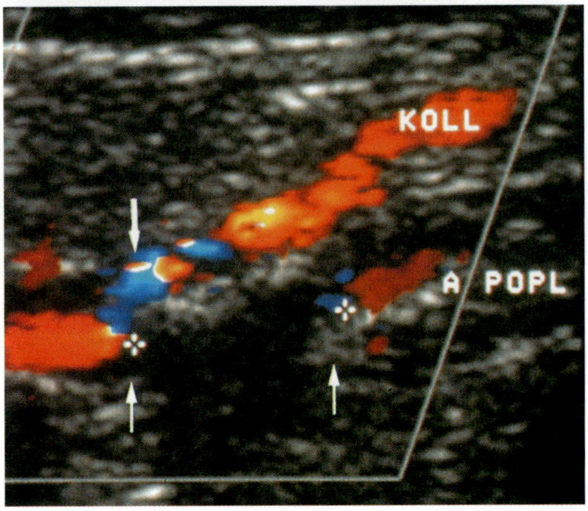

b

Color flow duplex sonogram. In a longitudinal section of the popliteal artery proceeding from a dorsal direction (with the patient in prone position), a short *occlusion of the artery over a distance of some 13.5 mm* (thin arrows) (**b**) is seen above the interarticular space of the knee. Hyperechoic structures in the (dorsal) section of the vascular occlusion near the transducer cause an ultrasound shadow.

Proximal to the beginning of the occlusion, a collateral blood vessel of the popliteal artery is given off. The aliasing of the collateral at the origin is greater than expected, due to a change in the ultrasound application angle. Because of a preselected high range for the average flow velocity of the color flow Doppler mode (92 cm/s), stenosis of the collateral blood vessel (thick arrow) can be deduced, solely based on the color coding.

Angiogram. Selective catheter angiography during percutaneous transluminal angioplasty shows the *short occlusion in the proximal segment of the popliteal artery* (thin arrows) (**a**). The collateral, which proceeds dorsally, also shows a stenosis at the origin (thick arrow) on the angiogram.

Comment. Clinically, a *stenosis* of the popliteal artery was suspected on the basis of a systolic bruit in the hollow of the knee. Color flow duplex sonography, however, detected an *occlusion* of the popliteal artery. The bruit was most probably caused by a stenosis of the collateral blood vessel.

In the *differential diagnosis,* this set of findings obtained with color flow duplex sonography, with massive ultrasound extinction over a short distance due to a hyperechoic plaque near the transducer, should also suggest consideration of a hemodynami-cally effective high-grade stenosis. The differentiation process must then involve a careful examination of this vascular region using the pulsed wave or continuous wave Doppler mode, searching for a "stenosis signal." Indirect criteria of a distal or perhaps proximal decrease in the flow velocity, as well as morphologi-cally and functionally detectable collateral circulation, is not always helpful in discriminating between stenosis and occlusion.

To provide orientation and assist in *localizing the level* of pathological vascular processes involving both the arteries and veins in the region of the popliteal artery, the *interarticular space of the knee* can be used as a reference point, since this bony structure usually presents a good image in sonographic examinations.

24 Aneurysm of the Femoral Artery

Patient: H.T., female, age 76.
Clinical diagnosis: **PAOD;** distal **aneurysm of the left femoral artery,** and **stenosis of the left femoral artery.**

CAD:	+	Vascular surgery:	+	Diabetes mellitus:	–
PAOD:	+	Heredity:	–	Hyperlipidemia:	+
TIA, stroke:	+	Hypertension:	–	Smoking:	+

Case report. Seven years previously, an open recanalization in the region of the *left* superficial femoral artery had been carried out, and a Dacron patch dilatation was conducted to alleviate a stenosis of the femoral artery. At presentation, there was reported increasing swelling in the area of the medial thigh, without any intermittent claudication.

Examination results. There is a well-healed, 15-cm long scar located distally at the medial side of the thigh.

	Pulse status		Auscultation	
	Right	Left	Right	Left
Abdomen			–	–
Common femoral artery	+	+	–	–
Superficial femoral artery		*	–	+
Popliteal artery	+	+	–	–
Posterior tibial artery	+	(+)		
Dorsal artery of the foot	(+)	(+)		

Systolic pressure measurement (Doppler ultrasound technique; mmHg):

	Right	Left
Brachial artery	160	
Posterior tibial artery	160	125
Anterior tibial artery	155	115

Color flow duplex sonogram. A sonographic longitudinal section through the left distal superficial femoral artery shows two aneurysms (**a**), in the distal section and in the femoropopliteal transition zone. In this plane, the larger *distal* aneurysm is approximately 5.9 cm × 7.5 cm in size. The lumen, through which there is still blood flowing, is partially shown in the image, and its diameter measures approximately 2 cm. A sonographically nonhomogeneous massive partial thrombosis of the aneurysmal cavity can be seen. There is a hemodynamically effective stenosis distal to the aneurysm in the femoropopliteal vascular segment (not shown).

A corresponding cross-section through the distal aneurysm, with the sectional plane perpendicular to the vascular axis, also shows the extreme wall dilatation (6.8 cm × 6.7 cm) and a lumen (2.3 cm × 2 cm) through which blood is still flowing (**b**).

In a cross-section *of the proximal* aneurysm, also partially thrombosed, its maximum size can be measured as 3.6 cm × 3.4 cm (**c**). Here the lumen with blood flowing through it shows a distension of 2.8 cm × 1.3 cm.

Angiogram. The intra-arterial angiographic image shows two aneurysms of the left superficial femoral artery, also involving the popliteal artery. The artery is already dilated proximal to the aneurysms (**d**).

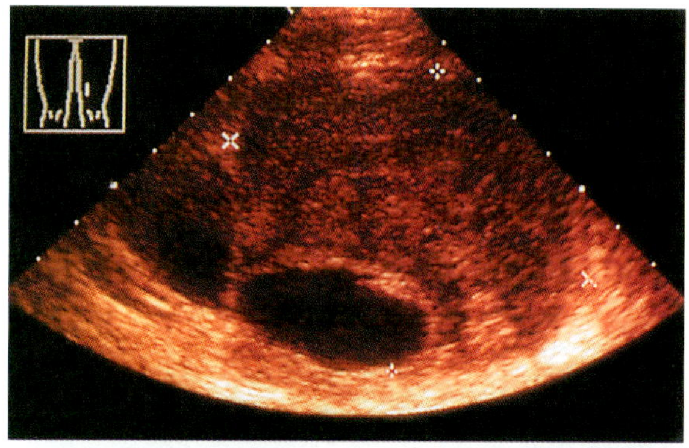
a

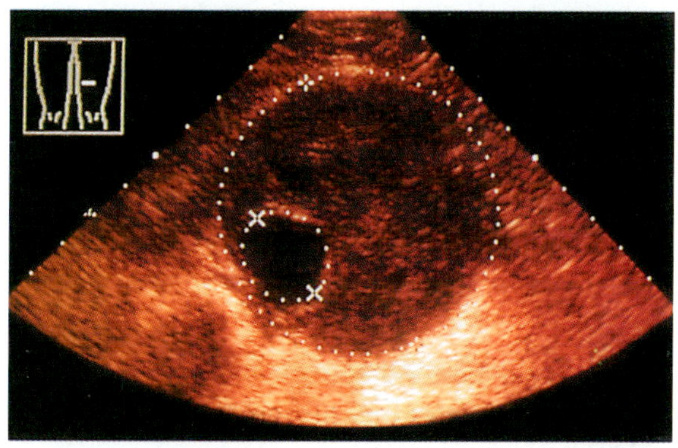
b

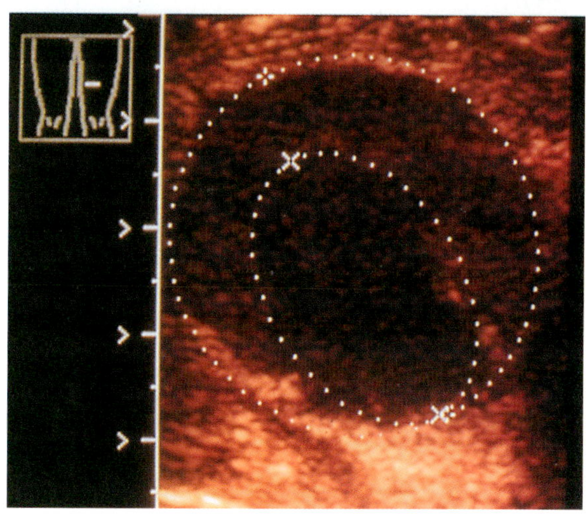
c

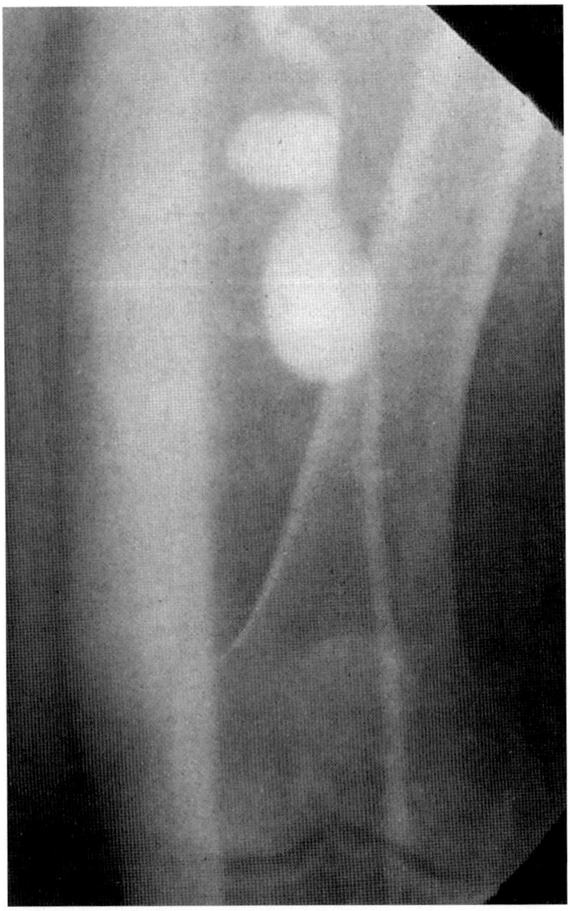

d

Doppler frequency spectrum. From the *right* proximal superficial femoral artery, an almost normal spectrum was registered, except for a missing forward flow component in the early diastole (**e**). The hemodynamic effects of an early *extrasystole* (second spectrum) can be seen, with a lowered systolic velocity (67 cm/s) and average flow velocity, corresponding to a lowered cardiac stroke volume and, assuming a constant vascular cross-section, also causing a lowered flow rate in the femoral artery. In the subsequent cardiac cycle (third spectrum), the systolic (111 cm/s) and average flow velocities are somewhat higher due to compensation, and then return to the steady prior values.

On the side of the aneurysms, the frequency spectrum in the *left* proximal femoral artery indicates pendular flow, with a physiological systolic flow direction and a clear holodiastolic backflow component, which corresponds to retrograde flow from the aneurysmal cavities (**f**).

Postoperatively, a triphasic Doppler frequency spectrum was again registered at the same location from the proximal superficial femoral artery (**g**).

A triphasic wave form was also registered in the flow passing through the Gore-Tex prosthesis in the distal thigh region (**h**).

Comment. In comparison with the angiogram (**d**), the B-mode findings (**a–c**) show the clear advantages of sonography in diagnosing aneurysms. While angiography is only able to depict the internal vascular lumen *through which blood is flowing,* sonography shows the full extent of the aneurysm, including its thrombosed segments. In this patient, the cross-sections of the thrombosed segments are several times the residual lumen (3605 mm^2 compared to 345 mm^2 in **b**).

Compared to the typical predilection points for aneurysms, the location here in the superficial femoral artery is quite rare (Case 22, p. 302). The cause of the wall dilatation in this vascular region is therefore very likely to be the earlier recanalization and patch operation.

The indication for surgery was based in particular on risk of embolism and thrombosis, as well as an increased risk of aneurysmal rupture.

During the electively conducted aneurysmal resection, with the implantation of a ring-stabilized Gore-Tex prosthesis, the sonographic findings were confirmed intraoperatively.

Postoperatively, all pulses in the left lower extremity were strongly palpable. The systolic pressure at the ankle arteries, above the anterior and posterior tibial arteries, corresponded at 190 mmHg to the pressure measured above the brachial artery.

We are grateful to Dr. Borchers of the Department of Radiological Diagnostics in Gummersbach Hospital for his kind permission to use the angiogram.

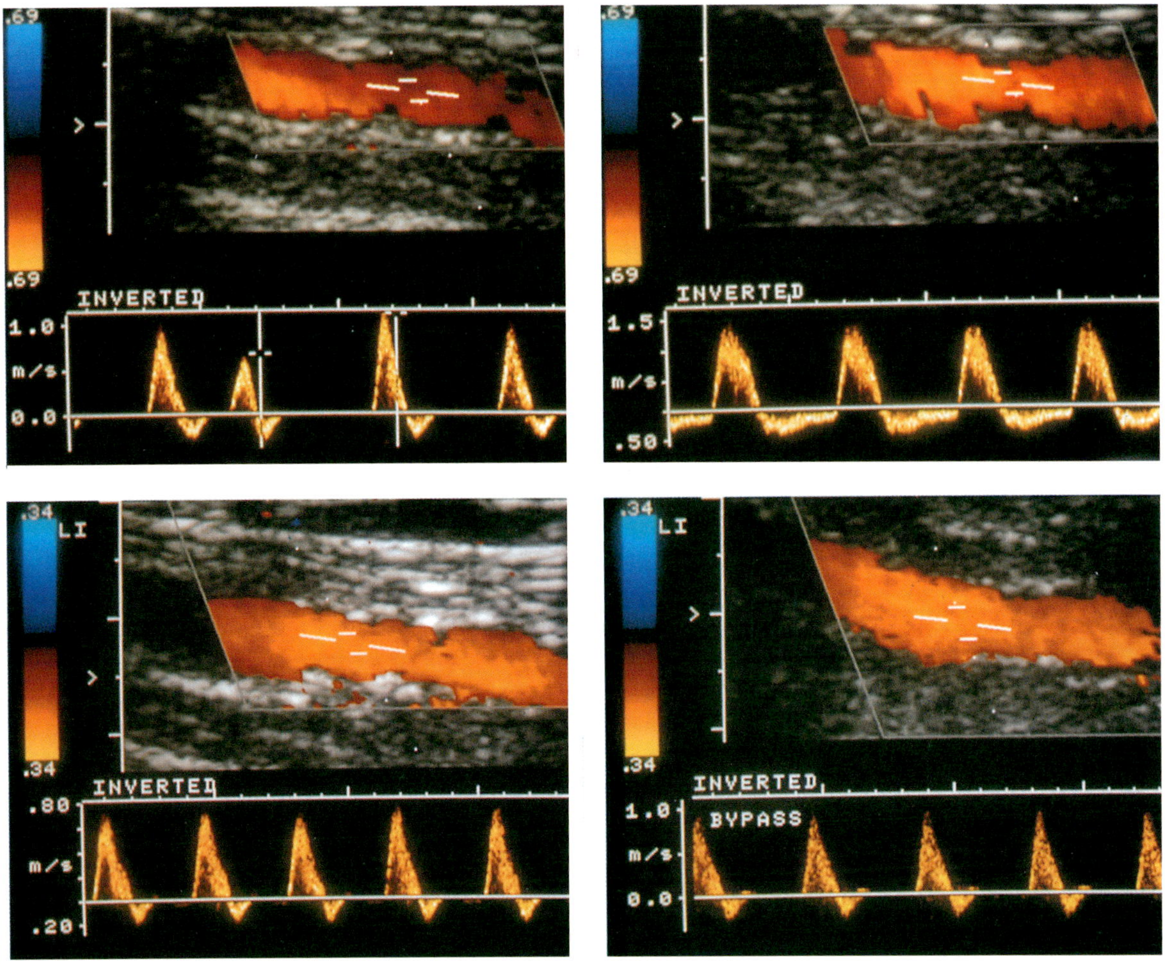

Case Figs. **24 e–h**

25 Aneurysm of the Popliteal Artery

Patient: M.B., male, age 70.
Clinical diagnosis: **dilatative/obliterative arteriopathy,** with previous bilateral **aneurysms of the popliteal arteries.** At presentation, there was an acutely appearing **occlusion of the right popliteal artery.**

CAD:	+	Vascular surgery:	+	Diabetes mellitus:	–
PAOD:	+	Heredity:	–	Hyperlipidemia:	+
TIA, stroke:	–	Hypertension:	+	Smoking:	+

Case report. Ten years previously, the first incident of exertion-dependent pains in the right calf region occurred. Two years later, successful catheter fibrinolysis and catheter dilatation of an acute occlusion in the *right* popliteal artery was carried out. There were also signs of incipient dilatative arteriopathy at this time. Eight years later, an angioplasty of a stenosis of the *left* popliteal artery was carried out, and *bilateral* aneurysms of the popliteal arteries were diagnosed, more pronounced on the right than on the left.

Three months earlier, there had been a sudden increase in calf claudication on the right side, without any pain while at rest.

Examination results:

	Pulse status		Auscultation	
	Right	Left	Right	Left
Abdomen			–	–
Common femoral artery	+	+	–	+
Superficial femoral artery			–	–
Popliteal artery	–	+	–	–
Posterior tibial artery	–	(+)		
Dorsal artery of the foot	–	(+)		

Systolic pressure measurement (Doppler ultrasound technique; mmHg):

	Right	Left
Brachial artery	145	
Posterior tibial artery	110	150
Anterior tibial artery	85	150

Angiogram. An image obtained using central venous digital subtraction angiography (DSA) two years earlier shows an aneurysm of the right popliteal artery, with wall changes in the femoropopliteal vascular segment (**a**). The current DSA image shows an occlusion of the right popliteal artery, beginning in the femoropopliteal transition zone (**d**). On the con-tralateral side, there is an incipient aneurysmal dilatation of the popliteal artery.

Color flow duplex sonogram. The color flow duplex sonogram corresponding to the first angiographic examination (**a**) shows the large aneurysm of the popliteal artery in longitudinal (**b**) and transverse section (**c**). The thrombosed lumen, with a residual lumen (coded red) of 5.4 mm through which blood is still flowing, measures 30.5 mm in its sagittal diameter (**b**) and 33.5 mm × 30.3 mm in its cross-sectional diameter (**c**).

Two years later, this true aneurysm is occluded. In the longitudinal section of the right popliteal artery proceeding from a dorsal orientation, hypoechoic internal structures can be recognized in the expanded vascular wall dilatation (**e**). In transverse section, the cross-sectional diameter 40 mm × 29 mm (**f**). Next to the aneurysm, color-coded flow is seen, belonging to the popliteal vein, which is displaced medially but not compressed.

Comment. The region of the popliteal arteries is a predilection point for aneurysmal vascular dilatations. In an occlusion of the popliteal artery, the possibility of a thrombosed popliteal aneurysm should always be included in the differential diagnosis.

The preferred method for detecting or excluding an aneurysm is B-mode imaging. Color flow Doppler sonography provides fast information about segments through which there is still blood flowing, and the extent of the thrombosis.

Since an aneurysmal structural deficiency of the arterial wall rarely occurs in isolation, when *one* aneurysm is found, additional arterial segments should always be examined. The contralateral vascular system, as well as the abdominal aorta and the iliac arteries, in particular, should be evaluated so as not to overlook possible dilatations of the lumen in these vessels.

In the case presented here, the patient declined to have the aneurysm of the popliteal artery corrected by vascular surgery. Despite the administration of oral anticoagulants, a thrombotic occlusion of the aneurysm occurred within two years.

Case Figs. **25 a–f** ▷

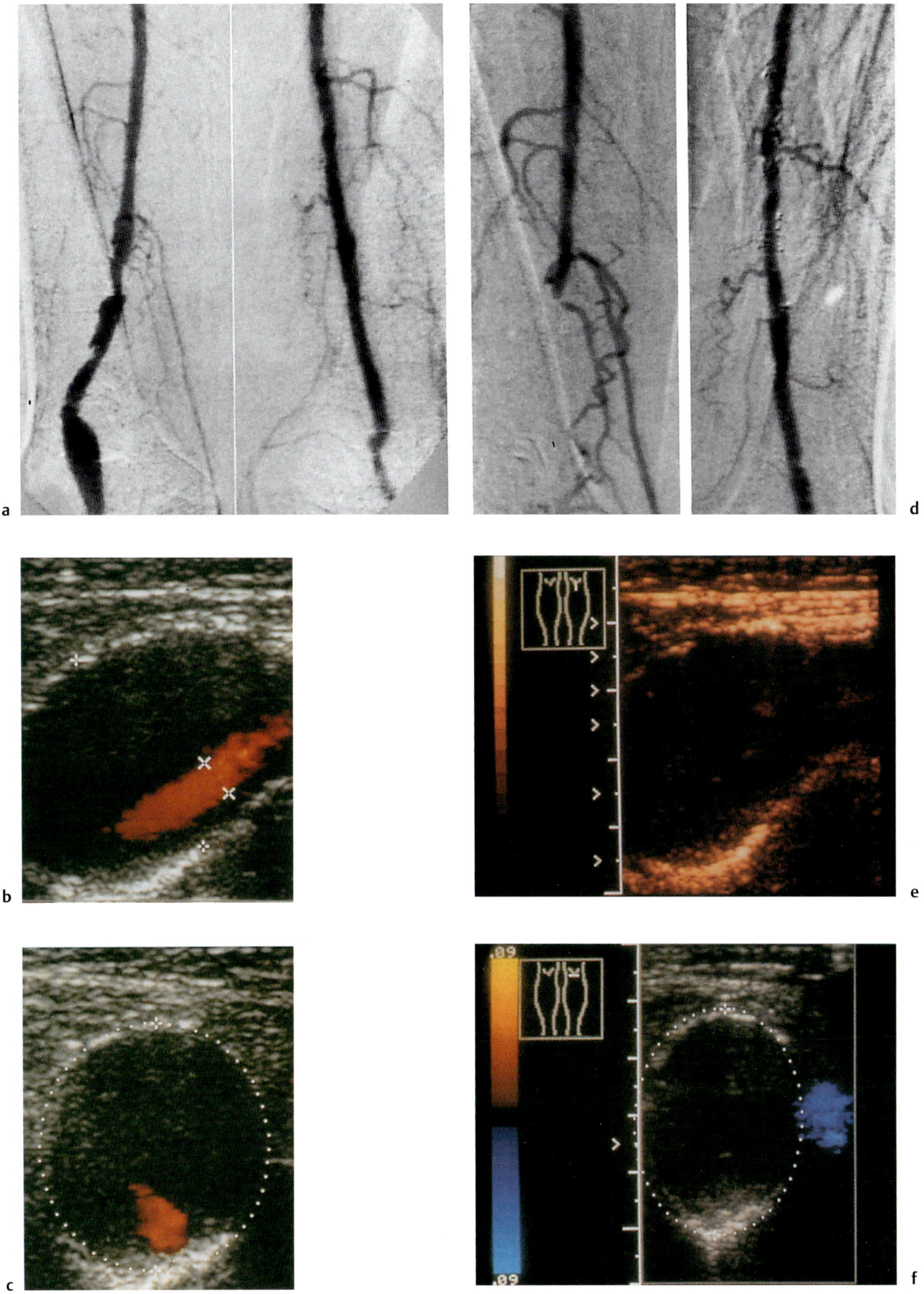

Case Figs. **25 a–f**

26 False Aneurysm of the Common Femoral Artery (Spontaneous Course)

Patient: B.D., female, age 67.
Clinical diagnosis: **PAOD stage IV on the right, amputation at the left thigh.** Stenosis of the right femoral artery and occlusion of the right popliteal artery. Following percutaneous transluminal angioplasty of the vascular obstructions on the right side, a **false aneurysm** formed in the right groin region.

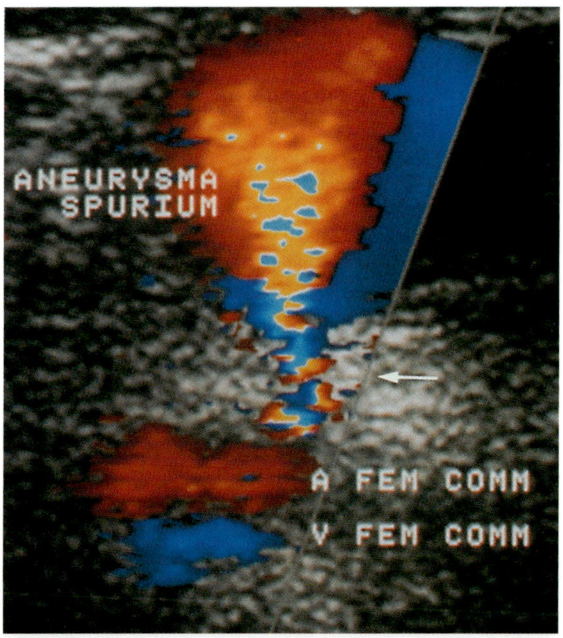

a

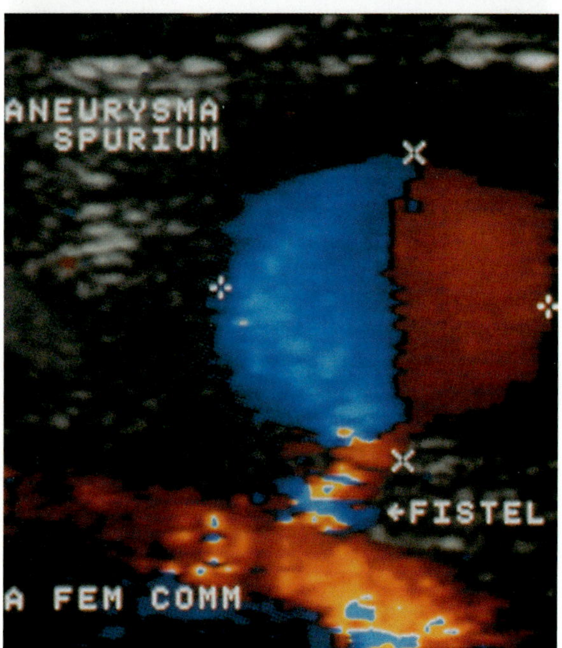

b

Case Figs. **26 c–e** ▷

CAD:	–	Vascular surgery:	+	Diabetes mellitus:	+
PAOD:	+	Heredity:	+	Hyperlipidemia:	+
TIA, stroke:	–	Hypertension:	+	Smoking:	–

Case report. The patient had a ten-year history of incidents of bilateral calf claudication. Two years previously, a left femoropopliteal bypass was placed due to a perforating ulcer of the foot on the left side. Postoperatively, an occlusion of the bypass occurred. Four months later, an amputation at the left thigh was carried out.

Ten months previously, a thromboendarterectomy of the right superficial femoral artery was carried out. For three months, there had been traumatic lesions (injuries due to chafing) in the region of the right Achilles' tendon and the right heel. A lumbar sympathectomy was carried out three months earlier.

Examination results at admission. A well-healed thigh stump was present on the left.

On the right side, there was a well-healed scar at the medial side of the thigh, a contraction of the knee joint (160°), and a stiffening of the talocalcaneal joint. Dry, mummified necroses were observed above the Achilles' tendon and above the heel, with a reddened marginal seam, a perforating ulcer above the metatarsal head both dorsally and ventrally, and hyperkeratoses.

	Pulse status		Auscultation	
	Right	Left	Right	Left
Abdomen			–	–
Common femoral artery	+	+	–	–
Superficial femoral artery		Amputation	+	Amputation
Popliteal artery	–		–	
Posterior tibial artery	–			
Dorsal artery of the foot	–			

Systolic pressure measurement (Doppler ultrasound technique; mmHg):

	Right	Left
Brachial artery	145	
Posterior tibial artery	120	Amputation
Anterior tibial artery	115	Amputation

After angioplasty of the stenoses of the femoral artery and the occlusion of the popliteal artery, there was a patent hematoma in the right groin, a pulsating tumor next to the puncture point that was painful on pressure, and a systolic sound above the groin, with a clinical suspicion *of false aneurysm.*

Color flow duplex sonogram. The course of spontaneous thrombosis of a false aneurysm in the region of the common femoral artery was observed over six weeks. All the images (**a–e**) represent sagittal longitudinal sections above the right groin.

Four days after PTA, a patent false aneurysm (**a**) was observed (only the proximal segment is shown here), located ventral to the common femoral artery and vein. It was approximately 60 mm × 20 mm × 20 mm in size, and was connected to the artery by a fistular canal (arrow).

Eleven days after catheter intervention, the section of the false aneurysm through which blood was flowing had decreased in size, and a thrombosed seam was observed at the margins (**b**). In the frequency spectrum of flow in the fistular canal (**c**), there was a shorter systolic inflow into the aneurysm (shown below the baseline) and a longer diastolic outflow (above the baseline), to-and-fro sign.

After a further ten days, increasing thrombosis of the aneurysmal cavity was seen (**d**). At the follow-up examination *39 days after angioplasty,* this cavity was completely occluded (**e**). The ellipse marks the region in the aneurysmal cavity corresponding to the lumen through which blood had recently flowed, which is still hypoechoic in this examination.

Comment. Due to her extensive necroses, the patient still remained hospitalized even after the successful angioplasty. It was therefore possible for the spontaneous thrombosis of the false aneurysm to be awaited on an in-patient basis.

A form of treatment which has shown good success rates involves compression of the fistular canal with the transducer, under the guidance of color flow Doppler sonography (see the following case report). If it persists for a longer period, a false aneurysm will require vascular surgery.

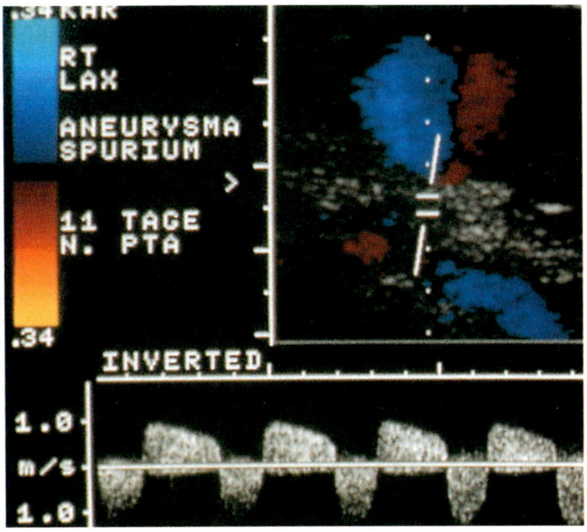

c

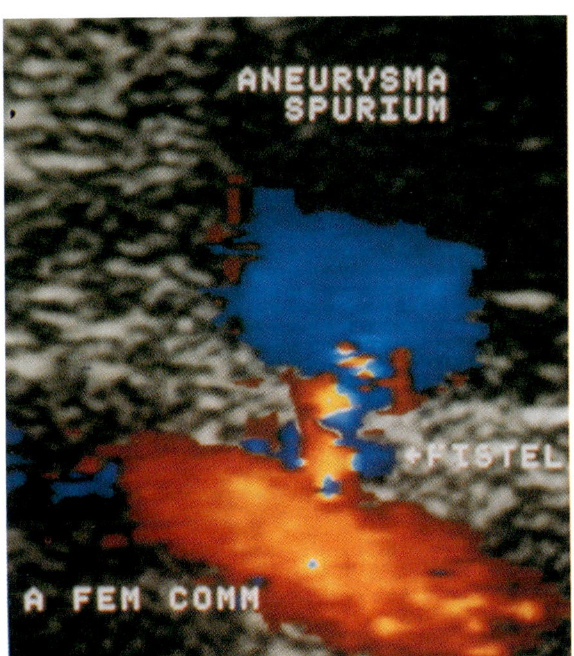

d

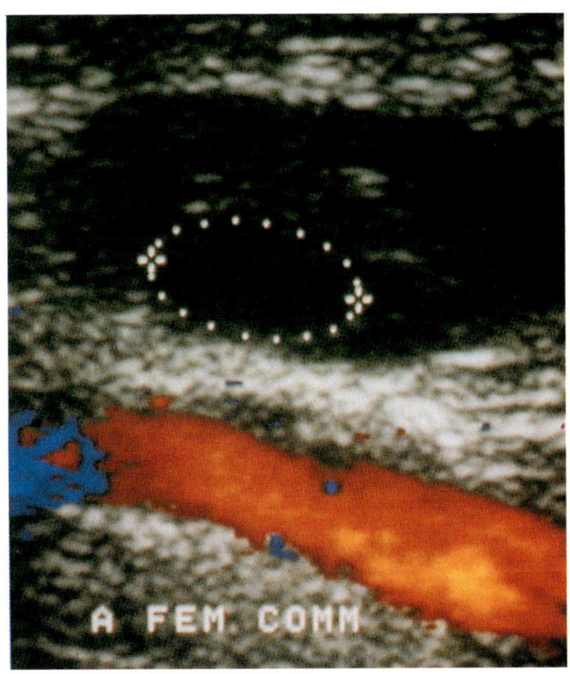

e

Case Figs. **26 c–e**

27 False Aneurysm (Compression Therapy)

Patient: J.E., male, age 63.
Clinical diagnosis: **PAOD stage IIa,** with a **stenosis of the left femoral artery** and occlusions of the arteries in the lower leg.

After percutaneous transluminal angioplasty (PTA) of a high-grade stenosis of the right renal artery, there was a suspicion of false aneurysm formation in the region of the right groin.

CAD:	–	PTA:	+	Diabetes mellitus:	–
PAOD:	+	Heredity:	–	Hyperlipidemia:	+
TIA, stroke:	–	Hypertension:	+	Smoking:	+

Case report. Four days previously, PTA of a stenosis in the right renal artery has been carried out in this patient suffering from arterial hypertension, renal failure (compensated retention), and bilateral stenoses of the renal arteries. After the patient had been on a long walk, a painful, pulsating tumor suddenly appeared, with a patent hematoma in the region of the right groin.

Examination results. There was a pulsating, full swelling as large as a walnut in the region of the right groin, which was painful on pressure, and a softer, fist-sized hematoma at the medial side of the thigh. Sys-tolic flow sound above the right common femoral artery was observed, with the following additional findings:

	Pulse status		Auscultation	
	Right	Left	Right	Left
Abdomen			+	+
Common femoral artery	+	+	+	–
Superficial femoral artery			–	+
Popliteal artery	+	(+)	–	–
Posterior tibial artery	+	+		
Dorsal artery of the foot	+	–		

Systolic pressure measurement (Doppler ultrasound technique; mmHg):

	Right	Left
Brachial artery	160	
Posterior tibial artery	174	186
Anterior tibial artery	196	148

Color flow duplex sonogram. The illustrations show *compression treatment of a false aneurysm* in the region of the common femoral artery. All the images **(a–d)** are sagittal longitudinal sections above the right groin at the level of the puncture point.

With B-mode sonography, the longitudinal section in the groin region depicts a pulsating, hypo-echoic, space-occupying lesion, approximately 45 mm × 30 mm × 14 mm in size (**a**). Dorsal to this, there is a connection, with a fluctuating caliber, to the common femoral artery (arrow), representing the reopened canal.

With color flow Doppler sonography, a fountain-like inflow of blood through the fistular canal (arrow) into the aneurysmal cavity during the systole is seen (**b**). Aliasing in the neck of the aneurysm reflects the relatively high flow velocity in the inflow jet. In this sectional plane, the jet can be readily followed into the cavity of the hematoma (light color mosaic), and has to be differentiated from the usual eddy-like flow, with slower velocities, that is found here.

After *compression of the aneurysm and the fistular canal* with the transducer lasting some 25 minutes, the cavity of the hematoma has clearly decreased in size, to 21.9 mm × 12.9 mm × 12 mm (**c**). The dorsally located fistular canal (arrow) still has a diameter of 2.6 mm.

With color flow Doppler sonography, occlusion of the false aneurysm and the fistular canal can be documented by the absent blood flow (**d**). It is important to use varying adjustments of the color parameters in this process, so as not to overlook fast as well as slow flow velocities.

After compression treatment, the common femoral artery (coded red) also shows a normal flow pattern at the puncture point.

Comment. With color flow duplex-sonographic guidance, targeted compression of the fistular canal and the aneurysmal cavity is possible. According to our own experience and communications we have received, compression therapy produces thrombosis of the false aneurysm in most patients. However, if an arteriovenous fistula is also present, this method is only rarely successful.

Case Figs. **27 a–d** ▷

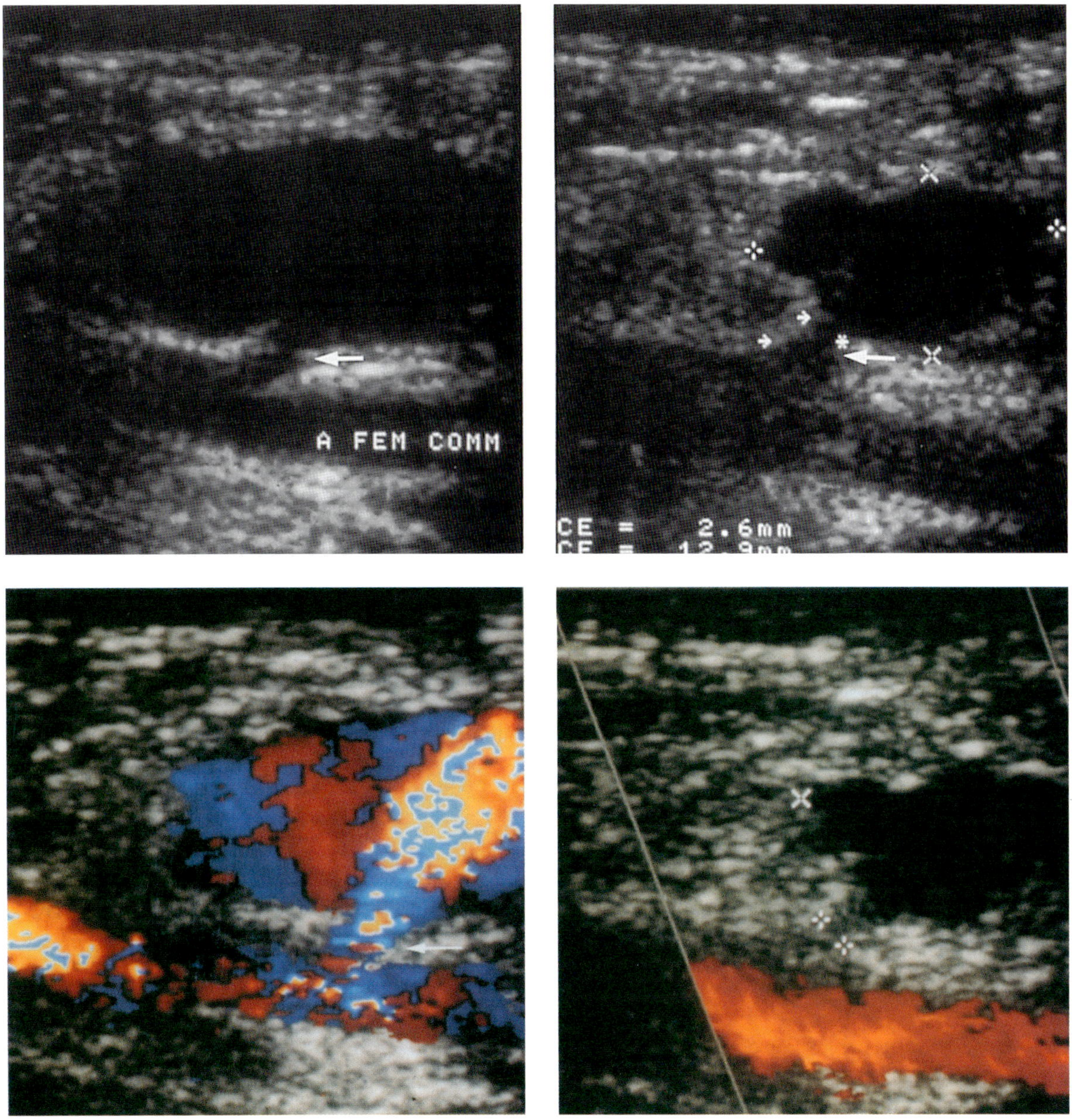

a c b d

Case Figs. **27 a–d**

28 Stenosis at a Bypass Insertion Point

Patient: W.U., male, age 45.
Clinical diagnosis: **PAOD stage IIb on the right, amputation at the left thigh,** suspicion of *right-sided* **stenoses in the pelvic vascular system and a stenosis at the bypass insertion point;** *on the left side,* occlusion at the origin of the iliac artery and occlusion at the origin of the femoral artery.

CAD:	+	Vascular surgery:	+	Diabetes mellitus:	–
PAOD:	+	Heredity:	+	Hyperlipidemia:	+
TIA, stroke:	–	Hypertension:	+	Smoking:	+

Case report. A laparotomy had been carried out 22 years earlier due to abdominal gunshot wounds, and a nephrectomy on the right followed somewhat later. The patient had been known to be suffering from PAOD for four years. Three years previously, an iliofemoral crossover bypass from left to right had been carried out, and four months later an amputation at the left thigh proved necessary. The year after that, a thromboendarterectomy of the left carotid vascular system and a femorocrural bypass of the right great saphenous vein followed. In the same year, a myocardial infarction occurred, and—with persistent pains at the stump remaining after amputation—a trial denudation of the arteries in the groin was conducted. Three months later, the stump was revised due to a suspicion of neurinoma, and a reamputation of the thigh was carried out. Two months previously, neurolysis had followed (femoral nerve and lateral cutaneous femoral nerve), and an arthroscopic operation on the lateral right meniscus was conducted. The patient was known to abuse alcohol.

In the region of the supra-aortal arteries, a high-grade stenosis of the left subclavian artery, with an ipsilateral occlusion of the vertebral artery, were detectable both clinically and with duplex sonography.

Examination results. A 45-year-old patient in a poor general state of health, with the following angiological findings:

	Pulse status		Auscultation	
	Right	Left	Right	Left
Abdomen			–	+
Common femoral artery	+	–	+	+
Superficial femoral artery		Amputation	+	Amputation
Popliteal artery	(+)		–	
Posterior tibial artery	+			
Dorsal artery of the foot	+			

Systolic pressure measurement (Doppler ultrasound technique; mmHg):

	Right	Left
Brachial artery	130	
Posterior tibial artery	130	Amputation
Anterior tibial artery	130	Amputation

Color flow duplex sonogram. In a sagittal longitudinal section of the right groin, the anastomotic region is seen, with a clear constriction of the lumen when in B-mode imaging (**a**).

Color flow Doppler sonography is able to image the high flow velocity in the stenosis and also displays (in shades of green) the different flow velocities that are simultaneously present intrastenotically, and here mainly poststenotically (**b**). The maximum systolic, end-diastolic, and time-standardized velocities were 306 cm/s, 93 cm/s, and 147 cm/s respectively. Poststenotically, in the dilated vascular segment, there is a flow jet (greenish–yellow), which is located mainly parallel to the vascular axis and secondary flow areas and has an eddy-like flow (dark blue–red).

Angiogram. Intra-arterial digital subtraction angiography (transaxillary access) shows the bypass stenosis on the right, with a poststenotic dilatation of the femorocrural bypass (**c**).

Comment. This patient, with similar systemic and ankle artery pressures, basically shows good compensation for the arterial obstruction. Planning appropriate treatment for the stenoses at the insertion point of a bypass has to take into account both the hemodynamic effect and also the possible progression of the stenosis, and the accompanying danger of a bypass occlusion, which would lead to a worse prognosis.

In the present case, the patient, who was not satisfactorily mobile due to significant fluctuations in the size of the stump, as well as an existing right-sided claudication, through compensatory additional exertion of the leg, declined a planned correction of the anastomotic stenosis.

Case Figs. **28 a–c** ▷

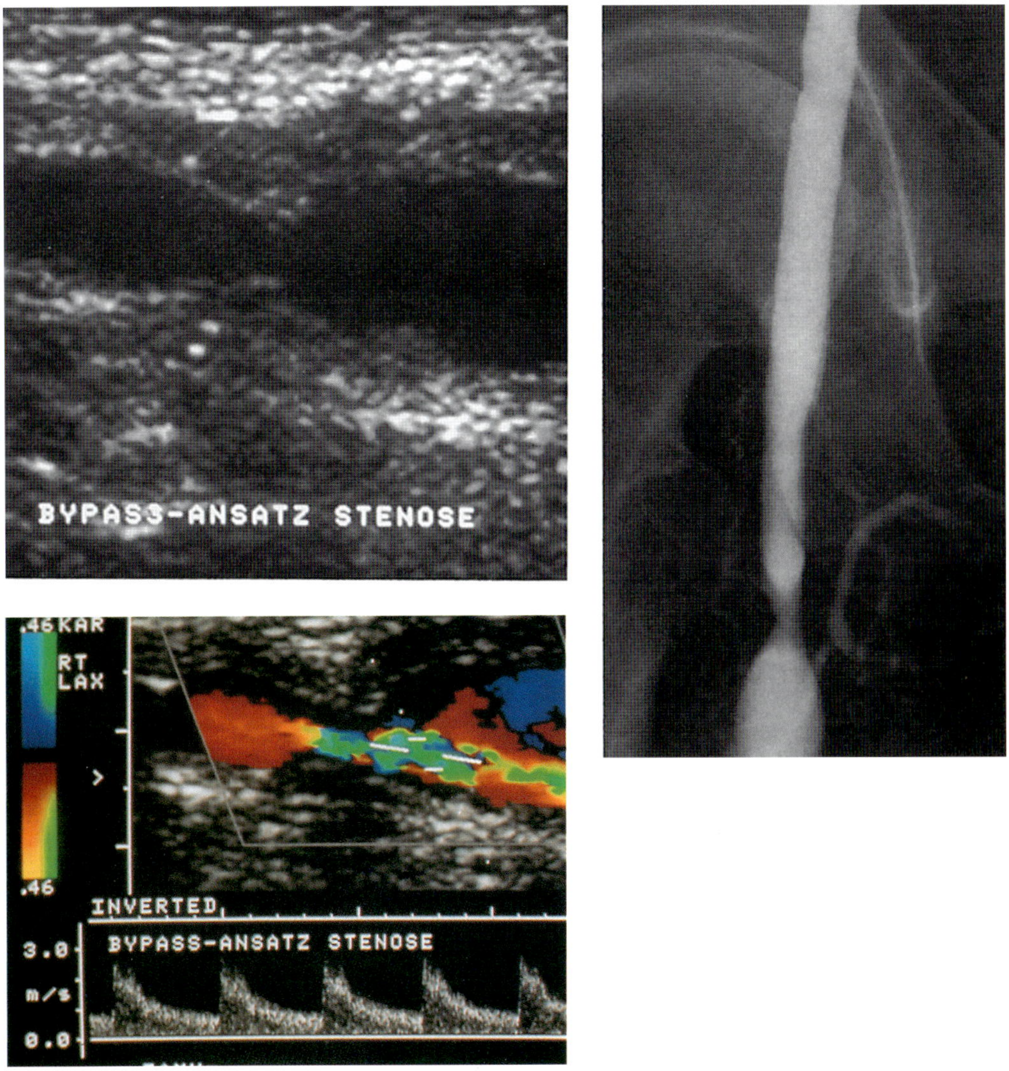

a
b
c

Case Figs. **28 a–c**

29 Stent Stenosis in the Superficial Femoral Artery (Percutaneous Transluminal Angioplasty)

Patient: D.J., male, age 53.
Clinical diagnosis: **PAOD stage IIa.** *Right:* **Stenosis of the femoral artery** following angioplasty of an occlusion of the superficial femoral artery, with stent implantation due to an early relapse. *Left:* stenosis of the femoral artery.

CAD:	–	PTA:	+	Diabetes mellitus:	+
PAOD:	+	Heredity:	–	Hyperlipidemia:	+
TIA, stroke:	–	Hypertension:	–	Smoking:	+

Case report. The patient had been suffering from claudication of the right calf for the previous four years. The symptoms had become worse approximately ten months earlier, and percutaneous transluminal angioplasty (PTA) of a distal occlusion in the right superficial femoral artery was carried out.

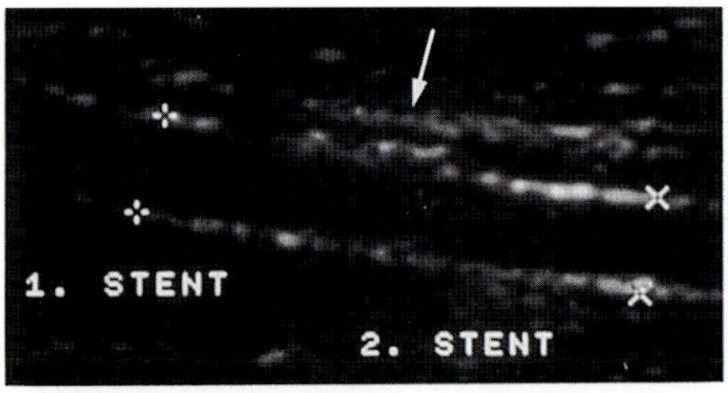

a

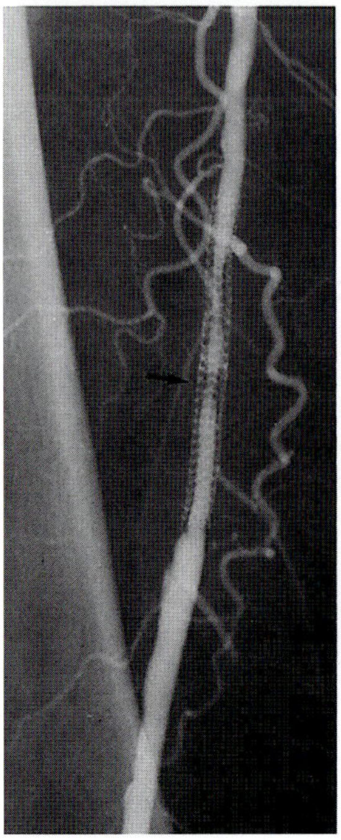

b

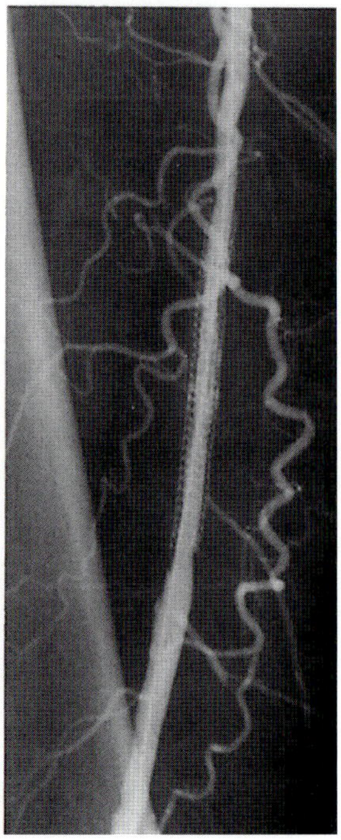

c

Reocclusion occurred the same day, and five days later, a renewed PTA was undertaken. Lysis was carried out with rt-PA (alteplase, 17.5 mg), and two overlapping stents were implanted, with a diameter of 5 mm and a length of 40 mm in each instance. The procedure led to good clinical success.

Four months later, right-sided calf claudication occurred again, and percutaneous transluminal angioplasty of a stenosis of the right superficial femoral artery in the stent region was carried out.

After a further six months, follow-up examinations with color flow duplex sonography detected a residual stenosis in the overlapping zone of the stents, which was again dilated using a balloon catheter. At the time of writing, the patient occasionally has pain and "paleness" in the region of the toes during longer walks.

Examination results:

	Pulse status		Auscultation	
	Right	Left	Right	Left
Abdomen			–	–
Common femoral artery	+	+	–	–
Superficial femoral artery			+	+
Popliteal artery	(+)	+	–	–
Posterior tibial artery	–	(+)		
Dorsal artery of the foot	(+)	+		

Systolic pressure measurement (Doppler ultrasound technique; mmHg):

	Right	Left
Brachial artery	120	
Posterior tibial artery	110	125
Anterior tibial artery	120	115

Color flow duplex sonogram. The illustrations show longitudinal sections through the right superficial femoral artery in its distal third, in the region of the implanted stents (**a, d–g**).

In B-mode imaging, the stents are depicted as hyperechoic structures within the vascular wall (**a**). The beginning of the distal stent (second stent) can be identified from a clearly recognizable thickening of the mesh (arrow). Intraluminal structures cannot be depicted with certainty.

Aliasing (white and light blue colors) in color flow Doppler sonography raised a suspicion of a hemodynamically effective stenosis in, or shortly distal to, the overlapping zone of the stents (**d**). In the Doppler frequency spectrum, there was an angle-corrected, elevated flow velocity in this region, which in the systole was maximally 323 cm/s, with a time-averaged maximal velocity of 89 cm/s (pulsatility index 3.84).

After percutaneous transluminal angioplasty (PTA), the maximum systolic flow velocity at the same location decreased to 189 cm/s (**e**). In B-mode imaging, the overlapping zone of the stents was again visualized by its structural thickening (arrow). With the color coding, a discretely elevated local flow velocity (light yellow) was still visible.

The outflow region of the distal stent also showed a higher average flow velocity in comparison with the subsequent (larger-caliber) vascular segment (**f**). The color in the stent segment here is lighter than in the native artery, and shows transitions to shades of yellow. The color reflects an increase in the luminal caliber of the artery distal to the stents, which is also evident in the angiograms (**b, c**).

Angiograms. These show the selective intra-arterial catheter angiography of the distal superficial femoral artery in the context of percutaneous transluminal angioplasty (PTA) of the stent stenosis (**b**). Slightly distal to the overlapping zone of the two stents, which had been implanted ten months previously, there is a concentric lumen constriction of the superficial femoral artery (arrow). Medial to the femoral artery, there is a corkscrew-like, winding collateral.

After angioplasty, the superficial femoral artery was patent (**c**), although the dilated stent region still showed wall irregularities. The course of the collateral vessels confirms that the projection is identical in both angiograms.

Comment. Two indications influenced the decision to use angioplasty in the recurring stenosis in the stent region. First, the patient presented both subjective and objective symptoms of peripheral arterial occlusive disease. A second, and no less important indication, was the prophylactic aspect of preventing a potential occlusion of the superficial femoral artery due to progression of the lumen constriction in the stent region. With regard to possible recanalization with angioplasty, an occlusion has a significantly poorer prognosis.

When the hemodynamic compensation for the vascular obstruction was estimated, pressure measurements at the right ankle arteries of 110 mmHg and 120 mmHg represented misleadingly high values, due to the probable media sclerosis caused by many years of diabetes mellitus. They did not reflect the true compensation or circulatory reserve.

In the clinical follow-up, the arterial vascular system in the thigh was still patent when examined with color flow duplex sonography 22 months after the last angioplasty of the stent stenoses on the right side. The only finding was a persistent locally elevated flow velocity in the overlapping region of the vascular prostheses, after PTA (**g**).

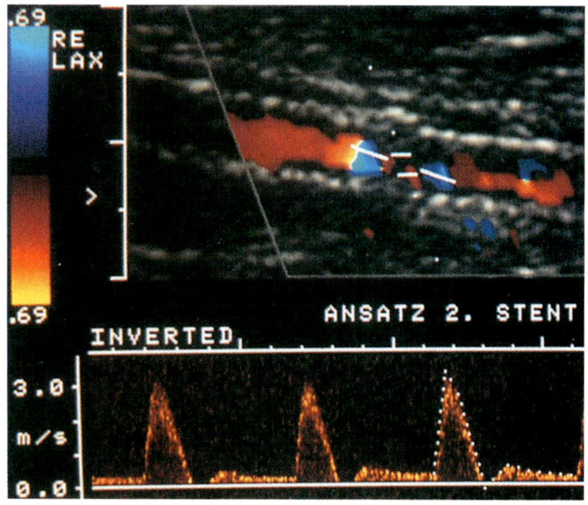

d

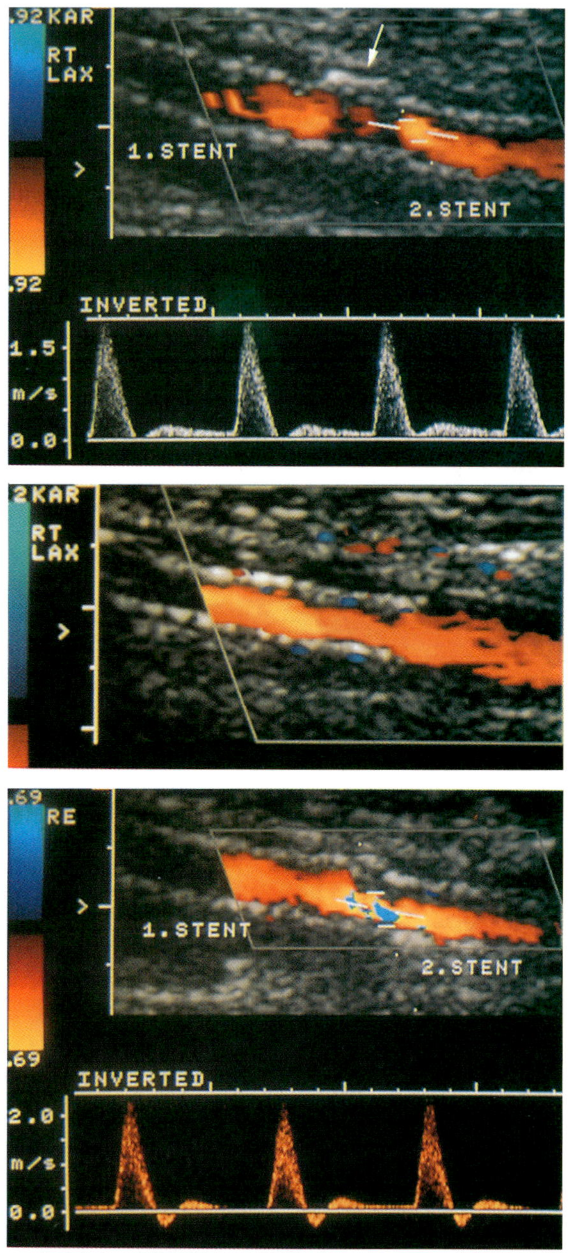

e

f

g

Case Figs. **29 e–g**

30 Bypass Rupture

Patient: E.-M.R., female, age 62.
Clinical diagnosis: **PAOD; restenosis of the right femoral artery** and a **bypass aneurysm on the left.**

CAD:	–	PTA:	+	Diabetes mellitus:	–
PAOD:	+	Heredity:	–	Hyperlipidemia:	+
TIA, stroke:	–	Hypertension:	+	Smoking:	+

Case report. Bilateral calf claudication had first occurred ten years previously. Five years after that, a thromboendarterectomy on the left and implantation of a femoropopliteal polytetrafluoroethylene (PTFE) bypass on the left was conducted, with good clinical success. Seven months previously, angioplasty of several stenoses of the right femoral artery was carried out, and a bypass rupture (aneurysm) was diagnosed with duplex sonography. The patient's current hospital admission was in order to implant a stent on the left.

Examination results. A pulsating, space-occupying lesion was observed in the middle of the left thigh.

	Pulse status Right	Pulse status Left	Auscultation Right	Auscultation Left
Abdomen			–	–
Common femoral artery	+	+	–	–
Superficial femoral artery			+	–
Popliteal artery	((+))	+	–	–
Posterior tibial artery	(+)	+		
Dorsal artery of the foot	(+)	+		

Systolic pressure measurement (Doppler ultrasound technique; mmHg):

	Right	Left
Brachial artery	130	
Posterior tibial artery	105	150
Anterior tibial artery	110	130

Angiogram. Using intra-arterial DSA, a bypass rupture on the left side can be seen, with the formation of a false aneurysm in the middle third of the thigh (**a**). In the same session, a stent was implanted in order to bridge the ruptured bypass in the aneurysmal dilatation.

Color flow duplex sonogram *after stent implantation.* In a *longitudinal section,* B-mode sonographic imaging can readily distinguish the false aneurysm, due to its stronger reflections (**b**). In *cross-section,* the vascular prosthesis can be seen in the middle of the aneurysmal cavity (**c**). The false aneurysm measured approximately 21 mm × 23 mm × 19 mm. The stent's sonographic diameter was approximately 6 mm.

In longitudinal section (**d**) and cross-section (**e**), the *color flow Doppler mode* demonstrates that blood is still flowing through the aneurysmal cavity, in addition to the stent. In the stent region, the flow is strong, while in the aneurysmal cavity there are slower components with eddy-like flow visible in the darker colors, with color changes passing through black.

Comment. As an alternative to immediate surgery, a therapeutic attempt to provide internal support and possibly cause the aneurysmal cavity to thrombose proved unsuccessful. It was ultimately necessary to resect the aneurysm and bridge the bypass using an end-to-end anastomosis with a polytetrafluoroethylene (PTFE) prosthesis.

With regard to the stenosis of the right femoral artery, conservative therapy in the form of interval training succeeded in increasing the patient's pain-free walking distance to over 1000 m.

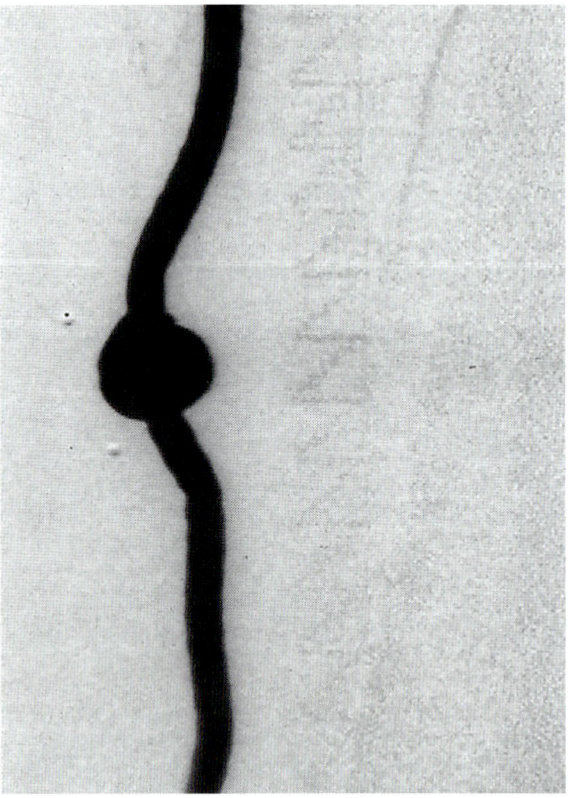

a

Case Figs. **30 b–e** ▷

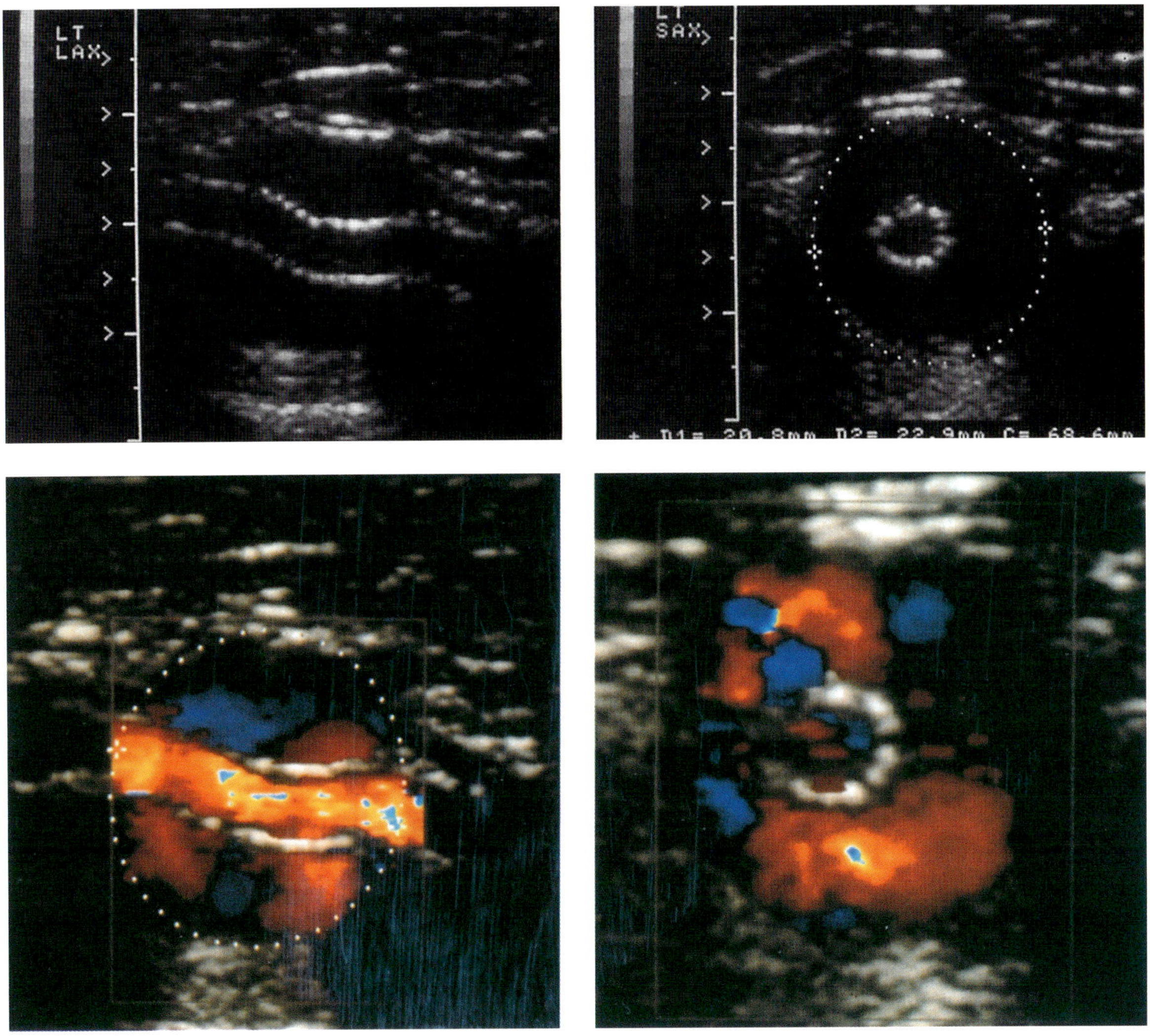

b c d e

Case Figs. **30 b–e**

31 Dissection of the Common Femoral Artery

Patient: R.H., male, age 45.
Clinical diagnosis: **PAOD bilateral stage IIa, stenosis of the right femoral artery,** and **stenosis of the left iliac artery.**

CAD:	–	PTA:	+	Diabetes mellitus:	–
PAOD:	+	Heredity:	+	Hyperlipidemia:	–
TIA, stroke:	–	Hypertension:	–	Smoking:	+

Case **report.** The patient had been suffering from calf claudication on the left over the previous three years. Two years previously, percutaneous transluminal angioplasty (PTA), initially of a stenosis in the left superficial femoral artery, was conducted, and two months after that, PTA of the ipsilateral external iliac artery. Initially, good clinical results were obtained in both instances. For the previous seven months, intermittent claudication had again been occurring in the region of the left calf, and three months after that also in the right calf. Currently, the chief complaint had been claudication of the right calf.

Examination results:

	Pulse status		Auscultation	
	Right	Left	Right	Left
Abdomen			–	+
Common femoral artery	+	(+)	–	+
Superficial femoral artery			+	–
Popliteal artery	(+)	(+)	–	–
Posterior tibial artery	(+)	(+)		
Dorsal artery of the foot	–	(+)		

Systolic pressure measurement (Doppler ultrasound technique; mmHg):

	Right	Left
Brachial artery	145	
Posterior tibial artery	120	115
Anterior tibial artery	–	105

Color flow duplex sonogram. *Left:* In a longitudinal section of the common femoral artery from a ventral orientation, with *poor delimitation in B-mode,* two hypoechoic pulsating structures are seen, which appear to be separated from one another by an oscillating membrane (arrows) (**a**).

Color flow Doppler sonography (**b**) shows two separate lumina with blood flowing through them. The flow velocity appears to be significantly elevated in the dorsal lumen; aliasing still occurs, although the color scale has been deliberately adjusted for high flow velocities (110 cm/s average velocity). Due to the color coding of the flow, inflow into the dissected lumen and reentry into the blood vessel can be recognized.

The registered *spectra* from both of the vascular channels (**c, d**) confirm both the slow and the fast flow velocity of around 100 cm/s systolic in the ventral lumen, and over 320 cm/s systolic in the dorsal lumen. This corresponds to a dissection with an overall hemodynamically effective constriction of the blood vessel.

Right: There is a high-grade effective stenosis of the superficial femoral artery at the transition zone from the proximal to the middle third of the blood vessel (not shown).

Comment. The clinical impression of the dissection was that of a moderate obstruction of the pelvic vascular system, most likely originating during the balloon dilatation of the left pelvic and leg vascular systems two years previously. There was currently no obligatory indication for surgical correction.

To alleviate the complaints, predominating on the right side, the ipsilateral stenosis of the femoral artery was dilated, allowing the patient's walking distance, at a rate of 120 steps/min, to increase to over 700 m.

Case Figs. **31 a–d** ▷

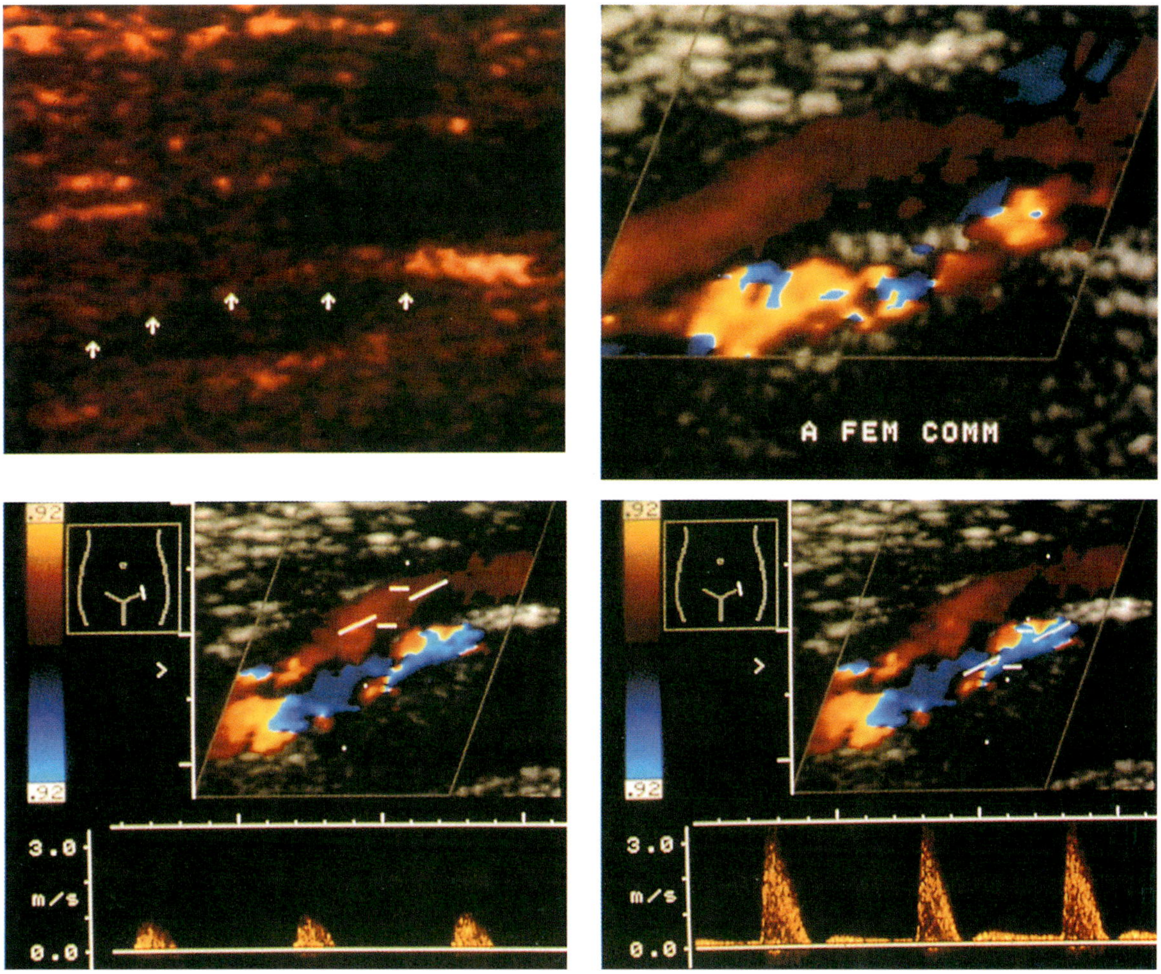

Case Figs. **31 a–d**

Arteries of the Lower Leg and Foot

32 Stenosis of the Tibiofibular Trunk (Percutaneous Transluminal Angioplasty)

Patient: H.R., male, age 66.
Clinical diagnosis: **PAOD bilateral stage IV,** suspicion of **low-grade bilateral stenoses of the superficial femoral artery and popliteal artery,** with **bilateral occlusions of the arteries in the lower leg.**

CAD:	+	Vascular surgery:	+	Diabetes mellitus:	–
PAOD:	+	Heredity:	–	Hyperlipidemia:	–
TIA, stroke:	–	Hypertension:	+	Smoking:	+

Case report. Ten years before, the right carotid artery was operated and five months previously, an asymptomatic occlusion of the ipsilateral common carotid artery occurred. An aortocoronary venous bypass (ACVB) was carried out three months previously, due to three-vessel disease of the coronary arteries. Occlusions of the arteries of the lower leg and toes had been occurring since approximately ten months before, and it was suspected that these were due to emboli associated with a dilatative and obliterative arteriopathy, in the presence a previously known saccular aneurysm of the abdominal aorta (70 mm × 41 mm × 39 mm).

Examination results. Edemas in the forefoot, which were more pronounced on the left than on the right, were seen. There were necroses of the tips of the toes, in the third digit on the right and on digits three and five on the left. There was inflammatory swelling and pain in the toes while at rest.

When the patient was admitted, the *pressure at the ankle arteries* could not be measured due to the edema in the feet. The brachial artery showed a pressure of 140 mmHg.

	Pulse status		Auscultation	
	Right	Left	Right	Left
Abdomen			–	–
Common femoral artery	+	+	+	+
Superficial femoral artery			+	+
Popliteal artery	+	+	+	+
Posterior tibial artery	–	–		
Dorsal artery of the foot	–	–		

Color flow duplex sonogram *before* (**a**) and *after* (**c**) percutaneous transluminal angioplasty. The illustrations show sagittal longitudinal sections from a dorsal orientation, which are adapted to the vascular axes in the region of the proximal lower leg, and the tibiofibular trunk.

Before angioplasty, color flow Doppler sonography showed aliasing in the trunk (**a**). The presence of a high-grade, hemodynamically effective stenosis, with maximum systolic and end-diastolic flow velocities of 338 cm/s and 43 cm/s, respectively, was confirmed.

After angioplasty, and using the *same* Doppler angle of 45° (!) with a still monophasic frequency spectrum, there was a clear decrease in the maximum systolic and end-diastolic flow velocities in the tibiofibular trunk to 83 cm/s and 11 cm/s, respectively (**c**). With a lower color scale, with a maximum of the average velocity at 34 cm/s compared to 69 cm/s, aliasing in the recanalized vascular segment still persisted.

Angiograms, before (**b**) and after (**d**) angioplasty. With selective application of an angiographic catheter, imaging (**b**) and simultaneous dilatation of the stenosis identified by duplex sonography in the tibiofibular trunk (arrow) was possible.

The second angiogram shows a satisfactory result, with clear lesions of the intima (**d**), and a persistent occlusion of the posterior and anterior tibial artery, both with peripheral refilling. There is also a stenosis of the fibular artery.

Comment. After angioplasty and conservative therapeutic measures (local treatment, antibiotics), the lesion in the third digit on the left healed. At the fifth toe, an amputation of the terminal joint was followed by satisfactory wound healing.

Five months after the above catheter intervention, the patency of the tibiofibular trunk was demonstrated during an angioplasty carried out due to fresh stenoses of the left superficial femoral artery.

Case Figs. **32 a–d** ▷

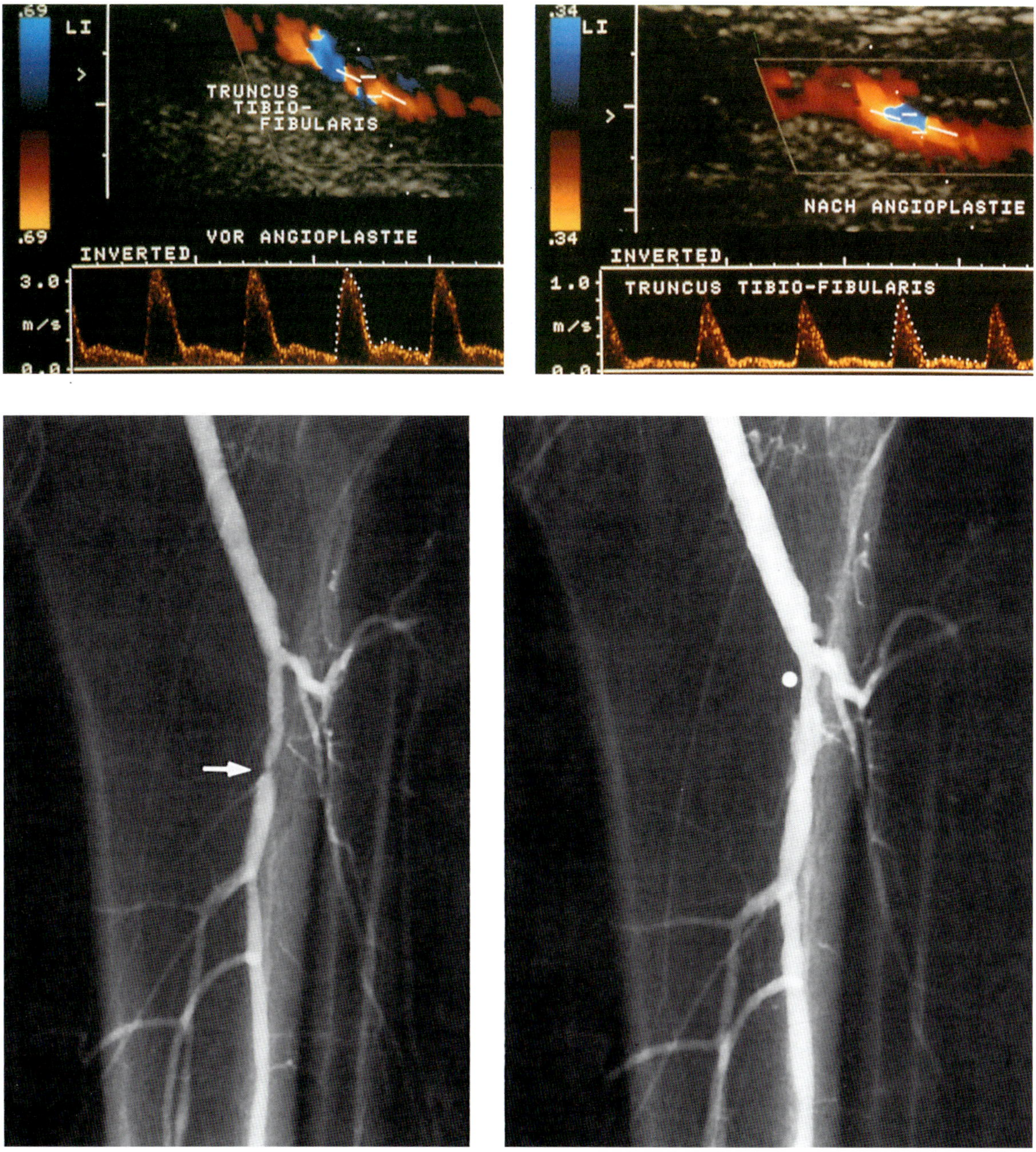

a b c d

Case Figs. **32 a–d**

33 Occlusions of the Lower Leg Arteries

Patient: H.M., male, age 51.
Clinical diagnosis: **PAOD** of the *acral* type, **occlusions of the lower leg arteries** (right anterior tibial artery and left anterior and posterior tibial artery).

CAD:	+	PTA:	–	Diabetes mellitus:	–
PAOD:	+	Heredity:	–	Hyperlipidemia:	+
TIA, stroke:	–	Hypertension:	–	Smoking:	+

Case report. For the previous five years, the patient had experienced exercise-dependent pain and paresthesia in the left foot. Prescribed foot supports had not led to any improvement in this condition.

Examination results. During Ratschow's test, the *left* forefoot and heel became pale, and when compared to the contralateral side showed delayed and excessive hyperemia.

	Pulse status		Auscultation	
	Right	Left	Right	Left
Abdomen			–	–
Common femoral artery	+	+	–	–
Superficial femoral artery			–	–
Popliteal artery	+	+	–	–
Posterior tibial artery	+	+		
Dorsal artery of the foot	–	–		

Systolic pressure measurement (Doppler ultrasound technique; mmHg):

	Right	Left
Brachial artery	135	
Posterior tibial artery	150	140
Anterior tibial artery	140	130

Angiogram. Intra-arterial needle angiography on the *left side* showed a normal image of the pelvic and thigh arteries. Stenoses of the posterior tibial artery are seen (upper arrow), and there is an occlusion of this vessel proximal to the talocrural articulation (**a**). There is strong refilling of the plantar artery of the foot and an occlusion of the anterior tibial artery at the level of the ankle.

Color flow duplex sonogram. The *left* posterior tibial artery (arrow in **a**) has a stenosis in its distal segment, showing local aliasing and a maximum systolic flow velocity of 103 cm/s; the pulsatility index was 2.8 (**b**). Distally, there was an occlusion of the blood vessel, with a hypoechoic image of the occluded lumen (arrows), and a collateral blood vessel given off preocclusively (**c**).

Comment. Particularly when atherogenic risk factors are present, the appearance of exercise-dependent pains in the sole of the foot should prompt consideration of a peripherally located obstruction of the arterial vascular system. A diagnosis indicating occlusions of the arteries in the distal lower leg or the foot rarely requires recanalization techniques. However, it is decisively important in relation to prognostic and prophylactic aspects (avoiding external noxa).

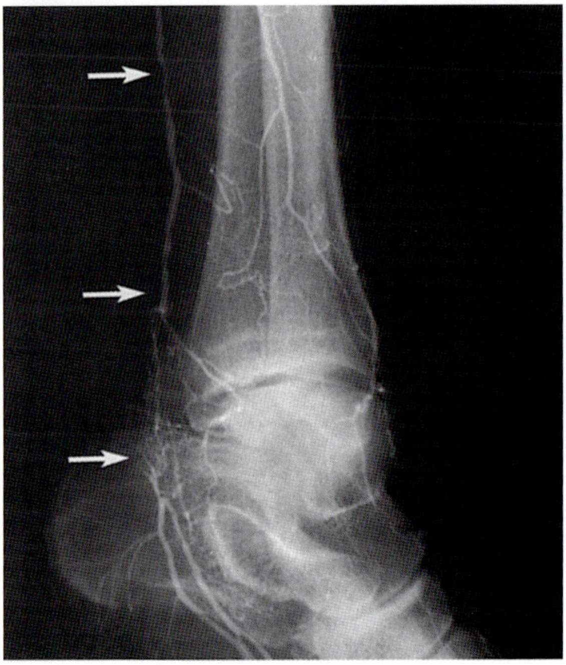

a

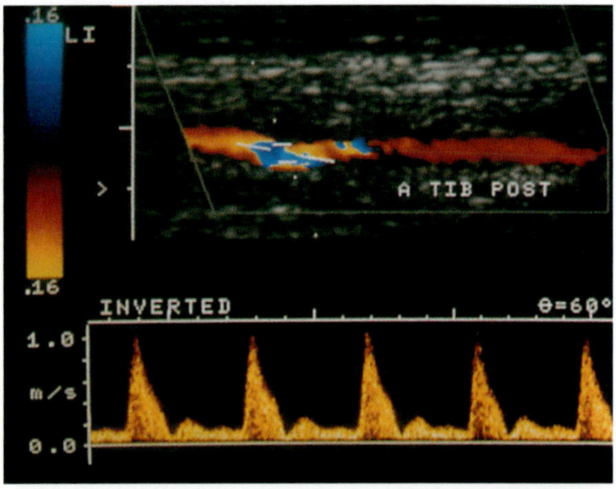

b

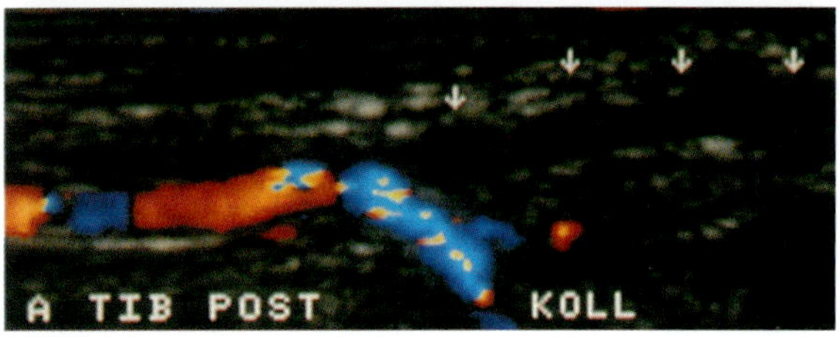

c

Shoulder and Arm Arteries

34 Stenosis of the Brachial Artery

Patient: P.F., male, age 57.
Clinical diagnosis: **PAOD** with **stenosis of the right brachial artery** and **bilateral occlusions of the ulnar artery.** Status after amputation at the right thigh, with occlusion of the iliac and femoral arteries; stenosis of the popliteal artery and occlusions of the arteries in the left lower leg. Aneurysm of the abdominal aorta.

CAD:	–	PTA:	+	Diabetes mellitus:	–
PAOD:	+	Heredity:	–	Hyperlipidemia:	+
TIA, stroke:	–	Hypertension:	–	Smoking:	+

Case report. Twenty-two years previously, calf claudication on the left had been accompanied by pallor of the toes when in a flat position. The diagnosis made at that time was occlusion of the left popliteal artery and bilateral occlusions of the lower leg arteries.

Since then, several percutaneous transluminal catheter treatments and recanalization attempts involving the leg arteries were conducted bilaterally. A lumbar sympathectomy on the left was completed 21 years ago. 4 years later, a right-sided lumbar and thoracic sympathectomy also followed. 1 year thereafter, an amputation at the level of the right thigh occurred.

Two years previously, a partially thrombosed infrarenal aneurysm of the abdominal aorta (70 mm × 35 mm × 30 mm), with a lumen of approximately 14 mm in its transverse diameter and with blood still flowing through it, had been detected sonographically.

There was no anamnestic arm claudication in the patient history.

Examination results:

	Pulse status Right	Pulse status Left	Auscultation Right	Auscultation Left
Common carotid artery	+	+	–	–
Superior temporal artery	+	+		
Facial artery	+	+		
Subclavian artery	+	+	–	–
Distal brachial artery	((+))	+	+	–
Radial artery	(+)	(+)		
Ulnar artery	–	–		
Abdomen	Broad pulse		–	–
Common femoral artery	+	+	–	–
Superficial femoral artery	Amputation		Amputation	–
Popliteal artery		+		+
Posterior tibial artery		–		+
Dorsal artery of the foot		–		

Systolic pressure measurement (Doppler ultrasound technique; mmHg):

	Right	Left
Brachial artery	115	150
Ulnar artery	80	110
Radial artery	85	140
Posterior tibial artery	Amputation	–
Anterior tibial artery	Amputation	90

Angiogram. In selective angiography of the brachial artery with hand enlargement, clear fluctuations in the lumen of the brachial artery are seen in its middle third. Stenosed and aneurysmally dilated segments are seen (**a**). In the subsequent course, there was an occlusion of the ulnar artery and the radial artery, each in the distal third of the lower arm, with peripheral refilling of the radial artery and occlusions of the digital arteries in the fourth and fifth digits, ulnar and radial (not shown).

Color flow duplex sonogram. In the *B-mode* image, a longitudinal section through the middle segment of the brachial artery shows the distension of the vascular lumen, with the vascular wall not continuously distinguishable (**b**).

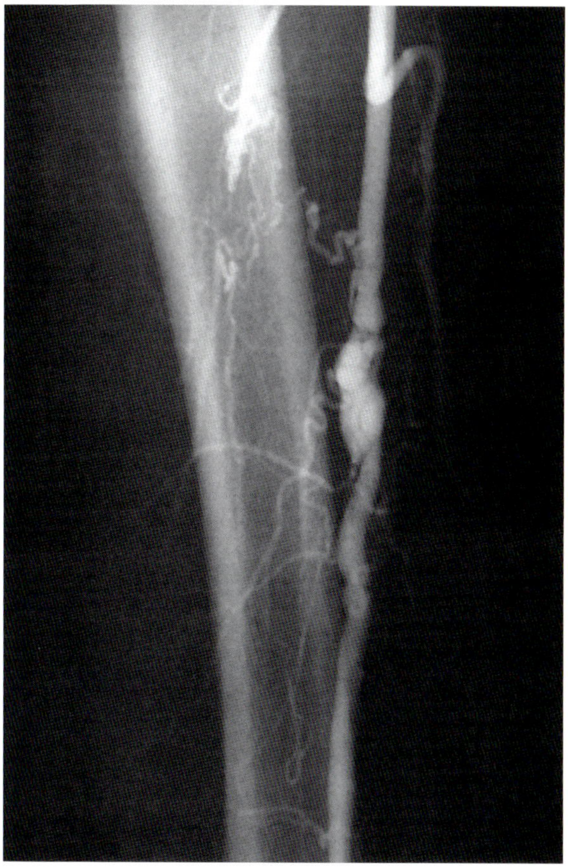

a

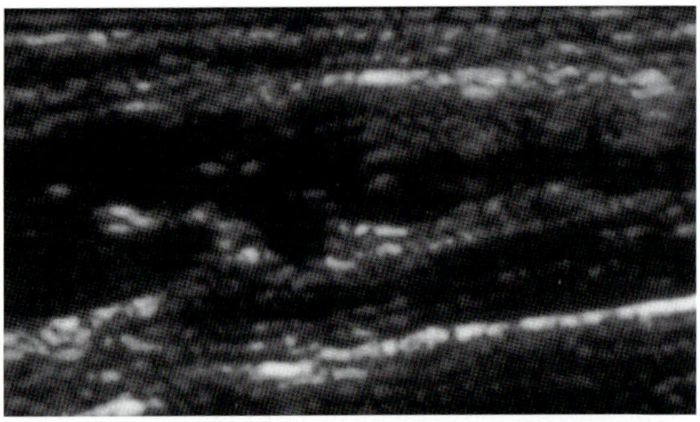

b

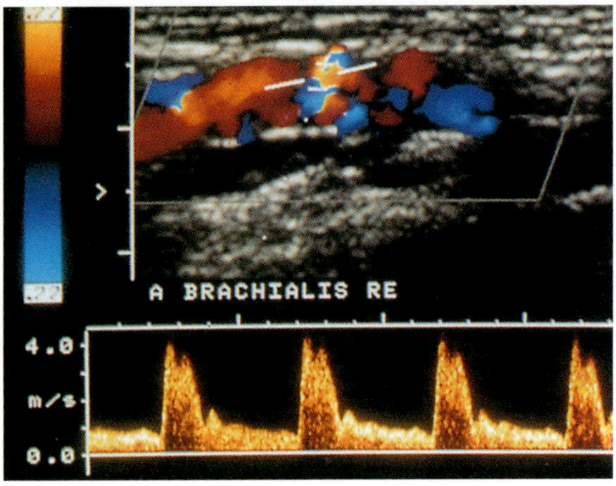

c

The lighter colors and aliasing at a high velocity scale indicate a stenosis in the *color flow Doppler sonogram* (**c**). The blood flow in the aneurysmal sac shows disturbed flow.

With the monophasically changed *Doppler frequency spectrum,* the clearly elevated systolic (411 cm/s) and end-diastolic (94 cm/s) flow velocities can be calculated.

Comment. In this patient, dilatative/obliterative arteriopathy can be assumed in addition to a previously known occurrence of obliterating endangiitis, and which has also led to the stenotic and aneurysmal phenomena in the region of the right brachial artery. The obstruction in the vascular system of the right arm was discovered simply through a routinely administered side-to-side comparison of the measured blood pressure. This led to the further diagnostic procedures. The aneurysm of the abdominal aorta had already been known of for the previous two years.

In principle, the arteries of the upper extremity are at least as readily accessible to examination using (color flow) duplex sonography as the arterial vascular system in the pelvis and the leg. However, occlusive arterial disease in the region of the arm arteries is a very much less frequent occurrence.

35 Occlusions of the Brachial Artery and Arteries of the Lower Arm (Neurovascular Compression Syndrome)

Patient: S.W., female, age 25.
Clinical diagnosis: **low-grade stenosis of the subclavian artery, proximal occlusions of the lower arm arteries on the right,** with suspicion of a **thoracic outlet syndrome,** suspicion of bilateral neck ribs.

CAD:	–	PTA:	–	Diabetes mellitus:	–
PAOD:	–	Heredity:	–	Hyperlipidemia:	–
TIA, stroke:	–	Hypertension:	–	Smoking:	+

Case report. The past medical history showed no significant prior illnesses. Three weeks previously, while lifting a heavy load (moving house), a tearing pain suddenly appeared in the right lower arm.

One week later, *pains in the right small finger and the hypothenar* appeared on slight exertion of the lower arm (rapid typing on a keyboard). The or-thopedist providing treatment diagnosed a capsule irritation and prescribed *cryotherapy* and dressings with ointment, which did not produce lasting mitigation of the pain. Only ice application briefly alleviated the pain during arm exertion.

The symptoms were then classified as tendovaginitis, and *vibration massage treatment of the connective tissue* of the lower arm with ultrasound was initiated. Subsequently, there were also pains in the thumb region on exertion. The patient did not notice any white or blue discoloration of the fingers.

In addition to pain while using the arm, a feeling of numbness and coldness had appeared for the first time in the area of the entire hand three days before presentation in our clinic.

Examination results. A 25-year-old patient with good nutritional status and in a good general state of health. The heart and lungs were normal on percussion and auscultation, and there were no cardiac arrhythmias. The blood pressure on the right was 100/80, and 120/85 mmHg on the left. The following angiological findings were obtained:

	Pulse status		Auscultation	
	Right	Left	Right	Left
Common carotid artery	+	+	–	–
Superior temporal artery	+	+		
Facial artery	+	+		
Subclavian artery	(+	+	+	–
Distal brachial artery	(+	+	–	–
Radial artery	–	+		
Ulnar artery	((+))	+		

The *Allen test* on the right proved to be pathological.

Bilaterally in the region of the supraclavicular fossa, a bony structure with a connection to the spinal column was palpated in the slim patient.

The *mechanical oscillogram* at rest showed no oscillations above the right lower arm, with normal findings on the left.

Measuring the pressure at the finger (strain gauge) yielded the following values (in mmHg):

In the lower extremities, the pulse status and auscultation findings were normal.

	Right	Left
Brachial artery	100	120
Digit II	33	120
Digit III	50	100
Digit IV	33	100

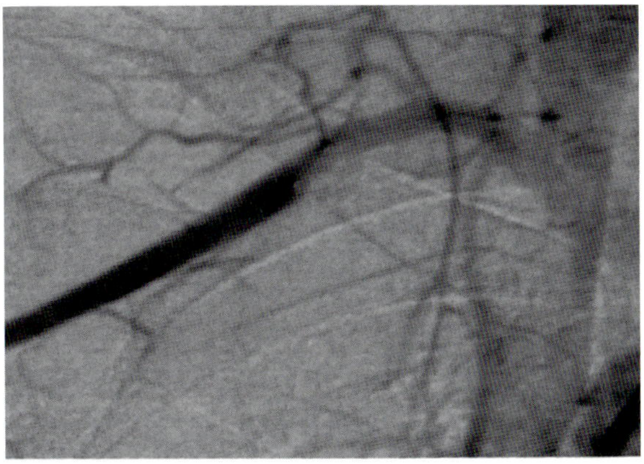
a

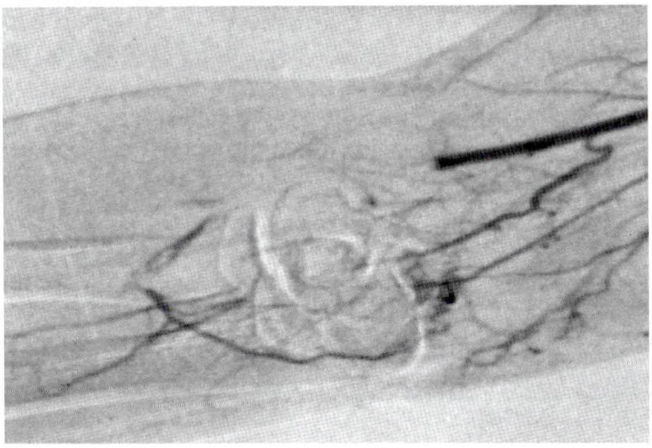
b

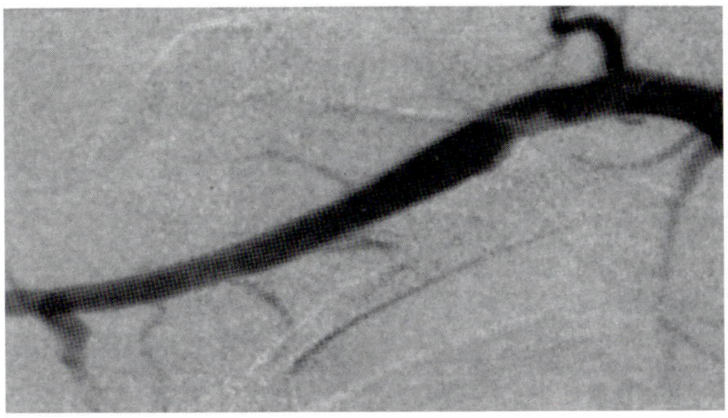
c

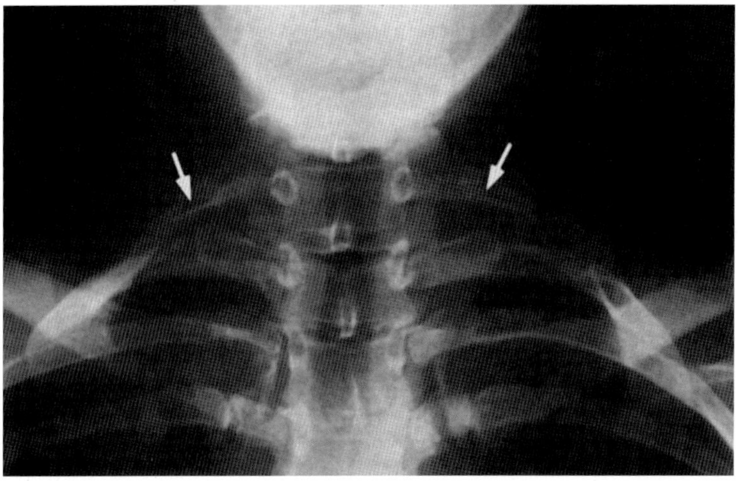

d

Case Figs. **35 e–i** ▷

Angiograms. Selective catheterization with intra-arterial digital subtraction angiography (DSA) shows the subclavian artery, axillary artery, and brachial artery (**a**). In the region of the subclavian artery, there is a poorly delimited stenosis with irregular wall changes.

The brachial artery is occluded approximately at the height of the elbow joint. There are collaterals poorly refilling the proximal radial artery, ulnar artery, and interosseous arteries (**b**). There are subsequent occlusions of the radial artery and ulnar artery, with only the distal radial artery being refilled.

After lysis with rt-PA (alteplase, 13 mg) *using a catheter,* a clear decrease in the stenosis of the subclavian artery, with persistent wall deposits, is seen. The occlusion in the brachial artery has decreased in length, although complete patency of the vessel segment has not been achieved (**c**).

Radiograph of the upper thoracic aperture. Regular transparency and a typical bone structure of the skeletal components depicted, with normal soft tissues, are seen (**d**). The transverse processes of the cervical spine appear normal up to C6.

At C7, there are approximately 5 cm–long *neck ribs* (arrows). The uncinate processes are normal.

Color flow duplex sonogram. With regard to the clinical diagnosis of proximal occlusions of the arteries in the lower arm, color flow duplex sonography diagnosed an occlusion of the distal brachial artery (not shown), which was followed by selective lysis using an angiographic catheter. The results of this lysis treatment, monitored using color flow duplex sonography, were:

After selective lysis with an angiographic catheter, a hemodynamically patent subclavian artery (**e**) and brachial artery (**f**) were seen.

Compared to the result obtained immediately after lysis with the catheter (**c**), an examination with color flow duplex sonography one day after lysis, due to a delayed effect of the fibrinolytic process, shows the bifurcation at the lower arm with a patent radial artery and ulnar artery (**g**). The elbow joint is seen at the lower edge of the image.

The ulnar artery then becomes occluded approximately 6 cm after its origin (**h**). The vascular occlusion can be recognized from the interruption in the color; the vascular structures (wall and internal lumen components) are clearly visible. The preocclusive Doppler frequency spectrum shows a flow pattern corresponding to high peripheral resistance. Distally, peripheral refilling of the artery (not shown) was observed with duplex sonography.

The patent radial artery can be followed as far as the joints of the hand (**i**). The Doppler frequency spectrum indicates lower peripheral resistance in the artery's area of distribution.

Comment. The past medical history, suggesting a progressive set of symptoms developing in phases, leads one to suspect that recurrent emboli from the stenosis of the subclavian artery were passing into the arteries of the lower arm. Heavy lifting three weeks earlier might have caused trauma to the subclavian artery, with a parietal thrombus that scattered over the following period.

The successful selective lysis with the angiographic catheter supports the suspicion that the illness, at least the occlusive symptoms, began in the relatively recent past. It can be assumed that there had already been long-term (micro-)traumatization of the subclavian artery due to the neck ribs.

The patient was treated by exarticulation of the right first rib and the neck rib from a transaxillary access point, in order to remove the cause of the traumatic damage to the subclavian artery. Macroscopically, the neck rib and first rib showed a synostotic connection, with a club-like protrusion pointing in a ventral direction.

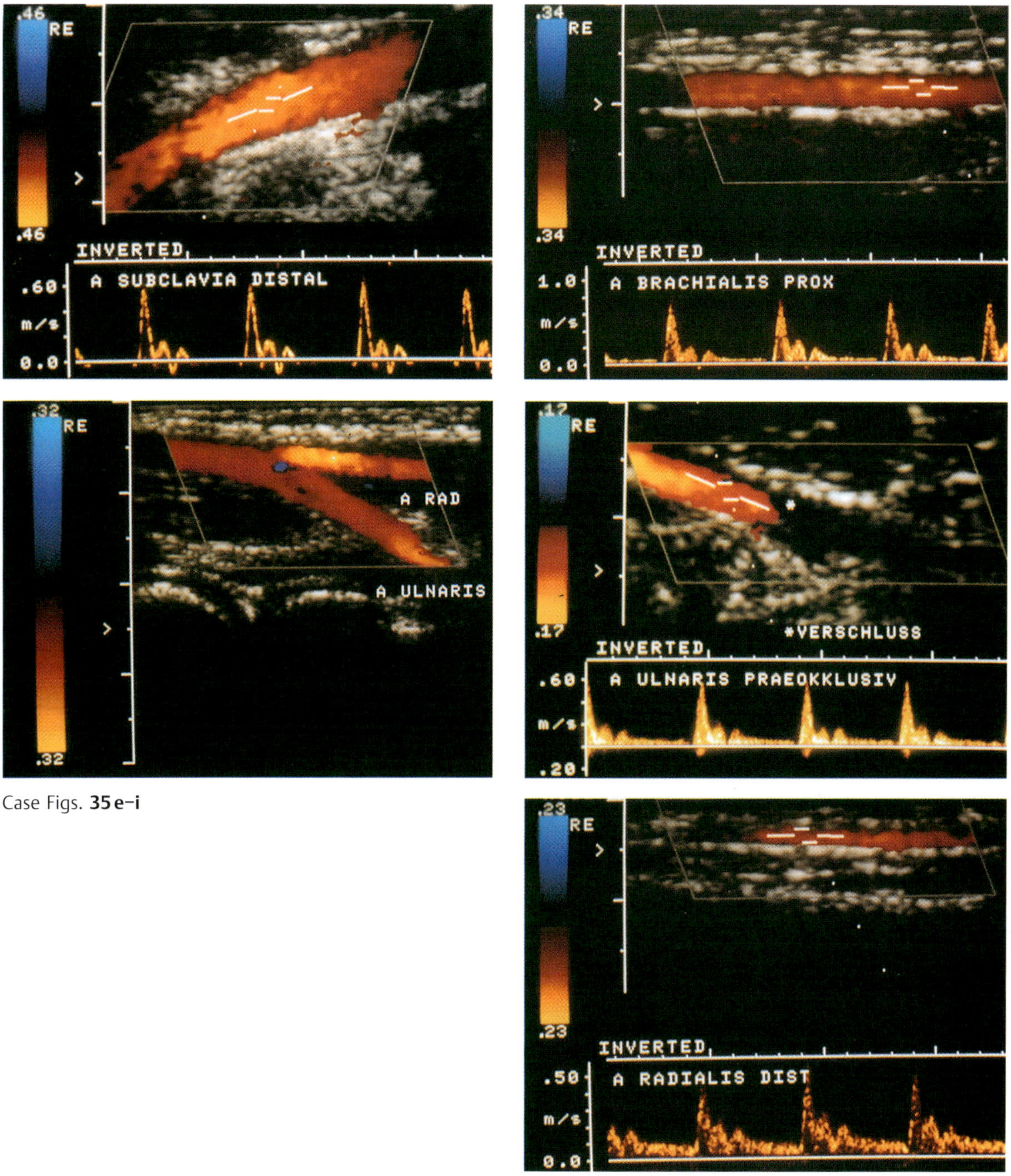

Case Figs. **35 e–i**

36 Occlusions of the Digital Arteries

Patient: H.M., male, age 47. Clinical diagnosis: **occlusions of the digital arteries**
in the feet and the right hand, with an atypical rheumatoid arthritis as an underlying condition.

CAD:	–	Vascular surgery:	–	Diabetes mellitus:	–
PAOD:	–	Heredity:	+	Hyperlipidemia:	–
TIA, stroke:	–	Hypertension:	–	Smoking:	+

Case report. Since 1971, the patient had had painful swelling in the region of the right talocalcaneal joint, later also involving the right joints of the hand. From 1975, the joints of the left hand were also affected. In 1982, a spontaneous ulcer of the lower leg appeared in the pretibial region on the right. In 1985, livid discolorations of the first and second right toes occurred. Gangrene developed. Since 1986, there had been a blue discoloration of the fourth right finger.

Laboratory results indicated the presence of a highly inflammatory process. Rheumatoid factors were negative. An atypical rheumatoid arthritis was diagnosed. Vascular occlusions were confirmed in 1987.

Examination results:

	Pulse status Right	Pulse status Left	Auscultation Right	Auscultation Left
Subclavian artery	+	+	–	–
Brachial artery	+	+		
Radial artery	+	+		
Ulnar artery	+	+		

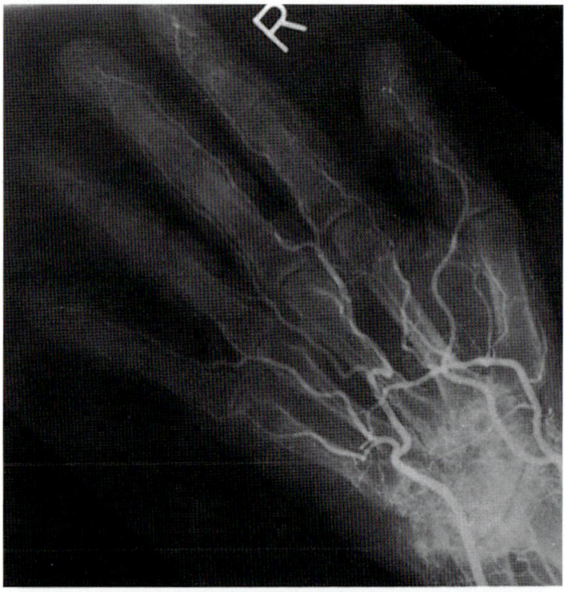

a

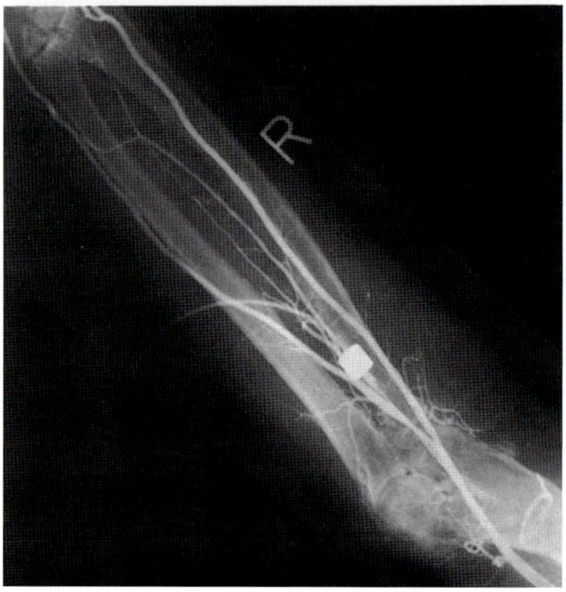

b

Doppler sonogram *(analogue registration)* (**c**). The Doppler sonograms of the radial artery and ulnar artery show a high diastolic resting flow rate and an absent reverse flow component. This is caused by the "warm hand" phenomenon, and partly by inflammation and the resultant peripheral hyperperfusion. The digital arteries could only be located on the radial side of the second, third, and fifth fingers.

Angiograms (**a, b**). *Angiography of the right brachial artery:* There are occlusions of the radial digital artery of the second digit, and also of the ulnar digital artery of the fifth digit. Stenoses of both the radial and the ulnar digital arteries of the fourth digit were noted. There was significant arthrosis in the metacarpal region.

Comment. Subsequent to an underlying case of atypical rheumatoid arthritis, the patient developed an acral occlusive disease, leading the ulcer formation in the feet and hand. A suspicion of crystalline emboli was excluded by examining a resected piece of skin from an ulcer on the tibia. Pressure measurement of the digital arteries also confirmed reduced pressures on the right. The blue, discolored finger (fourth digit), in particular, had a pressure of only 50 mmHg, indicating that both digital arteries were affected. The patient received sympatholytics and an antibiotic. With this therapy and careful treatment of the wound, remission of the necroses was achieved.

Case Figs. **36c** ▷

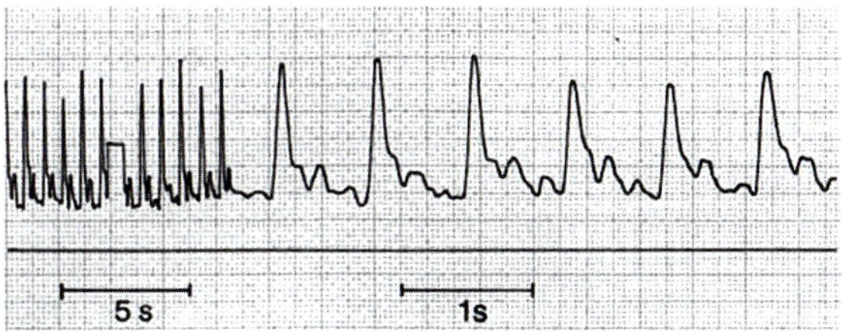

Right radial artery

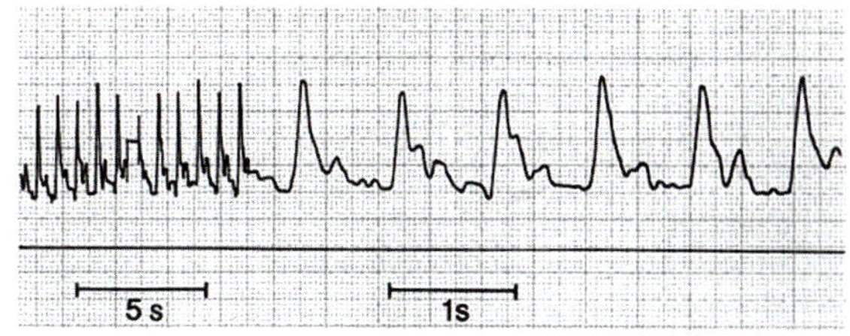

Right ulnar artery

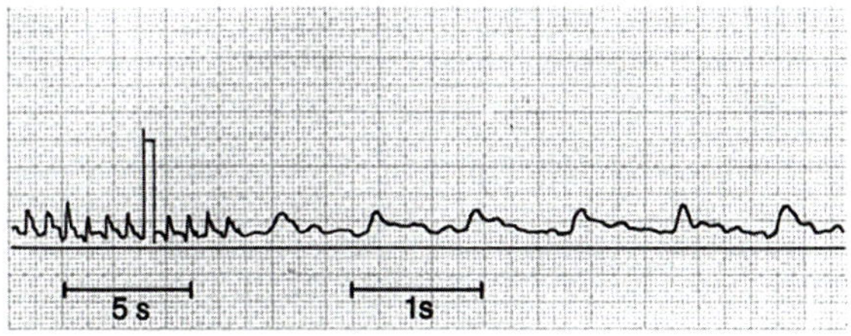

Digital artery, second right digit, radial side

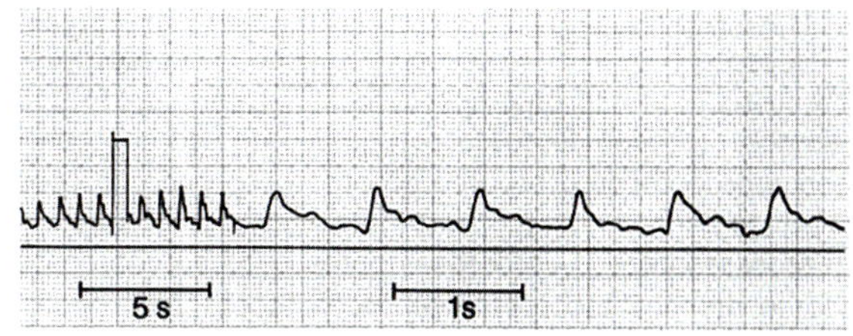

Digital artery, third right digit, radial side

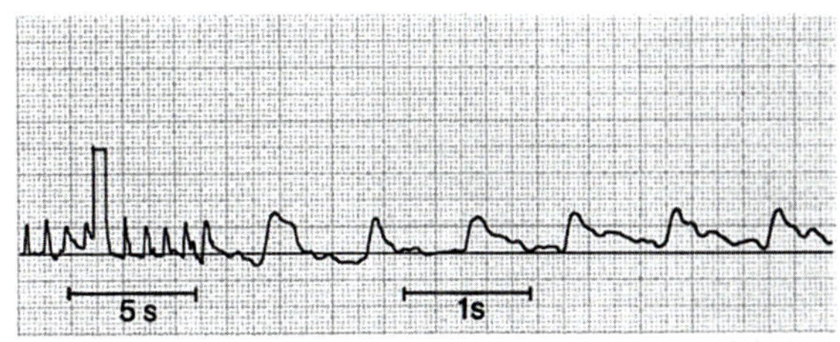

Digital artery, fifth right digit, radial side

c CW Doppler sonograms

Case Figs. **36c**

Systolic pressure measurement (Doppler ultrasound and strain gauge technique; mmHg)

Cuff position	Right radial artery	Left radial artery
Upper arm	115	120
Lower arm	130	135

Cuff position	Right radial artery	Left radial artery
2nd basal joint of the finger	90	90
3rd basal joint of the finger	90	110
4th basal joint of the finger	50	110
5th basal joint of the finger	110	115

Peripheral Veins

Thrombotic Processes

37 Iliac Vein Thrombosis

Patient: P.S., male, age 50.
Clinical diagnosis: **postthrombotic syndrome of the left extremity.**

Case report. In 1978, the patient had suffered a deep venous thrombosis on the left following a meniscus operation. An ulcer of the lower leg developed in 1979, which opened repeatedly. A venous exeresis at the left leg was carried out. Currently, there was hyperpigmentation of the left lower leg and an induration of the calf. The patient presented with status after a healed ulcer of the lower leg. He was substantially obese (107 kg with a height of 169 cm). Treatment was given with intermittent compression and fast compression bandages.

Examination results. The left lower leg showed hyperpigmentation, induration, and ulcer scars. There are also scars present due to venous stripping.

Doppler sonogram *(analogue registration)* (**c**): *Right:* There is spontaneous modulation of the respiration-dependent venous wave in the groin, and reflux during forced inspiration and during the Valsalva maneuver. There are pronounced flow peaks (A sounds) on compression in the thigh. *Left:* Continuous flow is seen during normal respiration with slight modulation during deepened respiration and a cessation of flow during the Valsalva maneuver. When the thigh is compressed, there are no flow peaks in the direction of the heart (central), due to the proximal flow obstruction. On decompression, there is suction in a peripheral direction due to the pressure gradient associated with valvular insufficiency.

Phlebogram (**a, b**). Ascending leg phlebography was carried out bilaterally.

Right: The deep venous system is intact, both in the pelvis and the leg, and there is no indication of a vari-cosis. *Left:* The image shows the status after a thrombosis of the proximal and middle superficial femoral vein, with poor recanalization, as well as a thrombosis of the external iliac vein, which is also poorly recanal-ized. The common femoral vein is patent.

Comment. As a consequence of the venous thrombosis, the patient has developed a marked postthrombotic syndrome on the left side. The Doppler-sonographic findings, showing continuous flow above the left groin, reflect the elevated pressure in the venous system of the left leg.

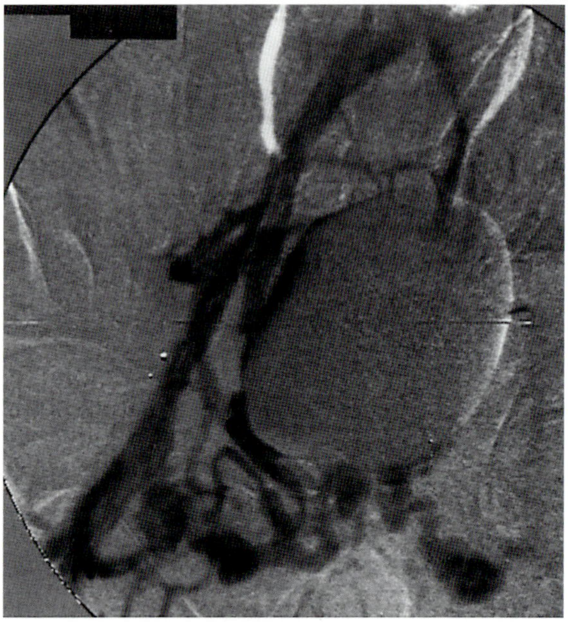

a

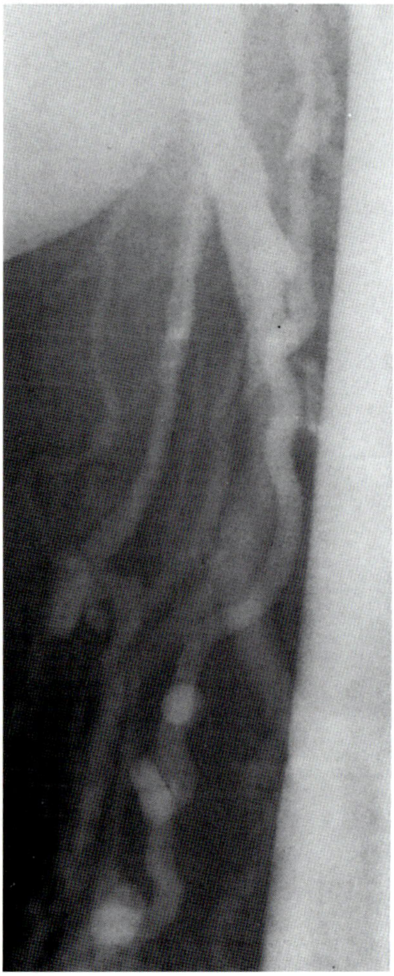

b

Case Figs. **37 c** ▷

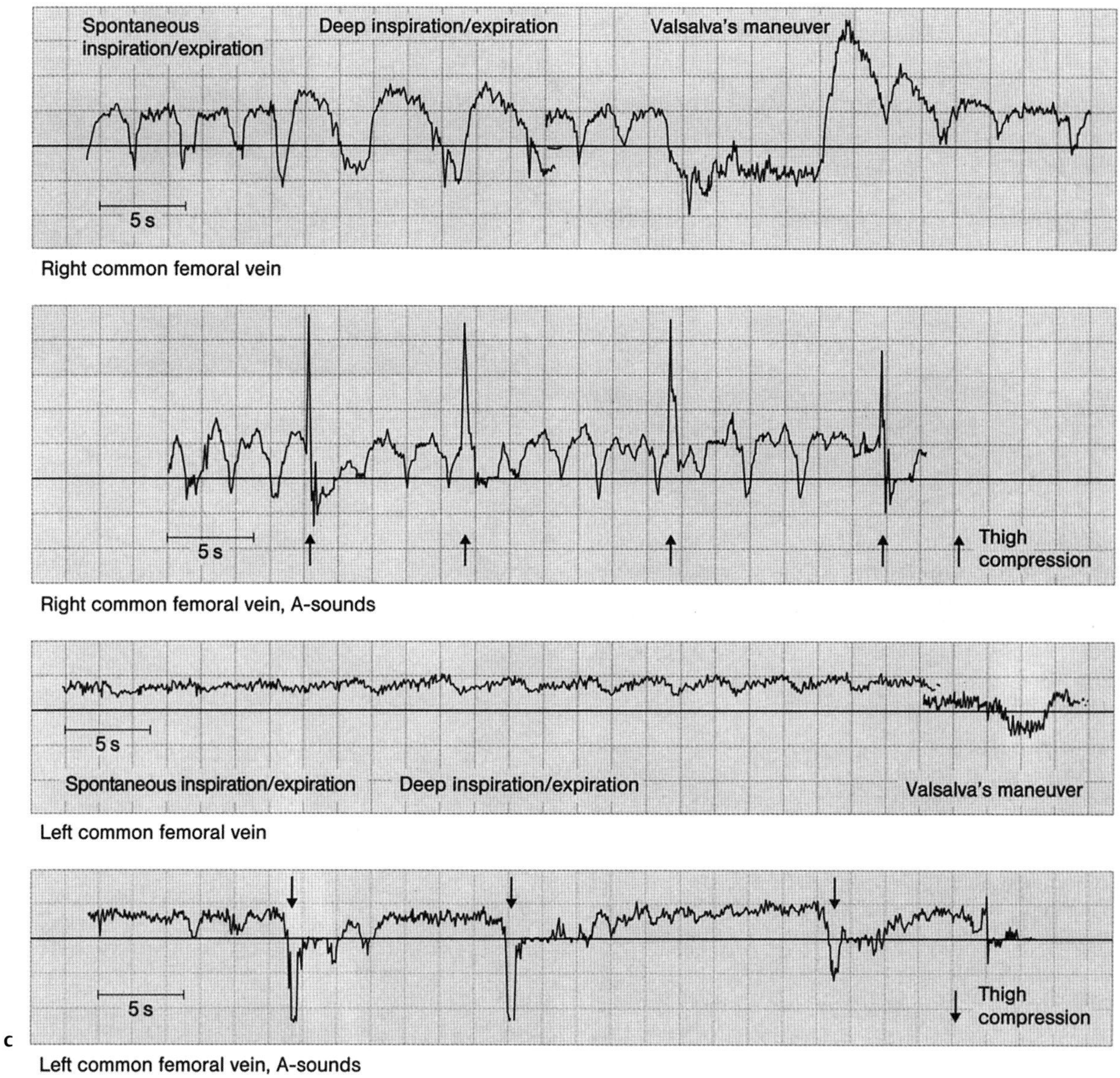

Case Figs. **37 c**

The phlebogram confirms the poor recanalization of the thrombosis in the thigh and pelvic region. A well-developed suprapubic collateral has formed to compensate (spontaneous Palma-like shunt. Case 40, p. 338).

For the imaging of iliac vein thromboses using color flow duplex sonography.

38 Thrombosis of the Femoral Vein

Patient: H.K., male, age 63.
Clinical diagnosis: **acute thrombosis of the right deep leg veins.** Suspected rethrombosis after previous thrombosis of the right leg and pelvic veins.

Case report. Three years previously, a thrombosis of the deep leg and iliac veins occurred on the *right*, involving the common femoral vein, the superficial femoral vein, the popliteal vein, and also all three pairs of veins in the lower leg. At that time, successful fibrinolytic therapy was given, followed by oral anticoagulation treatment for 12 months, without any compression therapy. At presentation, an acutely appearing swelling of the *right* thigh and lower leg had been noticed four days before.

Examination results. On the *right*, there was disseminated hyperpigmentation in the lower leg region, with pretibial edema. In comparison with the left side, the circumference of the thigh and the lower leg was enlarged by approximately 1 cm. On palpation, the thigh was soft to the touch; the lower leg was somewhat firmer. There were no visible varices, and no suprapubic collateral vascular system.

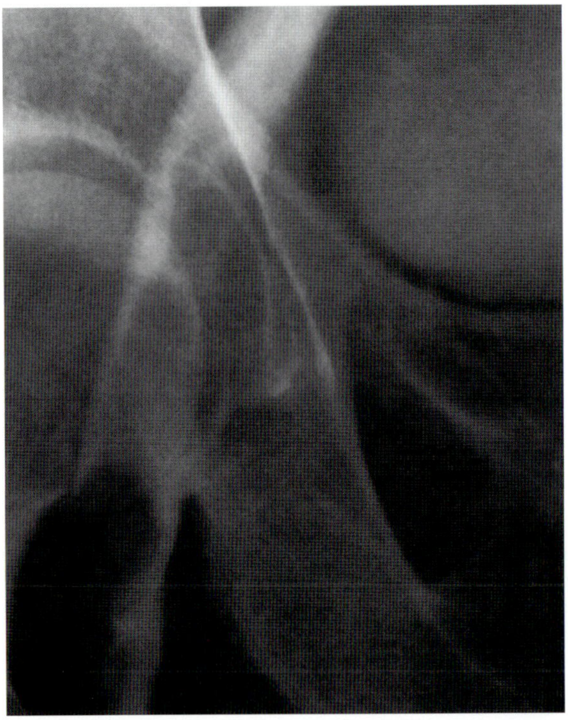

a

Color flow duplex sonogram. Initially, *B-mode* imaging shows the right common femoral artery and vein in transverse section in the groin region (**c**). Compared to the artery shown laterally, the vein has a larger caliber. Even when pressure was applied with the transducer (right section of the image), hardly any compression of the venous lumen was possible.

In the same sectional plane, *color flow Doppler sonography* shows blood flow in the artery in red (**b**). Within the medially located vein, there is only sparse, marginal flow in the dorsal vascular segment (coded blue).

The corresponding longitudinal section through the right groin shows dorsal marginal flow in the segment of the common femoral vein furthest removed from the transducer (**d**).

In a slightly medial section, the confluence of the great saphenous vein (VSM) and common femoral vein (**e**) is seen. In this segment, the partially throm-bosed common femoral vein has respiration-modulated physiological blood flow. The superficial femoral vein (VFS) is completely thrombosed. Further examination showed absent compressibility of the deep veins in the thigh and the lower leg (not shown).

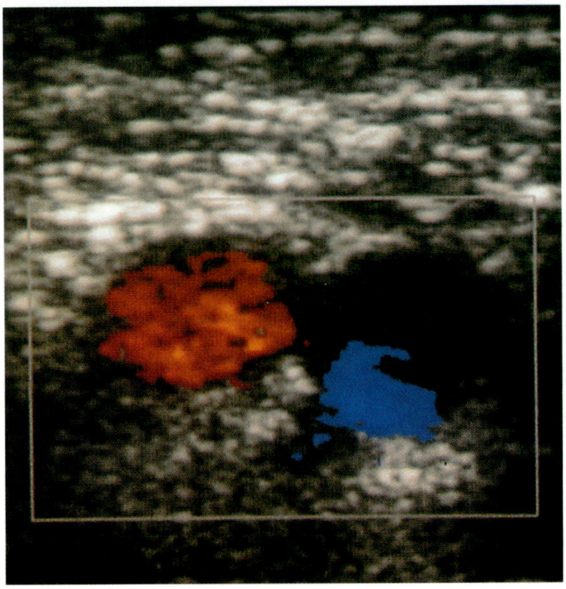

b

Phlebogram. Imaging of the deep veins of the lower leg or thigh on the right side was not possible. Contrast drainage was exclusively through the epifascial venous system. Contrast is seen in the great saphenous vein, with drainage through the common femoral vein, presenting a clear, cone-shaped gap (**a**). Drainage continues freely through the iliac veins into the vena cava. There is a *venous thrombosis* of the right lower leg and thigh, with a cone-shaped thrombus in the common femoral vein.

Comment. Using B-mode imaging alone, it is often difficult to detect a rethrombosis in the context of a postthrombotically altered venous system. In this set of findings, decreased or absent compressibility of the vein is not a definite indication of acute thrombosis. If detectable, the clearly dilated venous lumen is a more reliable sign. Color flow duplex sonography provides an additional margin of diagnostic certainty when evaluating absent or only sparse marginal flow near thrombi that have blood flowing round them. It is helpful to carry out comparisons with findings precisely documented earlier using phlebography or duplex sonography, to provide additional criteria for classifying the current findings on the basis of the location and extent of prior postthrombotic changes.

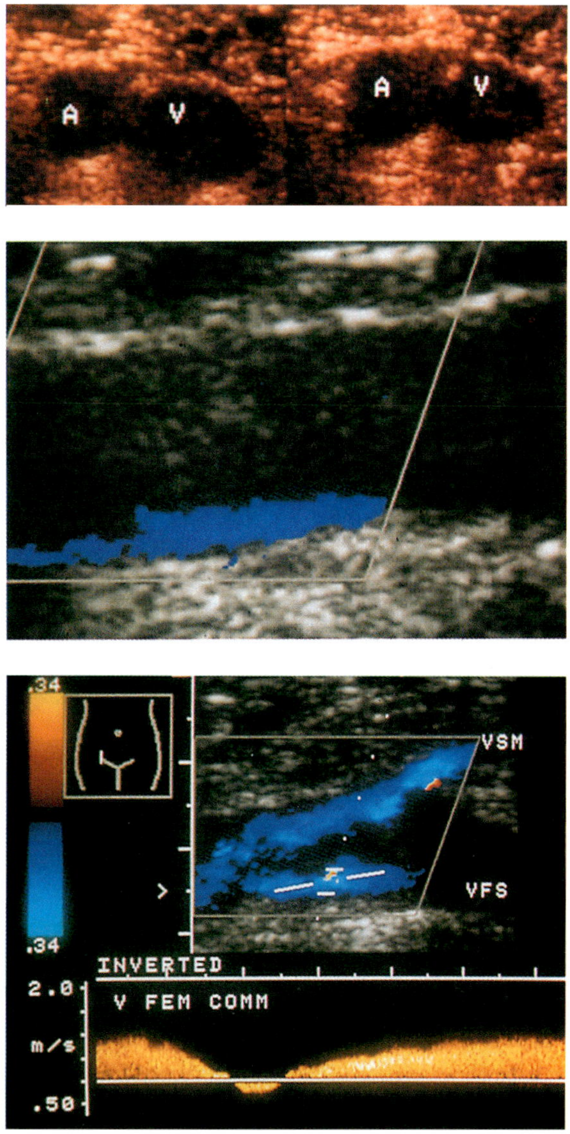

c

d

e

39 Partially Occluded Thrombosis of the Popliteal Vein

Patient: M.P., male, age 54.
Clinical diagnosis: **bilateral primary varicosis, insufficient perforating veins** (blow-outs), and **lymphedema** of the left lower leg, significant obesity.

Case report. The patient had been suffering from bilateral primary varicosis for the previous ten years. Four weeks previously, erysipelas developed in the left lower leg, with lymphangitis extending to the groin. Due to the difficult examination conditions when using continuous wave Doppler sonography (obesity), color flow duplex sonography was used, particularly to evaluate the valvular function of the superficial and deep venous system.

Examination results. The patient, with a height of 179 cm and weight of 110 kg, was obese.

The *left lower leg* appeared reddened, scaly, and swollen; the swelling was more pronounced distally (back of the foot) than proximally. Congestive dermatosis was observed, as a symptom of chronic venous insufficiency. There was a varicose, crown-like formation of the veins (cockpit varices) at the margins of the foot. All three Cockett groups in the left lower leg showed blow-outs.

Color flow duplex sonogram. The sagittal longitudinal section (**b**), proceeding from a dorsal direction (hollow of the knee) with color-coded flow visualization, depicts a partially occluded thrombus in the proximal popliteal vein. Using B-mode imaging in this sectional plane only allows moderate delimitation of the thrombotic material.

In the corresponding transverse section (**c**), the base of the thrombus, which is adherent to the wall, can be seen. In this plane, too, the round thrombus is recognizable through the marginal blood flow; the flow is modulated by respiration, and there is *no* hemodynamically effective venous flow obstruction at this location.

In a longitudinal section (**e**) and cross-section (**f**) from a dorsal direction in the same segment of the popliteal vein *16 days later,* color flow duplex sonography shows complete (spontaneous) lysis of the thrombotic material.

Phlebogram. Ascending leg phlebography on the left shows the floating thrombus (**a**) protruding into the lumen of the popliteal vein, with a paired distal superficial femoral vein.

In addition, varicosis of the trunk of the great saphenous vein was seen, with insufficient perforating veins of the lower, middle, and upper Cockett groups (not shown).

The *follow-up phlebography* after heparin-supported fibrinolysis (**d**) confirms the findings obtained with color flow duplex sonography (**e, f**), i.e., complete recanalization of the popliteal vein.

Comment. The underlying condition in this patient was chronic venous insufficiency, superimposed on insufficiencies of the perforating veins. A previously undiscovered lymphedema was complicated by erysipelas, leading to a clear increase in swelling of the left lower leg, so that hospital admission was required.

The fresh thrombus confirmed in the left popliteal vein is an *accidental finding* with color flow duplex sonography.

The patient was initially treated with intravenous full heparinization, with a subsequent, overlapping period of transfer to an oral anticoagulant, and the course was typical for heparin-supported spontaneous lysis of the thrombus.

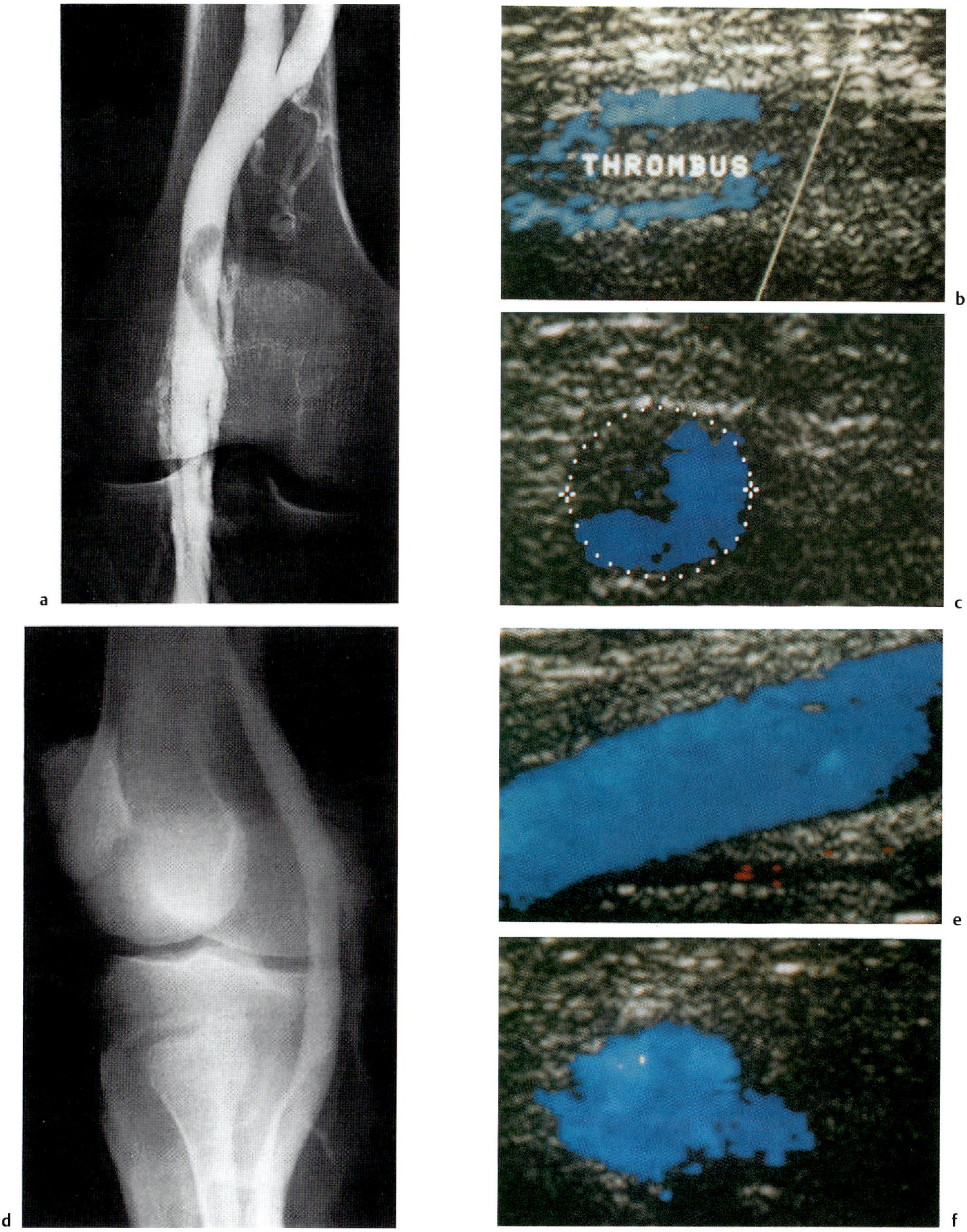
THROMBUS
a
b
c
d
e
f

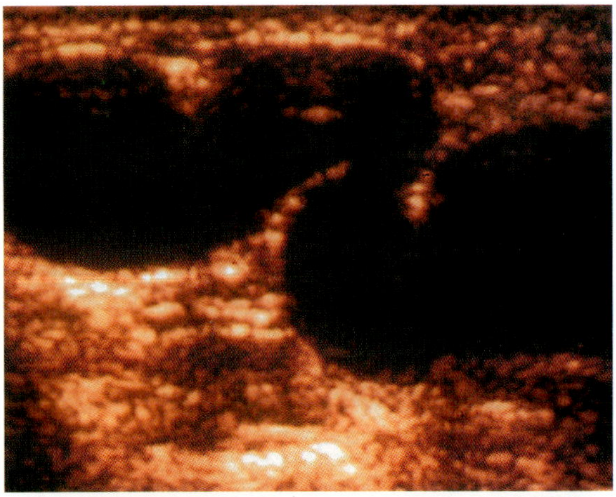

a

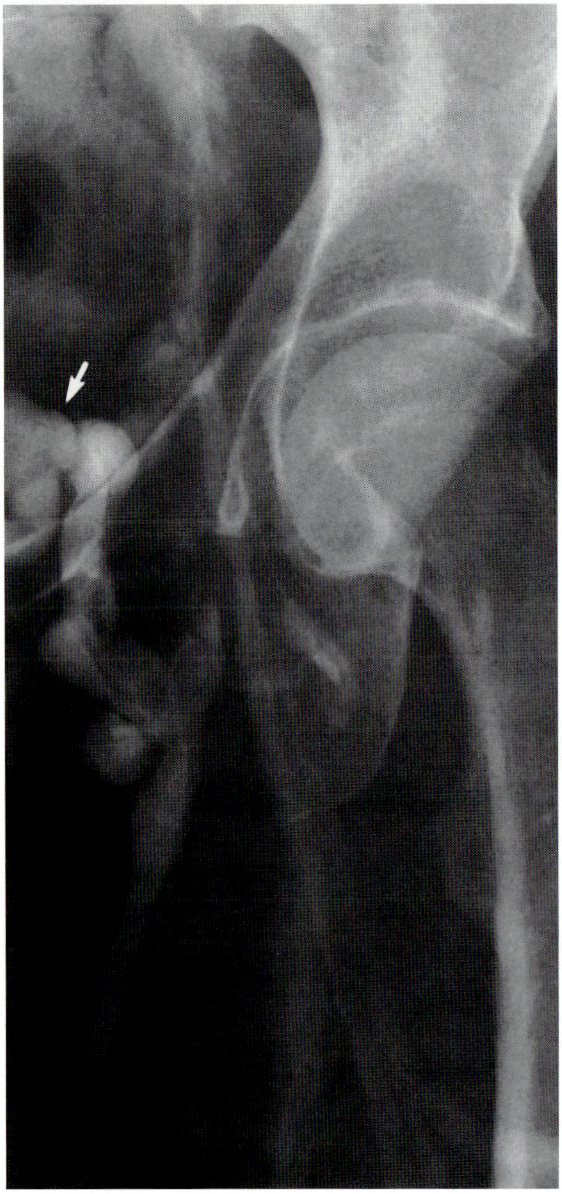

b

40 Thrombosis of the Iliac and Leg Veins (Spontaneous Palma-like Shunt)

Patient: R.F., female, age 38.
Clinical diagnosis: status after postpartal **thrombosis of the left pelvic, thigh, and lower leg veins** 15 years previously, with a postthrombotic syndrome. There was a primary varicosis of the right great saphenous vein.

Case report. Fifteen years previously, a thrombosis of the *left* leg and iliac veins developed postpartally, and conservative treatment was given. Compression therapy was not consistently followed subsequently. Two years previously, a herniotomy was carried out on the left side, followed by a surgical neurolysis six months later. The patient now presented with pains in the left groin area, radiating into the thigh, and with the lower leg showing a tendency toward swelling.

Examination results. There was a slight increase in the circumference of the *left* leg and a suprapubic bypass circulation system had developed. In addition, varices of the great saphenous vein and pain when exerting pressure on the scar in the left groin were noted. Varicosis of the great saphenous vein was present in the *right* thigh and lower leg.

B-mode imaging. A nearly sagittal suprapubic section shows a dilated collateral vein, with blood flowing through it very slowly from left to right (**a**). On the left, color flow duplex sonography (not shown) demonstrated a recanalized external iliac vein, common femoral vein, and superficial femoral vein, as well as recanalized veins of the lower leg, with an occluded common iliac vein accompanied by iliac vein collaterals.

Phlebogram. On the left, there are partially recanalized lower leg veins, superficial femoral vein and iliac veins, and an occlusion of the common iliac vein. Contrast medium is draining via a dilated great saphenous vein into a dilated complex of suprapubic collateral veins, and passing to the contralateral side (**b**).

Comment. Due to the complete obstruction of the left pelvic vascular system, the patient has developed dilated suprapubic collateral veins (spontaneous Palma-like collateralization) toward the contralateral side. These provide substantial venous drainage for the left leg, and should therefore not be removed.

The patient was treated by an initial use of intensive methods to counteract the venous congestion (tight wrapping of the legs, intermittent compression treatment), and these were subsequently supported by a custom-made knee-length compression stocking.

We are grateful to Mr. G.G. Anding, Dr. H.-J. Walter, Mr. M. Schubert and Dr. H. Fallenski of Lüdenscheid for their kind permission to use the phlebogram.

Aneurysm

41 Aneurysm of a Great Saphenous Vein Bypass

Patient: I.R., male, age 51.
Clinical diagnosis: **PAOD on the right,** status after **multiple bypass implantations** and **revisions.** Currently, **patent great saphenous vein bypass on the right.**

CAD:	–	Vascular surgery:	+	Diabetes mellitus:	–
PAOD:	+	Heredity:	–	Hyperlipidemia:	–
TIA, stroke:	–	Hypertension:	+	Smoking:	+

Case **report.** Three years previously, claudication in the right calf had occurred for the first time. Nine months earlier, a femoropopliteal Gore-Tex bypass was implanted. An occlusion of the bypass occurred two months previously, and fibrinolysis proved unsuccessful. A renewed implantation of a femoropopliteal bypass was carried out, requiring a triple revision (open recanalization, thrombectomies, venous patch surgery of the anastomoses). Finally, a femoropopliteal venous bypass (autologous left great saphenous vein) with an infraglenoid anastomosis was implanted. Currently, the patient was experiencing paresthesias and dysesthesias in the right lower leg region. There was no claudication.

Examination results. There was a pulsating resistance in the region of the right proximal thigh, located medially, that was painful on pressure. The scars at the right and left thigh and lower leg were mainly free of any irritation. The right lower leg was swollen due to postoperative lymphedema.

Color flow duplex sonogram. With *B-mode imaging,* and using a longitudinal section adjusted to the course of the bypass, an aneurysmal dilatation of the venous bypass with a diameter of approximately 8 × 11 mm, without any thrombotic covering of the aneurysmal wall, is seen in the proximal right thigh (**a**). At the bypass wall located closest to the transducer, there is a small, hyperechoic structure, representing a remaining venous valve.

Color flow Doppler sonography (**b**) shows that there is blood flowing freely through the bypass. In the *aneurysmal cavity,* the color change from blue through black to red, using a lower preselected scale, indicates slow, eddy-like flow. In this sectional plane, the main flow direction is marked by a curved arrow.

Angiogram. Central venous digital subtraction angiography (DSA) shows the bypass aneurysm in the proximal right thigh in two planes (**c**). There is an occlusion at the origin of the superficial femoral artery.

Comment. Due to the small size of the bypass aneurysm and the multiple operations the patient had previously undergone, it initially seemed justified to wait cautiously while providing oral anticoagulation, closely monitoring the patient's status with color flow duplex sonography. If the size of the aneurysm increases, vascular surgical intervention may eventually become necessary.

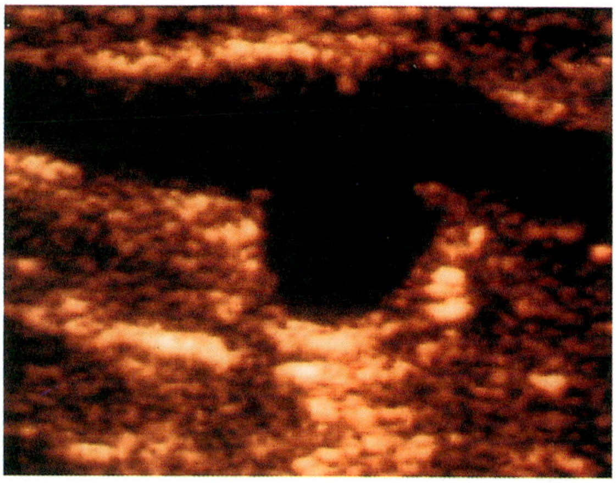
a

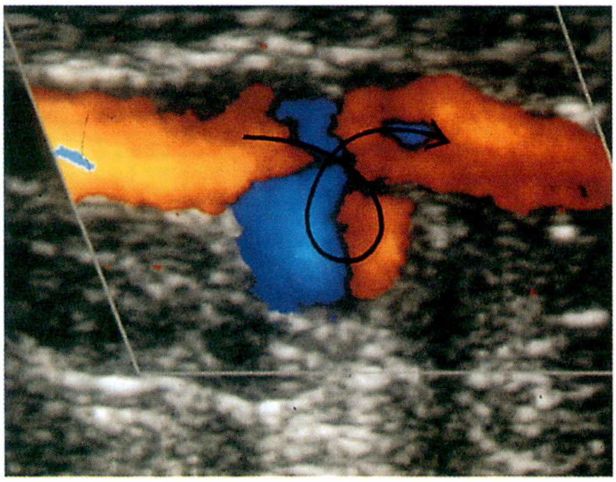
b

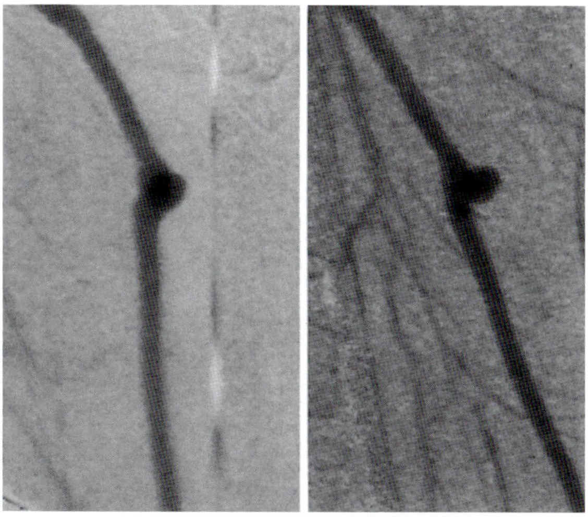
c

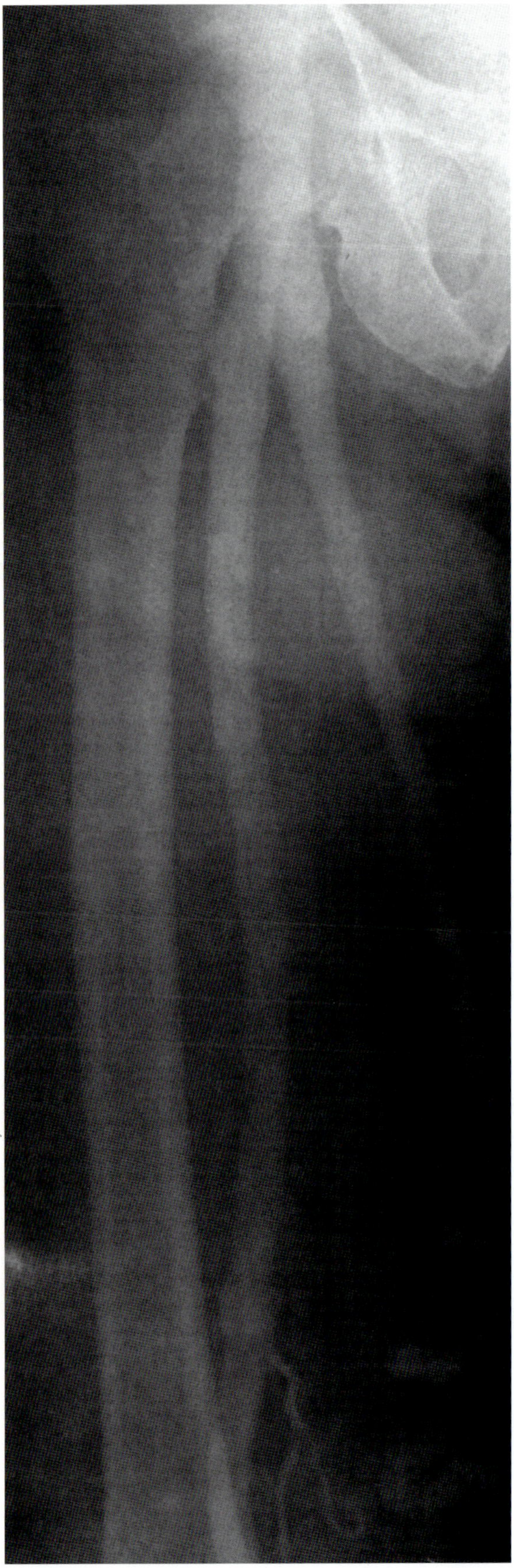
a

Valvular Insufficiency

42 Varicosis of the Great Saphenous Vein

Patient: A.Q., male, age 49.
Clinical diagnosis: **varicosis of the main trunk and side branch of the right great saphenous vein.** Chronic venous insufficiency.

Case report. For the previous ten years, the patient had been suffering increasing varicosis, which was more pronounced on the right than on the left. One year previously, a swelling of the right leg occurred, and there was a suspicion of thrombosis or thrombophlebitis, as there was a sense of tightness in both calves. Clinically, the right leg presented with an increased circumference and signs of chronic venous congestion, induration, and brown pigmentation. Varicosis of the right great saphenous vein was observed.

Examination results. There was an increase in the circumference of the right leg, with chronic signs of venous congestion, and a slight increase in consistency was detected.

Doppler sonogram *(analogue registration)* (**b**). *Right:* Registration above the great saphenous vein, 7 cm below the confluence: there is respiratory modulation during both normal and deep inspiration, with pendular flow present as a sign of valvular insufficiency. The Valsalva maneuver was positive, and showed a reverse flow phenomenon.

Phlebogram (**a**). *Right:* Ascending leg phlebography depicts an intact deep venous system in the pelvis and leg, and varicosis of the trunk of the great saphenous vein, with insufficient lower and middle Cockett perforating veins. During the Valsalva maneuver, clearly insufficient venous valves were detected at the origin and in the course of the great saphenous vein.

Comment. The findings are typical of a pronounced trunk varicosis of the great saphenous vein. Surgery was proposed, but was delayed due to the patient's significant obesity.

The clinical suspicion of a deep vein thrombosis was not confirmed. The congestion is a symptom of chronic venous insufficiency, caused by pronounced primary varicosis, with insufficient venous valves and insufficiency of the perforating veins.

Example using color flow duplex sonography. The findings in a *different* patient (H.C., male, age 51) with varicosis of the great saphenous vein on the right, and with an insufficient valve at the vessel's origin were documented with color flow duplex sonography.

During spontaneous, quiet respiration, the normal blood flow in the proximal great saphenous vein directed toward the heart is coded blue in a longitudi-

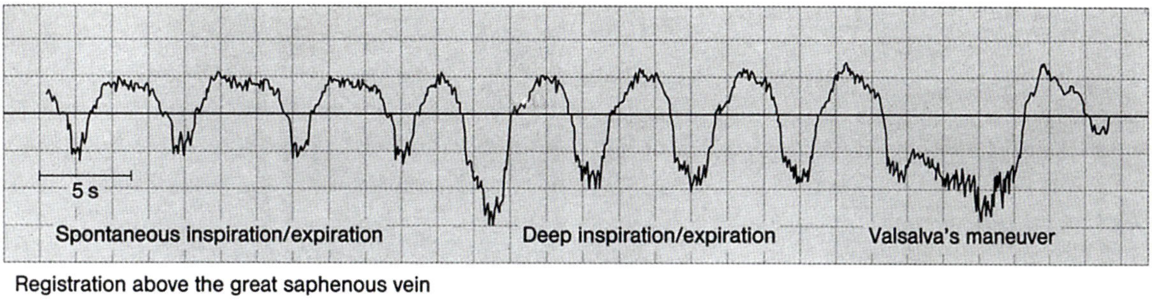

b

nal section (**c**). The frequency spectrum shows good respiratory modulation.

During the Valsalva maneuver, the red color coding in the vein shows a reversal in the flow direction, with retrograde flow (**d**). Also typical is the local appearance of red and blue pixels, representing eddies, e.g., at the valvular structures. The retrograde blood flow during the Valsalva maneuver is also shown below the baseline in the frequency spectrum.

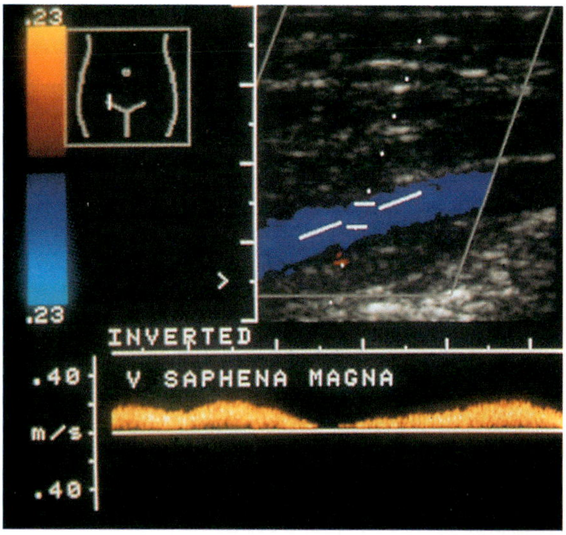

c

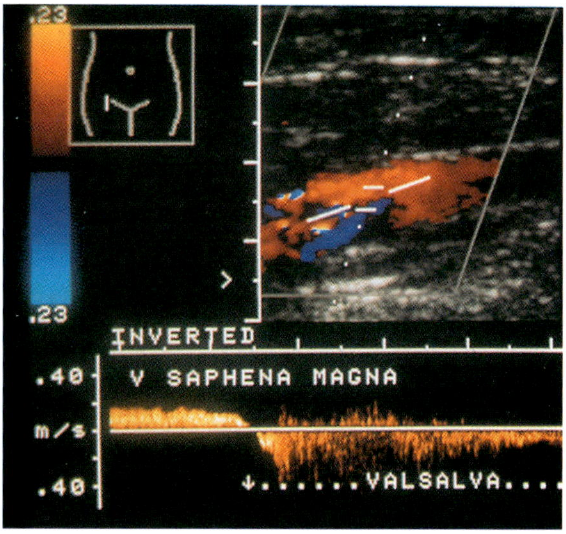

d

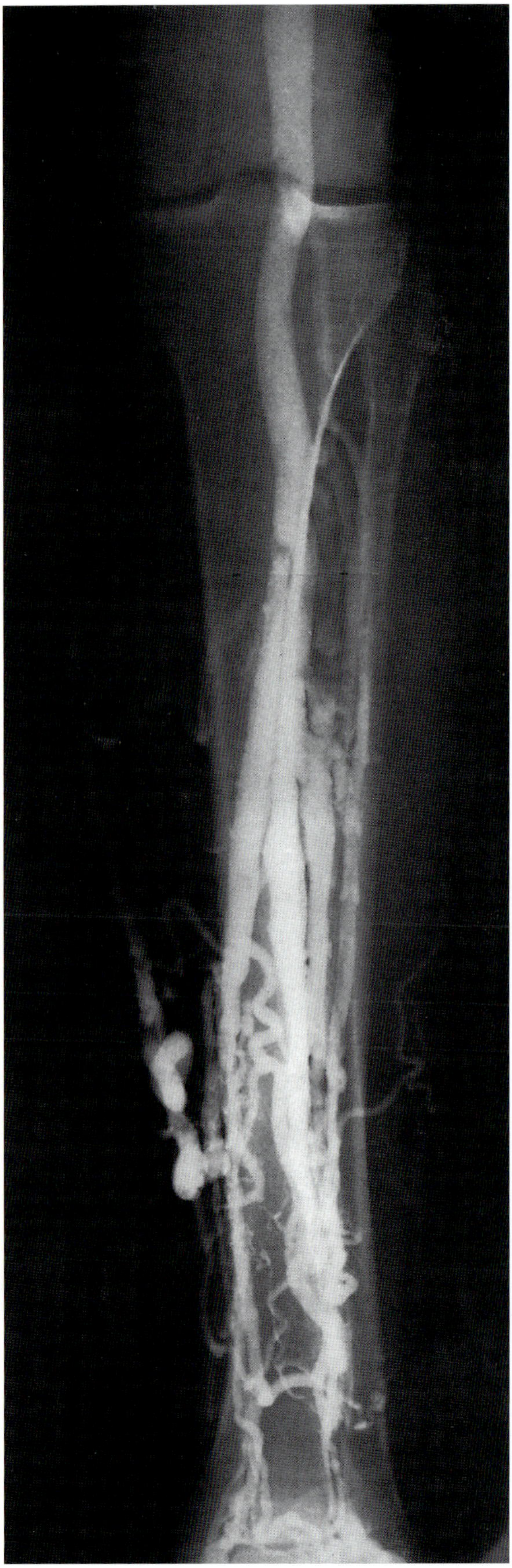

a

43 Insufficient Perforating Veins

Patient: K.G., male, age 51.
Clinical diagnosis: **chronic venous insufficiency with an insufficiency of perforating veins of the left leg.**

Case report. The patient had been suffering varicosis of the left lower leg since the age of 40. Initially, there were minimal symptoms. There was a discrete supramalleolar tendency toward venous congestion. Three years previously, brown spots had developed on the left distal medial lower leg, with a small lesion exuding pus. As yet, neither phlebography, sclerotherapy, nor surgery had been carried out. There was a familial predisposition to varicosis.

Examination results. In the left ankle region, there was a clear increase in the circumference of about 3 cm in comparison with the right side. The patient presented with status following a healed ulcer of the lower leg, with pigmented skin discoloration and moderate varicosis in the thigh region and in the course of the left great saphenous vein.

Doppler sonogram *(analogue registration)* (**b**). Above the left groin, there is a slight reflux during the Valsalva maneuver: normal, steep A sounds with a narrow peak were noted indicating an intact deep venous system.

The probe was placed above the blow-out of the perforating vein. Quick compression of the *distal* calf caused forward flow toward the probe, and retrograde suction when the compression was released. During this maneuver, afferent flow from the superficial varices was ligated using tourniquets.

Compression *proximal* to the registration point only resulted in minimal flow toward the probe, and on decompression there was reflux moving away from the probe. The slight reaction to proximal compression indicates that the valves of the deep venous system are intact.

Phlebogram. Ascending phlebography of the left leg (**a**) shows a freely patent venous system in the deep lower leg veins, popliteal vein, superficial and deep femoral veins, and iliac vein. Under the Valsalva maneuver, retrograde low-contrast imaging of the great saphenous vein was seen, but not of the small saphenous vein. There are insufficient perforating veins of the upper Cockett type, Sherman type, and Boyd type.

Comment. The patient was suffering from chronic venous insufficiency, superimposed on a slight insufficiency of the great saphenous vein and a pronounced insufficiency of the perforating veins.

During standing exercise, blood flowed out of the deep venous system through the insufficient valves of the perforating veins into the varicose superficial venous region. The result was chronic venous insufficiency, and clearly damaged skin with brown discoloration had formed due to the chronic venous congestion. This represents an indication for ligation the perforating veins.

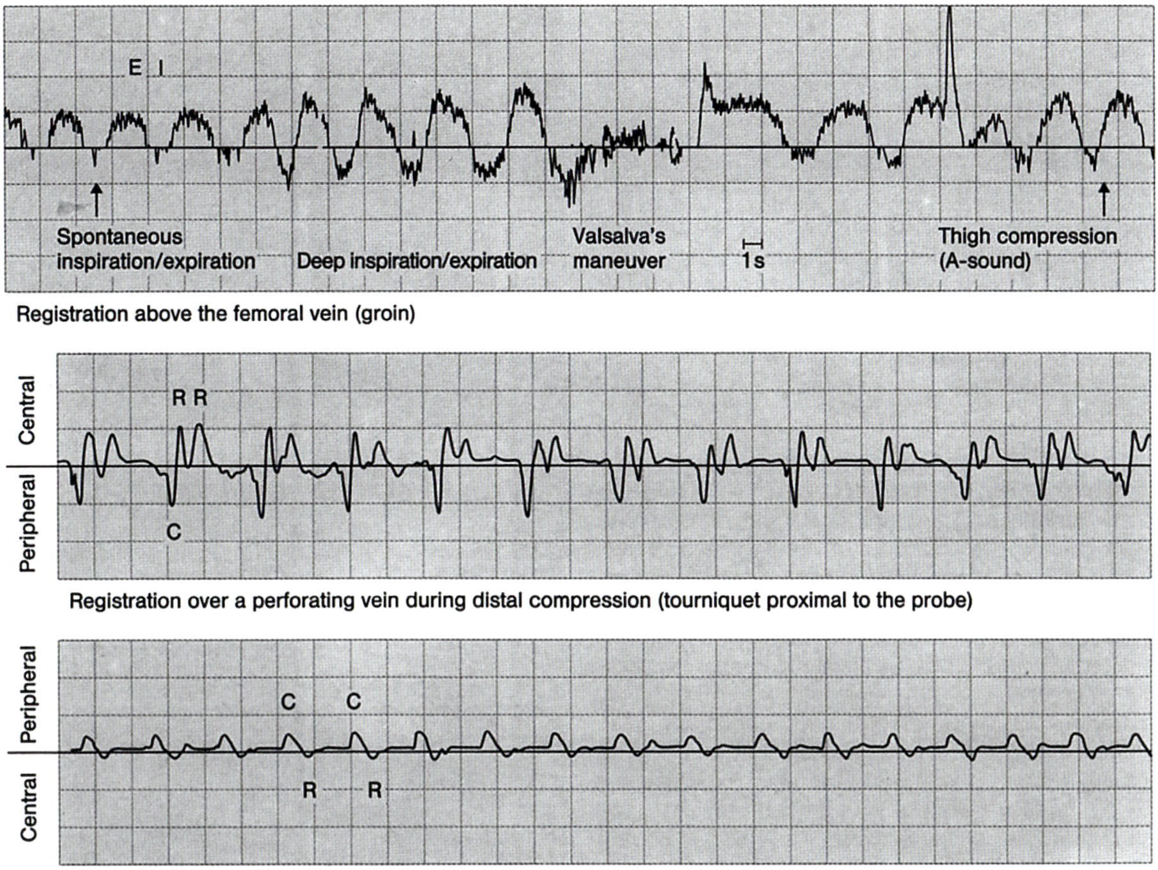

Registration above a perforating vein during proximal compression (tourniquet distal to the probe) (C: compression, R: reflux)

b

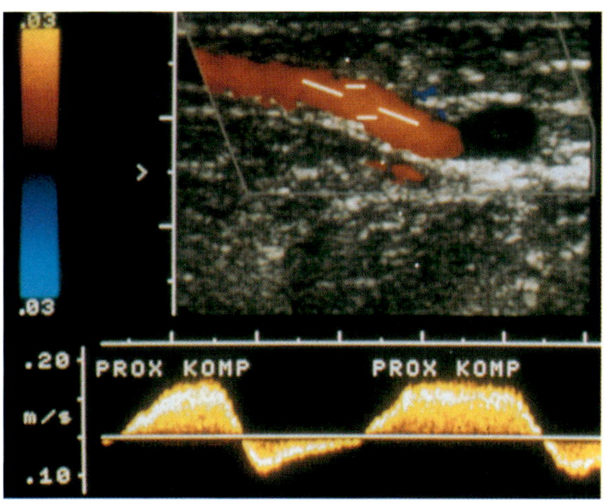

c

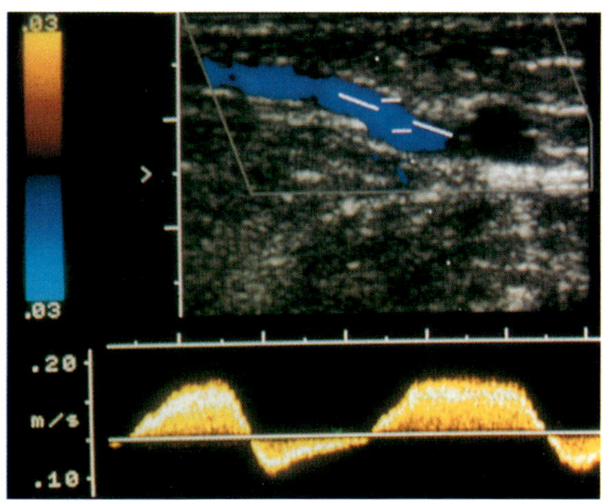

d

Example using color flow duplex sonography. In a *different* patient (R.F., female, age 38) with an insufficiency of the perforating veins in the right lower leg region, pendular flow was demonstrated visually by the varying color coding of the opposite flow directions.

In adapted sectional planes along the anatomical course of a perforating vein above a *blow-out,* flow from the depths proceeding to the surface was recorded (coded red) during *proximal compression* of the calf above the registration point (**c**). After the pressure on the calf had been *released* (**d**), inflow into the deep vein proceeding toward the interior followed (coded blue), suggesting decompression. In both spectra flow above the baseline is coded red, and the reverse flow after release of compression is coded blue.

(The proximal compression is only effective if there are insufficient valves in the deep peripheral veins or when there is no valve in the region between the compression and perforating veins.)

Upper Extremities

44 Subclavian Vein Thrombosis

Patient: W.L., male, age 36.
Clinical diagnosis: **thrombosis of the right subclavian vein.**

Case report. For the previous five days, there had been swelling of the right arm, which had begun in the hand. The patient was unable to recall any possible cause of the condition, such as a vigorous or unusual exertion of the arm, or trauma.

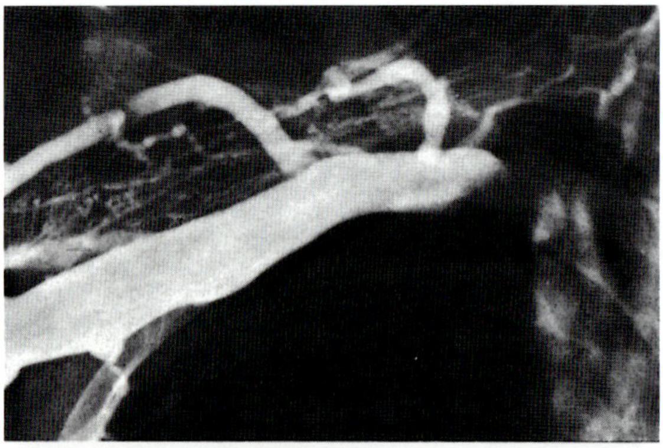
a

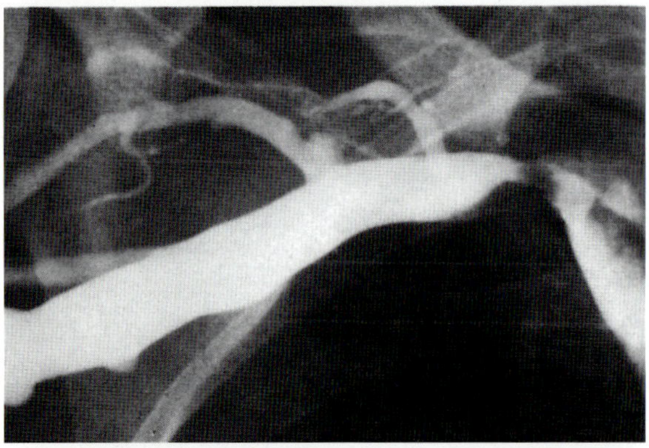
b

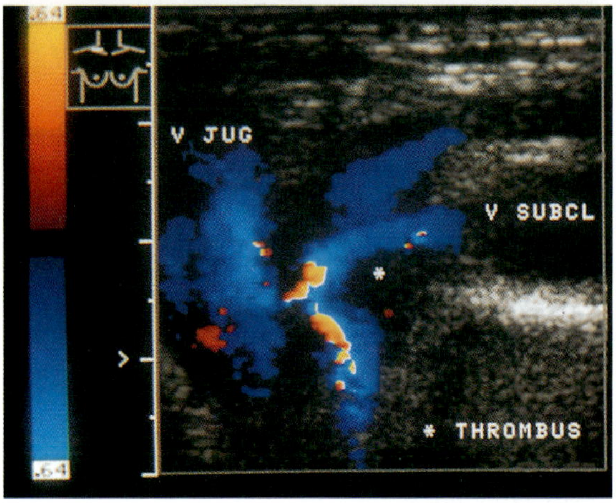

c

Case Figs. **44 d–h** ▷

Examination results. A bluish, livid discoloration of the *right* hand and forearm was seen, which became pronounced when the arms were allowed to hang at the sides. Emptying of the veins was delayed on arm elevation. The right upper arm and forearm were approximately 2–3 cm larger in circumference in comparison with the contralateral side. Palpation detected somewhat firmer subfascial tissue, with an increase in consistency in comparison with the left side. Collateral veins were not visible in the thoracic region.

Phlebogram of the arm. There is normal venous drainage in the axillary vein, and an occlusion over a short distance in the proximal right subclavian vein (**a**).

The follow-up phlebogram after fibrinolysis documents continuing unobstructed venous drainage through the axillary vein, and a recanalization of the subclavian vein with a residual thrombotic fragment in the proximal vascular segment (**b**).

Color flow duplex sonogram *during the first fibrinolytic cycle.* The deep and superficial veins of the forearm and upper arm were completely compressible (not shown). In an almost frontal sectional plane proceeding from a supraclavicular direction, the proximal right subclavian vein and the jugular vein can be seen.
Blood is already flowing round the thrombus in the proximal subclavian vein (asterisk). The high flow velocity in the partially recanalized vein is reflected by the light shades of blue (**c**).

The *Doppler frequency spectrum* from the re-canalized venous segment (**d**) documents a high flow velocity of maximally 100–130 cm/s, which is not modulated by respiration.

The right *jugular vein* is not thrombosed, and shows flow affected by both respiratory and cardiac modulation (**e**).

The *M-mode* confirms the absence of a collapse of the right axillary vein during forced panting in the *sniff test* (**f**). The fine indentations in the venous wall reflections above and below the anechoic lumen correspond to the high respiration frequency, and are caused by concomitant transducer movements due to respiration-dependent skin and tissue movement above the supraclavicular fossa.

In the contralateral *left* axillary vein, the M-mode image shows a decrease in the venous caliber during rapid, panting respiration (**g**).

After fibrinolytic therapy, the right subclavian vein is functionally patent again. In the same sectional plane as in Figures **c** and **d,** the patent proximal segment of the subclavian vein is seen with color flow duplex sonography (**h**). The *Doppler frequency spectrum* once again shows a normal cardiac modulation of the flow velocity, whereas a continuous flow was registered previously (**d**). Also, the flow velocity has decreased in comparison to the earlier findings due to the wider venous lumen through which the blood is flowing.

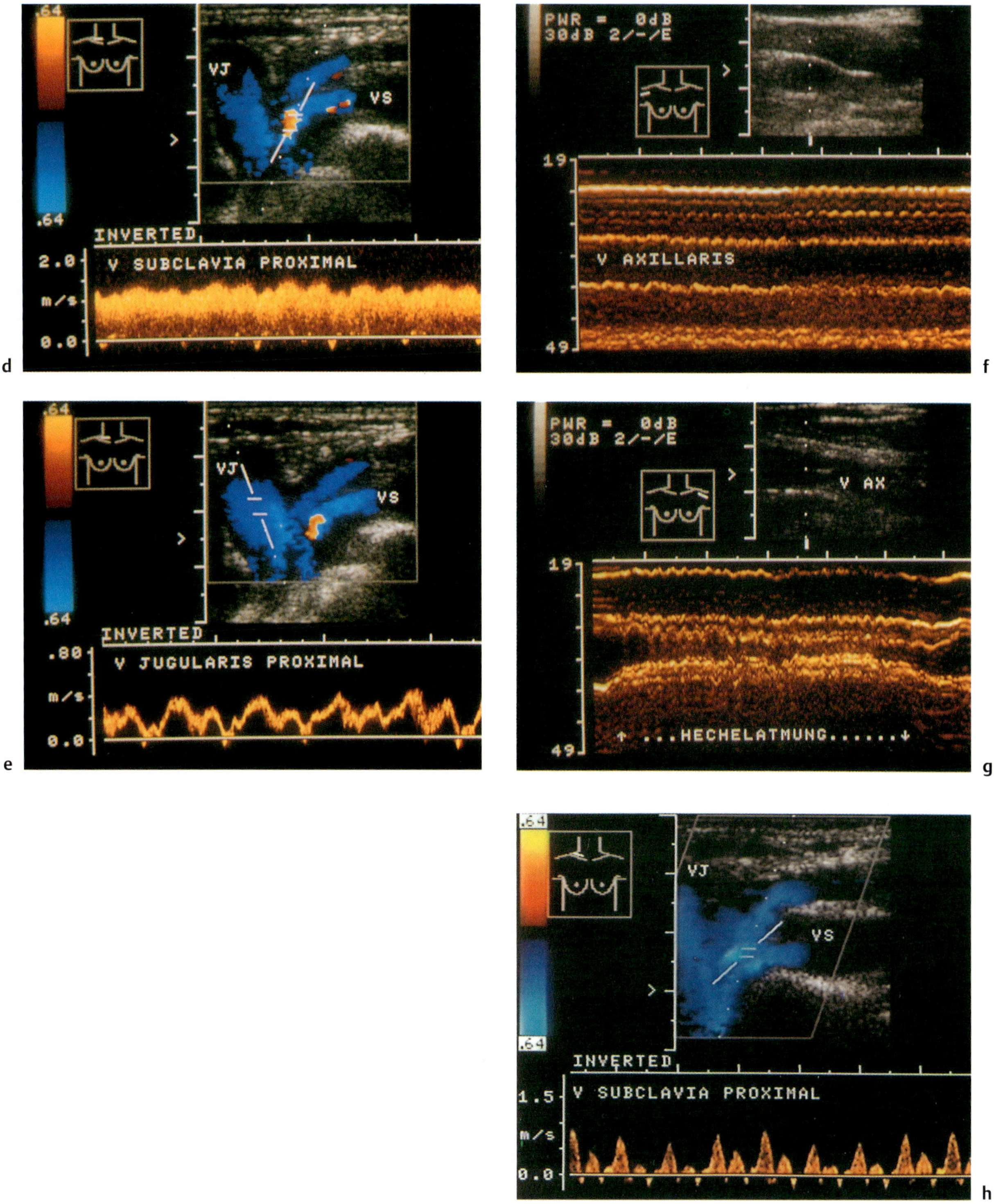

Case Figs. **44d–h**

Comment. With two fibrinolytic cycles administered over four hours using six million units of streptokinase in each instance (ultrahigh short-time lysis), successful thrombolysis of the subclavian vein occlusion was achieved. Oral anticoagulation treatment followed.

The cause of the subclavian vein thrombosis remained unclear. The patient history showed no evidence of an effort thrombosis, nor were there any indications of neurovascular compression or a paraneoplastic syndrome.

8 Glossary

The following terminological overview attempts to summarize brief definitions of selected important specialist terms that are relevant to angiological ultrasound procedures. The selection is mainly restricted to equipment-related technical concepts, to meet users' actual needs. Cross-references are given in italics.

Acoustic energy: mechanical energy that is transported in the form of an acoustic wave.
Unit: joule = watt × second.

Acoustic impedance: the product formed from the tissue density and the ultrasound velocity. Changing the acoustic resistance leads to a reflection of acoustic energy at an interface.

Acoustic power: acoustic energy that is transported per unit of time in the form of a wave.
Unit: joule/second = watt.

Acoustic shadow: the region of limited echo reflection located behind a strongly reflecting interface (e.g., calcified plaque).

Acoustic wave: the transport of energy by the propagation of mechanical oscillations. Both spatially and with respect to time, this is a periodic process, in which neighboring locations in the same phase define the acoustic wavelength. In an acoustic wave front, all the points are in the same phase.

Aliasing: the situation that arises when the *Nyquist* sampling limit is exceeded by the frequency of the input signal.

A-mode (amplitude mode): depicting the amplitudes of the echo signals as a function of the transit time between echogenic structures and the transducer: abscissa = time axis, ordinate = axis showing the amplitude of the echo signal. Corresponding to the transit time of the echo, the distance on the time axis indicates the location of the reflection.

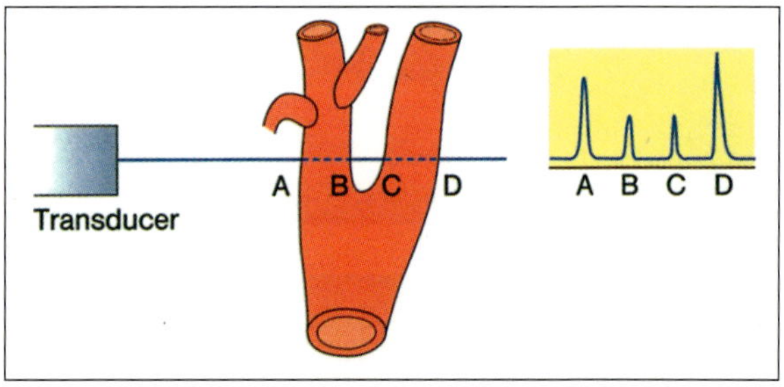

Amplitude: the magnitude of the wave variable, such as velocity, displacement, or acceleration.

Analogue: the transfer methodology of a system in which the initial value is a continuous function of the entered value. This is in contrast to digital.

Analysis, three-dimensional: the depiction of ultrasound sectional images (B-mode or Doppler mode) in three spatial dimensions.

Analysis, four-dimensional: the depiction of ultrasound sectional images (B-mode or Doppler mode) in spatial (three-dimensional) and time dimensions in order to show wall and plaque movements.

Angle: The important angle is the one formed by the ultrasound beam and the vascular axis. According to the *Doppler formula,* the frequency of the reflected ultrasound signal is reduced proportionately to the cosine of the angle. For various technical and physical reasons, it can be difficult to measure this angle, e.g., in a *duplex system.* In addition, inaccurate estimates are also made at vascular bends.

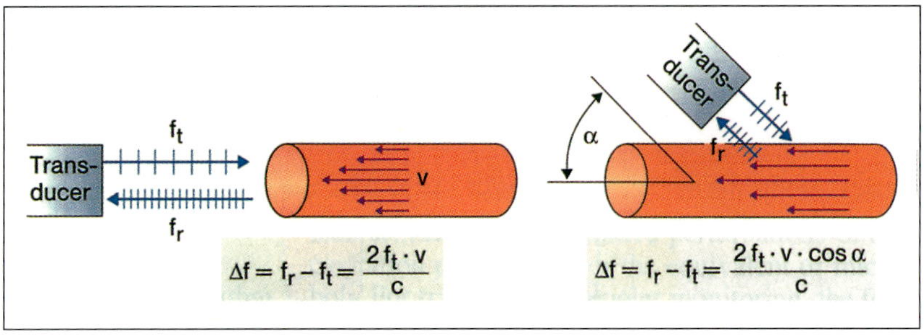

◁ to **Angle**

Angle of incidence: angle between the ultrasound axis and the structure to be examined, e.g., axis of a blood vessel.

Array: a spatial arrangement of several transducer components:

- Linear
- Annular
- Circular

Artifact: an interfering echo that does not belong to a true reflector, and is usually caused by multiple reflection or scattering.

Audio analysis: the simplest, purely acoustic reproduction of the amplified Doppler signal, using a loudspeaker or headphones.

Autocorrelation: multiplication of a wave by a time-shifted section of the same wave.

Axial resolution: the minimum distance between two reflectors, in the direction of ultrasound propagation, at which they are still shown as separate entities. This is also called depth resolution or range resolution, and among other factors depends on the transmission frequency (e.g., at 8 MHz: 0.3 mm in muscle).

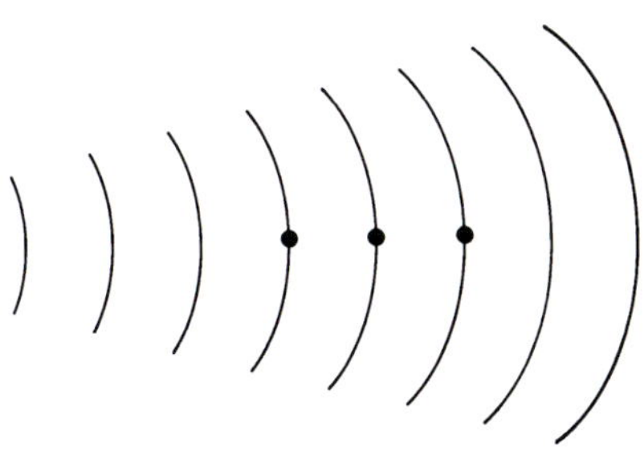

Backscatter: the energy reradiated by a scatterer in a direction opposite to that of the incident wave.

Beam: an acoustic field produced by the transducer with the axis located at the point of maximal acoustic pressure (far-field). The transducer form also determines the form of the beam.

Beam angle: the angle between the direction of propagation of ultrasound and the direction of anatomical structures reflecting ultrasound. For B-mode imaging, perpendicular insonation is optimal; for Doppler mode studies, 30–60° are optimal.

Bernouilli, Daniel (1700–1782): Swiss mathematician, sometimes referred to as the founder of mathematical physics, who unified the study of hydrodynamics under what became known later as the principle of the conservation of energy. Daniel Bernouilli was one of an extraordinary dynasty of eight mathematicians spanning three generations of the same family. Among his contributions to hemodynamics are the *Bernouilli effect* and the *Bernouilli equation.*

Bernouilli effect: the reduction in pressure that accompanies an increase in velocity of fluid flow.

Bernouilli equation: the equation that states that the total fluid energy along a streamline of fluid flow is constant.

Bidirectional: This refers to the ability of Doppler equipment to distinguish between positive and negative Doppler shifts (flow toward the probe or away from the probe).

B-mode (brightness mode): a two-dimensional ultrasound image produced by depicting echo signals on the screen. The amplitude is indicated by the brightness modulation (gray scales). The point of origin of the image is located on the abscissa, and, depending on the echo's transit time (= depth), indicates the reflection's location. Placing single scans one next to the other creates a two-dimensional image (echotomogram = B-scan).

Boundary layer: the thin layer of stationary fluid adjacent to the walls of the containing vessel.

Bruit: the name given to sounds sometimes associated with disturbed and turbulent flow. They arise from the periodic variation in shear stress on the vessel wall, which causes it to vibrate.

Cathode-ray oscilloscope: electronic registration equipment used to record a rapidly occurring fluctuation in voltage. It mainly consists of a cathode-ray tube with x–y deflection and an amplification unit. The electrode beam is made visible on a fluorescent screen, with the abscissa corresponding to the time and the

to **B-mode** ▷

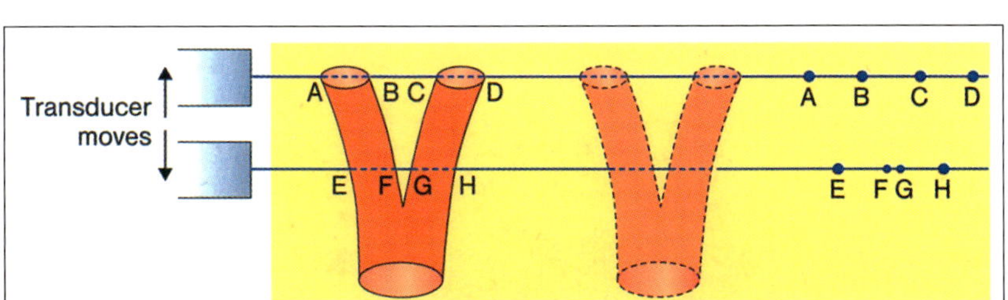

ordinate representing the amplified applied voltage. The beam's movement without any inertia also makes it possible to register brief or high-frequency fluctuations in voltage (see *M-mode).*

Color flow duplex sonography: a new procedure for imaging structural and hemodynamic relationships within tissue in a two-dimensional, almost real-time image.

Compliance: the rate of change of cross-sectional area with pressure ("area" compliance); or the rate of change of diameter with pressure ("diameter" compliance).

Continuous wave (CW) Doppler: a Doppler method in which ultrasound waves are continuously transmitted. It functions simultaneously with two piezoelectric crystals (transmitter/receiver). Its advantage is that there is no physical upper limit (Nyquist frequency) for clearly measuring the Doppler shift. High flow velocities can therefore be measured. Its disadvantage is that no information about the location of a measurement is possible.

Critical Reynolds number: the Reynolds number around which the transition from laminar to turbulent flow takes place.

Critical stenosis: a stenosis of sufficient diameter to reduce flow rate and pressure significantly. Sometimes called "hemodynamically significant" stenosis.

Curved-array transducer: a transducer consisting of a curved arrangement of piezoelectric crystals lying next to one another. The image field that results is trapezoid in form (see pp. 17–19).

DC: direct current.

Decibel (dB): the relative measurement for the signal intensity. For the intensity (I), or intensity ratio, and amplitude (A), or the amplitude ratio, the following formulas apply:

$$x\ (dB) = 10 \times \log I_1/I_2$$

$$x\ (dB) = 20 \times \log A_1/A_2$$

Density: specific weight. Tissue that is sonographically "dense" is strongly echogenic, due to the presence of structures that have large differences in impedance.

Digital: numerical, having to do with numbers. In contrast to analogue methods, a continuously modulated input signal is transformed into a series of discrete symbols that lie next to one another or one after another, e.g., the conversion of spoken language into Morse code.

Directionality:

- Unidirectional equipment functions without detecting the flow direction.
- Using a phase detector in bidirectional systems, the flow direction is indicated in relation to the transducer as a positive or negative Doppler shift value.

Disturbed flow: deviations from laminar flow consisting of oscillating variations in the direction of the formation of vortices. Disturbance of blood flow may be caused by various reasons, e.g., high peak velocities, by curving, branching, and divergence of vessels, or by projections into the vessel lumen.

Doppler, Christian Andreas (1803–1853): a mathematician and physicist, who was born in Salzburg and died in Venice. In 1842, Doppler published "Concerning the Colored Light of the Double Stars and Several Other Celestial Bodies," in which he described alterations in stellar light that were dependent on the movement of a star either toward the Earth or away from it. The frequency shift that forms the basis of this phenomenon is called the Doppler effect, and it applies to all types of waves. He enunciated his principle in 1842, but unfortunately confused its interpretation. The acoustic Doppler effect was demonstrated in 1845 by Buys Ballot using a trumpeter riding on a steam locomotive.

Doppler angle: the angle between the direction of propagation of the ultrasound and the direction of vessel flow. As an approximation, the angle between the axis of the ultrasound beam and the axis of the vessel lumen is generally used (see *angle).*

Doppler effect: a frequency change in ultrasound waves due to the relative movement of the receiver or the transmitter, when one component of this movement lies in the ultrasound axis.

Doppler formula:

$$\Delta f = \frac{2 f_t \times v \times \cos\alpha}{c}$$

where:

c = ultrasound velocity in tissue (1540 m/s)
v = velocity of red blood cells in blood vessels
f_t = transmission frequency
f_r = reception frequency
Δf = Doppler shift = $f_r - f_t$
α = the angle between the ultrasound and vascular axes

Example:

c = 1540 m/s
v = 1 m/s
f_t = 2.5 MHz
α = 0°; cos = 1
Δf = 3247 Hz
(see *angle*)

Doppler shift: the difference between the transmission frequency f_t and the reception frequency f_r:

$$\Delta f = f_r - f_t$$

Doppler shift frequency, peak: the highest Doppler shift frequency at a moment in time or in an individual Doppler spectrum, corresponding to the fastest moving target in the Doppler sample volume.

Doppler shift frequency, mean: the average Doppler shift frequency above and below which half of the total power in the spectrum resides.

Doppler shift frequency, mode: the Doppler shift frequency with the greatest power in a given spectrum.

Doppler spectrum: In the Doppler frequency spectrum, the frequencies are indicated that appear within a measurement interval, along with how often they occur, i.e., their amplitudes. The Doppler spectrum thus reflects the number of corpuscular elements in a measured vascular segment that can be classified as having certain flow velocities. The imaging is usually accomplished using a *filter bank analysis* or *fast Fourier transform analysis.* Three forms of graphic representation are commonly used:

- Abscissa = time
- Ordinate = frequency (flow velocity)
- Point density = (brightness or color level) = amplitude

In the second form of representation, the frequency is indicated on the abscissa, the amplitude on the ordinate, and the time on the Z axis. In the *power spectrum,* the square of the amplitude is plotted.

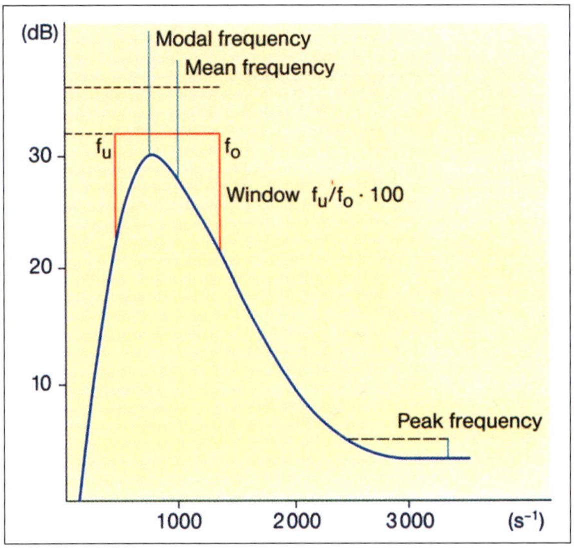

Duplex system: the combination of echo impulse and Doppler procedures to locate the Doppler *sample volume* in the B-mode image in real time, using a single *transducer.* In equipment that has a separately mounted Doppler probe on the side of the transducer, it should be noted that the refractive relationships differ in B-mode (perpendicular) and in Doppler mode (oblique). This leads to errors in the location of the sample volume shown on the monitor. The true position of the sample volume is in reality usually displaced to the side or ahead of the location indicated on the monitor.

Dynamic focusing: This is used to improve the lateral resolution in linear and phased-array systems. While the echo pulse is being received, the effective focal zone is displaced (receiving focus following a single transmitted pulse).

Dynamic range:

- *For amplifiers:* maximal transmittable signal range above the noise level (in dB).
- *For displays:* the range of the input signal (in dB) which lies between the saturation level and the noise level.

Echo: an acoustic signal resulting from scattering or reflection of acoustic waves at an interface.

Envelope: the curve that joins the peaks of oscillations following one another in a wave.

Far field (Fraunhofer zone): the region of the acoustic field beyond the focus.

Fast Fourier transform (FFT): a quick version of the *Fourier transformation,* adapted for computers.

Filter: an electronic circuit designed to allow signals of certain frequencies to pass and to stop signals of other frequencies.

Filter, high-pass: a device that allows high-frequency but not low-frequency variations to pass through. An example is the electrical filter used in Doppler devices to eliminate low-frequency Doppler shifts caused by clutter.

Filter, low-pass: a device that allows low-frequency but not high-frequency variations to pass through. An example is a stenosis, which has the effect of damping rapid variations in the pressure and flow waveforms.

Filter bank analysis: This is often used in the spectral analysis of a Doppler signal. Through a certain number of high-pass and low-pass filters that are switched in parallel and have the same bandwidth, the range of Doppler frequencies appearing (0–20 kHz) is covered. Since the filters more or less represent large frequency ranges overlapping at their limits, artificial changes in the spectrum result.

Flow separation: If a body of fluid within a vessel has a particularly high momentum (because it has entered the vessel as a jet through a tight stenosis, for example), its boundary will separate from the laminae of surrounding fluid. The region of flow separation is marked by a high velocity gradient and hence shear stress.

Focal zone: the area of an ultrasound beam in which the lateral resolution is the best. Continuous wave Doppler systems are usually not very focused. In systems that use a mechanical transducer, the focal zone is fixed between the near field and the far field, while it is variable (usually selectable in steps) in systems with an electronic transducer. Depending on the focusing, different segments of a blood vessel with a constant diameter are faded out: e.g., flow regions near the wall may be in the center of focus, but not those in the far field.

Fourier, Jean Baptiste (1768–1830): son of a French tailor who was persuaded to leave his Benedictine monastery to become a professor of mathematics at the age of 21. Among his many original and outstanding contributions to mathematics and physics was heat analysis. Some say that his death was speeded by his insistence that heat was good for the health and his later habit of living in rooms "hotter than the Sahara desert."

Fourier transformation: a mathematical algorithm that is used to represent a periodic function as a sum of sinusoidal functions of different frequencies, named after Jean Baptiste Fourier (1768–1830); French physicist and mathematician. The calculation allows the presentation of the Doppler signal in terms of the relative power of the various Doppler shift frequencies.

Freeze mode: a method of displaying images that allows functions to be depicted in real time as individual images.

Frequency: in periodic signals, the number of oscillations per unit of time (unit: Hz, Hertz; kHz kilohertz, MHz megahertz; 1 cycle/s = 1 Hz, 1000 Hz = 1 kHz, 1,000,000 Hz = 1 MHz).

Frequency modulation: the principle of information transfer by modulating the signal. Usually, groups of impulses are used that are successive over time, in the form of impulse frequency modulation or pulse frequency modulation.

Gray scale: the brightness-modulated depiction of echo signal amplitudes in several gray scales ranging between black and white (16–128). Alternatively, color scales can also be used. The adjustment of brightness differences in an gray scale can be carried out in either a linear or a logarithmic fashion.

Harmonic imaging: the possibility of improving the registration of weak Doppler signals from reflections occurring at interfaces by using a separate analysis of signals, which, after the application of an *ultrasound contrast medium* (oscillating MAB), reflect harmonic frequencies (*2f*, i.e., at a transmission frequency of 4 MHz, harmonic frequencies of 8 MHz are received and recorded in so-called sonoangiograms) (see p. 12). This method is also used to analyze movement phenomena and flow structures (unresolved flow), which would otherwise not be detectable in small blood vessels. Practical experience with this methodology is still scarce. In comparison with the standard procedure, disadvantageous effects may occur due to examination times that can take several times longer.

Heat effect: A portion of the acoustic energy that impinges on the tissue is converted into heat by absorption, the extent of the temperature elevation depending on the ultrasound intensity, the duration of the ultrasound application, its frequency, and also the characteristics of the tissue affected. This is an important parameter, to which special attention needs to be given during long monitoring and follow-up examinations.

High intensity transient signal (HITS): frequency bands of short duration with a high signal amplitude in the Doppler spectrum. A controversial interpretation has recently suggested that such signals are little more than artifacts, i.e. transmitted perturbations or microcavitation phenomena, or the equivalent of microemboli. For transcranial Doppler monitoring, the following criteria can be used to differentiate a true signal from an artifact:

- Duration < 0.1 s
- Amplitude > 3 dB above background
- Unidirectional
- Variable distribution in subsequent cardiac cycles
- A discrete, metallic audio signal

Impedance: density × acoustic velocity (also termed wave resistance).

Intensity: the acoustic power in the propagation direction. Units:

I_{SATA} [mW/cm²]	=	spatial average and temporal average intensity
I_{SPTP} [mW/cm²]	=	spatial peak and temporal peak intensity
I_{SPTA} [mW/cm²]	=	spatial peak and temporal average intensity

According to the American Institute for Ultrasound in Medicine (AIUM), $I_{SPTA} < 100$ mW/cm² is not harmful. According to studies conducted by the AIUM, there is no proof of any biological changes associated with the MHz spectrum commonly used in medical diagnostics.

Intima media thickness (IMT): a statistical measurement parameter used to define the extent of atherogenesis in its early phase (discrimination threshold in the resolution area of ultrasound equipment: $\geq 100\,\mu m$).

Intravascular sonography: the introduction of suitable transducers into blood vessels for B-mode and Doppler mode imaging.

Laminar flow: ideal flow, with its highest velocity in the center and its lowest velocity at the wall of the blood vessel (see *velocity profile).*

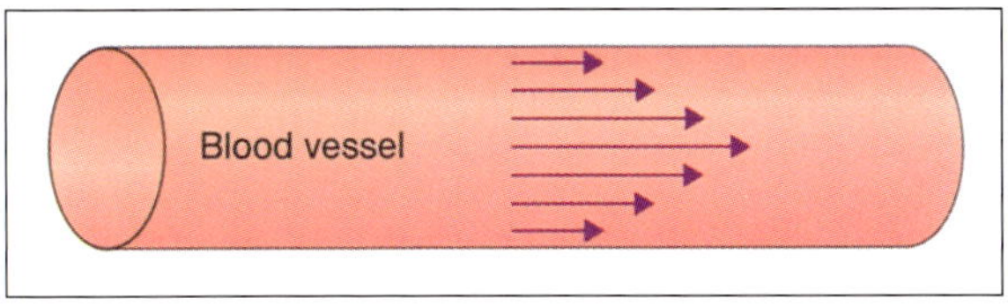

Lateral resolution: minimal distance between two reflecting structures perpendicular to the ultrasound beam at which they can still be depicted as separate entities. The lateral resolution improves at a higher transmission frequency, with a shorter focal zone and with a transducer that covers a greater surface area (important when applying the echo impulse technique to the carotid artery).

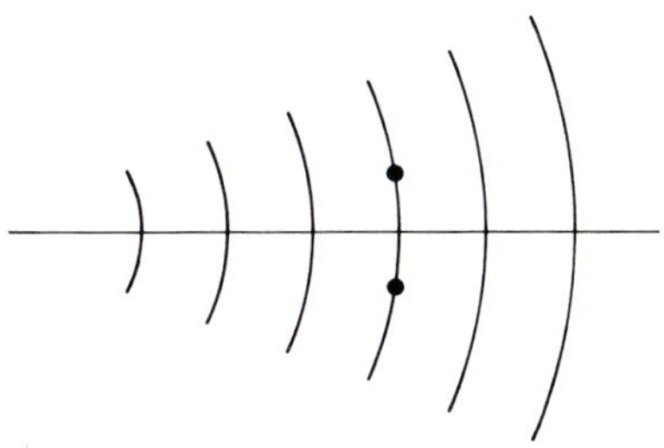

Linear-array transducer: a transducer that has a straight arrangement of the piezoelectric elements, which results in a rectangular-shaped image field.

Mean frequency: the average value of all the frequency components in a spectrum, calculated according to varying formulas (see the illustration under *Doppler spectrum,* above). (This value can be determined for every time interval analyzed).

M-mode (motion mode): continuous depiction of changes in the *A-mode* or *B-mode* (ordinate) over time (abscissa) in order to document moving (pulsatile) echo structures (vascular wall, heart valves). The signal intensity can be documented on paper using a gray scale, or directly with a *cathode-ray oscilloscope.* It is important for measurements of the systolic and diastolic arterial diameter.

Mode frequency: indicates the frequency with the highest amplitude, and is the value that can be measured most reliably (see the illustration under *Doppler spectrum,* above). (This value can be determined for every interval analyzed).

Multi-gate pulsed wave (PW) Doppler system: Several sample volumes are placed along the length of the ultrasound beam simultaneously, and are analyzed independently of each other. This is necessary to register a *velocity profile.*

Near field (Fresnel zone): the region of the acoustic field from the transducer to the focus.

Newtonian fluid: a fluid in which the viscous force opposing flow is proportional to the velocity gradient.

Nyquist theorem (aliasing): In order to classify one wave unambiguously with the assistance of another, the applied frequency must be at least twice as large, so that stroboscopic effects are avoided (F_{max} = PRF/2). This can be important, for example, when measuring high Doppler frequencies (F_{max}) with pulsed wave Doppler procedures in which the *pulse repetition frequency (PRF)* determines the upper limit of the velocity measurement. According to the *Doppler formula,* the maximum flow velocity (V_{max}) is calculated as follows:

$$V_{max} = \frac{PRF}{2} \cdot \frac{c}{2 \cdot f_t \cdot \cos \alpha}$$

This means that higher flow velocities can still be measured by increasing the PRF, or, if this is no longer feasible, by optimizing the angle at which ultrasound is applied (without *aliasing).*

Peak frequency: maximum frequency 3–6 dB above the noise level, or the frequency which borders 95–98% of the spectrum integral in a positive (upward) direction (see the illustration under *Doppler spectrum,* above). It is methodologically restricted by artifacts resulting from inaccurate parameters that have measurement errors of a magnitude of 1 kHz (can be determined for every analyzed time-interval).

Phased-array transducer: transducer with delayed regulation of the crystal elements in order to form a sector-shaped image (see p. 18).

Piezoelectric elements: crystals (e.g., quartz, barium titanate) that change their form in the presence of an electric voltage (transmission mode) or, vice versa, convert changes in their form into an electric voltage (receiving mode).

Pixel: area elements, or also volume elements, that result from the spatial analysis of object and image during imaging with digital systems.

Poiseuille, Jean Leonard M. (1797–1869): French physician and physicist who performed the first experiments demonstrating viscosity in a fluid and its relationship to the pressure gradient in a tube. Poiseuille also refined the techniques of Hales for measuring pressure in the arterial circulation.

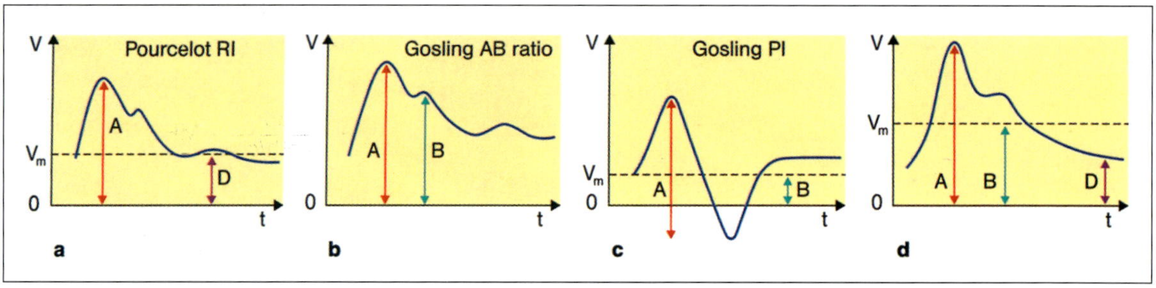

to **Pulsatility index**

Power: the energy delivered by a wave in unit time, measured in watts.

Power mode (intensity imaging, amplitude-mode imaging): application of various imaging techniques to enhance signal-noise reduction in the color image.

Power spectrum: a graph using the square of the amplitude in order to show the intensity component of the flowing corpuscular elements in the blood.

Pulsatility index (PI): Several pulsatility indices (PI), or resistance indices (RI) are suggested in the literature as methods of describing peripheral vascular resistance:

- (a) According to Pourcelot, the RI = (A – D)/A
- (b) According to Gosling, the ratio A/B
- (c) According to Gosling, PI = A/B
- (d) Or, for monophasic curves, PI = (A – D)/B

Pulse duration: interval (usually 0.5–2.0 μs) after which the absolute value of the acoustic pressure exceeds 32% (–10 dB) of the maximum value until the point in time at which the acoustic pressure again falls below this value.

Pulsed wave (PW) Doppler (range-gated Doppler): Ultrasound pulses are discontinuously transmitted at a *pulse repetition frequency* (PRF). A *sample volume* from which the Doppler shift is to be measured can be defined. This method has the advantage of topographically specifying the region from which a Doppler signal will be registered. It has a disadvantage in that the upper boundary for clear measurement of the Doppler shift is restricted by an upper velocity range that can be only partially evaluated (see *Nyquist theorem).*

Pulse repetition frequency (PRF): an important parameter in pulsed wave Doppler procedures. The sample volume depth determines the maximal PRF due to the specific transit time (T) of sound through tissue.

$$PRF = 1/T.$$

The maximum penetration depth Z is calculated by the formula:

$$Z = \frac{c}{2} \times T$$

(c = velocity of sound in tissue). The PRF, in turn, determines the maximum measurable Doppler frequency, amounting to half of the PRF, due to the outward and return path. If the PRF is not twice as large as the Doppler shift that is to be measured, a stroboscopic effect occurs (see *Nyquist theorem).* This means that the true frequency can no longer be ascertained. Modern systems automatically calculate the PRF for every adjusted depth.

Quadrature detector: equipment designed to determine the *Doppler shift,* including a directional analysis. In order to separate the two components (toward the

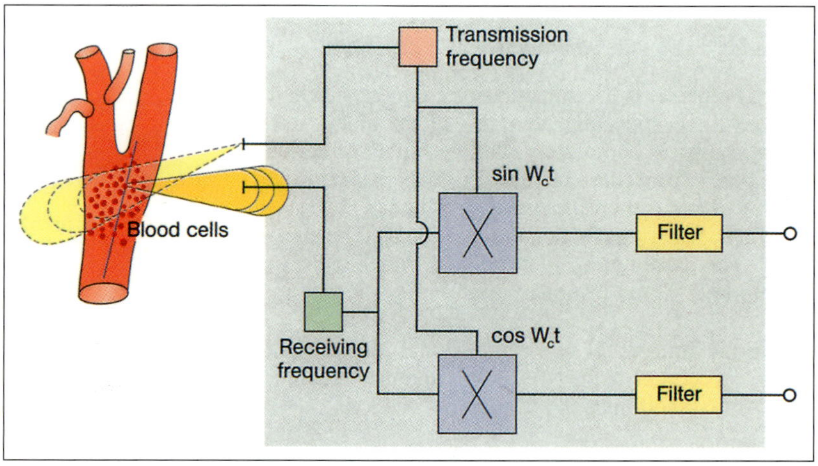

to **Quadrature detector**

probe, away from the probe) the received signal is split. One part is multiplied by the sine component of the transmission frequency, and the other by the cosine component. Both signals are filtered. At the exit point of the quadrature detector, both channels contain all the information from the received signal, but differ from one another in the phase. With a 90° phase shift, both voltages can be examined in such a manner that during further evaluation in spectral analysis the forward flow and return flow components of the blood circulation can be separated in relation to the probe.

Real-time ultrasound: displaying the ultrasound image at an appropriate rate above the flicker-fusion threshold (> 15–20 images/s).

Red blood cell density imaging: Instead of depicting the frequency information in the color-coded Doppler sonogram, echo amplitudes obtained from erythrocyte densities are shown in the sample volume. This means that signals can also be recognized with perpendicular application of the ultrasound.

Reflection (echo): an acoustic signal caused by a reflected beam. The incident angle is equal to the angle of reflection.

Reverberation (multiple reflections): In multiple interface reflections, repeating echoes can cause an untrue image, or poor *axial resolution.* Due to a longer transit time, these disturbing echoes appear displaced in relation to the true echo reflections in the B-mode image, and can complicate the interpretation of the image.

Reynolds, Sir Osborne (1842–1912): English mechanical civil engineer who pioneered the study of vortical and turbulent flow in liquids and laid the theoretical basis for subsequent study of the behavior of viscous fluids. Reynolds also built a steam engine to determine the mechanical equivalent of heat and held patents for the design of marine turbines.

Reynolds number: a number expressing the balance of inertial and viscous forces acting on a flowing fluid. Reynolds numbers higher than a critical value result in disturbed or turbulent flow.

Sample volume: the region in which a Doppler shift is measured. The axial and lateral dimensions (the size and form of the sample volume) are dependent on the duration of the pulse and on the geometry of the ultrasound beam. A variable sample volume can be selected by changing the time of measurement or the number of oscillations per transmitted impulse, or both. The greater the number of oscillations per transmitted impulse, the better the sensitivity, but the poorer the axial resolution.

Scattering: In contrast to interface reflections, which depict a directed transmission of ultrasound, scattering describes a diffuse directional distribution of the ultrasound energy subsequent to its interaction with the tissue structure:

$$f = \frac{30}{\text{depth/cm}} \cdot \text{MHz}$$

where f represents the transmission frequency. When it reaches the corpuscular elements of the blood, ultrasound energy is completely scattered in all directions. The energy from these reflected signals that returns to the transducer is therefore much less than in the case of interface reflections (e.g., a vascular wall). The signals need to be more intensely amplified, which causes an unfavorable signal-to-noise ratio.

Sectional image: a two-dimensional depiction of tissue structures using the echo impulse technique (e.g., cross-sections and longitudinal sections).

Sector scan: a transducer with electronic or mechanical regulation used to generate a sector-shaped sectional image (see p. 18).

Signal movement artifacts (clutter, degradation): superimpositions of low-amplitude Doppler signals by signal-intense interface reflections of the vascular walls (e.g., poststenotic) or capillary/tissue movements. Methods available to compensate for this are ultrasound contrast media and *harmonic imaging.*

Single-gate pulsed wave (PW) Doppler: Along the length of the ultrasound beam, a sample volume can be placed at varying depths, and can be adjusted in its width to match the vascular segments to be analyzed. It is not adequate for obtaining a *velocity profile.*

Sound velocity:

$c = f \times \lambda$
f = sound frequency (Hz)
λ = wavelength (m)

Spectral broadening (window): as an indication of turbulence, this is defined differently by various authors. It is usually calculated over a certain amplitude distance from the modal frequency (see the illustration under *Doppler spectrum,* above).

Spectrum, see *Doppler spectrum.*

Stenosis evaluation (carotid artery): Various angiographic classification schemes are used for stenoses of the carotid artery (local, distal), as well as the hemodynamic parameters used in duplex analysis.

Local stenosis: % stenosis = [1– (c/b)] × 100%
Distal stenosis: % stenosis = [1– (c/a)] × 100%

a = poststenotic lumen width, b = suspected original width of the lumen in the stenotic region, c = lumen width in the stenosis.

The mathematically calculated and experimentally confirmed relationship between these parameters—from conventional angiograms—is as follows:

Distal stenosis (%) = 1.67 × local stenosis (%)

Time gain compensation (TGC): electronic compensation for weakened reflections from deeper tissue layers.

Tone: sound with a pure sinus oscillation of a single frequency in the audible frequency range.

Transducer: using the piezoelectric crystals (= transducer elements) in the transducer of an ultrasound device (= transmitting-receiving unit), electrical energy is transformed into mechanical energy, and vice versa. Various arrangements are available for the B-mode procedure (see pp. 17–19).

- Annular array: elements arranged in a ring shape.
- Linear array: a series of crystal elements which are simultaneously controlled, apart from the accompanying focusing, and which construct wave fronts consisting of acoustic pulses moving in a forward direction.
- Phased array: a series of crystal elements that are individually regulated. Using a time delay in the impulses, acoustic wave fronts are emitted that allow the construction of a sector-shaped image (swiveling ultrasound beam).

Transmission frequency: the exact frequency of the acoustic waves emitted from a transducer, the so-called *mean frequency,* lies in the range of the manufacturer's indicated transmission frequency, allowing for small deviations upward and downward. For example, a designated transmission frequency of 3.0 MHz may apply when the actual middle frequency is 2.8 MHz. (The term is synonymous with nominal frequency).

Turbulence: disorganized flow chaotically oriented in many directions simultaneously.

Turbulent flow: According to Reynolds, flow that deviates from Poiseuille's law can be defined as a dimensionless number (Reynolds number; RN) that depends on the liquid's density and its viscosity. When the RN > 2000, turbulent flow in the moving liquid's components appears.

Ultrasound: Frequencies above the range of human hearing (16,000 Hz to 1 GHz) are called ultrasound. In examining biological tissues, frequencies of approximately 0.5–20.0 MHz are used. The acoustic properties of tissue are exploited diagnostically in this process, e.g.:

- Acoustic quantities (conduction of sound waves, attenuation of sound waves, reflection, scattering, Doppler shift)
- Intrinsic quantities (density, compressibility, viscosity)
- Quantities related to the state (temperature and flow velocity)

Ultrasound beam: In addition to the main ultrasound beam within the acoustic field of a probe, there are many subsidiary beams of varying magnitudes that are dependent on the focusing *(focal zone).* The width of the ultrasound beam determines the diameter of a sample volume, and therefore varies in the *near field* and the *far field.* Usually, the manufacturers of the equipment only provide insufficient information on this aspect.

Ultrasound contrast media: These are used to improve the Doppler signal analysis (see also *harmonic imaging, signal movement artifacts),* using:

- Microscopic air bubbles of varying sizes (12–250 μm)
- Air-bubble microspheres
- Microparticle suspensions (e.g., Echovist, Levovist) consisting of water-soluble microparti-cles as carriers for the microbubbles—depending on their size, these are stable either when passing through the lung (2–8 μm) or not (8–20 μm)

Possible areas of application lie in the differential diagnosis of subtotal stenoses versus occlusions of large blood vessels, analyzing flow structures in smaller blood vessels and tissue capillaries, diagnosing tumors, and in regions that strongly absorb ultrasound (transcranial sonography).

Ultrasound frequency: frequencies used in the field of vascular diagnostics:

- 1–2 MHz: for deep blood vessels (e.g., transcranial Doppler)
- 3–5 MHz: extracranial blood vessels, large arteries
- 8–10 MHz: for superficially located arteries (orbital arteries, arteries of the hands and feet)

Ultrasound probe: with one piezoelectric crystal *(pulsed wave Doppler)* or two *(continuous wave Doppler)* in the ultrasound probe, Doppler systems transform electrical energy into mechanical energy, and vice versa.

Vector: quantity defined by both magnitude and direction. Velocity is a vector quantity; the Doppler shift frequency is determined by the magnitude of the component of the velocity vector along a line between the source and the receiver of sound.

Velocity profile: With the varying flow velocities that occur in the cross-section of a blood vessel, measurements have to be carried out simultaneously at several measuring points in order to obtain the flow velocity profile. Physiologically, the flow velocity is at its lowest near the vessel wall, and at it is highest in the center (see *laminar flow*).

Viscosity: the tendency of a fluid to resist deformation, such as that required to maintain laminar flow. Viscous forces have their origin in the internal cohesion of the fluid.

Vortices: elements of rotational flow often seen with flow separation and disturbance. Rotating flow comprises a wide range of velocities aligned in both directions along a line passing through its center. Doppler shifts from vortical flow are thus characterized by spectral broadening and simultaneous forward and reverse flow.

Wall motion filter: a high-pass filter that distinguishes Doppler shifts in the low-frequency range, caused by wall movements, from the signals caused by the corpuscular flow elements that are actually of interest, and are often more than 20 dB weaker (see *scattering*). This allows the lower measurable flow velocity to be identified. However, it also leads to errors in the calculated parameters of the spectrum, e.g., overestimation of the *mean frequency*.

Wave, wavelength: a periodic process involving oscillation or vibration (λ).

Zero-crossing counter: transformation of the Doppler input signal into a sequence of rectangular pulses, according to the frequency (number) of times that zero is crossed (0 volt comparator).

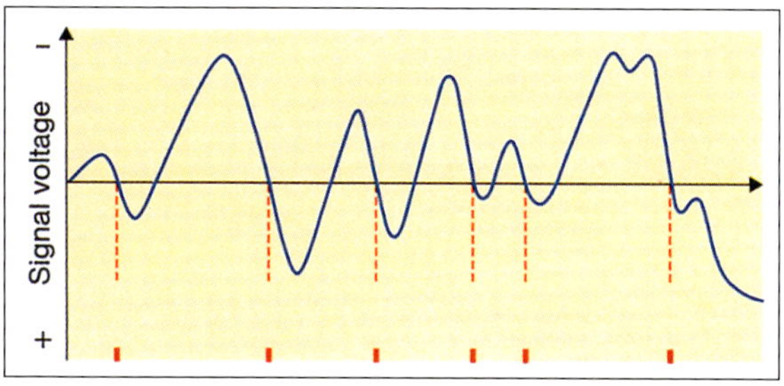

Index

Page numbers in *italics* refer to illustrations or tables where these are separate from the text reference

G

H

I

J

K

L

Q

R

S

T

U

V

W

Z